Multiplication Rule	Given two activities, A_1 and A_2, that can be performed in N_1 and N_2 different ways, respectively, the total number of ways A_1 followed by A_2 can be performed is $N_1 \times N_2$.
Permutations	$P(n, r) = \dfrac{n!}{(n-r)!}$
Combinations	$C(n, r) = \dfrac{n!}{(n-r)!\,r!}$
Union	$n(A \cup B) = n(A) + n(B) - n(A \cap B)$ $P(E \cup F) = P(E) + P(F) - P(E \cap F)$
Probability of Equally Likely Events	If an event E contains s simple outcomes (successes) and the sample space S contains n simple outcomes, and if the outcomes are equally likely, then the probability of E is $P(E) = \dfrac{s}{n} = \dfrac{n(E)}{n(S)}$
Complement Theorem	For an event E, $P(E') = 1 - P(E)$ $P(E) = 1 - P(E')$ $P(E) + P(E') = 1$ where E' is the complement of E in the sample space S.
Conditional Probability	If $P(F) \neq 0$, $P(E \mid F) = \dfrac{P(E \cap F)}{P(F)}$
Product Rule	$P(E \cap F) = P(F) \cdot P(E \mid F)$
Independent Events	E and F are independent if $P(E \mid F) = P(E)$, $P(F \mid E) = P(F)$, or $P(E \cap F) = P(E) \cdot P(F)$
Bayes' Formula	$P(E_i \mid F) = \dfrac{P(E_i)P(F \mid E_i)}{P(E_1)P(F \mid E_1) + P(E_2)P(F \mid E_2) + \cdots + P(E_n)P(F \mid E_n)}$
Binomial Experiments	Probability of x successes in n trials $= C(n, x) \cdot p^x \cdot (1-p)^{n-x}$ $E(X) = np$ $\sigma^2(X) = npq$ $\sigma(X) = \sqrt{npq}$
Expected Value	$E(x) = x_1 p_1 + x_2 p_2 + \cdots + x_n p_n$

Finite Mathematics

SIXTH EDITION

Howard L. Rolf

Baylor University

THOMSON

BROOKS/COLE

Australia • Canada • Mexico • Singapore • Spain
United Kingdom • United States

THOMSON

BROOKS/COLE

Executive Editor: Curt Hinrichs
Development Editor: Cheryll Linthicum
Assistant Editor: Ann Day
Editorial Assistant: Katherine Brayton
Technology Project Manager: Earl Perry
Marketing Manager: Joseph Rogove
Advertising Project Manager: Nathaniel Michelson
Project Manager, Editorial Production: Belinda Krohmer
Art Director: Rob Hugel
Print/Media Buyer: Barbara Britton

Permissions Editor: Sarah Harkrader
Production Service: G&S Typesetters, Inc.
Text Designer: Lisa Henry
Copy Editor: Jan Six
Cover Designer: Lisa Henry
Cover Image: david bishop/san francisco
Cover Printer: Coral Graphic Services
Compositor: G&S Typesetters, Inc.
Printer: Transcontinental

Printed in Canada
2   3   4   5   6   7   08   07   06   05

For more information about our products, contact us at:
Thomson Learning Academic Resource Center
1-800-423-0563

For permission to use material from this text, contact us at:
http://www.thomsonrights.com

Library of Congress Control Number: 2003114133

Student Edition: ISBN 0-534-46539-0
Instructor's Edition: ISBN 0-534-46540-4

Brooks/Cole—Thomson Learning
10 Davis Drive
Belmont, CA 94002
USA

Asia
Thomson Learning
5 Shenton Way #01-01
UIC Building
Singapore 068808

Australia/New Zealand
Thomson Learning
102 Dodds Street
Southbank, Victoria 3006
Australia

Canada
Nelson
1120 Birchmount Road
Toronto, Ontario M1K 5G4
Canada

Europe/Middle East/Africa
Thomson Learning
High Holborn House
50/51 Bedford Row
London WC1R 4LR
United Kingdom

Latin America
Thomson Learning
Seneca, 53
Colonia Polanco
11560 Mexico D.F.
Mexico

Spain/Portugal
Paraninfo
Calle Magallanes, 25
28015 Madrid, Spain

To Some Special People

Lauren Elizabeth McClintock

Anna Lynn McClintock

Elizabeth Kate Dietze

Contents

3   LINEAR PROGRAMMING   195

4   LINEAR PROGRAMMING: THE SIMPLEX METHOD   249

5   MATHEMATICS OF FINANCE   342

9 GAME THEORY   697

10 LOGIC   717

A REVIEW TOPICS   740

B USING A TI-83 GRAPHING CALCULATOR   756

C USING EXCEL   766

ANSWERS TO SELECTED ODD-NUMBERED EXERCISES   A-1

INDEX   I-1

Preface

To the Student

Here is your first quiz in Finite Mathematics.
What do the following have in common?

A banker.

A sociologist studying a culture.

A person planning for retirement.

A proud new parent.

A young couple buying their first house.

A politician assessing the chances of winning an election.

A gambling casino.

The feedlot operator caught in a highly competitive cattle market.

The marketing manager of a corporation who wants to know if the company should invest in marketing a new product.

Here is the answer. You get to grade your own quiz.

All of these persons directly or indirectly use or can use some area of *Finite Mathematics* to help determine the best course of action. *Finite Mathematics* helps to analyze problems in business and the social sciences and provides methods that help determine the implications and consequences of various choices available.

This book gives an introduction to mathematics that is useful to a variety of disciplines. Mathematics can help you understand the underlying concepts of a discipline. It can help you organize information into a more useful form. Predictions and trends can be obtained from mathematical models. A mathematical analysis can provide a basis for making a good decision.

How can this course benefit you? First, you must understand your discipline and second, you must understand this course. It is up to you to learn your discipline. This book is written to help you understand the mathematics. Perhaps these suggestions will be helpful.

1. You must study the material on a regular basis. Do your homework.
2. Study to understand the concepts. Read the explanations and study the examples.
3. Work the exercises and relate them to the concepts presented. Selected exercises refer to specific examples to help you get started.

4. After you work an exercise, take a minute to review what you have done and make sure you understand how you worked the problem.

Your general problem solving skills can improve because of this course. As you analyze problems, consider which method to use, and work through the solution, you are experiencing a simple form of the kind of process you will use throughout your life in your job or in daily living when you respond to the question "Hey, we have a problem, what should we do?" And most of life's problems are word problems, so do those in this course.

Student Aids

Several features in the text assist you in your study of the concepts.

Boldface words indicate new terms.

Boxes emphasize definitions, theorems, procedures, and summaries.

Notes and **Warnings** show typical problem areas and provide reminders of concepts introduced earlier.

Review exercises at the end of each chapter help you to review the concepts.

Answers to odd-numbered exercises are provided.

A **Student's Solution Manual** contains worked-out solutions to the odd-numbered exercises. It is available as a separate item.

Important Terms are summarized at the end of each chapter.

Selected exercises are **cross-referenced** with examples to help you get started with homework.

To the Instructor

Mathematics as we know it came into existence through an evolutionary process, and that process continues today. Occasionally a mathematical idea will fade away as it is replaced by a better idea. Old ideas become modified, or new and significant concepts are born and take their place. Some mathematical concepts have been developed in an attempt to solve problems in a particular discipline. Many disciplines have found that mathematical concepts are useful in understanding and applying the ideas of that discipline.

As technology impacts more and more areas of the workplace and our culture, mathematical proficiency increases in importance. Science and engineering traditionally rely heavily on mathematics for analyzing and solving problems. Disciplines in business, life sciences, and the social sciences have more recently applied and developed mathematical concepts in an attempt to solve problems in those disciplines. This book deals with topics, such as functions, linear systems, and matrices, that serve as useful tools in expressing and analyzing problems in areas using mathematics. Other topics, such as linear programming, probability, and mathematics of finance, have more direct applications to problems in business and industry.

Prerequisites

This book assumes at least three semesters of high school algebra. *Appendix A* provides a brief review for those who may need to refresh their memory.

Audience

This book is designed for students majoring in business, the social sciences, and some areas of the life sciences who wish to develop mathematical and quantitative skills that will be of value in their discipline. Liberal arts students and prospective teachers also can profit from the study of several areas of mathematics that apply to familiar areas of life.

Philosophy

Because the potential audience includes a wide range of students with different interests, the author is sensitive to the needs of students and heeds the advice he received years ago, "Write for the student."

The purpose of *Finite Mathematics* is the learning of mathematical concepts and techniques with applications of these concepts as the reason for studying them. For this reason, the examples and exercises in the text are designed to point toward these applications. Even so, the application of mathematical concepts requires an understanding of the mathematics and an expertise in the area of application. Generally, a straightforward application of mathematics does not occur because there is a certain amount of "fuzziness" due to complications, exceptions, and variations. Thus, an application may be more difficult to accomplish than it appears on the surface. In spite of the difficulties in analyzing real-world problems, we can obtain an idea of the usefulness of finite mathematics using examples and exercises that are greatly simplified versions of actual applications.

While the author assumes only a background in high school algebra, more challenging exercises are included for students who are ready to dig deeper. Some exercises involve the use of a graphing calculator or spreadsheet, but the course is designed to be taught independent of them.

Changes in the Sixth Edition

New and updated exercises have been added to increase the relevance of the text. More than 3200 exercises and 500 examples throughout the text provide instructors and students with an abundance of homework and practice problems.

For those who wish to use spreadsheets in the course, instructions on the use of EXCEL have been added. At the end of appropriate sections we show the use of a TI-83 calculator or an EXCEL spreadsheet as they relate to that section. In addition, a new Appendix C provides guidance on the use of EXCEL.

Coverage of probability has been revised and expanded to include empirical probability.

A new optional section on partitions has been added to Chapter 6.

THE INVERSE OF A MATRIX

The inverse of a square matrix can be obtained by using the x^{-1} key. For example, to find the inverse of a matrix stored in [A] where

$$[A] = \begin{bmatrix} 3 & 2 & 1 \\ 0 & 4 & 1 \\ 1 & 2 & 1 \end{bmatrix}$$

use [A] x^{-1} ENTER, and the screen will show

Some matrices such as

$$[A] = \begin{bmatrix} 3 & 2 & 2 \\ 0 & 4 & 1 \\ 1 & 2 & 1 \end{bmatrix}$$

have no inverse. In this case, an error message is given:

EXERCISES

Find the inverse of each of the following.

1. $\begin{bmatrix} 2 & -1 & 3 \\ 3 & 1 & 2 \\ 4 & 1 & 4 \end{bmatrix}$
2. $\begin{bmatrix} 1 & 0 & 1 \\ 0 & 1 & 1 \\ 1 & 1 & 0 \end{bmatrix}$
3. $\begin{bmatrix} 4 & -2 & 1 \\ 1 & -4 & -1 \\ 3 & 2 & 2 \end{bmatrix}$

EXCEL has the **MINVERSE** command that calculates the inverse of a square matrix. To find the inverse of

$$A = \begin{bmatrix} 1 & 2 & 1 \\ 2 & 4 & 1 \\ 1 & 3 & 2 \end{bmatrix}$$ enter the matrix in cells A2:C4. Next, select the cells E2:G4 for the location of the inverse

of A, type =MINVERSE(A2:C4) and simultaneously press CTRL + SHIFT + ENTER. The inverse of A then appears in E2:G4.

	A	B	C	D	E	F	G
1	Matrix A in A2:C4				Inverse of A in E2:G4		
2	1	2	1		5	-1	-2
3	2	4	1		-3	1	1
4	1	3	2		2	-1	0
5							
6	If A has no inverse, you will get a matrix like this						
7	#NUM!	#NUM!	#NUM!				
8	#NUM!	#NUM!	#NUM!				
9	#NUM!	#NUM!	#NUM!				
10							

EXERCISES

Find the inverse of A in the following exercises.

1. $A = \begin{bmatrix} -2 & 6 & 3 \\ 7 & -3 & 1 \\ 9 & 2 & 5 \end{bmatrix}$
2. $A = \begin{bmatrix} 1 & 2 & 1 \\ 2 & 5 & 1 \\ 1 & 3 & 2 \end{bmatrix}$
3. $A = \begin{bmatrix} -0.4 & 6 & 3.75 \\ 1.4 & -3 & 1.25 \\ 1.8 & 2 & 6.25 \end{bmatrix}$

4. $A = \begin{bmatrix} 2 & -1 & 5 & 5 \\ 3 & 5 & 1 & -1 \\ 1 & 3 & 6 & -2 \\ 2 & 2 & -1 & 1 \end{bmatrix}$
5. $A = \begin{bmatrix} -2 & 3 & 4 \\ 7 & -3 & 1 \\ 1 & 2 & 5 \end{bmatrix}$

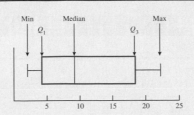

NOTE

The box contains the *middle 50%* of the data.

Notice how the box plot is constructed. The ends of the box are Q_1 and Q_3, with the location of the median shown in the box. Whiskers are attached to each end of the box. The whisker on the left extends to the minimum value, and the whisker on the right extends to the maximum value.

■ Now You Are Ready to Work Exercise 25

Stem-and-Leaf Plots

To more orderly organize data into categories, a **stem-and-leaf plot** can be used. We do so by breaking the scores into two parts, the *stem,* consisting of the first one or two digits, and the *leaf,* consisting of the other digits.

EXAMPLE 3 ➤   Make a stem-and-leaf plot of the following scores: 21, 13, 17, 24, 48, 7, 31, 46, 44, 39, 9, 15, 10, 41, 46, 33, 24

SOLUTION

We use the first digits 0, 1, 2, 3, 4 for the stems, which will divide the data into intervals 0–9, 10–19, 20–29, 30–39, and 40–49.

Stem	Leaves
0	79
1	3750
2	144
3	193
4	86416

We can now easily count the frequency of each category.
    Notice that the digits for the leaves are not in order. They can be listed as they occur in the list.

The addition of material on boxplots and stem-and-leaf analysis to Chapter 8 provides new ways to look at data.

A free IText pincode card comes with every student copy of the text. This code will allow students to access an online version of the previous edition of the text for additional online practice.

Continuing Features

Exposition

The author has concentrated on writing that is lucid, friendly, and considerate of the student. As a result, the text offers a clear explanation of concepts, and the computations are detailed enough that students can easily follow successive steps in the problem-solving process.

Exercises

The text contains more than 3200 exercises, including an abundance of word problems, providing students ample means to apply mathematical concepts to problems and giving the instructor a variety of choices in assigning problems.

▦ 1.2 EXERCISES

LEVEL 1

Draw the graphs of the lines in Exercises 1 through 6.

1. *(See Example 2)*
$f(x) = 3x + 8$

2. $f(x) = 4x - 2$

3. $f(x) = x + 7$

4. $f(x) = -2x + 5$

5. $f(x) = -3x - 1$

6. $f(x) = \dfrac{2}{3}x + 4$

Find the slope and *y*-intercept of the lines in Exercises 7 through 10.

7. *(See Example 3)*
$y = 7x + 22$

8. $y = 13x - 4$

9. $y = \dfrac{-2}{5}x + 6$

10. $y = \dfrac{-1}{4}x - \dfrac{1}{3}$

Find the slope and *y*-intercept of the lines in Exercises 11 through 14.

11. *(See Example 4)*
$2x + 5y - 3 = 0$

12. $4x +$

13. $x - 3y + 6 = 0$

14. $5x -$

**Determine the slopes of the straight lin[e]
pairs of points in Exercises 15 through 1[**

15. *(See Example 5)*
$(1, 2), (3, 4)$

16. $(2, 3), (-3, 1)$

17. $(-4, -1), (-1, -5)$

18. $(2, -4), (6, -3)$

For the graphs shown in Exercises 19 through 22 indicate whether the lines have positive, negative, or zero slope.

19. *(See Example 6)*

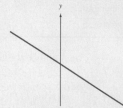

EXPLORATIONS

107. The birth rate in Japan has declined for a number of years. The birth rates per 1000 population for three different years are

1980	13.7
1985	11.9
2002	10.03

(a) Use the 1980 and 2002 figures to find birth rate as a linear function of number of years since 1980.

(b) Use the linear function to estimate the birth rate for 1985. Does it give a realistic rate?

(c) Use the linear function to determine when the birth rate will reach zero. Comment on the reasonableness of the result.

108. The U.S. Census Bureau reports a wide variety of population information. One bit of information is the percentage of the population that never marries.

(a) In 1980, the percentage of males in the 20–24 age group who had never married was 68.8%. In 2000 the percentage rose to 83.7%.

  (i) Based on this information, find the percentage of unmarried males as a linear function of time in years.

  (ii) Use the linear function you found to estimate the percentage of males, ages 20–24, who never married for the year 1998.

  (iii) The actual percentage given by the Census Bureau for 1998 was 83.4%. How does this compare with your estimate? Does the linear function found in part (i) seem to be a reasonably good representation of the growth of the percentage?

(b) In 1980, 50.2% of the female population in the 20–24 age range had never married. The percentage rose to 72.8% in 2000.

  (i) Based on this information, find the percentage of females who had never married as a linear function of time in years since 1980.

  (ii) Use the linear function just found to estimate the percentage of females, ages 20–24, who had never married for the year 1998.

  (iii) The actual percentage reported by the Census Bureau for 1998 was 70.3%. Compare the estimate obtained in part (ii). Does the linear function seem to provide a reasonable estimate of the growth of the percentage of females who never married?

109. The manager of the Ivy Square Cinema observed that Friday night estimated ticket sales were 185 when the admission was $5. When the admission was increased to $6, the estimated attendance fell to 140.

(a) Use this information to find estimated attendance as a linear function of admission price.

(b) Use the function found in part (a) to estimate attendance if admission is increased to $7.

(c) Use the function found in part (a) to determine the admission price that will yield an estimated attendance of 250.

(d) The manager had no desire to set an admission price that would drive away all customers, but she was curious to know when that would occur. Based on the function in part (a), find the admission price that would result in a zero attendance. (Does this seem reasonable to you?)

Exercises are graded by level of difficulty: level 1 for routine problems, level 2 for elementary word problems and somewhat more challenging problems, and level 3 for the most difficult problems.

"Explorations" exercises encourage students to think more deeply about mathematical concepts, often providing an opportunity to use the graphing calculator. Many of these exercises may be used for group projects or writing assignments.

EXAMPLE 2 ➤   A company made a cost study and found that it cost $10,170 to produce 800 pairs of running shoes and $13,810 to produce 1150 pairs.

(a) Determine the cost–volume function.
(b) Find the fixed cost and the unit cost.

SOLUTION

(a) Let x = the number of pairs. The information gives two points on the cost–volume line: $(800, 10170)$ and $(1150, 13810)$. The slope of the line is

$$m = \frac{13{,}810 - 10{,}170}{1150 - 800} = \frac{3640}{350} = 10.40$$

Using the point $(800, 10170)$ in the point-slope equation, we have

$$y - 10{,}170 = 10.40(x - 800)$$
$$y = 10.40x + 1850$$

Therefore, $C(x) = 10.40x + 1850$.

(b) From the equation $C(x) = 10.40x + 1850$, the fixed cost is $1850 per week, and the unit cost is $10.40.

■ **Now You Are Ready to Work Exercise 13**

Cross-Referencing of Examples and Exercises

The exercise sets form an integral part of any mathematics textbook, but they are not the only part. The exercises are structured to encourage students to read the body of the text for explanations and examples. We have done this by cross-referencing some examples with exercises. After reading a particular example, students can go directly to the referenced exercise to test their understanding. Conversely, selected exercises refer to an example that illustrates the concepts needed to work them.

More than 500 examples — many of them based on real-world data — illustrate the useful application of the mathematics studied.

Discussion questions provide the option of using the graphing calculator or spreadsheet technology to solve the problem.

Technology boxes focus on concepts and walk students through the exercises using the graphing calculator and EXCEL spreadsheets.

Appendix B provides guidance on use of the TI-83 graphing calculator, whereas the "Graphing Calculator Manual" offers more in-depth coverage.

Use of Technology

A graphing calculator or spreadsheet is not required to use the textbook. However, for those who wish to incorporate a graphing calculator or spreadsheet in the course, some exploration exercises show a graphing calculator icon ⬤ or a spreadsheet icon ▦. These exercises may require the use of a graphing calculator or spreadsheet, they may be more accessible with the use of a graphing calculator or spreadsheet, or they may illustrate a use of technology. Depending on the extent to which the instructor wishes to integrate the graphing calculator or spreadsheet into the course, the student may use a graphing calculator or spreadsheet on other exercises not marked by an icon.

In addition to the *Using Your TI-83* features at the end of some sections, *Appendix B* gives some guidance on the use of the TI-83 graphing calculator.

Flexibility

This book was written to provide flexibility in the choice and order of topics. Some topics must necessarily be covered in sequence. The chapters that are prerequisites are shown in the following diagram.

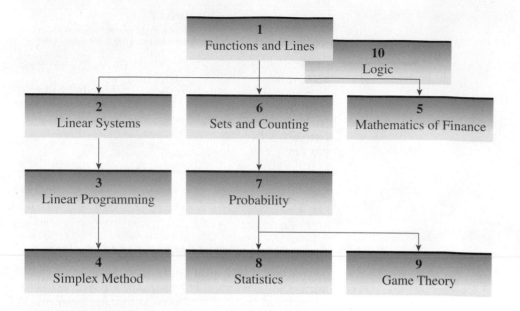

Use of Calculators

Students are encouraged to use a calculator in working problems. A calculator with an exponential function is needed for the section on the binomial distribution and for the chapter on mathematics of finance. A graphing calculator is optional unless the instructor uses the graphing calculator exercises. Students are expected to use a graphing calculator for those exercises.

Learning Aids

Material at the end of each chapter includes important terms referenced to the section in which they are defined, and review exercises for the chapter.

- Definitions, key formulas, procedures, summaries, and theorems are boxed to emphasize their importance to the student.
- Figures use two colors to reinforce learning.
- Occasional **Caution** or **Note** statements are included as warnings of typical problem areas or reminders of concepts introduced earlier.
- Answers to selected odd-numbered exercises provide feedback to students.

Accuracy Check

Examples and exercises have been checked by the author and a graduate student. Brooks/Cole Publishers also arranges for an independent accuracy check. Any errors that remain are the work of gremlins.

Supplements

Student Solutions Manual (0534491626)

The *Student Solutions Manual,* written by the author, provides worked-out solutions to the odd-numbered exercises in the text.

Instructor's Solutions Manual (053449160X)

The *Instructor's Solutions Manual,* written by the author, provides worked-out solutions to all of the exercises (with the exception of some Exploration exercises). It also contains transparency masters.

Instructor's Suite CD (0534491642)

The Instructor's Suite CD contains the *Instructor's Solutions Manual,* Test Bank, and PowerPoint Presentation.

Microsoft EXCEL Spreadsheet Manual (0534491618)

The *Microsoft EXCEL Spreadsheet Manual* parallels the textbook, providing instruction on the use of EXCEL and offering exercises that correspond to the material in the text.

Graphing Calculator Manual (0534491650)

The *Graphing Calculator Manual,* written by the author, parallels the textbook and provides hints on the use of the TI-83 graphing calculator, exercises for the graphing calculator, and special programs that are useful in some topics.

Test Bank (0534491634)

A *Test Bank,* written by the author, is available both in print and on the Instructor's Suite CD. It contains approximately 1600 multiple-choice and short-answer test questions by section.

Finite Mathematics *Video Lecture CD* (0534491596)

Packaged free of charge with the textbook, the CD provides video tutorials, demonstrating how to work out selected exercises. Students select compressed videos of the exercises based on a chapter and section menu.

ILrn/BCA Instructor Version (0534491669)

iLrn Testing provides testing and course-management software using correct statistical notation and delivered in a browser-based format. Results flow automatically to a grade book for tracking so that instructors are can better able to assess student understanding of the material before class or an actual test.

Book-specific Web site (http://mathematics.brookscole.com)

When you adopt a Thomson–Brooks/Cole mathematics text, you and your students will have access to a variety of teaching and learning resources. This Web site features everything from book-specific resources to newsgroups.

Acknowledgments

A number of people have contributed to the writing of this book. I am most grateful to users of the book and the following reviewers who provided constructive criticism, suggestions on clarifying ideas, and improved presentation of the concepts.

Thanks especially are due to these reviewers, whose careful attention to detail greatly improved the manuscript: Pedro Barquero, Santa Monica College; Sipra Eko, University of Maine; Jaclyn LeFebvre, Illinois Central College; Scott Martin, Arizona State University; Thomas Porter, Southern Illinois University; and Danny Turner, Winthrop University.

During galley and page proof stages, this edition was checked for accuracy by Jo Cannon. Her careful attention to detail is appreciated.

Many thanks to the staff of Brooks/Cole for their support of this book, especially to Curt Hinrichs, Mathematics Editor; Cheryll Linthicum, Developmental Editor; Ann Day, Assistant Editor, Statistics & Applied Mathematics; Katherine Brayton, Senior Editorial Assistant; Jessica Perry, Marketing Assistant; and Belinda Krohmer, Production Project Manager.

Thanks to the teachers and students who called attention to errors and to examples and exercises that could be stated more clearly.

Howard L. Rolf
November 2003

CHAPTER

1

Functions and Lines

Mathematics is a powerful tool used in the design of automobiles, electronic equipment, and buildings. It helps in solving problems of business, industry, environment, science, and social sciences. Mathematics helps to predict sales, population growth, the outcome of elections, and the location of black holes. Some techniques of mathematics are straightforward; others are complicated and difficult. Nearly all practical problems involve two or more quantities that are related in some manner. For example, the amount withheld from a paycheck for FICA is related to an employee's salary; the area of a rectangle is related to the length of its sides (area = length × width); the amount charged for sales tax depends on the price of an item; and UPS shipping costs depend on the weight of the package and the distance shipped.

One of the simplest relationships between variables can be represented by a straight line, a basic geometric concept encountered by people of different ages, cultures, and times in history. Few people likely contemplate the common property held by the shaft of an arrow, the tightly stretched rope used by a sailor, the fold of a blanket, the crease formed by folding a paper, a string between two stakes to plant a straight row of vegetables, the stripes on a parking lot, the boundaries of a basketball court, and the representation of streets and highways on a map.

We realize that the "lines" mentioned are not really lines. They are at best an approximation of a line segment. They may not be exactly straight, a basketball court boundary is really a stripe, and the most carefully drawn line has "ragged edges" when enlarged sufficiently.

The ancient Greeks are credited with the generalization and abstraction of geometric concepts such as the ideal line. We mention this because the ideal line — the line that expresses the essence of the stretched rope, the mark in the sand, or the artist's finely drawn line — can be extended to certain relationships between two variables and can be expressed mathematically and used to give insight into the behavior of phenomena and activities in science, business, and some daily activities. In this chapter we discuss linear equations and their applications. ■

1.1 Functions

Mathematicians formalize certain kinds of relationships between quantities and call them **functions.** Nearly always in this book, the quantities involved are measured by real numbers.

A function consists of three parts: two sets and a rule. The **rule of a function** describes the relationship between a number in the first set (called the **domain**) and a number in the second set (called the **range**). The rule is often stated in the form of an equation like $A = \text{length}^2$ for the relationship between the area of a square and the length of a side.

In this case, the domain consists of the set of numbers representing lengths and the range consists of the set of numbers representing areas.

DEFINITION
Function

> A function is a rule that assigns to each number from the first set (domain) exactly one number from the second set (range).

We generally use the letter x to represent a number from the first set and the letter y to represent a number from the second set. Thus, for each value of x there is exactly one value of y assigned to x.

Generally, a number may be arbitrarily selected from the domain, so x is called an **independent variable.** Once a value of x is selected, the rule determines the corresponding value of y. Because y depends on the value of x, y is called a **dependent variable.**

EXAMPLE 1 ➤

Mr. Riggs is a consultant for a trailer manufacturing company. His fee is $300 for miscellaneous expenses plus $50 per hour. What is the domain of this function? The number of hours worked determines the fee, so the domain consists of the number of hours worked and the range consists of the fees charged. It makes sense to say that hours worked must be a positive number, and the minimum fee is $300. Thus, positive numbers make the domain, and numbers larger than 300 make the range. The rule that determines the fee that corresponds to a certain number of consulting hours is given by the formula

$$y = 50x + 300$$

where x represents the number of hours consulted and y represents the total fee in dollars. Notice that the formula gives exactly one fee for each number of consulting hours. So, the domain consists of the set of numbers representing hours worked, and the range consists of the set of numbers representing the dollar amount of fees.

■ **Now You Are Ready to Work Exercise 1**

In some cases, the quantities in the domain and range of a function may be limited to a few values, as illustrated in the next example.

EXAMPLE 2 ➤

When a family goes to a concert, the amount paid depends on the number attending; that is, the total admission is a function of the number attending. The ticket office usually has a chart giving total admission, so for them the rule is a chart something like this:

Number of Tickets	Total Admission
1	$ 6.50
2	$13.00
3	$19.50
4	$26.00
5	$32.50
6	$39.00

From the chart, the domain is the set of numbers representing the number of tickets sold. Because we never sell a fractional number of tickets, the domain is restricted to positive integers. In this case, the domain is the set $\{1, 2, 3, 4, 5, 6\}$.

■ **Now You Are Ready to Work Exercise 3**

Mathematicians have a standard notation for functions. For example, the equation $y = 50x + 300$ is often written as

$$f(x) = 50x + 300$$

$f(x)$ is read "f of x," indicating "f is a function of x." The notation $f(x)$ is a way of naming a function f and indicating that the variable used is x. The notation $g(t)$ indicates another function named g using the variable t. The $f(x)$ notation is especially useful to indicate the substitution of a number for x. $f(3)$ looks as though 3 has been put in place of x in $f(x)$. This is the correct interpretation. $f(3)$ represents the **value** of the function when 3 is substituted for x in

$$f(x) = 50x + 300$$

That is,

$$f(3) = 50(3) + 300$$
$$= 150 + 300$$
$$= 450$$

The next three examples illustrate some uses of the $f(x)$ type notation.

EXAMPLE 3 ➤   **(a)** If $f(x) = -7x + 22$, then

$$f(2) = -7(2) + 22 = 8$$
$$f(-1) = -7(-1) + 22$$
$$= 7 + 22$$
$$= 29$$

(b) If $f(x) = 4x - 11$, then

$$f(5) = 4(5) - 11$$
$$= 20 - 11$$
$$= 9$$
$$f(0) = 4(0) - 11$$
$$= 0 - 11$$
$$= -11$$

(c) If $f(x) = x(4 - 2x)$, then

$$f(6) = 6(4 - 2(6))$$
$$= 6(4 - 12)$$
$$= 6(-8)$$
$$= -48$$
$$f(a) = a(4 - 2a)$$
$$f(a + 3) = (a + 3)(4 - 2(a + 3))$$
$$= (a + 3)(4 - 2a - 6)$$
$$= (a + 3)(-2a - 2)$$
$$= -2a^2 - 8a - 6$$

■ **Now You Are Ready to Work Exercise 5**

Note: Sometimes the equation that defines a function, the rule that relates numbers in the domain to those in the range, is given, but the domain and range are not specified. In such cases, we define the domain to be all real numbers that can be substituted for x and that yield a real number for y. The range is the set of values of y so obtained. For example, $x = 9$ is in the domain of $y = \sqrt{x}$ because it yields the real number $y = 3$. However, $x = -4$ is not in the domain because $\sqrt{-4}$ is not a real number.

For $f(x) = \dfrac{3x + 2}{x - 5}$, 2 is in the domain because $f(2) = \frac{8}{-3}$ but 5 is not in the domain because $f(5) = \frac{17}{0}$, which is undefined. In fact, the domain consists of all real numbers except $x = 5$.

When applying a function, the nature of the application may restrict the domain or range. For example, the function that determines the amount of postage depends on the weight of the letter. It makes no sense for the domain to contain negative values of weight

We conclude this section with some applications.

EXAMPLE 4 ➤   From 1980 to 2000, the population of the United States can be estimated with the function

$$p(t) = 2.745t - 5210.35$$

where

$$t = \text{the year}$$
$$p(t) = \text{the population in millions}$$

(a) Based on this function, find $p(1970)$. Find $p(2010)$.
(b) Estimate when the population will reach 300 million.

SOLUTION

(a) $p(1970) = 2.745(1970) - 5210.35 = 5407.65 - 5210.35 = 197.3$ million.
$p(2010) = 2.745(2010) - 5210.35 = 307.1$ million
(b) If $p(t) = 300$, then

$$300 - 2.745t - 5210.35$$
$$5510.35 = 2.745t$$
$$t = \frac{5510.35}{2.745} = 2007.4$$

The function estimates that the population will reach 300 million in the year 2007.

■ **Now You Are Ready to Work Exercise 9**

EXAMPLE 5 ➤ Andy works at Papa Rolla's Pizza Parlor. He makes $8 per hour and time-and-a-half for all hours over 40 in a week. Thus, his weekly salary is $S(h) = 12h + 320$, where h is the number of hours overtime and $S(h)$ is his weekly salary.

(a) What is $S(5.25)$?
(b) Find his weekly salary when he works 44.5 hours.
(c) One week Andy's salary was $362. How many hours overtime did he work?

SOLUTION

(a) $S(5.25) = 12(5.25) + 320 = 383$
(b) In this case, $h = 44.5 - 40 = 4.5$, so $S(h) = 12(4.5) + 320 = 374$.
His salary was $374.
(c) $S(h) = 362$, so
$$362 - 12h + 320$$
$$12h = 362 - 320 = 42$$
$$h = \frac{42}{12} = 3.5$$
Andy worked 3.5 hours overtime.

■ **Now You Are Ready to Work Exercise 23**

Observe that we use letters other than f and x to represent functions and variables. We might refer to the cost of producing x items as $C(x) = 5x + 540$; the price of x pounds of steak as $p(x) = 5.19x$; the area of a circle of radius r as $A(r) = \pi r^2$; and the distance in feet an object falls in t seconds as $d(t) = 16t^2$.

A function requires that each number in the domain be associated with *exactly* one number in the range. There are times when a rule assigns more than one number in the range to a number in the domain. In such a case, the relationship is *not* a function. Here is an example.

EXAMPLE 6 ➤ A grocery store sells apples for $1.29 a pound. When Sarah buys six apples ($x = 6$), the checker determines the weight in order to know the cost. The six apples of the customer behind Sarah likely will correspond to a different weight. Thus, the weight "function" of x apples is not a function of the number of apples because a number in the domain (number of apples) may be related to more than one number in the range (weight of x apples).

If we let $x =$ the weight of the apples, then the cost relation to x pounds is a function because there is a unique cost for a given weight. ■

▪▪ 1.1 EXERCISES

These exercises are designed to help you understand the concept of a function and the use of the function notation.

LEVEL 1

Exercises 1 through 4 give information that describes the relationship between two variables. You are to convert this information into a rule that relates the variables. The rule may take the form of an equation or some other form.

1. *(See Example 1)* Barber's Tree Service charges $20 plus $15 per hour to trim trees. Write the rule relating the fee and hours worked using x for the number of hours worked and y for the fee. What do the numbers in the domain represent? What do the numbers in the range represent?

2. An appliance repairman charges $30 plus $20 per hour for house calls. Write the rule that relates hours worked and his fee.

3. *(See Example 2)* The price of movie tickets is given by the following chart:

Number of Tickets	Total Admission
1	$4.75
2	$9.50
3	$14.25
4	$19.00
5	$23.75
6	$28.50
7	$33.25

   (a) What is $f(5)$?   **(b)** What is $f(3)$?

4. Tickets to a football game cost $14 each. Make a chart showing the total cost function for the purchase of one, two, three, four, five, and six tickets.

Be sure you understand the use of functional notation by working Exercises 5 through 11.

5. *(See Example 3)* $f(x) = 4x - 3$. Determine:
   (a) $f(1)$ **(b)** $f(-2)$
   (c) $f\left(\dfrac{1}{2}\right)$ **(d)** $f(a)$

6. $f(x) = x(2x - 1)$. Determine:
   (a) $f(3)$ **(b)** $f(-2)$
   (c) $f(0)$ **(d)** $f(b)$

7. $f(x) = \dfrac{x + 1}{x - 1}$. Determine:
   (a) $f(5)$ **(b)** $f(-6)$
   (c) $f(0)$ **(d)** $f(2c)$

LEVEL 2

8. $f(x) = -4x + 7$. Determine:
   (a) $f(a)$ **(b)** $f(y)$
   (c) $f(a + 1)$ **(d)** $f(a + h)$
   (e) $f(3a)$ **(f)** $f(2b + 1)$

9. *(See Example 4)* In recent years, the number of people in the United States who are 100 years old or older can be estimated by the function

$$p(t) = 1.32t - 2589.5$$

where

$t =$ the year
$p(t) =$ the number of people, in thousands, who are 100 or older.

(a) Find $p(1995)$. Find $p(2010)$.
(b) Use the function to estimate when the number of people 100 years or older will reach 75,000.

10. A small cheese pizza is cut into four pieces. The calcium content of pizza is given by the function

$$f(x) = 221x$$

where x is the number of pieces of pizza and $f(x)$ is the quantity of calcium in milligrams.

(a) How many milligrams of calcium are contained in one pizza?

(b) The recommended daily requirement for calcium is 1000 mg. How many pieces should be eaten to get that amount?

11. Swimming requires 9 calories per minute of energy, so the function for calories used in swimming is given by

$$f(x) = 9x$$

where x is the number of minutes and $f(x)$ is the number of calories used.

(a) How many calories are used in swimming for one hour?
(b) A swimmer wants to use 750 calories. How long should she swim?

LEVEL 3

Each of the following statements describes a function. Write an equation of the function.

12. The cost of grapes at the Corner Grocery is 49¢ per pound.

13. The cost of catering a hamburger cookout is a $25 service charge plus $2.40 per hamburger.

14. The monthly income of a salesman is $500 plus 5% of sales.

15. The sale price of all items in The Men's Clothing Store is 20% off the regular price.

16. The monthly sales of Pappa's Pizza are $1200 plus $3 for each dollar spent on advertising.

17. Dion hauls sand and gravel. His hauling costs (per load) are overhead costs of $12.00 per load and operating costs of $0.60 per mile.

18. Becky has a lawn-mowing service. She charges a base price of $10 plus $7 per hour.

19. Paloma University found that a good estimate of its operating budget is $5,000,000 plus $3500 per student.

20. Peoples Bank collects a monthly service charge of $2.00 plus $0.10 per check.

21. An automobile dealer's invoice cost is 0.88 of the list price of an automobile.

22. A telephone company provides measured phone service. The rate is $7.60 per month plus $0.05 per call.

23. (See Example 5)   Maradee works at Cox's Jewelers for $7.50 per hour with time-and-a-half for all hours over 40 in a week. Thus, her weekly salary is given by

$$S(h) = 11.25h + 300$$

where h is the number of overtime hours and $S(h)$ is her weekly salary.

(a) Find $S(2.5)$.
(b) One week, Maradee worked 45 hours. Find her salary for that week.
(c) Maradee's Valentine week salary was $395.63. How many hours overtime did she work?

24. If

$$f(x) = (x + 2)(x - 1)$$

and

$$g(x) = \frac{7x + 4}{x + 1}$$

find

$$f(3) + g(2)$$

Exercises 25 through 36 describe two variables. In each case determine the following:

(a) Does the relationship between the two variables define y as a function of x (or a function of the variables indicated)?

(b) If the relationship is a function, indicate what would be a typical domain and range for the function.

25. r = the radius of a circle, A = the area of a circle.

26. L = the length of a side of a square, P = the perimeter of a square.

27. w = the weight of a package of hamburger meat, p = the price of the package.

28. N = a Social Security number of a student at Midtown College, GPA = a student's grade point average.

29. x = a real number, y = square of the number.

30. x = a real number, y = cube of the number.

31. x = a family name, y = a person with that family name. (Notice that the variables are not numbers in this case.)

32. x = a positive real number, y = a number whose square is x.

33. x = the number of boys in a kindergarten class, y = the combined weights of the boys.

34. x = the age of an elementary school girl, y = the height of the girl.

35. x = the number of children in a family, y = the number of boys in a family.

36. x = the price of a book in the bookstore, y = the price of the book rounded to the nearest dollar.

In Exercises 37 through 40, determine the domain and range of the function shown by a graph.

37.

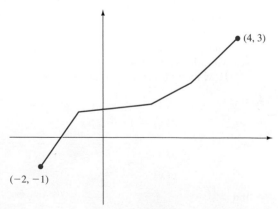

38.

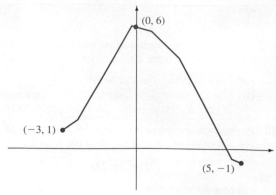

39.

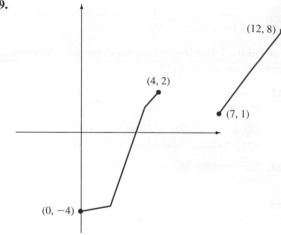

40.

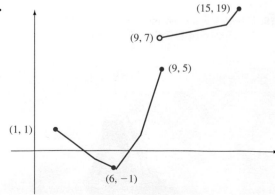

EXPLORATIONS

41. The domain and range of functions can be sets other than sets of numbers. For example, the Fingerprint Function has the set of all people for the domain and the set of all individual's fingerprints for the range. Since each person has fingerprints unique to him or her, there is exactly one set of fingerprints in the range for each person in the domain. Thus, this is a function assuming each person has at least one finger. Would this be a function if someone in the set of people had no hands?

If we reverse the role and use the set of all fingerprints as the domain, then each set of fingerprints belongs to just one person, so it is a function with the range consisting of those people who have fingerprints.

Give three examples of a function for which either the domain or the range is not a set of numbers.

42. The strength of a concrete structure increases with time as the concrete "cures." The percentage of maximum strength attained (p) after d days is given by

$$p = 1.9d + 24$$

(a) Find the percentage of strength attained after five days.

(b) When does p reach 100?

(c) What does your answer in part (b) mean?

43. Bivin is the fitness director at Lake Air Fitness Center. He recommends that the intensity of aerobic activities be measured by pulse rate. Following generally accepted practice, he recommends that the clients achieve a pulse rate depending on their age and level of activity. The target pulse rate is determined as follows: Subtract the person's age from 220 to obtain a "maximum pulse rate." Those beginning aerobic activities should exercise at a level so that their pulse rate (per minute) is 40% of their maximum rate. For burning calories, the person should exercise at a level that achieves 70% of his or her maximum rate.

Write the equations that give the relationship between a person's age and desired pulse rate for

(a) the beginner, and

(b) the person desiring to burn calories.

44. Here is a game to determine a person's age and month of birth. Take an example of an 18-year-old who was born in November. The person does the following computations without telling you anything except the final answer. Have the person start with the number of the month of birth and proceed as follows.

Number of month of birth	11
Multiply by 2	22
Add 5	27
Multiply by 50	1350
Then add the person's age, 18	1368
Subtract 365, giving	1003
Next, add 115	1118

The person gives their result, 1118, whereupon you announce the person's age is 18 (from the last two digits) and the birth month is November (from the first two digits).

Now you are to show how this procedure works in general. Let M represent the number of the birth month and A the age of the person. Write the process described above as a mathematical expression in terms of M and A. From that expression conclude how the first two digits will be the number of the birth month and the last two digits will be the person's age.

45. Beginning drivers soon learn that bringing an automobile to a stop requires greater distances for higher speeds. A formula that approximates the stopping distance under normal driving conditions is

$$d = 1.1v + 0.055v^2$$

where v = speed of the auto in miles per hour and d = distance in feet required to bring the auto to a stop.

(a) Find the stopping distance of an auto traveling 30 mph.

(b) Find the stopping distance of an auto traveling 60 mph.

46. 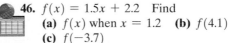 $f(x) = 1.5x + 2.2$ Find
(a) $f(x)$ when $x = 1.2$ (b) $f(4.1)$
(c) $f(-3.7)$

47. 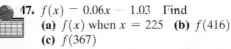 $f(x) = 0.06x - 1.03$ Find
(a) $f(x)$ when $x = 225$ (b) $f(416)$
(c) $f(367)$

48. 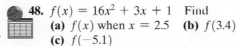 $f(x) = 16x^2 + 3x + 1$ Find
(a) $f(x)$ when $x = 2.5$ (b) $f(3.4)$
(c) $f(-5.1)$

49. 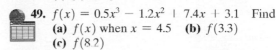 $f(x) = 0.5x^3 - 1.2x^2 + 7.4x + 3.1$ Find
(a) $f(x)$ when $x = 4.5$ (b) $f(3.3)$
(c) $f(8.2)$

50. 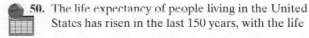 The life expectancy of people living in the United States has risen in the last 150 years, with the life expectancy of females consistently greater than that of males. One approximation, based on data from 1940 through 2000, of life expectancy is given by the linear equation

For males: $y = 0.1958x - 317.846$

For females: $y = 0.2174x - 354.092$

where x is the year a person is born and y is the expected age at death.
(a) Find the approximate life expectancy of a male born in 1980, 1950, 1900, 1850, 2000, 2025, and 2050.
(b) Find the life expectancy of a female born in the same years.
(c) Find your life expectancy.

 51. Based on data from 1981 through 2000, women's earnings as a percentage of men's can be approximated by the linear equation

$$y = 0.69x - 1305$$

where x is the year and y is the percentage. Based on this equation, find the approximate percentage of women's earnings for
(a) 1990   (b) 1950
(c) 2000   (d) 2010

 52. Based on data from 1950 through 1999, the number of deaths (per 100,000 population) due to suicide in the 15- to 24-year-old age range can be approximated by the equation

$$y = 0.1448x - 277.23$$

where x is the year and y is the number of suicides per 100,000 population. Based on this equation, find the approximate suicide rate in
(a) 1950   (b) 1965   (c) 1990
(d) 2000   (e) 2025

 53. Based on the U.S. censuses taken from 1800 through 1990, the population of the United States can be approximated by the equation

$$y = 0.00677x^2 - 24.379x + 21950.4$$

where x is the year and y is the approximate U.S. population, in millions. Using this equation, find the approximate population for
(a) 2000   (b) 1950   (c) 1900   (d) 1850
(e) 1800   (f) 2020   (g) 2050
(h) According to the equation, when will the U.S. population reach 400 million? 500 million?

In Exercises 54 through 60, calculate y for the given values of x. (Use EXCEL.)

54. $y = 4x + 7$, $x = 1, 5, 6, 14$

55. $y = -3x + 21$, $x = 5, 6, 9, 13, 22$

56. $y = x^2 + 2x - 8$, $x = -3, 5, 4.5, 6$

57. $y = 3.2x + 31.6$, $x = 2.4, 4.65, 22.7$

58. $y = 5.4(2x - 54.2)^3 - 119.7$, $x = 29.5$

59. $y = 3x + 7$, $x = 3, 4, 5, 6, 7$

60. $y = \dfrac{2.1x - 3.3}{0.5x + 4.6}$, $x = 1.6, 5.9, 7.1, 8.0, 11.2$

USING YOUR TI-83

The TI-83 can be used to calculate values of a function $y = f(x)$ for a single or several values of x.

Note: The notation using a box such as ENTER *indicates the key to be pressed.*

EXAMPLE

For $y = 7x - 5$, calculate y for $x = 3, 7, 2,$ and -12. Here's how:

1. Select Y= and enter $7x - 5$ as the Y1 function.

    Next, we set up a table that will calculate the values of y for values of x listed in the table.

2. Press TblSet (i.e. 2nd WINDOW) to display the TABLE SETUP screen.

    Enter 0 for **TblStart** and press ENTER .

    Enter 1 for **ΔTbl** and press ENTER .

    Select the **Ask** option for **Indpnt** and press ENTER .

    Select **Auto** for **Depend** and press ENTER .

    You will then have the first screen shown below.

3. Enter the values of x in the table:

    Press TABLE (i.e. 2nd GRAPH) and enter the values of x in the list headed X. As you enter the values of x, the values of $7x - 5$ will appear in the list under Y_1 as shown in the second screen below.

EXERCISES

1. Calculate $y = 6x - 3$ for $x = 1, 5,$ and 9.
2. Calculate $y = 17x + 4$ for $x = 23$.
3. Calculate $y = x^2 + 5$ for $x = -2, 2, 3,$ and 5.
4. Calculate $y = (x - 2)^2$ for $2, 3, 5,$ and 6.
5. Calculate $y = \dfrac{2x + 1}{x - 4}$ for $x = -1/2, 0, 1.5,$ and 5.3.
6. Calculate $y = 1.98x - 3.11$ for $x = 6.16, 8.25,$ and 9.80.

USING EXCEL

A cell in a spreadsheet may be used to record either a number or alphabetic information. For example, to enter 17.5 in cell B3, select the cell and type 17.5.

To enter the current date in cell C2, select the cell and type the current date.

	A	B	C
1			
2			Sept. 18
3		17.5	
4			
5			

FORMULAS

A cell may contain a number or a formula that uses numbers from other cells.

Here's how a spreadsheet adds the numbers in cells B3 and C3 with the answer stored in cell D3.

1. Select the cell D3.
2. Type an = (or click on = in the top bar).
3. Type B3+C3

Click on this when formula is complete or press the return key

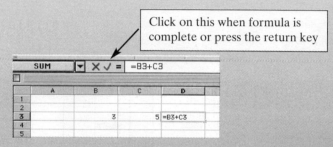

4. Press the return key or click on the check mark in the top bar. The value 8 (3 + 5 in this case) appears in D3. If you change the numbers in B3 or C3, the new result will appear in D3.

EXAMPLE

Calculate $y = 3x + 4$ for x found in A2. Store the result in B2.

1. Type =3*A2+4 in B2

2. Press return or click on the check mark in the top bar.
3. For $x = 5$ in A2, the result $3 \times 5 + 4$ appears in B2.

You may calculate values for $y = 3x + 4$ using several values of x as follows:

1. Enter the values of x (5 values in this case) in A2 through A6.
2. Select cell B2 where =3*A2+4 is stored. Notice that the dark rectangle outlining B2 has a small square hanging on the lower right corner.
3. Place the cursor on the small square, click, and hold down while dragging the cell down to B6.

	A	B	C
1			
2	5	19	=3*A2+4
3	7	25	=3*A3+4
4	-3	-5	=3*A4+4
5	13	43	=3*A5+4
6	44	136	=3*A6+4
7			

Dragging B2 puts these formulas in B2 through B6

The B column now shows the values of y corresponding to the x values in column A.

The next screen shows cell B3 selected and the bar at the top shows the formula in B3 is the same as the one entered in B2, except it uses cell A3 instead of A2. The formulas in B4 through B6 use cells A4 through A6.

	A	B	C
1			
2	5	19	
3	7	25	
4	-3	-5	
5	13	43	
6	44	136	

A formula may be written in standard mathematical notation using +, -, *, and / for addition, subtraction, multiplication, and division.

Example: =(A2-B2)/C2.

Exponentiation is indicated with ^.

Example: =A1^3 indicates x^3 where x is in A1.

EXERCISES

Write the EXCEL formulas for the calculations described in the exercises.

1. Add the numbers in A4 and B4 with the result in C4.
2. Add the numbers in A1, B1, and C1 with the result in D1.
3. Add the numbers in C4 and C5 with the result in C6.
4. Subtract the number in B4 from the number in A4 with the result in C4.
5. Multiply the numbers in B2 and B3 with the result in B4.
6. Divide the number in C2 by the number in D2 with the result in E2.
7. Divide the sum of the numbers in B1 and B2 by 2 with the result in B3.
8. Calculate $2x + 6$ where x is in B3 and the result is in C3.
9. Calculate $2.1x   1.8$ where x is in A5 and the result is in B5.
10. For each of these values of x: 2, 5, -1, and 8 (x's in A1 through A4), calculate $2x - 3$ with the results in B1 through B4.

1.2   Graphs and Lines

- Definition of a Graph
- Linear Functions and Straight Lines
- Slope and Intercept
- Horizontal and Vertical Lines
- Slope-Intercept Equation

- Point-Slope Equation
- Two-Point Equation
- The *x*-Intercept
- Parallel Lines
- Perpendicular Lines

Definition of a Graph

"A picture is worth a thousand words" may be an overworked phrase, but it does convey an important idea. You may even occasionally use the expression "Oh, I see!" when you really grasp a difficult concept. A **graph** of a function shows a picture of a function and can help you to understand the behavior of the function.

A graph often makes it easier to notice trends and to draw conclusions. Let's look at an elementary example. Jason kept a record of the number of e-mail messages he received. He told his roommate that he averaged 12 messages per day over a two-week period. "Tell me more," was his roommate's response. "Were there any days when you received none? What's the most you got? How many times did you get 12 messages? On what days, if any, did you receive 20 messages?"

To satisfy his roommate's curiosity and obtain the information of interest, Jason made a graph showing the number of messages for each day. Figure 1–1 shows the graph.

The graph shows that Jason received at least 7 messages each day. The largest number of messages was 26 on the tenth day. The graph shows no consistent pattern of the number of messages but indicates that there is considerable variation from day to day. He never received exactly 12 messages, although the graph does coincide with 12 messages at about days 4.3, 7.5, 8.5, and 11.5. This doesn't make much sense, but it does illustrate an important point.

The daily number of messages really should be unconnected dots. However, it is much harder to get information from a graph drawn with just dots. Remember those drawings you made as a child by connecting the dots? The picture made a lot more sense after you drew in the connecting lines. In Figure 1–1, we sacrificed some technical accuracy by "connecting the dots" but got a better picture of what happened by doing so. Professional users of mathematics do the same thing; an accountant may let $C(x)$ represent the cost of manufacturing x items, or a manager may let $E(x)$ represent an efficiency index when x people are involved in a large project. In reality, the domains of these functions involve only positive integers. But many methods of mathematics require the domain of the function to be an interval or intervals, rather than isolated points. These methods have proven so powerful in solving problems that people set up their functions using such domains. They then use some common sense in interpreting their answer. If the manager finds that the most efficient number of people to assign to a project is 54.87, she will probably end up using either 54 or 55 people.

Because the picture of a function might help convey the information the function represents, we might ask how the picture of a function, its graph, relates to the rule, or equation, of the function. We obtain a point on a graph from the value

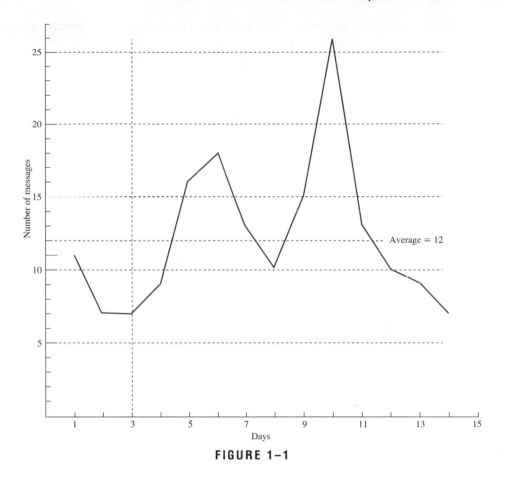

FIGURE 1–1

of a number in the domain, x, and the corresponding value of the function, $f(x)$. We obtain the complete graph by using all numbers in the domain. Here is the definition.

DEFINITION
Graph of a Function

The **graph of a function** f is the set of points (x, y) in the plane that satisfy the equation $y = f(x)$.

Generally, it is impossible to plot *all* points of the graph of a function. Sometimes we can find several points on the graph, and that suffices to give us the general shape of the graph.

EXAMPLE 1 ➤

We write the function

$$f(x) = 2x^2 - 3 \quad as$$
$$y = 2x^2 - 3$$

so that we can relate the x-coordinates and y-coordinates of points on its graph to the equation of the function.

When $x = 2$, we find from $y = 2(2^2) - 3$ that $y = 5$, so the ordered pair $(2, 5)$, is a **solution** to $y = 2x^2 - 3$.

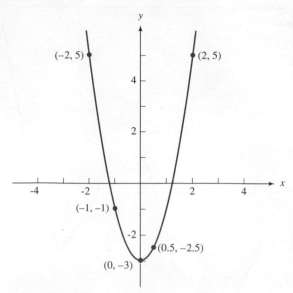

FIGURE 1-2   Graph of $y = 2x^2 - 3$.

Thus, the point $(2, 5)$ is a point on the graph of $y = 2x^2 - 3$. Other solutions include the ordered pairs (points on the graph) $(-2, 5)$, $(0.5, -2.5)$, $(3, 15)$, $(0, -3)$, $(-1, -1)$, and $(5, 47)$. When we plot these points and all other points that are solutions, we have the graph of the function $y = 2x^2 - 3$ (Figure 1–2). Since the graph of $y = 2x^2 - 3$ extends upward indefinitely, we cannot show all points on the graph. We can, however, show enough to convey the shape and location of the graph. ■

As we develop different mathematical techniques throughout this text, we will use some concrete applications. This in turn will require some familiarity with the functions involved and some idea of the shape of their graphs. We start with the simplest functions and graphs.

Linear Functions and Straight Lines

DEFINITION
Linear Function

A function is called a **linear function** if its rule — its defining equation — can be written $f(x) = mx + b$. Such a function is called linear because its graph is a straight line.

From geometry we learned that two points determine a line. One point and the direction of a line also determine a line. We will learn how to find the equation of a line in each of these situations.

EXAMPLE 2 ➤

Let's see how we can draw the graph of $f(x) = 2x + 5$.

SOLUTION

The graph will be a straight line, and it takes just two points to determine a straight line. If we let $x = 1$, then we have $f(1) = 7$; if we let $x = 4$, then $f(4) = 13$. This means that the points $(1, 7)$ and $(4, 13)$ are on the graph of

$f(x) = 2x + 5$. Because we also use y for $f(x)$, we could also say that these points are on the line $y = 2x + 5$. By plotting the points $(1, 7)$ and $(4, 13)$ and drawing the line through them, we obtain Figure 1–3. (We can use any pair of x-values to get two points on the line.) It is usually a good idea to plot a third point to help catch any error. Because $f(0) = 5$, the point $(0, 5)$ is also on the graph of f.

■ **Now You Are Ready to Work Exercise 1**

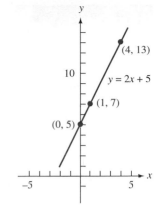

FIGURE 1–3   Graph of $f(x) = 2x + 5$.

> **CAUTION**
>
> The coefficient of x in a general linear equation does not automatically give you the slope of the line. When the equation of the line is in the form $y = mx + b$, the coefficient of x is the slope of the line, and the constant term is the y-intercept. If the equation is in another form, it is a good idea to change to this form to determine the slope and y-intercept.

Slope and Intercept

The equations such as $y = 3x + 8$, $y = -2.5x + 17$, and $y = 12.1x - 62$ are equations of lines, and all are of the form $y = mx + b$. In that form, the constants m and b give key information about the line. The constant b is the value of y that corresponds to $x = 0$, so $(0, b)$ is a point on the line. We call $(0, b)$ the **y-intercept** of the line, because it tells where the line intercepts the y-axis. It is a common practice to shorten the y-intercept notation of $(0, b)$ to just the letter b. Thus, in the linear equation $y = mx + b$, b is called the y-intercept of the line. The other constant, m, determines the direction, or slant, of a line. We call m the **slope** of the line. It measures the relative steepness of a line and will be discussed in detail later.

Notice the equation of the line in Example 2 can be written $y = 2x + 5$, so it has slope $m = 2$, a y-intercept of 5, and passes through $(0, 5)$.

$y = mx + b$

> For a linear function written in the form $y = mx + b$, b is called the **y-intercept** of the line and $(0, b)$ is on the line.
>
> m is called the **slope** of the line and determines the direction and steepness of the line.

EXAMPLE 3 ➤

Find the slope and y-intercept of each of the following lines.

(a) $y = 3x - 5$ **(b)** $y = -6x + 15$

SOLUTION

(a) For the line $y = 3x - 5$, the slope $m = 3$ and the y-intercept $b = -5$.
(b) For the line $y = -6x + 15$, the slope is -6 and the y-intercept is 15.

■ **Now You Are Ready to Work Exercise 7**

Some equations may represent a line even though they are not in the form $y = mx + b$. The next example illustrates how we can still find the slope and y-intercept of those lines.

EXAMPLE 4 ➤

Find the slope and y-intercept of the line $3x + 2y - 4 = 0$.

SOLUTION

We rewrite the equation $3x + 2y - 4 = 0$ in the slope-intercept form, $y = mx + b$, by solving the given equation for y:

$$3x + 2y - 4 = 0$$
$$2y = -3x + 4$$
$$y = -\frac{3}{2}x + 2$$

Thus, the slope-intercept form is $y = -\frac{3}{2}x + 2$. Now we can say that the slope is $-\frac{3}{2}$ and the y-intercept is 2.

■ **Now You Are Ready to Work Exercise 11**

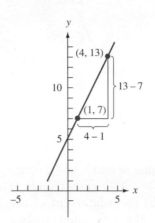

FIGURE 1–4   Graph of $y = 2x + 5$.

Now let's use the linear function $y = 2x + 5$ to illustrate the way the slope relates to the direction of a line. Select two points on the line $y = 2x + 5$, such as $(1, 7)$ and $(4, 13)$. (See Figure 1–4.) Compute the difference in the y-coordinates of the two points: $13 - 7 = 6$. Now compute the difference in x-coordinates: $4 - 1 = 3$. The quotient $\frac{6}{3} = 2$ is m, and the slope of the line $y = 2x + 5$. Following this procedure with any other two points on the line $y = 2x + 5$ will also yield the answer 2.

Examples 3 and 4 illustrate the following general formula that shows how to compute the slope of a line.

Slope Formula

> Choose two points P and Q on the line. Let (x_1, y_1) be the coordinates of P and (x_2, y_2) be the coordinates of Q. The **slope** of the line, m, is given by the equation
>
> $$m = \frac{y_2 - y_1}{x_2 - x_1} = \frac{\text{change in } y}{\text{change in } x} \qquad \text{where} \qquad x_2 \neq x_1$$

 **N O T E**

Notice that it doesn't matter which point we call (x_1, y_1) and which one we call (x_2, y_2); it doesn't affect the computation of m. If we label the points differently in Example 5, the computation becomes

$$m = \frac{1-5}{6-2} = \frac{-4}{4} = -1$$

The answer is the same. Just be sure to subtract the x- and y-coordinates in the same order.

The slope is the difference in the y-coordinates divided by the difference in the x-coordinates.

Figure 1–5 shows the geometric meaning of this quotient.

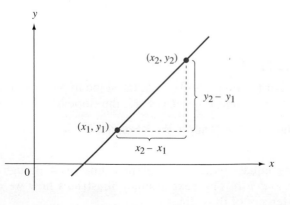

FIGURE 1–5   Geometric meaning of slope quotient.

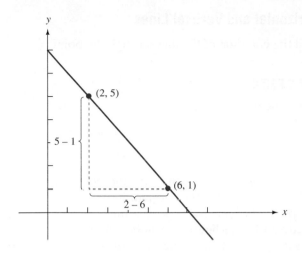

FIGURE 1–6 Geometric relationship of the slope between the points $(2, 5)$ and $(6, 1)$.

We now use the slope formula to obtain the slope of a line through two points.

EXAMPLE 5 ➤

Find the slope of a line through the points $(2, 5)$ and $(6, 1)$.

SOLUTION

Let the point (x_1, y_1) be $(2, 5)$ and (x_2, y_2) be $(6, 1)$. Substituting these values into the definition of m,

$$m = \frac{5 - 1}{2 - 6} = \frac{4}{-4} = -1$$

Figure 1–6 shows the geometric relationship.

■ **Now You Are Ready to Work Exercise 15**

EXAMPLE 6 ➤

Determine the slope of a line through the points $(2, 2)$ and $(7, 10)$.

SOLUTION

For these points

$$m = \frac{2 - 10}{2 - 7} = \frac{-8}{-5} = \frac{8}{5}$$

The geometric relationship is shown in Figure 1–7.

■ **Now You Are Ready to Work Exercise 19**

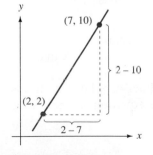

FIGURE 1–7
Geometric relationship of the slope between the points $(2, 2)$ and $(7, 10)$.

The two preceding examples suggest the direction a line takes when the slope is positive and when it is negative.

Now we consider situations when the slope is neither positive nor negative.

Horizontal and Vertical Lines

Find the equation of the line through the points $(3, 4)$ and $(7, 4)$.

SOLUTION

The slope of the line is

$$m = \frac{4 - 4}{7 - 3} = \frac{0}{4} = 0$$

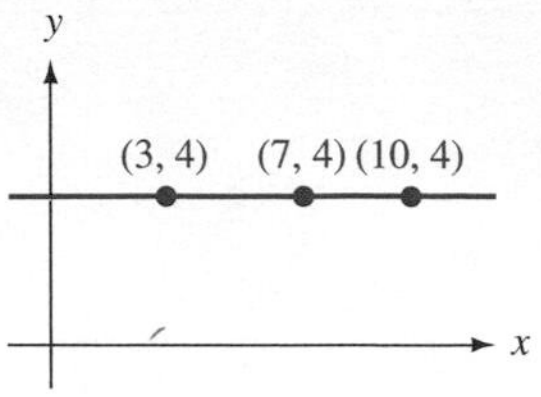

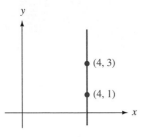

FIGURE 1–8 The horizontal line $y = 4$. Notice the y-coordinate of a point on the line is always 4.

Whenever $m = 0$, we can write the equation $f(x) = 0x + b$ more simply as $f(x) = b$; f is called the **constant function.** The graph of a constant function is a line parallel to the x-axis; such a line has an equation of the form $y = b$ and is called a **horizontal line.** (See Figure 1–8.) Because all points on a horizontal line have the same y-coordinates, the value of b can be determined from the y-coordinate of any point on the line. The equation of the line in this example is $y = 4$.

■ **Now You Are Ready to Work Exercise 23**

When the slope of a line is zero, the line is a horizontal line. Conversely, a horizontal line has slope zero.

Horizontal Line | A **horizontal line** has slope zero.

The next example illustrates a line that has no slope.

Determine the equation of the line through the points $(4, 1)$ and $(4, 3)$.

SOLUTION

We can try to use the rule for computing the slope, but we obtain the quotient

$$\frac{3 - 1}{4 - 4} = \frac{2}{0}$$

FIGURE 1–9 The vertical line $x = 4$. Notice the x-coordinate of a point on the line is always 4.

which doesn't make sense because division by zero is not defined. The slope is not defined. When we plot the two points, however, we have no difficulty in drawing the line through them. See Figure 1–9.

The line, parallel to the y-axis, is called a **vertical line.** A point lies on this line when the first coordinate of the point is 4, so the equation of the line is $x = 4$.

■ **Now You Are Ready to Work Exercise 31**

When the slope of a line is undefined, the line is a vertical line. Conversely, a vertical line has undefined slope.

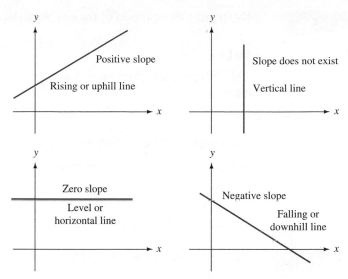

FIGURE 1–10 Relationship of the direction of a line to its slope (reading from left to right).

Vertical Line

The slope of a **vertical line** is undefined.

CAUTION

A vertical line does *not* have a slope, but it does have an equation.

Whenever $x_2 = x_1$, you get a 0 in the denominator when computing the slope, so we say that the *slope does not exist* for such a line.

The slope of a line can be positive, negative, zero, or even not exist. These situations are depicted in Figure 1–10.

This figure shows the relationship between the slope and the slant of the line. If $m > 0$, the graph slants up as x moves to the right. If $m < 0$, the graph slants down as x moves to the right. If $m = 0$, the graph remains at the same height. If m does not exist, the line is vertical, and the line is *not* the graph of a linear function. The equation of a vertical line cannot be written in the form $f(x) = mx + b$ because there is no m.

We conclude this section by showing how to find equations of lines. Linear functions arise in many applied settings. When an application provides the appropriate two pieces of information, you can find an equation of the corresponding line. This information may be given either as:

1. the slope and a point on the line or
2. two points on the line.

We now show each form in a particular application and then give the general method of solving the problem.

Slope-Intercept Equation

When the slope of a line is known and the point given is the y-intercept, the equation of the line is given directly by the $y = mx + b$ form.

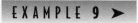

Determine an equation of the line with slope 3 and y-intercept 2.

SOLUTION

This information gives $m = 3$ and $b = 2$ in the equation $y = mx + b$, so the equation is

$$y = 3x + 2$$

The graph of this equation is shown in Figure 1–11.

■ **Now You Are Ready to Work Exercise 39**

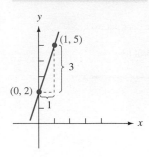

FIGURE 1–11 Graph of a line with slope 3 and y-intercept 2.

When the slope of a line and a point on the line are given, we can find the values of m and b in the slope-intercept form $y = mx + b$.

Determine an equation of the line that has slope -2 and passes through the point $(-3, 5)$.

SOLUTION

The value of $m = -2$, so the line has an equation of form

$$y = -2x + b$$

To complete the solution, we find the value of b. Since the point $(-3, 5)$ lies on the line, $x = -3$ and $y = 5$ must be a solution to $y = -2x + b$. Just substitute those values into $y = -2x + b$ to obtain

$$5 = (-2)(-3) + b$$
$$5 = 6 + b$$
$$5 - 6 = b$$

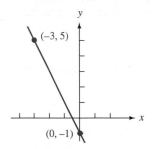

FIGURE 1–12 Graph of a line with slope -2 passing through $(-3, 5)$.

Thus, $b = -1$ and the equation of the line is $y = -2x - 1$. The graph of this equation is shown in Figure 1–12.

■ **Now You Are Ready to Work Exercise 45**

The following table displays some values of x and the corresponding values of y for the line $y = -2x - 1$ from Example 10. The values of x listed increase in steps of 2 in order to point out that y also changes in equal steps (-4). Although the step size for y differs from the step size of x, they are related. Observe that the change in y equals the slope times the change in $x(-4 = -2(2))$. This relationship holds in general for linear functions.

x	y
-2	3
0	-1
2	-5
4	-9
6	-13

We can use another method for writing the equation of a line when the slope and a point are given. This method is called the **point-slope equation.**

Point-Slope Equation

We will work with the information in Example 10, $m = -2$ and the point $(-3, 5)$. Use the slope formula with $(-3, 5)$ as (x_1, y_1) and use an arbitrary point, (x, y), as (x_2, y_2). Because $m = -2$, we can write

$$-2 = \frac{y - 5}{x - (-3)} = \frac{y - 5}{x + 3}$$

Multiply both sides by $x + 3$ to obtain

$$-2(x + 3) = y - 5$$

The formula is usually written as

$$y - 5 = -2(x + 3)$$

Check that this has the same slope-intercept form as we obtained in Example 10.

Notice what we did in this example and observe that the procedure generally holds. When we are given the slope of a line, m, and a specific point on the line, (x_1, y_1), then for any other point on the line, (x, y), the slope of the line can be obtained as $\frac{y - y_1}{x - x_1}$, so we write

$$\frac{y - y_1}{x - x_1} = m$$

Now multiply through by $x - x_1$ to obtain

$$y - y_1 = m(x - x_1)$$

an equation of a line with a given slope and point.

Point-Slope Equation

If a line has slope m and passes through (x_1, y_1), an equation of the line is
$$y - y_1 = m(x - x_1)$$

Now we use the point-slope equation to find the equation of a line.

EXAMPLE 11 ➤

Find an equation of the line with slope 4 that passes through $(-1, 5)$.

SOLUTION

$x_1 = -1$ and $y_1 = 5$; $m = 4$, so the point-slope formula gives us

$$y - 5 = 4(x - (-1)) = 4(x + 1)$$

which can be simplified to $y = 4x + 9$.

■ **Now You Are Ready to Work Exercise 49**

Now let's look at a simple application of a linear function that illustrates the role of the slope and the y-intercept.

EXAMPLE 12 ➤ Carlos started a paper route and decided to add $7 each week to his savings account. By the eighth week, he had $242 in savings. Write his total savings as a linear function of the number of weeks since he started his paper route.

SOLUTION

Let x be the number of weeks and y be the total in savings. Each time x increases by 1 (one week), y increases by 7 ($7 deposit). Thus, the slope equals

$$\frac{\text{change in } y}{\text{change in } x} = \frac{7}{1} = 7$$

NOTE

The slope of the line is the rate of change of y per unit change in x.

At 8 weeks, the savings totaled $242, so the point $(8, 242)$ lies on the line. Using the point-slope formula, we have

$$y - 242 = 7(x - 8)$$

Solving for y, we have $y = 7x + 186$.

Notice that the y-intercept, 186, indicates that Carlos had $186 when he decided to start his periodic savings.

■ **Now You Are Ready to Work Exercise 84**

Two-Point Equation

Just as we have a point-slope formula, we have a formula for finding an equation of a line using the coordinates of two points on the line.

Two-Point Equation

If a line passes through the points (x_1, y_1) and (x_2, y_2), with $x_1 \neq x_2$, an equation of the line is

$$y - y_1 = m(x - x_1)$$

where

$$m = \frac{y_2 - y_1}{x_2 - x_1}$$

Notice that the **two-point equation** is a variation of the point-slope equation. The two given points enable us to find the slope. Then use the slope and one of the given points in the point-slope equation. (It doesn't matter which one of the two points is used.)

EXAMPLE 13 ➤ Determine an equation of the straight line through the points $(1, 3)$ and $(4, 7)$.

SOLUTION

Let (x_1, y_1) be the point $(1, 3)$ and let (x_2, y_2) be the point $(4, 7)$; then

$$m = \frac{7 - 3}{4 - 1} = \frac{4}{3}$$

The point-slope formula gives $y - 3 = \frac{4}{3}(x - 1)$ as an equation of this line, whose graph is shown in Figure 1–13.

■ **Now You Are Ready to Work Exercise 53**

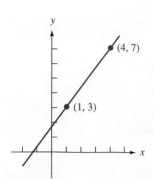

FIGURE 1–13 Graph of a line through $(1, 3)$ and $(4, 7)$.

EXAMPLE 14 ➤

An electric utility computes the monthly electric bill for residential customers with a linear function of the number of kilowatt hours (kWh) used. One month, a customer used 1560 kWh, and the bill was $118.82. The next month, the bill was $102.26 for 1330 kWh used. Find the equation relating kWh used and the monthly bill.

SOLUTION

Let x = the number of kWh used and y = the monthly bill. The information provided gives two points on the line, $(1560, 118.82)$ and $(1330, 102.26)$. The slope of the line is

$$\frac{118.82 - 102.26}{1560 - 1330} = \frac{16.56}{230} = 0.072$$

The equation can now be written as

$$y - 102.26 = 0.072(x - 1330)$$
$$y - 102.26 = 0.072x - 95.76$$
$$y = 0.072x + 6.50$$

■ **Now You Are Ready to Work Exercise 86**

> **NOTE**
>
> There is no one way the equation of a line must be written. This equation can be simplified to $y = \frac{4}{3}x + \frac{5}{3}$. Another form is $4x - 3y = -5$. Each form is correct, but one form may be preferred depending on the situation.

The *x*-Intercept

The **x-intercept** of a line is similar to the y-intercept. It is the x-coordinate of the point where the line crosses the x-axis. To find the x-intercept, set $y = 0$ in the linear equation and solve for x.

EXAMPLE 15 ➤

Find the x-intercept of the line $4x - 9y = 30$.

SOLUTION

Let $y = 0$. We have $4x - 9(0) = 30$ and $x = \frac{30}{4} = 7.5$. The x-intercept is 7.5, and the line crosses the x-axis at $(7.5, 0)$. (See Figure 1–14.)

■ **Now You Are Ready to Work Exercise 59**

We conclude this section with a discussion of parallel and perpendicular lines.

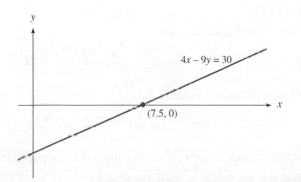

FIGURE 1–14   The x-intercept of $4x - 9y = 30$.

Parallel Lines

Two lines are parallel if they have the same slope or if they are both vertical lines.

EXAMPLE 16 ➤   Determine if the line through $(2, 3)$ and $(6, 8)$ is parallel to the line through $(3, 1)$ and $(11, 11)$.

SOLUTION

Let L_1 be the line through $(2, 3)$ and $(6, 8)$ and let m_1 be its slope. Then,

$$m_1 = \frac{8 - 3}{6 - 2} = \frac{5}{4}$$

Let L_2 be the line through $(3, 1)$ and $(11, 11)$ and let m_2 be its slope. Then,

$$m_2 = \frac{11 - 1}{11 - 3} = \frac{10}{8} = \frac{5}{4}$$

The slopes are identical, so the lines are parallel. (See Figure 1–15.)

FIGURE 1–15   Lines that have the same slope are parallel.

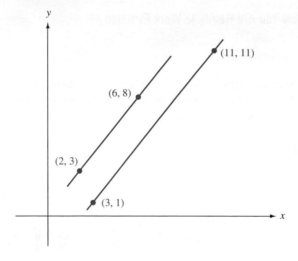

■ **Now You Are Ready to Work Exercise 63**

EXAMPLE 17 ➤   Is the line through $(5, 4)$ and $(-1, 2)$ parallel to the line through $(3, -2)$ and $(6, 4)$?

SOLUTION

The slope of the first line is

$$m_1 = \frac{4 - 2}{5 - (-1)} = \frac{2}{6} = \frac{1}{3}$$

The slope of the second line is

$$m_2 = \frac{-2 - 4}{3 - 6} = \frac{-6}{-3} = 2$$

Since the slopes are not equal, the lines are not parallel.

■ **Now You Are Ready to Work Exercise 65**

EXAMPLE 18 ➤

When we have the equation of a line, we can find the equation of a parallel line, provided we know a point on the second line. We illustrate this by finding the line through the point $(8, 5)$ that is parallel to the line $4x - 3y = 12$. We know a point on the line — namely, $(8, 5)$ — and we can find the slope from the given line $4x - 3y = 12$ by putting it in the slope-intercept form:

$$4x - 3y = 12$$
$$y = \frac{4}{3}x - 4$$

The desired slope is $\frac{4}{3}$, so the equation of the parallel line is found from

$$y - 5 = \frac{4}{3}(x - 8)$$
$$y = \frac{4}{3}x - \frac{17}{3}$$

We can put this in another standard form by multiplying by 3

$$3y = 4x - 17 \quad \text{or} \quad 4x - 3y = 17$$

Notice that the coefficients of x and y are the same as those in the given line. This generally holds, giving us another way to find the equation of the parallel line using the knowledge that its equation is of the form

$$4x - 3y = some\ constant$$

Since the point $(8, 5)$ lies on the line, it must give the desired constant when the coordinates are used for x and y; that is,

$$4(8) - 3(5) = the\ constant$$

and the equation is

$$4x - 3y = 17 \quad \blacksquare$$

Perpendicular Lines

When two lines are **perpendicular,** their slopes always relate in the following way.

Perpendicular Lines

Two lines with slopes m_1 and m_2 are perpendicular if and only if

$$m_1 m_2 = -1$$

or in another form

$$m_2 = -\frac{1}{m_1}$$

EXAMPLE 19 ➤

The lines

$$y = -\frac{2}{3}x + 17 \quad \text{and} \quad y = \frac{3}{2}x - 9$$

are perpendicular because the product of their slopes is

$$-\frac{2}{3} \times \frac{3}{2} = -1$$

The lines

$$3x + 5y = 7 \quad and \quad 11x - 3y = 15$$

are not perpendicular because the product of their slopes

$$-\frac{3}{5} \times \frac{11}{3} = -\frac{11}{5}$$

is not -1.

■ **Now You Are Ready to Work Exercise 71**

Summary of Equations of a Linear Function

> **Slope:** $m = \dfrac{y_2 - y_1}{x_2 - x_1}$, where (x_1, y_1) and (x_2, y_2) are points on the line with $x_1 \neq x_2$.
>
> **Standard Equation:** $Ax + By = C$ where (x, y) is any point on the line and at least one of A, B is not zero.
>
> **Slope-Intercept Equation:** $y = mx + b$, where m is the slope and b is the y-intercept.
>
> **Point-Slope Equation:** $y - y_1 = m(x - x_1)$, where m is the slope and (x_1, y_1) is a given point on the line.
>
> **Horizontal Line:** $y = k$, where (h, k) is a point on the line and all points on the line have the same y-coordinate, k.
>
> **Vertical Line:** $x = h$, where (h, k) is a point on the line and all points on the line have the same x-coordinate, h.

⬛ 1.2 EXERCISES

LEVEL 1

Draw the graphs of the lines in Exercises 1 through 6.

1. *(See Example 2)*
$f(x) = 3x + 8$

2. $f(x) = 4x - 2$

3. $f(x) = x + 7$

4. $f(x) = -2x + 5$

5. $f(x) = -3x - 1$

6. $f(x) = \dfrac{2}{3}x + 4$

Find the slope and y-intercept of the lines in Exercises 7 through 10.

7. *(See Example 3)*
$y = 7x + 22$

8. $y = 13x - 4$

9. $y = \dfrac{-2}{5}x + 6$

10. $y = \dfrac{-1}{4}x - \dfrac{1}{3}$

Find the slope and y-intercept of the lines in Exercises 11 through 14.

11. *(See Example 4)*
$2x + 5y - 3 = 0$

12. $4x + y - 3 = 0$

13. $x - 3y + 6 = 0$

14. $5x - 2y = 7$

Determine the slopes of the straight lines through the pairs of points in Exercises 15 through 18.

15. *(See Example 5)*
$(1, 2), (3, 4)$

16. $(2, 3), (-3, 1)$

17. $(-4, -1), (-1, -5)$

18. $(2, -4), (6, -3)$

For the graphs shown in Exercises 19 through 22 indicate whether the lines have positive, negative, or zero slope.

19. *(See Example 6)*

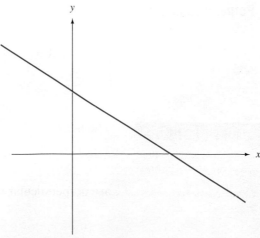

20.

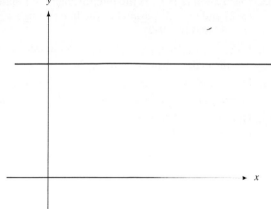

21.

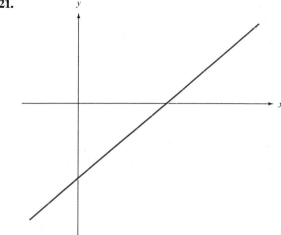

22.

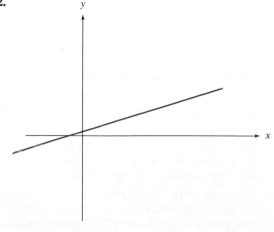

Find the equations of the lines in Exercises 23 through 26 through the given points.

23. *(See Example 7)*
(5, −2), (−3, −2)

24. (8, 3), (1, 3)

25. (−1, 0), (−4, 0)

26. (0, 0), (17, 0)

Graph the lines in Exercises 27 through 30.

27. $y = 6$

28. $y = -2$

29. $y = 4.5$

30. $y = -3.5$

Find an equation of the line through the given points in Exercises 31 through 34.

31. *(See Example 8)*
(3, 2), (3, 5)

32. (−4, 6), (−4, 9)

33. (10, 0), (10, 7)

34. (−6, −1), (−6, 13)

Graph the lines in Exercises 35 through 38.

35. $x = 8$

36. $x = -1$

37. $x = 1.5$

38. $x = -4.25$

Find the equations of the lines in Exercises 39 through 44 with the given slope and y-intercept.

39. *(See Example 9)* Slope 4 and y-intercept 3

40. Slope −2 and y-intercept 5

41. $m = -1, b = 6$

42. $m = \dfrac{-3}{4}, b = 7$

43. $m = \dfrac{1}{2}, b = 0$

44. $m = 3.5, b = -1.5$

Find the equations of the lines in Exercises 45 through 48 with the given slope and passing through the given point.

45. *(See Example 10)* Slope −4 and point (2, 1)

46. Slope 6 and point (−1, −1)

47. Slope $\dfrac{1}{2}$ and point (5, 4)

48. Slope −1.5 and point (2.6, 5.2)

Use the point-slope formula to find the equations of the lines with given point and slope in Exercises 49 through 52.

49. *(See Example 11)* Slope 7 and point (1, 5)

50. Slope −2 and point (3, 1)

51. Slope $\dfrac{1}{5}$ and point (9, 6)

52. Slope $\dfrac{-2}{3}$ and passing through (−1, 4)

In Exercises 53 through 58, determine the equations of the straight lines through the pairs of points, and sketch the lines.

53. *(See Example 13)*
(−1, 0), (2, 1)

54. (3, 0), (1, −1)

55. (0, 0), (1, 2)

56. (1, 3), (1, 5)

57. (2, 4), (5, 4)

58. (0, 3), (7, 0)

Find the x- and y-intercepts of the lines in Exercises 59 through 62 and sketch the graph.

59. *(See Example 15)*
$5x - 3y = 15$

60. $6x + 5y = 30$

61. $2x - 5y = 25$

62. $3x + 4y = 15$

Exercises 63 through 83 deal with parallel and perpendicular lines to help you understand the concept and properties of parallel and perpendicular lines.

63. *(See Example 16)* Is the line through the points $(8, 2)$ and $(3, -3)$ parallel to the line through $(6, -1)$ and $(16, 9)$?

64. Is the line through $(9, -1)$ and $(2, 8)$ parallel to the line through $(3, 5)$ and $(10, -4)$?

65. *(See Example 17)* Is the line through $(5, 4)$ and $(1, -2)$ parallel to the line through $(1, 2)$ and $(6, 8)$?

66. Determine whether the line through $(6, 2)$ and $(-3, 5)$ is parallel to the line through $(4, 1)$ and $(0, 5)$.

LEVEL 2

Determine whether the pairs of lines in Exercises 67 through 70 are parallel.

67. $y = 6x + 22$
$y = 6x - 17$

68. $3x + 2y = 5$
$6x + 4y = 15$

69. $x - 2y = 3$
$2x + y = 1$

70. $3x - 5y = 4$
$-6x + 10y = -8$

Determine if the pairs of lines in Exercises 71 through 74 are perpendicular.

71. *(See Example 19)*
$y = -2x + 5$
$y = 0.5x - 8$

72. $6x + 5y = 17$
$10x - 12y = 22$

73. $y = 3x + 14$
$y = 5x - 2$

74. $2x - 7y = 13$
$7x - 2y = 21$

75. Write the equation of the line through $(-1, 5)$ that is parallel to $y = 3x + 4$.

76. Write the equation of the line through $(2, 6)$ that is parallel to $3x + 2y = 17$.

77. Write the equation of the line with y-intercept 8 that is parallel to $5x + 7y = -2$.

78. Write the equation of the line through $(6, 2)$ that is parallel to $5x - 2y = 20$.

79. Write the equations of the parallel lines obtained in Exercises 76 through 78 in the form $ax + by = c$. How do the values of a and b compare with the coefficients of x and y in the given equation?

80. Find an equation of the line passing through the point $(0, 0)$ with slope 2.3.

81. Find the y-intercept of the line passing through $(2, 5)$ and having slope $\frac{2}{3}$.

82. Write the equation of the line through $(5, 7)$ and perpendicular to $y = 0.25x + 16$.

83. Write the equation of the line through $(2, 3)$ and perpendicular to the line $5x + 3y = 22$.

LEVEL 3

Many of the following problems can be modeled with a linear equation. Analyze them and identify what quantities the variables x and y represent. The information provided in the problems is adequate to determine a point and a slope or two points on the line.

84. *(See Example 12)* Jane got a regular job and started adding $9 per week to her savings account. At the end of 11 weeks, she has $315 in savings. Write her savings as a linear function of the number of weeks since she started the job.

85. Raul is on a carefully supervised diet that causes a weight loss of 3 pounds per week. After 14 weeks, his weight is 196 pounds.

(a) Does the information provided give two points or a point and a slope?

(b) Write Raul's weight as a linear function of weeks on the diet.

(c) What was his weight when the diet started?

86. *(See Example 14)* Valley Electric Utility computes the monthly electric bill for residential customers with a linear function of the number of kilowatt hours (kWh) used. One month, a customer used 1170 kWh, and the bill was $100.02. Another month, the bill was $120.27 for 1420 kWh used.

(a) Does the information provided give two points or a point and a slope?

(b) Find the equation relating kWh used and the monthly bill.

87. Robinson Wholesale Company paid $1340 for 500 items. Later they bought 800 items for $1760. Assume that the cost is a linear function of the number of items. Write the equation.

88. The average cost of attending a four-year public college or university was $5138 in 1991 and was $7621 in 2000.
    (a) Find the average cost as a linear function of time since 1991.
    (b) Use the linear function to estimate the average cost in 2005.

89. The slope and one point on a line are given. A coordinate of a second point is missing. Find the missing coordinate so that the second point is on the line.
    (a) $m = 3$, $(2, 3)$, $(-1,\quad)$
    (b) $m = -4$, $(1, 2)$, $(\quad, 3)$
    (c) $m = \dfrac{3}{4}$, $(-2, 0)$, $(5,\quad)$
    (d) $m = -\dfrac{1}{2}$, $(-1, -3)$, $(\quad, 4)$

90. Brian requires 3000 calories per day to maintain his daily activities and to maintain a constant weight. He wants to gain weight. Each pound of body weight requires an additional 3500 calories. Write his daily calorie intake as a linear function of the number of pounds gained per day.

91. The average cost of attending a four-year private college or university in 1991 was $13,892 and $21,423 in 2000.
    (a) Find the average cost as a linear function of time since 1991.
    (b) Use the linear function to estimate the average cost in 2005.

92. Temperature measured in degrees Celsius (C) is related to temperature measured in degrees Fahrenheit (F) by a linear equation. Water boils at 100°C and 212°F. Water freezes at 0°C and 32°F. Find the linear equation relating Celsius and Fahrenheit temperatures.

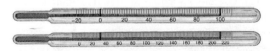

93. Alpine College plans to increase tuition $50 per semester hour each year. In 2001, the tuition was $375 per semester hour. Find a linear function that describes tuition as a function of time since 2001.

94. In 1995, Angie paid $48 for a textbook for English 2302. In 2002, her brother paid $69 for a textbook for the same course. Assume the price of a textbook

increases linearly as a function of time since 1995. Write a linear equation that describes this relation.

95. City Electric Company calculates a customer's monthly electricity bill as $5.00 plus 7.8 cents for each kWh of electricity used.
    (a) Does the information provided give two points or a point and a slope?
    (b) Write the linear equation relating kWh used and the monthly bill.

96. An electric company calculates a customer's monthly electricity bill as $7.50 plus 8.2 cents for each kWh of electricity used over 50 kWh. Write the linear equation relating kWh used and the monthly bill. (Assume that at least 50 kWh is used.) Explain why you should use at least 50 kWh.

97. The slope of a line is the amount of change in y when x increases by 1 unit. For $y = 3x + 7$, y increases by 3 when x increases by 1. For $y = -2x + 4$, y decreases by 2 when x increases by 1. (The negative value indicates a decrease in y.) Find the change in y when x increases by 1:
    (a) $y = 4x - 5$ (b) $y = -3x + 4$
    (c) $y = \dfrac{2}{3}x + 7$ (d) $y = \dfrac{-1}{2}x + \dfrac{2}{5}$
    (e) $3y + 2x - 4 = 0$ (f) $y = 17$

98. Linda needs 2100 calories per day to maintain her present weight and daily activities. She wants to lose x pounds per day. Each pound of body weight is equivalent to 3500 calories. Write her daily calorie intake as a linear function of the number of pounds of weight lost per day.

99. Executive Auto Rental charges a fixed daily rate and a mileage charge. One customer rents a car for one day and drives it 125 miles. His bill is $35.75. Another customer rents a car for one day and drives it 265 miles. Her bill is $51.15. Write the linear equation relating miles driven and total cost.

100. When their daughter Rhonda was studying in Japan, the Ward family enrolled in their phone company's International Call Plan with a monthly fee of $4.95 and a cost of 12 cents per minute called. Find the linear equation that relates total monthly cost to the number of minutes called.

101. The American Automobile Manufacturers' Association estimated that 536,000 passenger cars were exported in 1997.
    (a) One analyst expected exports to decrease by 15,000 passenger cars per year. Find the equation for this linear trend.
    (b) Another analyst expected exports to increase by 4500 passenger cars per year. Find the equation for this linear trend.

102. The U.S. Department of Commerce reported in a *Survey of Current Business* that the per capita income in Alabama was $7465 in 1980 and was $23,521 in 2000.

    (a) Assume this is a linear trend and write the linear equation relating years and per capita income.

    (b) The per capita income in Alabama was $19,683 in 1995. Does this conform to the linear trend found in (a)?

103. The IRS Form 1040 for 2002 shows that the income tax on a taxable income in the $27,950 to $67,700 range is $3892.50 plus 27% of the taxable income over $27,950. (This is for a single person.) Let x be the taxable income and y the tax paid. Write the linear equation relating taxable income and tax in that income range.

104. Central Waste Management has a fleet of 35 trucks that are able to haul 178 tons of trash per day. When they increase the number of trucks to 47, they can haul 230 tons of trash per day. When the company contracts to pick up trash for the city of Bellmead, the amount of trash is expected to increase to 255 tons per day. Estimate the number of trucks that will be required.

105. The IRS Form 1040 for 2002 shows, for a married couple filing jointly, that the income tax on a taxable income in the $12,000 to $46,700 range is $1200 plus 15% of the taxable income over $12,000. Let x be the taxable income and y the tax paid. Write the linear equation relating taxable income and tax in that income range.

106. The Colorado 2002 state income tax for taxable income over $25,000 is $1157 plus 4.67% of taxable income over $25,000. Write the linear equation relating taxable income and tax in that income range.

EXPLORATIONS

107. The birth rate in Japan has declined for a number of years. The birth rates per 1000 population for three different years are

        1980   13.7

        1985   11.9

        2002   10.03

    (a) Use the 1980 and 2002 figures to find birth rate as a linear function of number of years since 1980.

    (b) Use the linear function to estimate the birth rate for 1985. Does it give a realistic rate?

    (c) Use the linear function to determine when the birth rate will reach zero. Comment on the reasonableness of the result.

108. The U.S. Census Bureau reports a wide variety of population information. One bit of information is the percentage of the population that never marries.

    (a) In 1980, the percentage of males in the 20–24 age group who had never married was 68.8%. In 2000 the percentage rose to 83.7%.

        (i) Based on this information, find the percentage of unmarried males as a linear function of time in years.

        (ii) Use the linear function you found to estimate the percentage of males, ages 20–24, who never married for the year 1998.

        (iii) The actual percentage given by the Census Bureau for 1998 was 83.4%. How does this compare with your estimate? Does the linear function found in part (i) seem to be a reasonably good representation of the growth of the percentage?

    (b) In 1980, 50.2% of the female population in the 20–24 age range had never married. The percentage rose to 72.8% in 2000.

        (i) Based on this information, find the percentage of females who had never married as a linear function of time in years since 1980.

        (ii) Use the linear function just found to estimate the percentage of females, ages 20–24, who had never married for the year 1998.

        (iii) The actual percentage reported by the Census Bureau for 1998 was 70.3%. Compare the estimate obtained in part (ii). Does the linear function seem to provide a reasonable estimate of the growth of the percentage of females who never married?

109. The manager of the Ivy Square Cinema observed that Friday night estimated ticket sales were 185 when the admission was $5. When the admission was increased to $6, the estimated attendance fell to 140.

    (a) Use this information to find estimated attendance as a linear function of admission price.

    (b) Use the function found in part (a) to estimate attendance if admission is increased to $7.

    (c) Use the function found in part (a) to determine the admission price that will yield an estimated attendance of 250.

    (d) The manager had no desire to set an admission price that would drive away all customers, but she was curious to know when that would occur. Based on the function in part (a), find the admission price that would result in a zero attendance. (Does this seem reasonable to you?)

(e) The manager was also curious to know what kind of attendance would result from free admission to all. (Popcorn sales should zoom upward.) Based on the function in part (a) find the estimated attendance for the manager. (Does this seem reasonable to you?)

110. The linear trend of the median age of men when they first married has been approximated from Census Bureau data for the years 1950–2000. The linear equation is

$$y = 0.099x + 22.1$$

where x is the number of years since 1950 ($x = 0$ for 1950) and y is the median age.
   (a) Use this to estimate the median age of marriage in
      (i) 1975
      (ii) 2000
      (iii) 2050
   (b) Do you think this is a realistic model over a long period of time?

111. On October 14, 2002, Audrey Mestre, a French woman, died attempting to break the world record in free diving. In free diving, a diver, without oxygen, uses a 200-pound weight attached to a cable to pull them into the ocean depths. Ms. Mestre's goal was a depth of 561 feet, which would have set a new record. As a diver descends, the pressure increases due to the weight of the water above. At 18 feet, the water pressure is approximately 8 pounds per square inch and increases to approximately 40 pounds per square inch at 90 feet. Use a linear function to approximate the water pressure at 561 feet depth.

112. Find a linear function so that as x changes in equal increasing steps, the y-coordinates of the points on the line *increase* in the same size steps as x. In general, what form must this function take?

113. Find a linear function so that as x changes in equal increasing steps, the y-coordinates of the points on the line *decrease* in the same size steps as x. In general, what form must this function take?

114. One standard form of an equation of a straight line is $Ax + By = C$. In this form show that:

   (a) The slope of the line is $-\dfrac{A}{B}$.

   (b) The y-intercept is $\dfrac{C}{B}$.

   (c) The x-intercept is $\dfrac{C}{A}$.

115. The 10 o'clock evening news forecast steadily falling snow later that night. Sure enough, 2 inches had fallen at 6 A.M. when Andy went out to get the pa-

per. When he left for work at 8 o'clock, $4\frac{1}{2}$ inches had fallen. By noon, the snow was 10 inches deep. Plot a graph of these points and connect them to form a graph. Use the graph to estimate the time the snow started falling.

116. Compare the graph of $y = 1.5x + 2$ with the graphs of $y = 1.5x + b$ using $b = -1, 3, 4.5$, and 7. How are the graphs related?

117. Compare the graphs of $y = 2x + 4$ and $y = mx + 4$ using $m = 1, -2, -\frac{1}{2}, 0$, and 3. What do the graphs have in common and how do they differ?

118. Compare the graphs of $y = mx + b$ using $m = 0$ and $b = 1, -5, 6$, and 4. What is a common feature of the graphs?

119. Compare the graphs of $y = mx + b$ using $b = 0$ and $m = 3, -2$, and 0.25. How are the graphs related?

120. Graph each of the following lines and observe the direction of the line.
   (a) $y = 4x + 5$ **(b)** $y = -2x + 3$
   (c) $y = 0.2x - 4$ **(d)** $y = -0.5x + 3$
   (e) $y = 7$ **(f)** $y = 2.6x + 1.9$
   (g) $y = -3.4x - 2.1$
   (h) What information in the equation of a line determines if the line is an "uphill" line or a "downhill" line?

121. Find the slope of the line through the points $(3, 8)$ and $(6, -2)$.

122. Find the slope of the line through the points $(5.3, 1.9)$ and $(9.7, 13.5)$.

123. Find the slope of the line through the points $(156.2, 298.5)$ and $(644.4, 903.8)$.

124. Graph the line through the two points $(2, 3)$ and $(8, -1)$ and find its equation.

125. Graph the line through the two points $(5, 4)$ and $(10, 16)$ and find its equation.

126. Graph the line through the two points $(1.5, 3.4)$ and $(4.7, 1.6)$ and find its equation.

127. Graph the line through the two points $(21.65, 13.42)$ and $(5.75, 7.84)$ and find its equation.

128. The linear trend of the median age of women when they first married has been approximated from Census Bureau data for the years 1950 to 2000. The linear equation is

$$y = 0.114x + 19.4$$

where x is the number of years since 1950 ($x = 0$ for 1950) and y is the median age. Use this to estimate the median age of marriage in
   (a) 1975   **(b)** 2025
   (c) 2050   **(d)** 2075

USING YOUR TI-83

Graphs of Lines and Evaluating Points on a Line

The graphing of a line, or lines, on a graphics calculator requires these steps:

A. Set the x and y ranges for the window.

B. Enter the linear equation(s).

C. Select the GRAPH command.

Here are examples of the TI-83 screens resulting from these steps using the linear equations

$$y = 2x + 5$$
$$3x + 4y = 12$$

Note: the second equation must be solved for y to enter it. It may be written as

$$y = (12 - 3x)/4 \qquad \text{or} \qquad y = -.75x + 3$$

A. Settings for the window to show x and y ranges from -10 to 10, with tick marks 2 units apart. The WINDOW key will select the screen.

B. The y= key selects the screen for entering the equations $y = 2x + 5$ and $y = (12 - 3x)/4$.

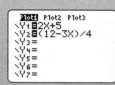

C. The GRAPH key activates the graphing of the equations.

EXERCISES

Graph the lines in Exercises 1 through 8.

1. $y = 5x + 4$ **2.** $y = 0.5x - 6$ **3.** $y = -1.4x + 8.2$

4. $y = -0.65x + 7.3$ **5.** $5x - 2y = 12$ **6.** $3x + y = 8$

7. $2.4x + 5.3y = 15.6$ **8.** $3.3x - 7.2y = 22.8$

You can make the cursor trace the graph of the line by pressing TRACE and using the < and > keys to move the cursor to the left or right along the line. The x and y coordinates of the point where the cursor lies will show on the screen. For example, for the line $y = 2x - 3$, the value of y where $x = 4.68$ (to two decimals) is shown here.

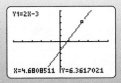

9. For the line $y = 3.86x + 1.22$, use ⌐TRACE⌐ to find the value of y when $x = 4.5$. If the cursor does not fall exactly on 4.5, use the nearest value of x.

10. For the line $y = 1.98x - 3.11$, use ⌐TRACE⌐ to find the value of x for which $y = 6.16$.

11. Use ⌐TRACE⌐ to find the x- and y-intercepts of the line $y = 4.6x + 7.2$.

12. Find the x- and y-intercepts for the line $3.8x + 5.4y = 29.7$.

USING EXCEL

We now illustrate how to find the slope of a line through two points and the slope-intercept equation of the line.

EXAMPLE 1

Find the slope of the line through the points $(3, 8)$ and $(4, 5)$.

Solution

We enter the coordinates of the first point on line 2, and the coordinates of the second point on line 3, and the slope in cell C3. Line 1 is used to identify the entries in the columns. Here is the screen that we get.

	A	B	C	D	E
1	X	Y	Slope		
2	3	8			The formula
3	4	5	-3		=(B3-B2)/(A3-A2)
4					is entered in C3
5					
6					
7					

EXAMPLE 2

Find the slope-intercept equation of the line through the points $(2, 5)$ and $(4, 9)$.

Solution

The y-intercept of a line, b, can be found from the slope of the line, m, and a point on the line, (x_1, y_1) using

$$b = y_1 - mx_1$$

(continued)

Here is the spreadsheet using Line 1 to identify the entries in each column, entering the coordinates of the first point in Line 2, the coordinates of the second point in Line 3, the calculated slope in C3, and the y-intercept in D3.

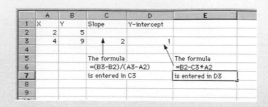

The slope-intercept line is

$$y = 2x + 1$$

EXERCISES

1. Find the slope of the line through the points $(6, 5)$ and $(-2, 3)$.

2. Find the slope of the line through the points $(3, 8)$ and $(6, -2)$.

3. Find the slope of the line through the points $(5.3, 1.9)$ and $(9.7, 13.9)$.

4. Find the slope of the line through the points $(156.2, 298.5)$ and $(644.4, 903.8)$.

5. Find the equation of the line through the points $(2, 3)$ and $(8, -1)$.

6. Find the equation of the line through the points $(5, 4)$ and $(10, 16)$.

7. Find the equation of the line through the points $(1.5, 3.4)$ and $(4.7, 1.6)$.

Finding the Graph and Equation of a Line Through Two Points

A series of EXCEL menu selections will give the graph of a line through two points and find the equation of the line.

EXAMPLE 3

Graph the line through the points $(1, 4)$ and $(3, 8)$.

Solution

Here are the steps to find the graph and equation.

- Enter the first point in cells A2:B2 (x in A2, y in B2) and the second point in A3:B3.
- Select the cells, A2:B3, containing the points.
- Click on the **Chart Wizard** icon in the Tool Bar.
- Under **Chart type,** select **xy(Scatter).**
- Under **Chart Sub-type,** select the lower left pattern.
- Click **Next.**
- On the **Data range** tab, select **Columns** for **Series in:**
- Click **Next.**

- Select the **Legend** tab and make sure there is no check for **Show legend.**
- Click **Finish.**
- Under **Chart** in the tool bar, select **Add Trendline.**
- Select the Options tab and check **Display equation on chart.**
- Click **OK.**

Here is the result of the above sequence.

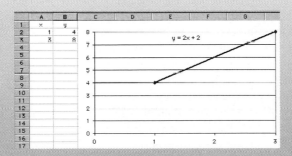

Once you have this sequence set up on your spreadsheet, you can change the points, and the graph and equation will automatically give the new graph and equation.

EXERCISES

1. Find the graph and equation of the line through the points (3, 5) and (8, 12).

2. Find the graph and equation of the line through the points (1, 3) and (5, –11).

3. Find the graph and equation of the line through the points (–2.2, –1.4) and (4.2, 6.6).

1.3   Mathematical Models and Applications of Linear Functions

- Applications
- Cost–Volume Function
- Revenue Function

- Break-Even Analysis
- Straight-Line Depreciation
- Applications of Linear Inequalities

Nat Hambrick is a contractor in a competitive construction market. He needs to answer such questions as, "We want to add another floor to the top of the Amico Building for an elite restaurant. Will the present building support it?" "Will this roof design support a 2-foot snowfall?" "Will the proposed office tower withstand 100-mile-per-hour winds?"

The building trade knows how to handle many problems from past experiences of successes and failures. A builder may avoid a failure by deliberately overdesigning, but that may drive up costs until the builder cannot compete. Because a trial-and-error approach may result in costly failures or expensive successes, an approach that uses proper mathematical analyses can provide valuable answers. Engineers have developed mathematical equations and formulas, called **mathematical models,** that can be used to answer "what if?" questions of building design.

Mathematical models abound in numerous other disciplines. They describe scientific phenomena, population growth, sociological trends, economic growth, and product costs. Politicians depend on mathematically based opinion polls to determine campaign strategy.

One of the more famous mathematical models is the equation $E = mc^2$, which was chalked on sidewalks of campuses across the nation the day after Albert Einstein died. Written in tribute to his profound work on the theory of relativity, this equation relates energy, E, the mass of a body, m, and the speed of light, c. This equation does not tell all about energy, but it describes the energy released when matter is fully transformed into energy. It helps predict the energy released from uranium and the energy stored in the Sun.

Although mathematical models generally only approximate the real situations, they can be useful in making decisions and estimating the consequences of these decisions: A mathematical analysis can help a hospital administrator determine the most economical order for the quantity of X-ray films, can give a proud new grandparent an estimate of the worth of an investment when the grandchild begins college, or can help the phone company plan for the growing demand for cell phones.

Applications

In practical problems, the relationship between the **variables** can be quite complicated. For example, the variables and their relationship that affect the stock market still defy the best analysts. However, many times a linear relationship can be used to provide a reasonable and useful model for solving practical problems. We now look at several applications of the linear function.

Cost–Volume Function

The manufacturer of a home theater system conducted a study of production costs and found that fixed costs averaged $5600 per week and component costs averaged $359 per system. This information can be stated as

$$C = 359x + 5600$$

where x represents the number of systems produced per week, also called the **volume,** and C is the total weekly cost of producing x systems. A linear function like this is used when:

1. there are **fixed costs** — such as rent, utilities, and salaries — that are the same each week, independent of the number of items produced;
2. there are **variable costs** that depend on the number of items produced, such as the cost of materials for the items, packaging, and shipping costs.

The home theater system example illustrates a linear **cost–volume function** (often simply called the *cost function*). A linear function is appropriate when the general form of the cost function C is given by

$$C(x) = ax + b$$

where

x is the number of items produced (**volume**),

b is the **fixed cost** in dollars,

a is the **unit cost** (the cost per item) in dollars,

$C(x)$ is the **total cost** in dollars of producing x items.

Notice the form of the cost function. It is essentially the slope-intercept form of a line, where the slope is the unit cost and the intercept is the fixed cost.

EXAMPLE 1 ➤

If the cost of manufacturing x home theater systems per week is given by

$$C(x) = 415x + 5200$$

then:

(a) Determine the unit cost and the fixed cost.
(b) Determine the cost of producing 700 systems per week.
(c) Determine how many systems were produced if the production cost for one week was $230,130.

SOLUTION

(a) The unit cost is $415, and the fixed cost is $5200 per week.
(b) Substitute $x = 700$ into the cost equation to obtain

$$
\begin{aligned}
C(700) &= 415(700) + 5200 \\
&= 290,500 + 5200 \\
&= 295,700
\end{aligned}
$$

So, the total cost is $295,700. (See Figure 1–16.)

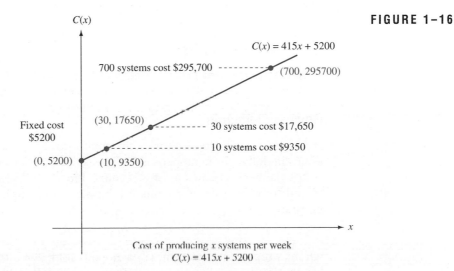

FIGURE 1–16

Cost of producing x systems per week
$C(x) = 415x + 5200$

(c) This information gives $C(x) = 230{,}130$, so we have

$$230{,}130 = 415x + 5200$$

We need to solve this for x:

$$230{,}130 - 5200 = 415x$$
$$224{,}930 = 415x$$
$$x = \frac{224{,}930}{415} = 542$$

So, 542 systems were produced that week.

■ **Now You Are Ready to Work Exercise 1**

The unit cost and fixed costs may not be known directly, but the cost function can be obtained if the information gives two points on the line, as illustrated in the next example.

EXAMPLE 2 ▶   A company made a cost study and found that it cost $10,170 to produce 800 pairs of running shoes and $13,810 to produce 1150 pairs.

(a) Determine the cost–volume function.
(b) Find the fixed cost and the unit cost.

SOLUTION

(a) Let $x =$ the number of pairs. The information gives two points on the cost–volume line: (800, 10170) and (1150, 13810). The slope of the line is

$$m = \frac{13{,}810 - 10{,}170}{1150 - 800} = \frac{3640}{350} = 10.40$$

Using the point (800, 10170) in the point-slope equation, we have

$$y - 10{,}170 = 10.40(x - 800)$$
$$y = 10.40x + 1850$$

Therefore, $C(x) = 10.40x + 1850$.
(b) From the equation $C(x) = 10.40x + 1850$, the fixed cost is $1850 per week, and the unit cost is $10.40.

■ **Now You Are Ready to Work Exercise 13**

Revenue Function

If Joan's Sporting Goods Store sells mopeds for $798 each, the total income (**revenue**) in dollars from mopeds is 798 times the number of mopeds sold. This illustrates the general concept of a **revenue function;** it gives the total revenue obtained from the sale of x items. In the moped example the revenue function is given by

$$R(x) = 798x$$

where x represents the number of mopeds sold, 798 is the selling price in dollars for each item, and $R(x)$ is the total revenue in dollars from x items.

EXAMPLE 3 ➤ The sporting goods store has a sale on mopeds at $725 each.

(a) Give the revenue function.

(b) The store sold 23 mopeds. What was the total revenue?

(c) One salesperson sold $5075 worth of mopeds. How many did she sell?

SOLUTION

(a) The revenue function is given by

$$R(x) = 725x$$

(b) The revenue for 23 mopeds is obtained from the revenue function when $x = 23$:

$$R(23) = 725(23) = 16,675$$

The revenue in this case is $16,675.

(c) This gives $R(x) = 5075$, so

$$725x = 5075$$
$$x = \frac{5075}{725} = 7$$

So, she sold 7 mopeds.

■ **Now You Are Ready to Work Exercise 5**

Break-Even Analysis

Break-even analysis answers a common management question: At what sales volume will we break even? When do revenues equal costs? Greater sales will induce a profit, whereas lesser sales will show a loss.

The **break-even point** occurs when the cost equals the revenue, so the cost and revenue functions can be used to determine the break-even point. Using function notation, we write this as $C(x) = R(x)$.

Break-Even Point

The point at which cost equals revenue,

$$C(x) = R(x)$$

EXAMPLE 4 ➤ Cox's Department Store pays $99 each for DVD players. The store's monthly fixed costs are $1250. The store sells the DVD players for $189.95 each.

(a) What is the cost–volume function?

(b) What is the revenue function?

(c) What is the break-even point?

SOLUTION

Let x represent the number of DVD players sold.

(a) The cost function is given by

$$C(x) = 99x + 1250$$

(b) The revenue function is defined by

$$R(x) = 189.95x$$

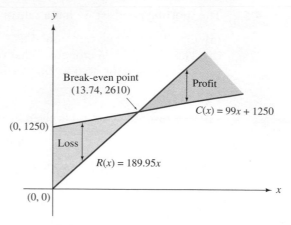

FIGURE 1–17   A graph showing

Cost function: $C(x) = 99x + 1250$
Revenue function: $R(x) = 189.95x$
Break-even point: $(13.74, 2610)$

(c) The break-even point occurs when cost equals revenue,

$$C(x) = R(x)$$

Writing out the function gives

$$99x + 1250 = 189.95x$$

The solution of this equation gives the break-even point:

$$99x + 1250 = 189.95x$$
$$1250 = 189.95x - 99x$$
$$1250 = 90.95x$$
$$x = \frac{1250}{90.95}$$
$$= 13.74$$

Because x represents the number of DVD players, we use the next integer, 14, as the number sold per month to break even. If more than 14 are sold, there will be a profit. If fewer than 14 are sold, there will be a loss.

The break-even point is at $x = 14$ DVD players, with a revenue of \$2659.30.

Geometrically (See Figure 1–17), the break-even point occurs where the cost function and revenue function intersect. A profit occurs for the values of x when the graph of the revenue function is above the cost function. At a given value of x, the profit is the vertical distance between the graphs. Similarly, a loss occurs when the graph of the revenue function is below the graph of the cost function and the amount of the loss is represented by the vertical distance between the graphs.

■ **Now You Are Ready to Work Exercise 9**

EXAMPLE 5 ➤ A temporary secretarial service has a fixed weekly cost of \$896. The wages and benefits of the secretaries amount to \$7.65 per hour. A firm that employs a secretary pays Temporary Service \$10.40 per hour. How many hours per week of secretarial service must Temporary Service place to break even?

SOLUTION

First, write the cost and revenue functions. The fixed cost is $896, and the unit cost is $7.65, so the cost function is given by

$$C(x) = 7.65x + 896$$

where x is the number of hours placed each week. The revenue function is given by

$$R(x) = 10.40x$$

Equating cost and revenue, we have

$$7.65x + 896 = 10.40x$$

This equation reduces to

$$10.40x - 7.65x = 896$$
$$2.75x = 896$$
$$x = \frac{896}{2.75}$$
$$= 325.8$$

Temporary Service must place secretaries for a total of 326 hours per week (rounded up) to break even.

■ **Now You Are Ready to Work Exercise 15**

Straight-Line Depreciation

To prepare a tax return and report the financial condition of a company to its stockholders, a company needs to estimate the value of buildings, equipment, and so on. Part of this procedure sometimes uses a linear function to estimate the value of equipment. For instance, when a corporation buys a fleet of cars, it expects them to decline in value because of wear and tear. For example, if the corporation purchases new cars for $17,500 each, they may be worth only $5500 each three years later, due to normal wear and tear. This decline in value is called **depreciation.** The value of an item after deducting depreciation is called its **book value.** In three years, each car depreciated $12,000, and its book value at the end of three years was $5500. For tax and accounting purposes, a company will report depreciation and book value each year during the life of an item. The Internal Revenue Service allows several methods of depreciation. The simplest is **straight-line depreciation.** This method assumes that the book value is a linear function of time — that is,

$$B = mx + b$$

where B is the book value and x is the number of years. For example, each car had a book value of $17,500 when $x = 0$ ("brand new" occurs at zero years). When $x = 3$, the car's book value declined to $5500. This information is equivalent to giving two points $(0, 17500)$ and $(3, 5500)$ on the straight line representing book value. (See Figure 1–18.)

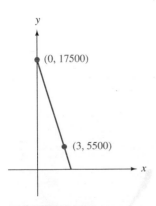

FIGURE 1–18   Straight-line depreciation.

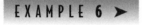

Over time an auto declines in value (depreciates).

We obtain the linear equation of the book value by finding the equation of a line through these two points. The slope of the line is

$$m = \frac{y_2 - y_1}{x_2 - x_1} = \frac{5500 - 17,500}{3 - 0} = \frac{-12,000}{3} = -4000$$

and the y-intercept is 17,500, so the equation is

$$B = -4000x + 17,500$$

The book value at the end of two years is

$$B = -4000(2) + 17,500 = -8000 + 17,500 = 9500$$

The *negative* value of the slope indicates a *decrease* in the book value of $4000 each year. This annual decrease is the **annual depreciation.**

Generally, a company will estimate the number of years of useful life of an item. The estimated value of the item at the end of its useful life is called its **scrap value.** The values of x are restricted to $0 \le x \le n$, where n is the number of years of useful life.

Here is a simple example.

EXAMPLE 6 ▶   Acme Manufacturing Co. purchases a piece of equipment for $28,300 and estimates its useful life as eight years. At the end of its useful life, its scrap value is estimated at $900.

(a) Find the linear equation expressing the relationship between book value and time.
(b) Find the annual depreciation.
(c) Find the book value at the end of the first, fifth, and seventh years.

SOLUTION

(a) The line passes through the two points $(0, 28300)$ and $(8, 900)$. Therefore, the slope is

$$m = \frac{900 - 28,300}{8 - 0} = -3425$$

and the y-intercept is 28,300, giving the equation

$$B = -3425x + 28,300$$

(See Figure 1–19.)

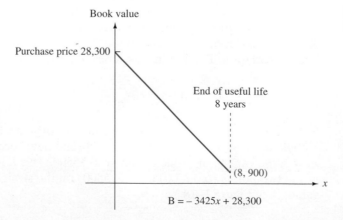

Book value

Purchase price 28,300

End of useful life
8 years

(8, 900)

x

$B = -3425x + 28,300$

FIGURE 1–19
Book value.

(b) The annual depreciation is obtained from the slope and is $3425.

(c) For year 1, $B = -3425(1) + 28,300 = 24,875$. For year 5, $B = -3425(5) + 28,300 = 11,175$. For year 7, $B = -3425(7) + 28,300 = 4325$. Thus, after seven years the book value of the equipment is $4325.

■ **Now You Are Ready to Work Exercise 17**

Applications of Linear Inequalities

For some applications, an interval of values — instead of one specific value — gives an appropriate answer to the problem. These situations can often be represented by inequalities. Here are two examples using linear inequalities. (See Appendix A.4 for a review of the properties of linear inequalities.)

EXAMPLE 7 ➤   A doughnut shop sells doughnuts for $3.29 per dozen. The shop has fixed weekly costs of $650 and unit costs of $1.55 per dozen. How many dozens of doughnuts must be sold weekly for the shop to make a profit?

SOLUTION

Let x = number of dozens of doughnuts sold per week. The revenue and cost functions are

$$R(x) = 3.29x$$
$$C(x) = 1.55x + 650$$

The shop makes a profit when the revenue exceeds the costs — that is, when $R(x) > C(x)$. Therefore, we want to solve

$$3.29x > 1.55x + 650$$
$$3.29x - 1.55x > 650$$
$$1.74x > 650$$
$$x > \frac{650}{1.74} = 373.56$$

Therefore, at least 374 dozen doughnuts must be sold in order to make a profit. The interval notation for this solution is $[374, \infty)$. (See Figure 1–20.)

■ **Now You Are Ready to Work Exercise 21**

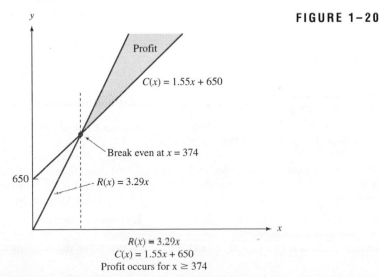

FIGURE 1–20

$R(x) = 3.29x$
$C(x) = 1.55x + 650$
Profit occurs for x ≥ 374

The following example illustrates how an analysis of an inequality can help decide on the best course of action.

EXAMPLE 8 ➤ A quick-copy store can select from two plans to lease a copy machine. Plan A costs $75 per month plus five cents per copy. Plan B costs $200 per month plus two cents per copy. When will it be to the copy shop's advantage to lease under plan A?

SOLUTION

Let x = number of copies per month. Then the monthly costs are as follows:

$$\text{Plan A:} \quad CA(x) = 75 + 0.05x$$
$$\text{Plan B:} \quad CB(x) = 200 + 0.02x$$

Plan A is better when $CA(x) < CB(x)$. We find the solution by solving

$$75 + 0.05x < 200 + 0.02x$$
$$0.05x - 0.02x < 200 - 75$$
$$0.03x < 125$$
$$x < \frac{125}{0.03} = 4166.7$$

Plan A is better when the number of copies per month is fewer than 4167 copies. The interval notation for this answer is $(0, 4167)$.

■ **Now You Are Ready to Work Exercise 27**

▦ 1.3 EXERCISES

LEVEL 1

Exercises 1 through 4 help you understand cost functions.

1. *(See Example 1)* The weekly cost function of manufacturing x mountain bikes is given by $C(x) = 43x + 2300$.
   (a) Determine the cost of producing 180 bikes per week.
   (b) One week the total production cost was $11,889. How many bikes were produced that week?
   (c) Find the unit cost and fixed cost.

2. A software company produces a home accounting system. The company's cost function for producing x systems per month is $C(x) = 16.25x + 28,300$.
   (a) Determine the cost of producing 2500 systems per month.
   (b) One month the production costs were $63,010. How many systems did the company produce?
   (c) Find the company's unit cost and fixed cost.

3. T-Shirts and More has determined that the relationship between daily cost and volume (the cost–volume formula) for T-shirts is $C(x) = 3x + 400$.

   (a) Determine the fixed cost and the unit cost.
   (b) Find the total daily costs when the production is 600 units per day and 1000 units per day.

4. The cost–volume formula for men's belts is $y = 2.5x + 750$, where x is the number of belts produced per week. Determine the fixed cost and the unit cost. Find the total costs when the weekly production is
   (a) 100 units   (b) 300 units   (c) 650 units

Exercises 5 through 8 focus on the revenue function.

5. *(See Example 3)* A store sells jogging shoes for $32 per pair.
   (a) Write the revenue function.
   (b) The store sold 78 pairs. What was the revenue?
   (c) One day the store sold $672 worth of jogging shoes. How many pairs did it sell?

6. Tony's Cassette Warehouse sells cassettes for $6.25 each.
   (a) Write the revenue function.

(b) What is the revenue from selling 265 cassettes?

7. Cold Pizza sells frozen pizzas for $3.39 each.
   (a) Write the revenue function.
   (b) What is the revenue from selling 834 pizzas?

8. A club sells cookies for 60 cents each.
   (a) Write the rule that gives the revenue function.
   (b) The revenue one day was $343.80. How many cookies were sold?

Exercises 9 through 16 give you experience in using the cost and revenue functions to find the break-even point.

9. *(See Example 4)*   Austin Avenue Clothiers pays $57 each for sports coats and has a fixed monthly cost of $780. The store sells the coats for $79 each.
   (a) What is the linear cost–volume function?
   (b) What is the linear revenue function?
   (c) What is the break-even number of coats?

10. Find the linear cost equation if the fixed cost is $700 and the cost per unit volume is $2.50. What are the total costs when the volume produced is 400 units?

11. Find the cost–volume formula if the fixed cost is $500 and the cost per unit volume is $4. What are the total costs when the volume produced is 800 units?

12. A company has a cost function of
$$C(x) = 22x + 870$$
and a revenue function of
$$R(x) = 37.50x$$
Find the break-even point.

13. *(See Example 2)*   The Academic T-Shirt Company did a cost study and found that it costs $1400 to produce 600 "I Love Math" T-shirts. The total cost is $1600 for a volume of 700 T-shirts.
   (a) Determine the linear cost–volume function.
   (b) What is the fixed cost?
   (c) What is the unit cost?

14. The monthly expenses of The Campus Copy Shop are given by the cost equation
$$C(x) = 3690 + 0.025x$$
where x is the number of pages copied in a month. The revenue function is
$$R(x) = 0.055x$$
Find the break-even point of The Campus Copy Shop.

15. *(See Example 5)*   The Computer Shop sells computers. The shop has fixed costs of $1500 per week.

Its average cost per computer is $649 each, and the average selling price is $899 each.
   (a) Write the linear cost function.
   (b) Write the linear revenue function.
   (c) Find the cost of selling 37 computers per week.
   (d) Find the revenue from selling 37 computers.
   (e) Find the break-even point.

16. A Toy Co. estimates that total costs are $1000 when its volume is 500 Fastback cars and $1200 when its volume is 900 Fastback cars.
   (a) Determine the linear cost–volume function.
   (b) Find the fixed cost and unit cost.
   (c) What are the estimated total costs when the volume is 1200 units?

The linear function provides one model of depreciation as illustrated in Exercises 17 through 20.

17. *(See Example 6)*   A TV costs $425, has a scrap value of $25, and has a useful life of eight years. Find
   (a) the linear equation relating book value and number of years.
   (b) the annual depreciation.
   (c) the book value at the end of year 3.

18. A fax machine costs $1500 and has a useful life of ten years. If it has a scrap value of $200, find
   (a) the linear equation relating book value and number of years.
   (b) the annual depreciation.
   (c) the book value at the end of year 7.

19. An automobile costs $9750, has a useful life of six years, and has a scrap value of $300. Find
   (a) the linear equation relating book value and number of years.
   (b) the annual depreciation.
   (c) the book value at the end of years 2 and 5.
   (d) Explain why an auto might have significantly lower value after two years than the value indicated by the linear equation found in part (a).

20. A machine costs $13,500, has a useful life of 12 years, and has no scrap value. Find the linear equation relating book value and number of years.

The following exercises demonstrate applications of linear functions. Analyze the information given to determine the equation of the relevant line or lines. Use the linear equation to determine the requested information.

21. *(See Example 7)*   The Cookie Store sells cookies for 45 cents each. The store has monthly fixed costs of $475 per month and unit costs of 23 cents per cookie. How many cookies must it sell each month to make a profit?

22. A manufacturing company produces an item that sells for $145 and has a unit cost of $70. If fixed manufacturing costs are $225,000 per year, how many must the company sell to make a profit?

23. The Tie Shop sells its ties for $21 each. The shop has weekly fixed costs of $845, and each tie costs $12. How many ties must be sold to make a profit?

24. The Pretzel Place sells pretzels for $1.75 each. The cost of materials to make the pretzels (unit cost) is 33 cents per pretzel. The fixed costs of rent, salaries, and so on are $8200 per month. How many pretzels must they sell to make a profit?

25. Tony plans to open a bagel shop. He estimates the weekly costs are: rent, $300; utilities, $130; salaries, $1800; insurance, $90; and bagel ingredients, 32 cents per bagel. At the price of $0.95 per bagel, how many must be sold to make a profit?

26. Tina investigates the possibility of opening a donut shop. She estimates weekly costs of $275 for rent, $2100 for employee costs, and $125 for miscellaneous costs. The ingredients for the donuts cost 17 cents per donut. She believes she can sell 4400 donuts per week. How much should she charge to make a profit?

LEVEL 2

27. *(See Example 8)* The Cannota family plans to rent a car for a one-week vacation. Company A rents a car for $105 per week plus 14 cents per mile. Company B rents a car for $23 per day plus 10 cents per mile. Find the number of miles traveled that makes the car from company A the better deal.

28. A graduating senior has a choice of two salary plans from a company. The first plan will pay $1000 per month plus 2% of the senior's monthly sales. The second plan will pay a straight 7.5% commission. How much must the senior sell per month to do better under the second plan?

29. The *Times-Herald* is planning a special-edition magazine. The publishing expenses include fixed costs of $1400 and printing costs of 40 cents per magazine. The magazines will sell for 85 cents each, and advertising revenue is 20 cents per magazine. How many magazines must be sold to make a profit?

30. The Student Center plans to bring a musical group for a concert. It expects a sellout crowd of 1200. The group will come for a flat fee of $3000 or for a fee of $1000 plus 30% of gate receipts. The Student Center decides to pay $1000 plus 30% of the gate receipts.
(a) What admission prices should the Student Center charge in order for this cost to be less than the other choice of $3000?
(b) If other costs are $2500, what admission price should the Student Center charge in order to make a profit?

31. Davis Printing finds that the unit cost of printing a book is $12.65. The total cost for printing 2700 books is $36,295. Find the linear cost–volume function.

32. A company makes microwave stands at a cost of $48 each and sells them for $62 each. The company spent $28,000 to begin making the stands. How many stands must it sell to break even?

LEVEL 3

33. The Beta Club plans a dance as a fund-raiser. The band costs $650, decorations cost $45, and the refreshments cost $2.20 per person. The admission tickets are $6 each.
(a) How many tickets must be sold to break even?
(b) How many tickets must be sold to clear $700?
(c) If admission tickets are $7.50 each, how many must be sold to clear $700?

34. A store handles a specialty item that has a unit cost of $24 and a fixed cost of $360. The sales required to break even are 75 items.
(a) Find the linear cost function.
(b) Find the linear revenue function.

35. A health club membership costs $35 per month.
(a) What is the linear monthly revenue function?
(b) What is the monthly revenue if there are 1238 members?
(c) One month the revenue increased by $595. What was the increase in membership?

36. A company's records showed that the daily fixed costs for one of its production lines was $1850 and the total cost of one day's production of 320 items was $3178. What is the linear cost–volume function?

37. Brooks Bros. Company purchased a piece of equipment and used the straight-line depreciation method. The company books showed that the book value was $14,175 at the end of the third year and $8475 at the end of the seventh year.
  (a) Find the linear equation relating book value and number of years.
  (b) Find the annual depreciation.
  (c) Find the purchase price.

38. The Unique Shoppe sells personalized telephones. Its weekly cost function is
$$C(x) = 28x + 650$$
At the break-even point, $x = 65$ phones per week. What is the revenue function?

39. The break-even point for a tanning salon is 260 memberships that will produce $3120 in monthly revenue. If the salon sells only 200 memberships, it will lose $330 per month.
  (a) What is the linear revenue function?
  (b) What is the linear cost function?

40. The profit function is revenue minus cost; that is,
$$P(x) = R(x) - C(x)$$
  (a) The cost and revenue functions for Acme Manufacturing are
$$C(x) = 28x + 465$$
$$R(x) = 52x$$
  (i) Write the profit function.
  (ii) What is the profit from selling 25 items?
  (b) The weekly expenses of selling x bicycles in The Bike Shop are given by the cost function
$$C(x) = 1200 + 130x$$

and revenue is given by
$$R(x) = 210x$$
  (i) Write the profit function.
  (ii) Find the profit from selling 18 bicycles in a week.
  (c) Another Bike Shop has monthly fixed costs of $5200 and unit costs of $145. Its bicycles sell for $225 each.
  (i) Write the linear profit function.
  (ii) What is the profit from selling 75 bicycles per month?

41. A specialty shop owner used a revenue function and a cost–volume function to analyze his monthly sales. One month he found that with a sales volume of 1465 items he had revenues of $32,962.50 and a total cost of $26,405.50. Another month he had total costs of $17,638 on a sales volume of 940 items.
  (a) Find the linear revenue function.
  (b) Find the linear cost function.
  (c) Find the break-even point.

42. Midwest Office Supply sold 27 briefcases for a total revenue of $1134. The cost is a linear function, and the break-even point occurs when 88 briefcases are sold. If the company sells 100 briefcases, it makes a profit of $132. Find the cost and revenue functions.

43. The Simpson family has two long-distance billing options. The first option has a monthly fee of $7.95 and 6 cents per minute called. The second option costs 9 cents per minute with no monthly fee.
  (a) Find the number of minutes for which the two options have the same cost.
  (b) When does the first option cost less?

EXPLORATIONS

44. If a function $C(x)$ gives the cost of x items, then the average cost is given by $\dfrac{C(x)}{x}$. For each of the linear cost functions given, find the average cost of the indicated number of items.
  (a) The monthly cost of producing x cookies at The Cookie Store is
$$C(x) = 0.23x + 475$$
  One month it produced 12,500 cookies. Find the average cost.
  (b) The weekly cost function of Common Ground Coffeehouse is
$$C(x) = 0.35x + 255$$
  where x is the number of cups of coffee. Find the average cost of 700 cups of coffee.

  (c) The cost of x T-shirts for the Eta Pi Spring Fling is quoted as $C(x) = 7.85x + 82.5$. Find the average cost of 120 T-shirts.
  (d) The daily cost of baking x bagels at the Bagel Bakery is
$$C(x) = 0.125x + 382$$
  Find the average cost of baking x bagels a day.

45. The Neighborhood Bank has three checking account plans:
  Plan 1: A $15 monthly charge for which a customer can write an unlimited number of checks.
  Plan 2: A $5 monthly fee plus a charge of 8 cents per check.
  Plan 3: No monthly fee and a 16-cent charge for each check written.

Find the conditions under which plan 2 is the most economical for the customer.

46. When Anita's Ford Explorer was a year old it had a Blue Book value of $14,340. Two years later the Blue Book value was $8665.

   (a) Using straight-line depreciation, estimate the cost when it was new.

   (b) Estimate when the Blue Book value will reach $1000.

   (c) Explain why your answer in parts (a) and (b) are, or are not, realistic.

47. The Crofford Cutlery Company manufactures a supersharp knife that sells for $14.50. The current process has fixed costs of $1400 per day and unit costs of $3.85. The vice-president of the company recommends leasing new equipment that will reduce unit costs to $2.70. The president of the company opposes this because it will increase fixed costs to $1725 per day. The vice-president claims the decrease in unit costs will more than offset the increase in fixed costs. Determine if the vice-president is correct.

48. A bank usually has more than one checking account plan. Call a local bank to determine the cost to the customer for a checking account. Compare two of the plans and determine which is the better plan.

49. Let x represent the demand for DVD players and y represent the price. Then the demand equation for DVD players is

$$y = -25x + 460$$

and the supply equation is

$$y = 17x + 100$$

   (a) Graph the two equations on the [0, 25] [0, 500] screen.

   (b) Is there a shortage or surplus of DVD players when the price is $200?

   (c) Is there a shortage or surplus of DVD players when the price is $300?

   (d) Is there a shortage or surplus of DVD players when the demand = 20?

50. The supply and demand equations for in-line skates are

$$\text{Demand:} \quad y = -0.07x + 375$$
$$\text{Supply:} \quad y = 0.03x + 76$$

   (a) Graph these equations in the window X [0, 3500], Y [0, 400].

   (b) Determine the price for which there is a surplus.

   (c) Determine the price for which there is a shortage.

51. The cost function of producing x VCRs is

$$C(x) = 16,700 + 140x$$

Use a graph to determine the maximum production level if the costs cannot exceed $100,000.

52. One electric utility company charges its residential customers 7.5 cents per kWh in addition to a base charge of $8 per month. A second electric utility charges 6.6 cents per kWh in addition to a base charge of $20 per month.

   (a) For each company, write and graph the equation of the monthly charge y in terms of x, the number of kWh used.

   (b) For what amounts of use are the first company's rates better for the consumer?

53. Juan is a salesman for L. L. Bowers, Inc. He has a choice of three compensation plans. Plan 1 pays $2500 per month. Plan 2 pays $2000 per month plus 17% commission. Plan 3 pays $1700 per month plus 25% commission. Graph the three plans and determine the best plan.

54. After many years of service, a politician has offers from three publishing companies to publish her memoirs. Publisher A offers a flat payment of $750,000. Publisher B offers a payment of $300,000 plus 17% of gross sales. Publisher C offers a payment of $100,000 plus 28% of gross sales. Write and graph the equation of each plan. Determine the best offer based on gross sales.

55. The Cameron Art Center is planning a fund-raising dinner. The costs include a speaker's fee of $2000 and publicity costs of $700. If they hold the dinner at the Convention Center, it will cost $600 for the use of the center plus $20 per plate. If they hold the dinner at the Ferrell Center, it will cost $1300 for the use of the building plus $17 per plate. The admission is $35 per plate.

   (a) How many tickets must be sold to break even at the Convention Center?

   (b) How many tickets must be sold to break even at the Ferrell Center?

   (c) Graph the profit function in each case to determine the better place to hold the dinner.

56. The Butler family plans to rent a car for a two-week vacation trip. They have obtained prices from three rental companies. They are

   $35 per day with unlimited mileage.

   $20 per day plus 10¢ per mile.

   $25 per day plus 8¢ per mile.

Graph each of these and determine which plan is the best.

57. Air North provides special fares for international visitors. The special fare is good for two months. The visitor can select one of three plans:

Plan 1:   Purchase a pass for $2500 that allows unlimited trips for the two-month period.

Plan 2:   Purchase a discount coupon for $300 that allows the visitor to purchase any trip for $130 during the 60 days.

Plan 3:   The visitor may purchase any trip for $165 during the 60 days.

(a) Which plan is best for 10 trips? For 15 trips?

(b) Graph each plan and determine the number of trips for which each plan is the best.

58. T-Shirts and More has determined the relationship between daily cost and volume (the cost–volume formula) for T-shirts is

$$C(x) = 4x + 600.$$

Find the total daily costs when the production is 700 units per day and 1200 units per day.

59. The cost–volume formula for men's belts is

$$y = 3.1x + 950,$$

where x is the number of belts produced per week. Find the total costs when the weekly production is
(a) 1500 units.
(b) 2250 units.

60. The profit function is revenue minus cost; that is,

$$P(x) = R(x) - C(x).$$

(a) The cost and revenue functions for Acme Manufacturing are

$$C(x) = 97x + 785$$
$$R(x) = 215x$$

(i) Write the profit function.
(ii) What is the profit from selling 30 items? From 426 items?

(b) The weekly expenses of selling x bicycles in The Bike Shop are given by the cost function

$$C(x) = 1620 + 143x$$

and revenue is given by

$$R(x) = 234x$$

(i) Write the profit function.
(ii) What is the profit from selling 27 bicycles in a week? From 42 bicycles?

61. For the cost and revenue functions

$$C(x) = 42x + 1380$$
$$R(x) = 96x$$

Find the profit for $x = 10, 30, 45$, and 62.

62. For the cost and revenue functions

$$C(x) = 37x + 2470$$
$$R(x) = 99x$$

Find the profit for $x = 33, 47$, and 74.

63. Find the break-even point for the cost and revenue functions

$$C(x) = 8.7x + 350.88$$
$$R(x) = 21.6x$$

64. Find the break-even point for the cost and revenue functions

$$C(x) = 49.5x + 2167$$
$$R(x) = 98.75x$$

65. (a) Find the break-even point for the cost and revenue functions

$$C(x) = 39.25x + 2576$$
$$R(x) = 79.50x$$

(b) Find the value of x so that profit $= 2300$
(c) Find the value of x so that loss $= 675$

66. (a) Find the break-even point for the cost and revenue functions

$$C(x) = 136x + 12,312$$
$$R(x) = 244x$$

(b) Find the value of x so that profit $= 16,500$
(c) Find the value of x so that loss $= 4000$

USING YOUR TI-83

INTERSECTION OF LINES

Finding the break-even point requires finding the intersection of two lines. This can be done with a TI graphing calculator. We illustrate by finding the intersection of the lines

$$y = 2x + 5$$
$$3x + 4y = 12$$

Graph the two equations using $\boxed{\text{Y=}}$ and $\boxed{\text{GRAPH}}$:

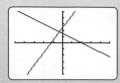

The intersection of the two lines can then be found using the **intersect** command. With the graph of the two lines on the screen, press $\boxed{\text{CALC}}$ ($\boxed{\text{2nd}}$ $\boxed{\text{TRACE}}$) and select <5 : intersect> from the menu. Press $\boxed{\text{ENTER}}$. three times, and the coordinates of the intersection will appear as in the following window:

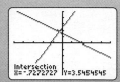

Note: Notation such as <5 : intersect> indicates a command to be selected from a menu.

EXERCISES

Use the **intersect** command to find the intersection of the following pairs of lines.

1. $y = 2x - 1$
  $y = 1.75x$

2. $y = 3x - 11.5$
  $y = 2x - 5$

3. $y = -4x + 18$
  $y = 1.5x - 5.5$

4. $y = 6.2x + 1.4$
  $y = -2.1x + 9.3$

5. $3x + 4y = 30$
  $5x - 2y = -2$

6. $3.1x + 1.4y = 19.46$
  $5.7x - 2.3y = 6.46$

USING EXCEL

EXAMPLE 1

Let's use EXCEL to find the profit for the cost and revenue functions

$$C(x) = 38x + 648$$
$$R(x) = 95x$$

where x = number of items. Find the profit for selling 10, 58, 72, 109, and 223 items.
     Here is the EXCEL template to calculate $C(x)$, $R(x)$, and the profit, $P(x)$.

	A	B	C	D
1	X	C(x)	R(x)	P(x)
2	10	1028	950	70
3	58	2852	5510	2658
4	72	3384	6840	3456
5	109	4790	10355	5565
6	223	9122	21185	12063
7				
8		Enter the formula	Enter the formula	Enter the formula
9		=38*A2+648	=95*A2	=C2-B2
10		in B2 and drag	in C2 and drag	in D2 and drag
11		down to B6	down to C6	down to D6
12				

Notice that for $x = 10$ there is a loss of 78.
     EXCEL has a tool called **Goal Seek** that can be used to find the intersection of two linear functions such as when you want to find the break-even point.

EXAMPLE 2

Find the break-even point for the cost and revenue functions

$$C(x) = 3.6x + 9895$$
$$R(x) = 12.6x$$

     We set up the spreadsheet as shown below with entries in Line 1 to identify the entries in Line 2 that are used in the analysis. The entry of 10 for x is an arbitrary value.

	A	B	C	D
1	X	C(X)	R(X)	R(X)-C(X)
2	10	=3.6*A2+9895	=12.6*A2	=C2-B2
3				
4				

The calculations for this are

	A	B	C	D
1	X	C(X)	R(X)	R(X)-C(X)
2	10	9931	126	-9805
3				
4				
5				

     We will use the formula in D2, $R(x) - C(x)$, to find the intersection, because this difference is zero at the intersection. Now we are ready to use this spreadsheet to find the intersection of the cost and revenue lines.

(*continued*)

1. Select **Goal Seek** under the Tools menu. This will bring up the following screen:

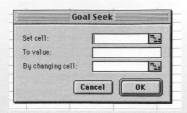

2. Place the cursor in the blank to the right of **Set cell:** and then select the cell D2 on the spreadsheet.
3. Move the cursor to the blank to the right of **To value:** and enter the number zero.
4. Move the cursor to the blank to the right of **By changing cell:** and then select cell A2 in the spreadsheet. This gives you the following screen:

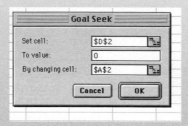

5. Click on OK and the spreadsheet shows

	A	B	C	D
1	X	C(X)	R(X)	R(X)–C(X)
2	1099.44444	13853	13853	0
3				
4				

Notice that the value of x, 1099.4444, gives the same cost and revenue value, 13,853, so we have found the break-even point (1099.4444, 13853). When x represents a number of items, we would round it to 1099.

The **Goal Seek** procedure we have just used seeks the value of x that makes the value of D2, $R(x) - C(x)$, equal to the value placed in "To value," namely zero. If you use the same procedure and enter 1000 in "To value," **Goal Seek** will seek the value of x that makes $R(x) - C(x)$ equal to 1000, that is, find the x that yields a profit of 1000. Try it. You should get $x = 1210.55556$.

You can use the same spreadsheet to find the loss when $x = 900$. Enter 900 in A2, do not use **Goal Seek,** and D2 shows -1795, a loss of $1795.

EXERCISES

1. $C(x) = 10x + 535$ and $R(x) = 35x$
   Calculate profit for $x = 20, 30, 45, 62,$ and 81.

2. $C(x) = 54x + 1250$ and $R(x) = 97.50x$
   Calculate profit for $x = 225, 315, 450,$ and 620.

3. $C(x) = 537.60x + 7431$ and $R(x) = 779x$
   Calculate profit for $x = 25, 100, 250, 350,$ and 560.

4. Find the break-even point for the cost and revenue functions.

$$C(x) = 5x + 450 \text{ and } R(x) = 12.5x$$

5. Find the break-even point for the cost and revenue functions.

$$C(x) = 22.2x + 1165 \text{ and } R(x) = 48.6x.$$

6. (a) Find the break-even point for the cost and revenue functions.

$$C(x) = 3.6x + 224.64 \text{ and } R(x) = 18x$$

(b) Find the value of x for which profit = 1500.
(c) Find the profit, or loss, when $x = 10$, when $x = 33$.

■■ IMPORTANT TERMS

1.1

Function
Rule of a Function
Domain
Range
Independent Variable
Dependent Variable
Function Value

1.2

Graph
Graph of a Function
Solution
Linear Function
y-Intercept

Slope
Constant Function
Horizontal Line
Vertical Line
Slope-Intercept Equation
Point-Slope Equation
Two-Point Equation
x-Intercept
Parallel Lines
Perpendicular Lines

1.3

Mathematical Model
Variables
Volume

Fixed Costs
Variable Costs
Cost–Volume Function
Unit Costs
Total Cost
Revenue
Revenue Function
Break-Even Analysis
Break-Even Point
Depreciation
Book Value
Straight-Line Depreciation
Annual Depreciation
Scrap Value

■■ REVIEW EXERCISES

1. If $f(x) = \dfrac{7x - 3}{2}$, find

   (a) $f(5)$ (b) $f(1)$
   (c) $f(4)$ (d) $f(b)$

2. If $f(x) = 8x - 4$, find
   (a) $f(2)$ (b) $f(-3)$
   (c) $f\left(\dfrac{1}{2}\right)$ (d) $f(c)$

3. If $f(x) = \dfrac{x + 2}{x - 1}$ and $g(x) = 5x + 3$, find
   $f(2) + g(3)$.

4. If $f(x) = (x + 5)(2x - 1)$, find
   (a) $f(1)$ (b) $f(0)$
   (c) $f(-5)$ (d) $f(a - 5)$

5. Apples cost $1.20 per pound, so the price of a bag of apples is $f(x) = 1.20x$, where x is the weight in pounds and $f(x)$ is the purchase price in dollars.
   (a) What is $f(3.5)$?
   (b) A bag of apples cost $3.30. How much did it weigh?

6. Tuition and fees charges at a university are given by

$$f(x) = 135x + 450$$

where x is the number of semester hours enrolled and $f(x)$ is the total cost of tuition and fees.
  (a) Find $f(15)$.
  (b) A student's bill for tuition and fees was $2205; for how many semester hours was she enrolled?

7. Write an equation of the function described by the following statements.
  (a) All the shoes on this table are $29.95 per pair.
  (b) A catering service charges $40 plus $1.25 per person to cater a reception.

8. Sketch the graph of
  (a) $f(x) = 2x - 5$ (b) $6x + 10y = 30$

9. Graph the following lines:
  (a) $y = 3x - 5$ (b) $y = -7$
  (c) $x = 5.5$ (d) $y = x$

10. Graph the following lines:
  (a) $y = 6.5$ (b) $x = -4.75$
  (c) $y = -1.3$ (d) $x = 7$

11. Find the slope and y-intercept for the following lines:
  (a) $y = -2x + 3$ (b) $y = \dfrac{2}{3}x - 4$
  (c) $4y = 5x + 6$ (d) $6x + 7y + 5 = 0$

12. Find the slope of the line through the following pairs of points:
  (a) $(2, 7)$ and $(-3, 4)$ (b) $(6, 8)$ and $(-11, 8)$
  (c) $(4, 2)$ and $(4, 6)$

13. For the line $6x + 5y = 15$, find
  (a) the slope   (b) the y-intercept
  (c) the x-intercept

14. For the line $-2x + 9y = 6$, find
  (a) the slope   (b) the y-intercept
  (c) the x-intercept

15. Find an equation of the following lines:
  (a) With slope $-\dfrac{3}{4}$ and y-intercept 5
  (b) With slope 8 and y-intercept -3
  (c) With slope -2 and passing through $(5, -1)$
  (d) With slope 0 and passing through $(11, 6)$
  (e) Passing through $(5, 3)$ and $(-1, 4)$
  (f) Passing through $(-2, 5)$ and $(-2, -2)$
  (g) Passing through $(2, 7)$ and parallel to $4x - 3y = 22$

16. Find an equation of the line with the given slope and passing through the given point:
  (a) $m = 5$ and point $(2, -1)$
  (b) With slope $-\dfrac{2}{3}$ and point $(5, 4)$
  (c) With $m = 0$ and point $(7, 6)$
  (d) With slope 1 and point $(-2, -2)$

17. Find an equation of the line through the given points:
  (a) $(6, 2)$ and $(-3, 2)$
  (b) $(-4, 5)$ and $(-4, -2)$
  (c) $(5, 0)$ and $(5, 10)$
  (d) $(-7, 6)$ and $(7, 6)$

18. Determine whether the following pairs of lines are parallel:
  (a) $7x - 4y = 12$ and $-21x + 12y = 17$
  (b) $3x + 2y = 13$ and $2x - 3y = 28$

19. Is the line through $(5, 19)$ and $(-2, 7)$ parallel to the line through $(11, 3)$ and $(-1, -5)$?

20. Is the line through $(4, 0)$ and $(7, -2)$ parallel to the line through $(7, 4)$ and $(10, 2)$?

21. Determine whether the line through $(8, 6)$ and $(-3, 14)$ is parallel to the line $8x + 4y = 34$.

22. Determine whether the line through $(-2.5, 0)$ and $(-1, 4.5)$ is parallel to the line $3x - y = 19$.

23. Determine whether the line through $(9, 10)$ and $(5, 6)$ is parallel to the line $3x - 2y = 14$.

24. Determine whether the following pairs of lines are parallel:
  (a) $y = 5x + 13$ (b) $6x + 2y = 15$
      $y = 5x - 24$ $15x + 5y = -27$
  (c) $-8x + 9y = 41$ (d) $12x - 5y = 60$
      $9x - 8y = 13$ $6x + \ \ y = 15$

25. A manufacturer has fixed costs of $12,800 per month and a unit cost of $36 per item produced. What is the cost function?

26. The weekly cost function of a manufacturer is

$$C(x) = 83x + 960$$

  (a) What are the weekly fixed costs?
  (b) What is the unit cost?

27. The cost function of producing x bags of Hi-Gro fertilizer per week is

$$C(x) = 3.60x + 2850$$

  (a) What is the cost of producing 580 bags per week?
  (b) If the production costs for one week amounted to $5208, how many bags were produced?

28. The Shoe Center has a special sale in which all jogging shoes are $28.50 per pair. Write the revenue function for jogging shoes.

29. A T-shirt shop pays $6.50 each for T-shirts. The shop's weekly fixed expenses are $675. It sells the T-shirts for $11.00 each.
  (a) What is the revenue function?
  (b) What is the cost function?
  (c) What is the break-even point?

30. Midstate Manufacturing sells calculators for $17.45 each. The unit cost is $9.30, and the fixed cost is $17,604. Find the quantity that must be sold to break even.

31. The comptroller of Southern Watch Company wants to find the company's break-even point. She has the following information: The company sells the watches for $19.50 each. One week it produced 1840 watches at a production cost of $25,260.00. Another week it produced 2315 watches at a production cost of $31,102.50. Find the weekly volume of watches the company must produce to break even.

32. Find an equation of the line described in the following:
   (a) Through the point $(5, 7)$ and parallel to $6x - y = 15$
   (b) Through the point $(\ 2,\ \ 5)$ and parallel to the line through $(4, -2)$ and $(9, 5)$
   (c) With y-intercept 6 and passing through $(2, -5)$

33. Norton's, Inc. purchases a piece of equipment for $17,500. The useful life is eight years, and the scrap value at the end of eight years is $900.
   (a) Find the equation relating book value and its age using straight-line depreciation.
   (b) What is the annual depreciation?
   (c) What is the book value for the fifth year?

34. The function for the book value of a truck is

$$f(x) = -2300x + 16,500$$

where x is its age and $f(x)$ is its book value.
   (a) What did the truck cost?
   (b) If its useful life is seven years, what is its scrap value?

35. An item cost $1540, has a useful life of five years, and has a scrap value of $60. Find the equation relating book value and number of years using straight-line depreciation.

36. Find the x- and y-intercepts of the line $8x + 6y = 24$ and sketch its graph.

37. A line passes through the point $(2, 9)$ and is parallel to $4x - 5y = 10$. Find the value of k so that $(-3, k)$ is on the line.

38. Lawn Care Manufacturing estimates that the material for each lawnmower costs $85. The manufacturer's fixed operating costs are $4250 per week. Find the company's weekly cost function.

39. Nguyen's Auto Parts bought a new delivery van for $22,000. Nguyen expects the van to be worth $3000 in five years. Find the value of the van as a linear function of its age.

40. A manufacturer found that it cost $48,840 to produce 940 items in one week. The next week it cost $42,535 to produce 810 items. Find the linear cost–volume function.

41. A hamburger place estimates that the materials for each hamburger cost $0.67. One day it made 1150 hamburgers, and the total operating costs were $1250.50. Find the cost function.

42. A fast-food franchise owner must pay the parent company $1200 per month plus 4.1% of receipts. Find the monthly franchise cost function.

43. A company offers an inventor two royalty options for her product. The first is a one-time-only payment of $17,000. The second is a payment of $2000 plus 75 cents for each item sold. Determine when the second option is better than the first.

44. A university estimates that 92% of the applicants who pay the admissions deposit will enroll. How many applicants who pay their admissions deposit are required in order to have a class of 2300 students?

45. Find the value of k so that the points $(9, 4)$ and $(-2, k)$ lie on a line with slope -2.

46. A caterer's fee to cater a wedding reception is a linear function based on a fixed fee and an amount per person. Stephanie and Roy's wedding was planned for 350 guests and cost $2950. Jennifer and Brett's wedding was planned for 290 guests and cost $2530. Find the cost function.

47. A manufacturer of videotapes uses a linear cost function. One week, the company produced 1730 tapes for a total cost of $12,813.60. The unit cost is known to be $6.82. Find the fixed cost.

48. Jones established a small business with a reserve of $12,000 to cover operating expenses in the early stages when a loss is expected. The reserve is reduced $620 each week to cover operating costs.
   (a) Write the function giving the amount remaining in the reserve fund.
   (b) How much is in the fund after eight weeks?
   (c) How long will it take to deplete the fund?

49. Juan is a salesman for L. L. Bowers Corp. He has a choice of three compensation plans. Plan 1 pays $2500 per month. Plan 2 pays $2000 per month plus 15% commission. Plan 3 pays $1700 per month plus 30% commission. Graph the three plans and determine which is best.

50. The Grounds, a coffee shop, has average daily fixed costs of $215 per day and unit costs of $0.65 per cup of coffee. The coffee sells for $1.75 per cup.
   (a) Find the daily average number of cups of coffee that must be sold to break even.
   (b) After several months fixed costs rise to $265 per day. At that time, the manager introduces a

small cup of coffee that contains half as much coffee and is priced at $1.00 per cup. The regular size cup remains at $1.75. At the break-even point, find the relationship between the number of small and regular sizes that must be sold.

51. The Homewood branch library opened with 8400 books. The budget provides for 120 additional books each month.
   (a) Write the equation that gives the number of books x months after the library opened.
   (b) When will the library have 10,000 books?

52. Richard Fenton spent $4.5 million in an unsuccessful campaign for the U.S. Senate. He obtained 32% of the votes. Political analysts estimate that if he runs again, he can gain an additional 5% of the votes for each additional million dollars spent on the campaign.
   (a) Assuming a linear relationship, write the equation relating expenditures, x, to percent of votes obtained, y.
   (b) Based on this relationship, estimate the cost of a campaign that will yield 51% of the votes.

53. Pete sells used cars at a profit of $500 per car, not taking into account monthly overhead expenses of $1700. Taking overhead expense into account, write the equation relating monthly profit (y) to monthly sales of cars (x).

CHAPTER

2

Linear Systems

In Chapter 1, you learned that a linear equation may represent or model a situation reasonably well. You also learned that an important business concept, the break-even point, is determined from two equations, a cost function and a revenue function. The intersection of the two lines determines the break-even point.

Other, more complex situations may require two or more linear equations to describe. As with the break-even problem, the goal is to find a point common to all equations, their intersection. We illustrate with a simple example.

The Bluebonnet Campfire Kids sold cookies and candy to raise money for summer camp. They sold a total of 325 boxes of candy and cookies. The candy sold for $4 per box, and the cookies sold for $3 per box. They made a profit of $2.10 for each box of candy and $1.80 for each box of cookies. Their sales totaled $1165, with a profit of $642. This information can be represented mathematically in the following way. Let $x = $ the number of boxes of candy sold and let $y = $ the number of boxes of cookies sold. Then, $x + y = 325$ represents the total number of boxes sold. Because each box of candy sells for $4 and each box of cookies sells for $3, $4x + 3y = 1165$ represents total sales. The profit from sales is

$$2.10x + 1.80y = 642$$

To answer the question of the number of boxes of candy and the number of boxes of cookies sold, we want the solution to the system

$$\begin{aligned} x + y &= 325 \\ 4x + 3y &= 1165 \\ 2.10x + 1.80y &= 642 \end{aligned}$$

To find a solution to the system, we want to find a value of x and a value of y that makes *all three* equations true. We can easily find a solution to the first equation, such as $x = 100$ and $y = 225$, but these values make neither of the last two equations true. The values $x = 265$ and $y = 35$ make the second equation true but not the other two. Thus, the point $(265, 35)$ lies on the second line but not on the other two. We have not found the solution of the system until we find a point common to all three lines, their point of intersection. We do not solve this system here. The purpose of this chapter is to demonstrate some methods of finding a solution to the system of equations; then you will be able to solve this system. Some of the methods used can be applied to systems with a large number of variables and equations. ■

2.1   Systems of Two Equations

- Solution by Graphing
- Substitution Method
- Elimination Method
- Inconsistent Systems
- Systems That Have Many Solutions
- Application: Supply-and-Demand Analysis

The equations used in the introduction are called a **system of equations.** A pair of numbers, one a value of x and the other a value of y, that makes *all* equations true is called a **solution of the system of equations.**

We now look at some methods used to find the solution to a system of equations.

Solution by Graphing

We use the following simple system to illustrate the geometric meaning of solving a system of linear equations.

EXAMPLE 1 ➤   Solve the system

$$\begin{aligned} 2x - y &= 3 \\ x + 2y &= 4 \end{aligned}$$

NOTE

Remember that the pair of numbers $x = 2$ and $y = 1$ forms *one* solution.

SOLUTION

Geometrically, each of these equations represents a line. When the lines are graphed with the same coordinate axes, they intersect at the point $(2, 1)$. (Figure 2–1) The values $x = 2$ and $y = 1$ satisfy both equations (check to be sure). Thus, they form a solution to the system. ■

We can estimate the solution to a system by graphing the lines and noting the intersection of the lines. However, the graph must be accurate, and even so the precision of the solution may be in doubt.

The precision of a graphical solution depends on the accuracy of the graph. Generally, a pencil-and-paper graph may lack the desired precision. A graphing calculator with a zoom feature can often yield quite an accurate solution. Algebraic techniques can be used to obtain precision solutions. One algebraic method is the **substitution method.**

Substitution Method

The Substitution Method

To solve a system of two linear equations in two variables by the substitution method:

1. Solve for a variable in one of the equations (say x in terms of y).
2. Substitute for x in the other equation.
3. You now have an equation in one variable (say y). Solve for that variable.
4. Substitute the value of the variable (y) just obtained into the first equation and solve for the other variable (x).

EXAMPLE 2 ➤

Solve the system

$$2x - y = 3$$
$$x + 2y = 4$$

by substitution.

SOLUTION

In this case, it is easy to solve for x in the second equation because its coefficient is 1. We obtain

$$x = 4 - 2y$$

Substitute this expression for x in the first equation, $2x - y = 3$:

$$2(4 - 2y) - y = 3$$
$$8 - 4y - y = 3$$
$$8 - 5y = 3$$
$$8 = 3 + 5y$$
$$5 = 5y$$
$$y = 1$$

Now substitute 1 for y in $x = 4 - 2y$ to obtain $x = 4 - 2 = 2$. You could also substitute in $2x - y = 3$. Thus, the solution to the system is $(2, 1)$. You may also solve for y in the first equation and substitute it in the second equation. This will yield the same solution. (Try it.) To be sure you have made no error, you should check your solution in *both* equations. Figure 2–1 shows the graph of the solution.

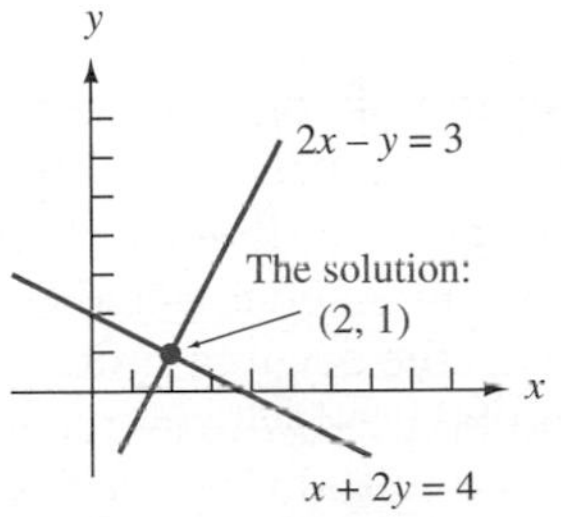

■ **Now You Are Ready to Work Exercise 1**

FIGURE 2–1
$2x - y = 3$ and $x + 2y = 4$ intersect at the point $(2, 1)$.

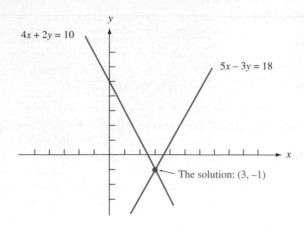

FIGURE 2–2 $5x - 3y = 18$ and $4x + 2y = 10$ intersect at the point $(3, -1)$.

EXAMPLE 3 ➤

Solve the system

$$5x - 3y = 18$$
$$4x + 2y = 10$$

SOLUTION

In this case, it is a little easier to solve for y in the second equation.

$$2y = 10 - 4x$$
$$y = 5 - 2x$$

Substitute the expression for y in the first equation:

$$5x - 3(5 - 2x) = 18$$
$$5x - 15 + 6x = 18$$
$$11x = 33$$
$$x = 3$$

Now substitute $x = 3$ into one of the equations. We use $y = 5 - 2x$.

$$y = 5 - 2(3) = -1$$

The pair $x = 3$, $y = -1$ gives the solution to the system. Check it in both equations. Figure 2–2 shows the graph of the solution.

■ **Now You Are Ready to Work Exercise 13**

Elimination Method

The **elimination method** finds the solution by systematically modifying the system to simpler systems. It does so in a manner that algebraically modifies the system without disturbing the solution. The goal is to modify the system until one of the equations contains just one unknown. The simpler system so obtained is called an **equivalent system** because it has exactly the same solution as the original system. The elimination method is especially useful because it can be used with systems with several variables and equations. The elimination method produces a series of systems of equations by eliminating a variable from an equation or equations, to obtain a simpler system. As you study the examples, observe that the operations used to transform the system (eliminate a variable) into a simpler yet equivalent system are the following:

Equivalent Linear Systems

> To transform one system of linear equations into an **equivalent linear system,** use one or more of the following:
>
> **1.** Interchange two equations.
> **2.** Multiply or divide one or more equations by a nonzero constant.
> **3.** Multiply one equation by a constant and add the result to or subtract it from another equation.

The procedure of converting a linear system to an equivalent linear system will be used again later in this chapter and in Chapter 4, so be sure you understand the process.

We illustrate this method with the system

$$3x - y = 3$$
$$x + 2y = 8$$

The arithmetic is a little easier if we eliminate a variable that has 1 as a coefficient. It is usually more convenient to use the top equation to eliminate x, so we interchange the two equations to get

$$x + 2y = 8$$
$$3x - y = 3$$

Now eliminate x from the bottom equation as follows:

$$-3x - 6y = -24 \quad \text{(Multiply the top equation by } -3 \text{ because it gives } -3x,$$
$$\text{the negative of the } x\text{-term in the bottom equation)}$$

$$\underline{3x - y = 3} \quad \text{(Now add it to the bottom equation)}$$
$$-7y = -21 \quad \text{(The new second equation)}$$

Since the new equation came from equations in the system, it is true whenever the system is true. It replaces the equation $3x - y = 3$ to give the system

$$x + 2y = 8$$
$$-7y = -21$$

Notice that the bottom equation has been modified so that the variable x has been *eliminated* from it. Simplify the bottom equation further by dividing by -7:

$$x + 2y = 8$$
$$y = 3$$

This system has the same solution as the original system, but it has the advantage of giving the value of y at the common solution, namely, 3. Now substitute 3 for y in the top equation to obtain

$$x + 2(3) = 8$$

(Actually, you can substitute y into either of the original equations.) This simplifies to $x = 2$, so the solution to the system is $(2, 3)$.

When a system has no coefficient equal to 1, you can still solve such a system by elimination, as illustrated in the next example.

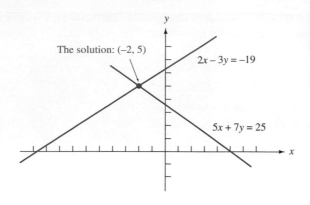

The solution: (–2, 5)

$2x - 3y = -19$

$5x + 7y = 25$

FIGURE 2–3 $2x - 3y = -19$ and $5x + 7y = 25$ intersect at the point $(-2, 5)$.

EXAMPLE 4 ➤

Solve this system by elimination:

$$2x - 3y = -19$$
$$5x + 7y = 25$$

SOLUTION

We want to eliminate x from the second equation and find the value of y for the common solution. We modify the system of equations as follows:

$$
\begin{aligned}
-10x + 15y &= 95 \quad \text{(Multiply the top equation by } -5\text{)} \\
10x + 14y &= 50 \quad \text{(Multiply the bottom equation by 2)} \\
29y &= 145 \quad \text{(Now add)} \\
y &= 5 \quad \text{(Divide by 29 to find } y\text{)}
\end{aligned}
$$

Replace the bottom equation to obtain the modified, equivalent system:

$$2x - 3y = -19$$
$$y = 5$$

Now substitute this value of y into one of the equations. We use the first to obtain

$$
\begin{aligned}
2x - 3(5) &= -19 \\
2x &= -4 \\
x &= -2
\end{aligned}
$$

The solution to the original system is $(-2, 5)$. Figure 2–3 shows the graph of the solution.

■ **Now You Are Ready to Work Exercise 17**

> ### NOTE
>
> Each of the original equations was multiplied by a number so that the resulting equations have the same coefficients of x, except for the sign. Then it was easy to eliminate x from the second equation by adding.
>
> It is a good practice to check your solutions because errors in arithmetic sometimes occur. To check, substitute your solution into *all* equations in the original system. If your solution fails to satisfy one or more of the equations, an error has occurred.

Observe that the elimination method follows this general procedure.

Elimination Method

To solve a system of two linear equations in two unknowns by the **elimination method:**

1. Multiply one or both equations by appropriate constants so that a variable in one equation has a coefficient c and the same variable in the other equation has a coefficient $-c$.
2. Add the equations to eliminate this variable.
3. The resulting equation replaces one of the two equations.

Before mathematics can be used to solve an application, the information given must be converted to a mathematical form. Depending on the application, one of many forms may be appropriate. For our purposes in this chapter, we look at some applications that can be represented with a system of equations and solved by elimination.

EXAMPLE 5 ➤

A woman must control her diet. She selects milk and bagels for breakfast. How much of each should she serve in order to consume 700 calories and 28 grams of protein? Each cup of milk contains 170 calories and 8 grams of protein. Each bagel contains 138 calories and 4 grams of protein.

SOLUTION

Let m be the number of cups of milk and b the number of bagels. Then the total number of calories is

$$170m + 138b$$

and the total protein is

$$8m + 4b$$

So, we need to solve the system

$$\begin{aligned} 8m + \quad 4b &= 28 \\ 170m + 138b &= 700 \end{aligned}$$

Divide the top equation by 4 and the bottom equation by 2 to simplify somewhat:

$$\begin{aligned} 2m + \quad b &= 7 \\ 85m + 69b &= 350 \end{aligned}$$

Next, multiply the top equation by -69 and add it to the bottom equation in order to eliminate b from the second equation:

$$\begin{aligned} -138m - 69b &= -483 \\ \underline{85m + 69b} &= \underline{\quad 350} \\ -53m \qquad\quad &= -133 \\ m = \frac{133}{53} &= 2.509 \quad \text{(rounded)} \end{aligned}$$

Now substitute and solve for b:

$$\begin{aligned} 2(2.509) + b &= 7 \\ 5.018 + b &= 7 \\ b &= 1.982 \end{aligned}$$

It is reasonable to round these answers to 2.5 cups of milk and 2 bagels.

■ **Now You Are Ready to Work Exercise 43**

EXAMPLE 6 ➤

A health food company has two nutritional drinks prepared. One contains 6.25 grams of carbohydrates per ounce, and the second contains 5.125 grams of carbohydrate per ounce. A customer wants 400 ounces of a nutritional drink containing 5.625 grams of carbohydrate per ounce. The company dietitian will mix the two drinks on hand to provide the requested drink. How much of each should be used?

SOLUTION

Let x = number of ounces of the first drink and let y = number of ounces of the second drink. Then,

$$x + y = 400 \quad \text{(Number of ounces of the mixture)}$$
$$6.25x = \text{grams of carbohydrate from the first drink}$$
$$5.125y = \text{grams of carbohydrate from the second drink}$$
$$5.625(400) = 2250 = \text{grams of carbohydrate in the mixture}$$

We summarize this information with the system

$$x + \quad y = \quad 400$$
$$6.25x + 5.125y = 2250$$

We can use the substitution method to solve the system. Substitute $x = 400 - y$ into the second equation:

$$6.25(400 - y) + 5.125y = 2250$$
$$2500 - 6.25y + 5.125y = 2250$$
$$-1.125y = -250$$
$$1.125y = 250$$
$$y = 222.22$$

Rounded to the nearest ounce, the dietitian uses 222 ounces of the second drink and $400 - 222 = 178$ ounces of the first drink.

■ **Now You Are Ready to Work Exercise 49**

EXAMPLE 7 ➤

An electronics firm manufactures DVD players at its plants in Jonesboro and Smithville. The firm receives an order for 1200 players. To support the operation of both plants, the vice-president will use both plants to fill the order. At the Jonesboro plant, the unit cost is $6.60, and the fixed cost is $9360. At the Smithville plant, the unit cost is $7.40, and the fixed cost is $8400. The budget for combined production costs totals $26,200. How many DVD players should each plant make to provide the 1200 DVD players and stay within the budget?

SOLUTION

Let x = the number of players produced at Jonesboro and let y = the number of players produced at Smithville. The total number of players produced is

$$x + y = 1200$$

The total cost at Jonesboro is $C_J = 6.60x + 9360$. The total cost at Smithville is $C_S = 7.40y + 8400$. Because the combined costs are $26,200,

$$(6.60x + 9360) + (7.40y + 8400) = 26{,}200$$

which we can reduce to

$$6.60x + 7.40y = 8440$$

We seek the solution of the system

$$x + \quad y = 1200$$
$$6.60x + 7.40y = 8440$$

(See Figure 2–4.)

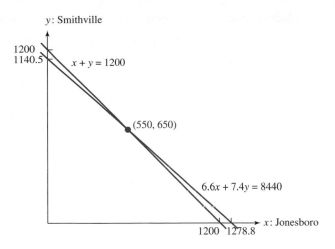

FIGURE 2-4 For combined production costs to be $26,200, 550 DVD players should be produced at Jonesboro, and 650 players should be produced at Smithville.

Substitute $x = 1200 - y$ into the second equation:

$$6.60(1200 - y) + 7.40y = 8440$$
$$0.8y = 520$$
$$y = 650$$
$$x = 1200 - 650 = 550$$

If 550 DVD players are produced at Jonesboro and 650 at Smithville, the 1200 DVD players will be produced within budget.

■ **Now You Are Ready to Work Exercise 51**

Inconsistent Systems

Each system in the preceding examples has exactly one solution. Do not expect this always to be the case. If the equations represent two parallel lines, they have no points in common, and a solution to the system does not exist. We say that this is an **inconsistent system.** Figure 2–5 shows the graph of the two lines

$$-2x + y = 3$$
$$-4x + 2y = 2$$

Each line has slope 2, and they do not intersect, so no solution exists for the system.

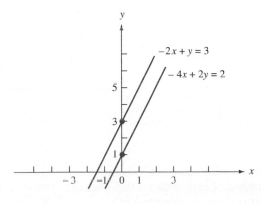

FIGURE 2-5 Each line has a slope 2, and they do not intersect, so no solution exists for this system.

Notice that the equations in this system have the slope-intercept forms

$$y = 2x + 3 \quad and \quad y = 2x + 1$$

In an inconsistent system of two equations, the lines have the same slope but different y-intercepts.

EXAMPLE 8 ➤ The equations of two parallel lines give a system of equations. Let's see what happens when we try to solve such a system:

$$3x - 2y = 5$$
$$6x - 4y = -6$$

We can eliminate x from an equation by multiplying the first equation by -2 and adding the two equations:

$$\begin{array}{r} -6x + 4y = -10 \\ \underline{6x - 4y = -6} \\ 0x + 0y = -16 \quad \text{or} \\ 0 = -16 \end{array}$$

The process of solving the system leads to an inconsistency, $0 = -16$, so the system has *no* solution. You may expect such an inconsistency when attempting to solve a system that represents two different parallel lines.

■ **Now You Are Ready to Work Exercise 29**

Systems That Have Many Solutions

When you graph the lines

$$12x + 9y = 24$$
$$8x + 6y = 16$$

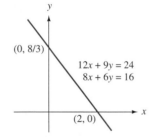

FIGURE 2–6   The graphs of $12x + 9y = 24$ and $8x + 6y = 16$ coincide.

you will find that the graphs coincide, so they represent the same line. When you put both equations in the slope-intercept form, you find that both have slopes $-\frac{4}{3}$ and y-intercept $\frac{8}{3}$, so the lines and their graphs are identical. Because the lines coincide, *every* point on this line is a solution to the given system (see Figure 2–6). Let's look at an example that illustrates what happens when we try to solve such a system.

EXAMPLE 9 ➤ Solve the system

$$12x + 9y = 24$$
$$8x + 6y = 16$$

First, eliminate x from an equation:

$$\begin{array}{rl} 24x + 18y = 48 & \text{(Multiply the first equation by 2)} \\ \underline{-24x - 18y = -48} & \text{(Multiply the second equation by } -3) \\ 0 = 0 & \text{(Add the equations)} \end{array}$$

This transforms the system into

$$12x + 9y = 24$$
$$0 = 0$$

Any solution to the equation $12x + 9y = 24$ is a solution to the system. Since an infinite number of points lie on this line, the system has an infinite number of solutions that may be represented as

$$y = -\frac{4}{3}x + \frac{8}{3}$$

where x may be any number. For example, when $x = 1, 5,$ and -2, we have the solutions

$$x = 1, \qquad y = \frac{4}{3}$$
$$x = 5, \qquad y = -4$$
$$x = -2, \qquad y = \frac{16}{3}$$

■ **Now You Are Ready to Work Exercise 31**

NOTE

Because k can be any real number, an infinite number of solutions exist.

In this example, the variable y can be expressed in terms of x. When this occurs, we call x a **parameter.** In practice, we often use another letter — say, k — to represent an arbitrary value of x. We obtain the corresponding value of y when $x = k$ (k — any real number) is substituted into the equation. When doing so, we can express the infinity of solutions as

$$x = k \qquad \text{and} \qquad y = -\frac{4}{3}k + \frac{8}{3}$$

We call this the **parametric form** of the solution, which we can also write as $\left(k, -\frac{4}{3}k + \frac{8}{3}\right)$.

Application: Supply-and-Demand Analysis

Department stores are well aware that they can sell large quantities of goods if they advertise a reduction in price. The lower the price, the more they sell. Retailers understand this relationship between the price of a commodity and the consumer **demand** (the amount consumers buy). They also know that there may be more than one relationship between price and demand, depending on circumstances. In times of shortages, a different psychology takes effect, and prices tend to *increase* when demand *increases.*

In the competitive situation, a decrease in price can cause an increase in demand (when a store has a sale). This suggests that demand is a function of price. On other occasions, a store may lower prices because an item is in great demand and it expects to increase profits by a greater volume. This suggests that price is a function of demand. Because the cause-and-effect relationship between price and demand can go either way — a change in price causes a change in demand or a change in demand can cause a change in price — we need to decide how to write the demand equation. The analysis is easier if we write the demand equation (and the supply equation in the next paragraph) so that the price is a function of demand (or supply).

EXAMPLE 10 ➤

The Bike Shop held an annual sale. The consumer price and demand relationship for the Ten-Speed Special was

$$p = -2x + 179$$

where x is the number of bikes in demand at the price p. The negative slope, -2, indicates that when an *increase* occurs in one of the variables, price or demand, a decrease occurs in the other. This relationship between price and demand is a linear function. Its graph illustrates the decrease in price with an increase in demand

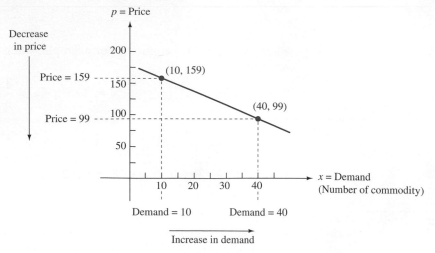

The price-demand function $p = -2x + 179$

FIGURE 2–7 A decrease in price leads to a higher demand.

(see Figure 2–7). When demand increases from 10 to 40, prices drop from $159 to $99.

The Bike Shop cannot lower prices indefinitely because the supplier wants to make a profit also. In a competitive situation, a price increase gives the supplier incentive to produce more. When prices fall, the supplier tends to produce less. The quantity produced by the supplier is called **supply.** Suppose Bike Manufacturing produces the Ten-Speed Special, and the relationship between supply and price is given by the linear function

$$p = 1.5x + 53$$

The graph of this equation illustrates that an increase in price leads to a higher supply (see Figure 2–8). When the price increases from $83 to $128, the supply increases from 20 to 50 units.

Supply and demand are two sides of a perfect competitive market. They interact to determine the price of a commodity. The price of a commodity settles

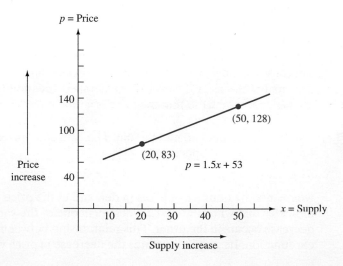

FIGURE 2–8 An increase in price leads to a higher supply.

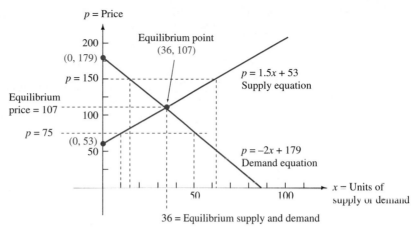

FIGURE 2–9 Equilibrium solution of supply-and-demand equations.

down in the market to one at which the amount willingly supplied and the amount willingly demanded are equal. This price is called the **equilibrium price.** The equilibrium price may be determined by solving a system of equations. In our example, the system is

$$p = -2x + 179$$
$$p = 1.5x + 53$$

Solve the system to obtain the equilibrium solution by either the substitution method or the elimination method. The solution is $x = 36$ and $p = 107$. (Be sure you can find this solution.) The equilibrium price is \$107 when the supply and demand are 36 bikes (see Figure 2–9).

Figure 2–9 also helps us see when there is a surplus or shortage of bikes. For example, draw a horizontal line at $p = 150$. Notice that it intersects the demand equation at about $x = 15$ and it intersects the supply equation at about $x = 65$. Thus, when the price is \$150, there is a demand for 15 bikes, but there are 65 supplied, so a surplus exists. Similarly, when $p = 75$, the demand is about 50, and the supply is about 10, so a shortage exists.

Observe that the p-intercept for the demand equation is \$179. This indicates that the price must be less than \$179 if the shop expects to sell any bikes. Similarly, the p-intercept for the supply equation is 53, so the price must be more than \$53 for the supplier to manufacture bikes.

■ **Now You Are Ready to Work Exercise 41**

▚ 2.1 EXERCISES

LEVEL 1

Solve the systems of equations in Exercises 1 through 16 by substitution.

1. *(See Example 2)*
$4x - y = 5$
$x + 2y = 8$

2. $2x + y = 7$
$x - 3y = 7$

3. $5x - y = -15$
$x + y = -3$

4. $2x - y = -1$
$2x + y = -3$

5. $y = 5x$
$6x - 2y = 12$

6. $x = 5 - 2y$
$3x - y = 15$

7. $7x - y = 32$
$2x + 3y = 19$

8. $x - 2y = -16$
$3x - 4y = -34$

9. $5x + 2y = 14$
$x - 3y = 30$

10. $4x - y = 13$
$3x - 5y = 31$

11. $22x + y = 81$
$8x - 3y = 16$

12. $3x - y = -47$
$x + 2y = 17$

13. *(See Example 3)*
$6x - 3y = 9$
$9x - 15y = 31$

14. $0.06x - y + 1.4 = 0$
$0.07x - 0.04y - 0.62 = 0$

15. $y = 3x - 5$
$8x - 4y - 30 = 0$

16. $5x - y = 8$
$8x - 1.5y = 6.5$

Solve the systems of equations in Exercises 17 through 28 by elimination.

17. *(See Example 4)*
$3x - 4y = 22$
$2x + 5y = 7$

18. $2x + y = 13$
$3x + 5y = 16$

19. $6x - y = 18$
$2x + y = 2$

20. $2x - 3y = -14$
$3x + 4y = -4$

21. $-2x + y = 7$
$6x + 12y = 24$

22. $7x - 2y = 14$
$-14x + 6y = -18$

23. $2x + y = -9$
$4x + 3y = 1$

24. $2x + 3y = 237$
$6x - 2y = 370$

25. $7x + 3y = -1.5$
$2x - 5y = -30.3$

26. $11x - 2y = 387$
$3x + 6y = -189$

27. $2x - 3y = -0.27$
$5x - 2y = 0.04$

28. $49x - 27y + 47 = 0$
$14x - 45y + 30 = 0$

The systems of equations in Exercises 29 through 34 do not have unique solutions. Determine whether each system has no solution or an infinite number of solutions.

29. *(See Example 8)*
$6x - 9y = 8$
$10x - 15y = -20$

30. $8x + 10y = 18$
$20x + 25y = 45$

31. *(See Example 9)*
$8x + 10y = 2$
$12x + 15y = 3$

32. $3x - 7y = 20$
$-6x + 14y = 15$

33. $x - 6y = 4$
$5x - 30y = 20$

34. $4x + 14y = 15$
$6x + 21y = 26$

Determine the equilibrium solutions in Exercises 35 through 40. The demand equation is given first and the supply equation second.

35. $p = -3x + 15$
$p = 2x - 5$

36. $p = -8x + 200$
$p = 3x - 20$

37. $p = -4x + 130$
$p = x - 20$

38. $p = -6x + 68$
$p = 3x - 4$

39. $p = -5x + 83$
$p = 4x - 52$

40. $p = -8x + 2000$
$p = 6x - 800$

LEVEL 2

41. *(See Example 10)* The demand equation for portable television sets is $p = -2.5x + 148$, where x is the demand quantity and p is the price. The supply equation is $p = 1.7x + 43$, where x is the supply quantity and p is the price. Find the equilibrium solution.

42. The research department of a major corporation keeps careful records on its product. It finds that the demand equation is

$$p = -18x + 970$$

where x is the demand and p is the price. The supply equation is

$$p = 15x + 640$$

Find the equilibrium solution.

43. *(See Example 5)* An orange contains 50 mg (milligrams) of calcium and 0.5 mg of iron. An apple contains 8 mg of calcium and 0.4 mg of iron. How many of each are required to obtain 151 mg of calcium and 2.55 mg of iron?

44. Jim has $440 in his savings account and adds $12 per week. At the same time, Rhonda has $260 in her savings account and adds $18 per week.

(a) How long will it take for Rhonda to have the same amount as Jim?
(b) How much will each have?

45. Belmont Records produces CD records. The fixed costs for producing a record are $160,000, and unit costs amount to 85 cents per record. The revenue is $5 per CD.
(a) How many CDs must be sold in order to break even?
(b) Find the profit or loss if 20,000 CDs are sold.
(c) Find the profit or loss if 50,000 CDs are sold.

46. The supply and demand for a CD player in The Discount Store are given for two prices:

Demand	Supply	Price
10	40	$130
25	10	$100

(a) Find the linear demand equation for the CD player.
(b) Find the linear supply equation for the CD player.
(c) Find the equilibrium quantity and price.

LEVEL 3

47. A child has $14.35 worth of nickels and dimes in her piggy bank. There is a total of 165 coins. How many of each does she have?

48. A child has 34 nickels and dimes. Their total value is $2.35. How many of each does he have?

49. *(See Example 6)* A health club asked the dietitian to prepare 600 ounces of a nutritional drink containing 5.5 grams of carbohydrates per ounce. The dietitian has a nutritional drink containing 5.0 grams per ounce and one that contains 5.8 grams of carbohydrate per ounce. How much of each should be used to provide the requested drink?

50. Mrs. Alford invested $5000 in securities. Part of the money was invested at 8% and part at 9%. The total annual income was $415. How much was invested at each rate?

51. *(See Example 7)* An electronics firm manufactures calculators at the McGregor plant and the Ennis plant. At McGregor, the unit cost is $8.40, and the fixed cost is $7480. At Ennis, the unit cost is $7.80, and the fixed cost is $5419. The company wants the two plants to produce a combined total of 1500 calculators, at combined costs of $24,956. How many should be produced at each plant?

52. Action, Inc. makes videotapes at two locations, Atlanta and Baltimore. At Atlanta, the unit cost is $4.30, and the fixed cost is $1840. At Baltimore, the unit cost is $4.70, and the fixed cost is $2200. The company wishes to produce a combined total of 900 tapes at the two locations at combined costs of $8066. How many tapes should be produced at each location?

53. The Kiwanis Club sold citrus fruit to raise money for their scholarship fund. A box of oranges cost $14 and a box of grapefruit cost $16. They sold 502 boxes of citrus for a total of $7570. How many boxes of each fruit did they sell?

54. Wholesale Jewelry receives an order for a total of 800 tie tacks and lapel pins. The tie tacks cost $5.65 each, and the lapel pins cost $7.42 each. A check for $4897.01 is enclosed, but the order fails to specify the number of each. Help the distributor determine how to fill the order.

55. Zest Fruit Juices makes two kinds of fruit punch from apple juice and pineapple juice. The company has 142 gallons of apple juice and 108 gallons of pineapple juice. Each case of Golden Punch requires 4 gallons of apple juice and 6 gallons of pineapple juice. Each case of Light Punch requires 7 gallons of apple juice and 3 gallons of pineapple juice. How many cases of each punch should be made in order to use all the apple and pineapple juice?

56. Urban Developers has two sizes of lots in its development. One sells for $2500 and the other for $3000. One month, the developer sold 22 lots for a total of $61,500. How many of each did it sell?

57. Danny's Bagel Shop has tables that seat two people and tables that seat four people. There are a total of 20 tables, and 66 people can be seated. How many tables of each size are there?

58. Pet Products has two production lines, I and II. Line I can produce 5 tons of regular dog food per hour and 3 tons of premium per hour. Line II can produce 3 tons of regular dog food per hour and 6 tons of premium. How many hours of production should be scheduled in order to produce 360 tons of premium and 460 tons of regular dog food?

59. Mr. Hamilton has $50,000 to invest in a tax-free fund and in a money market fund. The tax-free fund pays 7.4%, and the money market pays 8.8%. How much should he invest in each to get a return of $4071 per year?

60. Cutter and Andrew bring soft drinks and chips for a club party. They buy them at the same store. Cutter buys four six-packs of soft drinks and three bags of chips for a total price of $21.90. Andrew buys two six-packs of soft drinks and five bags of chips for a total price off $22.50. At the end of the party, they have one six-pack of drinks and two bags of chips left. Madeline offers to buy them for the purchase price. How much should she pay?

61. Home Service Corp. has taxable income of $198,000. The federal tax is 20% of taxable income after state taxes have been paid. The state tax is 5% of taxable income after federal taxes have been paid. Find the amount of each tax.

62. The population of two cities, Woodward and Kingman, in 1980 and 1990 were

	1980	1990
Woodward	15,000	19,500
Kingman	12,000	18,000

Assume each continues to grow at a linear rate.
(a) Write the equation of the population of Woodward.
(b) Write the equation of the population of Kingman.
(c) Find when they will have equal populations.

63. Charlie and Susi were planning their annual budget. They agreed they should give $5000 to charitable causes. They could not agree on the charities. Charlie wanted to give some to Habitat for Humanity, Susi wanted to give some to the Family Abuse Center, and both wanted to give to their church. They finally negotiated a rather odd agreement:

- They would give to Habitat for Humanity 60% of what was left after donating to the Family Abuse Center.
- The would give to the Family Abuse Center 40% of what was left after donating to Habitat for Humanity.
- They would give what was left to their church.

(a) Determine, to the nearest $10, the amounts thus budgeted for Habitat for Humanity and for the Family Abuse Center.

(b) Determine the amount left for their church.

64. Professor Hyden, a retired mathematics teacher, had tried to impress upon his granddaughters the importance of mathematics, so in his will he specified that they should share an $800,000 inheritance as follows: Anna and Lauren, the twins, would each receive 40% of the $800,000 *after* the other's share was deducted. The third granddaughter, Beth, would receive the remainder after the twins got their share. Find each girl's share.

65. Discuss why a system of two linear equations in two variables can have exactly one solution but cannot have exactly two solutions.

66. The graph of a system of three linear equations (two variables) has the following form:

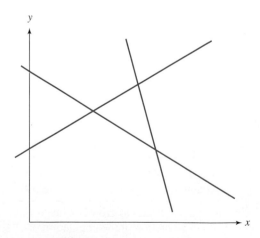

Discuss why the system can or cannot have three solutions.

67. Central Bank is located in the suburbs, has deposits totaling $3,000,000, and the deposits are growing at the rate of $250,000 per year. Citizens Bank is located downtown, has deposits totaling $9,000,000, and the deposits are declining at the rate of $300,000 per year. If these rates continue, find the number of years until the total deposits of Central Bank equal those of Citizens Bank.

68. The table gives the monthly supply, demand, and prices of a new frozen pizza in a chain grocery.

Month	Demand	Supply	Price
1	50,000	20,000	3.00
2	40,000	21,000	3.25
3	29,000	24,000	3.50
4	20,000	30,000	3.75
5	5,000	50,000	4.00

(a) In which months was there a surplus?
(b) In which months was there a shortage?
(c) Give an estimate of the equilibrium quantity and price.

69. Graph each of the following systems and estimate the solution (to two decimals) from the graph.
(a) $y = -2.13x + 17.4$
   $y = 1.19x + 2.34$
(b) $y = 3x + 7$
   $y = -5x + 13$
(c) $y = -3.1x + 8.7$
   $y = 2.3x + 1.9$
(d) $5x - 11y = 23$
   $7x + 3y = 19$

70. Graph each of the following systems of equations. Based on the graph, identify each system as having one solution, no solution, or many solutions.
(a) $4x + 5y = 20$
   $3x - 2y = 6$
(b) $4x - 2y = 9$
   $-6x + 3y = 12$
(c) $x + 3y = 8$
   $2x + 6y = 16$
(d) $y = -4x + 1$
   $y = -4x + 5$
(e) $y = -2x + 1$
   $y = 2x + 3$
(f) $y = -x + 12$
   $y = 2x - 5$
   $y = -4x + 8$

 71. The demand equation for tricycles is $p = -4.5x + 210$, and the supply equation is $p = 2.8x + 60$. Estimate the equilibrium quantity and price.

 72. The demand equation for toasters is $p = -7x + 160$, and the supply equation is $p = 4x + 75$. Estimate the equilibrium quantity and price.

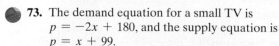 **73.** The demand equation for a small TV is $p = -2x + 180$, and the supply equation is $p = x + 99$.
(a) Draw graphs of the supply and demand equations.

(b) Is there a shortage or a surplus when the price is $110? How much?

(c) Is there a shortage or a surplus when the price is $135? How much?

74. The demand equation for a computer desk is $p = -4x + 240$, and the supply equation is $p = 2x + 60$.

(a) Graph the supply and demand lines.

(b) Find the equilibrium quantity and price.

(c) Find the price at which the buyer stops buying.

(d) Find the price at which the supplier stops supplying.

(e) Is there a shortage or surplus when the price is $100? How much?

(f) Is there a shortage or surplus when the price is $130? How much?

In Exercises 75 through 78, graph both lines on the same sheet and estimate the point of intersection from the graph.

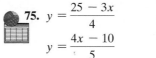
75. $y = \dfrac{25 - 3x}{4}$

$y = \dfrac{4x - 10}{5}$

76. $y = \dfrac{25 - 2x}{3}$

$y = \dfrac{42 - 7x}{4}$

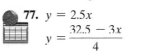
77. $y = 2.5x$

$y = \dfrac{32.5 - 3x}{4}$

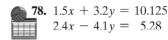
78. $1.5x + 3.2y = 10.125$

$2.4x - 4.1y = 5.28$

USING EXCEL

See **Using EXCEL** in Section 1.3 for the procedure to graph two lines and estimate their intersection.

2.2   Systems with Three Variables: An Introduction to a Matrix Representation of a Linear System of Equations

- Elimination Method
- Matrices
- Matrices and Systems of Equations
- Gauss-Jordan Method
- Application

The tuition–fee cost of courses at Mountainview Community College is determined by a general fee of $250 and tuition of $38 per semester hour and can be represented by a linear equation in two variables:

$$y = 38x + 250$$

where x is the number of semester hours and y is the total tuition–fee cost.

At Valley Junior College, the tuition–fee cost is determined by a general fee of $225, a building use fee of $25 per course, and tuition of $41 per semester hour. The tuition–fee cost can be represented by a linear equation

$$y = 41h + 25x + 225$$

where h is the number of semester hours, x is the number of courses, and y is the total tuition–fee cost.

In the first case, the tuition–fee costs can be represented with a linear equation with two variables. For Valley Junior College, the tuition–fee function is also linear, but it requires the use of three variables.

This illustrates that, as in Section 2.1, many applications can be modeled using a linear equation with two variables, but some linear models require more than two variables.

Elimination Method

Some applications require the solution of two or more linear equations in three variables (or more), so we will use some examples with "nice" numbers to illustrate the procedure for solving such a system.

EXAMPLE 1 ➤

Cutter's mother opened Amy's Online Store so he could purchase some books from the Bargain Baskets for his cousin Andrew's birthday. He was allowed to select from the $1 basket, the $2 basket, and the $3 basket. Based on the following information, determine how many books Cutter selected from each basket:

(a) He selected five books at a total cost of $10.
(b) Shipping costs were $2.00 for each $1 book and $1.00 for each $2 and $3 book.
(c) The total shipping cost was $6.00

SOLUTION

Let's state this information in mathematical form and determine the number of books chosen from each table. Let

$$x = \text{number of books from the \$1 basket}$$
$$y = \text{number of books from the \$2 basket}$$
$$z = \text{number of books from the \$3 basket}$$

The given information may be written as follows:

$$
\begin{aligned}
x + \;\;y + \;\;z &= \;\;5 \quad \text{(Total number of books)} \\
x + 2y + 3z &= 10 \quad \text{(Total cost of books)} \\
2x + \;\;y + \;\;z &= \;\;6 \quad \text{(Total shipping costs)}
\end{aligned}
$$

The solution to this system of three equations in three variables gives the number of books chosen from each basket. Before we solve this system, let's look at some basic ideas.

A set of values for x, y, z that satisfies all three equations is called a **solution** to the system. *Be sure* you understand that a solution consists of three numbers, one each for x, y, and z.

You know that a linear equation in two variables represents a line in two-dimensional space. However, a linear equation in three variables does not represent a line, it represents a plane in three-dimensional space. A solution to a system of three equations in three variables corresponds to a point that lies in all three planes. Figure 2–10 illustrates some possible ways in which the planes might intersect. If the three planes have just one point in common, the solution will be unique. If the planes have no points in common, there will be no solution to the system. If the planes have many points in common, the system will have many solutions.

We need not stop with a system of three variables. Applications exist that require larger systems with more variables. Larger systems are more difficult to interpret geometrically and are more tedious to solve, but they also can have a unique solution, no solution, or many solutions.

The method of elimination used to solve systems of two equations can be adapted to larger systems. We now solve the system in Example 1. First, we make

Unique solution:

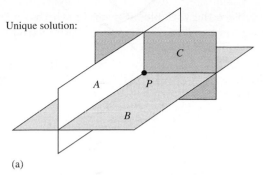

(a)

Three planes A, B, C intersect at a single point P;
P corresponds to a unique solution.

No solutions:

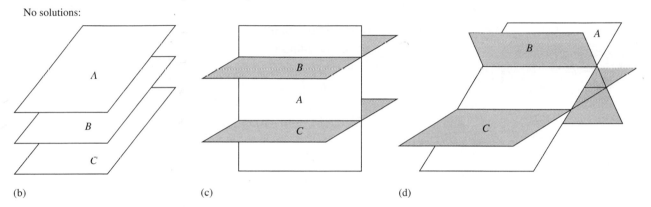

(b)　　　　　　　　　　　　(c)　　　　　　　　　　　　(d)

Planes A, B, C have no point of common intersection, no solution.

Many solutions:

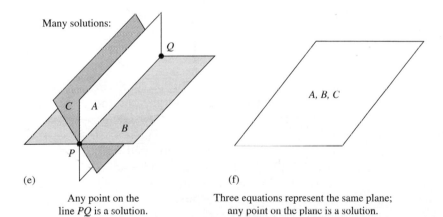

(e)　　　　　　　　　　　　(f)

Any point on the
line PQ is a solution.

Three equations represent the same plane;
any point on the plane is a solution.

FIGURE 2–10　Six possible ways three planes can intersect. (a) Unique Solution: Planes A, B, and C intersect at a single point P; P corresponds to a unique solution. (b, c, and d) No Solutions: Planes A, B, and C have no point of common intersection; no solution. (e and f) Many Solutions: (e) Any point on the line PQ is a solution. (f) Three equations represent the same plane, and any point on the plane is a solution.

a change in notation. Replace the variables x, y, and z with x_1, x_2, and x_3. We do this to emphasize that we can run out of letters when more variables are needed. (Some applications use dozens of variables.) The use of x_1, x_2, x_3, x_4, ... for variables is another way to identify different variables, and it has the advantage of providing a notation for any number of variables. So, here is the system of Example 1 using x_1, x_2, and x_3 :

$$\begin{aligned} x_1 + x_2 + x_3 &= 5 \\ x_1 + 2x_2 + 3x_3 &= 10 \\ 2x_1 + x_2 + x_3 &= 6 \end{aligned}$$

Now let's proceed with the solution. First, eliminate x_1 from the second equation by subtracting the first equation from the second equation to obtain a new second equation:

$$\begin{array}{ll} x_1 + 2x_2 + 3x_3 = 10 & \text{(Equation 2)} \\ \underline{x_1 + x_2 + x_3 = 5} & \text{(Equation 1)} \\ x_2 + 2x_3 = 5 & \text{(This will replace the second equation in the system)} \end{array}$$

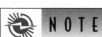

N O T E

Observe that the reason we multiplied the first equation by -2 was to give a coefficient of x_1 that was the negative of the coefficient of x_1 in the third equation. Adding the two equations then eliminated x_1 from the third equation.

As we work through the example, notice that we usually multiply one equation by a constant before adding it to another. The choice of the multiple is determined by the coefficients of the variable we want to eliminate.

Now eliminate x_1 from the third equation by multiplying the first equation by -2 and adding it to the third:

$$\begin{array}{ll} 2x_1 + x_2 + x_3 = 6 & \text{(Equation 3)} \\ \underline{-2x_1 - 2x_2 - 2x_3 = -10} & (-2 \times \text{Equation 1}) \\ -x_2 - x_3 = -4 & \text{(This will replace the third equation in the system)} \end{array}$$

This gives the new and simpler equivalent system

$$\begin{aligned} x_1 + x_2 + x_3 &= 5 \\ x_2 + 2x_3 &= 5 \\ -x_2 - x_3 &= -4 \end{aligned}$$

Observe that the variable x_1 has been eliminated from both the second and third equations. This system of three equations becomes the current system from which we eliminate another variable.

To describe the operations performed in a concise manner, we will use the notation Eq. 2 − Eq. 1 to mean the first equation is subtracted from the second. $-2(\text{Eq. 1}) + \text{Eq. 3}$ means to multiply the first equation by -2 and add it to the third. We continue the process by eliminating more variables from the new system.

Now we can eliminate the variable x_2 from the first and third equations of the current system by using the second equation:

Current System		**Next System**
$x_1 + x_2 + x_3 = 5$	Eq. 1 − Eq. 2 yields	$x_1 - x_3 = 0$
$x_2 + 2x_3 = 5$	remains	$x_2 + 2x_3 = 5$
$-x_2 - x_3 = -4$	Eq. 3 + Eq. 2 yields	$x_3 = 1$

The system on the right is equivalent to the original system and is now the current system. We can complete the solution by eliminating x_3 from the first and second equations of this latest system:

Current System		**Next System**
$x_1 \quad - \quad x_3 = 0$	Eq. 1 + Eq. 3 yields	$x_1 = 1$
$x_2 + 2x_3 = 5$	Eq. 2 − 2(Eq. 3) yields	$x_2 = 3$
$x_3 = 1$	remains	$x_3 = 1$

The solution to this system and, thus, the solution to the original system is $x_1 = 1$, $x_2 = 3$, $x_3 = 1$, which is also written $(1, 3, 1)$. This solution tells us that Cutter chose 1 book from the $1 basket, 3 books from the $2 basket, and 1 book from the $3 basket. (Check the solution in the original system.)

■ **Now You Are Ready to Work Exercise 5**

Matrices

The method of elimination may be used to solve larger systems, but it is tedious, with errors easily made. To make the elimination method more efficient, we convert it to an equivalent procedure that can still be tedious, but less so. The good news is that it can be used to obtain computer solutions of a system. We will learn the **Gauss-Jordan Method,** which uses matrices, a notation that keeps track of the variables in the system but does not write them.

When you worked some solution by elimination exercises, you likely noticed that the coefficients of the variables determined the multipliers used in the operations. Thus, some labor can be saved if we can avoid writing the variables and concentrate on the coefficients. The matrix notation allows that.

Before we show you the Gauss-Jordan Method, we introduce you to matrices. The formal definition of a matrix is the following:

DEFINITION

Matrix

> A **matrix** is a rectangular array of numbers. The numbers in the array are called the **elements of the matrix.** The array is enclosed with brackets.
>
> An array composed of a single row of numbers is called a **row matrix.**
> An array composed of a single column of numbers is called a **column matrix.**

Some examples of matrices are

$$\begin{bmatrix} 1 & 2 & 3 \\ 0 & -1 & 1 \end{bmatrix} \qquad \begin{bmatrix} 2 & 3 \\ 1 & 1 \\ 4 & 1 \end{bmatrix} \qquad \begin{bmatrix} 1 & 2 & 3 \\ 4 & 5 & 6 \\ 0 & 1 & 2 \end{bmatrix} \qquad \begin{bmatrix} 2.5 & 8.3 \end{bmatrix}$$

The location of each element in a matrix is described by the row and column in which it lies. Count the rows from the top of the matrix and the columns from the left.

	Col. 1	Col. 2	Col. 3	Col. 4
Row 1	4	6	−3	2
Row 2	1	5	9	−2
Row 3	7	8	3	4

The element 9 is in row 2 and column 3 of the matrix. We call this location the $(2, 3)$ location; the row is indicated first and the column second. A standard notation for the number in that location is a_{23} (read as "a sub two-three"), so

$a_{23} = 9$. The element 7, designated $a_{31} = 7$, is in the $(3, 1)$ location; and -3, designated $a_{13} = -3$, is in the $(1, 3)$ location.

EXAMPLE 2 ➤

For the matrix

$$\begin{bmatrix} 2 & -6 & -5 & -1 & 0 \\ 1 & 7 & 6 & -4 & 4 \\ 9 & 5 & -8 & 3 & -2 \end{bmatrix}$$

find the following:

(a) The $(1, 1)$ element (a_{11})
(b) The $(2, 5)$ element (a_{25})
(c) The $(3, 3)$ element (a_{33})
(d) The location of -4
(e) The location of 0

SOLUTION

(a) 2 is the $(1, 1)$ element $(a_{11} = 2)$.
(b) 4 is the $(2, 5)$ element $(a_{25} = 4)$.
(c) The $(3, 3)$ element is -8 $(a_{33} = -8)$.
(d) -4 is in the $(2, 4)$ location $(a_{24} = -4)$.
(e) 0 is in the $(1, 5)$ location $(a_{15} = 0)$.

■ **Now You Are Ready to Work Exercise 11**

Let's pause to point out that you encounter matrices more often than you might realize. When you go to a fast-food restaurant, the cashier may record your order on a device that resembles Figure 2–11. When you order three Big Burgers, the cashier presses the Big Burger location, the $(1, 2)$ location, to record the number

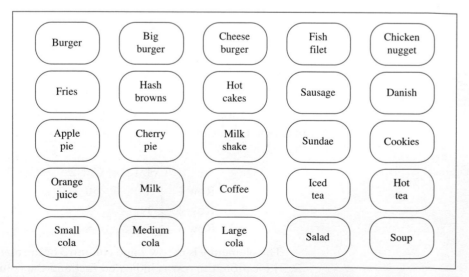

FIGURE 2–11   An example of how matrices are used in the real world.

ordered. Your completed order forms a matrix, with some entries zero, which the computer uses to compute your bill. Both the cashier and the company computer understand that a number in a location counts the number of the item represented by the location.

The rows and columns of a matrix often represent categories. A grade book shows a rectangular array of numbers. Each column represents a test, and each row represents a student.

Matrices and Systems of Equations

Let's use the following system to see how matrices relate to systems of equations.

$$\begin{aligned} 2x_1 + x_2 - x_3 &= 5 \\ 3x_1 + 5x_2 + 2x_3 &= 11 \\ x_1 - 2x_2 + x_3 &= -1 \end{aligned}$$

We form one matrix by using only the coefficients of the system. This gives the **coefficient matrix:**

$$\begin{bmatrix} 2 & 1 & -1 \\ 3 & 5 & 2 \\ 1 & -2 & 1 \end{bmatrix}$$

Column 1 lists the coefficients of x_1, column 2 lists the coefficients of x_2, and column 3 lists the coefficients of x_3. Notice that the entries in each column are listed in the order first, second, and third equations, so the first row represents the left-hand side of the first equation, and so on.

A matrix that also includes the numbers on the right-hand side of the equation is called the **augmented matrix** of the system:

$$\left[\begin{array}{ccc|c} 2 & 1 & -1 & 5 \\ 3 & 5 & 2 & 11 \\ 1 & -2 & 1 & -1 \end{array}\right]$$

The augmented matrix gives complete information, in a compact form, about a system of equations, provided that we agree that each row represents an equation and each column, except the last, consists of the coefficients of a variable.

Generally, we place a vertical line between the coefficients and the column of constant terms. This gives a visual reminder of the location of the equal sign in the equations.

EXAMPLE 3 ➤

Write the coefficient matrix and the augmented matrix of the system

$$\begin{aligned} 5x_1 - 7x_2 + 2x_3 &= 17 \\ -x_1 + 3x_2 + 8x_3 &= 12 \\ 6x_1 + 9x_2 - 4x_3 &= -23 \end{aligned}$$

SOLUTION

The coefficient matrix is

$$\begin{bmatrix} 5 & -7 & 2 \\ -1 & 3 & 8 \\ 6 & 9 & -4 \end{bmatrix}$$

and the augmented matrix is

$$\left[\begin{array}{ccc|c} 5 & -7 & 2 & 17 \\ -1 & 3 & 8 & 12 \\ 6 & 9 & -4 & -23 \end{array}\right]$$

■ **Now You Are Ready to Work Exercise 13**

Write the system of linear equations represented by the augmented matrix

$$\left[\begin{array}{cccc|c} 3 & 7 & 2 & -3 & 8 \\ 4 & 0 & -5 & 7 & -2 \end{array}\right]$$

SOLUTION

The system is

$$3x_1 + 7x_2 + 2x_3 - 3x_4 = 8$$
$$4x_1 \qquad - 5x_3 + 7x_4 = -2$$

■ **Now You Are Ready to Work Exercise 19**

We can solve a system of linear equations by using its augmented matrix. Since each row in the matrix represents an equation, we perform the same kinds of operations on rows of the matrix as we do on the equations in the system. These **row operations** are as follows:

Row Operations

1. Interchange two rows.
2. Multiply or divide a row by a nonzero constant.
3. Multiply a row by a constant and add it to or subtract from another row.

Two **augmented matrices are equivalent** if one is obtained from the other by using row operations.

These row operations are the same kind of operations that we used on the equations in Section 2.1 to convert a system of equations to an equivalent system.

Gauss-Jordan Method

To illustrate the matrix technique, we solve a system of linear equations using the augmented matrix and row operations. This technique is basically a simplification of the elimination method. You will soon find that even this "simplified" method can be tedious and subject to arithmetic errors. However, a method such as this is widely used on computers to solve systems of equations. Whereas the examples and exercises use equations that yield relatively simple answers, actual applications don't always have nice, neat, unique solutions.

We cannot give you a manual method for solving systems of equations that is quick, simple, and foolproof. We want you to understand and apply systems of equations. When a "messy" system needs to be solved, you need to be aware that solutions are available using computers or a TI graphing calculator.

First, we solve a system of two equations in two variables and then a system of three equations in three variables. In each case, we show the solutions of the systems by both the elimination method and the method using matrices. They are shown in parallel so that you can see the relationship between the two methods.

EXAMPLE 5 ➤

Solve the system of equations

$$x + 3y = 11$$
$$2x - 5y = -22$$

Sequence of Equivalent Systems of Equations	**Corresponding Equivalent Augmented Matrices**

Original system:

$$x + 3y = 11$$
$$2x - 5y = -22$$

Original augmented matrix:

$$\begin{bmatrix} 1 & 3 & | & 11 \\ 2 & -5 & | & -22 \end{bmatrix}$$

Eliminate x from the second equation by multiplying the first equation by -2 and adding to the second:

$$x + 3y = 11$$
$$-11y - -44$$

Get a 0 in the second row, first column by multiplying the first row by -2 and adding it to the second row:

$$\begin{bmatrix} 1 & 3 & | & 11 \\ 0 & -11 & | & -44 \end{bmatrix}$$

Simplify the second equation by dividing by -11:

$$x + 3y = 11$$
$$y = 4$$

Simplify the second row by dividing by -11:

$$\begin{bmatrix} 1 & 3 & | & 11 \\ 0 & 1 & | & 4 \end{bmatrix}$$

Eliminate y from the first equation by multiplying the second equation by -3 and adding it to the first:

$$x = -1$$
$$y = 4$$

Get a 0 in the first row, second column by multiplying the second row by -3 and adding it to the first:

$$\begin{bmatrix} 1 & 0 & | & -1 \\ 0 & 1 & | & 4 \end{bmatrix}$$

Read the solution from this augmented matrix. The first row gives $x = -1$, and the second row gives $y = 4$.

■ **Now You Are Ready to Work Exercise 31**

Some of the details that arise in larger systems do not show up in a system of two equations with two variables, so we now solve a system of three equations with three variables.

EXAMPLE 6 ➤ Solve the system of equations.

$$\begin{aligned} x_1 + x_2 + x_3 &= 5 \\ x_1 + 2x_2 + 3x_3 &= 10 \\ 2x_1 + x_2 + x_3 &= 6 \end{aligned}$$

SOLUTION

This system was solved earlier in Example 1. We use it again so that you can concentrate on the procedure. We show the solution of this system by both the elimination method and the method using matrices. They are shown in parallel so that you can see the relationship between the two methods.

Sequence of Equivalent Systems of Equations	**Corresponding Equivalent Augmented Matrices**
Original system:	*Original augmented matrix:*

$$\begin{aligned} x_1 + x_2 + x_3 &= 5 \\ x_1 + 2x_2 + 3x_3 &= 10 \\ 2x_1 + x_2 + x_3 &= 6 \end{aligned} \qquad \begin{bmatrix} 1 & 1 & 1 & 5 \\ 1 & 2 & 3 & 10 \\ 2 & 1 & 1 & 6 \end{bmatrix}$$

Eliminate x_1 from the second equation by multiplying the first equation by -1 and adding to the second.

Get 0 in the second row, first column by multiplying the first row by -1 and adding to the second.

Eliminate x_1 from the third equation by multiplying the first equation by -2 and adding to the third:

Get 0 in the third row, first column by multiplying the first row by -2 and adding to the third:

$$\begin{aligned} x_1 + x_2 + x_3 &= 5 \\ x_2 + 2x_3 &= 5 \\ -x_2 - x_3 &= -4 \end{aligned} \qquad \begin{bmatrix} 1 & 1 & 1 & 5 \\ 0 & 1 & 2 & 5 \\ 0 & -1 & -1 & -4 \end{bmatrix}$$

Eliminate x_2 from the first equation by multiplying the second equation by -1 and adding to the first.

Get 0 in the first row, second column by multiplying the second row by -1 and adding to the first.

Eliminate x_2 from the third equation by adding the second equation to the third:

Get 0 in the third row, second column by adding the second row to the third:

$$\begin{aligned} x_1 \quad - x_3 &= 0 \\ x_2 + 2x_3 &= 5 \\ x_3 &= 1 \end{aligned} \qquad \begin{bmatrix} 1 & 0 & -1 & 0 \\ 0 & 1 & 2 & 5 \\ 0 & 0 & 1 & 1 \end{bmatrix}$$

Eliminate x_3 from the first equation by adding the third equation to the first.

Get 0 in the first row, third column by adding the third row to the first.

Eliminate x_3 from the second equation by multiplying the third equation by -2 and adding to the second:

$$\begin{aligned} x_1 \quad\quad &= 1 \\ x_2 \quad &= 3 \\ x_3 &= 1 \end{aligned}$$

Get 0 in the second row, third column by multiplying the third row by -2 and adding to the second:

$$\begin{bmatrix} 1 & 0 & 0 & 1 \\ 0 & 1 & 0 & 3 \\ 0 & 0 & 1 & 1 \end{bmatrix}$$

Read the solution from this augmented matrix. The first row gives $x_1 = 1$, the second row gives $x_2 = 3$, and the third row gives $x_3 = 1$.

■ **Now You Are Ready to Work Exercise 37**

This technique of using row operations to reduce an augmented matrix to a simple matrix is called the **Gauss-Jordan Method.** The form of the final matrix is such that the solution to the original system can easily be read from the matrix. Notice that the final matrix in the example preceding was

$$\begin{bmatrix} 1 & 0 & 0 & 1 \\ 0 & 1 & 0 & 3 \\ 0 & 0 & 1 & 1 \end{bmatrix}$$

For the moment, ignore the last column of the matrix. The remaining columns have zeros everywhere except in the $(1, 1)$, $(2, 2)$, and $(3, 3)$ locations. These are called the **diagonal locations.** The Gauss-Jordan Method attempts to reduce the augmented matrix until there are 1's in the diagonal locations and 0's elsewhere (except in the last column). This procedure of obtaining a 1 in one position of a column and making all other entries in that column equal to 0 is called **pivoting.**

When a matrix is reduced to this diagonal form, each row easily shows the value of a variable in the solution. In the matrix above, the rows represent

$$\begin{aligned} x_1 &= 1 \\ x_2 &= 3 \\ x_3 &= 1 \end{aligned}$$

which gives the solution.

We now look at another example and focus attention on the procedure for arriving at this desired diagonal form. In this section, we focus our attention on augmented matrices that can be reduced to this diagonal form. The cases in which the diagonal form is not possible are studied in the next section.

A new notation is introduced in the next example to reduce the writing involved. When we are reducing a matrix and you see

$$\frac{1}{4}\text{R1} \quad \textit{gives} \quad [1 \quad 2 \quad -3 \quad 11] \rightarrow \text{R1}$$

this means that row 1 of the current matrix is divided by 4 and gives the new row $[1 \quad 2 \quad -3 \quad 11]$, which is placed in row 1 of the next matrix. The notation

$-2R2 + R3 \rightarrow R3$ means that row 2 of the current matrix is multiplied by -2 and added to row 3. The result becomes row 3 of the next matrix.

EXAMPLE 7 ➤ Solve the following system by reducing the augmented matrix to the diagonal form.

$$2x_1 - 4x_2 + 6x_3 = 20$$
$$3x_1 - 6x_2 + x_3 = 22$$
$$-2x_1 + 5x_2 - 2x_3 = -18$$

The augmented matrix of this system is

$$\begin{bmatrix} 2 & -4 & 6 & 20 \\ 3 & -6 & 1 & 22 \\ -2 & 5 & -2 & -18 \end{bmatrix}$$

We now use row operations to find the solution to the system.

Matrix	This Operation on Present Matrix	Put in New Row

Need 1 here
$$\begin{bmatrix} ② & -4 & 6 & 20 \\ 3 & -6 & 1 & 22 \\ -2 & 5 & -2 & -18 \end{bmatrix}$$
$(\frac{1}{2})$R1 gives $[1 \quad -2 \quad 3 \quad 10] \rightarrow$ R1

Need 0 here
$$\begin{bmatrix} 1 & -2 & 3 & 10 \\ ③ & -6 & 1 & 22 \\ ⊖2 & 5 & -2 & -18 \end{bmatrix}$$
-3R1 + R2 gives $[0 \quad 0 \quad -8 \quad -8] \rightarrow$ R2
2R1 + R3 gives $[0 \quad 1 \quad 4 \quad 2] \rightarrow$ R3

$$\begin{bmatrix} 1 & -2 & 3 & 10 \\ 0 & 0 & -8 & -8 \\ 0 & 1 & 4 & 2 \end{bmatrix}$$
Interchange R2 and R3

Need 0 here
$$\begin{bmatrix} 1 & ⊖2 & 3 & 10 \\ 0 & 1 & 4 & 2 \\ 0 & 0 & -8 & -8 \end{bmatrix}$$
2R2 + R1 gives $[1 \quad 0 \quad 11 \quad 14] \rightarrow$ R1

Need 1 here
$$\begin{bmatrix} 1 & 0 & 11 & 14 \\ 0 & 1 & 4 & 2 \\ 0 & 0 & ⊖8 & -8 \end{bmatrix}$$
$(-\frac{1}{8})$R3 gives $[0 \quad 0 \quad 1 \quad 1] \rightarrow$ R3

Need 0 here
$$\begin{bmatrix} 1 & 0 & ⑪ & 14 \\ 0 & 1 & ④ & 2 \\ 0 & 0 & 1 & 1 \end{bmatrix}$$
-11R3 + R1 gives $[1 \quad 0 \quad 0 \quad 3] \rightarrow$ R1
-4R3 + R2 gives $[0 \quad 1 \quad 0 \quad -2] \rightarrow$ R2

$$\begin{bmatrix} 1 & 0 & 0 & 3 \\ 0 & 1 & 0 & -2 \\ 0 & 0 & 1 & 1 \end{bmatrix}$$

The last matrix is in diagonal form and represents the system

$$x_1 = 3 \qquad x_2 = -2 \qquad x_3 = 1$$

so the solution is $(3, -2, 1)$.

■ **Now You Are Ready to Work Exercise 41**

We point out that a system of linear equations can be solved without reducing the augmented matrix to diagonal form. Reducing the matrix to a form with zeros below the main diagonal suffices. For example, we use the same system of Example 7 and reduce as follows:

$$\begin{bmatrix} 2 & -4 & 6 & | & 20 \\ 3 & -6 & 1 & | & 22 \\ -2 & 5 & -2 & | & -18 \end{bmatrix}$$

$$\begin{bmatrix} 1 & -2 & 3 & | & 10 \\ 0 & 0 & -8 & | & -8 \\ 0 & 1 & 4 & | & 2 \end{bmatrix}$$

$$\begin{bmatrix} 1 & -2 & 3 & | & 10 \\ 0 & 1 & 4 & | & 2 \\ 0 & 0 & -8 & | & -8 \end{bmatrix}$$

From the last row of this matrix, we have

$$-8x_3 = -8 \quad or \quad x_3 = 1$$

From the second row, we have

$$x_2 + 4x_3 = 2$$

Since $x_3 = 1$, we have

$$x_2 + 4 = 2$$
$$x_2 = -2$$

From the first row, we have

$$x_1 - 2x_2 + 3x_3 = 10$$

Substituting $x_2 = -2$ and $x_3 = 1$, we have

$$x_1 + 4 + 3 = 10$$
$$x_1 = 3$$

Thus, the solution is $(3, -2, 1)$.

Generally, you can reduce the matrix to a form with zeros below the diagonal and then work from the bottom row up substituting the values of the variables.

EXAMPLE 8 ➤   Solve the system

$$\begin{aligned} x_1 - x_2 + x_3 + 2x_4 &= 1 \\ 2x_1 - x_2 \quad\quad + 3x_4 &= 0 \\ -x_1 + x_2 + x_3 + x_4 &= -1 \\ x_2 \quad\quad + x_4 &= 1 \end{aligned}$$

SOLUTION

The augmented matrix of this system is

$$\begin{bmatrix} 1 & -1 & 1 & 2 & | & 1 \\ 2 & -1 & 0 & 3 & | & 0 \\ -1 & 1 & 1 & 1 & | & -1 \\ 0 & 1 & 0 & 1 & | & 1 \end{bmatrix}$$

Performing the indicated row operations produces the following sequence of equivalent augmented matrices:

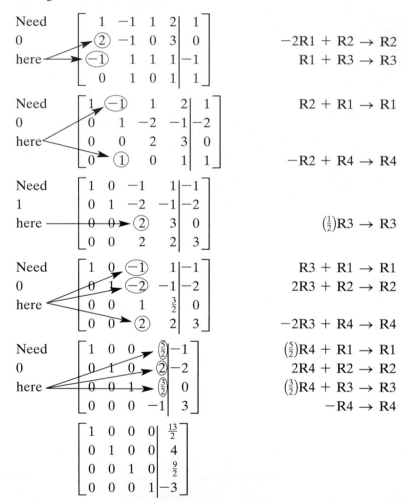

Need
0
here $-2R1 + R2 \rightarrow R2$
     $R1 + R3 \rightarrow R3$

Need $R2 + R1 \rightarrow R1$
0
here
     $-R2 + R4 \rightarrow R4$

Need
1
here $\left(\frac{1}{2}\right)R3 \rightarrow R3$

Need $R3 + R1 \rightarrow R1$
0      $2R3 + R2 \rightarrow R2$
here
     $-2R3 + R4 \rightarrow R4$

Need $\left(\frac{5}{2}\right)R4 + R1 \rightarrow R1$
0      $2R4 + R2 \rightarrow R2$
here $\left(\frac{3}{2}\right)R4 + R3 \rightarrow R3$
     $-R4 \rightarrow R4$

$$\begin{bmatrix} 1 & 0 & 0 & 0 & | & \frac{13}{2} \\ 0 & 1 & 0 & 0 & | & 4 \\ 0 & 0 & 1 & 0 & | & \frac{9}{2} \\ 0 & 0 & 0 & 1 & | & -3 \end{bmatrix}$$

NOTE

In some systems, it might be impossible to get a 1 in some of the diagonal positions. This next section deals with that situation.

This matrix is in the reduced diagonal form, so we can read the solution to the system:

$$x_1 = \frac{13}{2} \qquad x_2 = 4 \qquad x_3 = \frac{9}{2} \qquad x_4 = -3$$

■ **Now You Are Ready to Work Exercise 49**

The examples in this section all have unique solutions. In such cases, the last column of the diagonal form gives the unique solution.

Summary

> ### Solve a System with the Gauss-Jordan Method Using Augmented Matrices
>
> Solve a system of n equations in n variables by using row operations on the augmented matrix.
>
> 1. Write the augmented matrix of the system.
> 2. Get a 1 in the $(1, 1)$ position of the matrix by
>   (a) rearranging rows or
>   (b) dividing row 1 by the $(1, 1)$ entry, a_{11}.
> 3. Get a zero in the other positions of column 1.
>   (a) Take the negative of the number in the position to be made zero and multiply row 1 by it.
>   (b) Add this multiple of row 1 to the row where the zero is needed. The result becomes a new row there.
> 4. Get a 1 in the $(2, 2)$ location by
>   (a) rearranging the rows below row 1 or
>   (b) dividing row 2 by the $(2, 2)$ entry, a_{22}.
> 5. Get a zero in all other positions of column 2 (leave 1 in the $(2, 2)$ position).
>   (a) Take the negative of the number in the position to be made zero and multiply row 2 by it.
>   (b) Add this multiple of row 2 to the row where the zero is needed. The result becomes the new row there.
> 6. Get a 1 in the $(3, 3)$, $(4, 4)$, … positions, and in each case get zeros in the other positions of that column.
> 7. Each row now gives the value of a variable $x_1 = c_1$, $x_2 = c_2$, etc. in the solution.
> 8. If some step in the process produces a row with all zeros except the last column, the system has produced an equation of the form $0 = c$ $(c \neq 0)$, which is a contradiction. Thus, the system has no solution.

Application

EXAMPLE 9 ➤

The Midway High School band sold citrus fruit for a fund-raiser. They sold grapefruit for $16 per box, oranges for $13 per box, and a grapefruit–orange combination for $15 per box. Rhonda, a French horn player, was pleased with her sales. "I sold 57 boxes, and here is the money, $845." The band director insisted he needed to know the number of boxes of each kind of fruit to process the order. "I don't remember the number of each, but I did notice that if we increased the price of a box of grapefruit by $2 and the price of a box of oranges by $1, I would have an even $900."

"That's odd information to notice, but perhaps we can use it to determine the number of each box."

Help Rhonda and the band director determine the number of boxes of each kind of fruit sold.

SOLUTION

Let $x_1 = $ number of boxes of grapefruit
$x_2 = $ number of boxes of oranges
$x_3 = $ number of boxes of grapefruit–orange

We can state the given information as follows:

$$\begin{aligned}
x_1 + x_2 + x_3 &= 57 \quad \text{(Number of boxes sold)} \\
16x_1 + 13x_2 + 15x_3 &= 845 \quad \text{(Amount Rhonda received)} \\
18x_1 + 14x_2 + 15x_3 &= 900 \quad \text{(Amount if prices were increased)}
\end{aligned}$$

The solution to this system gives the desired number of boxes for each kind of fruit. Find the solution by using augmented matrices. The following sequence of matrix reductions leads to the solution:

$$\begin{bmatrix} 1 & 1 & 1 & | & 57 \\ 16 & 13 & 15 & | & 845 \\ 18 & 14 & 15 & | & 900 \end{bmatrix} \quad \begin{matrix} \\ -16R1 + R2 \rightarrow R2 \\ -18R1 + R3 \rightarrow R3 \end{matrix}$$

$$\begin{bmatrix} 1 & 1 & 1 & | & 57 \\ 0 & -3 & -1 & | & -67 \\ 0 & -4 & -3 & | & -126 \end{bmatrix} \quad (-\tfrac{1}{3})R2 \rightarrow R2$$

$$\begin{bmatrix} 1 & 1 & 1 & | & 57 \\ 0 & 1 & \tfrac{1}{3} & | & \tfrac{67}{3} \\ 0 & 4 & 3 & | & 126 \end{bmatrix} \quad \begin{matrix} -R2 + R1 \rightarrow R1 \\ \\ -4R2 + R3 \rightarrow R3 \end{matrix}$$

$$\begin{bmatrix} 1 & 0 & \tfrac{2}{3} & | & \tfrac{104}{3} \\ 0 & 1 & \tfrac{1}{3} & | & \tfrac{67}{3} \\ 0 & 0 & \tfrac{5}{3} & | & \tfrac{110}{3} \end{bmatrix} \quad (\tfrac{3}{5})R3 \rightarrow R3$$

$$\begin{bmatrix} 1 & 0 & \tfrac{2}{3} & | & \tfrac{104}{3} \\ 0 & 1 & \tfrac{1}{3} & | & \tfrac{67}{3} \\ 0 & 0 & 1 & | & 22 \end{bmatrix} \quad \begin{matrix} (-\tfrac{2}{3})R3 + R1 \rightarrow R1 \\ (-\tfrac{1}{3})R3 + R2 \rightarrow R2 \end{matrix}$$

$$\begin{bmatrix} 1 & 0 & 0 & | & 20 \\ 0 & 1 & 0 & | & 15 \\ 0 & 0 & 1 & | & 22 \end{bmatrix}$$

The last matrix gives $x_1 = 20$, $x_2 = 15$, and $x_3 = 22$.

Rhonda sold 20 boxes of grapefruit, 15 boxes of oranges, and 22 boxes of grapefruit–orange.

■ **Now You Are Ready to Work Exercise 59**

▨ 2.2 EXERCISES

LEVEL 1

Use the elimination method to solve the systems in Exercises 1 through 10.

1. $x + 2y = 7$
$3x + 5y = 19$

2. $x - 4y = 0$
$2x + 6y = 14$

3. $2x + 5y = -1$
$6x - 4y = 16$

4. $3x + 7y = -44$
$12x + 5y = -61$

5. *(See Example 1)*
$x + y - z = -1$
$x - y + z = 5$
$x - y - z = 1$

6. $x + y + 2z = 12$
$2x + 3y - z = -2$
$-3x + 4y + z = -8$

7. $\begin{aligned} x + 4y - 2z &= 21 \\ 3x - 6y - 3z &= -18 \\ 2x + 4y + z &= 37 \end{aligned}$ **8.** $\begin{aligned} x + y + z &= 2 \\ 4x + 3y + 2z &= -4 \\ -2x + y - z &= 2 \end{aligned}$

9. $\begin{aligned} 6x + 4y - 6z &= -2 \\ 4x - 3y + z &= 11 \\ 3x + 2y - 2z &= 7 \end{aligned}$ **10.** $\begin{aligned} x + 2y &= 3 \\ y + z &= 3 \\ x + 3z &= 2 \end{aligned}$

11. *(See Example 2)* For the matrix

$$\begin{bmatrix} 2 & 4 & 6 \\ -1 & 3 & 5 \\ 7 & 0 & 6 \\ 9 & 8 & 11 \end{bmatrix}$$

(a) Give the $(1,1)$, $(2,2)$, $(3,3)$, and $(4,3)$ elements.
(b) Determine the location of 5.
(c) Find a_{12}, a_{32}, and a_{41}.

12. For the matrix

$$\begin{bmatrix} 6 & -2 & 3 & 1 & 9 \\ -7 & 5 & 0 & -5 & 4 \\ 8 & -2 & 12 & -2 & 11 \\ -1 & -9 & 15 & -4 & 0 \end{bmatrix}$$

(a) Give the $(1,1)$, $(2,2)$, $(3,4)$, $(4,3)$, $(2,5)$ elements.
(b) Determine the locations that contain -2.
(c) Find a_{35}, a_{24}, and a_{13}.

Determine the matrix of coefficients and the augmented matrix for each of the systems of linear equations in Exercises 13 through 18.

13. *(See Example 3)*
$\begin{aligned} 5x_1 - 2x_2 &= 1 \\ 3x_1 + x_2 &= 7 \end{aligned}$

14. $\begin{aligned} x_1 + x_2 + x_3 &= 9 \\ 5x_1 + 3x_2 + 6x_3 &= 21 \\ x_1 - 4x_2 + x_3 &= 7 \end{aligned}$

15. $\begin{aligned} x_1 + x_2 - x_3 &= 14 \\ 3x_1 + 4x_2 - 2x_3 &= 9 \\ 2x_1 + x_3 &= 7 \end{aligned}$

16. $\begin{aligned} 6x_1 + 4x_2 - x_3 &= 0 \\ -x_1 + x_2 - x_3 &= 7 \\ 5x_1 + &= 15 \end{aligned}$

17. $\begin{aligned} x_1 + 5x_2 - 2x_3 + x_4 &= 12 \\ x_1 - x_2 + 2x_3 + 4x_4 &= -5 \\ 6x_1 + 3x_2 - 11x_3 + x_4 &= 14 \\ 5x_1 - 3x_2 - 7x_3 + x_4 &= 22 \end{aligned}$

18. $\begin{aligned} x_1 - 3x_2 + x_3 + 5x_4 &= 10 \\ 6x_1 - x_2 - x_3 - x_4 &= -18 \\ x_1 + 7x_2 &= 1 \\ x_1 + x_2 + x_3 + x_4 &= 9 \end{aligned}$

Write the systems of equations represented by the augmented matrices in Exercises 19 through 23.

19. *(See Example 4)*
$$\begin{bmatrix} 5 & 3 & -2 \\ -1 & 4 & 4 \end{bmatrix}$$

20. $\begin{bmatrix} 1 & 4 & -1 & 5 \\ 2 & 1 & 2 & 0 \\ -3 & 5 & 1 & 6 \end{bmatrix}$

21. $\begin{bmatrix} 5 & 2 & -1 & 3 \\ -2 & 7 & 8 & 7 \\ 3 & 0 & 1 & 5 \end{bmatrix}$

22. $\begin{bmatrix} 1 & 3 & 2 & 1 \\ 0 & 1 & 5 & 2 \\ 5 & 4 & 0 & 9 \end{bmatrix}$

23. $\begin{bmatrix} 3 & 0 & 2 & 6 & 4 \\ -4 & 5 & 7 & 2 & 2 \\ 1 & 3 & 2 & 5 & 0 \\ -2 & 6 & -5 & 3 & 4 \end{bmatrix}$

24. List the diagonal elements of the following matrices:

(a) $\begin{bmatrix} 2 & 1 & 7 & -2 \\ 5 & 0 & 1 & 4 \\ 7 & 2 & -3 & 14 \end{bmatrix}$

(b) $\begin{bmatrix} 1 & 2 & 3 & 4 & 5 \\ -2 & -6 & 5 & 9 & 14 \\ 8 & 0 & -7 & 6 & 19 \\ 1 & 1 & -5 & 13 & 21 \end{bmatrix}$

Perform the row operations indicated beside each matrix in Exercises 25 through 30. Write the matrix obtained.

25. $\begin{bmatrix} 3 & 6 & -12 & 18 \\ 4 & 2 & 5 & 7 \\ 1 & -1 & 0 & 4 \end{bmatrix}$ $(\tfrac{1}{3})$R1 $\rightarrow$ R1

26. $\begin{bmatrix} 5 & 1 & 4 & 6 \\ 3 & 2 & 1 & 2 \\ 4 & 2 & -3 & 1 \end{bmatrix}$ $-$R3 $+$ R1 $\rightarrow$ R1

27. $\begin{bmatrix} 1 & 3 & 2 & -4 \\ 2 & -1 & 3 & 5 \\ 4 & 6 & -2 & 3 \end{bmatrix}$ $\begin{aligned} -2R1 + R2 &\rightarrow R2 \\ -4R1 + R3 &\rightarrow R3 \end{aligned}$

28. $\begin{bmatrix} 1 & 3 & -2 & 0 \\ 0 & 1 & 1 & 4 \\ 0 & 5 & 2 & 13 \end{bmatrix}$ $\begin{aligned} -3R2 + R1 &\rightarrow R1 \\ -5R2 + R3 &\rightarrow R3 \end{aligned}$

29. $\begin{bmatrix} 1 & -3 & 2 & | & -6 \\ 0 & 5 & -10 & | & 20 \\ 0 & 4 & 3 & | & 8 \end{bmatrix}$ $\left(\frac{1}{5}\right)$R2 $\rightarrow$ R2

30. $\begin{bmatrix} 1 & 0 & 3 & | & 4 \\ 0 & 1 & -6 & | & -1 \\ 0 & 0 & 1 & | & -2 \end{bmatrix}$ $\begin{array}{l} -3\text{R3} + \text{R1} \rightarrow \text{R1} \\ 6\text{R3} + \text{R2} \rightarrow \text{R2} \end{array}$

Solve the following systems of equations by reducing the augmented matrix.

31. *(See Example 5)*
$$2x + 3y = 5$$
$$x - 2y = -1$$

32. $\quad x + 2y = -3$
$$12x + 4y = -12$$

33. $\quad x - 3y = -1$
$$4x + 5y = 30$$

34. $\quad 3x + 7y = -10$
$$15x - 2y = 98$$

35. $2x + 4y = -7$
$$x - 3y = 9$$

36. $\quad x + 8y = 13$
$$-5x + 2y = 19$$

37. *(See Example 6)*
$$x_1 + 2x_2 - x_3 = 3$$
$$x_1 + 3x_2 - x_3 = 4$$
$$x_1 - x_2 + x_3 = 4$$

38. $\quad x_1 + x_2 + 2x_3 = 5$
$$3x_1 + 2x_2 = 4$$
$$2x_1 - x_3 = 2$$

39. $2x_1 + 4x_2 + 2x_3 = 6$
$$2x_1 + x_2 + x_3 = 16$$
$$x_1 + x_2 + 2x_3 = 9$$

40. $x_1 + 2x_3 = 5$
$$x_2 - 30x_3 = -17$$
$$x_1 - 2x_2 + 4x_3 = 10$$

41. *(See Example 7)*
$$x_1 + 2x_2 - x_3 = -1$$
$$2x_1 - 3x_2 + 2x_3 = 15$$
$$x_2 + 4x_3 = -7$$

42. $x_1 + 2x_2 + x_3 = 3$
$$x_2 - x_3 = -1$$
$$x_2 + 5x_3 = 14$$

43. $\quad x_1 + 5x_2 - x_3 = 1$
$$4x_1 - 2x_2 - 3x_3 = 6$$
$$-3x_1 + x_2 + 3x_3 = -3$$

44. $\quad x_1 - 2x_2 + x_3 = -4$
$$3x_2 + 6x_3 = 12$$
$$1.5x_1 + x_2 - 2.5x_3 = -2$$

45. $\quad 2x_1 + 2x_2 + 4x_3 = 16$
$$4x_1 + 2x_2 - x_3 = 6$$
$$-3x_1 + x_2 - 2x_3 = 0$$

46. $\quad x_1 + 5x_2 - 2x_3 = 4$
$$4x_1 - 2x_2 + 3x_3 = 5$$
$$2x_1 + 5x_3 = 7$$

47. $x_1 + 2x_2 - 3x_3 = -6$
$$x_1 - 3x_2 - 7x_3 = 10$$
$$x_1 - x_2 + x_3 = 10$$

48. $\quad x_1 - x_2 + 4x_3 = -3$
$$2x_1 + x_2 + 2x_3 = 1.5$$
$$4x_1 + 3x_2 + x_3 = 6$$

49. *(See Example 8)*
$$x_1 + x_2 + x_3 + x_4 = 4$$
$$x_1 + 2x_2 - x_3 - x_4 = 7$$
$$2x_1 - x_2 - x_3 - x_4 = 8$$
$$x_1 - x_2 + 2x_3 - 2x_4 = -7$$

50. $x_1 + x_2 + 2x_3 + x_4 = 3$
$$x_1 + 2x_2 + x_3 + x_4 = 2$$
$$x_1 + x_2 + x_3 + 2x_4 = 1$$
$$2x_1 + x_2 + x_3 + x_4 = 4$$

51. $2x_1 + 6x_2 + 4x_3 - 2x_4 = 18$
$$x_1 + 4x_2 - 2x_3 - x_4 = -1$$
$$3x_1 - x_2 - x_3 + 2x_4 = 6$$
$$-x_1 - 2x_2 - 5x_3 = -20$$

52. $\quad x_1 + 3x_2 - x_3 + 2x_4 = -2$
$$-3x_1 + x_2 + x_3 + 3x_4 = -1$$
$$2x_1 - 4x_2 - 2x_3 + x_4 = 3$$
$$2x_2 - 4x_4 = -6$$

Set up the systems of equations needed to solve Exercises 53 through 58. Do *not* solve the systems.

53. Fruit Products makes three kinds of fruit punch, Regular, Premium, and Classic. Each case of Regular punch uses 4 gallons of apple juice, 5 gallons of pineapple juice, and 1 gallon of cranberry juice.

Each case of Premium punch uses 4 gallons of apple juice, 4 gallons of pineapple juice, and 2 gallons of cranberry juice. Each case of Classic punch uses 5 gallons of apple juice, 2 gallons of pineapple juice, and 3 gallons of cranberry juice. The firm has 142 gallons of cranberry juice, 292 gallons of pineapple juice, and 316 gallons of apple juice in stock. Set up the equations that determine how many cases of each kind of punch should be made in order to use all the juices.

54. Dealer's Electric makes three kinds of monitors, A, B, and C, which must be manufactured and tested. A requires 3 hours of manufacturing and 1 hour of testing; B requires 4 hours of manufacturing and 2 hours of testing; C requires 6 hours of manufacturing and 2 hours of testing. The company has 285 hours for manufacturing and 115 hours for testing available. Set up the equations that determine how many of each kind should be made so that a total of 70 monitors is produced.

55. A student club sponsored a jazz concert and charged $3 admission for students, $5 for faculty, and $8 for the general public. The total ticket sales amounted to $2542. Three times as many students bought tickets as faculty. The general public bought twice as many tickets as did the students. Set up the equations that determine how many tickets were sold to each group.

56. The Snack Shop makes three mixes of nuts in the following proportions:

   Mix I: 7 pounds peanuts, 5 pounds cashews, and 1 pound pecans

   Mix II: 3 pounds peanuts, 2 pounds cashews, and 2 pounds pecans

   Mix III: 4 pounds peanuts, 3 pounds cashews, and 3 pounds pecans

The shop has 67 pounds of peanuts, 48 pounds of cashews, and 32 pounds of pecans in stock. How much of each mix can it produce?

57. An investor owns a portfolio of three stocks, X, Y, and Z. On March 1, the closing prices were X, $44; Y, $22; and Z, $64. The value of the investor's portfolio amounted to $20,480. Three weeks later, the closing prices were X, $42; Y, $28; and Z, $62. The investor's portfolio was valued at $20,720. Two weeks later, the closing prices were X, $42; Y, $30; and Z, $60 — with a total portfolio value of $20,580. How many shares of each stock did the investor hold in the portfolio?

58. Watson Electric has production facilities in Valley Mills, Marlin, and Hillsboro. Each one produces radios, stereos, and TV sets. Their production capacities are

   Valley Mills: 10 radios, 12 stereos, and 6 TV sets per hour

Marlin: 7 radios, 10 stereos, and 8 TV sets per hour

Hillsboro: 5 radios, 4 stereos, and 13 TV sets per hour

(a) The firm receives an order for 1365 radios, 1530 stereos, and 1890 TV sets. How many hours should each plant be scheduled in order to produce these amounts?

(b) How many hours should each plant be scheduled to fill an order of 1095 radios, 1230 stereos, and 1490 TV sets?

Solve Exercises 59 through 64.

59. (See Example 9) Andrew and Cutter were assigned to buy refreshments for the History Club meeting. They bought the same items in the following amounts:

   Andrew: 3 six-packs of drinks, 2 bags of chips, and 4 packages of cookies, for a total cost of $22.00.

   Cutter: 2 six-packs of drinks, 4 bags of chips, and 5 packages of cookies, for a total cost of $26.30.

   After the meeting, Madeline agreed to buy the unused items at the purchase cost. She got 1 six-pack of drinks, 2 bags of chips, and 4 packages of cookies, for a total cost of $17.20.
   Find the cost of each item.

60. As part of a promotional campaign, The T-Shirt Company packaged thousands of cartons of T-shirts for its retail outlets. Each carton contained small, medium, and large sizes. Three types of cartons were packed according to the quantities shown in the table. The entries in the table give the number (in dozens) of each size of T-shirt in the carton:

Size	Carton		
	A	B	C
Small	2	5	4
Medium	5	8	6
Large	3	2	10

The promotion was a flop, and the company's warehouse was full of the packed cartons. When the company received an order from T-Shirt Orient for 45 dozen small, 78 dozen medium, and 52 dozen large, it wanted to fill the order with the packed cartons in order to save repacking costs. How many of each carton should it send to fill the order?

61. Ms. Hardin invested $40,000 in three stocks. The first year, stock A paid 6% dividends and increased

3% in value; stock B paid 7% dividends and increased 4% in value; stock C paid 8% dividends and increased 2% in value. If the total dividends were $2730 and the total increase in value was $1080, how much was invested in each stock?

62. Curt has one hour to spend at the athletic club, where he will jog, play handball, and ride a bicycle. Jogging uses 13 calories per minute; handball, 11; and cycling, 7. He jogs as long as he rides the bicycle. How long should he participate in each of these activities in order to use 640 calories?

63. Concert Special has a package deal for three concerts, the Rock Stars, the Smooth Sounds, and the Baroque Band. The customer must buy a ticket to all three concerts in order to get the special prices. This chart shows how ticket prices are divided among the musical groups:

	Ticket Prices		
	High School Students	**College Students**	**Adults**
Rock Stars	$4	$6	$8
Smooth Sounds	$4	$5	$9
Baroque Band	$2	$7	$7

The total ticket sales were $14,980 for the Rock Stars, $14,430 for the Smooth Sounds, and $14,450 for the Baroque Band. Determine the number of tickets sold to high school students, college students, and adults.

64. A study showed that for young women a breakfast containing approximately 23 g (grams) of protein, 49 g of carbohydrate, 20 g of fat, and 460 calories is a nutritious breakfast that prevents a hungry feeling before lunch. The following table shows the content of four breakfast foods. How much of each should

be served to obtain the desired amounts of protein, carbohydrates, fat, and calories?

	Protein	Carbo-hydrates	Fat	Calories
1 cup orange juice	2	24	0	110
1 scrambled egg	8	4	8	120
1 slice bread	2	10	6	100
1 cup skim milk	10	13	0	85

65. Solve Exercise 53.

66. Solve Exercise 54.

67. Solve Exercise 55.

68. Solve Exercise 56.

69. Solve Exercise 57.

70. Solve Exercise 58.

71. Brent Industries manufactures three kinds of plywood, regular interior, regular exterior, and rough cedar, at three different mills. The chart below shows the amounts of each plywood produced at each mill. Find the equations that determine the number of days each mill must operate in order to supply the plywood needed to meet the demand shown.

	Mill A Sheets/day	**Mill B** Sheets/day	**Mill C** Sheets/day	**Demand** Sheets
Regular interior	300	700	400	39,500
Regular exterior	500	900	400	52,500
Rough cedar	200	100	800	12,500

EXPLORATIONS

72. Explain why a system of two equations in three variables cannot have exactly one solution.

73. Sketch a graph to illustrate that a system of three linear equations with two variables can have exactly one solution.

74. In this section, the Gauss-Jordan Method solves a system of equations by reducing the coefficient part of the augmented matrix so that 1's appear in the diagonal positions and 0's elsewhere. A solution can

also be found if the reduced matrix has 0's only below the diagonal, such as

$$\begin{bmatrix} 1 & 3 & -2 & | & 1 \\ 0 & 1 & 4 & | & 2 \\ 0 & 0 & 1 & | & 5 \end{bmatrix}$$

Explain how the solution to the system can be found without reducing the matrix to one with 0's above and below the diagonal.

75. Madeline, a free spirit, reduced an augmented matrix in a different order and obtained

$$\begin{bmatrix} 0 & 0 & 1 & 3 & | & 5 \\ 1 & 0 & 0 & -1 & | & 4 \\ 0 & 0 & 0 & 1 & | & 3 \\ 0 & 1 & 0 & 2 & | & 2 \end{bmatrix}$$

She claimed the solution could easily be obtained from this matrix. Is she correct? If so, find the solution. If not, explain why.

Find the intersection of the lines given in Exercises 76 through 79.

76. $y = 3x + 5$
$y = -2x + 15$

77. $y = 4x - 2$
$y = x + 7$

78. $y = 2.5x - 7.5$
$y = -1.4x - 1.65$

79. $2x + 5y = 10$
$6x + 3y = 18$

80. (a) Set up the augmented matrix of this system of equations:

$$x_1 + x_2 + x_3 = 10$$
$$5x_1 + 3x_2 + 6x_2 = 25.8$$
$$x_1 - 4x_2 + x_3 = 4$$

(b) Use matrix row operations to solve the system.

81. (a) Set up the augmented matrix of the system:

$$1.1x_1 + 1.2x_2 - 1.3x_3 = 29.76$$
$$3.5x_1 + 4.1x_2 - 2.2x_3 = 81.94$$
$$2.3x_1 \qquad + 1.4x_3 = 17.70$$

(b) Use row operations to solve the system.

82. (a) Solve the following system using **intersect:**

$$1.2x - 0.5y = 20$$
$$0.3x + 4y = 12$$

(b) Enter the augmented matrix of the system and solve using row operations.

(c) Compare the solutions obtained in parts (a) and (b).

83. (a) Solve the following system using **intersect:**

$$1.28x - 3.44y = 9.995$$
$$3.80x + 9.22y = 10.44$$

(b) Enter the augmented matrix and solve using row operations.

(c) Compare the answers obtained in parts (a) and (b).

USING YOUR TI-83

Solving a System of Equations Using Row Operations

Let's see how we can solve the following system of equations using row operations on the TI-83 calculator.

$$2x_1 - 3x_2 + x_3 = 25$$
$$6x_1 + x_2 + 5x_3 = 33$$
$$x_1 + 4x_2 - 3x_3 = -29$$

The augmented matrix is entered in matrix [A], using $\boxed{\text{MATRX}}$ <EDIT>, which displays the following screens:

```
MATRIX[A] 3 ×4        MATRIX[A] 3 ×4
[ 2    -3   1   :      -3   1    25   ]
[ 6    1    5   :      1    5    33   ]
[ 1    4   -3   :      4   -3    -29  ]

3,1=1                 3,4=-29
```

Notice that the screen does not show the entire matrix. The left screen shows the left-hand portion of the matrix. The rest of the matrix is viewed by scrolling the screen to the right. Use the $\boxed{>}$ key.

At any time, you can view matrix [A] by the sequence $\boxed{\text{MATRX}}$ <1:[A]3x4> $\boxed{\text{ENTER}}$.

```
[A]
[[2  -3  1   25 ]
 [6  1   5   33 ]
 [1  4  -3  -29]]
```

(continued)

We now solve the system using the following sequence of row operations. Multiply row 1 by 0.5: $\boxed{\text{MATRX}}$ <MATH> <E:*row(.5,[A],1)> $\boxed{\text{ENTER}}$, which displays the following screens:

```
*row(.5,[A],1)          *row(.5,[A],1)
[[1  -1.5  .5  12...     ...  -1.5  .5  12.5]
 [6   1     5  33...     ...   1     5  33. ]
 [1   4    -3 -29...     ...   4    -3 -29 ]]
                         ■
```

The matrix shown is in working memory, not in [A]. To place it in [A], use $\boxed{\text{ANS}}$ $\boxed{\text{STO}}$ [A] $\boxed{\text{ENTER}}$.

Next, obtain a zero in row 2, column 1 by multiplying row 1 by −6 and adding to row 2. Use $\boxed{\text{MATRX}}$ <MATH> <F:*row+(-6,[A],1,2)> $\boxed{\text{ENTER}}$. You should get the following:

```
*row+(-6,[A],1,2        *row+(-6,[A],1,2
[[1  -1.5  .5  12...     ...  -1.5  .5  12.5]
 [0  10    2  -42...     ...  10    2  -42 ]
 [1   4   -3  -29...     ...   4   -3  -29 ]]
■                       ■
```

$\boxed{\text{ANS}}$ $\boxed{\text{STO}}$ [A] $\boxed{\text{ENTER}}$ stores the result in [A].

To get a zero in the (3, 1) position, multiply the first row by −1 and add to the third row. You should get the following:

```
*row+(-1,[A],1,3        *row+(-1,[A],1,3
[[1  -1.5  .5   1...     ....5  .5   12.5 ]
 [0  10    2   -...      ...   2   -42  ]
 [0  5.5  -3.5  -...     ...5  -3.5 -41.5]]
■
```

$\boxed{\text{ANS}}$ $\boxed{\text{STO}}$ [A] $\boxed{\text{ENTER}}$ stores the result in [A].

Now proceed to obtain a 1 in row 2, column 2, and zero in the rest of column 2. Then obtain a 1 in row 3, column 3, and zero in the rest of column 3. The final matrix is

```
*row+(-.2,[A],3,
2
     [[1 0 0  3 ]
      [0 1 0 -5]
      [0 0 1  4 ]]
```

This matrix shows the solution of the system to be $(3, -5, 4)$.

EXERCISES

Use row operations to solve the following systems:

1. $2x_1 + x_2 + 4x_3 = 12$
$-x_1 + 3x_2 + 5x_3 = 8$
$3x_1 + 2x_3 = 9$

2. $x_1 + 3x_2 - 2x_3 = 19$
$2x_1 + x_2 + x_3 = 13$
$5x_1 - 2x_2 + 4x_3 = 7$

3. $4x_1 + 3x_2 + x_3 = 26$
$-5x_1 + 2x_2 + 2x_3 = -22$
$3x_1 + x_2 + 5x_3 = 42$

4. $5x_1 - 3x_2 + x_3 = 31$
$2x_1 + 4x_2 - 3x_3 = 33$
$3x_1 + x_2 + x_3 = 23$

USING EXCEL

SOLVING SYSTEMS OF EQUATIONS USING ROW OPERATIONS

Let's see how we can solve the following system of equations using row operations in EXCEL.

$$2x_1 - 3x_2 + x_3 = 25$$
$$6x_1 + x_2 + 5x_3 = 33$$
$$x_1 + 4x_2 - 3x_3 = -29$$

We enter the augmented matrix on the worksheet in cells A2:D4.

	A	B	C	D
1	Matrix A	in A2:D4		
2	2	-3	1	25
3	6	1	5	33
4	1	4	-3	-29

We will use row operations to obtain a sequence of matrices that eventually give the solution to the system.

First Pivot

Obtain the first matrix by pivoting on the (1,1) entry, 2. We place the matrix obtained by this pivot in cells A7:D9. The first step of the pivot divides the first row by 2 to get a 1 in the (1, 1) position, then we use row operations to get zeros in the rest of the first column.

To obtain a 1 in the (1, 1) position, enter =A2/2 in cell A7 (then press the return key). Then copy A7 in cells B7 through D7 by selecting A7 and placing the cursor on the little square at the lower right corner of A7. See the next figure.

	A	B	C	D
1	Matrix A	in A2:D4		
2	2	-3	1	25
3	6	1	5	33
4	1	4	-3	-29
5				
6	Pivot on the	1, 1 entry	Result in A7:D9	
7	=A2/2			

Place cursor here

Drag across to D7. The next figure shows that this creates the formulas =B2/2, =C2/2, and =D2/2 in B7 through D7.

	A	B	C	D
1	Matrix A	in A2:D4		
2	2	-3	1	25
3	6	1	5	33
4	1	4	-3	-29
5				
6	Pivot on the	1, 1 entry	Result in A7:D9	
7	=A2/2	=B2/2	=C2/2	=D2/2

Here we see the results of these formulas.

	A	B	C	D
1	Matrix A	in A2:D4		
2	2	-3	1	25
3	6	1	5	33
4	1	4	-3	-29
5				
6	Pivot on the	1, 1 entry	Result in A7:D9	
7	1	-1.5	0.5	12.5
8				
9				

(continued)

To complete the pivot, now use the following row operations:

> In cell A8 enter: =-6*A7+A3 and copy it in cells B8 through D8.
> In cell A9 enter: =-1*A7+A4 and copy it in cells B9 through D9.
> The following figure shows the matrix in A7:D9.

Second Pivot

We pivot on the $(2, 2)$ entry, 10, in the matrix in A7:D9 to obtain the next matrix, which we place in A12:D14. Observe that the row operations will operate on rows in the matrix A7:D9 and place the results in the matrix located in A12:D14.

Pivot on the $(2, 2)$ entry with the following formulas (*Note:* I usually enter the pivot row formula first, row A13 here):

> In A12: =1.5*A13+A7
> In A13: =A8/10
> In A14: =-5.5*A13+A9

Copy each formula across the row to the D column. The resulting matrix is shown in A12:D14 in the following figure.

Third Pivot

Pivot on the $(3, 3)$ entry in the A12:D14 matrix and place the result in A17:D19 using the formulas:

> In A17: =-.8*A19+A12
> In A18: =-.2*A19+A13
> In A19: =A14/-4.6

The resulting matrix is the last matrix shown in the sequence of matrices in the following figure, which gives the solution $x_1 = 3$, $x_2 = -5$, $x_3 = 4$.

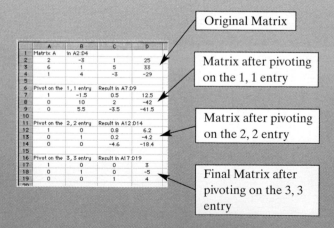

The preceding process can be used in general after we observe the way the row operations are constructed. Let's look at the row operations used to pivot on the $(2, 2)$ entry.

First, we point out that the pivot was made on the matrix in A7:D9 (let's call it the Current Matrix). The matrix formed by pivoting (let's call it the Next Matrix) was located in A12:D14. The following figure identifies the row operations in terms of the Current Matrix and the Next Matrix. Study the figure and then describe the row operations used to pivot on the $(1, 1)$ and the $(3, 3)$ entries in terms of Current Matrix and Next Matrix. Then work the exercises using row operations.

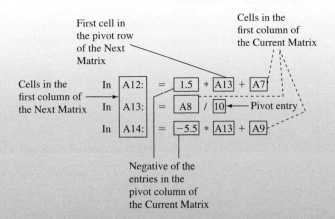

Cells in the first column of the Next Matrix

First cell in the pivot row of the Next Matrix

Cells in the first column of the Current Matrix

In A12: = 1.5 * A13 + A7

In A13: = A8 / 10 ← Pivot entry

In A14: = −5.5 * A13 + A9

Negative of the entries in the pivot column of the Current Matrix

Row 2 of the Current Matrix is the pivot row. Column 2 of the Current Matrix is the pivot column.

EXERCISES

Use row operations to solve the following systems.

1.
$$2x_1 + x_2 + 4x_3 = 12$$
$$-x_1 + 3x_2 + 5x_3 = 8$$
$$3x_1 \qquad + 2x_3 = 9$$

2.
$$x_1 + 3x_2 - 2x_3 = 19$$
$$2x_1 + x_2 + x_3 = 13$$
$$5x_1 - 2x_2 + 4x_3 = 7$$

3.
$$4x_1 + 3x_2 + x_3 = 26$$
$$-5x_1 + 2x_2 + 2x_3 = -22$$
$$3x_1 + x_2 + 5x_3 = 42$$

4.
$$5x_1 - 3x_2 + x_3 = 31$$
$$2x_1 + 4x_2 - 3x_3 = 33$$
$$3x_1 + x_2 + x_3 = 23$$

5.
$$x_1 + 2x_2 - x_3 + 3x_4 = 12$$
$$2x_1 - x_2 + 2x_3 - x_4 = 10$$
$$x_1 + 3x_2 - 2x_3 + 2x_4 = -3$$
$$3x_1 - 2x_2 + 3x_3 - 2x_4 = 13$$

2.3 Gauss-Jordan Method for General Systems of Equations

- Reduced Echelon Form
- Application

In this section, we expand on the Gauss-Jordan Method presented in Section 2.2. In that section, the systems had the same number of equations as variables, dealt mainly with three variables, and had a unique solution. In general, a system may have many variables, it may have more, or fewer, equations than variables, and it may have many solutions or no solution at all. In any case, the Gauss-Jordan Method can be used to solve the system by starting with an augmented matrix as before. Using a sequence of row operations eventually gives a simpler form of the matrix, which yields the solutions. In Section 2.2, the simplified matrices reduced to a diagonal form that gave a unique solution. Some augmented matrices will not reduce to a diagonal form, but they can always be reduced to another standard form called the **reduced echelon form.**

Reduced Echelon Form

We give the general definition of a reduced echelon form. The diagonal forms of the preceding section also conform to this definition.

DEFINITION
Reduced Echelon Form

A matrix is in **reduced echelon form** if all the following are true:

1. All rows consisting entirely of zeros are grouped at the bottom of the matrix.
2. The leftmost nonzero number in each row is 1. This element is called the *leading 1 of the row.*
3. The leading 1 of a row is to the right of the leading 1 of the rows above.
4. All entries above and below a leading 1 are zeros.

The following matrices are all in reduced echelon form. Check the conditions in the definition to make sure you understand why.

$$\begin{bmatrix} 1 & 0 & 0 & 5 \\ 0 & 1 & 0 & -3 \\ 0 & 0 & 1 & 7 \end{bmatrix} \qquad \begin{bmatrix} 1 & 0 & 3 & 0 & 8 \\ 0 & 1 & -1 & 0 & 2 \\ 0 & 0 & 0 & 1 & 7 \end{bmatrix}$$

$$\begin{bmatrix} 1 & 5 & 0 & 2 \\ 0 & 0 & 1 & 3 \\ 0 & 0 & 0 & 0 \end{bmatrix} \qquad \begin{bmatrix} 1 & 0 & -2 & 0 & 0 & 3 & 7 \\ 0 & 1 & 4 & 0 & 0 & -5 & 6 \\ 0 & 0 & 0 & 1 & 0 & 4 & 11 \\ 0 & 0 & 0 & 0 & 1 & 2 & 2 \end{bmatrix}$$

The following matrices are *not* in reduced echelon form.

$$\begin{bmatrix} 1 & 0 & 0 & 2 & 1 \\ 0 & 1 & 0 & 2 & 3 \\ 0 & 0 & 0 & 0 & 0 \\ 0 & 0 & 1 & 0 & 2 \end{bmatrix} \qquad\qquad \begin{bmatrix} 1 & 0 & 0 & 6 \\ 0 & 1 & 0 & 5 \\ 0 & 0 & 4 & 7 \end{bmatrix}$$

The row of zeros is not at the bottom of the matrix.

The leftmost nonzero entry in row 3 (4) is not 1.

$$\begin{bmatrix} 1 & 0 & 0 & 0 & 5 \\ 0 & 1 & 0 & 0 & 3 \\ 0 & 0 & 0 & 1 & 4 \\ 0 & 0 & 1 & 0 & -2 \end{bmatrix} \qquad\qquad \begin{bmatrix} 1 & 0 & 0 & 0 & 7 \\ 0 & 1 & 3 & 0 & -2 \\ 0 & 0 & 1 & 0 & 5 \\ 0 & 0 & 0 & 1 & -1 \end{bmatrix}$$

The leading 1 in row 4 is not to the right of the leading 1 in row 3.

The entry in row 2 above the leading 1 in row 3 is not zero.

We now work through the details of modifying a matrix until we obtain the reduced echelon form. As we work through it, notice how we use row operations to obtain the leading 1 in row 1, row 2, and so on, and then get zeros in the rest of a column with a leading 1.

We use the same row operations that are used in reducing an augmented matrix to obtain a solution to a linear system. Generally, we want to obtain a reduced matrix with 1's on the diagonal.

EXAMPLE 1 ➤ Find the reduced echelon form of the matrix

$$\begin{bmatrix} 0 & 1 & -3 & 2 \\ 2 & 4 & 6 & -4 \\ 3 & 5 & 2 & 2 \end{bmatrix}$$

SOLUTION

	Matrices	Row Operations	Comments
Need 1 here	$\begin{bmatrix} ⓪ & 1 & -3 & 2 \\ 2 & 4 & 6 & -4 \\ 3 & 5 & 2 & 2 \end{bmatrix}$	R2 ↔ R1	Interchange row 1 and row 2 to get nonzero number at top of column 1.
Need 1 here	$\begin{bmatrix} ② & 4 & 6 & -4 \\ 0 & 1 & -3 & 2 \\ 3 & 5 & 2 & 2 \end{bmatrix}$	$(\frac{1}{2})$R1 → R1	Divide row 1 by 2 to get a 1.
Need 0 here	$\begin{bmatrix} 1 & 2 & 3 & -2 \\ 0 & 1 & -3 & 2 \\ ③ & 5 & 2 & 2 \end{bmatrix}$	−3R1 + R3 → R3	Get zeros in rest of column 1.
Leading 1 here	$\begin{bmatrix} 1 & 2 & 3 & -2 \\ 0 & ① & -3 & 2 \\ 0 & -1 & -7 & 8 \end{bmatrix}$		Now get leading 1 in row 2. No changes necessary this time.
Need 0 here	$\begin{bmatrix} 1 & ② & 3 & -2 \\ 0 & 1 & -3 & 2 \\ 0 & ⊖① & -7 & 8 \end{bmatrix}$	−2R2 + R1 → R1 R2 + R3 → R3	Zero entries above and below leading 1 of row 2.
Need leading 1 here	$\begin{bmatrix} 1 & 0 & 9 & -6 \\ 0 & 1 & -3 & 2 \\ 0 & 0 & ⊖⑩ & 10 \end{bmatrix}$	$(-\frac{1}{10})$R3 → R3	Get a leading 1 in the next row.
Need 0 here	$\begin{bmatrix} 1 & 0 & ⑨ & -6 \\ 0 & 1 & ⊖③ & 2 \\ 0 & 0 & 1 & -1 \end{bmatrix}$	−9R3 + R1 → R1 3R3 + R2 → R2	Zero entries above, leading 1 in row 3.
	$\begin{bmatrix} 1 & 0 & 0 & 3 \\ 0 & 1 & 0 & -1 \\ 0 & 0 & 1 & 1 \end{bmatrix}$		This is the reduced echelon form.

■ **Now You Are Ready to Work Exercise 15**

EXAMPLE 2 ➤ Find the reduced echelon form of this matrix:

$$\begin{bmatrix} 0 & 0 & 2 & -2 & 2 \\ 3 & 3 & -3 & 9 & 12 \\ 4 & 4 & -2 & 11 & 12 \end{bmatrix}$$

SOLUTION

Again we show much of the detailed row operations.

	Matrices	**Row Operations**	**Comments**

Need 1 here
$$\left[\begin{array}{cccc|c} ⓪ & 0 & 2 & -2 & 2 \\ 3 & 3 & -3 & 9 & 12 \\ 4 & 4 & -2 & 11 & 12 \end{array}\right]$$
R1 ↔ R2

Need 1 here
$$\left[\begin{array}{cccc|c} ③ & 3 & -3 & 9 & 12 \\ 0 & 0 & 2 & -2 & 2 \\ 4 & 4 & -2 & 11 & 12 \end{array}\right]$$
$(\frac{1}{3})$R1 → R1

Need 0 here
$$\left[\begin{array}{cccc|c} 1 & 1 & -1 & 3 & 4 \\ 0 & 0 & 2 & -2 & 2 \\ ④ & 4 & -2 & 11 & 12 \end{array}\right]$$
$(\frac{1}{2})$R2 → R2
−4R1 + R3 → R3

Leading 1 row 2
$$\left[\begin{array}{cccc|c} 1 & 1 & -1 & 3 & 4 \\ 0 & 0 & ① & -1 & 1 \\ 0 & 0 & 2 & -1 & -4 \end{array}\right]$$

The leading 1 of row 2 must come from row 2 or below. Since all entries in column 2 are zero in rows 2 and 3, go to column 3 for leading 1.

Need 0 here
$$\left[\begin{array}{cccc|c} 1 & 1 & ⑴-1 & 3 & 4 \\ 0 & 0 & 1 & -1 & 1 \\ 0 & 0 & ② & -1 & -4 \end{array}\right]$$
R2 + R1 → R1

−2R2 + R3 → R3

Need 0 here
$$\left[\begin{array}{cccc|c} 1 & 1 & 0 & ② & 5 \\ 0 & 0 & 1 & -1 & 1 \\ 0 & 0 & 0 & 1 & -6 \end{array}\right]$$
−2R3 + R1 → R1
R3 + R2 → R2

$$\left[\begin{array}{cccc|c} 1 & 1 & 0 & 0 & 17 \\ 0 & 0 & 1 & 0 & -5 \\ 0 & 0 & 0 & 1 & -6 \end{array}\right]$$

This is the reduced echelon form.

■

We now solve various systems of equations to illustrate the Gauss-Jordan Method of elimination. As you work through the examples, notice that solving a system of equations using the Gauss-Jordan Method is basically a process of manipulating the augmented matrix to its reduced echelon form.

EXAMPLE 3 ➤   Solve, if possible, the system

$$\begin{array}{rcl} 2x_1 - 4x_2 + 12x_3 &=& 20 \\ -x_1 + 3x_2 + 5x_3 &-& 15 \\ 3x_1 - 7x_2 + 7x_3 &=& 5 \end{array}$$

SOLUTION

We start with the augmented matrix and convert it to reduced echelon form:

$$\begin{bmatrix} 2 & -4 & 12 & | & 20 \\ -1 & 3 & 5 & | & 15 \\ 3 & -7 & 7 & | & 5 \end{bmatrix} \qquad \tfrac{1}{2}\text{R1} \to \text{R1}$$

$$\begin{bmatrix} 1 & -2 & 6 & | & 10 \\ -1 & 3 & 5 & | & 15 \\ 3 & -7 & 7 & | & 5 \end{bmatrix} \qquad \begin{array}{l} \text{R1} + \text{R2} \to \text{R2} \\ -3\text{R1} + \text{R3} \to \text{R3} \end{array}$$

$$\begin{bmatrix} 1 & -2 & 6 & | & 10 \\ 0 & 1 & 11 & | & 25 \\ 0 & -1 & -11 & | & -25 \end{bmatrix} \qquad \begin{array}{l} 2\text{R2} + \text{R1} \to \text{R1} \\ \\ \text{R2} + \text{R3} \to \text{R3} \end{array}$$

$$\begin{bmatrix} 1 & 0 & 28 & | & 60 \\ 0 & 1 & 11 & | & 25 \\ 0 & 0 & 0 & | & 0 \end{bmatrix}$$

This matrix is the reduced echelon form of the augmented matrix. It represents the system of equations

$$\begin{aligned} x_1 \quad + 28x_3 &= 60 \\ x_2 + 11x_3 &= 25 \end{aligned}$$

When the reduced echelon form gives equations containing more than one variable, such as $x_1 + 28x_3 = 60$, the system has many solutions. Many sets of x_1, x_2, and x_3 satisfy these equations. Usually, we solve the first equation for x_1 and the second for x_2 to get

$$\begin{aligned} x_1 &= 60 - 28x_3 \\ x_2 &= 25 - 11x_3 \end{aligned}$$

This represents the general solution with x_1 and x_2 expressed in terms of x_3. When some variables are expressed in terms of another variable, x_3 in this case, we call x_3 a **parameter.** We find specific solutions to the system by substituting values for x_3. For example, one specific solution is found by assigning $x_3 = 1$. Then $x_1 = 60 - 28 = 32$, and $x_2 = 25 - 11 = 14$. In general, we assign the arbitrary value k to x_3 and solve for x_1 and x_2. The arbitrary solution can then be expressed as $x_1 = 60 - 28k$, $x_2 = 25 - 11k$, and $x_3 = k$. As k ranges over the real numbers, we get all solutions. In such a case, k is called a *parameter*. For example, when $k = 2$, we get $x_1 = 4$, $x_2 = 3$, and $x_3 = 2$. When $k = -1$, we get the solution $x_1 = 88$, $x_2 = 36$, and $x_3 = -1$. In summary, the solutions to this example may be written in two ways:

$$\begin{aligned} x_1 &= 60 - 28x_3 \\ x_2 &= 25 - 1 + 1x_3 \end{aligned}$$

or

$$\begin{aligned} x_1 &= 60 - 28k \\ x_2 &= 25 - 11k \\ x_3 &= k \end{aligned}$$

The latter is sometimes written as $(60 - 28k, 25 - 11k, k)$.

■ **Now You Are Ready to Work Exercise 29**

The reduction of an augmented matrix can be tedious. However, this method reduces the solution of a system of equations to a routine. This routine can be carried out by a computer or a graphing calculator. When dozens of variables are involved, a computer is the only practical way to solve such a system. We want you to be able to perform this routine, so we have two more examples to help you.

EXAMPLE 4 ➤   Solve the system by reducing the augmented matrix.

$$
\begin{aligned}
x_1 + 3x_2 - 5x_3 + 2x_4 &= -10 \\
-2x_1 + x_2 + 3x_3 - 4x_4 + 7x_5 &= -22 \\
3x_1 - 7x_2 - 3x_3 - 2x_4 + 4x_5 &= -18
\end{aligned}
$$

We write the augmented matrix and start the process of reducing to echelon form:

$$
\left[\begin{array}{ccccc|c}
1 & 3 & -5 & 2 & 0 & -10 \\
-2 & 1 & 3 & -4 & 7 & -22 \\
3 & -7 & -3 & -2 & 4 & -18
\end{array}\right]
\begin{array}{l}
\\
2\text{R1} + \text{R2} \to \text{R2} \\
-3\text{R1} + \text{R3} \to \text{R3}
\end{array}
$$

$$
\left[\begin{array}{ccccc|c}
1 & 3 & -5 & 2 & 0 & -10 \\
0 & 7 & -7 & 0 & 7 & -42 \\
0 & -16 & 12 & -8 & 4 & 12
\end{array}\right]
\begin{array}{l}
\\
\frac{1}{7}\text{R2} \to \text{R2} \\
-\frac{1}{4}\text{R3} \to \text{R3}
\end{array}
$$

Simplify rows 2 and 3:

$$
\left[\begin{array}{ccccc|c}
1 & 3 & -5 & 2 & 0 & -10 \\
0 & 1 & -1 & 0 & 1 & -6 \\
0 & 4 & -3 & 2 & -1 & -3
\end{array}\right]
\begin{array}{l}
-3\text{R2} + \text{R1} \to \text{R1} \\
\\
-4\text{R2} + \text{R3} \to \text{R3}
\end{array}
$$

$$
\left[\begin{array}{ccccc|c}
1 & 0 & -2 & 2 & -3 & 8 \\
0 & 1 & -1 & 0 & 1 & -6 \\
0 & 0 & 1 & 2 & -5 & 21
\end{array}\right]
\begin{array}{l}
2\text{R3} + \text{R1} \to \text{R1} \\
\text{R3} + \text{R2} \to \text{R2}
\end{array}
$$

$$
\left[\begin{array}{ccccc|c}
1 & 0 & 0 & 6 & -13 & 50 \\
0 & 1 & 0 & 2 & -4 & 15 \\
0 & 0 & 1 & 2 & -5 & 21
\end{array}\right]
$$

We now have the reduced echelon form of the augmented matrix. This matrix represents the system:

$$
\begin{aligned}
x_1 + 6x_4 - 13x_5 &= 50 \\
x_2 + 2x_4 - 4x_5 &= 15 \\
x_3 + 2x_4 - 5x_5 &= 21
\end{aligned}
$$

Solving for x_1, x_2, and x_3 in terms of x_4 and x_5, we get

$$
\begin{aligned}
x_1 &= 50 - 6x_4 + 13x_5 \\
x_2 &= 15 - 2x_4 + 4x_5 \\
x_3 &= 21 - 2x_4 + 5x_5
\end{aligned}
$$

Here, x_1, x_2, and x_3 are solved in terms of x_4 and x_5, so this solution has two parameters, x_4 and x_5. We can select arbitrary values for x_4 and x_5 and use them to obtain values for x_1, x_2, and x_3. We denote this by assigning the arbitrary value k to x_4 and m to x_5. We use them to obtain solutions

$$x_1 = 50 - 6k + 13m$$
$$x_2 = 15 - 2k + 4m$$
$$x_3 = 21 - 2k + 5m$$
$$x_4 = k$$
$$x_5 = m$$

which we can also write as $(50 - 6k + 13m, 15 - 2k + 4m, 21 - 2k + 5m, k, m)$. Specific solutions are obtained when specific values for k and m are selected, such as $k = 5, m = 1$, which yields the solution $(33, 9, 16, 5, 1)$. Because we can choose any real numbers for k and m, this system has an infinite number of solutions.

■ **Now You Are Ready to Work Exercise 33**

It is possible for a system to have **no solution.** We illustrate this in the following example.

EXAMPLE 5 ➤   The following system has no solution. Let's see what happens when we try to solve it.

$$x_1 + 3x_2 - 2x_3 = 5$$
$$4x_1 - x_2 + 3x_3 = 7$$
$$2x_1 - 7x_2 + 7x_3 = 4$$

SOLUTION

$$\begin{bmatrix} 1 & 3 & -2 & | & 5 \\ 4 & -1 & 3 & | & 7 \\ 2 & -7 & 7 & | & 4 \end{bmatrix} \quad \begin{array}{l} -4R1 + R2 \to R2 \\ -2R1 + R3 \to R3 \end{array}$$

$$\begin{bmatrix} 1 & 3 & -2 & | & 5 \\ 0 & -13 & 11 & | & -13 \\ 0 & -13 & 11 & | & -6 \end{bmatrix} \quad -R2 + R3 \to R3$$

$$\begin{bmatrix} 1 & 3 & -2 & | & 5 \\ 0 & -13 & 11 & | & -13 \\ 0 & 0 & 0 & | & 7 \end{bmatrix}$$

The last matrix is not yet in reduced echelon form. However, we need proceed no further because the last row represents the equation $0 = 7$. When we reach an inconsistency like this, we know that the system has no solution.

■ **Now You Are Ready to Work Exercise 35**

Usually, you cannot look at a system of equations and tell whether there is no solution, just one solution, or many solutions. When a system has fewer equations than variables, we generally expect many solutions. Here is such an example.

EXAMPLE 6 ➤   Solve the system

$$x_1 + 2x_2 - x_3 = -3$$
$$4x_1 + 3x_2 + x_3 = 13$$

SOLUTION

Set up the augmented matrix and solve:

$$\begin{bmatrix} 1 & 2 & -1 & | & -3 \\ 4 & 3 & 1 & | & 13 \end{bmatrix} \quad -4R1 + R2 \rightarrow R2$$

$$\begin{bmatrix} 1 & 2 & -1 & | & -3 \\ 0 & -5 & 5 & | & 25 \end{bmatrix} \quad -\tfrac{1}{5}R2 \rightarrow R2$$

$$\begin{bmatrix} 1 & 2 & -1 & | & -3 \\ 0 & 1 & -1 & | & -5 \end{bmatrix} \quad -2R2 + R1 \rightarrow R1$$

$$\begin{bmatrix} 1 & 0 & 1 & | & 7 \\ 0 & 1 & -1 & | & -5 \end{bmatrix}$$

This matrix is in reduced echelon form and represents the equations

$$x_1 = 7 - x_3$$
$$x_2 = -5 + x_3$$

Since x_3 can be chosen arbitrarily, this system has many solutions. Letting $x_3 = k$, the parametric form of this solution is

$$x_1 = 7 - k$$
$$x_2 = -5 + k$$
$$x_3 = k$$

which may be written $(7 - k, -5 + k, k)$.

■ **Now You Are Ready to Work Exercise 39**

You should not conclude from the preceding example that a system with fewer equations than variables will always yield **many solutions.** In some cases, the system contains an inconsistency, and no solution is possible. Here is such an example.

EXAMPLE 7 ➤　Attempt to solve the following system:

$$x_1 + x_2 - x_3 + 2x_4 = 4$$
$$-2x_1 + x_2 + 3x_3 + x_4 = 5$$
$$-x_1 + 2x_2 + 2x_3 + 3x_4 = 6$$

SOLUTION

$$\begin{bmatrix} 1 & 1 & -1 & 2 & | & 4 \\ -2 & 1 & 3 & 1 & | & 5 \\ -1 & 2 & 2 & 3 & | & 6 \end{bmatrix} \quad \begin{array}{l} 2R1 + R2 \rightarrow R2 \\ R1 + R3 \rightarrow R3 \end{array}$$

$$\begin{bmatrix} 1 & 1 & -1 & 2 & | & 4 \\ 0 & 3 & 1 & 5 & | & 13 \\ 0 & 3 & 1 & 5 & | & 10 \end{bmatrix} \quad -R2 + R3 \rightarrow R3$$

$$\begin{bmatrix} 1 & 1 & -1 & 2 & | & 4 \\ 0 & 3 & 1 & 5 & | & 13 \\ 0 & 0 & 0 & 0 & | & -3 \end{bmatrix}$$

The last row of this matrix represents $0 = -3$, an inconsistency, so the system has **no solution.**

■ **Now You Are Ready to Work Exercise 43**

A system with more equations than variables may have a **unique solution,** no solution, or many solutions. The following examples illustrate these cases.

EXAMPLE 8 ➤   If possible, solve the system

$$\begin{aligned} x + 3y &= 11 \\ 3x - 4y &= -6 \\ 2x - 7y &= -17 \end{aligned}$$

SOLUTION

$$\left[\begin{array}{cc|c} 1 & 3 & 11 \\ 3 & -4 & -6 \\ 2 & -7 & -17 \end{array}\right] \quad \begin{array}{l} -3R1 + R2 \to R2 \\ -2R1 + R3 \to R3 \end{array}$$

$$\left[\begin{array}{cc|c} 1 & 3 & 11 \\ 0 & -13 & -39 \\ 0 & -13 & -39 \end{array}\right] \quad -R2 + R3 \to R3$$

$$\left[\begin{array}{cc|c} 1 & 3 & 11 \\ 0 & -13 & -39 \\ 0 & 0 & 0 \end{array}\right] \quad \left(-\tfrac{1}{13}\right)R2 \to R2$$

$$\left[\begin{array}{cc|c} 1 & 3 & 11 \\ 0 & 1 & 3 \\ 0 & 0 & 0 \end{array}\right] \quad -3R2 + R1 \to R1$$

$$\left[\begin{array}{cc|c} 1 & 0 & 2 \\ 0 & 1 & 3 \\ 0 & 0 & 0 \end{array}\right]$$

This reduced echelon matrix gives the solution $x = 2$, $y = 3$, a unique solution.

■ **Now You Are Ready to Work Exercise 47**

EXAMPLE 9 ➤   If possible, solve the system

$$\begin{aligned} x_1 - x_2 + 2x_3 &= 2 \\ 2x_1 + 3x_2 - x_3 &= 14 \\ 3x_1 + 2x_2 + x_3 &= 16 \\ x_1 + 4x_2 - 3x_3 &= 12 \end{aligned}$$

SOLUTION

$$\begin{bmatrix} 1 & -1 & 2 & | & 2 \\ 2 & 3 & -1 & | & 14 \\ 3 & 2 & 1 & | & 16 \\ 1 & 4 & -3 & | & 12 \end{bmatrix} \quad \begin{array}{l} \\ -2R1 + R2 \rightarrow R2 \\ -3R1 + R3 \rightarrow R3 \\ -R1 + R4 \rightarrow R4 \end{array}$$

$$\begin{bmatrix} 1 & -1 & 2 & | & 2 \\ 0 & 5 & -5 & | & 10 \\ 0 & 5 & -5 & | & 10 \\ 0 & 5 & -5 & | & 10 \end{bmatrix} \quad \begin{array}{l} \\ \\ -R2 + R3 \rightarrow R3 \\ -R2 + R4 \rightarrow R4 \end{array}$$

$$\begin{bmatrix} 1 & -1 & 2 & | & 2 \\ 0 & 5 & -5 & | & 10 \\ 0 & 0 & 0 & | & 0 \\ 0 & 0 & 0 & | & 0 \end{bmatrix} \quad \tfrac{1}{5}R2 \rightarrow R2$$

$$\begin{bmatrix} 1 & -1 & 2 & | & 2 \\ 0 & 1 & -1 & | & 2 \\ 0 & 0 & 0 & | & 0 \\ 0 & 0 & 0 & | & 0 \end{bmatrix} \quad R2 + R1 \rightarrow R1$$

$$\begin{bmatrix} 1 & 0 & 1 & | & 4 \\ 0 & 1 & -1 & | & 2 \\ 0 & 0 & 0 & | & 0 \\ 0 & 0 & 0 & | & 0 \end{bmatrix}$$

This matrix represents the system

$$\begin{array}{l} x_1 \quad + x_3 = 4 \\ x_2 - x_3 = 2 \end{array}$$

so there are an infinite number of solutions of the form

$$\begin{array}{l} x_1 = 4 - x_3 \\ x_2 = 2 + x_3 \end{array}$$

or $(4 - k, 2 + k, k)$ in parametric form.

■ **Now You Are Ready to Work Exercise 51**

EXAMPLE 10 ➤

We now attempt to solve a system having no solutions, to observe the effect on the reduced matrix.

We use the system

$$\begin{array}{r} x_1 + 2x_2 - 2x_3 = 5 \\ 3x_1 + x_2 + 4x_3 = 10 \\ x_1 - 2x_2 - 2x_3 = -7 \\ 2x_1 \quad - 4x_3 = 9 \end{array}$$

SOLUTION

The augmented matrix of the system is

$$\left[\begin{array}{ccc|c} 1 & 2 & -2 & 5 \\ 3 & 1 & 4 & 10 \\ 1 & -2 & -2 & -7 \\ 2 & 0 & -4 & 9 \end{array}\right]$$

We now use row operations to reduce the matrix.

$$\left[\begin{array}{ccc|c} 1 & 2 & -2 & 5 \\ 3 & 1 & 4 & 10 \\ 1 & -2 & -2 & -7 \\ 2 & 0 & -4 & 9 \end{array}\right] \qquad \begin{array}{l} \\ -3R1 + R2 \rightarrow R2 \\ -R1 + R3 \rightarrow R3 \\ -2R1 + R4 \rightarrow R4 \end{array}$$

$$\left[\begin{array}{ccc|c} 1 & 2 & -2 & 5 \\ 0 & -5 & 10 & -5 \\ 0 & -4 & 0 & -12 \\ 0 & -4 & 0 & -1 \end{array}\right] \qquad \begin{array}{l} \\ -\frac{1}{5}R2 \rightarrow R2 \\ -\frac{1}{4}R3 \rightarrow R3 \\ \\ \end{array}$$

$$\left[\begin{array}{ccc|c} 1 & 2 & -2 & 5 \\ 0 & 1 & -2 & 1 \\ 0 & 1 & 0 & 3 \\ 0 & -4 & 0 & -1 \end{array}\right] \qquad \begin{array}{l} -2R2 + R1 \rightarrow R1 \\ \\ -R2 + R3 \rightarrow R3 \\ 4R2 + R4 \rightarrow R4 \end{array}$$

$$\left[\begin{array}{ccc|c} 1 & 0 & 2 & 3 \\ 0 & 1 & -2 & 1 \\ 0 & 0 & 2 & 2 \\ 0 & 0 & -8 & 3 \end{array}\right] \qquad \begin{array}{l} -R3 + R1 \rightarrow R1 \\ R3 + R2 \rightarrow R2 \\ \frac{1}{2}R3 \rightarrow R3 \\ 4R3 + R4 \rightarrow R4 \end{array}$$

$$\left[\begin{array}{ccc|c} 1 & 0 & 0 & 1 \\ 0 & 1 & 0 & 3 \\ 0 & 0 & 1 & 1 \\ 0 & 0 & 0 & 11 \end{array}\right]$$

The last matrix is not quite in the reduced echelon form because the last column has not been reduced to a 1 in the last row and zeros elsewhere. However, we need not proceed any further because the matrix in the present form represents the system

$$\begin{array}{rcl} x_1 & = & 1 \\ x_2 & = & 3 \\ x_3 & = & 1 \\ 0 & = & 11 \end{array}$$

Because the system includes a false equation, $0 = 11$, the system cannot be satisfied by any values of x_1, x_2, and x_3. Thus, we must conclude that the original system has no solution.

■ **Now You Are Ready to Work Exercise 55**

Each nonzero row in the reduced echelon matrix gives the value of one variable — either as a number or expressed in terms of another variable. When the system reduces to fewer equations than variables, not enough rows in the matrix exist to give a row for each variable. This means that you can solve only for some of the variables, and they will be expressed in terms of the remaining variables (parameters). Examples 3, 4, and 6 illustrate the relationship between the number of variables solved in terms of parameters. Notice the following:

Example 3 reduces to two equations and three variables. Two of the variables, x_1 and x_2, were solved in terms of one variable, x_3. We call x_3 a parameter.

In Example 4, which has three equations and five variables, three of the variables, x_1, x_2, and x_3, were solved in terms of two variables, x_4 and x_5. We have two parameters in this case, x_4 and x_5.

In Example 6, which has two equations and three variables, two variables, x_1 and x_2, were solved in terms of one variable, x_3. This solution has one parameter, x_3.

The relationship is as follows: If there are k equations with n variables in the reduced echelon matrix, and $n > k$, then k of the variables can be solved in terms of $n - k$ parameters. The system has many solutions.

Whenever a row in a reduced matrix becomes all zeros, then an equation is eliminated from the system, and the number of equations is reduced by one. In the reduced echelon matrix, count the nonzero rows to count the number of equations in the solution.

Application

EXAMPLE 11 ➤

A brokerage firm packaged blocks of common stocks, bonds, and preferred stocks into three different portfolios. The portfolios contained the following:

Portfolio	Common	Bonds	Preferred
I	3 blocks	2 blocks	5 blocks
II	2 blocks	6 blocks	8 blocks
III	5 blocks	8 blocks	13 blocks

A customer wants to buy 110 blocks of common stock, 190 blocks of bonds, and 300 blocks of preferred stock. How many of each portfolio should be purchased to accomplish this?

SOLUTION

Let x, y, and z represent the number of portfolios I, II, and III used. The information given can be stated as a system of equations

$$\begin{aligned} \text{Common stock:} \quad & 3x + 2y + 5z = 110 \\ \text{Bonds:} \quad & 2x + 6y + 8z = 190 \\ \text{Preferred stock:} \quad & 5x + 8y + 13z = 300 \end{aligned}$$

The augmented matrix is

$$\begin{bmatrix} 3 & 2 & 5 & | & 110 \\ 2 & 6 & 8 & | & 190 \\ 5 & 8 & 13 & | & 300 \end{bmatrix}$$

This matrix reduces to the following matrix. (Show that it does.)

$$\begin{bmatrix} 1 & 0 & 1 & | & 20 \\ 0 & 1 & 1 & | & 25 \\ 0 & 0 & 0 & | & 0 \end{bmatrix}$$

The system has an infinite number of solutions

$$x = 20 - z$$
$$y = 25 - z$$

Common sense tells us that x, y, and z cannot be negative, so z can be an integer zero through 20. A larger value of z makes x or y negative. To fill the customer's order, the brokerage firm can use the following number of each portfolio:

Portfolio III: From 0 through 20

Portfolio II: 25 reduced by the number of III used

Portfolio I: 20 reduced by the number of III used

■ **Now You Are Ready to Work Exercise 63**

Summary

The nonzero rows of the reduced echelon matrix give the needed information about the solutions to a system of equations. Three situations are possible.

1. **No solution.**   At least one row has all zeros in the coefficient portion of the matrix (the portion to the left of the vertical line) and a nonzero entry to the right of the vertical line.

$$\begin{bmatrix} 1 & 0 & 0 & | & 3 \\ 0 & 1 & 0 & | & 2 \\ 0 & 0 & 0 & | & 5 \end{bmatrix} \quad \text{(No solution)}$$

Two more possibilities arise when a solution exists.

2. **The solution is unique.**   The number of nonzero rows equals the number of variables in the system.

$$\begin{bmatrix} 1 & 0 & | & 5 \\ 0 & 1 & | & -2 \end{bmatrix} \qquad \begin{bmatrix} 1 & 0 & 0 & | & -4 \\ 0 & 1 & 0 & | & 3 \\ 0 & 0 & 1 & | & 2 \\ 0 & 0 & 0 & | & 0 \end{bmatrix} \quad \text{(Unique solution)}$$

3. **Infinite number of solutions.**   The number of nonzero rows is less than the number of variables in the system.

$$\begin{bmatrix} 1 & 0 & 0 & 2 & | & 3 \\ 0 & 1 & 0 & 5 & | & 2 \\ 0 & 0 & 1 & 3 & | & 4 \end{bmatrix} \qquad \begin{bmatrix} 1 & 0 & 1 & | & 2 \\ 0 & 1 & 2 & | & 4 \\ 0 & 0 & 0 & | & 0 \end{bmatrix} \quad \text{(Infinite number of solutions)}$$

Notice that we solve for the variable in each row where a leading 1 occurs, and that variable is written in terms of the other variables in that row.

NOTE

In the Gauss-Jordan Method, we start by obtaining a 1 in the (1, 1) position of the augmented matrix. Actually, this is not essential. We do not need to conform exactly to this sequence, and we could begin with another column.

▪▪ 2.3 EXERCISES

LEVEL 1

You are expected to recognize when a matrix is in reduced echelon form. Thus, state whether or not the matrices in Exercises 1 through 8 are in reduced echelon form. If a matrix is not in reduced echelon form, explain why it is not.

1. $\begin{bmatrix} 1 & 0 & 0 & 0 & | & 0 \\ 0 & 1 & 0 & 2 & | & 4 \\ 0 & 0 & 0 & 0 & | & 0 \end{bmatrix}$

2. $\begin{bmatrix} 1 & 0 & 0 & 0 & | & -3 \\ 0 & 0 & 1 & 0 & | & 5 \\ 0 & 0 & 0 & 3 & | & 7 \end{bmatrix}$

3. $\begin{bmatrix} 1 & 0 & 0 & | & 11 \\ 0 & 1 & 3 & | & 6 \\ 0 & 0 & 1 & | & 5 \end{bmatrix}$

4. $\begin{bmatrix} 1 & 2 & 0 & 0 & -11 & | & 8 \\ 0 & 0 & 1 & 0 & 13 & | & 10 \\ 0 & 0 & 0 & 1 & 37 & | & 12 \end{bmatrix}$

5. $\begin{bmatrix} 1 & 0 & 0 & | & 4 \\ 0 & 1 & 0 & | & 3 \\ 0 & 0 & 1 & | & 6 \\ 0 & 0 & 0 & | & 0 \\ 0 & 0 & 0 & | & 0 \end{bmatrix}$

6. $\begin{bmatrix} 1 & 0 & 0 & 1 & | & 4 \\ 0 & 1 & 0 & 2 & | & 6 \\ 0 & 0 & 0 & 0 & | & 0 \\ 0 & 0 & 1 & 3 & | & 1 \end{bmatrix}$

7. $\begin{bmatrix} 1 & 0 & 0 & | & 8 \\ 0 & 0 & 1 & | & 5 \\ 0 & 1 & 3 & | & 2 \end{bmatrix}$

8. $\begin{bmatrix} 1 & 0 & 0 & | & 0 \\ 0 & 1 & 0 & | & 0 \\ 0 & 0 & 1 & | & 0 \\ 0 & 0 & 0 & | & 1 \end{bmatrix}$

Each of the matrices in Exercises 9 through 14 is a matrix from a sequence of matrices obtained when reducing a matrix to echelon form. The next step in the sequence is to reduce another column to the appropriate form. Find the next step in the sequence in each exercise.

9. $\begin{bmatrix} 1 & 0 & 2 & | & 5 \\ 0 & 1 & 3 & | & -2 \\ 0 & 0 & 4 & | & 8 \end{bmatrix}$

10. $\begin{bmatrix} 1 & -3 & | & 5 \\ 0 & 2 & | & 4 \\ 0 & -5 & | & 1 \end{bmatrix}$

11. $\begin{bmatrix} 1 & 2 & 3 & | & 4 \\ 0 & 0 & 5 & | & -3 \\ 0 & 1 & 2 & | & 6 \end{bmatrix}$

12. $\begin{bmatrix} 0 & 1 & 2 & | & 1 \\ 2 & 8 & -6 & | & 10 \\ 3 & 4 & -2 & | & 6 \end{bmatrix}$

13. $\begin{bmatrix} 1 & 3 & -1 & 4 & | & 2 \\ 0 & 0 & 2 & 3 & | & 5 \\ 0 & 1 & 4 & -2 & | & 7 \end{bmatrix}$

14. $\begin{bmatrix} 1 & 0 & 0 & 2 & | & 5 \\ 0 & 1 & -3 & 1 & | & 3 \\ 0 & 0 & 0 & 0 & | & 1 \\ 0 & 0 & 0 & 1 & | & 4 \end{bmatrix}$

In Exercises 15 through 20, reduce each matrix to its reduced echelon form.

15. *(See Example 1)*
$\begin{bmatrix} 1 & 2 & -3 & | & 2 \\ 1 & 0 & 3 & | & -2 \\ 3 & 5 & -7 & | & 2 \end{bmatrix}$

16. $\begin{bmatrix} 1 & 4 & -2 & | & 7 \\ 2 & 5 & 2 & | & 2 \\ 0 & 3 & 2 & | & 5 \end{bmatrix}$

17. $\begin{bmatrix} 1 & 3 & 2 & | & 1 \\ 3 & -1 & 4 & | & 9 \\ 2 & -4 & 2 & | & 8 \end{bmatrix}$

18. $\begin{bmatrix} 2 & 6 & | & 4 \\ 3 & 5 & | & 12 \\ -2 & 0 & | & 3 \end{bmatrix}$

19. $\begin{bmatrix} 1 & -1 & 3 & 1 & | & 0 \\ 2 & -2 & 7 & 0 & | & -5 \\ 1 & -1 & 2 & 1 & | & -1 \\ -2 & 2 & 6 & 2 & | & -1 \end{bmatrix}$

20. $\begin{bmatrix} 1 & 4 & 3 & -1 & | & 2 \\ -3 & -5 & 7 & 0 & | & 6 \\ 2 & 1 & -10 & 1 & | & -8 \\ -1 & 3 & 13 & -2 & | & 10 \end{bmatrix}$

Each of the matrices in Exercises 21 through 28 is in reduced echelon form. Write the system of equations represented by each and find the solution, if possible.

21. $\begin{bmatrix} 1 & 0 & 0 & | & 3 \\ 0 & 1 & 0 & | & -2 \\ 0 & 0 & 1 & | & 5 \end{bmatrix}$

22. $\begin{bmatrix} 1 & 0 & 3 & | & -1 \\ 0 & 1 & 2 & | & 5 \\ 0 & 0 & 0 & | & 0 \end{bmatrix}$

23. $\begin{bmatrix} 1 & 0 & 3 & 0 & | & 4 \\ 0 & 1 & 1 & 0 & | & -6 \\ 0 & 0 & 0 & 1 & | & 2 \end{bmatrix}$

24. $\begin{bmatrix} 1 & 0 & 2 & 0 & 5 & | & 3 \\ 0 & 1 & -1 & 0 & 4 & | & 9 \\ 0 & 0 & 0 & 1 & 3 & | & 7 \end{bmatrix}$

25. $\begin{bmatrix} 1 & 0 & | & 0 \\ 0 & 1 & | & 0 \\ 0 & 0 & | & 1 \end{bmatrix}$

26. $\begin{bmatrix} 1 & 0 & 0 & 2 & | & 0 \\ 0 & 1 & 0 & -3 & | & 0 \\ 0 & 0 & 1 & 5 & | & 0 \\ 0 & 0 & 0 & 0 & | & 1 \end{bmatrix}$

27. $\begin{bmatrix} 1 & 0 & 0 & 3 & | & 0 \\ 0 & 1 & 0 & -2 & | & 0 \\ 0 & 0 & 1 & 7 & | & 0 \\ 0 & 0 & 0 & 0 & | & 0 \end{bmatrix}$

28. $\begin{bmatrix} 1 & 0 & 2 & 3 & | & 0 \\ 0 & 1 & -1 & 5 & | & 0 \\ 0 & 0 & 0 & 0 & | & 0 \\ 0 & 0 & 0 & 0 & | & 0 \end{bmatrix}$

Solve each system of equations in Exercises 29 through 61 (if possible).

29. *(See Example 3)*
$$x_1 + 4x_2 - 2x_3 = 13$$
$$3x_1 - x_2 + 4x_3 = 6$$
$$2x_1 - 5x_2 + 6x_3 = -4$$

30. $2x_1 + 4x_2 - x_3 = -4$
$$x_1 + 3x_2 + 6x_3 = -15$$
$$3x_1 + 5x_2 - 8x_3 = 7$$

31. $3x_1 - 2x_2 + 2x_3 = 10$
$$2x_1 + x_2 + 3x_3 = 3$$
$$x_1 + x_2 - x_3 = 5$$

32.
$$\begin{aligned} x_1 + 3x_2 + 6x_3 - 2x_4 &= -7 \\ -2x_1 - 5x_2 - 10x_3 + 3x_4 &= 10 \\ x_1 + 2x_2 + 4x_3 \quad &= 0 \\ x_2 + 2x_3 - 3x_4 &= -10 \end{aligned}$$

33. *(See Example 4)*
$$\begin{aligned} x_1 + x_2 + x_3 - x_4 &= -3 \\ 2x_1 + 3x_2 + x_3 - 5x_4 &= 9 \\ x_1 + 3x_2 - x_3 - 6x_4 &= 7 \end{aligned}$$

34.
$$\begin{aligned} x_1 + 6x_2 - x_3 - 4x_4 &= 0 \\ -2x_1 - 12x_2 + 5x_3 + 17x_4 &= 0 \\ 3x_1 + 18x_2 - x_3 - 6x_4 &= 0 \end{aligned}$$

35. *(See Example 5)*
$$\begin{aligned} x_1 - x_2 + x_3 &= 3 \\ -2x_1 + 3x_2 + x_3 &= -8 \\ 4x_1 - 2x_2 + 10x_3 &= 10 \end{aligned}$$

36.
$$\begin{aligned} x_1 + 4x_2 - 2x_3 &= 10 \\ 3x_1 - x_2 + 4x_3 &= 6 \\ 2x_1 - 5x_2 + 6x_3 &= 7 \end{aligned}$$

37.
$$\begin{aligned} x_1 + x_2 + x_3 &= 6 \\ x_1 - 3x_2 + 2x_3 &= 1 \\ 3x_1 - x_2 + 4x_3 &= 5 \end{aligned}$$

38.
$$\begin{aligned} 2x_1 + 4x_2 + 2x_3 &= 4 \\ 6x_1 + 3x_2 - x_3 &= -5 \\ 7x_1 + 5x_2 \quad &= 8 \end{aligned}$$

39. *(See Example 6)*
$$\begin{aligned} x_1 + 2x_2 - x_3 &= -13 \\ 2x_1 + 5x_2 + 3x_3 &= -3 \end{aligned}$$

40.
$$\begin{aligned} 2x_1 + 3x_2 - 4x_3 &= 1 \\ 4x_1 + x_2 + 2x_3 &= -3 \end{aligned}$$

41.
$$\begin{aligned} x_1 - 2x_2 + x_3 + x_4 - 2x_5 &= -9 \\ 5x_1 + x_2 - 6x_3 - 6x_4 + x_5 &= 21 \end{aligned}$$

42.
$$\begin{aligned} x_1 - 3x_2 + 4x_3 &= 6 \\ 2x_1 - 5x_2 - 6x_3 &= 11 \end{aligned}$$

43. *(See Example 7)*
$$\begin{aligned} x_1 + x_2 - 3x_3 + x_4 &= 4 \\ -2x_1 - 2x_2 + 6x_3 - 2x_4 &= 3 \end{aligned}$$

44.
$$\begin{aligned} 2x_1 - 4x_2 + 16x_3 - 14x_4 &= 12 \\ -x_1 + 5x_2 - 17x_3 + 19x_4 &= -2 \\ x_1 - 3x_2 + 11x_3 - 11x_4 &= 4 \end{aligned}$$

45.
$$\begin{aligned} x_1 + x_2 - x_3 - x_4 &= -1 \\ 3x_1 - 2x_2 - 4x_3 + 2x_4 &= 1 \\ 4x_1 - x_2 - 5x_3 + x_4 &= 5 \end{aligned}$$

46.
$$\begin{aligned} 2x_1 - 3x_2 + x_3 &= 5 \\ -4x_1 + 6x_2 - 2x_3 &= 4 \end{aligned}$$

47. *(See Example 8)*
$$\begin{aligned} x + 4y &= -10 \\ -2x + 3y &= -13 \\ 5x - 2y &= 16 \end{aligned}$$

48.
$$\begin{aligned} x + y &= -5 \\ 4x + 5y &= 2 \\ 3x + y &= 7 \end{aligned}$$

49.
$$\begin{aligned} x - y &= -7 \\ x + y &= -3 \\ 3x - y &= -17 \end{aligned}$$

50.
$$\begin{aligned} 3x + 4y &= 14 \\ 6x - y &= 10 \\ 3x - 5y &= -4 \end{aligned}$$

LEVEL 2

51. *(See Example 9)*
$$\begin{aligned} x_2 + 2x_3 &= 7 \\ x_1 - 2x_2 - 6x_3 &= -18 \\ x_1 - x_2 - 2x_3 &= -5 \\ 2x_1 - 5x_2 - 15x_3 &= -46 \end{aligned}$$

52.
$$\begin{aligned} x_1 + x_2 - x_3 &= 3 \\ x_1 + 2x_2 + 2x_3 &= 10 \\ 2x_1 + 3x_2 + x_3 &= 13 \\ x_1 - 4x_3 &= -7 \end{aligned}$$

53.
$$\begin{aligned} 3x_1 - 2x_2 + 4x_3 &= 4 \\ 2x_1 + 5x_2 - x_3 &= -2 \\ x_1 - 7x_2 + 5x_3 &= 6 \\ 5x_1 + 3x_2 + 3x_3 &= 3 \end{aligned}$$

54.
$$\begin{aligned} x_1 - x_2 - 5x_3 &= 4 \\ x_1 + 2x_2 - x_3 &= 7 \\ 3x_1 + 3x_2 - 7x_3 &= 18 \\ 3x_2 + 4x_3 &= 3 \end{aligned}$$

55. *(See Example 10)*
$$\begin{aligned} 2x - 5y &= 5 \\ 6x + y &= 31 \\ 2x + 11y &= 18 \end{aligned}$$

56.
$$\begin{aligned} 2x_1 - 4x_2 - 14x_3 &= 6 \\ x_1 - x_2 - 5x_3 &= 4 \\ 2x_1 - 4x_2 - 17x_3 &= 9 \\ -x_1 + 3x_2 + 10x_3 &= -3 \\ 2x_2 + 2x_3 &= 4 \end{aligned}$$

57.
$$\begin{aligned} x_1 - x_2 + 2x_3 \quad &= 7 \\ 3x_1 - 4x_2 + 18x_3 - 13x_4 &= 17 \\ 2x_1 - 2x_2 + 2x_3 - 4x_4 &= 12 \\ -x_1 + x_2 - x_3 + 2x_4 &= -6 \\ -3x_1 + x_2 - 8x_3 - 10x_4 &= -21 \end{aligned}$$

58.
$$\begin{aligned} 4x_1 + 8x_2 - 12x_3 &= 28 \\ -x_1 - 2x_2 + 3x_3 &= -7 \\ 2x_1 + 4x_2 - 6x_3 &= 14 \\ -3x_1 - 6x_2 + 9x_3 &= 15 \end{aligned}$$

LEVEL 3

59.
$$x_1 + 2x_2 - x_3 - x_4 = 0$$
$$x_1 + 2x_2 \quad\quad + x_4 = 0$$
$$-x_1 - 2x_2 + 2x_3 + 4x_4 = 0$$
$$-x_1 - x_2 - x_3 \quad\quad = 0$$

60.
$$x_1 + 2x_2 - 2x_3 + 7x_4 + 7x_5 = 0$$
$$-x_1 - 2x_2 + 2x_3 - 9x_4 - 11x_5 = 4$$
$$x_3 - 2x_4 - x_5 = 5$$
$$2x_1 + 4x_2 - 3x_3 + 12x_4 + 13x_5 = 1$$

61.
$$x_1 + 2x_2 - 3x_3 + 2x_4 + 5x_5 - x_6 = 0$$
$$-2x_1 - 4x_2 + 6x_3 - x_4 - 4x_5 + 5x_6 = 0$$
$$3x_1 + 6x_2 - 9x_3 + 5x_4 + 13x_5 - 4x_6 = 0$$

62. A brokerage firm packaged blocks of common stocks, bonds, and preferred stocks into three different portfolios. They contained the following:

> Portfolio I: 3 blocks of common stock, 2 blocks of bonds, and 1 block of preferred stock
>
> Portfolio II: 1 block of common stock, 4 blocks of bonds, and 1 block of preferred stock
>
> Portfolio III: 5 blocks of common stock, 10 blocks of bonds, and 3 blocks of preferred stock

A customer wants to buy 50 blocks of common stock, 160 blocks of bonds, and 25 blocks of preferred stock. Show that it is impossible to fill this order with the portfolios described.

63. *(See Example 11)*   An investor bought $45,000 in stocks, bonds, and money market funds. The total invested in bonds and money market funds was twice the amount invested in stocks. The return on the stocks, bonds, and money market funds was 10%, 7%, and 7.5%, respectively. The total return was $3660. How much was purchased of each?

64. A contractor builds houses, duplexes, and apartment units. He has financial backing to build 250 units. He makes a profit of $4500 on each house, $4000 on each duplex, and $3000 on each apartment unit. Each house requires 10 person-months of labor, each duplex requires 12 person-months, and each apartment requires 6 person-months. How many of each should the contractor build if he has 2050 person-months of labor available and wishes to make a total profit of $875,000?

65. Celia had one hour to spend at the athletic club, where she will jog, play handball, and ride a bicycle. Jogging uses 13 calories per minute; handball, 11; and cycling, 7. She jogs twice as long as she rides the bicycle. How long should she participate in each of these activities in order to use 660 calories?

66. Here is a problem that has been making the rounds of offices and stores. It has most employees stumped. Use your skill to solve it.

A farmer has $100 to buy 100 chickens. Roosters cost $5 each, hens $3 each, and baby chicks 5 cents each. How many of each does the farmer buy if he must buy at least one of each and pay exactly $100 for exactly 100 chickens?

67. At the beginning of a new semester, Andy makes plans for a successful semester. He allocates 42 hours per week for study time for the four courses he is taking: math, English, chemistry, and history. He decides to allocate half of his time to math and English and twice as much time to math as to English. He decides to allocate twice as much time to English as to history.
   (a) Find a system of equations that represents this information.
   (b) Solve the system to determine the number of hours allocated to each subject.

68. Ansel and his friends went to their favorite fast-food place for a late-night snack. They ordered Big Burgers, French fries, and soft drinks. Ansel studied a sheet giving nutrition information about the food items. "Guess what, guys, we just ordered a total of 9360 calories, 477 grams of fat, and 352 grams of sugar." The others questioned him about the calories, fat, and sugar content of their food. He summarized the information in a table.

	Calories	**Fat**	**Sugar**
Big Burger	710	45 g	9 g
French fries	360	18 g	1 g
Soft drink	230	0	56 g

About that time, Basil came in and they told him about their food order. "Gee, how many of those things did you eat?" Their response was for him to figure it out. Help Basil by determining the number of Big Burgers, orders of French fries, and drinks they ordered.

69. Mr. Oliver's income is subject to federal, state, and city taxes. The tax rates are:

> Federal: 40% of taxable income after deducting state and city taxes.
>
> State: 20% of taxable income after deducting federal and city taxes.
>
> City: 10% of taxable income after deducting federal and state taxes.

His taxable income is $58,400. Find the amount of each tax.

70. For their grand opening, Super Sound had special prices on their CDs, their DVD movies, and their videocassettes. Joshua bought 4 CDs, 1 DVD movie, and 2 videocassettes for a total cost of $47. Shilpa bought 3 CDs, 2 DVD movies, and 1 videocassette for a total cost of $45.

(a) From this information, can you find the sale price of each item? Why?

(b) A little later, Lilia bought 1 CD, 3 DVD movies, and 1 videocassette for a total cost of $46. Using the information of the three customers, can you find the cost of each item?

71. The Sound Source has a going-out-of-business sale. Amy, Bill, and Carlton each purchase some CDs, some DVD movies, and some cassette tapes. Amy buys 6 CDs, 2 DVD movies, and 4 cassette tapes for a total cost of $40. Bill buys 3 CDs, 6 DVD movies, and 1 cassette tape for a total cost of $53. Carlton buys 6 CDs, 7 DVD movies, and 3 cassette tapes for a total cost of $73. From the information given, can you find the cost of each item? Why?

EXPLORATIONS

72. The public library budgeted for fiction, nonfiction, and reference books as follows:

Purchase 500 books each month at a total cost of $19,500.

Purchase 50 more fiction than reference books.

The average cost of fiction books is $30, of nonfiction is $40, and of reference is $50.

(a) Find the number of books the library should buy in each category.

(b) Because the library cannot buy a negative number of books, use the solution to find the maximum and minimum number in each category.

73. The Spirit Shop has stores at three locations, and they obtain Homecoming sweatshirts from two suppliers, Sweats-Plus and Imprint-Sweats. During Homecoming Week, Spirit Shop makes the following order:

	No. Sweatshirts Ordered
Spirit Shop 1	15
Spirit Shop 2	20
Spirit Shop 3	30

The suppliers have the following number of the desired sweatshirts in stock:

	No. Sweatshirts in Stock
Sweats-Plus	40
Imprint-Sweats	25

(a) Find a system of equations that represents how the orders may be filled and solve the system.

(b) Give three different ways the orders can be filled. From the solution found in part (a), determine the following and justify your answer:

(c) What is the maximum number of sweatshirts that Imprint-Sweats supplies to Spirit Shop 2?

(d) Can the order be filled if Imprint-Sweats supplies no sweatshirts to Spirit Shop 2 and Spirit Shop 3?

(e) What is the minimum total number of sweatshirts supplied to Spirit Shop 2 and Spirit Shop 3 by Imprint-Sweats?

(f) What is the maximum number of sweatshirts supplied to Spirit Shop 2 by Sweats-Plus?

74. The system

$$x - 2y = 3$$
$$-2x + 4y = 5$$

has no solution. Change the constant in the second equation to give a system that has a solution.

75. The system

$$x + 2y + z = 3$$
$$2x - y + 3z = 2$$
$$3x + y + 4z = 2$$

has no solution. Change the constant in the third equation to give a system that has a solution.

76. The system

$$x - y + z = 2$$
$$3x + 2y - z = 5$$
$$x + 4y - 3z = 4$$

has no solution. Change the constant in the third equation to give a system that has a solution.

77. A group of students were discussing systems of linear equations. The following claims were made by two of the students:

(a) If the system has the same number of equations as variables, then the system has a unique solution.

(b) If the system has fewer equations than variables, then the system has an infinite number of solutions.

(c) If the system has more equations than variables, then the system has no solution.

Which, if any, of these claims are correct? Explain and support your response with examples.

78. Make a system of three linear equations with two variables that has no solution and for which no pair of lines is parallel.

79. Make a system of three linear equations with two variables that has a unique solution.

80. Make a system of three linear equations in three variables that has no solution.

81. A system has four linear equations with four variables. The reduced echelon form of the augmented matrix has two rows of all zeros. What does this information tell about the solutions of the system, if it is known that the system has at least one solution?

82. The city of Tulsa conducts a traffic flow study at the area bounded by Clay Ave., Webster Ave., and 4th and 5th Streets. All of the streets are one-way streets. The map shows the number of vehicles entering or leaving the area. For example, 800 vehicles per hour enter intersection A from Clay Ave., and 300 vehicles per hour leave intersection D on Webster Ave. You are to find the traffic flow on the four blocks between intersections A, B, C, and D.

 (a) Set up a system of equations that represents the relationships between the given traffic flows and the traffic between intersection A and B, B and C, C and D, and D and A.

 (b) Solve the system.

 (c) Determine the maximum and minimum number of vehicles per hour on each street in the block ABCD.

 (d) What is the largest traffic flow on Clay Ave. so that the traffic flow on each of the other three streets is nonnegative? How do you interpret this value?

83. The map shown represents a network of five one-way streets. and we want to analyze the traffic flow within the network bounded by the intersections A, B, C, D, E, and F. The numbers shown indicate the number of vehicles per hour that enter or leave that intersection on the indicated street.

 (a) Write a system of equations that represents the traffic flow in each of the six blocks within the network.

 (b) Solve the system.

 (c) The street department needs to make repairs on 10th St. between Colcord and Blair, so they want the minimum traffic flow. Find the minimum traffic flow for that block and the traffic flow in the other six blocks as a result of cutting back traffic on that block.

84. Uncle Dan's Bar-B-Que has peak crowds at lunch (11:30–1:30) and at evening (5:30–7:00). He schedules his employees in three shifts, 7:00–2:00, 11:00–7:30, and 5:00–10:00, in order to have an overlap of shifts at peak times. He needs 12 employees for the lunch peak and 8 for the evening peak. The early morning is busier than late evening, so he wants the morning schedule (7:00–2:00) to have 5 more employees than the evening schedule (5:00–10:00). Help Uncle Dan write the equations that represent these conditions and determine what you should recommend for the number in each shift.

85. Find the values of c so the system represented by the augmented matrix has no solution.

$$\begin{bmatrix} 1 & 2 & -1 & 3 & | & 5 \\ 2 & 8 & 2 & -2 & | & 6 \\ 1 & 0 & 3 & 1 & | & 3 \\ 4 & 10 & 4 & 2 & | & c \end{bmatrix}$$

86. Use a graphing calculator to find the reduced echelon form of the following matrices.

(a) $\begin{bmatrix} 1 & 3 & 2 & | & 4 \\ 2 & 1 & -1 & | & 6 \\ -2 & 3 & 2 & | & 1 \end{bmatrix}$ (b) $\begin{bmatrix} 1 & 4 & -2 & | & 7 \\ 3 & 0 & 2 & | & 5 \\ 5 & 8 & -2 & | & 19 \end{bmatrix}$

(c) $\begin{bmatrix} 2 & 3 & 2 & | & -3 \\ 1 & 1 & -3 & | & 4 \\ 3 & 4 & -1 & | & 1 \\ 4 & 5 & -4 & | & 5 \\ 1 & 2 & 5 & | & -7 \end{bmatrix}$ (d) $\begin{bmatrix} 2 & 1 & | & 2 \\ -1 & 0 & | & 3 \\ 0 & 1 & | & 8 \\ 3 & 2 & | & 7 \end{bmatrix}$

(e) $\begin{bmatrix} 1 & 3 & 0 & 1 & | & 4 \\ 2 & 1 & 4 & -3 & | & 5 \\ 5 & 8 & 9 & 6 & | & 2 \end{bmatrix}$

87. Solve the following system.

$$2x_1 - 3x_2 + x_3 = 1$$
$$4x_1 - x_2 - 5x_3 = 1$$
$$x_1 + x_2 - 3x_3 = 0$$

88. Solve the following system.

$$2x_1 - 3x_2 + x_3 = 1$$
$$4x_1 - x_2 - 5x_3 = 3$$
$$x_1 + x_2 - 3x_3 = 0$$

89. Solve the following system.

$$x_1 + 2x_2 + 4x_3 = -3$$
$$4x_1 + 7x_2 + 13x_3 = -10$$
$$2x_1 + 7x_2 + 15x_3 = 8$$

90. Solve the following system.

$$2x_1 + 2x_2 - x_3 - x_4 = 2$$
$$x_1 - x_2 - x_3 - 3x_4 = 0$$
$$x_1 + x_2 + x_3 - 4x_4 = 1$$
$$x_1 + 5x_2 + 2x_3 + x_4 = 1$$

91. (a) Find the solution to the following system.

$$2x_1 + 2x_2 - x_3 - x_4 = 2$$
$$x_1 - x_2 - x_3 - 3x_4 = 0$$
$$x_1 + x_2 + x_3 - 4x_4 = 1$$
$$x_1 + 5x_2 + 2x_3 + x_4 = 3$$

(b) If the solutions must contain only positive numbers, find the permissible range of each variable in the solution.

92. The Student Store bought a total of 1000 homecoming T-shirts and sweatshirts. The T-shirts cost $7.80 each and sold for $13.50 each. The white sweatshirts cost $16.50 each and sold for $28.00 each. The gold sweatshirts cost $18.00 each and sold for $30.00 each. The total cost was $11,505. The Store sold all the shirts for $19,600 revenue. How many of each kind did they buy?

USING YOUR TI-83

OBTAINING THE REDUCED ECHELON FORM OF A MATRIX

The process of solving a system of equations by using row operations on the augmented matrix seeks to reduce the columns to a single entry of 1 and 0's in the rest of the column. The same process is used to obtain the reduced echelon form of a matrix.

This process of modifying a matrix so a column contains a single entry with 1 and 0's in the rest of the column is called *pivoting*. A system of three equations can require up to nine row operations to solve the system. These row operations can be accomplished by the TI-83 program called PIVOT, which can be used to pivot on specified entries of the matrix stored in [A]. The pivot row and column determine where the pivot, the 1 entry, is located.

```
PIVOT
: [A] → [B]                        : Goto 4
: 2 → dim(L1)                      : *Row+(-[B](K,J),[B],I,K) → [B]
: dim([A]) → L1                    : Lbl 4
: Lbl 1                            : 1 + K → K
: Disp "PIVOT ROW"                 : If K ≤ L1(1)
: Input I                          : Goto 5
: Disp "PIVOT COL"                 : round([B],2) → [C]
: Input J                          : Pause [C]
: *row(1/[B](I,J),[B],I) → [B]     : Goto 1
: 1 → K                            : End
: Lbl 5
: If K = I
```

(continued)

We now show some screens that occur when using the PIVOT program. The matrix used is

$$[A] = \begin{bmatrix} 2 & 4 & 6 & -4 \\ 0 & 1 & -3 & 2 \\ 3 & 5 & 2 & 2 \end{bmatrix}$$

Begin the program with $\boxed{\text{PRGM}}$ <PIVOT> $\boxed{\text{ENTER}}$ $\boxed{\text{ENTER}}$
and enter the row and column numbers where the pivot occurs
(row 1, column 1, here). The screen shows the result of the pivot:

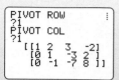

```
PIVOT ROW        :
?1
PIVOT COL
?1
   [[1  2   3  -2]
    [0  1  -3  2 ]
    [0 -1  -7  8 ]]
```

Press $\boxed{\text{ENTER}}$ to enter next pivot
(row 2, column 2).

```
PIVOT ROW        :
?2
PIVOT COL
?2
   [[1  0  9   -6]
    [0  1 -3   2 ]
    [0  0 -10 10]]
```

Finally, pivot using row 3, column 3, to
obtain the reduced echelon matrix:

```
PIVOT ROW        :
?3
PIVOT COL
?3
   [[1 0 0  3 ]
    [0 1 0 -1]
    [0 0 1 -1]]
```

If the original matrix represents the system

$$\begin{aligned} 2x_1 + 4x_2 + 6x_3 &= -4 \\ x_2 - 3x_3 &= 2 \\ 3x_1 + 5x_2 + 2x_3 &= 2 \end{aligned}$$

then the reduced matrix above gives the solution $x_1 = 3$, $x_2 = -1$, and $x_3 = -1$.

EXERCISES

Use the PIVOT program to solve the following systems:

1. $\quad x_1 + 3x_2 + 2x_3 = 1$
$\quad 6x_1 - x_2 + 4x_3 = 31$
$\quad 2x_1 + x_2 + 2x_3 = 9$

2. $\quad x_1 + 4x_2 + 2x_3 = 15$
$\quad 3x_1 + x_2 - x_3 = 4.7$
$\quad 2x_1 - 2x_2 + 3x_3 = 11.6$

3. $\quad x_1 - 2x_2 + x_3 = 1$
$\quad 2x_1 + x_2 - 2x_3 = -2$
$\quad -x_1 + x_2 + 2x_3 = 13$

4. $1.2x_1 + 3.1x_2 - 4.5x_3 = -6.71$
$\quad 2.3x_1 - 1.8x_2 + 2.5x_3 = 6.82$
$\quad 4.1x_1 + 2.6x_2 - 3.4x_3 = -0.99$

THE RREF COMMAND

We now show you a better way to obtain the reduced echelon matrix for solving a system. Use a command that finds the reduced echelon form of a matrix, the **rref** command.

Let's illustrate **rref** using the matrix [A]
shown on the screen:

```
[A]
  [[1 2 1 0 9 ]
   [2 3 0 1 7 ]
   [3 5 2 1 20]]
```

Access the rref command with $\boxed{\text{MATRX}}$ <MATH>
<B:rref (> and **rref** will show on the screen). Enter the
matrix name, [A], and press $\boxed{\text{ENTER}}$. The reduced
echelon matrix is

```
rref([A])
  [[1 0 0 2  -1]
   [0 1 0 -1 3 ]
   [0 0 1 0  4 ]]
```

EXERCISES

1. Use **rref** to find the reduced echelon form of the matrix

$$\begin{bmatrix} 1 & 3 & 2 & 1 & 5 \\ 2 & 2 & 0 & 4 & 6 \\ 4 & -4 & 2 & 0 & 9 \end{bmatrix}$$

Use **rref** to solve the following systems:

2.
$$\begin{aligned} 5x_1 + 4x_2 - \ x_3 &= \ 2 \\ 2x_1 + \ x_2 + 3x_3 &= 19 \\ 4x_1 + 7x_2 + 2x_3 &= \ 8 \end{aligned}$$

3.
$$\begin{aligned} x_1 + 3x_2 + 2x_3 &= \ \ \ 7 \\ -4x_1 + 7x_2 + 5x_3 &= \ \ 28 \\ 2x_1 + \ x_2 + 2x_3 &= -5 \end{aligned}$$

4.
$$\begin{aligned} 1.2x_1 + 3.6x_2 + 2.4x_3 &= 24.0 \\ 2.3x_1 + 1.1x_2 + 4.3x_3 &= 32.9 \\ 3.5x_1 + 2.2x_2 + 5.1x_3 &= 43.9 \end{aligned}$$

USING EXCEL

GENERAL PROCEDURE FOR PIVOTING

Using EXCEL in Section 2.2 showed how to pivot on a matrix entry. For different matrices, you need to change the formulas in the template using the entries in the new matrix. In this section, we show a template for pivoting on the diagonal entries of a 3 × 4 matrix. In this template, the original matrix is placed in cells A2:D4 and the sequence of matrices obtained by pivoting are located in cells A7:D9, A12:D14, and A17:D19. As in Section 2.2, it suffices to enter the formulas in the A column. The rest of the formulas are obtained by dragging the A column formula across the row to column D. For our example, we again use the matrix

$$\left[\begin{array}{ccc|c} 2 & -3 & 1 & 25 \\ 6 & 1 & 5 & 33 \\ 1 & 4 & -3 & -29 \end{array}\right]$$

Below, we show the formulas as entered in the A column of each matrix. The spreadsheet with the completed calculations is shown on the right.

	A
1	Augmented matrix in A2:D4
2	
3	
4	
5	
6	Pivot on the (1, 1) entry in A2
7	=A2/$A$2
8	=-$A$3*A2/$A$2+A3
9	=-$A$4*A2/$A$2+A4
10	
11	Pivot on the (2, 2) entry in B8
12	=-$B$7*A8/$B$8+A7
13	=A8/$B$8
14	=-$B$9*A8/$B$8+A9
15	
16	
17	Pivot on the (3, 3) entry in C14
18	=-$C$12*A14/$C$14+A12
19	=-$C$13*A14/$C$14+A13
20	=A14/$C$14
21	

	A	B	C	D
1	Augmented	matrix in	A2:D4	
2	2	-3	1	25
3	6	1	5	33
4	1	4	-3	-29
5				
6	Pivot on the	(1, 1) entry	in A2	
7	1	-1.5	0.5	12.5
8	0	10	2	-42
9	0	5.5	-3.5	-41.5
10				
11	Pivot on the	(2, 2) entry	in B8	
12	1	0	0.8	6.2
13	0	1	0.2	-4.2
14	0	0	-4.6	-18.4
15				
16	Pivot on the	(3, 3) entry	in C14	
17	1	0	0	3
18	0	1	0	-5
19	0	0	1	4
20				

(*continued*)

Warning: This will work only on 3×4 matrices. If a pivot element is zero, this will not work.

Enter this template on your spreadsheet and use it to work some of the exercises. Notice that the formulas include cell addresses like $A$2, $A$3, $B$8, and $C$14. Let's explain the reason for this.

When you drag the formula in A7, =A2/$A$2, across to the D7 cell, the A2 address changes to B2, C2, and D2. The $A$2 cell address *does not change*. This is correct, because you want every entry in the matrix row to be divided by the pivot element in A2. All cell addresses in the template with a $ sign before the column and row number do not change when the formula is dragged across to the D column.

Study the formulas here and then make a template to solve a system with four equations and four variables.

EXERCISES

Solve the following systems using an EXCEL template.

1.
$$4x_1 - 2x_2 + x_3 = -18$$
$$x_1 + x_2 + 3x_3 = 15$$
$$3x_1 + 7x_2 + 2x_3 = 11$$

2.
$$x_1 + 2x_2 - x_3 = 13$$
$$4x_1 - x_2 + 3x_3 = 10$$
$$2x_1 - 5x_2 + 6x_3 = -13$$

3.
$$2.1x_1 + 1.5x_2 - 3.4x_3 = -8.2$$
$$1.7x_1 + 4.1x_2 + 0.4x_3 = 21.2$$
$$2.6x_1 + 5.3x_2 + 0.8x_3 = 10.6$$

4.
$$x_1 + 2x_2 - x_3 + 3x_4 = 7$$
$$2x_1 + 3x_2 + x_3 - 5x_4 = 8$$
$$5x_1 + 2x_2 - 4x_3 + x_4 = 15$$
$$x_1 + x_2 + x_3 + x_4 = 8$$

2.4 Matrix Operations

- Additional Uses of Matrices
- Equal Matrices
- Addition of Matrices
- Scalar Multiplication

Additional Uses of Matrices

You have used augmented matrices to represent a system of linear equations (Section 2.3) and then used row operations to solve the system of equations. This method reduces the procedure to a more straightforward sequence of steps. The procedure can also be performed on a computer that handles the computational drudgery and reduces errors.

A casual observer might view an augmented matrix as simply a table of numbers. In one sense that is correct. The interpretation of the table and an understanding of the row operations gives significant meaning to "just a table of numbers." Matrices are mathematically interesting because they can be used to represent more than a system of equations. Many applications in business, social, and biological sciences represent information in tables or rectangular arrays of numbers. The analyses of such data can often be done using matrices and matrix operations. We add, multiply, and perform other operations with matrices. We will study these operations and suggest ways in which they are useful.

We begin with a simple application using a matrix to summarize information in tabular forms. For example, a mathematics professor records grade information for two sections of a mathematics course in the following matrix:

$$\begin{array}{c} \\ \\ \text{Homework} \\ \text{Quizzes} \\ \text{Exams} \end{array} \begin{array}{c} \text{Section} \\ \begin{array}{cc} 1 & 2 \end{array} \\ \begin{bmatrix} 76 & 79 \\ 71 & 70 \\ 73 & 74 \end{bmatrix} \end{array}$$

Each row represents a different type of grade, and each column represents a section.

Because there is an endless variety of ways in which someone might want to break information into categories and summarize it, matrices come in various shapes and sizes. Independent of the source of the information summarized in a matrix, we can classify matrices by the number of rows and columns they have. For example,

$$\begin{bmatrix} 3 & -1 & 4 \\ 2 & 1 & 5 \end{bmatrix}$$

is a 2 × 3 matrix because it has two rows and three columns.

$$\begin{bmatrix} 5 & 0 & 1 \\ 2 & 1 & 4 \\ 3 & 2 & 2 \\ -1 & 6 & -2 \end{bmatrix}$$

is a 4 × 3 matrix, having four rows and three columns. The convention used in describing the size of a matrix states the number of rows first, followed by the number of columns.

Two matrices are the **same size** if they have the same dimensions; that is, the number of rows is the same for both matrices, and the number of columns is also the same. For example,

$$\begin{bmatrix} -2 & 8 & -3 \\ 1 & 0 & 1 \end{bmatrix} \quad \text{and} \quad \begin{bmatrix} 2 & 5 & 9 \\ 3 & 6 & 7 \end{bmatrix}$$

are matrices of the same size; they are both 2 × 3 matrices.

Here is another example of using a matrix to summarize information.

EXAMPLE 1 ➤

The Campus Bookstore carries spirit shirts in white, green, and gold. In September it sold 238 white, 317 green, and 176 gold shirts. In October it sold 149 white, 342 green, and 369 gold shirts. In November it sold 184 white, 164 green, and 201 gold shirts. Summarize this information in a matrix.

SOLUTION

Let each of three columns represent a month and each of three rows represent a color of shirts. Label the columns and rows.

$$\begin{array}{c} \\ \text{White} \\ \text{Green} \\ \text{Gold} \end{array} \begin{array}{c} \begin{array}{ccc} \text{Sept.} & \text{Oct.} & \text{Nov.} \end{array} \\ \begin{bmatrix} 238 & 149 & 184 \\ 317 & 342 & 164 \\ 176 & 369 & 201 \end{bmatrix} \end{array}$$

■ **Now You Are Ready to Work Exercise 1**

A matrix n in which the number of rows equals the number of columns is called a **square matrix.**

Equal Matrices

Two matrices of the same size are **equal matrices** if and only if their corresponding components are equal. If the matrices are not the same size, they are not equal.

EXAMPLE 2 ➤

$$\begin{bmatrix} 3 & 7 \\ 5-1 & 4 \times 4 \end{bmatrix} = \begin{bmatrix} \frac{6}{2} & 7 \\ 4 & 16 \end{bmatrix}$$

because corresponding components are equal.

$$\begin{bmatrix} 1 & 2 & 5 \\ 3 & 6 & 4 \end{bmatrix} \neq \begin{bmatrix} 1 & 2 & 5 \\ 3 & -1 & 4 \end{bmatrix}$$

because the entries in row 2, column 2 are different; that is, the $(2, 2)$ entries are not equal.

■ **Now You Are Ready to Work Exercise 17**

EXAMPLE 3 ➤   Find the value of x such that

$$\begin{bmatrix} 3 & 4x \\ 2.1 & 7 \end{bmatrix} = \begin{bmatrix} 3 & 9 \\ 2.1 & 7 \end{bmatrix}$$

SOLUTION

For the matrices to be equal, the corresponding components must be equal, so $4x = 9$ and $x = \frac{9}{4}$.

■ **Now You Are Ready to Work Exercise 43**

Addition of Matrices

A businesswoman has two stores. She is interested in the daily sales of the regular size and the giant economy size of laundry soap. She can use matrices to record this information. These two matrices show sales for two days.

	Store 1	Store 2		Store 1	Store 2
Regular	8	12		6	5
Giant	9	7		11	4
	Day 1			Day 2	

CAUTION

To add two matrices, they must be the same size.

The position in the matrix identifies the store and package size. For example, store 2 sold 7 giant sizes and 12 regular on day 1, and so on.

The total sales, by store and package size, can be obtained by adding the sales in each individual category to get the total in that category — that is, by adding corresponding entries of the two matrices. We indicate this procedure with the notation

$$\begin{bmatrix} 8 & 12 \\ 9 & 7 \end{bmatrix} + \begin{bmatrix} 6 & 5 \\ 11 & 4 \end{bmatrix} = \begin{bmatrix} 14 & 17 \\ 20 & 11 \end{bmatrix}$$

This procedure applies generally to the **addition of matrices.**

DEFINITION
Matrix Addition

The **sum** of two matrices of the same size is obtained by adding corresponding elements. If two matrices are not of the same size, they cannot be added; we say that their sum does not exist. **Subtraction** is performed on matrices of the same size by subtracting corresponding elements.

Now we apply the definition of matrix addition in the next two examples.

EXAMPLE 4 ➤

For the following matrices:

$$A = \begin{bmatrix} 2 & 1 & -1 \\ 0 & 5 & 2 \end{bmatrix} \qquad B = \begin{bmatrix} 1 & 3 & 1 \\ 2 & 1 & 4 \end{bmatrix} \qquad C = \begin{bmatrix} 4 & 1 \\ -1 & 2 \end{bmatrix}$$

determine the sums $A + B$ and $B + C$ if possible.

SOLUTION

$$A + B = \begin{bmatrix} 2 & 1 & -1 \\ 0 & 5 & 2 \end{bmatrix} + \begin{bmatrix} 1 & 3 & 1 \\ 2 & 1 & 4 \end{bmatrix}$$

$$= \begin{bmatrix} 2+1 & 1+3 & -1+1 \\ 0+2 & 5+1 & 2+4 \end{bmatrix} = \begin{bmatrix} 3 & 4 & 0 \\ 2 & 6 & 6 \end{bmatrix}$$

Neither the sum $A + C$ nor $B + C$ exists, because matrices A and C, and matrices B and C are not of the same size. (Try adding these matrices using the rule.)

■ **Now You Are Ready to Work Exercise 23**

Let's extend our definition to enable us to add more than just two matrices. For example, define the sum of three matrices as

$$\begin{bmatrix} 1 & 2 \\ 0 & -1 \end{bmatrix} + \begin{bmatrix} 3 & 4 \\ 2 & 1 \end{bmatrix} + \begin{bmatrix} 5 & 2 \\ -1 & 0 \end{bmatrix} = \begin{bmatrix} 1+3+5 & 2+4+2 \\ 0+2-1 & -1+1+0 \end{bmatrix}$$

$$= \begin{bmatrix} 9 & 8 \\ 1 & 0 \end{bmatrix}$$

We add a string of matrices that are the same size by adding corresponding elements. The following example illustrates a use of this rule.

EXAMPLE 5 ➤

The Green Earth Recycling Center has three locations. They recycle aluminum, plastic, and newspapers. Each location keeps a daily record in matrix form. Here is an illustration of one week's records. The entries represent the number of pounds collected.

Location 1

	Alum.	Plastic	Paper
Mon.	920	140	1840
Tue.	640	96	1260
Wed.	535	80	955
Thu.	768	32	1030
Fri.	420	55	1320
Sat.	1590	205	2340

Location 2

	Alum.	Plastic	Paper
Mon.	435	60	2840
Tue.	620	45	2665
Wed.	240	22	3450
Thu.	195	38	1892
Fri.	530	52	1965
Sat.	895	74	3460

Location 3

	Alum.	Plastic	Paper
Mon.	634	110	1565
Tue.	423	86	948
Wed.	555	142	1142
Thu.	740	93	1328
Fri.	883	135	1476
Sat.	976	234	1928

A summary of the total collected at the three locations can be obtained by matrix addition.

All Locations

Location 1 + Location 2 + Location 3 =

	Alum.	Plastic	Paper
Mon.	1989	310	6245
Tue.	1683	227	4873
Wed.	1330	244	5547
Thu.	1703	163	4250
Fri.	1833	242	4761
Sat.	3461	513	7728

Thus, the total aluminum collected on Monday was 1989 pounds, the amount of newspaper collected on Saturday was 7728 pounds, and so on.

■ **Now You Are Ready to Work Exercise 41**

Although this analysis and others like it can be carried out without the use of matrices, the handling of large quantities of data is often most efficiently done on computers using matrix techniques.

Scalar Multiplication

Another matrix operation multiplies a matrix by a number like

$$4\begin{bmatrix} 3 & 2 \\ 1 & 7 \end{bmatrix}$$

This product is defined to be

$$4\begin{bmatrix} 3 & 2 \\ 1 & 7 \end{bmatrix} = \begin{bmatrix} 12 & 8 \\ 4 & 28 \end{bmatrix}$$

Notice that this operation multiplies each entry in the matrix by 4. This illustrates the procedure for **scalar multiplication,** so called because mathematicians often use the term *scalar* to refer to a *number.*

DEFINITION
Scalar Multiplication

Scalar multiplication is the operation of multiplying a matrix by a number (scalar). Each entry in the matrix is multiplied by the scalar.

EXAMPLE 6 ➤

$$-3\begin{bmatrix} 5 & 2 & 1 \\ 0 & 1 & 4 \\ -1 & 3 & 6 \end{bmatrix} = \begin{bmatrix} -15 & -6 & -3 \\ 0 & -3 & -12 \\ 3 & -9 & -18 \end{bmatrix}$$

■ **Now You Are Ready to Work Exercise 31**

EXAMPLE 7 ➤ A class of ten students had five tests during the quarter. A perfect score on each of the tests is 50. The scores are listed in this table.

	Test 1	Test 2	Test 3	Test 4	Test 5
Anderson	40	45	30	48	42
Boggs	20	15	30	25	10
Chittar	40	35	25	45	46
Diessner	25	40	45	40	38
Farnam	35	35	38	37	39
Gill	50	46	45	48	47
Homes	22	24	30	32	29
Johnson	35	27	20	41	30
Schomer	28	31	25	27	31
Wong	40	35	36	32	38

We can express these scores as column matrices:

$$\begin{bmatrix} 40 \\ 20 \\ 40 \\ 25 \\ 35 \\ 50 \\ 22 \\ 35 \\ 28 \\ 40 \end{bmatrix} \begin{bmatrix} 45 \\ 15 \\ 35 \\ 40 \\ 35 \\ 46 \\ 24 \\ 27 \\ 31 \\ 35 \end{bmatrix} \begin{bmatrix} 30 \\ 30 \\ 25 \\ 45 \\ 38 \\ 45 \\ 30 \\ 20 \\ 25 \\ 36 \end{bmatrix} \begin{bmatrix} 48 \\ 25 \\ 45 \\ 40 \\ 37 \\ 48 \\ 32 \\ 41 \\ 27 \\ 32 \end{bmatrix} \begin{bmatrix} 42 \\ 10 \\ 46 \\ 38 \\ 39 \\ 47 \\ 29 \\ 30 \\ 31 \\ 38 \end{bmatrix}$$

To obtain each person's average, we use matrix addition to add the matrices and then scalar multiplication to multiply by $\frac{1}{5}$ (dividing by the number of tests). We get

$$\frac{1}{5} \begin{bmatrix} 205 \\ 100 \\ 191 \\ 188 \\ 184 \\ 236 \\ 137 \\ 153 \\ 142 \\ 181 \end{bmatrix} = \begin{bmatrix} 41.0 \\ 20.0 \\ 38.2 \\ 37.6 \\ 36.8 \\ 47.2 \\ 27.4 \\ 30.6 \\ 28.4 \\ 36.2 \end{bmatrix} \quad \text{(Column matrix giving each person's average score)}$$

■ **Now You Are Ready to Work Exercise 51**

Row matrices are also useful; a person's complete set of scores corresponds to a row matrix. For example,

$$[25 \quad 40 \quad 45 \quad 40 \quad 38]$$

is a row matrix giving Diessner's scores.

This approach to analyzing test scores has the advantage of lending itself to implementation on the computer. A computer program can be written that will perform the desired matrix additions and scalar multiplications.

2.4 EXERCISES

LEVEL 1

In Exercises 1 through 4, summarize the given information in matrix form.

1. (*See Example 1*) The Alpha Club and the Beta Club perform service work for the Salvation Army, the Boys' Club, and the Girl Scouts. The Alpha Club performs 50 hours at the Salvation Army, 85 hours at the Boys' Club, and 68 hours for the Girl Scouts. The Beta Club performs 65 hours at the Salvation Army, 32 hours at the Boys' Club, and 94 hours for the Girl Scouts.

2. An appliance saleswoman sold 15 washers, 8 dryers, and 13 microwave ovens in March. She sold 12 washers, 11 dryers, and 6 microwave ovens in April.

3. Citizen's Bank awards prizes to the employees who sign up the largest number of new customers. In October, Joe signed up 12 new checking accounts, 15 savings accounts, and 8 safe deposit boxes. Jane signed up 11 new checking accounts, 18 savings accounts, and 9 safe deposit boxes. Judy signed up 5 new checking accounts, 8 savings accounts, and 21 safe deposit boxes.

4. Tom scored 78, 82, and 72 on the first three biology exams. Dick scored 62, 71, and 76. Harriet scored 98, 70, and 81.

State the size of each matrix in Exercises 5 through 16.

5. $\begin{bmatrix} 7 & 13 \\ -4 & 8 \end{bmatrix}$

6. $\begin{bmatrix} 4 & 1 \\ 5 & -2 \\ 0 & 0 \end{bmatrix}$

7. $\begin{bmatrix} 4 & 1 & 9 \\ 8 & 4 & 8 \\ 2 & -1 & -5 \end{bmatrix}$

8. $\begin{bmatrix} 4 & 8 & 8 & 7 \\ 4 & 5 & 3 & 9 \\ 3 & 5 & 6 & 1 \end{bmatrix}$

9. $\begin{bmatrix} 1 \\ 2 \\ 3 \\ 4 \end{bmatrix}$

10. $[-6 \quad -5 \quad 0 \quad 5 \quad 6]$

11. $\begin{bmatrix} 4 & 2 & -1 & 6 \\ 3 & 8 & 9 & 2 \end{bmatrix}$

12. $\begin{bmatrix} 1 \\ 1 \end{bmatrix}$

13. $\begin{bmatrix} 4 & 4 & 2 \\ 1 & 0 & 1 \end{bmatrix}$

14. $[1 \quad 1 \quad 0 \quad 1 \quad 1 \quad 2]$

15. $\begin{bmatrix} 1 & 0 \\ 0 & 1 \end{bmatrix}$

16. $\begin{bmatrix} 0 & 0 & 0 \\ 0 & 0 & 0 \\ 0 & 0 & 0 \end{bmatrix}$

In Exercises 17 through 22, determine which of the pairs of matrices are equal.

17. *(See Example 2)*

$\begin{bmatrix} 2 & 1 & 3 \\ 4 & 0 & 2 \end{bmatrix}, \begin{bmatrix} 4 & 0 & 2 \\ 2 & 1 & 3 \end{bmatrix}$

18. $\begin{bmatrix} \frac{3}{4} & 16 \\ 8 & \frac{1}{2} \end{bmatrix}, \begin{bmatrix} 0.75 & 16 \\ 8 & 0.5 \end{bmatrix}$

19. $\begin{bmatrix} 5+2 & 5-2 \\ \frac{5}{2} & \frac{2}{5} \end{bmatrix}, \begin{bmatrix} 7 & 3 \\ 2.5 & 0.4 \end{bmatrix}$

20. $\begin{bmatrix} 2 & 1 & 3 & 6 \\ 5 & 9 & 4 & 1 \end{bmatrix}, \begin{bmatrix} 2 & 3 & 6 \\ 5 & 4 & 1 \end{bmatrix}$

21. $\begin{bmatrix} -1 & 4 & 1 \\ 3 & 2 & 0 \end{bmatrix}, \begin{bmatrix} -1 & 3 \\ 4 & 2 \\ 1 & 0 \end{bmatrix}$

22. $\begin{bmatrix} 1 & 0 \\ 0 & 2 \end{bmatrix}, \begin{bmatrix} 2 & 0 \\ 0 & 1 \end{bmatrix}$

If possible, add the matrices in Exercises 23 through 30. We say that the sum does *not* exist if the matrices cannot be added.

23. *(See Example 4)*

$\begin{bmatrix} 1 & -1 & 3 \\ 2 & 4 & 1 \end{bmatrix} + \begin{bmatrix} 2 & 4 & -1 \\ 5 & 0 & 2 \end{bmatrix}$

24. $\begin{bmatrix} 0 & 1 \\ 2 & 3 \\ -1 & 4 \end{bmatrix} + \begin{bmatrix} 3 & -1 \\ 2 & 0 \\ 4 & -6 \end{bmatrix}$

25. $\begin{bmatrix} 2 \\ 5 \end{bmatrix} + \begin{bmatrix} -1 \\ 4 \end{bmatrix} + \begin{bmatrix} 1 \\ 49 \end{bmatrix}$

26. $\begin{bmatrix} 5 & 1 \\ -2 & 0 \end{bmatrix} + \begin{bmatrix} 2 & 8 \\ 4 & 6 \end{bmatrix}$

27. $[4 \quad 5 \quad 2] + [7 \quad -1]$

28. $\begin{bmatrix} 4 & 1 & 5 \\ 6 & -3 & 9 \end{bmatrix} + \begin{bmatrix} -1 & 3 \\ 6 & 8 \\ 7 & 2 \end{bmatrix}$

29. $\begin{bmatrix} 1 & 4 & 2 \\ 3 & 0 & 1 \\ 6 & -1 & 5 \end{bmatrix} + \begin{bmatrix} 5 & 1 & 6 \\ 2 & 1 & 2 \\ -1 & -2 & -1 \end{bmatrix}$

30. $\begin{bmatrix} 1 & 0 & 1 & 0 \\ 0 & 1 & 0 & 1 \\ 2 & 1 & 2 & 1 \\ 1 & 2 & 1 & 2 \end{bmatrix} + \begin{bmatrix} 3 & 4 & 3 & 4 \\ 2 & 5 & 2 & 5 \\ 3 & 0 & 3 & 0 \\ 2 & 2 & 3 & 3 \end{bmatrix}$

Perform the scalar multiplication in Exercises 31 through 37.

31. *(See Example 6)*

$3\begin{bmatrix} 4 & 1 \\ 2 & 5 \end{bmatrix}$

32. $2\begin{bmatrix} 3 & -2 & 1 \\ 6 & 0 & -3 \end{bmatrix}$

33. $5\begin{bmatrix} 4 \\ 3 \\ 1 \\ 2 \end{bmatrix}$

34. $-5\begin{bmatrix} 1 & 2 & 3 \\ 4 & 1 & -2 \end{bmatrix}$

35. $-3[4 \quad -2 \quad 5]$

36. $\frac{1}{2}\begin{bmatrix} 4 & 5 \\ 8 & 6 \end{bmatrix}$

37. $0\begin{bmatrix} 3 & 5 \\ 0 & 2 \end{bmatrix}$

LEVEL 2

38. The Music Store's inventory of recordings is:

Popular Music: cassettes, 848; LPs, 145; CDs, 969

Classical Music: cassettes, 159; LPs, 37; CDs, 246

The Sound Shop's inventory of recordings is:

Popular Music: cassettes, 753; LPs, 252; CDs, 639

Classical Music: cassettes, 342; LPs, 19; CDs, 113

(a) Represent the Music Store's inventory in a matrix A and the Sound Shop's inventory in a matrix B.

(b) If the stores merge, represent the merged inventory by adding the matrices.

39. Let $A = \begin{bmatrix} 1 & 4 \\ -2 & 3 \end{bmatrix}$, $B = \begin{bmatrix} 0 & 2 \\ 4 & 1 \end{bmatrix}$, and $C = \begin{bmatrix} 1 & -2 \\ 1 & -3 \end{bmatrix}$. Find these matrices:

(a) $3A, -2B, 5C$ **(b)** $A + C$
(c) $3A - 2B$ **(d)** $A - 2B + 5C$

40. Let $A = \begin{bmatrix} 1 & -2 & 4 \\ 3 & 1 & 0 \end{bmatrix}$, $B = \begin{bmatrix} -2 & 2 & 5 \\ 0 & 1 & 1 \end{bmatrix}$, and $C = \begin{bmatrix} 7 & -3 & 9 \\ -2 & 4 & 6 \end{bmatrix}$. Find these matrices:

(a) $A - 2B$ **(b)** $-C$ and $B - C$
(c) $2A + 3B - 4C$

41. *(See Example 5)* A distributor furnishes PC computers, printers, and disks to three retail stores. He summarizes monthly sales in a matrix.

	Customer								
	I	II	III	I	II	III	I	II	III
PC	5	7	8	8	6	5	10	4	7
Printer	6	4	5	10	9	4	3	9	2
Disk	45	52	35	52	60	42	54	39	28
	June			July			August		

Find the three-month total by item and by store.

In Exercises 42 through 46, find the value of x that makes the pairs of matrices equal.

42. $\begin{bmatrix} 3 & x \\ 2 & 1 \end{bmatrix} = \begin{bmatrix} 3 & 9 \\ 2 & 1 \end{bmatrix}$

43. *(See Example 3)*
$\begin{bmatrix} 5x & 7 \\ 2 & 4 \end{bmatrix} = \begin{bmatrix} 15 & 7 \\ 2 & 4 \end{bmatrix}$

44. $\begin{bmatrix} 2x + 3 & -2 \\ 6 & 1 \end{bmatrix} = \begin{bmatrix} 3x - 1 & -2 \\ 6 & 1 \end{bmatrix}$

45. $\begin{bmatrix} 17 & 6x + 4 \\ 94 & -39 \end{bmatrix} = \begin{bmatrix} 17 & 14x - 13 \\ 94 & -39 \end{bmatrix}$

46. $\begin{bmatrix} 2x + 1 & 6 \\ 5 & -4 \end{bmatrix} = \begin{bmatrix} 3x + 5 & 6 \\ 5 & -4 \end{bmatrix}$

47. A firm has three plants, all of which produce small, regular, and giant size boxes of detergent. The annual report shows the total production (in thousands of boxes), broken down by plant and size, in the following matrix:

	Plant		
	A	B	C
Small	65	110	80
Regular	90	135	60
Giant	75	112	84

Find the average monthly production by plant and size.

LEVEL 3

48. A cafeteria manager estimates that the amount of food needed to serve one person is

Meat:	4 oz
Peas:	2 oz
Rice:	3 oz
Bread:	1 slice
Milk:	1 cup

Use matrix arithmetic to find the amount needed to serve 114 people.

49. A manufacturer has plants at Fairfield and Tyler that produce the pollutants sulfur dioxide, nitric oxide, and particulate matter in the amounts shown below. The amounts are in kilograms and represent the average per month.

	Sulfur dioxide	Nitric oxide	Particulate matter
Fairfield	230	90	140
Tyler	260	115	166

Find the annual totals by plants and pollutant.

50. The total sales of regular and giant size boxes of detergent at two stores is given for three months.

	Store		Store		Store	
	1	2	1	2	1	2
Regular	85	46	80	61	50	42
Giant	77	93	93	47	61	38
	March		April		May	

Use matrix arithmetic to find the average monthly sales by store and size.

51. (*See Example 7*) The test scores for five students are given in the following table:

	Test 1	Test 2	Test 3
A	90	88	91
B	62	69	73
C	76	78	72
D	82	80	84
E	74	76	77

Use matrix arithmetic to find each student's average.

52. Use a matrix to display the following information about students at City College.

645 freshmen had GPAs of 3.0 or higher.
982 freshmen had GPAs of less than 3.0.
569 sophomores had GPAs of 3.0 or higher.
722 sophomores had GPAs of less than 3.0.
531 juniors had GPAs of 3.0 or higher.
562 juniors had GPAs of less than 3.0.
478 seniors had GPAs of 3.0 or higher.
493 seniors had GPAs of less than 3.0.

53. While on a study break at the library, Alissa came upon information guides to private colleges and universities. She found the following information for six of the schools:

For the year 2002:

School	Number of students	Tuition	Room/Board
Bowdoin	1609	$25,345	$6760
Caltech	929	$20,904	$6543
Marquette	7496	$17,080	$6090
Rhodes	1537	$19,303	$5671
Samford	2870	$10,738	$4720
Xavier (OH)	4019	$15,680	$6940

For the year 1994:

School	Number of students	Tuition	Room/Board
Bowdoin	1410	$17,355	$5855
Caltech	862	$15,160	$4676
Marquette	8409	$ 9,900	$4350
Rhodes	1414	$14,916	$4708
Samford	3194	$ 6,074	$3436
Xavier (OH)	3996	$10,970	$4470

(a) Represent the given information for each year with a matrix. Use a matrix operation to find the change in number of students, tuition, and room and board from 1994 to 2002.

(b) Which school had the largest increase in tuition?

(c) Which schools, if any, had a decrease in students?

54. The Department of Veteran Affairs keeps records of surviving veterans, their surviving dependent children, and surviving spouses. The tables below show, as of July 1997 and May 2001, the number surviving for the Civil War, World War I, and World War II.

As of July 1997:

War	Veterans	Children	Spouses
Civil War	0	17	2
World War I	1,269	7,419	56,267
World War II	855,049	20,792	308,675

As of May 2001:

War	Veterans	Children	Spouses
Civil War	0	12	1
World War I	144	5,810	25,573
World War II	647,205	18,707	272,793

Use matrices to represent the information in these tables and use a matrix operation to find the decrease, from 1997 to 2001, in each category. (*Trivia note:* The last Revolutionary War dependent died in 1911 at the age of 90. The last Civil War veteran died in 1958 at the age of 112.)

EXPLORATIONS

55. Discuss the similarities of repeated addition and scalar multiplication of matrices to repeated addition and multiplication of numbers.

56. Let $\begin{bmatrix} a & b \\ c & d \end{bmatrix}$ be an arbitrary 2×2 matrix and $A = \begin{bmatrix} 0 & 0 \\ 0 & 0 \end{bmatrix}$. Discuss the similarity of the relationship of A to all 2×2 matrices to the relationship of the number zero to all numbers. Does A have the same relationship to all 3×3 matrices?

57. $A = \begin{bmatrix} 3 & 1 & 4 \\ 2 & 0 & 2 \\ 1 & 1 & -1 \end{bmatrix}$ and $B = \begin{bmatrix} 2 & 0 & 5 \\ 8 & 4 & 8 \\ 2 & 5 & 4 \end{bmatrix}$. Find these matrices:
(a) $A + B$ **(b)** $A - B$
(c) $2A$ **(d)** $4A + 3B$

58. $A = \begin{bmatrix} 2 & 2 & 2 \\ 3 & 0 & 1 \end{bmatrix}$ and $B = \begin{bmatrix} 1 & 5 & 9 \\ 2 & -1 & 4 \end{bmatrix}$. Find these matrices:
(a) $A + B$ **(b)** $A - B$
(c) $-5A$ **(d)** $2A - B$

59. Use matrix operations to calculate the following.
(a) $\begin{bmatrix} 3 & 1 & 5 \\ 6 & 0 & 2 \\ 2 & 4 & 4 \end{bmatrix} + \begin{bmatrix} 6 & -1 & 3 \\ 0 & 2 & 5 \\ 2 & 4 & 4 \end{bmatrix}$

(b) $\begin{bmatrix} 1.3 & 2.1 \\ 4.4 & 8.4 \end{bmatrix} + \begin{bmatrix} -2.7 & 6.2 \\ 3.3 & -1.8 \end{bmatrix}$

(c) $\begin{bmatrix} 2.1 & 3.0 & -1.1 \\ 5.4 & 2.2 & 4.7 \\ -1.5 & 1.0 & -2.6 \end{bmatrix} + \begin{bmatrix} 3.3 & 0 & 4.2 \\ -1.6 & 2.0 & -3.1 \\ 4.6 & 5.2 & -2.4 \end{bmatrix}$

(d) $3\begin{bmatrix} 1 & 5 & 1 \\ 2 & 1 & 2 \\ 3 & 0 & 4 \end{bmatrix}$ **(e)** $5.4\begin{bmatrix} 3.0 & 2.2 \\ 6.1 & 7.3 \\ -1.4 & -2.5 \end{bmatrix}$

(f) $\begin{bmatrix} 3 & 5 & -2 \\ 4 & 1 & 7 \\ 2 & 6 & -4 \end{bmatrix} + \begin{bmatrix} 3 & 4 & 2 \\ 5 & 1 & 6 \\ -2 & 7 & -4 \end{bmatrix}$

60. Try to perform the following matrix operations on your graphing calculator or spreadsheet. What happens?
(a) $\begin{bmatrix} 1 & 2 \\ 3 & 4 \end{bmatrix} + \begin{bmatrix} 4 & 5 & 6 \\ 1 & -1 & 0 \end{bmatrix}$

(b) $\begin{bmatrix} 3 & 2 & 1 \\ 4 & 5 & 4 \end{bmatrix} + \begin{bmatrix} 6 & 7 \\ -5 & 4 \end{bmatrix}$

USING YOUR TI-83

MATRIX OPERATIONS

The notation used to add matrices and to do scalar multiplication resembles that used for the addition and multiplication of numbers. The following matrices are stored in [A] and [B].

$$[A] = \begin{bmatrix} 1 & 2 & 3 \\ 4 & 5 & 6 \\ 7 & 8 & 9 \end{bmatrix} \text{ and } [B] = \begin{bmatrix} 10 & 11 & 12 \\ 13 & 14 & 15 \\ 16 & 17 & 18 \end{bmatrix}$$

Their sum is obtained by
[A] + [B] ENTER,

```
[A]+[B]
      [[11 13 15]
       [17 19 21]
       [23 25 27]]
```

and 3[A] is obtained by
3 × [A] ENTER

```
3*[A]
      [[3  6  9 ]
       [12 15 18]
       [21 24 27]]
```

EXERCISES

1. $[A] = \begin{bmatrix} 1 & 3 & 2 \\ 4 & 5 & 7 \end{bmatrix}$ and $[B] = \begin{bmatrix} 3 & -2 & 1 \\ 6 & 8 & -5 \end{bmatrix}$. Find the matrices:

  (a) $[A] + [B]$　　　　**(b)** $2[A]$　　　　**(c)** $[A] - [B]$　　　　**(d)** $3[A] + 2[B]$

2. $[A] = \begin{bmatrix} 4 & 2 \\ 3 & 9 \end{bmatrix}$ and $[B] = \begin{bmatrix} 2 & 6 \\ 5 & 12 \end{bmatrix}$. Find the matrices:

  (a) $[A] + [B]$　　　　**(b)** $-4[A]$　　　　**(c)** $[A] - [B]$　　　　**(d)** $2[A] - 5[B]$

USING EXCEL

MATRIX ADDITION AND SCALAR MULTIPLICATION

We now show how to perform matrix addition and scalar multiplication using EXCEL.

EXAMPLE

$A = \begin{bmatrix} 1 & 2 & 3 \\ 4 & 5 & 6 \end{bmatrix}$　$B = \begin{bmatrix} -1 & 3 & 1 \\ 2 & -3 & 4 \end{bmatrix}$

Find $A + B, 5A, 3A - 2B$

Solution

We enter A in cells A2:C3 and B in cells E2:G3. We will put $A + B$ in cells A6:C7, 5A in cells E6:G7, and $3A - 2B$ in A10:C11.

The formulas to perform the desired operations follow much like the same operations on numbers.

For $A + B$: Enter =A2+E2 in A6. Then copy the formulas in the rest of the matrix by dragging A6. Press RETURN to activate the operation.

For 5A: Enter =5*A2 in E6 and drag to other cells in E6:G7.

For $3A - 2B$: Enter =3*A2-2*E2 in A10 and drag to the other cells in A10:C11.

Here is the spreadsheet with the results of the operation.

	A	B	C	D	E	F	G
1	Matrix A	in A2:C3			Matrix B	in E2:G3	
2	1	2	3		-1	3	1
3	4	5	6		2	-3	4
4							
5	A + B	in A6:C7			5A	in E6:G7	
6	0	5	4		5	10	15
7	6	2	10		20	25	30
8							
9	3A - 2B	in A10:C11					
10	5	0	7				
11	8	21	10				

(*continued*)

EXERCISES

For Exercises 1 through 5 use

$$A = \begin{bmatrix} 1 & 3 & -2 \\ 5 & 9 & 7 \\ -4 & 0 & 6 \end{bmatrix} \qquad B = \begin{bmatrix} 8 & 2 & 0 \\ -3 & 5 & 4 \\ 2 & 9 & 1 \end{bmatrix}$$

1. Find $A + B$ **2.** Find $A - B$ **3.** Find $4A$

4. Find $-0.35B$ **5.** Find $4A + 6B$

6. Repeat Exercises 1 through 5 using

$$A = \begin{bmatrix} 2 & 3 & 2 \\ 5 & 1 & 5 \\ 6 & 9 & 3 \end{bmatrix} \qquad B = \begin{bmatrix} 4 & 1 & 7 \\ 2 & -3 & 6 \\ 3 & 3 & 2 \end{bmatrix}$$

2.5 Multiplication of Matrices

- Dot Product
- Matrix Multiplication
- Identity Matrix
- Row Operations Using Matrix Multiplication

You have learned to add matrices and multiply a matrix by a number. You may naturally ask whether one can multiply two matrices together and whether this helps to solve problems. Mathematicians have devised a way of multiplying two matrices. It might seem rather complicated, but it has many useful applications. For example, you will learn how to use matrix multiplication to solve a problem like the following.

A manufacturer makes tables and chairs. The time, in hours, required to assemble and finish the items is given by the matrix

$$\begin{array}{cc} & \text{Chair} \quad \text{Table} \\ \begin{array}{c} \text{Assemble} \\ \text{Finish} \end{array} & \begin{bmatrix} 2 & 3 \\ 2.5 & 4.75 \end{bmatrix} \end{array}$$

The total assembly and finishing time required to produce 950 chairs and 635 tables can be obtained by an appropriate matrix multiplication. The procedure of multiplying matrices may appear strange. Following a description of how to multiply two matrices are some examples of the uses of matrix multiplication. But first, here's an overview of the process:

1. For two matrices A and B, we will find their product AB. Their product is a matrix called $C (AB = C)$.
2. C is a matrix. The problem is to find the entries of C.
3. Each entry in C will depend on a *row* from matrix A and a *column* from matrix B.

We call the entry in row i and column j the (i, j) entry of a matrix. The (i, j) entry in C is a number obtained using all entries of row i in A and using all entries of column j in B. For example, the $(2, 3)$ entry in C depends on row 2 of A and column 3 of B. We show you how to find an entry in C by using the *dot product* of a row in A and a column in B.

Dot Product

We use the two matrices

$$A = \begin{bmatrix} 1 & 3 \\ 2 & -1 \end{bmatrix} \quad \text{and} \quad B = \begin{bmatrix} 4 & -5 \\ 1 & 6 \end{bmatrix}$$

to illustrate matrix multiplication AB. We use the first row of A, $R1 = [1 \quad 3]$, and the first column of B,

$$C1 = \begin{bmatrix} 4 \\ 1 \end{bmatrix}$$

to find the $(1, 1)$ entry of the product. To do so, we need to find what is called the **dot product,** $R1 \cdot C1$, of the row and column. It is

$$R1 \cdot C1 = [1 \quad 3] \cdot \begin{bmatrix} 4 \\ 1 \end{bmatrix} = 1(4) + 3(1) = 7$$

Notice the following:

1. The dot product of a row and a column is a single number.
2. Obtain the dot product by multiplying the first numbers from both the row and column, then the second numbers from both, and so on, and then adding the results.

There are three other dot products possible using a row from A and a column from B. They are:

$$R1 \cdot C2 = [1 \quad 3] \cdot \begin{bmatrix} -5 \\ 6 \end{bmatrix} = 1(-5) + 3(6) = 13$$

$$R2 \cdot C1 = [2 \quad -1] \cdot \begin{bmatrix} 4 \\ 1 \end{bmatrix} = 2(4) + (-1)(1) = 7$$

$$R2 \cdot C2 = [2 \quad -1] \cdot \begin{bmatrix} -5 \\ 6 \end{bmatrix} = 2(-5) + (-1)(6) = -16$$

The general form of the dot product of a row and column is

$$[a_1 \quad a_2 \cdots a_n] \cdot \begin{bmatrix} b_1 \\ b_2 \\ \vdots \\ b_n \end{bmatrix} = a_1 b_1 + a_2 b_2 + \cdots + a_n b_n$$

> **NOTE**
>
> The dot product is defined only when the row and column matrices have the same number of entries.

The total cost of a purchase at the grocery store can be determined by using the dot product of a price matrix and a quantity matrix as illustrated in the next example.

EXAMPLE 1 ➤ Let the row matrix $[0.95 \quad 1.75 \quad 2.15]$ represent the prices of a loaf of bread, a six-pack of soft drinks, and a package of granola bars, in that order. Let

$$\begin{bmatrix} 5 \\ 3 \\ 4 \end{bmatrix}$$

represent the quantity of bread (5), soft drinks (3), and granola bars (4) purchased in that order. Then the dot product

$$[0.95 \quad 1.75 \quad 2.15] \cdot \begin{bmatrix} 5 \\ 3 \\ 4 \end{bmatrix} = 0.95(5) + 1.75(3) + 2.15(4)$$

$$= 4.75 + 5.25 + 8.60 = 18.60$$

gives the total cost of the purchase.

■ **Now You Are Ready to Work Exercise 7**

Matrix Multiplication

Recall that we said the entries in $C = AB$ depend on a row of A and a column of B. The entries are actually the dot product of a row and column. In the product

$$C = AB = \begin{bmatrix} 1 & 3 \\ 2 & -1 \end{bmatrix} \begin{bmatrix} 4 & -5 \\ 1 & 6 \end{bmatrix}$$

the $(1, 2)$ entry in C is the dot product R1 · C2, for example. In the product AB,

$$C = \begin{bmatrix} \text{R1} \cdot \text{C1} & \text{R1} \cdot \text{C2} \\ \text{R2} \cdot \text{C1} & \text{R2} \cdot \text{C2} \end{bmatrix}$$

$$= \begin{bmatrix} [1 \quad 3] \cdot \begin{bmatrix} 4 \\ 1 \end{bmatrix} & [1 \quad 3] \cdot \begin{bmatrix} -5 \\ 6 \end{bmatrix} \\ [2 \quad -1] \cdot \begin{bmatrix} 4 \\ 1 \end{bmatrix} & [2 \quad -1] \cdot \begin{bmatrix} -5 \\ 6 \end{bmatrix} \end{bmatrix}$$

$$= \begin{bmatrix} 1(4) + 3(1) & 1(-5) + 3(6) \\ 2(4) + (-1)(1) & 2(-5) + (-1)(6) \end{bmatrix}$$

$$= \begin{bmatrix} 7 & 13 \\ 7 & -16 \end{bmatrix}$$

EXAMPLE 2 ➤ Find the product AB of

$$A = \begin{bmatrix} 1 & 3 & 2 \\ -1 & 0 & 4 \end{bmatrix} \quad \text{and} \quad B = \begin{bmatrix} 7 & 5 \\ -2 & 6 \\ -3 & -4 \end{bmatrix}$$

SOLUTION

$$AB = \begin{bmatrix} \text{R1} \cdot \text{C1} & \text{R1} \cdot \text{C2} \\ \text{R2} \cdot \text{C1} & \text{R2} \cdot \text{C2} \end{bmatrix}$$

$$= \begin{bmatrix} 1(7) + 3(-2) + 2(-3) & 1(5) + 3(6) + 2(-4) \\ (-1)(7) + 0(-2) + 4(-3) & -1(5) + 0(6) + 4(-4) \end{bmatrix}$$

$$= \begin{bmatrix} -5 & 15 \\ -19 & -21 \end{bmatrix}$$

■ **Now You Are Ready to Work Exercise 11**

EXAMPLE 3 ➤   Multiply the matrices

$$A = \begin{bmatrix} 1 & 3 \\ 5 & 4 \end{bmatrix} \quad \text{and} \quad B = \begin{bmatrix} -1 & 6 \\ 2 & 7 \\ 0 & 8 \end{bmatrix}$$

SOLUTION

The product AB is not possible because a row–column dot product can occur only when the rows of A and the columns of B have the same number of entries.

■ **Now You Are Ready to Work Exercise 26**

This example illustrates that two matrices may or may not have a product. There must be the same number of columns in the first matrix as there are rows in the second in order for multiplication to be possible.

As you work with the product of two matrices, AB, notice how the size of AB relates to the size of A and the size of B. You will find that

- The number of rows of A equals the number of rows of AB.
- The number of columns of B equals the number of columns of AB.
- The number of columns of A must equal the number of rows of B.

Multiplication of Matrices

> Given matrices A and B, to find $AB = C$ **(matrix multiplication):**
>
> 1. Check the number of columns of A and the number of rows of B. If they are equal, the product is possible. If they are not equal, no product is possible.
> 2. Form all possible dot products using a row from A and a column from B. The dot product of row i with column j gives the entry for the (i, j) position in C.
> 3. The number of rows in C is the same as the number of rows in A. The number of columns in C is the same as the number of columns in B.

We now return to the problem at the beginning of the section and show a simple use of matrix multiplication.

EXAMPLE 4 ➤   The time, in hours, required to assemble and finish a table and a chair is given by the matrix

$$\begin{array}{c} \\ \text{Assemble} \\ \text{Finish} \end{array} \begin{array}{cc} \text{Chair} & \text{Table} \\ \begin{bmatrix} 2 & 3 \\ 2.5 & 4.75 \end{bmatrix} \end{array}$$

How long will it take to assemble and finish 950 chairs and 635 tables?

SOLUTION

Matrix multiplication gives the answer when we let

$$\begin{bmatrix} 950 \\ 635 \end{bmatrix}$$

be the column matrix that specifies the number of chairs and tables produced. Multiply the matrices:

$$\begin{bmatrix} 2 & 3 \\ 2.5 & 4.75 \end{bmatrix} \begin{bmatrix} 950 \\ 635 \end{bmatrix} = \begin{bmatrix} 2(950) & + 3(635) \\ 2.5(950) & + 4.75(635) \end{bmatrix} = \begin{bmatrix} 3805 \\ 5391.25 \end{bmatrix}$$

The rows of the result correspond to the rows in the *first* matrix; the first row in each represents assembly time, and the second represents finishing time. In the final matrix, 3805 is the *total* number of hours of assembly, and 5391.25 is the *total* number of hours for finishing required for 950 chairs and 635 tables.

■ **Now You Are Ready to Work Exercise 63**

The next example illustrates that AB and BA may both exist but are not equal.

EXAMPLE 5 ➤   Find AB and BA:

$$A = \begin{bmatrix} 1 & 3 \\ 5 & -2 \end{bmatrix} \quad \text{and} \quad B = \begin{bmatrix} 2 & 1 \\ 3 & -4 \end{bmatrix}$$

SOLUTION

$$AB = \begin{bmatrix} 1 & 3 \\ 5 & -2 \end{bmatrix}\begin{bmatrix} 2 & 1 \\ 3 & -4 \end{bmatrix} = \begin{bmatrix} 11 & -11 \\ 4 & 13 \end{bmatrix}$$

$$BA = \begin{bmatrix} 2 & 1 \\ 3 & -4 \end{bmatrix}\begin{bmatrix} 1 & 3 \\ 5 & -2 \end{bmatrix} = \begin{bmatrix} 7 & 4 \\ -17 & 17 \end{bmatrix} \quad ■$$

This example shows that AB and BA are not always equal. In fact, sometimes one of them may exist and the other not. The following example illustrates this.

EXAMPLE 6 ➤   Find AB and BA, if possible.

$$A = \begin{bmatrix} 1 & 2 & 3 \\ -4 & 0 & -2 \\ 1 & 1 & 1 \end{bmatrix} \quad \text{and} \quad B = \begin{bmatrix} 5 & -2 \\ 1 & 4 \\ 2 & 3 \end{bmatrix}$$

SOLUTION

$$AB = \begin{bmatrix} 1 & 2 & 3 \\ -4 & 0 & -2 \\ 1 & 1 & 1 \end{bmatrix}\begin{bmatrix} 5 & -2 \\ 1 & 4 \\ 2 & 3 \end{bmatrix} = \begin{bmatrix} 13 & 15 \\ -24 & 2 \\ 8 & 5 \end{bmatrix}$$

$$BA = \begin{bmatrix} 5 & -2 \\ 1 & 4 \\ 2 & 3 \end{bmatrix}\begin{bmatrix} 1 & 2 & 3 \\ -4 & 0 & -2 \\ 1 & 1 & 1 \end{bmatrix} = 5(1) + (-2)(-4) + ?(1)$$

When we attempt to use row 1 from B and column 1 from A to find the $(1, 1)$ entry of BA, we find no entry in row 1 of matrix B to multiply by the bottom entry, 1, of column 1 of A. Therefore, we, cannot complete the computation. BA does not exist.

■ **Now You Are Ready to Work Exercise 37**

Matrix multiplication can be used in a variety of applications, as illustrated in the next two examples.

EXAMPLE 7 ➤ The Kaplans have 150 shares of Acme Corp., 100 shares of High Tech, and 240 shares of ABC in an investment portfolio. The closing prices of these stocks one week were:

Monday:	Acme, $56; High Tech, $132; ABC, $19
Tuesday:	Acme, $55; High Tech, $133; ABC, $19
Wednesday:	Acme, $55; High Tech, $131; ABC, $20
Thursday:	Acme, $54; High Tech, $130; ABC, $22
Friday:	Acme, $53; High Tech, $128; ABC, $21

Summarize the closing prices in a matrix. Write the number of shares in a matrix and find the value of the Kaplans' portfolio each day by matrix multiplication.

SOLUTION

Set up the matrix of closing prices by letting each column represent a stock and each row a day:

$$
\begin{array}{c}
 \\
\text{Mon.} \\
\text{Tue.} \\
\text{Wed.} \\
\text{Thur.} \\
\text{Fri.}
\end{array}
\begin{array}{ccc}
\text{Acme} & \overset{\text{High}}{\text{Tech}} & \text{ABC} \\
\end{array}
\begin{bmatrix}
56 & 132 & 19 \\
55 & 133 & 19 \\
55 & 131 & 20 \\
54 & 130 & 22 \\
53 & 128 & 21
\end{bmatrix}
$$

We point out that using rows to represent stocks and columns to represent days is also acceptable. The matrix showing the number of shares of each company could be either a row matrix or a column matrix. Which of the matrices giving the number of shares,

$$
[150 \quad 100 \quad 240] \qquad or \qquad \begin{bmatrix} 150 \\ 100 \\ 240 \end{bmatrix}
$$

should be used to find the daily value of the portfolio? First of all, notice that the products

$$
[150 \quad 100 \quad 240]\begin{bmatrix}
56 & 132 & 19 \\
55 & 133 & 19 \\
55 & 131 & 20 \\
54 & 130 & 22 \\
53 & 128 & 21
\end{bmatrix}
$$

$$
\begin{bmatrix} 150 \\ 100 \\ 240 \end{bmatrix}\begin{bmatrix}
56 & 132 & 19 \\
55 & 133 & 19 \\
55 & 131 & 20 \\
54 & 130 & 22 \\
53 & 128 & 21
\end{bmatrix}
$$

$$
\begin{bmatrix}
56 & 132 & 19 \\
55 & 133 & 19 \\
55 & 131 & 20 \\
54 & 130 & 22 \\
53 & 128 & 21
\end{bmatrix}[150 \quad 100 \quad 240]
$$

are not possible.

The product

$$
\begin{bmatrix}
56 & 132 & 19 \\
55 & 133 & 19 \\
55 & 131 & 20 \\
54 & 130 & 22 \\
53 & 128 & 21
\end{bmatrix}
\begin{bmatrix}
150 \\
100 \\
240
\end{bmatrix}
=
\begin{bmatrix}
26{,}160 \\
26{,}110 \\
26{,}150 \\
26{,}380 \\
25{,}790
\end{bmatrix}
$$

is possible. Does it give the desired result? Notice that the first entry in the answer, 26,160, is obtained by

$$
56(150) + 132(100) + 19(240)
$$

which is

(price of Acme) $\times$ (no. shares of Acme)
 $+$ (price of High Tech) $\times$ (no. shares of High Tech)
 $+$ (price of ABC) $\times$ (no. shares of ABC)

This is exactly what is needed to find the total value of the portfolio on Monday. The other entries are correct for the other days.

■ **Now You Are Ready to Work Exercise 65**

When you use a matrix product in an application, check to see which order of multiplication makes sense.

If you ask a person to set up the stock data in matrix form, they might well let the columns represent days of the week and the rows the stocks. The matrix then takes the form

	Mon	Tue	Wed	Thu	Fri
Acme	56	55	55	54	53
High Tech	132	133	131	130	128
ABC	19	19	20	22	21

In this case, the matrix representing the number of shares must assume a form that makes sense when the matrices are multiplied to obtain the portfolio value. Use the form [150   100   240], and the product

$$
\begin{bmatrix} 150 & 100 & 240 \end{bmatrix}
\begin{bmatrix}
56 & 55 & 55 & 54 & 53 \\
132 & 133 & 131 & 130 & 128 \\
19 & 19 & 20 & 22 & 21
\end{bmatrix}
$$
$$
= \begin{bmatrix} 21{,}160 & 26{,}110 & 26{,}150 & 26{,}380 & 25{,}790 \end{bmatrix}
$$

gives daily portfolio values.

Thus, a matrix may be written in different ways as long as it is understood what the rows and columns represent, and the rows and columns of each matrix are arranged so the multiplication makes sense.

EXAMPLE 8 ➤ Five people attend a party. They are Alan, Bob, Cindy, Dana, and Eric. The diagram on the next page shows which ones are acquainted. A line joins two people's initials if they are acquainted. This acquaintance relationship can also be represented by the matrix

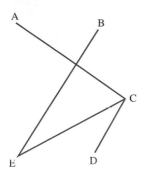

$$
\begin{array}{c}
 \\
A \\ B \\ C \\ D \\ E
\end{array}
\begin{array}{cc}
\begin{array}{ccccc} A & B & C & D & E \end{array} \\
\left[\begin{array}{ccccc}
1 & 0 & 1 & 0 & 0 \\
0 & 1 & 0 & 0 & 1 \\
1 & 0 & 1 & 1 & 1 \\
0 & 0 & 1 & 1 & 0 \\
0 & 1 & 1 & 0 & 1
\end{array}\right]
\end{array} = M
$$

where each entry is associated with two people, the person associated with the entry row and the person associated with the entry column. The entry is a 1 if the two people are acquainted and a 0 otherwise. We use the rather redundant convention that each person is acquainted with himself or herself. Thus, a 1 appears in row A column A, row B column B, and so on.

Now that we have an acquaintance matrix, we can multiply M by itself to find its square, which is

$$
\begin{array}{c}
 \\
A \\ B \\ C \\ D \\ E
\end{array}
\begin{array}{cc}
\begin{array}{ccccc} A & B & C & D & E \end{array} \\
\left[\begin{array}{ccccc}
2 & 0 & 2 & 1 & 1 \\
0 & 2 & 1 & 0 & 2 \\
2 & 1 & 4 & 2 & 2 \\
1 & 0 & 2 & 2 & 1 \\
1 & 2 & 2 & 1 & 3
\end{array}\right]
\end{array} = M^2
$$

We show you this matrix because of the information it contains. Observe for each position in M that contains a 1, the same position in M^2 contains a 2 or larger, and some positions that had a zero in M now have a positive integer in M^2.

Look at the A row and E column of M^2 and you find a 1, whereas the same position in M has a zero. Matrix M tells us that A and E are not acquainted. The 1 in M^2 tells us that A and E are acquainted through another person, they are "two-step" acquaintances. Look at M and you will see that A and C are acquainted and C and E are acquainted. Thus, A and E are "acquainted" through C. All the entries that are zero in M and positive in M^2 indicate the two persons of that row and column are acquainted through a third party and so are "two-step" acquaintances. The 1 in row A and column D of M^2 indicates that A and D are two-step acquaintances.

In a similar fashion, an entry of zero in M^2 that is positive in M^3 indicates the two people are "three-step" acquaintances. For the matrix M in this example, M^3 has no zero entries, which indicates that everyone is acquainted in three steps or less. ■

Identity Matrix

You are familiar with the number fact

$$1 \times a = a \times 1 = a$$

where a is any real number. We call 1 the **identity** for multiplication.

In general, we have no similar property for multiplication of matrices; there is no one matrix I such that $AI = IA = A$ for all matrices A. However, there is such a matrix for square matrices of a given size. For example, if

$$
A = \begin{bmatrix} 4 & 3 \\ 7 & 2 \end{bmatrix} \quad \text{and} \quad I = \begin{bmatrix} 1 & 0 \\ 0 & 1 \end{bmatrix}
$$

then

$$AI = \begin{bmatrix} 4 & 3 \\ 7 & 2 \end{bmatrix} \begin{bmatrix} 1 & 0 \\ 0 & 1 \end{bmatrix} = \begin{bmatrix} 4 & 3 \\ 7 & 2 \end{bmatrix} = A$$

and

$$IA = \begin{bmatrix} 1 & 0 \\ 0 & 1 \end{bmatrix} \begin{bmatrix} 4 & 3 \\ 7 & 2 \end{bmatrix} = \begin{bmatrix} 4 & 3 \\ 7 & 2 \end{bmatrix} = A$$

Furthermore, for any 2×2 matrix A, the matrix I has the property that $AI = A$ and $IA = A$. This can be justified by using a 2×2 matrix with arbitrary entries

$$A = \begin{bmatrix} a & b \\ c & d \end{bmatrix}$$

Now

$$AI = \begin{bmatrix} a & b \\ c & d \end{bmatrix} \begin{bmatrix} 1 & 0 \\ 0 & 1 \end{bmatrix} \begin{bmatrix} a \times 1 + b \times 0 & a \times 0 + b \times 1 \\ c \times 1 + d \times 0 & c \times 0 + d \times 1 \end{bmatrix}$$

$$= \begin{bmatrix} a & b \\ c & d \end{bmatrix} = A$$

You should now multiply IA to verify that it is indeed A. Thus,

$$\begin{bmatrix} 1 & 0 \\ 0 & 1 \end{bmatrix}$$

is the **identity matrix** for all 2×2 matrices. If we try to multiply the 3×3 matrix

$$A = \begin{bmatrix} 1 & 2 & 3 \\ 5 & 7 & 12 \\ 8 & 4 & -2 \end{bmatrix} \quad \text{by} \quad \begin{bmatrix} 1 & 0 \\ 0 & 1 \end{bmatrix}$$

we find we are unable to multiply at all because A has 3 columns and I has only two rows. So,

$$\begin{bmatrix} 1 & 0 \\ 0 & 1 \end{bmatrix}$$

is *not* the identity matrix for 3×3 matrices. However, the matrix

$$I = \begin{bmatrix} 1 & 0 & 0 \\ 0 & 1 & 0 \\ 0 & 0 & 1 \end{bmatrix}$$

is an identity matrix for the set of all 3×3 matrices:

$$\begin{bmatrix} a & b & c \\ d & e & f \\ g & h & i \end{bmatrix} \begin{bmatrix} 1 & 0 & 0 \\ 0 & 1 & 0 \\ 0 & 0 & 1 \end{bmatrix} = \begin{bmatrix} a & b & c \\ d & e & f \\ g & h & i \end{bmatrix}$$

$$\begin{bmatrix} 1 & 0 & 0 \\ 0 & 1 & 0 \\ 0 & 0 & 1 \end{bmatrix} \begin{bmatrix} a & b & c \\ d & e & f \\ g & h & i \end{bmatrix} = \begin{bmatrix} a & b & c \\ d & e & f \\ g & h & i \end{bmatrix}$$

In general, if we let I be the $n \times n$ matrix with ones on the *main diagonal* and zeros elsewhere, it is the identity matrix for the class of all $n \times n$ matrices. (The **main diagonal** runs from the upper left to the lower right corner.)

Row Operations Using Matrix Multiplication

We can perform row operations on a matrix by multiplying by a modified identity matrix. Let's illustrate.

$$\text{Let } A = \begin{bmatrix} 1 & 2 & 3 & 4 \\ 5 & 6 & 7 & 8 \\ 9 & 10 & 11 & 12 \end{bmatrix} \text{ and } I = \begin{bmatrix} 1 & 0 & 0 \\ 0 & 1 & 0 \\ 0 & 0 & 1 \end{bmatrix}$$

We know that $IA = A$. Now interchange rows 1 and 2 of I and multiply times A.

$$\begin{bmatrix} 0 & 1 & 0 \\ 1 & 0 & 0 \\ 0 & 0 & 1 \end{bmatrix} \begin{bmatrix} 1 & 2 & 3 & 4 \\ 5 & 6 & 7 & 8 \\ 9 & 10 & 11 & 12 \end{bmatrix} = \begin{bmatrix} 5 & 6 & 7 & 8 \\ 1 & 2 & 3 & 4 \\ 9 & 10 & 11 & 12 \end{bmatrix}$$

Notice that this multiplication interchanges rows 1 and 2 of A. This illustrates a general property.

Interchange Rows by Matrix Multiplication

If a matrix A has n rows and I is the $n \times n$ identity matrix, then modify I by interchanging two rows, giving matrix I_M. The product $I_M A$ interchanges the corresponding rows of A.

Next, modify I by adding row 3 to row 1 giving $I_M = \begin{bmatrix} 1 & 0 & 1 \\ 0 & 1 & 0 \\ 0 & 0 & 1 \end{bmatrix}$.

$$\text{The product } I_M A \text{ is } \begin{bmatrix} 1 & 0 & 1 \\ 0 & 1 & 0 \\ 0 & 0 & 1 \end{bmatrix} \begin{bmatrix} 1 & 2 & 3 & 4 \\ 5 & 6 & 7 & 8 \\ 9 & 10 & 11 & 12 \end{bmatrix} = \begin{bmatrix} 10 & 12 & 14 & 16 \\ 5 & 6 & 7 & 8 \\ 9 & 10 & 11 & 12 \end{bmatrix}.$$

Notice that rows 2 and 3 are unchanged, but row 1 is now the sum of rows 1 and 3. In the notation we have used for row operations, this product gives R1 + R3 → R1. Next we add row 1 to row 3 (R1 + R3 → R3) and multiply.

$$\begin{bmatrix} 1 & 0 & 0 \\ 0 & 1 & 0 \\ 1 & 0 & 1 \end{bmatrix} \begin{bmatrix} 1 & 2 & 3 & 4 \\ 5 & 6 & 7 & 8 \\ 9 & 10 & 11 & 12 \end{bmatrix} = \begin{bmatrix} 1 & 2 & 3 & 4 \\ 5 & 6 & 7 & 8 \\ 10 & 12 & 14 & 16 \end{bmatrix}$$

Row 1 has been added to row 3 and the result stored in row 3.

Let's give one more example. Modify I by the row operation 2R2 + R1 → R1,

$$\text{giving } I_M = \begin{bmatrix} 1 & 2 & 0 \\ 0 & 1 & 0 \\ 0 & 0 & 1 \end{bmatrix}.$$

$$\text{Now } I_M A = \begin{bmatrix} 1 & 2 & 0 \\ 0 & 1 & 0 \\ 0 & 0 & 1 \end{bmatrix} \begin{bmatrix} 1 & 2 & 3 & 4 \\ 5 & 6 & 7 & 8 \\ 9 & 10 & 11 & 12 \end{bmatrix} = \begin{bmatrix} 11 & 14 & 17 & 20 \\ 5 & 6 & 7 & 8 \\ 9 & 10 & 11 & 12 \end{bmatrix}.$$

The result is the row operation 2R2 + R1 → R1 on A.

We summarize these examples with a general property they represent.

Row Operations Using Matrix Multiplication

If a matrix A has n rows and I is the $n \times n$ identity matrix, then modify I by a row operation like $5R2 + R4 \rightarrow R4$, giving matrix I_M. The product $I_M A$ performs the same row operation on A.

2.5 EXERCISES

LEVEL 1

Find the dot products in Exercises 1 through 6.

1. $[1 \quad 3] \cdot \begin{bmatrix} 2 \\ 4 \end{bmatrix}$

2. $[-2 \quad 5] \cdot \begin{bmatrix} 4 \\ 1 \end{bmatrix}$

3. $[6 \quad 5] \cdot \begin{bmatrix} 2 \\ 0 \end{bmatrix}$

4. $[3 \quad -1 \quad 2] \cdot \begin{bmatrix} -2 \\ 5 \\ 3 \end{bmatrix}$

5. $[1 \quad 0 \quad 1] \cdot \begin{bmatrix} 6 \\ 7 \\ 8 \end{bmatrix}$

6. $[2 \quad 1 \quad 3 \quad -2] \cdot \begin{bmatrix} 5 \\ 5 \\ -1 \\ 3 \end{bmatrix}$

7. *(See Example 1)* The price matrix of bread, milk, and cheese is, in that order, $[0.90 \quad 1.85 \quad 0.65]$. The quantity of each purchased, in the same order, is given by the column matrix

$$\begin{bmatrix} 2 \\ 1 \\ 4 \end{bmatrix}$$

Find the total bill for the purchases.

8. Find the total bill for the purchase of hamburgers, fries, and drink where the price and quantities are listed in that order in the matrices

$$\text{Price matrix} = [3.25 \quad 1.09 \quad 1.19]$$

$$\text{Quantity matrix} = \begin{bmatrix} 10 \\ 15 \\ 8 \end{bmatrix}$$

Find the products in Exercises 9 through 12.

9. $\begin{bmatrix} 3 & 1 \\ 2 & 4 \end{bmatrix} \begin{bmatrix} -2 & 3 \\ 1 & 2 \end{bmatrix}$

10. $\begin{bmatrix} -6 & 2 \\ 0 & 4 \end{bmatrix} \begin{bmatrix} 1 & -1 \\ 3 & 5 \end{bmatrix}$

11. *(See Example 2)*
$\begin{bmatrix} 2 & 1 & 4 \\ 3 & -1 & 5 \end{bmatrix} \begin{bmatrix} 3 & -2 \\ 0 & 2 \\ 6 & 1 \end{bmatrix}$

12. $\begin{bmatrix} 1 & 2 & 3 \\ 5 & -1 & 2 \end{bmatrix} \begin{bmatrix} 1 & 0 \\ 3 & 2 \\ 4 & 5 \end{bmatrix}$

13. A is a 2×2 matrix and B is a 2×3 matrix. Determine if the following matrix operations are possible. If the operation is possible, give the size of the resulting matrix.
(a) $A + B$
(b) AB
(c) BA

14. A is a 2×3 matrix and B is a 3×2 matrix. Determine if the following matrix operations are possible. If the operation is possible, give the size of the resulting matrix.
(a) $3A$ (b) $A - B$
(c) AB (d) BA

15. A is a 3×4 matrix and B is a 4×5 matrix. Determine if the following matrix operations are possible. If the operation is possible, give the size of the resulting matrix.
(a) AB (b) BA

16. A is a 4×2 matrix, B is a 2×3 matrix, and C is a 3×5 matrix. Determine if the following matrix operations are possible. If the operation is possible, give the size of the resulting matrix.
(a) AB (b) BC
(c) AC (d) $(AB)C$

Find the matrix products in Exercises 17 through 32, if they exist.

17. $\begin{bmatrix} 2 & 3 \\ 1 & 4 \end{bmatrix} \begin{bmatrix} -5 & 2 \\ 1 & 1 \end{bmatrix}$

18. $\begin{bmatrix} 1 & 2 \\ 3 & 1 \end{bmatrix} \begin{bmatrix} -2 & 1 \\ 4 & 0 \end{bmatrix}$

19. $\begin{bmatrix} 1 & 3 \\ 2 & -2 \end{bmatrix} \begin{bmatrix} 3 \\ 4 \end{bmatrix}$

20. $\begin{bmatrix} 2 & -1 & 2 \\ 4 & 5 & 3 \end{bmatrix} \begin{bmatrix} 1 \\ 6 \end{bmatrix}$

21. $\begin{bmatrix} 2 \\ 5 \end{bmatrix} \begin{bmatrix} 3 & 1 \\ 0 & 2 \end{bmatrix}$

22. $\begin{bmatrix} 3 & 1 \\ 2 & 4 \\ 0 & -2 \end{bmatrix} \begin{bmatrix} 4 & -1 \\ 1 & 7 \end{bmatrix}$

23. $\begin{bmatrix} 4 & -1 \\ 1 & 7 \end{bmatrix} \begin{bmatrix} 3 & 1 \\ 2 & 4 \\ 0 & -2 \end{bmatrix}$

24. $\begin{bmatrix} 3 & 1 & 4 \\ 2 & 0 & 5 \\ 6 & 7 & -3 \end{bmatrix} \begin{bmatrix} 1 & 1 & 0 \\ 0 & 1 & 1 \\ 1 & 0 & 1 \end{bmatrix}$

25. $\begin{bmatrix} 2 & 0 \\ 1 & 5 \end{bmatrix}\begin{bmatrix} -2 & 2 \\ 1 & 5 \end{bmatrix}$ **26.** *(See Example 3)*

$$[4 \quad 1 \quad 3]\begin{bmatrix} 2 \\ 7 \\ 5 \end{bmatrix}$$

27. $\begin{bmatrix} 1 & 3 & 2 \\ 2 & 4 & 6 \end{bmatrix}\begin{bmatrix} 0 & 4 \\ 5 & 0 \end{bmatrix}$ **28.** $\begin{bmatrix} 2 & 4 & 6 \\ 3 & 1 & 3 \end{bmatrix}\begin{bmatrix} 0 & 5 & 5 \\ 2 & 1 & 3 \\ 6 & 0 & 0 \end{bmatrix}$

29. $\begin{bmatrix} 1 & 2 \\ 3 & 1 \end{bmatrix}\begin{bmatrix} 1 & 5 & 6 \\ 2 & 2 & 3 \end{bmatrix}$ **30.** $[1 \quad 2 \quad 3]\begin{bmatrix} 1 & 4 \\ 2 & 6 \\ 0 & 3 \end{bmatrix}$

31. $\begin{bmatrix} 1 & 2 & 3 \\ -1 & 4 & 6 \\ 2 & 1 & 3 \end{bmatrix}\begin{bmatrix} 1 \\ 2 \\ 1 \end{bmatrix}$ **32.** $\begin{bmatrix} 1 & 2 \\ 3 & 3 \end{bmatrix}\begin{bmatrix} 3 & 5 & 1 & 2 \\ -1 & 4 & 0 & -1 \end{bmatrix}$

Find AB and BA in Exercises 33 through 36.

33. $A = \begin{bmatrix} 1 & 2 \\ 0 & 1 \end{bmatrix}, \quad B = \begin{bmatrix} 3 & 4 \\ -1 & 2 \end{bmatrix}$

34. $A = \begin{bmatrix} 1 & 2 & 3 \\ 4 & 0 & -2 \end{bmatrix}, \quad B = \begin{bmatrix} 3 & 5 \\ 1 & -4 \\ 3 & 1 \end{bmatrix}$

35. $A = \begin{bmatrix} 1 & 3 \\ -1 & 2 \end{bmatrix}, \quad B = \begin{bmatrix} 0 & 1 \\ -1 & 3 \end{bmatrix}$

36. $A = [1 \quad 3 \quad 4], \quad B = \begin{bmatrix} 2 \\ 5 \\ 1 \end{bmatrix}$

In Exercises 37 through 42, find AB and BA, if possible.

37. *(See Example 6)*

$A = \begin{bmatrix} 1 & 2 \\ -3 & 1 \end{bmatrix}, \quad B = \begin{bmatrix} 4 & 2 & 3 \\ 1 & 0 & 5 \end{bmatrix}$

38. $A = \begin{bmatrix} 1 & 2 & 3 \\ 4 & 1 & 2 \\ 3 & -1 & 0 \end{bmatrix}, \quad B = \begin{bmatrix} 1 \\ 2 \\ 3 \end{bmatrix}$

39. $A = [1 \quad 3 \quad 5], \quad B = \begin{bmatrix} -1 & 4 \\ 6 & 3 \\ 2 & 5 \end{bmatrix}$

40. $A = [2 \quad 4 \quad 6], \quad B = \begin{bmatrix} 5 \\ 3 \end{bmatrix}$

41. $A = \begin{bmatrix} 1 & 2 \\ -3 & 1 \end{bmatrix}, \quad B = \begin{bmatrix} 4 & -2 \\ 3 & 4 \end{bmatrix}$

42. $A = \begin{bmatrix} 5 & 2 \\ 4 & 3 \end{bmatrix}, \quad B = \begin{bmatrix} 3 & 2 \\ 4 & 1 \end{bmatrix}$

LEVEL 2

Perform the indicated matrix operations in Exercises 43 through 58.

43. $\begin{bmatrix} 3 & 0 & 1 \\ 2 & 1 & 2 \end{bmatrix}\begin{bmatrix} 3 & -1 \\ 2 & 4 \\ 1 & 0 \end{bmatrix}\begin{bmatrix} -1 & 1 \\ 3 & -2 \end{bmatrix}$

44. $\begin{bmatrix} 3 & 1 \\ 1 & 2 \end{bmatrix}\begin{bmatrix} 1 & 0 & 4 \\ 2 & -1 & 5 \end{bmatrix}\begin{bmatrix} 3 \\ 1 \\ 1 \end{bmatrix}$

45. $\begin{bmatrix} 1 & 0 & 3 \\ 2 & 1 & 1 \\ 4 & -2 & -1 \end{bmatrix}\begin{bmatrix} 2 & 1 & 4 & -2 \\ -1 & 1 & 3 & -3 \\ 2 & 0 & -2 & -4 \end{bmatrix}$

46. $\begin{bmatrix} 4 & 1 \\ 3 & 2 \end{bmatrix} - \begin{bmatrix} 1 & 2 \\ 3 & 4 \end{bmatrix}\begin{bmatrix} 1 & 0 \\ 0 & 2 \end{bmatrix}$

47. $\begin{bmatrix} 5 & 2 \\ 1 & 3 \end{bmatrix}\begin{bmatrix} 4 & -2 \\ 1 & 1 \end{bmatrix} + \begin{bmatrix} 3 & -6 \\ 5 & 1 \end{bmatrix}\begin{bmatrix} 0 & 2 \\ 3 & -1 \end{bmatrix}$

48. $[3 \quad 1]\begin{bmatrix} 1 & 2 \\ 3 & 4 \end{bmatrix} + [2 \quad 0 \quad 3]\begin{bmatrix} 0 & 5 \\ 3 & 7 \\ -1 & 9 \end{bmatrix}$

49. $\begin{bmatrix} 0 & 1 \\ 1 & 0 \end{bmatrix}\begin{bmatrix} 1 & 2 \\ 3 & 4 \end{bmatrix}$

50. $\begin{bmatrix} 1 & 2 \\ 3 & 4 \end{bmatrix}\begin{bmatrix} 0 & 1 \\ 1 & 0 \end{bmatrix}$

51. $\begin{bmatrix} 1 & 0 \\ 0 & 1 \end{bmatrix}\begin{bmatrix} 2 & -10 \\ 3 & 7 \end{bmatrix}$

52. $\begin{bmatrix} 1 & 2 & 3 \\ 4 & 5 & 6 \\ 7 & 8 & 9 \end{bmatrix}\begin{bmatrix} 1 & 0 & 0 \\ 0 & 1 & 0 \\ 0 & 0 & 1 \end{bmatrix}$

53. $\begin{bmatrix} 3 & 1 \\ 2 & 4 \end{bmatrix}\begin{bmatrix} x \\ y \end{bmatrix}$

54. $\begin{bmatrix} 2 & 1 & 3 \\ 4 & -2 & 6 \\ 1 & 5 & -4 \end{bmatrix}\begin{bmatrix} x \\ y \\ z \end{bmatrix}$

55. $\begin{bmatrix} 1 & 2 & -1 \\ 3 & 1 & 4 \\ 2 & -1 & -1 \end{bmatrix}\begin{bmatrix} x_1 \\ x_2 \\ x_3 \end{bmatrix}$

56. $\begin{bmatrix} 1 & 5 & 9 \\ 2 & 1 & 6 \end{bmatrix}\begin{bmatrix} x_1 \\ x_2 \\ x_3 \end{bmatrix}$

57. $\begin{bmatrix} 1 & 3 & 5 & 6 \\ -2 & 9 & 6 & 1 \\ 8 & 0 & 17 & 5 \end{bmatrix}\begin{bmatrix} x_1 \\ x_2 \\ x_3 \\ x_4 \end{bmatrix}$

58. $\begin{bmatrix} 1 & 0 & 2 & -1 \\ 5 & 4 & 1 & 2 \\ 1 & 6 & -3 & -1 \\ 1 & -1 & 1 & -1 \end{bmatrix}\begin{bmatrix} x_1 \\ x_2 \\ x_3 \\ x_4 \end{bmatrix}$

59. A is a 3 × 3 matrix. For each I_M in (a) through (e), tell what row operations on A are the result of the matrix product $I_M A$.

(a) $I_M = \begin{bmatrix} 1 & 0 & 0 \\ 1 & 1 & 0 \\ 0 & 0 & 1 \end{bmatrix}$ **(b)** $I_M = \begin{bmatrix} 2 & 0 & 0 \\ 0 & 1 & 0 \\ 0 & 1 & 1 \end{bmatrix}$

(c) $I_M = \begin{bmatrix} 1 & 0 & -1 \\ 0 & 1 & 0 \\ 0 & 0 & 1 \end{bmatrix}$ **(d)** $I_M = \begin{bmatrix} 1 & 0 & 0 \\ 1 & 1 & 1 \\ 0 & 0 & 1 \end{bmatrix}$

(e) $I_M = \begin{bmatrix} 1 & 0 & 4 \\ 0 & 1 & 0 \\ 0 & 0 & 1 \end{bmatrix}$

60. $A = \begin{bmatrix} 2 & -1 & 5 & 8 \\ 3 & 7 & 0 & 9 \\ 6 & 12 & 8 & 4 \end{bmatrix}$

Find $I_M A$ where

(a) $I_M = \begin{bmatrix} 1 & 0 & 0 \\ 0 & 1 & 1 \\ 0 & 0 & 1 \end{bmatrix}$ **(b)** $I_M = \begin{bmatrix} 1 & 0 & 0 \\ 0 & 1 & 0 \\ 0 & 0 & 2 \end{bmatrix}$

(c) $I_M = \begin{bmatrix} 1 & 0 & 0 \\ 0 & 1 & 0 \\ 0 & 0 & 0.5 \end{bmatrix}$ **(d)** $I_M = \begin{bmatrix} 1 & 0 & 0 \\ 0 & 1 & 0 \\ -2 & 0 & 1 \end{bmatrix}$

61. $A = \begin{bmatrix} 1 & 5 & 3 & 9 \\ -2 & 7 & 4 & 11 \\ 9 & 0 & 2 & 5 \\ 6 & 3 & 3 & 2 \end{bmatrix}$

Find $I_M A$ where

(a) $I_M = \begin{bmatrix} 2 & 0 & 0 & 0 \\ 0 & 1 & 0 & 0 \\ 0 & 0 & -3 & 0 \\ 0 & 0 & 0 & 1 \end{bmatrix}$

(b) $I_M = \begin{bmatrix} 3 & 0 & 0 & 0 \\ 0 & 1 & 0 & 0 \\ 0 & 2 & 1 & 0 \\ 0 & 0 & 0 & 1 \end{bmatrix}$

(c) $I_M = \begin{bmatrix} 1 & 0 & 0 & 0 \\ 2 & 1 & 0 & 0 \\ 0 & 0 & 1 & 0 \\ 0 & 0 & 0 & 1 \end{bmatrix}$

(d) $I_M = \begin{bmatrix} 1 & 0 & 0 & 0 \\ 0 & 1 & 0 & 0 \\ -9 & 0 & 1 & 0 \\ 0 & 0 & 0 & 1 \end{bmatrix}$

(e) $I_M = \begin{bmatrix} 1 & 0 & 0 & 0 \\ 2 & 1 & 0 & 0 \\ -9 & 0 & 1 & 0 \\ -6 & 0 & 0 & 1 \end{bmatrix}$

62. $I = \begin{bmatrix} 1 & 0 & 0 \\ 0 & 1 & 0 \\ 0 & 0 & 1 \end{bmatrix}$ and $A = \begin{bmatrix} 1 & 2 & 3 \\ 4 & 5 & 6 \\ 7 & 8 & 9 \end{bmatrix}$

Perform the row operations indicated on I to obtain I_M. Then find $I_M A$.

(a) $4R2 \rightarrow R2$
(b) $3R1 + R2 \rightarrow R2$
(c) $0.25\, R2 \rightarrow R2$
(d) $-4R1 + R3 \rightarrow R3$

LEVEL 3

63. *(See Example 4)* The Home Entertainment Firm makes stereos and TV sets. The matrix below shows the time required for assembly and checking.

	Stereo	TV
Assembly	4	5.5
Check	1	2

Use matrices to determine the total assembly time and total checking time for 300 stereos and 450 TV sets.

64. Speed King has two production lines, I and II. Both produce ten-speed and three-speed bicycles. The number produced per hour is given by the matrix

	Line I	Line II
Three-speed	10	15
Ten-speed	12	20

Find the number of each type bicycle that is produced if line I operates 60 hours and line II 48 hours.

65. *(See Example 7)* An investment portfolio contains 60 shares of SCM and 140 shares of Apex Corp. The closing prices on three days were:

Monday:	SCM, $114; Apex, $85
Wednesday:	SCM, $118; Apex, $84
Friday:	SCM, $116; Apex, $86

Use matrix multiplication to find the value of the portfolio on each of the three days.

66. The following matrix gives the vitamin content of a typical breakfast in conveniently chosen units:

$$
\begin{array}{c}
\\
\\
\text{Orange juice} \\
\text{Oatmeal} \\
\text{Milk} \\
\text{Biscuit} \\
\text{Butter}
\end{array}
\begin{array}{c}
\text{Vitamin} \\
\begin{array}{cccc}
A & B_1 & B_2 & C
\end{array} \\
\left[
\begin{array}{cccc}
500 & 0.2 & 0 & 129 \\
0 & 0.2 & 0 & 0 \\
1560 & 0.32 & 1.7 & 6 \\
0 & 0 & 0 & 0 \\
460 & 0 & 0 & 0
\end{array}
\right]
\end{array}
$$

If you have 1 unit of orange juice, 1 unit of oatmeal, $\frac{1}{4}$ unit of milk, 2 units of biscuit, and 2 units of butter, find the matrix that tells how much of each type vitamin you have consumed. (Notice that you need to multiply a row matrix of the units times the given matrix.)

67. Use the vitamin content from Exercise 66. Two breakfast menus are summarized in the matrix

$$
\begin{array}{c}
\\
\\
\text{Orange juice} \\
\text{Oatmeal} \\
\text{Milk} \\
\text{Biscuit} \\
\text{Butter}
\end{array}
\begin{array}{c}
\text{Menus} \\
\begin{array}{cc}
\text{I} & \text{II}
\end{array} \\
\left[
\begin{array}{cc}
0.5 & 0 \\
1.5 & 1.0 \\
0.5 & 1.0 \\
1.0 & 3.0 \\
1.0 & 2.0
\end{array}
\right]
\end{array}
$$

Find the matrix that tells the amount of each vitamin consumed in each diet.

68. Data from three supermarkets are summarized in this matrix:

$$
\begin{array}{c}
\\
\text{Sugar (per pound)} \\
\text{Peaches (per can)} \\
\text{Chicken (per pound)} \\
\text{Bread (per loaf)}
\end{array}
\begin{array}{c}
\begin{array}{ccc}
\text{Store 1} & \text{Store 2} & \text{Store 3}
\end{array} \\
\left[
\begin{array}{ccc}
\$0.49 & \$0.47 & \$0.53 \\
\$1.39 & \$1.49 & \$1.54 \\
\$1.85 & \$1.79 & \$1.75 \\
\$1.19 & \$1.20 & \$1.15
\end{array}
\right]
\end{array}
$$

What is the total grocery bill at each store if the following purchase is made at each store: 5 pounds of sugar, 3 cans of peaches, 3 pounds of chicken, and 2 loaves of bread?

69. The Humidor blends regular coffee, High Mountain coffee, and chocolate to obtain three kinds of coffee: Early Riser, After Dinner, and Deluxe. The blends are:

	Blend		
	Early Riser	After Dinner	Deluxe
Regular	80%	75%	50%
High Mountain	20%	20%	40%
Chocolate	0%	5%	10%

Use matrix multiplication to determine the number of pounds of regular coffee, High Mountain coffee, and chocolate needed to fill an order of 400 pounds of Early Riser, 360 pounds of After Dinner, and 230 pounds of Deluxe coffees.

70. The Health Fare Cereal Company makes three cereals using wheat, oats, and raisins. The proportions of each cereal are:

	Proportion of Each Pound of Cereal		
Cereal	Wheat	Oats	Raisins
Lite	0.75	0.25	0
Trim	0.50	0.25	0.25
Health Fare	0.25	0.50	0.25

Use matrix multiplication to determine the number of pounds of wheat, oats, and raisins needed to fill an order of 1480 pounds of Lite, 1840 pounds of Trim, and 2050 pounds of Health Fare.

71. Professor Hurley gave four exams in his course. The grades of six students are shown in this matrix.

$$
\begin{array}{c}
\\
\\
\text{Amy} \\
\text{Bob} \\
\text{Cal} \\
\text{Dot} \\
\text{Eve} \\
\text{Fay}
\end{array}
\begin{array}{c}
\text{Exam} \\
\begin{array}{cccc}
1 & 2 & 3 & 4
\end{array} \\
\left[
\begin{array}{cccc}
78 & 83 & 81 & 86 \\
84 & 88 & 79 & 85 \\
70 & 72 & 77 & 73 \\
88 & 91 & 94 & 87 \\
96 & 95 & 98 & 92 \\
65 & 72 & 74 & 81
\end{array}
\right]
\end{array}
$$

(a) Use matrix multiplication to find each student's average if the exams are weighted the same.

(b) Use matrix multiplication to find each student's average if the exams are weighted as follows: Exam 1, 20%; Exam 2, 20%; Exam 3, 25%; and Exam 4, 35%.

72. Sedric works for a plumbing company at a wage of \$15 per hour for regular weekday hours. If he works on Saturday, he receives time-and-a-half. If he works on Sunday, he receives double-time. Matrix A gives the number of hours worked each day over a four-week pay period.

Hours Worked

	Days						
Week	M	Tu	W	Th	F	Sa	Su
1	8	7	8	8	6	4	0
2	8	8	8	7.5	8	6	2
3	8	8	7	8	8	3	1
4	8	8	8	8	8	8	4

$= A$

(a) $B = \begin{bmatrix} 1 \\ 1 \\ 1 \\ 1 \\ 1 \\ 1.5 \\ 2 \end{bmatrix}$ Find and interpret $15AB$.

(b) $C = [1\ \ 1\ \ 1\ \ 1]$ Find and interpret $C(15AB)$.

EXPLORATIONS

73. An epidemic hits a city. Each person is classified by the Health Department as either well, sick, or a carrier. The proportion of people in each category by age groups is given by the matrix

	Age		
	0–15	16–35	Over 35
Well	0.65	0.60	0.70
Sick	0.25	0.35	0.20
Carrier	0.10	0.05	0.10

$= A$

The population of the city by age and gender is

	Male	Female
0–15	35,000	30,000
16–35	55,000	50,000
Over 35	70,000	75,000

$= B$

(a) Compute AB. Interpret the meaning of the entries in AB.
(b) How many sick males are there?
(c) How many females are well?

74. The following figure shows airline routes connecting five cities. Two cities are connected with a line if there is a direct flight between them: This relationship can be represented by the matrix shown, where an entry of 1 indicates the two cities of that row and column are directly connected and a 0 indicates no direct connection

	PH	DAL	CHI	DC	ATL
PH	0	1	1	0	0
DAL	1	0	1	0	1
CHI	1	1	0	1	1
DC	0	0	1	0	1
ATL	0	1	1	1	0

$= A$

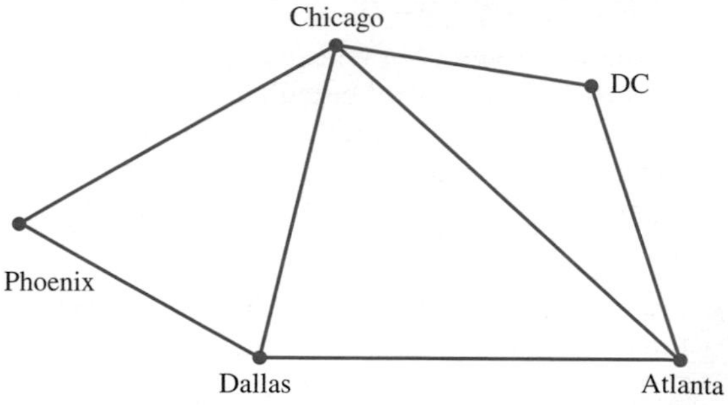

(a) Compute A^2.
(b) Verify from the figure that an entry in A^2 gives the number of two-flight routes between the cities for that row and column.

75. The following figure shows direct airline connections between five cities.
(a) Form the matrix that represents this figure.
(b) Find the matrix that gives the number of two-flight routes between cities.

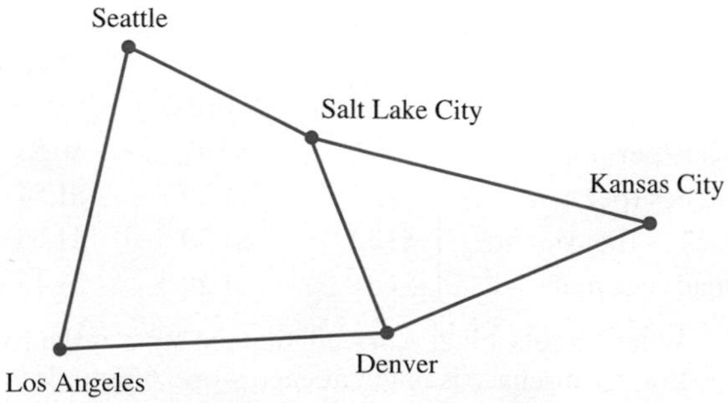

76. Computer Products makes three models of its laptop computer, the L100, the L150, and the L250. The computers require, among other components, circuit boards C1, C8, and C10. The number of each circuit board used per computer is given by the matrix

$$A = \begin{matrix} & \begin{matrix} \text{C1} & \text{C8} & \text{C10} \end{matrix} \\ \begin{matrix} \text{L100} \\ \text{L150} \\ \text{L250} \end{matrix} & \begin{bmatrix} 2 & 1 & 0 \\ 1 & 1 & 2 \\ 0 & 1 & 3 \end{bmatrix} \end{matrix}$$

Circuit Board

Each of the circuit boards in turn uses four microchips, M10, M11, M15, M16. The numbers of microchips per circuit board are given by the matrix.

$$B = \begin{matrix} & \begin{matrix} \text{M10} & \text{M11} & \text{M15} & \text{M16} \end{matrix} \\ \begin{matrix} \text{C1} \\ \text{C8} \\ \text{C10} \end{matrix} & \begin{bmatrix} 3 & 1 & 1 & 0 \\ 2 & 0 & 2 & 1 \\ 4 & 0 & 0 & 3 \end{bmatrix} \end{matrix}$$

Microchips

The number of each model laptop scheduled for next month is given by the matrix

$$C = \begin{matrix} \begin{matrix} \text{L100} & \text{L150} & \text{L250} \end{matrix} \\ [10,000 \quad 18,000 \quad 9,000] \end{matrix}$$

(a) Compute the matrix product AB and interpret the elements of the matrix.

(b) Compute the product $C(AB)$ and interpret the elements of the matrix.

77. What are the conditions on matrices A and B so that both AB and BA are defined?

78. What restrictions must be placed on the dimensions of matrices A and B so that $AB = BA$?

79. Review Exercises 33–42 and discuss the following: If A and B are matrices and AB exists, what are the possible outcomes for the product BA?

80. Matrix A shows the number of items purchased by a shopper, and matrix B represents the price per item.

$$A = \begin{matrix} \begin{matrix} \text{Chips} & \text{Water} & \text{Bread} & \text{Cereal} \end{matrix} \\ [4 \quad\quad 6 \quad\quad 2 \quad\quad 3] \end{matrix}$$

$$B = \begin{bmatrix} 0.59 \\ 0.85 \\ 0.65 \\ 2.15 \end{bmatrix} \begin{matrix} \text{Chips} \\ \text{Water} \\ \text{Bread} \\ \text{Cereal} \end{matrix}$$

Explain why AB gives the total bill for the purchase.

81. Matrix A shows the number of pieces shipped by three departments via overnight (ON), second (2nd) day, and regular (Reg) delivery. Matrix B shows the cost of shipping by two shipping companies.

$$A = \begin{matrix} & \begin{matrix} \text{No. pieces to} \\ \text{be shipped} \\ \begin{matrix} \text{Dept} & \text{ON} & \text{2nd} & \text{Reg} \end{matrix} \end{matrix} \\ \begin{matrix} 1 \\ 2 \\ 3 \end{matrix} & \begin{bmatrix} 3 & 5 & 9 \\ 1 & 6 & 4 \\ 8 & 4 & 0 \end{bmatrix} \end{matrix}$$

$$B = \begin{matrix} \begin{matrix} \text{Shipping Cost} \\ \begin{matrix} \text{Co. 1} & \text{Co. 2} \end{matrix} \end{matrix} \\ \begin{bmatrix} 12.50 & 14.40 \\ 8.25 & 7.20 \\ 2.35 & 2.00 \end{bmatrix} \begin{matrix} \text{ON} \\ \text{2nd} \\ \text{Reg} \end{matrix} \end{matrix}$$

Compute AB and tell what information this gives.

82. Matrix A represents the scores of five students on three exams.

$$A = \begin{matrix} & \begin{matrix} \text{Exam} & \text{Exam} & \text{Exam} \\ 1 & 2 & 3 \end{matrix} \\ \begin{matrix} \text{Al} \\ \text{Bea} \\ \text{Cindy} \\ \text{Dot} \\ \text{Ed} \end{matrix} & \begin{bmatrix} 88 & 76 & 81 \\ 92 & 84 & 79 \\ 72 & 78 & 76 \\ 94 & 90 & 95 \\ 68 & 73 & 78 \end{bmatrix} \end{matrix}$$

Professor Hubbs determines a grade by weighting the exams 25%, 40%, and 35%, which can be represented by the matrix

$$B = \begin{matrix} \text{Weight} \\ \begin{bmatrix} 0.25 \\ 0.40 \\ 0.35 \end{bmatrix} \end{matrix}$$

Compute and interpret the meaning of the product AB.

83. (a) The personnel office of Acme University determined the needs for administrative assistants and showed it in a matrix giving the number needed at each level for three schools in the university.

$$A = \begin{matrix} & \begin{matrix} \text{Level} & \text{Level} & \text{Level} \\ \text{I} & \text{II} & \text{III} \end{matrix} \\ \begin{bmatrix} 12 & 15 & 8 \\ 9 & 6 & 3 \\ 7 & 10 & 5 \end{bmatrix} & \begin{matrix} \text{Arts \& Sciences} \\ \text{Business} \\ \text{Engineering} \end{matrix} \end{matrix}$$

The average monthly cost of salary and fringe benefits for each level is

$$B = \begin{matrix} & \begin{matrix} \text{Salary} & \text{Benefits} \end{matrix} \\ \begin{bmatrix} \$1800 & \$360 \\ \$1600 & \$290 \\ \$1250 & \$110 \end{bmatrix} & \begin{matrix} \text{Level I} \\ \text{Level II} \\ \text{Level III} \end{matrix} \end{matrix}$$

Compute the matrix AB and interpret the meaning of its entries.

(b) The budget office wrote the salary–benefit matrix like this:

$$C = \begin{array}{c} \\ \end{array} \begin{array}{ccc} \text{Level} & \text{Level} & \text{Level} \\ \text{I} & \text{II} & \text{III} \end{array}$$
$$C = \begin{bmatrix} \$1800 & \$1600 & \$1250 \\ \$360 & \$290 & \$110 \end{bmatrix} \begin{array}{l} \text{Salary} \\ \text{Benefits} \end{array}$$

and wrote the school and number of employees as

$$\begin{array}{cccc} & \text{A\&S} & \text{Bus.} & \text{Engr.} \\ D = & \begin{bmatrix} 12 & 9 & 7 \\ 15 & 6 & 10 \\ 8 & 3 & 5 \end{bmatrix} & \begin{array}{l} \text{Level I} \\ \text{Level II} \\ \text{Level III} \end{array} \end{array}$$

Compute CD and interpret the meaning of its entries.

(c) The personnel office determined that the fraction of salary withheld for income tax and for the university's share of FICA is given by the matrix

$$\begin{array}{ccc} & \text{Income tax} & \text{FICA} \\ E = & \begin{bmatrix} 0.15 & 0.375 \\ 0.14 & 0.417 \\ 0.11 & 0.500 \end{bmatrix} & \begin{array}{l} \text{Level I} \\ \text{Level II} \\ \text{Level III} \end{array} \end{array}$$

Compute and interpret the meaning of the entries of the product FE where

$$\begin{array}{ccc} \text{Level} & \text{Level} & \text{Level} \\ \text{I} & \text{II} & \text{III} \end{array}$$
$$F = \begin{bmatrix} 1800 & 1600 & 1250 \end{bmatrix} \quad \text{Salary}$$

84. When a product of two numbers is zero, at least one of the numbers must be zero; that is, the product of two nonzero numbers is never zero. Symbolically, $ab = 0$ if and only if $a = 0$ or $b = 0$. This property does not hold for matrix multiplication. It is possible for two nonzero matrices (matrices with at least one entry not zero) to have a zero product. Find two nonzero 2×2 matrices whose product is the zero matrix (all entries zero).

85. $A = \begin{bmatrix} 2 & 1 \\ 3 & 5 \end{bmatrix}$ and $B = \begin{bmatrix} 1 & 4 \\ -1 & 2 \end{bmatrix}$. Find AB.

86. $A = \begin{bmatrix} 3 & 0 & 4 \\ 2 & 1 & 1 \end{bmatrix}$ and $B = \begin{bmatrix} 2 & -1 \\ 1 & 3 \\ 6 & 4 \end{bmatrix}$. Find AB.

87. $A = \begin{bmatrix} 1 & 2 & -1 \\ 3 & 1 & 5 \\ 2 & 0 & 2 \end{bmatrix}$ and $B = \begin{bmatrix} 4 & 1 & 6 \\ 1 & -3 & 2 \\ 2 & 2 & 3 \end{bmatrix}$.

Find AB and BA.

88. $A = \begin{bmatrix} 1 & -6 & 15 \\ 4 & 6 & -2 \\ 1 & 1 & 8 \end{bmatrix}$ and $B = \begin{bmatrix} 5 \\ 3 \\ -2 \end{bmatrix}$. Find AB.

89. $A = \begin{bmatrix} 2 & 1 & 5 \\ -2 & 4 & 2 \end{bmatrix}$ and $B = \begin{bmatrix} 4 & 1 \\ 3 & 0 \\ 1 & -3 \end{bmatrix}$. Find AB.

90. The survival pattern of an insect is the following:

On average, one-half of the insects survive their first year and live into a second year.

On average, one-fourth of the second-year insects survive and live into their third year.

On average, the third-year insects produce eight offspring.

By the end of the third year, all third-year insects die.

This survival pattern can be represented with a survival matrix

$$\begin{array}{cc} & \begin{array}{c} \text{Current Year of Life} \\ 1 \quad 2 \quad 3 \end{array} \\ \text{Age next year} \begin{array}{c} 1 \\ 2 \\ 3 \end{array} & \begin{bmatrix} 0 & 0 & 8 \\ \frac{1}{2} & 0 & 0 \\ 0 & \frac{1}{4} & 0 \end{bmatrix} = A \end{array}$$

Average proportion of insects that survive from current year into next year

The row 2, column 1 entry of $\frac{1}{2}$ indicates that $\frac{1}{2}$ of the insects survive from year 1 to year 2. The row 1, column 3 entry of 8 indicates that for each insect in year 3, 8 insects survive into year 1, that is, each insect produces 8 offspring in year 3.

Represent the current population of the three age groups with a column population matrix — say, 1000 — in each age group.

$$B = \begin{bmatrix} 1000 \\ 1000 \\ 1000 \end{bmatrix} \begin{array}{l} \text{Year 1} \\ \text{Year 1} \\ \text{Year 3} \end{array}$$

The product AB,

$$\begin{bmatrix} 0 & 0 & 8 \\ \frac{1}{2} & 0 & 0 \\ 0 & \frac{1}{4} & 0 \end{bmatrix} \begin{bmatrix} 1000 \\ 1000 \\ 1000 \end{bmatrix} = \begin{bmatrix} 8000 \\ 500 \\ 250 \end{bmatrix}$$

gives the population distribution for the next year. (Convince yourself this is true.)

(a) Fill in the population distributions for the next six years.

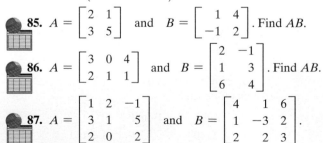

$$\begin{bmatrix} 1000 \\ 1000 \\ 1000 \end{bmatrix}, \begin{bmatrix} 8000 \\ 500 \\ 250 \end{bmatrix}, \begin{bmatrix} \\ \\ \end{bmatrix}, \dots$$

(b) Describe the population pattern.

(c) Compute

$$\begin{bmatrix} 0 & 0 & 8 \\ \frac{1}{2} & 0 & 0 \\ 0 & \frac{1}{4} & 0 \end{bmatrix}^2 \begin{bmatrix} 1000 \\ 1000 \\ 1000 \end{bmatrix}, \quad \begin{bmatrix} 0 & 0 & 8 \\ \frac{1}{2} & 0 & 0 \\ 0 & \frac{1}{4} & 0 \end{bmatrix}^3 \begin{bmatrix} 1000 \\ 1000 \\ 1000 \end{bmatrix},$$

and $\quad \begin{bmatrix} 0 & 0 & 8 \\ \frac{1}{2} & 0 & 0 \\ 0 & \frac{1}{4} & 0 \end{bmatrix}^7 \begin{bmatrix} 1000 \\ 1000 \\ 1000 \end{bmatrix}$

and compare the results with the 3rd, 4th, and 8th matrices in part (a). What conclusion do you draw regarding

$$A^n \begin{bmatrix} 1000 \\ 1000 \\ 1000 \end{bmatrix} ?$$

91. Suppose the insect population has a higher production rate, as in the survival matrix

$$A = \begin{bmatrix} 0 & 0 & 12 \\ \frac{1}{2} & 0 & 0 \\ 0 & \frac{1}{4} & 0 \end{bmatrix}$$

(a) Use the matrix giving the initial population for each of three age groups

$$B = \begin{bmatrix} 1000 \\ 1000 \\ 1000 \end{bmatrix}$$

and calculate the population matrices for the next seven years. Describe what is happening to the population.

(b) Calculate $A^{15}B$, $A^{16}B$, and $A^{17}B$. Do these population matrices follow the pattern described in part (a)?

92. Suppose the insect population has a lower reproduction rate, as given in the survival matrix

$$A = \begin{bmatrix} 0 & 0 & 4 \\ \frac{1}{2} & 0 & 0 \\ 0 & \frac{1}{4} & 0 \end{bmatrix}$$

Use the initial population matrix

$$B = \begin{bmatrix} 1000 \\ 1000 \\ 1000 \end{bmatrix}$$

and calculate the next seven population matrices. Describe what is happening to the population.

93. Use the survival matrix

$$A = \begin{bmatrix} 0 & 0 & 4 \\ 0.5 & 0 & 0 \\ 0 & 0.5 & 0 \end{bmatrix}$$

and the initial population matrix

$$B = \begin{bmatrix} 2000 \\ 1000 \\ 1000 \end{bmatrix}$$

and calculate the next five population matrices. What is the long-term population pattern?

94. Try to perform the following matrix multiplications on your graphing calculator or spreadsheet. What happens?

(a) $\begin{bmatrix} 1 & 2 & 3 \\ 4 & 5 & 6 \end{bmatrix} \begin{bmatrix} 6 & 3 \\ 4 & -1 \end{bmatrix}$

(b) $\begin{bmatrix} 1 & 0 & 1 \\ 2 & 1 & 2 \\ 3 & 4 & 3 \end{bmatrix} \begin{bmatrix} 6 \\ 2 \\ -1 \\ 5 \end{bmatrix}$

USING YOUR TI-83

MATRIX MULTIPLICATION

The multiplication of matrices is straightforward. When [A] and [B] contain the matrices

$$[A] = \begin{bmatrix} 2 & 1 & -3 \\ 1 & -1 & 1 \\ 4 & 2 & 5 \end{bmatrix} \quad [B] = \begin{bmatrix} 1 & 2 & 1 & 1 \\ 3 & -1 & 4 & 2 \\ 2 & 5 & 0 & 3 \end{bmatrix}$$

then their product is obtained by [A]$\times$[B] ENTER

(*continued*)

Here is the matrix [A] × [B]:

```
[A]*[B]
[[-1 -12  6  -5]
 [0   8  -3  2 ]
 [20 31  12 23]]
```

Use $\wedge$ to obtain powers of matrices. For example, compute $[A]^3$ with $[A]\boxed{\wedge}\boxed{3}\boxed{\text{ENTER}}$:

```
[A]^3
[[-99 -42 -84]
 [18   3  -6 ]
 [132 48  -3 ]]
■
```

Note: If you attempt to obtain the matrix product [A] × [B] when the number of columns of [A] does not equal the number of rows of [B], you will get the following error message: **ERR:DIM MISMATCH**

EXERCISES

1. $[A] = \begin{bmatrix} 2 & -4 \\ 3 & 7 \end{bmatrix}$ and $[B] = \begin{bmatrix} 5 & 1 \\ 2 & -2 \end{bmatrix}$. Find [A] [B].

2. $[A] = \begin{bmatrix} 2 & 1 & 3 \\ 4 & 6 & -2 \\ 5 & 9 & 1 \end{bmatrix}$ and $[B] = \begin{bmatrix} 1 & 0 & 2 \\ 6 & -2 & 1 \\ 3 & 1 & 1 \end{bmatrix}$. Find [A] [B].

3. $[A] = \begin{bmatrix} 8 & -2 & 1 & 4 \\ 3 & 0 & -1 & 5 \end{bmatrix}$ and $[B] = \begin{bmatrix} 1 & 3 \\ 2 & 0 \\ 1 & -1 \\ 2 & -4 \end{bmatrix}$. Find [A] [B].

4. $[A] = \begin{bmatrix} 3 & 1 \\ 1 & 2 \end{bmatrix}$. Find $[A]^2$ and $[A]^3$.

5. $[A] = \begin{bmatrix} 1 & 0 & 1 \\ 2 & 1 & 2 \\ 1 & 3 & 1 \end{bmatrix}$. Find $[A]^2$, $[A]^3$, and $[A]^4$.

USING EXCEL

EXCEL has a command, MMULT, that performs the multiplication of two matrices. To illustrate, let's use the matrices

$$A = \begin{bmatrix} 1 & 3 & 2 \\ 4 & 1 & 5 \end{bmatrix} \quad B = \begin{bmatrix} 2 & 1 & 1 & 3 \\ 5 & 0 & 2 & 4 \\ 3 & 2 & 2 & -1 \end{bmatrix}$$

Because A has 2 rows and B has 4 columns, the product AB has 2 rows and 4 columns.

Enter matrix A in cells A2:C3 and matrix B in E2:H4. We will put AB in cells A6:D7.

To calculate AB, select the cells A6:D7 and type =MMULT(A2:C3,E2:H4). Notice that this has the form =MMULT(Location of matrix A, Location of matrix B). The next step differs from the usual press RETURN. To activate matrix multiplication, simultaneously press the CTRL + SHIFT + ENTER keys. Then the product shows in cells A6:D7.

	A	B	C	D	E	F	G	H
1	Matrix A in	A2:C3			Matrix B in E2:H4			
2	1	3	2		2	1	1	3
3	4	1	5		5	0	2	4
4					3	2	2	-1
5	Matrix AB in	A6:D7						
6	23	5	11	13				
7	28	14	16	11				
8								

EXERCISES

Calculate AB in the following exercises.

1. $A = \begin{bmatrix} 2 & 1 & 1 \\ 3 & 4 & 2 \end{bmatrix}$ $B = \begin{bmatrix} 4 & 5 \\ 1 & 2 \\ 3 & 3 \end{bmatrix}$ **2.** $A = \begin{bmatrix} -1 & 1 & 2 \\ 5 & 0 & 3 \end{bmatrix}$ $R = \begin{bmatrix} 3 & 1 \\ -2 & 5 \\ 4 & 3 \end{bmatrix}$

3. $A = \begin{bmatrix} 1 & 5 & 1 \\ 2 & 3 & 2 \end{bmatrix}$ $B = \begin{bmatrix} 4 & 0 & 1 \\ -2 & 1 & 3 \\ 3 & 2 & 1 \end{bmatrix}$ **4.** $A = \begin{bmatrix} 1 & 2 & 1 \\ 2 & 1 & 2 \\ 3 & 3 & 1 \end{bmatrix}$ $B = \begin{bmatrix} 4 & 5 & 4 \\ 3 & 0 & 3 \\ 1 & 2 & 3 \end{bmatrix}$

2.6 The Inverse of a Matrix

- Inverse of a Square Matrix
- Matrix Equations
- Using A^{-1} to Solve a System

Inverse of a Square Matrix

We can extend another number fact to matrices. The simple multiplication facts

$$2 \times \frac{1}{2} = 1$$
$$\frac{3}{4} \times \frac{4}{3} = 1$$
$$1.25 \times 0.8 = 1$$

have a common property. Each of the numbers $2, \frac{3}{4}$, and 1.25 can be multiplied by another number to obtain 1. In general, for any real number a, except zero, there is a number b such that $a \times b = 1$. We call b the **inverse** of a. The standard notation for the inverse of a is a^{-1}.

EXAMPLE 1 ▶

$$3^{-1} = \frac{1}{3}, \qquad 2^{-1} = 0.5, \qquad \left(\frac{5}{8}\right)^{-1} = \frac{8}{5},$$
$$0.4^{-1} = 2.5, \qquad 625^{-1} = 0.0016$$

A similar property exists in terms of matrix multiplication. For example,

$$\begin{bmatrix} 1 & 1 \\ 1 & 2 \end{bmatrix} \begin{bmatrix} 2 & -1 \\ -1 & 1 \end{bmatrix} = \begin{bmatrix} 1 & 0 \\ 0 & 1 \end{bmatrix}$$

We can restate this equation as $AA^{-1} = I$, where

$$A = \begin{bmatrix} 1 & 1 \\ 1 & 2 \end{bmatrix} \quad \text{and} \quad A^{-1} = \begin{bmatrix} 2 & -1 \\ -1 & 1 \end{bmatrix}$$

We call A^{-1} the inverse of the matrix A.

■ **Now You Are Ready to Work Exercise 1**

DEFINITION
Inverse of a Matrix A

If A and B are square matrices such that $AB = BA = I$, then B is the **inverse matrix** of A. The inverse of A is denoted A^{-1}. If B is found so that $AB = I$, then a theorem from linear algebra states that $BA = I$, so it is sufficient to just check $AB = I$.

Only square matrices have inverses.
You can use the definition of an inverse matrix to check for an inverse.

EXAMPLE 2 ▶

(a) For the two matrices

$$A = \begin{bmatrix} 2 & 5 & 4 \\ 1 & 4 & 3 \\ 1 & -3 & -2 \end{bmatrix} \quad \text{and} \quad B = \begin{bmatrix} -1 & 2 & 1 \\ -5 & 8 & 2 \\ 7 & -11 & -3 \end{bmatrix}$$

determine whether B is the inverse of A.

(b) For the two matrices

$$A = \begin{bmatrix} 4 & 7 \\ 2 & 1 \end{bmatrix} \quad \text{and} \quad B = \begin{bmatrix} -\frac{1}{10} & \frac{7}{10} \\ \frac{1}{5} & -\frac{2}{5} \end{bmatrix}$$

determine whether $B = A^{-1}$.

(c) Determine whether B is the inverse of A for

$$A = \begin{bmatrix} 0 & 1 & 0 \\ 1 & 1 & 0 \\ 0 & 1 & 1 \end{bmatrix} \quad \text{and} \quad B = \begin{bmatrix} -1 & 1 & 0 \\ 1 & 0 & 1 \\ -1 & 1 & 0 \end{bmatrix}$$

SOLUTION

In each case it suffices to compute AB. If $AB = I$, then B is the inverse of A. If $AB \neq I$, then B is not the inverse of A.

(a) $AB = \begin{bmatrix} 2 & 5 & 4 \\ 1 & 4 & 3 \\ 1 & -3 & -2 \end{bmatrix} \begin{bmatrix} -1 & 2 & 1 \\ -5 & 8 & 2 \\ 7 & -11 & -3 \end{bmatrix}$

$= \begin{bmatrix} -2 - 25 + 28 & 4 + 40 - 44 & 2 + 10 - 12 \\ -1 - 20 + 21 & 2 + 32 - 33 & 1 + 8 - 9 \\ -1 + 15 - 14 & 2 - 24 + 22 & 1 - 6 + 6 \end{bmatrix}$

$= \begin{bmatrix} 1 & 0 & 0 \\ 0 & 1 & 0 \\ 0 & 0 & 1 \end{bmatrix} = I$

so B is the inverse of A.

(b) $AB = \begin{bmatrix} 4 & 7 \\ 2 & 1 \end{bmatrix} \begin{bmatrix} -\frac{1}{10} & \frac{7}{10} \\ \frac{1}{5} & -\frac{2}{5} \end{bmatrix}$

$= \begin{bmatrix} -\frac{4}{10} + \frac{7}{5} & \frac{28}{10} - \frac{14}{5} \\ -\frac{2}{10} + \frac{1}{5} & \frac{14}{10} - \frac{2}{5} \end{bmatrix} = \begin{bmatrix} 1 & 0 \\ 0 & 1 \end{bmatrix}$

so $B = A^{-1}$.

(c) $AB = \begin{bmatrix} 0 & 1 & 0 \\ 1 & 1 & 0 \\ 0 & 1 & 1 \end{bmatrix} \begin{bmatrix} -1 & 1 & 0 \\ 1 & 0 & 1 \\ -1 & 1 & 0 \end{bmatrix} = \begin{bmatrix} 1 & 0 & 1 \\ 0 & 1 & 1 \\ 0 & 1 & 1 \end{bmatrix} \neq I$

so B is not the inverse of A.

■ **Now You Are Ready to Work Exercise 3**

In general, a matrix A has an inverse if there is a matrix A^{-1} that fulfills the conditions that $AA^{-1} = A^{-1}A = I$. Not all matrices have inverses. In fact, a matrix must be square in order to have an inverse, and some square matrices have no inverse. We now come to the problem of deciding if a square matrix has an inverse. If it does, how do we find it? Let's approach this problem with a simple 2×2 example.

EXAMPLE 3 ➤

If we have the square matrix

$$A = \begin{bmatrix} 2 & 1 \\ 3 & 2 \end{bmatrix}$$

find its inverse, if possible.

SOLUTION

We want to find a 2×2 matrix A^{-1} such that $AA^{-1} = I$. Because we don't know the entries in A^{-1}, let's enter variables, x_1, x_2, y_1, and y_2 and attempt to find their values. Write

$$A^{-1} = \begin{bmatrix} x_1 & y_1 \\ x_2 & y_2 \end{bmatrix}$$

The condition $AA^{-1} = I$ can now be written

$$AA^{-1} = \begin{bmatrix} 2 & 1 \\ 3 & 2 \end{bmatrix} \begin{bmatrix} x_1 & y_1 \\ x_2 & y_2 \end{bmatrix} = \begin{bmatrix} 1 & 0 \\ 0 & 1 \end{bmatrix}$$

We want to find values of x_1, x_2, y_1, and y_2 so that the product on the left equals the identity matrix on the right. First, form the product AA^{-1}. We get

$$\overset{AA^{-1}}{\begin{bmatrix} (2x_1 + x_2) & (2y_1 + y_2) \\ (3x_1 + 2x_2) & (3y_1 + 2y_2) \end{bmatrix}} = \overset{I}{\begin{bmatrix} 1 & 0 \\ 0 & 1 \end{bmatrix}}$$

Recall that two matrices are equal only when they have equal entries in corresponding positions. So the matrix equality gives us the equations

$$\begin{array}{ccc} 2x_1 + x_2 = 1 & & 2y_1 + y_2 = 0 \\ 3x_1 + 2x_2 = 0 & \text{and} & 3y_1 + 2y_2 = 1 \end{array}$$

Notice that we have one system of two equations with variables x_1 and x_2:

1. $\begin{aligned} 2x_1 + x_2 &= 1 \\ 3x_1 + 2x_2 &= 0 \end{aligned}$ with augmented matrix $\begin{bmatrix} 2 & 1 & | & 1 \\ 3 & 2 & | & 0 \end{bmatrix}$

and a system with variables y_1 and y_2:

2. $\begin{aligned} 2y_1 + y_2 &= 0 \\ 3y_1 + 2y_2 &= 1 \end{aligned}$ with augmented matrix $\begin{bmatrix} 2 & 1 & | & 0 \\ 3 & 2 & | & 1 \end{bmatrix}$

The solution to system 1 gives $x_1 = 2$, $x_2 = -3$. The solution to system 2 gives $y_1 = -1$, $y_2 = 2$, so the inverse of

$$A = \begin{bmatrix} 2 & 1 \\ 3 & 2 \end{bmatrix} \quad \text{is} \quad A^{-1} = \begin{bmatrix} 2 & -1 \\ -3 & 2 \end{bmatrix}$$

We check our results by computing AA^{-1} and $A^{-1}A$:

$$AA^{-1} = \begin{bmatrix} 2 & 1 \\ 3 & 2 \end{bmatrix}\begin{bmatrix} 2 & -1 \\ -3 & 2 \end{bmatrix} = \begin{bmatrix} 1 & 0 \\ 0 & 1 \end{bmatrix}$$

$$A^{-1}A = \begin{bmatrix} 2 & -1 \\ -3 & 2 \end{bmatrix}\begin{bmatrix} 2 & 1 \\ 3 & 2 \end{bmatrix} = \begin{bmatrix} 1 & 0 \\ 0 & 1 \end{bmatrix}$$

It checks (*Note:* It suffices to check just one of these.)

■ **Now You Are Ready to Work Exercise 9**

Look at the two systems we just solved. The two systems have precisely the same coefficients; they differ only in the constant terms. The left-hand portions of the augmented matrices are exactly the same. In fact, each is the matrix A.

This means that when we solve each of the two systems using the Gauss-Jordan Method, we use precisely the same row operations. Thus, we can solve both systems using one matrix. Here's how: Combine the two augmented matrices into one using the common coefficient portion on the left, and list both columns from the right sides. This gives the matrix

$$\begin{bmatrix} 2 & 1 & | & 1 & 0 \\ 3 & 2 & | & 0 & 1 \end{bmatrix}$$

Notice that the left portion of the matrix is A and the right portion is the identity matrix.

Now proceed in the same way you do to solve a system of equations with an augmented matrix; that is, use row operations to reduce the left-hand portion to the identity matrix. This gives the following sequence:

$$\begin{bmatrix} 2 & 1 & | & 1 & 0 \\ 3 & 2 & | & 0 & 1 \end{bmatrix} \qquad \left(\tfrac{1}{2}\right)\text{R1} \to \text{R1}$$

$$\begin{bmatrix} 1 & \tfrac{1}{2} & | & \tfrac{1}{2} & 0 \\ 3 & 2 & | & 0 & 1 \end{bmatrix} \qquad -3\text{R1} + \text{R2} \to \text{R2}$$

$$\begin{bmatrix} 1 & \tfrac{1}{2} & | & \tfrac{1}{2} & 0 \\ 0 & \tfrac{1}{2} & | & -\tfrac{3}{2} & 1 \end{bmatrix} \qquad -\text{R2} + \text{R1} \to \text{R1}$$

$$\begin{bmatrix} 1 & 0 & | & 2 & -1 \\ 0 & \tfrac{1}{2} & | & -\tfrac{3}{2} & 1 \end{bmatrix} \qquad 2\text{R2} \to \text{R2}$$

$$\begin{bmatrix} 1 & 0 & | & 2 & -1 \\ 0 & 1 & | & -3 & 2 \end{bmatrix}$$

The final matrix has the identity matrix formed by the first two columns. The third column gives the solution to the first system, and the fourth column gives the solution to the second system. Notice that the last two columns form A^{-1}. This is no accident; one may find the inverse of a square matrix in this manner.

Method to Find the Inverse of a Square Matrix

1. To find the inverse of a matrix A, form an augmented matrix $[A \mid I]$ by writing down the matrix A and then writing the identity matrix to the right of A.
2. Perform a sequence of row operations that reduces the A portion of this matrix to reduced echelon form.
3. If the A portion of the reduced echelon form is the identity matrix, then the matrix found in the I portion is A^{-1}.
4. If the reduced echelon form produces a row in the A portion that is all zeros, then A has no inverse.

Now use this method to find the inverse of a matrix.

EXAMPLE 4 ➤

Find the inverse of the matrix

$$A = \begin{bmatrix} 1 & 3 & 2 \\ 2 & 4 & 2 \\ 1 & 2 & -1 \end{bmatrix}$$

SOLUTION

First, set up the augmented matrix $[A \mid I]$:

$$\left[\begin{array}{ccc|ccc} 1 & 3 & 2 & 1 & 0 & 0 \\ 2 & 4 & 2 & 0 & 1 & 0 \\ 1 & 2 & -1 & 0 & 0 & 1 \end{array}\right]$$

Next, use row operations to get zeros in column 1:

$$\left[\begin{array}{ccc|ccc} 1 & 3 & 2 & 1 & 0 & 0 \\ 0 & -2 & -2 & -2 & 1 & 0 \\ 0 & -1 & -3 & -1 & 0 & 1 \end{array}\right]$$

Now divide row 2 by -2:

$$\left[\begin{array}{ccc|ccc} 1 & 3 & 2 & 1 & 0 & 0 \\ 0 & 1 & 1 & 1 & -\frac{1}{2} & 0 \\ 0 & -1 & -3 & -1 & 0 & 1 \end{array}\right]$$

Next, get zeros in the second column:

$$\left[\begin{array}{ccc|ccc} 1 & 0 & -1 & -2 & \frac{3}{2} & 0 \\ 0 & 1 & 1 & 1 & -\frac{1}{2} & 0 \\ 0 & 0 & -2 & 0 & -\frac{1}{2} & 1 \end{array}\right]$$

Now divide row 3 by -2:

$$\left[\begin{array}{ccc|ccc} 1 & 0 & -1 & -2 & \frac{3}{2} & 0 \\ 0 & 1 & 1 & 1 & -\frac{1}{2} & 0 \\ 0 & 0 & 1 & 0 & \frac{1}{4} & -\frac{1}{2} \end{array}\right]$$

Finally, get zeros in the third column:

$$\left[\begin{array}{ccc|ccc} 1 & 0 & 0 & -2 & \frac{7}{4} & -\frac{1}{2} \\ 0 & 1 & 0 & 1 & -\frac{3}{4} & \frac{1}{2} \\ 0 & 0 & 1 & 0 & \frac{1}{4} & -\frac{1}{2} \end{array}\right]$$

When the left-hand portion of the augmented matrix reduces to the identity matrix, A^{-1} comes from the right-hand portion:

$$A^{-1} = \left[\begin{array}{ccc} -2 & \frac{7}{4} & -\frac{1}{2} \\ 1 & -\frac{3}{4} & \frac{1}{2} \\ 0 & \frac{1}{4} & -\frac{1}{2} \end{array}\right]$$

■ **Now You Are Ready to Work Exercise 13**

Now look at a case in which the matrix has no inverse.

EXAMPLE 5 ➤

Find the inverse of

$$A = \left[\begin{array}{cc} 1 & 3 \\ 3 & 9 \end{array}\right]$$

SOLUTION

Adjoin I to A to obtain

$$\left[\begin{array}{cc|cc} 1 & 3 & 1 & 0 \\ 3 & 9 & 0 & 1 \end{array}\right]$$

Now reduce this matrix using row operations:

$$\left[\begin{array}{cc|cc} 1 & 3 & 1 & 0 \\ 3 & 9 & 0 & 1 \end{array}\right] \quad -3R1 + R2 \rightarrow R2$$

$$\left[\begin{array}{cc|cc} 1 & 3 & 1 & 0 \\ 0 & 0 & -3 & 1 \end{array}\right]$$

The bottom row of the matrix represents two equations $0 = -3$ and $0 = 1$. Both of these are impossible, so in our attempt to find A^{-1} we reached an inconsistency. Whenever we reach an inconsistency in trying to solve a system of equations, we conclude that there is no solution. Therefore, in this case A has no inverse.

■ **Now You Are Ready to Work Exercise 17**

In general, when we use an augmented matrix $[A \mid I]$ to find the inverse of A and reach a step where a row of the A portion is all zeros, then A has no inverse.

Matrix Equations

We can write systems of equations using matrices and solve some systems using matrix inverses.

The **matrix equation**

$$\left[\begin{array}{cccc} 5 & 3 & -4 & 12 \\ 8 & -21 & 7 & -19 \\ 2 & 1 & -15 & 1 \end{array}\right] \left[\begin{array}{c} x_1 \\ x_2 \\ x_3 \\ x_4 \end{array}\right] = \left[\begin{array}{c} 7 \\ 16 \\ -22 \end{array}\right]$$

becomes the following when the multiplication on the left is performed:

$$\begin{bmatrix} 5x_1 + 3x_2 - 4x_3 + 12x_4 \\ 8x_1 - 21x_2 + 7x_3 - 19x_4 \\ 2x_1 + x_2 - 15x_3 + x_4 \end{bmatrix} = \begin{bmatrix} 7 \\ 16 \\ -22 \end{bmatrix}$$

These matrices are equal only when corresponding components are equal; that is,

$$\begin{array}{r} 5x_1 + 3x_2 - 4x_3 + 12x_4 = 7 \\ 8x_1 - 21x_2 + 7x_3 - 19x_4 = 16 \\ 2x_1 + x_2 - 15x_3 + x_4 = -22 \end{array}$$

In general, we can write a system of equations in the compact matrix form

$$AX = B$$

where A is a matrix formed from the coefficients of the variables

$$A = \begin{bmatrix} 5 & 3 & -4 & 12 \\ 8 & -21 & 7 & -19 \\ 2 & 1 & -15 & 1 \end{bmatrix}$$

X is a column matrix formed by listing the variables

$$X = \begin{bmatrix} x_1 \\ x_2 \\ x_3 \\ x_4 \end{bmatrix}$$

and B is the column matrix formed from the constants in the system

$$B = \begin{bmatrix} 7 \\ 16 \\ -22 \end{bmatrix}$$

EXAMPLE 6 ➤   Here is a system of equations.

$$\begin{array}{r} 4x_1 + 7x_2 - 2x_3 = 5 \\ 3x_1 - x_2 + 7x_3 = 8 \\ x_1 + 2x_2 - x_3 = 9 \end{array}$$

We can use matrices to represent this system in the following ways:

The coefficient matrix of this system is

$$\begin{bmatrix} 4 & 7 & -2 \\ 3 & -1 & 7 \\ 1 & 2 & -1 \end{bmatrix}$$

and the augmented matrix is

$$\begin{bmatrix} 4 & 7 & -2 & | & 5 \\ 3 & -1 & 7 & | & 8 \\ 1 & 2 & -1 & | & 9 \end{bmatrix}$$

The system of equations can be written in the matrix form, $AX = B$ as

$$\begin{bmatrix} 4 & 7 & -2 \\ 3 & -1 & 7 \\ 1 & 2 & -1 \end{bmatrix} \begin{bmatrix} x_1 \\ x_2 \\ x_3 \end{bmatrix} = \begin{bmatrix} 5 \\ 8 \\ 9 \end{bmatrix}$$

■ **Now You Are Ready to Work Exercise 27**

Using A^{-1} to Solve a System

Now we can illustrate the use of the inverse in solving a system of equations when the matrix of coefficients has an inverse. Sometimes it helps to be able to solve a system by using the inverse matrix. One such situation occurs when a number of systems need to be solved, and all have the same coefficients; that is, the constant terms change, but the coefficients don't. Here is a simple example.

A doctor treats patients who need adequate calcium and iron in their diet. The doctor has found that two foods, A and B, provide these. Each unit of food A has 0.5 milligram (mg) iron and 25 mg calcium. Each unit of food B has 0.3 mg iron and 7 mg calcium. Let x = number of units of food A eaten by the patient; let y = number of units of food B eaten by the patient. Then $0.5x + 0.3y$ gives the total milligrams of iron consumed by the patient and $25x + 7y$ gives the total milligrams of calcium. Suppose the doctor wants patient Jones to get 6 mg iron and 60 mg calcium. The amount of each food to be consumed is the solution to

$$0.5x + 0.3y = 6$$
$$25x + 7y = 60$$

If patient Smith requires 7 mg iron and 80 mg calcium, the amount of food required is found in the solution of the system

$$0.5x + 0.3y = 7$$
$$25x + 7y = 80$$

These two systems have the same coefficients; they differ only in the constant terms.

The inverse of the coefficient matrix

$$A = \begin{bmatrix} 0.5 & 0.3 \\ 25 & 7 \end{bmatrix}$$

may be used to avoid going through the Gauss-Jordan elimination process with each patient.

Here's how A^{-1} may be used to solve a system. Let $AX = B$ be a system for which A actually has an inverse. When both sides of $AX = B$ are multiplied by A^{-1}, the equation reduces to

$$A^{-1}AX = A^{-1}B$$
$$IX = A^{-1}B$$
$$X = A^{-1}B$$

The product $A^{-1}B$ gives the solution. The solution to such a system exists, and it is unique.

EXAMPLE 7 ▶   Use an inverse matrix to solve the system of equations:

$$x_1 + 3x_2 + 2x_3 = 3$$
$$2x_1 + 4x_2 + 2x_3 = 8$$
$$x_1 + 2x_2 - x_3 = 10$$

NOTE

Using the inverse of the coefficient matrix may not be the most efficient way to solve a *single* system of equations. However, some applications require the solution of several systems of equations in which *all* the systems have the same *coefficient matrix*. Using the inverse of the coefficient matrix can be more efficient in this situation.

Using the inverse of the coefficient matrix to solve a single system of equations may be the most efficient way when solving using a computer or graphing calculator.

SOLUTION

First, write the system in matrix form, $AX = B$:

$$\begin{bmatrix} 1 & 3 & 2 \\ 2 & 4 & 2 \\ 1 & 2 & -1 \end{bmatrix} \begin{bmatrix} x_1 \\ x_2 \\ x_3 \end{bmatrix} = \begin{bmatrix} 3 \\ 8 \\ 10 \end{bmatrix}$$

In matrix form the solution is

$$\begin{bmatrix} x_1 \\ x_2 \\ x_3 \end{bmatrix} = \begin{bmatrix} 1 & 3 & 2 \\ 2 & 4 & 2 \\ 1 & 2 & -1 \end{bmatrix}^{-1} \begin{bmatrix} 3 \\ 8 \\ 10 \end{bmatrix}$$

The inverse was found in Example 4. Substitute it and obtain

$$\begin{bmatrix} x_1 \\ x_2 \\ x_3 \end{bmatrix} = \begin{bmatrix} -2 & \frac{7}{4} & -\frac{1}{2} \\ 1 & -\frac{3}{4} & \frac{1}{2} \\ 0 & \frac{1}{4} & -\frac{1}{2} \end{bmatrix} \begin{bmatrix} 3 \\ 8 \\ 10 \end{bmatrix} = \begin{bmatrix} 3 \\ 2 \\ -3 \end{bmatrix}$$

The system has the unique solution $x_1 = 3$, $x_2 = 2$, $x_3 = -3$. (Check this solution in each of the original equations.)

■ **Now You Are Ready to Work Exercise 37**

EXAMPLE 8 ➤

Solve the systems

$$AX = B$$

where

$$A = \begin{bmatrix} 1 & 2 \\ 4 & 3 \end{bmatrix} \quad \text{and} \quad X = \begin{bmatrix} x \\ y \end{bmatrix}$$

using

$$B = \begin{bmatrix} 6 \\ 3 \end{bmatrix}, \quad \begin{bmatrix} 10 \\ 15 \end{bmatrix}, \quad \text{and} \quad \begin{bmatrix} 2 \\ 11 \end{bmatrix}$$

SOLUTION

First find A^{-1}. Adjoin the identity matrix of A:

$$\left[\begin{array}{cc|cc} 1 & 2 & 1 & 0 \\ 4 & 3 & 0 & 1 \end{array} \right]$$

This reduces to

$$\left[\begin{array}{cc|cc} 1 & 0 & -\frac{3}{5} & \frac{2}{5} \\ 0 & 1 & \frac{4}{5} & -\frac{1}{5} \end{array} \right]$$

so the inverse of A is

$$\begin{bmatrix} -\frac{3}{5} & \frac{2}{5} \\ \frac{4}{5} & -\frac{1}{5} \end{bmatrix}$$

For $B = \begin{bmatrix} 6 \\ 3 \end{bmatrix}$, the solution is

$$\begin{bmatrix} x \\ y \end{bmatrix} = \begin{bmatrix} -\frac{3}{5} & \frac{2}{5} \\ \frac{4}{5} & -\frac{1}{5} \end{bmatrix} \begin{bmatrix} 6 \\ 3 \end{bmatrix} = \begin{bmatrix} -\frac{12}{5} \\ \frac{21}{5} \end{bmatrix}$$

so $x = -\frac{12}{5}$, $y = \frac{21}{5}$ is the solution.

For $B = \begin{bmatrix} 10 \\ 15 \end{bmatrix}$,

$$\begin{bmatrix} x \\ y \end{bmatrix} = \begin{bmatrix} -\frac{3}{5} & \frac{2}{5} \\ \frac{4}{5} & -\frac{1}{5} \end{bmatrix} \begin{bmatrix} 10 \\ 15 \end{bmatrix} = \begin{bmatrix} 0 \\ 5 \end{bmatrix}$$

For $B = \begin{bmatrix} 2 \\ 11 \end{bmatrix}$,

$$\begin{bmatrix} x \\ y \end{bmatrix} = \begin{bmatrix} -\frac{3}{5} & \frac{2}{5} \\ \frac{4}{5} & -\frac{1}{5} \end{bmatrix} \begin{bmatrix} 2 \\ 11 \end{bmatrix} = \begin{bmatrix} \frac{16}{5} \\ -\frac{3}{5} \end{bmatrix}$$

■ **Now You Are Ready to Work Exercise 41**

Use the matrix solution $X = A^{-1}B$ to work the next example.

EXAMPLE 9 ➤   Let's return to the earlier example where a doctor prescribed foods containing calcium and iron. Let

$$x = \text{the number of units of food A}$$
$$y = \text{the number of units of food B}$$

where A contains 0.5 mg iron and 25 mg calcium and B contains 0.3 mg iron and 7 mg calcium per unit.

(a) Find the amount of each food for patient Jones, who needs 1.3 mg iron and 49 mg calcium.

(b) Find the amount of each food for patient Smith, who needs 2.6 mg iron and 106 mg calcium.

SOLUTION

(a) We need the solution to

$$0.5x + 0.3y = 1.3 \quad \text{(amount of iron)}$$
$$25x + 7y = 49 \quad \text{(amount of calcium)}$$

In matrix form this is

$$\begin{bmatrix} 0.5 & 0.3 \\ 25 & 7 \end{bmatrix} \begin{bmatrix} x \\ y \end{bmatrix} = \begin{bmatrix} 1.3 \\ 49 \end{bmatrix}$$

The inverse of

$$\begin{bmatrix} 0.5 & 0.3 \\ 25 & 7 \end{bmatrix} \quad \text{is} \quad \begin{bmatrix} -1.75 & 0.075 \\ 6.25 & -0.125 \end{bmatrix}$$

The solution to the system is

$$\begin{bmatrix} x \\ y \end{bmatrix} = \begin{bmatrix} -1.75 & 0.075 \\ 6.25 & -0.125 \end{bmatrix} \begin{bmatrix} 1.3 \\ 49 \end{bmatrix} = \begin{bmatrix} -1.75(1.3) + 0.075(49) \\ 6.25(1.3) - 0.125(49) \end{bmatrix}$$

$$= \begin{bmatrix} 1.4 \\ 2.0 \end{bmatrix}$$

so 1.4 units of food A and 2.0 units of food B are required.

(b) In this case, the solution is

$$\begin{bmatrix} x \\ y \end{bmatrix} = \begin{bmatrix} -1.75 & 0.075 \\ 6.25 & -0.125 \end{bmatrix} \begin{bmatrix} 2.6 \\ 106.0 \end{bmatrix} = \begin{bmatrix} 3.4 \\ 3.0 \end{bmatrix}$$

■ **Now You Are Ready to Work Exercise 45**

2.6 EXERCISES

LEVEL 1

1. *(See Example 1)* Find 25^{-1}, $\left(\dfrac{2}{3}\right)^{-1}$, $(-5)^{-1}$, 0.75^{-1}, and 11^{-1}.

Determine whether B is the inverse of A in each of Exercises 2 through 8.

2. $A = \begin{bmatrix} 4 & 7 \\ 1 & 2 \end{bmatrix}$, $B = \begin{bmatrix} 2 & -7 \\ -1 & 4 \end{bmatrix}$

3. *(See Example 2)*

$A = \begin{bmatrix} -2 & 1 & 3 \\ 2 & 4 & -1 \\ 3 & 0 & -4 \end{bmatrix}$, $B = \begin{bmatrix} -16 & 4 & -13 \\ 5 & -1 & 4 \\ -12 & 3 & -10 \end{bmatrix}$

4. $A = \begin{bmatrix} 1 & -2 & 3 \\ 2 & -2 & 0 \\ 4 & -5 & 6 \end{bmatrix}$, $B = \begin{bmatrix} -\frac{4}{6} & -\frac{1}{6} & \frac{2}{6} \\ \frac{4}{6} & -\frac{2}{6} & -\frac{2}{6} \\ 1 & -\frac{1}{6} & -\frac{2}{6} \end{bmatrix}$

5. $A = \begin{bmatrix} 2 & -1 \\ -6 & 2 \end{bmatrix}$, $B = \begin{bmatrix} -1 & -2 \\ -3 & -1 \end{bmatrix}$

6. $A = \begin{bmatrix} 2 & 1 & -1 \\ 1 & 1 & -1 \\ -1 & -2 & 3 \end{bmatrix}$, $B = \begin{bmatrix} 1 & -1 & 1 \\ -2 & 5 & 1 \\ -1 & 3 & 1 \end{bmatrix}$

7. $A = \begin{bmatrix} 2 & 0 & 0 \\ 0 & 3 & 0 \\ 0 & 0 & 5 \end{bmatrix}$, $B = \begin{bmatrix} \frac{1}{2} & 0 & 0 \\ 0 & \frac{1}{3} & 0 \\ 0 & 0 & \frac{1}{5} \end{bmatrix}$

8. $A = \begin{bmatrix} 3 & 2 \\ 0 & 0 \end{bmatrix}$, $B = \begin{bmatrix} \frac{1}{3} & 0 \\ 0 & \frac{1}{2} \end{bmatrix}$

Find the inverse of the matrices in Exercises 9 through 16.

9. *(See Example 3)*
$\begin{bmatrix} 1 & 2 \\ 3 & 5 \end{bmatrix}$

10. $\begin{bmatrix} 9 & 11 \\ 1 & 5 \end{bmatrix}$

11. $\begin{bmatrix} 3 & 2 \\ 4 & 3 \end{bmatrix}$

12. $\begin{bmatrix} 3 & 5 \\ 2 & 4 \end{bmatrix}$

13. *(See Example 4)*
$\begin{bmatrix} 1 & 3 & 9 \\ 0 & 1 & 4 \\ 3 & 2 & 3 \end{bmatrix}$

14. $\begin{bmatrix} 1 & 2 & 1 \\ 2 & -1 & 3 \\ 2 & 2 & 1 \end{bmatrix}$

15. $\begin{bmatrix} 0 & 4 & -2 \\ 1 & 3 & 5 \\ 1 & 4 & 2 \end{bmatrix}$

16. $\begin{bmatrix} 1 & 0 & 2 \\ 2 & -4 & 2 \\ 0 & 1 & -1 \end{bmatrix}$

Find the inverse, if possible, of the matrices in Exercises 17 through 19.

17. *(See Example 5)*
$\begin{bmatrix} 4 & -2 \\ -2 & 1 \end{bmatrix}$

18. $\begin{bmatrix} 1 & 0 & 1 \\ 1 & -1 & 2 \\ 3 & -1 & 4 \end{bmatrix}$

19. $\begin{bmatrix} 1 & 3 & 1 \\ 2 & 0 & -2 \\ 3 & 3 & -1 \end{bmatrix}$

26. $\begin{bmatrix} 1 & 2 & -1 \\ 2 & 4 & -3 \\ 1 & -2 & 0 \end{bmatrix}$

Determine the inverses (if they exist) of the matrices in Exercises 20 through 26.

20. $\begin{bmatrix} 1 & 2 & -1 \\ 3 & -1 & 0 \\ 2 & -3 & 1 \end{bmatrix}$

21. $\begin{bmatrix} 1 & 2 & 1 \\ 1 & -3 & 2 \\ 2 & -1 & 3 \end{bmatrix}$

22. $\begin{bmatrix} 1 & 0 \\ 2 & 1 \end{bmatrix}$

23. $\begin{bmatrix} 2 & 1 \\ 4 & 3 \end{bmatrix}$

24. $\begin{bmatrix} 0 & 2 \\ -\frac{1}{3} & \frac{1}{3} \end{bmatrix}$

25. $\begin{bmatrix} 1 & 2 & 3 \\ 2 & -1 & 4 \\ 0 & -1 & 1 \end{bmatrix}$

For each of the systems of equations in Exercises 27 through 30, write (a) the augmented matrix; (b) the co-efficient matrix; and (c) the system in the form $AX = B$.

27. *(See Example 6)*
$$\begin{aligned} 3x_1 + 4x_2 - 5x_3 &= 4 \\ 2x_1 - x_2 + 3x_3 &= -1 \\ x_1 + x_2 - x_3 &= 2 \end{aligned}$$

28.
$$\begin{aligned} x_1 - 4x_2 + 3x_3 &= 6 \\ x_1 \quad\quad + x_3 &= 2 \\ 2x_1 + 5x_2 - 6x_3 &= 1 \end{aligned}$$

29.
$$\begin{aligned} 4x + 5y &= 2 \\ 3x - 2y &= 7 \end{aligned}$$

30.
$$\begin{aligned} 7x_1 + 9x_2 - 5x_3 + x_4 &= 14 \\ 3x_1 + 5x_2 + 6x_3 - 8x_4 &= 23 \\ -2x_1 + x_2 \quad\quad + 17x_4 &= 12 \end{aligned}$$

LEVEL 2

Express each of the systems in Exercises 31 through 34 as a single matrix equation, $AX = B$.

31.
$$\begin{aligned} x_1 + 3x_2 &= 5 \\ 2x_1 - x_2 &= 6 \end{aligned}$$

32.
$$\begin{aligned} 2x_1 - 3x_2 + x_3 &= 4 \\ 4x_1 - x_2 + 2x_3 &= -1 \\ x_1 + x_2 - x_3 &= 2 \end{aligned}$$

33.
$$\begin{aligned} x_1 + 2x_2 - 3x_3 + 4x_4 &= 0 \\ x_1 + x_2 \quad\quad + x_4 &= 5 \\ 3x_1 + 2x_2 + x_3 + 2x_4 &= 4 \end{aligned}$$

34.
$$\begin{aligned} x_1 + 5x_2 - x_3 &= 7 \\ 4x_1 + 3x_2 + 6x_3 &= 15 \end{aligned}$$

Find the inverse of the matrices in Exercises 35 and 36.

35. $\begin{bmatrix} 1 & 1 & 0 & 0 \\ 0 & 1 & 1 & 0 \\ 1 & 0 & 0 & 1 \\ 0 & 0 & 1 & 1 \end{bmatrix}$

36. $\begin{bmatrix} -3 & -1 & 1 & -2 \\ -1 & 3 & 2 & 1 \\ 1 & 2 & 3 & -1 \\ -2 & 1 & -1 & -3 \end{bmatrix}$

Solve the systems of equations in Exercises 37 through 40 by determining the inverse of the matrix of coefficients and then using matrix multiplication.

37. *(See Example 7)*
$$\begin{aligned} x_1 + 2x_2 - x_3 &= 2 \\ x_1 + x_2 + 2x_3 &= 0 \\ x_1 - x_2 - x_3 &= 1 \end{aligned}$$

38.
$$\begin{aligned} x_1 + 3x_2 &= 5 \\ 2x_1 + x_2 &= 10 \end{aligned}$$

39.
$$\begin{aligned} x_1 + x_2 + 2x_3 + x_4 &= 4 \\ 2x_1 \quad\quad - x_3 + x_4 &= 6 \\ x_2 + 3x_3 - x_4 &= 3 \\ 3x_1 + 2x_2 \quad\quad + x_4 &= 9 \end{aligned}$$

40.
$$\begin{aligned} x_1 - x_2 \quad\quad &= 1 \\ x_1 + x_2 + 2x_3 &= 2 \\ x_1 + 2x_2 + x_3 &= 0 \end{aligned}$$

Using the inverse matrix method, solve the system of equations in Exercises 41 through 44 for each of the B matrices.

41. *(See Example 8)*
$$\begin{aligned} -2x_1 + x_2 + 3x_3 &= b_1 \\ 2x_1 + 4x_2 - x_3 &= b_2 \\ 3x_1 \quad\quad - 4x_3 &= b_3 \end{aligned}$$
$\begin{bmatrix} b_1 \\ b_2 \\ b_3 \end{bmatrix} = \begin{bmatrix} 1 \\ 5 \\ 2 \end{bmatrix}, \begin{bmatrix} -1 \\ 3 \\ 1 \end{bmatrix}, \begin{bmatrix} 0 \\ 1 \\ 2 \end{bmatrix}$

42.
$$\begin{aligned} x_1 + x_2 &= b_1 \\ 2x_1 + 3x_2 &= b_2 \end{aligned}$$
$\begin{bmatrix} b_1 \\ b_2 \end{bmatrix} = \begin{bmatrix} 0 \\ 1 \end{bmatrix}, \begin{bmatrix} 5 \\ 13 \end{bmatrix}, \begin{bmatrix} 1 \\ 2 \end{bmatrix}$

43.
$$\begin{aligned} x_1 + 2x_2 &= b_1 \\ 3x_1 + 5x_2 &= b_2 \end{aligned}$$
$\begin{bmatrix} b_1 \\ b_2 \end{bmatrix} = \begin{bmatrix} 3 \\ 8 \end{bmatrix}, \begin{bmatrix} 4 \\ 9 \end{bmatrix}, \begin{bmatrix} 3 \\ 7 \end{bmatrix}$

44.
$$\begin{aligned} x_1 + 3x_2 - x_3 &= b_1 \\ x_1 + x_2 + x_3 &= b_2 \\ 2x_1 + 5x_2 - 2x_3 &= b_3 \end{aligned}$$
$\begin{bmatrix} b_1 \\ b_2 \\ b_3 \end{bmatrix} = \begin{bmatrix} 2 \\ 0 \\ 2 \end{bmatrix}, \begin{bmatrix} 3 \\ 1 \\ -5 \end{bmatrix}, \begin{bmatrix} 4 \\ 6 \\ 0 \end{bmatrix}$

LEVEL 3

45. *(See Example 9)*   A doctor advises his patients to eat two foods for vitamins A and C. The contents per unit of food are given as follows:

$$\begin{array}{c} \quad\quad\quad\quad \text{Food} \\ \quad\quad\quad\quad \text{A} \quad\quad \text{B} \\ \begin{array}{l} \text{Vitamin C (mg)} \\ \text{Vitamin A (IU)} \end{array} \begin{bmatrix} 32 & 24 \\ 900 & 425 \end{bmatrix} = M \end{array}$$

It turns out that

$$M^{-1} = \begin{bmatrix} -0.053125 & 0.003 \\ 0.1125 & -0.004 \end{bmatrix}$$

Let

$x =$ number of units of food A
$y =$ number of units of food B
$b_1 =$ desired intake of vitamin C
$b_2 =$ desired intake of vitamin A

(a) Show that the $MX = B$ describes the relationship between units of food consumed and desired intake of vitamins.
(b) If a patient eats 3.2 units of food A and 2.5 units of food B, what is the vitamin C and vitamin A intake?
(c) If a patient eats 1.5 units of food A and 3.0 units of food B, what is the vitamin A and vitamin C intake?
(d) The doctor wants a patient to consume 107.2 mg of vitamin C and 2315 IU of vitamin A. How many units of each food should be eaten?
(e) The doctor wants a patient to consume 104 mg vitamin C and 2575 IU vitamin A. How many units of each food should be eaten?

46. The Restaurant Association sponsored a Taster's Choice evening in which restaurants set up booths and served food samples to attendees. The number of shrimp, steak bits, and cheese chunks given to each man, woman, and child by the Elite Cafe is summarized as follows.

	Man	Woman	Child
Shrimp	1	2	1
Steak	1	1	1
Cheese	3	1	1

Use

$$\begin{bmatrix} 0 & -\frac{1}{2} & \frac{1}{2} \\ 1 & -1 & 0 \\ -1 & \frac{5}{2} & -\frac{1}{2} \end{bmatrix} = \begin{bmatrix} 1 & 2 & 1 \\ 1 & 1 & 1 \\ 3 & 1 & 1 \end{bmatrix}^{-1}$$

(a) If the Elite Cafe served 614 shrimp, 404 steak bits, and 684 cheese chunks, how many men, women, and children were served?
(b) If the Elite Cafe served 740 shrimp, 510 steak bits, and 940 cheese chunks, how many men, women, and children were served?
(c) If the Elite Cafe served 409 shrimp, 278 steak bits, and 488 cheese chunks, how many men, women, and children were served?

47. A theater charges $4 for children and $8 for adults. One weekend, 900 people attended the theater, and the admission receipts totaled $5840. These can be represented by

$$\begin{bmatrix} 1 & 1 \\ 4 & 8 \end{bmatrix}\begin{bmatrix} x \\ y \end{bmatrix} = \begin{bmatrix} 900 \\ 5840 \end{bmatrix}$$

where $x =$ number of children and $y =$ number of adults. It is true that

$$\begin{bmatrix} 1 & 1 \\ 4 & 8 \end{bmatrix}^{-1} = \begin{bmatrix} 2 & -0.25 \\ -1 & 0.25 \end{bmatrix}$$

(a) Find the number of children and adults attending.
(b) If the total attendance is 1000 and the admission receipts total $6260, find the number of children and adults attending.
(c) If the attendance totals 750 and receipts total $5560, find the number of children and adults attending.

48. Carol Riggs is the plant manager of Health Farc Cereal Company, which makes three cereals using wheat, oats, and raisins. She receives a daily report on the amount of wheat, oats, and raisins used in production and the number of boxes of each kind of cereal produced. To determine the loss, if any, due to waste and other causes, she likes to compare the ingredients used to the amounts of cereals possible if there is no waste. She does not receive this information, so she must determine it from the information reported. She knows each box contains one pound of cereal and that the proportion of wheat, oats, and raisins of each cereal is the following:

Proportion of each pound	Ingredient		
	Wheat	Oats	Raisins
Lite	0.75	0.25	0
Trim	0.50	0.25	0.25
Health	0.25	0.50	0.25

One week the number of pounds of ingredients used were:

	Day				
	Mon.	Tue.	Wed.	Thu.	Fri.
Wheat	2320	2410	2260	2520	2150
Oats	1380	1400	1410	1640	1350
Raisins	700	760	680	830	740

(a) Represent the portions of wheat, oats, and raisins used in each cereal as a matrix A.
(b) Find A^{-1}.
(c) For each day, find the amount of each cereal that could be produced from the ingredients if there is no waste.

49. Wayne Lewis, an auditor for Humidor Coffees, wants to determine if there are any unexplained losses in the coffee blends produced. He obtains daily records showing the amounts of blends produced and the amounts of ingredients actually used. He wants to compare the amounts of ingredients used with the blends that could be produced if there is no waste. The composition of the three blends is:

| | **Ingredients** | | |
Blend	Regular	High Mountain	Chocolate
Early Riser	80%	20%	0
After Dinner	75%	20%	5%
Deluxe	50%	40%	10%

The records show the following amounts (in pounds) of ingredients used for a three-day period:

| | **Day** | | |
Ingredient	1	2	3
Regular	505	766	571
High Mt.	170	244	196
Chocolate	25	40	33

(a) Represent the portions of Regular, High Mountain, and chocolate used in the blends by a matrix A.
(b) Find A^{-1}.
(c) For each day, find the number of pounds of blends that could be produced if there is no waste.

50. Arnold Bowker is responsible for keeping the inventory of Arita China Company up to date. For a major regional sale, they package china in three combinations, which they label Basic, Plates Only, and Deluxe. The packages contain the following number of plates, cups and saucers, and salad plates.

	Basic	**Plates Only**	**Deluxe**
Plates	4	8	12
Cups, saucers	4	0	12
Salad plates	4	0	8

Each week, Mr. Bowker receives a daily report of the number of plates, cups and saucers, and salad plates taken from inventory to make the sales packages. One week, the daily number taken from inventory was:

	Mon.	**Tue.**	**Wed.**	**Thu.**	**Fri.**
Plates	1000	988	1252	1008	1776
Cups, saucers	760	764	916	576	1296
Salad plates	560	584	696	448	960

He needs to enter the number of packages produced into the inventory.

(a) Find the matrix A giving the number of items per package.
(b) Find A^{-1}.
(c) Use the inverse matrix to find the daily number of each kind of package produced.

EXPLORATIONS

51. (a) Find the inverse of

$$\begin{bmatrix} 4 & 0 \\ 0 & 5 \end{bmatrix}$$

(b) Find the inverse of

$$\begin{bmatrix} a & 0 \\ 0 & b \end{bmatrix} \quad (a \neq 0, b \neq 0)$$

(c) Find the inverse of

$$\begin{bmatrix} a & 0 & 0 \\ 0 & b & 0 \\ 0 & 0 & c \end{bmatrix} \quad (a \neq 0, b \neq 0, c \neq 0)$$

The matrices in parts (a)–(c) are diagonal matrices; all entries *not* on the main diagonal are zero.
(d) Based on the above, what is the form of the inverse of a diagonal matrix?

52. Can a diagonal matrix with a zero entry in the diagonal have an inverse? Explain. (For a matrix like

$$\begin{bmatrix} 2 & 0 & 0 \\ 0 & 0 & 0 \\ 0 & 0 & 4 \end{bmatrix}$$

try to find the inverse.)

53. Can a square matrix with a row of zeros have an inverse? Explain.

54. When reducing $[A \mid I]$, a row in the A part consists of all zeros. What does this indicate about A^{-1}?

55. $A = \begin{bmatrix} 3 & 1 \\ 2 & 1 \end{bmatrix}$. Find A^{-1}.

56. $A = \begin{bmatrix} 0.5 & 1 & 0.5 \\ 2 & 5 & 1 \\ 1 & 3 & 2 \end{bmatrix}$. Find A^{-1}.

57. $A = \begin{bmatrix} 2 & 0 & 0 \\ 0 & 4 & 0 \\ 0 & 0 & 5 \end{bmatrix}$. Find A^{-1}.

58. $A = \begin{bmatrix} 4 & -2 & 1 \\ 3 & 1 & 2 \\ 1 & 2 & 2 \end{bmatrix}$. Find A^{-1}.

59. $A = \begin{bmatrix} 2 & 1 & -1 \\ 1 & 1 & -1 \\ 1 & -2 & 3 \end{bmatrix}$.

Set up the $[A \mid I]$ matrix on your graphing calculator and find A^{-1} using row operations.

60. Find A^{-1} where
$$A = \begin{bmatrix} -0.80 & 0.25 & 0 \\ -0.40 & 0.75 & 0 \\ -0.25 & -0.20 & 1 \end{bmatrix}$$

61. Find the inverse of
$$\begin{bmatrix} 2 & -1 & 3 & 2 \\ 3 & 2 & -1 & 4 \\ 3 & 2 & 6 & 4 \\ 2 & 1 & -1 & 4 \end{bmatrix}$$

62. $A = \begin{bmatrix} 1 & -2 & -1 & -2 \\ 3 & -2 & -2 & -3 \\ 2 & -5 & -2 & -5 \\ -1 & 4 & 4 & 11 \end{bmatrix}$.

(a) Compute A^{-1} and $(A^{-1})^{-1}$.

(b) Based on these results, make a conjecture about $(A^{-1})^{-1}$ in general.

63. $A = \begin{bmatrix} 1 & 1 & 1 \\ 0 & 1 & 1 \\ 0 & 0 & 1 \end{bmatrix}$, $B = \begin{bmatrix} 1 & 0 & 0 \\ -1 & 1 & 0 \\ 1 & -1 & 1 \end{bmatrix}$.

(a) Find AB, A^{-1}, B^{-1}, $(AB)^{-1}$, $A^{-1}B^{-1}$, and $B^{-1}A^{-1}$.

(b) Which of the last three matrices are equal?

64. $A = \begin{bmatrix} 2 & 3 & -1 \\ 1 & 1 & 1 \\ 1 & 1 & 2 \end{bmatrix}$, $B = \begin{bmatrix} 1 & 1 & 1 \\ 1 & 2 & 3 \\ 3 & 2 & 0 \end{bmatrix}$.

(a) Find AB, A^{-1}, B^{-1}, $(AB)^{-1}$, $A^{-1}B^{-1}$, and $B^{-1}A^{-1}$.

(b) Which of the last three matrices are equal?

65. Based on the last two exercises, what do you conclude about the relationship of the matrices $(AB)^{-1}$, $A^{-1}B^{-1}$, and $B^{-1}A^{-1}$?

66. Find the inverse of
$$\begin{bmatrix} a & b \\ c & d \end{bmatrix}$$
Determine when the inverse matrix does not exist.

67. (a) For the matrix
$$A = \begin{bmatrix} 1 & -2 & -4 \\ -1 & 4 & -2 \\ -1 & 4 & 2 \end{bmatrix}$$
find A^{-1} and the reduced echelon form of A.

(b) For the matrix
$$A = \begin{bmatrix} 1 & -3 & 2 \\ 2 & 1 & 4 \\ 4 & -5 & 8 \end{bmatrix}$$
verify that A^{-1} does not exist and find the reduced echelon form of A.

68. (a) For the matrix
$$A = \begin{bmatrix} 1 & 2 & 4 & -1 \\ -1 & 4 & -2 & 6 \\ 1 & 4 & 2 & 0 \\ 3 & 6 & 2 & 8 \end{bmatrix}$$
find A^{-1} and the reduced echelon form of A.

(b) For the matrix
$$A = \begin{bmatrix} 3 & -1 & 0 & 2 \\ 4 & -1 & 0 & 5 \\ 1 & 2 & -2 & 3 \\ 0 & 2 & -2 & 0 \end{bmatrix}$$
verify that A^{-1} does not exist and find the reduced echelon form of A.

69. Based on the two previous exercises, make a conjecture on the relationship between the reduced echelon form of a matrix A and the inverse A^{-1}.

Solve Exercises 70 through 72 using the inverse of the coefficient matrix. (Use $X = A^{-1}B$.)

70. $\begin{aligned} x + y - 3z &= 4 \\ 2x + 4y - 4z &= 5 \\ -x + y + 4z &= -3 \end{aligned}$

71. $\begin{aligned} x + 2y + 3z &= -1 \\ 2x - 3y + 4z &= 2 \\ 3x - 5y + 6z &= -3 \end{aligned}$

72. $\begin{aligned} x_1 + x_2 - x_3 + x_4 &= 6 \\ 2x_1 + 3x_2 + x_3 - x_4 &= 4 \\ 3x_1 + 2x_2 - x_3 + x_4 &= 3 \\ x_1 + 2x_2 + x_3 + x_4 &= 9 \end{aligned}$

USING YOUR TI-83

THE INVERSE OF A MATRIX

The inverse of a square matrix can be obtained by using the $\boxed{x^{-1}}$ key. For example, to find the inverse of a matrix stored in [A] where

$$[A] = \begin{bmatrix} 3 & 2 & 1 \\ 0 & 4 & 1 \\ 1 & 2 & 1 \end{bmatrix}$$

use [A] $\boxed{x^{-1}}$ $\boxed{\text{ENTER}}$, and the screen will show

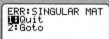

Some matrices such as

$$[A] = \begin{bmatrix} 3 & 2 & 2 \\ 0 & 4 & 1 \\ 1 & 2 & 1 \end{bmatrix}$$

have no inverse. In this case, an error message is given:

```
ERR:SINGULAR MAT
1:Quit
2:Goto
```

EXERCISES

Find the inverse of each of the following.

1. $\begin{bmatrix} 2 & -1 & 3 \\ 3 & 1 & 2 \\ 4 & 1 & 4 \end{bmatrix}$
2. $\begin{bmatrix} 1 & 0 & 1 \\ 0 & 1 & 1 \\ 1 & 1 & 0 \end{bmatrix}$
3. $\begin{bmatrix} 4 & -2 & 1 \\ 1 & -4 & -1 \\ 3 & 2 & 2 \end{bmatrix}$

USING EXCEL

EXCEL has the **MINVERSE** command that calculates the inverse of a square matrix. To find the inverse of

$$A = \begin{bmatrix} 1 & 2 & 1 \\ 2 & 4 & 1 \\ 1 & 3 & 2 \end{bmatrix}$$ enter the matrix in cells A2:C4. Next, select the cells E2:G4 for the location of the inverse

of A, type =MINVERSE(A2:C4) and simultaneously press CTRL + SHIFT + ENTER. The inverse of A then appears in E2:G4.

	A	B	C	D	E	F	G	H
1	Matrix A in A2:C4				Inverse of A in E2:G4			
2	1	2	1		5	-1	-2	
3	2	4	1		-3	1	1	
4	1	3	2		2	-1	0	
5								
6	If A has no inverse, you will get a matrix like this							
7	#NUM!	#NUM!	#NUM!					
8	#NUM!	#NUM!	#NUM!					
9	#NUM!	#NUM!	#NUM!					
10								

EXERCISES

Find the inverse of A in the following exercises.

1. $A = \begin{bmatrix} -2 & 6 & 3 \\ 7 & -3 & 1 \\ 9 & 2 & 5 \end{bmatrix}$

2. $A = \begin{bmatrix} 1 & 2 & 1 \\ 2 & 5 & 1 \\ 1 & 3 & 2 \end{bmatrix}$

3. $A = \begin{bmatrix} -0.4 & 6 & 3.75 \\ 1.4 & -3 & 1.25 \\ 1.8 & 2 & 6.25 \end{bmatrix}$

4. $A = \begin{bmatrix} 2 & -1 & 5 & 5 \\ 3 & 5 & 1 & -1 \\ 1 & 3 & 6 & -2 \\ 2 & 2 & -1 & 1 \end{bmatrix}$

5. $A = \begin{bmatrix} -2 & 3 & 4 \\ 7 & -3 & 1 \\ 1 & 2 & 5 \end{bmatrix}$

2.7   Leontief Input–Output Model in Economics

- The Leontief Economic Model

The economic health of our country affects each of us in some way. You want a good job upon graduation. The availability of a good job depends on the ability of an economic system to deal with problems that arise. Some problems are challenging indeed. How do we control inflation? How do we avoid a depression? How will a change in interest rates affect my options in buying a house or a car?

A better understanding of the interrelationships between prices, production, interest rates, consumer demand, and the like could improve our ability to deal with economic problems. Matrix theory has been successful in describing mathematical models used to analyze how industries depend on one another in an economic system. Wassily Leontief of Harvard University pioneered work in this area with a massive analysis of the U.S. economic system. As a result of the mathematical models he developed, he received the Nobel Prize in Economics in 1973. Since that time, we have seen the applications of the **Leontief input–output model** mushroom. The model is widely used to study the economic structure of businesses, corporations, and political units like cities, states, and countries. In practice, a large number of variables are required to describe an economic situation. Thus, the problems are quite complicated, so they can best be handled with computers using matrix techniques.

This section introduces the concept of the Leontief economic model, using relatively simple examples to suggest its use in more realistic situations.

The Leontief Economic Model

In the input–output model, we have an economic system consisting of a number of industries that both produce goods and use goods. Some of the goods produced are used in the industrial processes themselves, and some goods are available to outsiders, the consumers.

To illustrate the input–output model, imagine a simple economy with just two industries: electricity and steel. These industries exist to produce electricity and steel for the consumers. However, both production processes themselves use electricity and steel. The electricity industry uses steel in the generating equipment and uses electricity to light the plant and to heat and cool the buildings. The steel industry uses electricity to run some of its equipment, and that equipment in turn contains steel components.

In the economic model, the quantities in the input–output matrix can be approximate measures of the goods, such as weight, volume, or value. We will describe the quantities in terms of dollar values. We are interested in the quantities of each product needed to provide for the consumers (their demand) and to provide for that consumed internally.

EXAMPLE 1 ➤ The amount of electricity and steel consumed by the electric company and the steel company in producing their own products depends on the amount they produce. For example, an electric company may find that whatever the value of electricity produced, 15% of that goes to pay for the electricity used internally and 5% goes to pay for the steel used in production. Thus, if the electric company produces $200,000 worth of electricity, then $30,000 worth (15%) of electricity and $10,000 worth (5%) of steel are consumed by the electric company. Now let's state this concept in a more general form and bring in the cost of producing steel as well.

Let x be the value of electricity produced and y the value of steel produced. The cost of producing electricity includes the following: Of the value of the electricity produced, x, 15% of it, $0.15x$, pays for the electricity, and 5%, $0.05x$, pays for the steel consumed internally. Similarly, the cost of producing steel includes the following: Of the value of the steel produced, y, 40% of it, $0.40y$, pays for the electricity, and 10% of it, $0.10y$, pays for the steel consumed internally. We can express this information as

Total electricity consumed internally $= 0.15x + 0.40y$
Total steel consumed internally $= 0.05x + 0.10y$

Now note that this can also be expressed as

$$\begin{bmatrix} 0.15 & 0.40 \\ 0.05 & 0.10 \end{bmatrix} \begin{bmatrix} x \\ y \end{bmatrix}$$

The coefficient matrix used here is called the **input–output matrix** of the production model. The *column headings identify the output,* the amount produced, by each industry. The *row headings identify the input,* the amount used in production, by each industry.

$$
\text{Input} \begin{pmatrix} \text{Amount used} \\ \text{in production} \end{pmatrix} \quad \begin{array}{c} \\ \text{Electricity} \\ \text{Steel} \end{array} \overset{\begin{array}{cc} \text{Output} \\ \text{(Amount produced by)} \\ \text{Electricity} \quad \text{Steel} \end{array}}{\begin{bmatrix} 0.15 & 0.40 \\ 0.05 & 0.10 \end{bmatrix}} = A
$$

There is one row for each industry. The row labeled electricity gives the value of the electricity ($0.15) needed to produce $1 worth of electricity and the value of the electricity ($0.40) needed to produce $1 worth of steel. The second row shows the value of the steel ($0.05) needed to produce $1 worth of electricity and the value of the steel ($0.10) needed to produce $1 worth of steel. ∎

We have shown how the internal consumption of two industries can be represented with matrices. The input–output matrix is the central element of the Leontief economic model. We proceed to show how we can use matrices to answer other questions about an economy. But first, be sure you understand the makeup of the input–output matrix. The entries give the fraction of goods produced by one industry that are used in producing 1 unit of goods in another industry. The row heading of the input–output matrix identifies the industry that provides goods to the industry identified by the column heading.

The Electricity–Steel input–output matrix

$$
\begin{array}{c} E \\ S \end{array} \overset{\begin{array}{cc} E & S \end{array}}{\begin{bmatrix} 0.15 & 0.40 \\ 0.05 & 0.10 \end{bmatrix}}
$$

should be interpreted in the following way.

| | User | |
Supplier	**Electricity**	**Steel**
Electricity	*Electric* industry provides $0.15 worth of electricity to the **electric** industry to produce $1.00 worth of electricity.	*Electric* industry provides $0.40 worth of electricity to the **steel** industry to produce $1.00 worth of steel.
Steel	*Steel* industry provides $0.05 worth of steel to the **electric** industry to produce $1.00 worth of electricity.	*Steel* industry provides $0.10 worth of steel to the **steel** industry to produce $1.00 worth of steel.

WARNING

Don't make the error of mixing up the rows and columns of the input–output matrix. Remember that a row in the matrix represents the amounts of *one* material or good used by *all* industries.

You might be wondering why this information is presented in matrix form. It makes it easier to answer questions such as these:

1. The production capacity of industry is $9 million worth of electricity and $7 million worth of steel. How much of each is consumed internally by the production processes?
2. The consumers want $6 million worth of electricity and $8 million worth of steel for their use. How much of each should be produced to satisfy their demands and also to provide for the amounts consumed internally?

Before we learn how to answer these two questions, let's make some observations that will help set up the problems.

First, let x = the dollar value of electricity produced and y = the dollar value of steel produced. These values include that used internally for production and that available to the consumers. Then the total amounts consumed internally are

$$\text{Electricity consumed internally} = 0.15x + 0.40y$$
$$\text{Steel consumed internally} = 0.05x + 0.10y$$

which we expressed in matrix form as

$$\begin{bmatrix} \text{electricity consumed internally} \\ \text{steel consumed internally} \end{bmatrix} = \begin{bmatrix} 0.15 & 0.40 \\ 0.05 & 0.10 \end{bmatrix} \begin{bmatrix} x \\ y \end{bmatrix} = A \begin{bmatrix} x \\ y \end{bmatrix}$$

If production capacities are $9 million worth of electricity ($x = 9$) and $7 million worth of steel ($y = 7$), the amount consumed internally is

$$\begin{bmatrix} 0.15 & 0.40 \\ 0.05 & 0.10 \end{bmatrix} \begin{bmatrix} 9 \\ 7 \end{bmatrix} = \begin{bmatrix} 4.15 \\ 1.15 \end{bmatrix}$$

$4.15 million worth of electricity and $1.15 million worth of steel.

Another fact relates the amount of electricity and steel produced to that available to the consumer:

$$[amount\ produced] = \begin{bmatrix} \text{amount consumed} \\ \text{internally} \end{bmatrix} + \begin{bmatrix} \text{amount available} \\ \text{to consumer} \end{bmatrix}$$

We call these matrices **output, internal demand,** and **consumer demand** matrices, respectively.

In the case in which $9 million worth of electricity and $7 million worth of steel were produced with $4.15 and $1.15 million consumed internally, we have

$$\underset{\text{Output}}{\begin{bmatrix} 9 \\ 7 \end{bmatrix}} - \underset{\substack{\text{Internal} \\ \text{Demand}}}{\begin{bmatrix} 4.15 \\ 1.15 \end{bmatrix}} = \underset{\text{Consumer Demand}}{\begin{bmatrix} \text{electricity available to consumer} \\ \text{steel available to consumer} \end{bmatrix}}$$

We get $4.85 million and $5.85 million worth of electricity and steel available to the consumers.

If we call the output matrix X, the consumer demand matrix D, and the input–output matrix A, then the internal demand matrix is AX and $X - AX$ is the quantity of goods available for consumers, and so $X - AX = D$ expresses the relation between output, internal demand, and consumer demand.

We can look at the production problem from another perspective, the problem of determining the production needed to provide a known consumer demand. For example, the question, "What total output is necessary to supply consumers with $6 million worth of electricity and $8 million worth of steel?" asks for the output X when consumer demand D is given. Using the same input–output matrix, we want to find x and y (output) so that

$$X - AX = D$$
$$\begin{bmatrix} x \\ y \end{bmatrix} - \begin{bmatrix} 0.15 & 0.40 \\ 0.05 & 0.10 \end{bmatrix} \begin{bmatrix} x \\ y \end{bmatrix} = \begin{bmatrix} 6 \\ 8 \end{bmatrix}$$

Notice that the variables x and y appear in two matrices.

Let's use the equation $X - AX = D$ to apply some matrix algebra to find the matrix X.

You need to solve for the matrix X in

$$X - AX = D$$

This is equivalent to solving for X in the following:

$$
\begin{aligned}
X - AX &= D \\
IX - AX &= D \\
(I - A)X &= D \\
X &= (I - A)^{-1}D
\end{aligned}
$$

The last equation is the most helpful. To find the total production X that meets the final demand D and also provides the quantities needed to carry out the internal production processes, find the inverse of the matrix $I - A$ (as you did in Section 2.6) and multiply it by the matrix D. For example, using

$$A = \begin{bmatrix} 0.15 & 0.40 \\ 0.05 & 0.10 \end{bmatrix}$$

$$I - A = \begin{bmatrix} 0.85 & -0.40 \\ -0.05 & 0.90 \end{bmatrix}$$

and

$$(I - A)^{-1} = \begin{bmatrix} 1.208 & 0.537 \\ 0.0671 & 1.141 \end{bmatrix}$$

where the entries in $(I - A)^{-1}$ are rounded. For the demand matrix,

$$D = \begin{bmatrix} 6 \\ 8 \end{bmatrix}$$

$$X = \begin{bmatrix} 1.208 & 0.537 \\ 0.0671 & 1.141 \end{bmatrix} \begin{bmatrix} 6 \\ 8 \end{bmatrix} = \begin{bmatrix} 11.544 \\ 9.531 \end{bmatrix}$$

So, \$11.544 million worth of electricity and \$9.531 million worth of steel must be produced to provide \$6 million worth of electricity and \$8 million worth of steel to the consumers and to provide for the electricity and steel used internally in production.

Leontief Input–Output Model

The matrix equation for the Leontief input–output model that relates total production to the internal demands of the industries and to consumer demand is given by

$$X - AX = D$$

or the equivalent,

$$(I - A)X = D$$

where A is the input–output matrix giving information on internal demands, D represents consumer demands, and X represents the total goods produced.

The solution to $(I - A)X = D$ is

$$X = (I - A)^{-1}D \quad \text{[Provided } (I - A)^{-1} \text{ exists]}$$

EXAMPLE 2 ➤

An input–output matrix for electricity and steel is

$$A = \begin{bmatrix} 0.25 & 0.20 \\ 0.50 & 0.20 \end{bmatrix}$$

(a) If the production capacity of electricity is $15 million and the production capacity for steel is $20 million, how much of each is consumed internally for capacity production?

(b) How much electricity and steel must be produced to have $5 million worth of electricity and $8 million worth of steel available for consumer use?

SOLUTION

(a) We are given

$$A = \begin{bmatrix} 0.25 & 0.20 \\ 0.50 & 0.20 \end{bmatrix} \quad \text{and} \quad X = \begin{bmatrix} 15 \\ 20 \end{bmatrix}$$

We want to find AX:

$$AX = \begin{bmatrix} 0.25 & 0.20 \\ 0.50 & 0.20 \end{bmatrix}\begin{bmatrix} 15 \\ 20 \end{bmatrix} = \begin{bmatrix} 7.75 \\ 11.50 \end{bmatrix}$$

So, $7.75 million worth of electricity and $11.50 million worth of steel are consumed internally.

(b) We are given

$$A = \begin{bmatrix} 0.25 & 0.20 \\ 0.50 & 0.20 \end{bmatrix} \quad \text{and} \quad D = \begin{bmatrix} 5 \\ 8 \end{bmatrix}$$

and we need to solve for X in $(I - A)^{-1}D = X$ or in $(I - A)X = D$. We use the latter this time:

$$I - A = \begin{bmatrix} 0.75 & -0.20 \\ -0.50 & 0.80 \end{bmatrix}$$

Then the augmented matrix for $(I - A)X = D$ is

$$\begin{bmatrix} 0.75 & -0.20 & 5 \\ -0.50 & 0.80 & 8 \end{bmatrix}$$

which reduces to

$$\begin{bmatrix} 1 & 0 & 11.2 \\ 0 & 1 & 17.0 \end{bmatrix}$$ (Check it)

The two industries must produce $11.2 million worth of electricity and $17.0 million worth of steel to have $5 million worth of electricity and $8 million worth of steel available to the consumers.

■ **Now You Are Ready to Work Exercises 1 and 7**

EXAMPLE 3 ➤

Fantasy Island has an economy of three industries with the input–output matrix, A. Help the Industrial Planning Commission by computing the output levels of each industry to meet the demands of the consumers and the other industries for each of the two demand levels given. The units of D are millions of dollars.

$$A = \begin{bmatrix} 0.3 & 0.3 & 0.2 \\ 0.4 & 0.4 & 0 \\ 0 & 0 & 0.2 \end{bmatrix} \quad D = \begin{bmatrix} 6 \\ 9 \\ 12 \end{bmatrix}, \begin{bmatrix} 12 \\ 15 \\ 18 \end{bmatrix}$$

SOLUTION

We need to find the values of X that correspond to each D. That comes from the solution of

$$X = (I - A)^{-1}D$$

For the given matrix A,

$$I - A = \begin{bmatrix} 1 & 0 & 0 \\ 0 & 1 & 0 \\ 0 & 0 & 1 \end{bmatrix} - \begin{bmatrix} 0.3 & 0.3 & 0.2 \\ 0.4 & 0.4 & 0 \\ 0 & 0 & 0.2 \end{bmatrix} = \begin{bmatrix} 0.7 & -0.3 & -0.2 \\ -0.4 & 0.6 & 0 \\ 0 & 0 & 0.8 \end{bmatrix}$$

We can find $(I - A)^{-1}$ by using Gauss-Jordan elimination:

$$(I - A)^{-1} = \begin{bmatrix} 2 & 1 & \frac{1}{2} \\ \frac{4}{3} & \frac{7}{3} & \frac{1}{3} \\ 0 & 0 & \frac{5}{4} \end{bmatrix}$$

We can obtain $X = (I - A)^{-1}D$ with one matrix multiplication by forming a matrix of two columns using the two D matrices as columns.

$$x = \underbrace{\begin{bmatrix} 2 & 1 & \frac{1}{2} \\ \frac{4}{3} & \frac{7}{3} & \frac{1}{3} \\ 0 & 0 & \frac{5}{4} \end{bmatrix}}_{(I-A)^{-1}} \underbrace{\begin{bmatrix} 6 & 12 \\ 9 & 15 \\ 12 & 18 \end{bmatrix}}_{\substack{\text{Values} \\ \text{of } D}} = \underbrace{\begin{bmatrix} 27 & 48 \\ 33 & 57 \\ 15 & 22.5 \end{bmatrix}}_{\substack{\text{Corresponding} \\ \text{Outputs}}}$$

The output levels required to meet the demands $\begin{bmatrix} 6 \\ 9 \\ 12 \end{bmatrix}$ and $\begin{bmatrix} 12 \\ 15 \\ 18 \end{bmatrix}$ are $\begin{bmatrix} 27 \\ 33 \\ 15 \end{bmatrix}$

and $\begin{bmatrix} 48 \\ 57 \\ 22.5 \end{bmatrix}$, respectively, with the units being millions of dollars.

■ **Now You Are Ready to Work Exercise 8**

EXAMPLE 4 ➤

Hubbs, Inc., has three divisions: grain, lumber, and energy. An analysis of their operations reveals the following information. For each $1.00 worth of grain produced, they use $0.10 worth of grain, $0.20 worth of lumber, and $0.50 worth of energy. For each $1.00 worth of lumber produced, they use $0.10 worth of grain, $0.15 worth of lumber, and $0.40 worth of energy. For each $1.00 worth of energy produced, they use $0.05 worth of grain, $0.35 worth of lumber, and $0.15 worth of energy.

(a) How much energy is used in the production of $750,000 worth of lumber?
(b) Which division uses the largest amount of energy per unit of goods produced? The least?
(c) Set up the input–output matrix.
(d) Find the cost of producing $1.00 worth of lumber.
(e) Find the internal demand if the production levels are $640,000 worth of grain, $800,000 worth of lumber, and $980,000 worth of energy.
(f) Find the total production required in order to have the following available to consumers: $300,000 worth of grain, $500,000 worth of lumber, and $800,000 worth of energy.

SOLUTION

(a) Since $0.40 worth of energy is required to produce $1.00 worth of lumber, $0.40(750{,}000) = \$300{,}000$ worth of energy is required.

(b) The largest amount of energy per unit is $0.50 worth of energy used to produce $1.00 worth of grain. The least amount of energy per unit is $0.15 worth of energy used to produce $1.00 worth of energy.

(c) Let x_1 = total value of grain produced

x_2 = total value of lumber produced

x_3 = total value of energy produced

The value of grain used internally is

$$0.10x_1 + 0.10x_2 + 0.05x_3$$

The value of lumber used internally is

$$0.20x_1 + 0.15x_2 + 0.35x_3$$

The value of energy used internally is

$$0.50x_1 + 0.40x_2 + 0.15x_3$$

so the input–output matrix is

$$
\begin{array}{cc}
 & \text{User} \\
 & \begin{array}{ccc} \text{G} & \text{L} & \text{E} \end{array} \\
\text{Supplier}\ \begin{array}{c} \text{G} \\ \text{L} \\ \text{E} \end{array} &
\begin{bmatrix}
0.10 & 0.10 & 0.05 \\
0.20 & 0.15 & 0.35 \\
0.50 & 0.40 & 0.15
\end{bmatrix}
\end{array}
$$

(d) The total cost of producing $1.00 worth of lumber is found by totaling the entries in column 2 of the input–output matrix, because those entries represent the value of G, L, and E to produce $1.00 worth of lumber. Total cost = $0.65.

(e) The internal demand is given by

$$
\begin{bmatrix}
0.10 & 0.10 & 0.05 \\
0.20 & 0.15 & 0.35 \\
0.50 & 0.40 & 0.15
\end{bmatrix}
\begin{bmatrix}
640{,}000 \\
800{,}000 \\
980{,}000
\end{bmatrix}
=
\begin{bmatrix}
193{,}000 \\
591{,}000 \\
787{,}000
\end{bmatrix}
$$

The internal consumption of the system is $193,000 worth of grain, $591,000 worth of lumber, and $787,000 worth of energy to produce a total of $640,000 worth of grain, $800,000 worth of lumber, and $980,000 worth of energy.

(f) In this case,

$$
D =
\begin{bmatrix}
300{,}000 \\
500{,}000 \\
800{,}000
\end{bmatrix}
$$

so we need to solve $X = (I - A)^{-1}D$.

$$
\left(
\begin{bmatrix}
1 & 0 & 0 \\
0 & 1 & 0 \\
0 & 0 & 1
\end{bmatrix}
-
\begin{bmatrix}
0.10 & 0.10 & 0.05 \\
0.20 & 0.15 & 0.35 \\
0.50 & 0.40 & 0.15
\end{bmatrix}
\right)^{-1}
\begin{bmatrix}
300{,}000 \\
500{,}000 \\
800{,}000
\end{bmatrix}
=
\begin{bmatrix}
x \\
y \\
z
\end{bmatrix}
$$

$$
\begin{bmatrix}
0.90 & -0.10 & -0.05 \\
-0.20 & 0.85 & -0.35 \\
-0.50 & -0.40 & 0.85
\end{bmatrix}^{-1}
\begin{bmatrix}
300{,}000 \\
500{,}000 \\
800{,}000
\end{bmatrix}
=
\begin{bmatrix}
x \\
y \\
z
\end{bmatrix}
$$

With entries rounded to three decimal places, $(I - A)^{-1}$ is used. (See Exercise 31.)

$$\begin{bmatrix} 1.254 & 0.226 & 0.167 \\ 0.743 & 1.593 & 0.700 \\ 1.087 & 0.883 & 1.604 \end{bmatrix} \begin{bmatrix} 300{,}000 \\ 500{,}000 \\ 800{,}000 \end{bmatrix} = \begin{bmatrix} 622{,}800 \\ 1{,}579{,}400 \\ 2{,}050{,}800 \end{bmatrix}$$

Hubbs, Inc., needs to produce \$622,800 worth of grain, \$1,579,400 worth of lumber, and \$2,050,800 worth of energy to meet the consumer demand specified. ∎

Today the concept of a world economy has become a reality. In 1973, the United Nations commissioned an input–output model of the world economy. The aim of the model was to transform the vast collection of economic facts that describe the world economy into an organized system from which economic projections could and have been made. In the model, the world is divided into 15 distinct geographic regions, each one described by an individual input–output matrix. The regions are then linked by a larger matrix that is used in an input–output model. Overall, more than 200 variables enter into the model, and the computations are of course done on a computer. By feeding in projected values for certain variables, researchers use the model to create scenarios of future world economic possibilities.

We need to understand that this model is not a crystal ball that shows exactly what the future holds. It gives an indication of situations that can develop *if trends continue unchanged.* We can change the trends and thereby alter future conditions. For example, the model predicted energy problems of major negative consequences by 2025. Conservation, more efficient automobile design, and recycling have helped change the energy use patterns that were in place in 1973. Although energy problems have not vanished, they are different.

Mathematical models can provide an "early warning" of what might be and allow policymakers to make adjustments to avoid or soften potential problems.

2.7 EXERCISES

Exercises 1 through 4 give the input–output matrix and the output of some industries. Determine the amount consumed internally by the production processes.

1. *(See Example 2a)*

$$A = \begin{bmatrix} 0.15 & 0.08 \\ 0.30 & 0.20 \end{bmatrix}, X = \begin{bmatrix} 8 \\ 12 \end{bmatrix}$$

2. $A = \begin{bmatrix} 0.10 & 0.20 \\ 0.25 & 0.15 \end{bmatrix}, X = \begin{bmatrix} 20 \\ 15 \end{bmatrix}$

3. $A = \begin{bmatrix} 0.06 & 0.12 & 0.09 \\ 0.15 & 0.05 & 0.10 \\ 0.08 & 0.04 & 0.02 \end{bmatrix}, X = \begin{bmatrix} 8 \\ 14 \\ 10 \end{bmatrix}$

4. $A = \begin{bmatrix} 0.03 & 0 & 0.02 & 0.06 \\ 0.08 & 0.02 & 0 & 0.05 \\ 0.07 & 0.10 & 0.01 & 0.04 \\ 0.05 & 0.04 & 0.02 & 0.06 \end{bmatrix}, X = \begin{bmatrix} 10 \\ 30 \\ 20 \\ 40 \end{bmatrix}$

Compute $(I - A)^{-1}$ for the matrices in Exercises 5 and 6.

5. $A = \begin{bmatrix} 0.2 & 0.3 \\ 0.2 & 0.3 \end{bmatrix}$ **6.** $A = \begin{bmatrix} 0.32 & 0.16 \\ 0.22 & 0.36 \end{bmatrix}$

7. *(See Example 2b)* Find the output required to meet the consumer demand and internal demand for the following input–output matrix and consumer demand matrix:

$$A = \begin{bmatrix} 0.24 & 0.08 \\ 0.12 & 0.04 \end{bmatrix} \quad D = \begin{bmatrix} 15 \\ 12 \end{bmatrix}$$

The economies in Exercises 8 through 12 are either two or three industries. Determine the output levels required of each industry to meet the demands of the other industries and of the consumer. The units are millions of dollars.

8. *(See Example 3)*

$$A = \begin{bmatrix} 0.2 & 0.4 \\ 0.3 & 0.1 \end{bmatrix}, \quad D = \begin{bmatrix} 30 \\ 15 \end{bmatrix}, \quad \begin{bmatrix} 12 \\ 8 \end{bmatrix},$$

and $\begin{bmatrix} 15 \\ 15 \end{bmatrix}$

9. $A = \begin{bmatrix} 0.2 & 0.4 \\ 0.4 & 0.3 \end{bmatrix}, \quad D = \begin{bmatrix} 20 \\ 28 \end{bmatrix}$ and $\begin{bmatrix} 15 \\ 11 \end{bmatrix}$

10. $A = \begin{bmatrix} 0.4 & 0.2 \\ 0.1 & 0.3 \end{bmatrix}, \quad D = \begin{bmatrix} 22 \\ 18 \end{bmatrix}, \quad \begin{bmatrix} 16 \\ 20 \end{bmatrix},$

and $\begin{bmatrix} 9 \\ 7 \end{bmatrix}$

11. $A = \begin{bmatrix} 0.2 & 0.2 & 0.2 \\ 0.1 & 0.6 & 0.2 \\ 0.1 & 0.1 & 0.4 \end{bmatrix}, \quad D = \begin{bmatrix} 30 \\ 24 \\ 42 \end{bmatrix}$ and $\begin{bmatrix} 60 \\ 45 \\ 75 \end{bmatrix}$

12. $A = \begin{bmatrix} 0.2 & 0.1 & 0.2 \\ 0.2 & 0.2 & 0.4 \\ 0.25 & 0.2 & 0.4 \end{bmatrix}, \quad D = \begin{bmatrix} 200 \\ 360 \\ 180 \end{bmatrix}$

and $\begin{bmatrix} 640 \\ 850 \\ 900 \end{bmatrix}$

The economies in Exercises 13 through 15 are either two or three industries. The output level of each industry is given. Determine the amounts consumed internally and the amounts available for the consumer from each industry.

13. $A = \begin{bmatrix} 0.15 & 0.35 \\ 0.40 & 0.25 \end{bmatrix}, \quad X = \begin{bmatrix} 40 \\ 50 \end{bmatrix}$

14. $A = \begin{bmatrix} 0.25 & 0.15 & 0.20 \\ 0.30 & 0 & 0.40 \\ 0.20 & 0.30 & 0.25 \end{bmatrix}, \quad X = \begin{bmatrix} 660 \\ 720 \\ 540 \end{bmatrix}$

15. $A = \begin{bmatrix} 0.20 & 0.20 & 0 \\ 0.40 & 0.40 & 0.60 \\ 0.40 & 0.10 & 0.40 \end{bmatrix}, \quad X = \begin{bmatrix} 36 \\ 72 \\ 36 \end{bmatrix}$

16. An industrial system has two industries, coal and steel. To produce $1.00 worth of coal, the coal industry uses $0.30 worth of coal and $0.40 worth of

steel. To produce $1.00 worth of steel, the steel industry uses $0.40 worth of coal and $0.20 worth of steel.
 (a) Set up the input–output matrix for this system.
 (b) Find the output of each industry that will provide for $75,000 worth of coal and $45,000 worth of steel.

17. A small country's economy is based on agriculture and nonagriculture sectors. To produce $1.00 worth of agriculture products requires $0.10 worth of agriculture and $0.60 worth of nonagriculture products. To produce $1.00 worth of nonagriculture goods requires $0.30 worth of agriculture and $0.40 worth of nonagriculture products.
 (a) Write the input–output matrix of this economy.
 (b) One year the country produced $3.5 million in agriculture products and $5.2 million in nonagriculture products. Find the amount consumed internally and the amount left for export.
 (c) The export board has a goal of exporting $2 million worth of agriculture products and $2 million worth of nonagriculture products. Find the total production of each sector that is required to achieve this goal.
 (d) Find the total production required to export $2 million worth of agriculture and $3 million worth of nonagriculture products.

18. With the advent of NAFTA, Cranford Manufacturing has a U.S. division and a Mexico division. Each division builds specialty components that are used by both divisions. For each dollar in sales by the U.S. division, Mexico furnishes $0.20 in components, and for each dollar in sales of the Mexico division, the United States furnishes $0.05 in components. Each division uses $0.10 worth of their components in their own division.
 (a) Write the input–output matrix of this company.
 (b) Find the value of the components each division must produce to support external sales of $200 million by the U.S. division and $180 million by the Mexico division.

19. The Martinella Corporation has a digital electronics division and a plastics division. For each dollar's worth of plastics produced, the plastic division uses $0.10 worth of plastic and $0.20 worth of electronics. For each dollar's worth of electronic equipment produced by the electronics division, the electronics division uses $0.40 worth of plastics and $0.20 worth of electronics.
 (a) Write the input–output matrix of this corporation.
 (b) One month the plastics division produced $25 million worth of plastic products, and the elec-

tronics division produced $32 million worth of electronics. Find the value of plastics and electronics used internally.

(c) Find the value of plastics and electronics that must be produced for the corporation to provide external sales of $36 million worth of plastics and $44 million worth of electronics.

20. The economy of an island is based on industry, small business, and agriculture. For each $1.00 worth of goods produced by industry, industry uses $0.20 worth of its own products, $0.40 worth of the products of small business, and $0.20 worth of the products of agriculture. For each $1.00 worth of goods produced by small business, small business uses $0.40 worth of the products of industry, $0.20 worth of its own products, and $0.40 worth of the products of agriculture. For each $1.00 worth of its products, agriculture uses $0.20 worth of the products of industry, $0.20 worth of the products of small business, and $0.20 worth of its own products.

(a) Find the input–output matrix of this production model.

(b) Find the value of each of the products that must be produced to export $36,000 worth of industry's products, $40,000 worth of small business products, and $30,000 worth of agriculture products.

21. Enviro Transport is an international corporation that specializes in small pollution-free vehicles. They operate in Canada, Mexico, and the United States. The corporation has found it efficient for some com-

ponents to be produced by one country that then supplies facilities in all three countries. Here is the breakdown of components supplied to each country.

For each dollar's worth of vehicles produced in Canada, Canada supplies $0.20 worth of components, Mexico supplies $0.20, and the United States supplies $0.40.

For each dollar's worth of vehicles produced in Mexico, Canada supplies $0.10 worth of components, Mexico supplies $0.40 worth, and the United States supplies none.

For each dollar's worth of vehicles produced in the United States, Canada supplies $0.30 worth of components, Mexico supplies none, and the United States supplies $0.30 worth.

(a) Write the input–output matrix of the corporation.

(b) If Canada produces $10 million worth of vehicles, Mexico produces $18 million worth, and the United States produces $15 million worth, find the value of components used in each country.

(c) Find the total value of production required to produce external sales worth $24 million by Canada, $30 million by Mexico, and $20 million by the United States.

EXPLORATIONS

22. The entries in a column of an input–output matrix total 1, such as column 2 in the matrix

	Coal	Steel	Lumber
Coal	0.20	0.40	0.15
Steel	0.30	0.35	0.40
Lumber	0.10	0.25	0.10

Interpret the meaning of this.

23. Discuss why an input–output matrix should not contain negative entries nor entries greater than 1.

24. For a two-sector economy, the total production is

$$X = \begin{bmatrix} 50,000 \\ 65,000 \end{bmatrix}$$

and it provides for internal consumption as well as a consumer demand of

$$D = \begin{bmatrix} 50,000 \\ 60,000 \end{bmatrix}$$

What information does this provide about the input–output matrix A?

 25. A corporation has two divisions, service and retail. Each dollar value of service requires $0.20 of service and $0.50 of retail. Each dollar value of retail requires $0.60 of service and $0.60 of retail. Find the total production of service and retail that will provide to consumers $200,000 service and $100,000 retail. Comment on the production total. Does it suggest an efficient operation?

 26. An economy has three industries: grain, lumber, and energy. The input–output matrix of the economy is

$$A = \begin{bmatrix} 0.5 & 0.0 & 0.3 \\ 0.0 & 0.8 & 0.4 \\ 0.0 & 0.2 & 0.4 \end{bmatrix} \begin{matrix} \text{Grain} \\ \text{Lumber} \\ \text{Energy} \end{matrix}$$

Find all output levels that will provide equal values of grain, lumber, and energy for consumer demand.

EXPLORATIONS

35. The line $3x + 4y = 15$ divides the plane into two half planes with $(1, 3)$ as one point on the line.

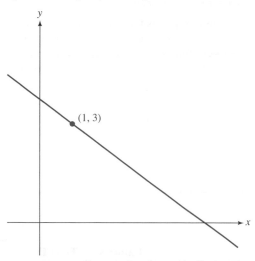

For the point $(1, 3)$, as well as any point on the line, we get a value of 15 when the coordinates are substituted in the expression $3x + 4y$.

(a) Discuss the change in the value of $3x + 4y$ when a point is chosen on a vertical line through $(1, 3)$ and *above* the line (that is, $x = 1, y > 3$). Does the value of $3x + 4y$ increase or decrease?

(b) Discuss the change in value of $3x + 4y$ when a point is chosen on a vertical line through $(1, 3)$ and *below* the line (that is, $x = 1, y < 3$). Does the value of $3x + 4y$ increase or decrease?

(c) For any point (x, y) on the line, $3x + 4y$ has the value 15. For points directly above or below (x, y), how does the value of $3x + 4y$ change?

(d) Do your observations in parts (a)–(c) support the claim that all points in one half plane determined by $3x + 4y = 15$ satisfy the inequality $3x + 4y < 15$ and all points in the other half plane satisfy $3x + 4y > 15$?

3.2   Solutions of Systems of Inequalities: A Geometric Picture

- Feasible Region
- Boundaries and Corners
- No Feasible Solution
- Bounded and Unbounded Feasible Solutions
- Graphing a System of Inequalities

Feasible Region

Linear programming problems are described by **systems of linear inequalities** rather than systems of linear equations. The solution to a linear programming problem depends on the ability to solve and graph such systems.

Because a linear inequality determines a region, we find it helpful to identify such a region; that is, we want to **graph the solution set of a system of inequalities.** Let's look at an example to illustrate.

EXAMPLE 1 ➤   Graph the region determined by the following system of inequalities.

$$2x + y \le 10$$
$$x + 3y \le 12$$

SOLUTION

We want to find and graph all points that make both inequalities true. We do so in the following manner.

1. We find the graph of $2x + y \le 10$.
   (a) Graph the line $2x + y = 10$ (Figure 3–9a). Recall that this line divides the plane into two half planes, one of which is included in the solution of the inequality.

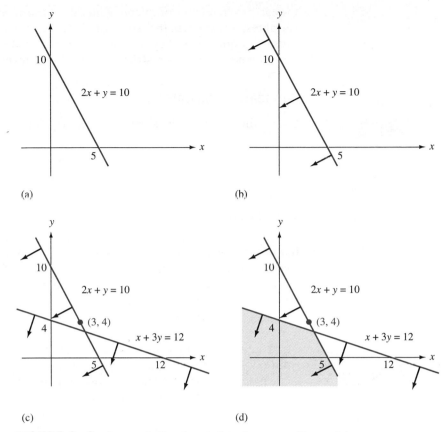

FIGURE 3–9   Steps to finding the solution of a system of inequalities.

    (b) Select a test point not on the line, say $(0, 0)$. If the test point satisfies $2x + y = 10$, it lies on the line, and you need to select another one.

    (c) Substitute $x = 0$ and $y = 0$ into $2x + y \le 10$, that is $2(0) + 0 \le 10$. As this is true, the point $(0, 0)$ lies in the half plane of the solution. The arrows attached to the graph of $2x + y = 10$ indicate the solution half plane (Figure 3–9b).

2. Next, graph $x + 3y \le 12$.

    (a) Graph the line $x + 3y = 12$.

    (b) Select a test point, say, $(3, 4)$.

    (c) Substitute $x = 3$, $y = 4$ into $x + 3y \le 12$; that is, $3 + 3(4) \le 12$. As this makes $x + 3y \le 12$ false, the point $(3, 4)$ is not in the solution half plane. The other half plane is the correct one. The arrows attached to $x + 3y = 12$ indicate the correct half plane (Figure 3–9c).

3. The region determined by the system of inequalities is the region of points that satisfy both inequalities. These points lie in the region where the two half planes overlap and is indicated by the shaded region in Figure 3–9d. This region of intersection is the **solution set of the system of linear inequalities,** and we call it the **feasible region.**

■ **Now You Are Ready to Work Exercise 1**

Linear programming problems generally have a **nonnegative condition** on the variables, which states that some or all of the variables can never be negative because the quantities they measure (number of items, weight of materials) can never be negative. This restriction is used in determining a feasible region.

Boundaries and Corners

EXAMPLE 2 ➤

Graph the solutions (feasible region) to the following system

$$\begin{aligned} x + y &\le 4 \\ -3x + 2y &\le 3 \\ x &\ge 0 \end{aligned}$$

SOLUTION

The lines $x + y = 4, -3x + 2y = 3$, and $x = 0$ determine **boundaries of the solution set.** The half plane of the solution to each inequality is indicated with arrows (Figure 3–10a). The intersection of these half planes forms the feasible region (Figure 3–10b). We call points A and B in Figure 3–10b **corners of the feasible region** because they are points in the feasible region where boundaries intersect. Even though C is a point of intersection of boundaries $x = 0$ and $x + y = 4$, it is not a corner because it is not in the feasible region.

You will learn that the corners of the feasible region determine the optimal solution to a linear programming problem. Thus, finding the corners becomes a critical step. We find corners by solving pairs of simultaneous equations, using equations of lines forming the boundary.

To find corner A, the point of intersection of lines $x = 0$ and $-3x + 2y = 3$, we solve the system

$$\begin{aligned} x &= 0 \\ -3x + 2y &= 3 \end{aligned}$$

and obtain $\left(0, \frac{3}{2}\right)$ for corner A.

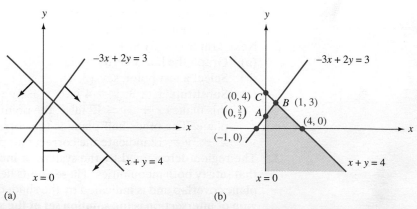

FIGURE 3–10   (a) Half planes determined by $x + y \le 4, -3x + 2y \le 3, x \ge 0$. (b) Solution to the system $x + y \le 4, -3x + 2y \le 3, x \ge 0$.

We find the point of intersection of $x + y = 4$ and $-3x + 2y = 3$ to obtain corner B. The solution of the system

$$\begin{aligned} x + y &= 4 \\ -3x + 2y &= 3 \end{aligned}$$

gives $(1, 3)$ for corner B.

The point C $(0, 4)$ is the solution of the system

$$\begin{aligned} x &= 0 \\ x + y &= 4 \end{aligned}$$

and so it is the intersection of two boundary lines. However, the point $(0, 4)$ does not satisfy the inequality $-3x + 2y \le 3$, so $(0, 4)$ lies outside the feasible region.

This illustrates that you cannot pick two boundary lines arbitrarily and expect their intersection to determine a corner point. You need to determine if the point lies in the feasible region.

■ **Now You Are Ready to Work Exercise 5**

Example 2 contained the nonnegative restriction, $x \ge 0$. Both conditions $x \ge 0$ and $y \ge 0$ generally enter into linear programming problems because the variables usually represent quantities such as the number of CD players. The other inequalities represent restrictions associated with these quantities that are imposed by equipment capacity, safety regulations, cost constraints, availability of materials, and so on.

The next example shows how the feasible region of Example 2 changes when both of the nonnegative conditions are included.

EXAMPLE 3 ➤   Sketch the feasible region (solution set) determined by the system

$$\begin{aligned} x + y &\le 4 \\ -3x + 2y &\le 3 \\ x \ge 0, \; y &\ge 0 \end{aligned}$$

SOLUTION

The feasible region of this system is bounded by the lines

$$\begin{aligned} x + y &= 4 \\ -3x + 2y &= 3 \\ x = 0, \; y &= 0 \end{aligned}$$

Notice that the nonnegative conditions restrict the feasible region to the first quadrant of the plane. As in Example 2, the points $(0, \frac{3}{2})$ and $(1, 3)$ are corners, but the condition $y \ge 0$ introduces the corners $(0, 0)$ and $(4, 0)$ (see Figure 3–11).

■ **Now You Are Ready to Work Exercise 18**

No Feasible Solution

Some systems of inequalities have no solution set, as the following example illustrates.

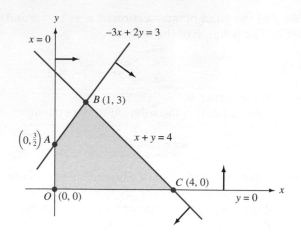

FIGURE 3-11 Corners of the feasible region: $A(0, \frac{3}{2})$, $B(1, 3)$, $C(4, 0)$, $O(0, 0)$.

EXAMPLE 4 ➤

Find the solution set (feasible region) of the system

$$5x + 7y \geq 35$$
$$3x + 4y \leq 12$$
$$x \geq 0$$
$$y \geq 0$$

SOLUTION

The inequalities $x \geq 0$ and $y \geq 0$ force the solutions to be in the first quadrant. The test point $(0, 0)$ shows that the points that satisfy $5x + 7y > 35$ lie above the line $5x + 7y = 35$ and the points that satisfy $3x + 4y < 12$ lie below the line $3x + 4y = 12$. As shown in Figure 3–12, these two regions do not intersect in the first quadrant. The system then has no solution. We sometimes say that there is **no feasible solution.**

■ **Now You Are Ready to Work Exercise 19**

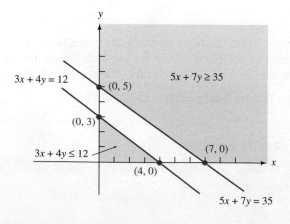

FIGURE 3-12 This system has no feasible solution.

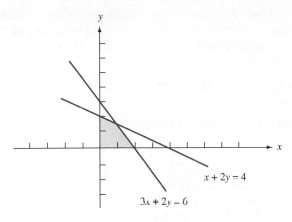

FIGURE 3–13 A system of inequalities with a bounded feasible region.

$x + 2y = 4$

$3x + 2y = 6$

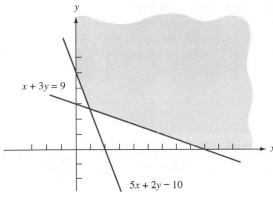

FIGURE 3–14 A system of inequalities with an unbounded feasible region.

$x + 3y = 9$

$5x + 2y = 10$

Bounded and Unbounded Feasible Solutions

A system of inequalities such as

$$x + 2y \le 4$$
$$3x + 2y \le 6$$
$$x \ge 0, y \ge 0$$

determines a feasible region, as shown in Figure 3–13. We say the system of inequalities has **bounded feasible solutions,** because the feasible region can be enclosed in a region where all the points are a finite distance apart.

On the other hand, a system of inequalities such as

$$x + 3y \ge 9$$
$$5x + 2y \ge 10$$
$$x \ge 0, y \ge 0$$

with a feasible region as shown in Figure 3–14, has **unbounded feasible solutions,** because some of the points in the feasible region are infinitely far apart.

EXAMPLE 5 ➤ A theater wishes to book a musical group that requires a guarantee of $7000. Ticket prices are $10 for students and $15 for adults, and the theater's maximum capacity is 550 seats. State the inequalities that represent this information.

SOLUTION

Let x equal the number of student tickets and let y equal the number of adult tickets. Then the total receipts is represented by $10x + 15y$, and the inequality

$10x + 15y \geq 7000$ states that total receipts must meet or exceed \$7000. The inequality $x + y \leq 550$ states that seating is limited to 550 people. Because a negative number of tickets makes no sense, the restrictions $x \geq 0$ and $y \geq 0$ are also needed.

■ **Now You Are Ready to Work Exercise 30**

Graphing a System of Inequalities

To Graph a System of Inequalities:

1. Replace each inequality symbol with an equals sign to obtain a linear equation.
2. Graph each line. Use a solid line if it is a part of the solution. Use a dotted line if it is not a part of the solution. The line is a part of the solution when $\leq$ or $\geq$ is used. The line is not a part of the solution when $<$ or $>$ is used.
3. Select a test point not on the line.
4. If the test point satisfies the original inequality, it is in the correct half plane. If it does not satisfy the inequality, the other half plane is the correct one.
5. Shade the correct half plane.
6. When the above steps are completed for each inequality, determine where the shaded half planes overlap. This region is the graph of the system of inequalities.

▪▪ 3.2 EXERCISES

LEVEL 1

Graph the systems of inequalities in Exercises 1 through 8.

1. *(See Example 1)*
$$x + y \leq 3$$
$$2x - y < -2$$

2. $2x + y \geq 4$
$$4x - y \geq 8$$

3. $x \geq 3$
$$y \geq 2$$
$$3x + 2y < 18$$

4. $-x + 3y < 6$
$$2x + y < 7$$
$$y \geq 0$$

5. *(See Example 2)*
$$4x + 6y \leq 18$$
$$x + 3y \leq 6$$
$$x \geq 0$$

6. $2x + y \leq 50$
$$4x + 5y \leq 160$$
$$x \geq 0, y \geq 0$$

7. $2x + y \leq 60$
$$2x + 3y \leq 120$$
$$x \geq 0, y \geq 0$$

8. $4x + 3y \leq 72$
$$4x + 9y \leq 144$$
$$x \geq 0, y \geq 0$$

Find the feasible regions of the systems of inequalities in Exercises 9 through 24. Determine the corners of each feasible region.

9. $x + y > 4$
$$2x - 3y \geq 8$$

10. $-x + y \leq 3$
$$2x + y \leq 6$$
$$x \geq 0, y \geq 0$$

11. $-3x + 10y \leq 15$
$$3x + 5y \geq 15$$

12. $-x + 3y \geq 5$
$$2x + y \geq 4$$

13. $x \leq 0$
$$y \geq 2$$

14. $x + y \leq 6$
$$-x + y \leq 2$$
$$x \geq 0, y \geq 0$$

15. $-x + 2y \geq 6$
$$3x + 2y \leq -2$$
$$y \geq 0$$

16. $x - y \leq 2$
$$2x + y \leq 4$$
$$x + 2y \leq 4$$
$$x \geq 0, y \geq 0$$

LEVEL 2

17. $3x + 4y > 24$
$$4x + 5y < 20$$
$$x > 0$$

18. *(See Example 3)*
$$x + y \leq 2$$
$$3x + y \leq 6$$
$$x \geq 0, y \geq 0$$

19. *(See Example 4)*
$$x + y \leq 2$$
$$2x + y \geq 6$$
$$y \geq 0$$

20. $2x + y \geq 2$
$$x - 3y \leq 6$$
$$-4x + y < -3$$

21. $x + y \geq 1$
$-x + y \geq 2$
$5x - y \leq 4$

22. $2x + y \leq 8$
$10x + y \leq 20$
$x \geq 0, y \geq 0$

23. $-x + y \leq 5$
$5x + 6y \geq -14$
$-4x + 6y \geq -32$
$6x + 7y \leq 48$

24. $4x - 3y \geq 60$
$x + y \leq 10$
$-x + 2y \geq -50$

In Exercises 25 through 28, sketch the feasible region and determine if the systems of inequalities have bounded or unbounded feasible regions.

25. $3x + 4y \leq 12$
$5x + 3y \leq 15$
$x \geq 0, y \geq 0$

26. $6x + 4y \geq 12$
$x + 2y \geq 4$
$x \geq 0, y \geq 0$

27. $x + y \geq 6$
$3x + 6y \geq 24$
$x \geq 0, y \geq 0$

28. $2x + 3y \leq 18$
$7x + 3y \leq 21$
$x \geq 0, y \geq 0$

LEVEL 3

29. Determine the feasible region and corners of this system:

$$-2x + y \leq 2$$
$$3x + y \leq 3$$
$$-x + y \geq -4$$
$$x + y \geq -3$$

30. *(See Example 5)* The seating capacity of a theater is 250. Tickets are \$3 for children and \$5 for adults. The theater must take in at least \$1000 per performance. Express this information as a system of inequalities.

31. High Fiber and Corn Bits cereals contain the following percentages of minimum daily requirements of vitamins A and D for each ounce of cereal.

	Vitamin A	Vitamin D
High Fiber	25%	4%
Corn Bits	2%	10%

Use inequalities to express how much of each Mrs. Smith should eat to obtain at least 40% of her minimum daily requirement of vitamin A and 25% of that for vitamin D.

32. High Fiber and Corn Bits cereals contain the following amounts of sodium and calories per ounce.

	Calories	Sodium
High Fiber	90	160
Corn Bits	120	200

Mr. Brown's breakfast should provide at least 600 calories but less than 800 milligrams of sodium. Find the constraints on the amount of each cereal.

33. The Musical Group has two admission prices for its concerts, \$12 for adults and \$6 for students. They will not book a concert unless they are assured an audience of at least 500 people and total ticket sales

of \$5400 or more. Write the constraints on the number of each kind of ticket sold.

34. To be eligible for a university scholarship, a student must score at least 600 on the SAT verbal test and 600 on the SAT mathematics test and must have a combined verbal–mathematics score of 1325 or more. Write this as a system of inequalities.

35. A test is scored by giving 4 points for each correct answer and -1 for each incorrect answer. To obtain an acceptable score, a student must answer at least 60 questions and attain a score of 200 points. Express this information with inequalities.

36. A plane has two kinds of packages to deliver; each of the first kind weighs 140 pounds and occupies 3 cubic feet of space. Each of the second kind weighs 185 pounds and occupies 5 cubic feet of space. The plane is limited to a total of 6000 pounds of cargo and has 300 cubic feet of cargo space available. Write the inequalities that describe the number of each kind of package that can be carried.

37. The Hippodrome Theater is to sell two types of tickets to a concert, balcony tickets at \$15 each and main-floor tickets at \$25 each. They must sell at least 3000 tickets and at least 1200 must be main-floor tickets. To break even, they must sell at least \$60,000 in tickets. Write the system of inequalities describing this information on the number of tickets to be sold of each kind.

38. A dietitian is helping a diabetic patient monitor intake of carbohydrates, protein, and fat. The number of grams of each in a serving of milk, vegetables, fruit, and meat is given by the following chart.

	Food			
	Milk	Vegetable	Fruit	Meat
Carbohydrates	12	7	10	0
Protein	8	2	0	7
Fat	10	0	0	5

The minimum requirements are carbohydrates: 30 grams; protein: 15 grams; and fat: 10 grams.

(a) Write a system of inequalities that must hold if a meal consists of two foods, vegetable and meat, and the minimum requirements are met.

(b) Write a system of inequalities that must hold if

the meal consists of milk and vegetable and the minimum requirements are met.

(c) Write a system of inequalities that must hold if the meal consists of vegetable and fruit and the minimum requirements are met.

EXPLORATIONS

Graph the boundary lines of the feasible region defined by the following inequalities and locate the corner points (to two decimal places). Then sketch the feasible region on your paper.

 39. $7x + 3y \leq 21$
$5x + 4y \leq 20$

 40. $3x + 8y \geq 24$
$6x + 5y \geq 30$

 41. $9x + 4y \leq 36$
$6x + 5y \leq 30$
$5x + 10y \leq 50$
$x \geq 0, y \geq 0$

 42. $4.5x + 9.5y \leq 42.75$
$8.8x + 6.2y \leq 54.56$
$x \geq 0, y \geq 0$

 43. $9.5x + 3.2y \geq 30.4$
$6.75x + 7.5y \geq 50.625$
$x \geq 0, y \geq 0$

44. $8.8x + 5.5y \geq 48.5$
$5.2x + 9.5y \geq 49.4$
$x + \quad y \leq 8.0$

45. $2.86x + 1.19y \leq 14.29$
$1.94x + 4.17y \leq 22.83$
$3.33x + 2.50y \leq 18.00$
$x \geq 0, y \geq 0$

USING YOUR TI-83

FINDING CORNER POINTS OF A FEASIBLE REGION

Use the TI-83 to find the corner points of the feasible region defined by the following constraints:

$$y + \frac{x}{2} \leq 15 \qquad y + \frac{4}{5}x \leq \frac{84}{5} \qquad y + \frac{5}{4}x \leq 24 \qquad x \geq 0, y \geq 0$$

We show the feasible region and the corners that we seek to find on the calculator:

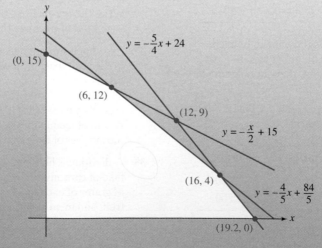

First, we graph the equations

$$y = \frac{x}{2} + 15 \qquad y = -\frac{4}{5}x + \frac{84}{5} \qquad y = -\frac{5}{4}x + 24$$

using $(0, 20)$ for both the x-range and y-range. We use the **intersect** command to find the corner points.

With the graph of the three lines on the screen, press the $\boxed{\text{CALC}}$ key ($\boxed{\text{2nd}}$ $\boxed{\text{TRACE}}$) and select <5:intersect> from the menu.

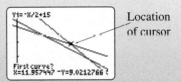

Press $\boxed{\text{ENTER}}$ and the following screen will appear:

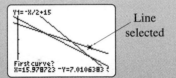

Location of cursor

Notice that the cursor lies at the intersection of two lines. Because the cursor lies on the line selected, we need to move it to determine which line is selected. Use the $\boxed{<}$ or $\boxed{>}$ key to move the cursor along a line. We move it and from the second screen we see the line selected.

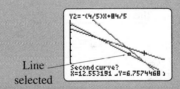

Line selected

Now press $\boxed{\text{ENTER}}$, and we are ready to select the second line. Use the $\boxed{\wedge}$ or $\boxed{\vee}$ key to move to another line.

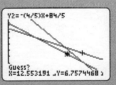

Line selected

Press $\boxed{\text{ENTER}}$ and obtain the following screen:

Press $\boxed{\text{ENTER}}$ again and the point of intersection and its coordinates, $(6, 12)$, are displayed:

Point of intersection $(6, 12)$

(continued)

To find other corners, follow the same procedure and select two lines that intersect at another corner. Other intersections are

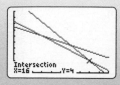

and

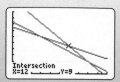

Notice the last intersection is not a corner because it is outside the feasible region.

EXERCISES

Find the corners of the feasible regions.

1. $1.5x + y \leq 14$
$0.8x + y \leq 8.4$
$x + 3y \leq 21$
$x \geq 0, y \geq 0$

2. $0.5x + y \leq 13$
$x + y \leq 15$
$2.5x + y \leq 33$
$x \geq 0, y \geq 0$

3. $5x + 11y \leq 149$
$x + 5y \leq 55$
$4x + 6y \leq 108$
$x \geq 0, y \geq 0$

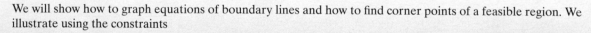

USING EXCEL

We will show how to graph equations of boundary lines and how to find corner points of a feasible region. We illustrate using the constraints

$$y + 3x \leq 30$$
$$y + \tfrac{7}{4}x \leq 20$$
$$y + \tfrac{1}{2}x \leq 15$$

To graph the boundary lines, we write the constraints as equations and put them in the form

$$y = -3x + 30$$
$$y = -\tfrac{7}{4}x + 20$$
$$y = -\tfrac{1}{2}x + 15$$

TO SET UP THE SPREADSHEET

We use line 1 of the spreadsheet to identify the contents of the columns. In columns A through D, enter x, $y = -3x + 30$, $y = -7x/4 + 20$, and $y = -0.5x + 15$ to identify the contents of those columns.

In cells A2 through A7, enter the numbers 0, 2, 4, 6, 8, 10. These values of x will be used to plot points on the graph of each line. When you use zero for a value of x, the graph will show the y-intercept of the boundary lines. Next enter the formulas:

In B2, enter =-3*A2+30
In C2, enter =-7*A2/4+20
In D2, enter =-.5*A2+15

Then drag each formula down to line 7. The values of y corresponding to each value of x appear for each function. This gives points used in plotting the graph.

	A	B	C	D
1	X	y=-3*x+30	y=-7*x/4+20	y=-.5*x+15
2	0	30.00	20.00	15.00
3	2	24.00	16.50	14.00
4	4	18.00	13.00	13.00
5	6	12.00	9.50	12.00
6	8	6.00	6.00	11.00
7	10	0.00	2.50	10.00
8				

TO SHOW THE GRAPHS

- Click on the **Chart Wizard** icon on the Ruler.
- Click on **XY (Scatter)** under **Chart Type.**
- Click on the second chart in the first column under **Chart Sub-type.**
- Click **Next.**
- Select cells A2:D7.
- Click **Next.**

You may now identify the graph by typing the title in **Chart title,** such as *Sec 3.2 Example*. You may also identify the x and y axes with a title in the next two boxes, if you like.

- Click on **Finish** to obtain the graphs of the boundary lines.

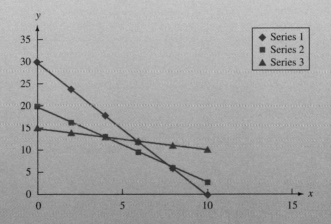

The Series 1, 2, and 3 legends shown refer to the three functions in the order they were entered in columns B, C, and D, respectively.

From the graph we see that the corners of the feasible region are $(0, 0)$, $(0, 15)$, $(10, 0)$, as well as the point where Series 2 and 3 intersect and the point where Series 1 and 2 intersect. Here's how we find those points of intersection.

POINTS OF INTERSECTION

The point of intersection of Series 1 and Series 2 is the point where $3x + 30 = -\frac{7}{4}x + 20$, or equivalently, $(-3x + 30) - (-\frac{7}{4}x + 20) = 0$. We use the second form and **Goal Seek** to find the point of intersection.

- In cell E2, enter =C2-D2, and in F2, enter =B2-C2. We will use E2 to find the intersection of Series 2 and 3, and we will use F2 to find the intersection of Series 1 and 2.
- Select cell E2.

(continued)

- Under **Tools** in the Tool Bar, select **Goal Seek.** In the menu that appears, fill the blanks with E2, 0, A2 as shown. The zero indicates the desired difference in y-values of the two lines, and A2 tells the location of the variable x.

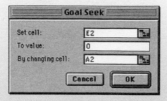

- Click **OK.**
  You will see that A2 is 4, and C2 and D2 are both 13. Thus, the point of intersection of the two lines is $(4, 13)$.
- Select cell F2.
- Under **Tools,** select **Goal Seek.** As you did for E2, fill the blanks in the menu that appears with F2, 0, A2.
- Click **OK.** Cell A2 shows $x = 8$ and cells B2 and C2 show $y = 6$. Thus, the two lines intersect at the point $(8, 6)$.

We have found the other two corners of the feasible region, so we can list all corners: $(0, 0)$, $(0, 15)$, $(4, 13)$, $(8, 6)$, and $(10, 0)$. You can use **Goal Seek** to find the x-intercept of a line, say $y = -\frac{7}{4}x + 20$, by entering C2, 0, A2 in the blanks of the Goal Seek menu that appears. The value of x in A2 is the x-intercept.

EXERCISES

1. Graph the following linear functions and find their point of intersection.
   $y = -2x + 19$
   $3x + 4y = 41$ (Put this in the form $y = (-3x + 41)/4$)

2. Graph the following linear functions and find their point of intersection.
   $y = 2.5x - 4.5$
   $6x + 5y = -4$

3. Graph the following linear functions and find their points of intersection.
   $2x + y = 32$
   $2x - 5y = -40$
   $2x + 3y = 40$

4. Graph the following linear functions and find their points of intersection.
   $x + y = 14$
   $x + 3y = 32$
   $2x + 3y = 40$

5. Graph the following linear functions and find their point of intersection.
   $4.2x + 2.2y = 24.4$
   $3.8x + 5.3y = 44.25$

3.3   Linear Programming: A Geometric Approach

- Constraints and Objective Function
- Geometric Solution
- Why Optimal Values Occur at Corner Points
- Unbounded Feasible Region
- Unusual Linear Programming Situations

Constraints and Objective Function

Managers in business and industry often make decisions in an effort to maximize or minimize some quantity. For example, a plant manager wants to minimize overtime pay for production workers, a store manager makes an effort to maximize revenue, and a stockbroker tries to maximize the return on investments. Most of these decisions are complicated by restrictions that limit choices. The plant manager might not be able to eliminate all overtime and still meet the contract deadline. A store might be swamped by customers because of its low prices, but the prices might be so low that the store is losing money.

A successful manager tries to make the right decision or to choose the best decision from several possible decisions. We will look at some simple examples where linear programming helps a manager make those decisions. First, an example that illustrates the form of a linear programming problem.

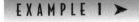

EXAMPLE 1 ➤   An appliance store manager plans to offer a special on washers and dryers. The storeroom capacity is limited to 50 items. Each washer requires 2 hours to unpack and set up, and each dryer requires 1 hour. The manager has 80 hours of employee time available for unpacking and setup. Washers sell for $300 each, and dryers sell for $200 each. How many of each should the manager order to obtain the maximum revenue?

SOLUTION

Convert the given information to mathematical statements:

$$\text{Let} \quad x = \text{the number of washers}$$
$$y = \text{the number of dryers}$$

The total number to be placed in the storeroom is

$$x + y$$

and the total setup time is

$$2x + 1y$$

Because the manager has space for only 50 items and setup time of 80 hours, we have the restrictions

$$x + y \leq 50$$
$$2x + y \leq 80$$

As x and y cannot be negative, we also have

$$x \geq 0 \quad \text{and} \quad y \geq 0$$

Because washers sell for \$300 and dryers sell for \$200, we want to find values of x and y that maximize the total revenue

$$z = 300x + 200y$$

Here is the problem stated in concise form:
Maximize z, where

$$z = 300x + 200y$$

is subject to

$$x + y \leq 50$$
$$2x + y \leq 80$$
$$x \geq 0$$
$$y \geq 0$$

■ **Now You Are Ready to Work Exercise 1**

We now discuss how you can solve a linear programming problem. In the preceding example, the inequalities

$$x + y \leq 50 \quad \text{and} \quad 2x + y \leq 80$$

impose restrictions on the problem, and we call them **constraints.** The restrictions $x \geq 0$, $y \geq 0$ are **nonnegative conditions.** We call the function $z = 300x + 200y$ the **objective function.** Find the values of x and y that satisfy the system of constraints (inequalities) by the methods from the last section. You should obtain the feasible region as shown in Figure 3–15a. Corner A is $(30, 20)$.

Each point in the feasible region determines a value for the objective function. We want to find the point in the feasible region that maximizes the objective function. The point $(10, 10)$ gives $z = 300x + 200y$ the value $z = 5000$, whereas the point $(20, 20)$ gives the value $z = 10,000$. Because the feasible region contains an infinite number of points, we will not easily find the maximum value of z by a haphazard trial-and-error process.

DEFINITION
**Constraints,
Nonnegative
Condition**

A linear inequality of the form

$$a_1 x + a_2 y \leq b$$

or

$$a_1 x + a_2 y \geq b$$

is called a **constraint** of a linear programming problem.
The restrictions

$$x \geq 0 \quad \text{and} \quad y \geq 0$$

are **nonnegative conditions.**

Geometric Solution

We now state a basic theorem that makes it easier to find a maximum or minimum value of an objective function. The term **optimal** refers to either maximum or minimum.

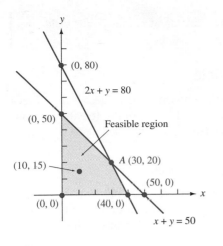

(a)

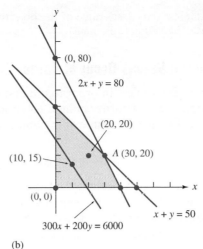

(b)

FIGURE 3–15 As an objective function increases, the objective function line moves toward a corner. In (e) the objective function moves out of the feasible region and out of consideration.

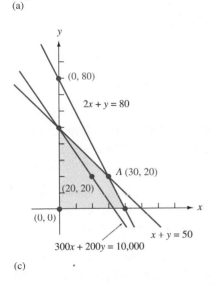

(c)

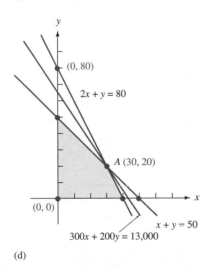

(d)

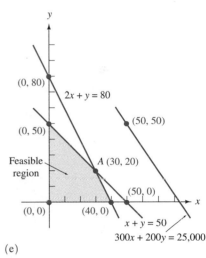

(e)

THEOREM **Optimal Values**	Given a linear **objective function** subject to linear inequality constraints, if the objective function has an **optimal value** (maximum or minimum), it must occur at a corner point of the feasible region.

The feasible region in Example 1 has corner points $(0, 0)$, $(0, 50)$, $(30, 20)$, and $(40, 0)$. The value of the objective function in each case is $z = 0$ at $(0, 0)$, $z = 10,000$ at $(0, 50)$, $z = 13,000$ at $(30, 20)$, and $z = 12,000$ at $(40, 0)$. Therefore, by the theorem, the maximum value of z within the feasible region is 13,000, and the minimum value is 0.

Graphical Solution to Linear Programming Problem	**1.** Use each constraint (linear inequality) in turn to sketch the boundary of the feasible region. **2.** Determine the corner points of the feasible region by solving pairs of linear equations (equations obtained from the constraints). **3.** Evaluate the objective function at each corner point.

> 4. The largest (smallest) value of the objective function at corner points yields the desired maximum (minimum).

Why Optimal Values Occur at Corner Points

Let's look at Example 1 in order to understand why optimal values of the objective function occur at corner points. The point $(10, 15)$ lies in the feasible region. (See Figure 3–15a.) The value of the objective function $z = 300x + 200y$ at that point is

$$z = 300(10) + 200(15) = 6000$$

This value of z can be obtained at other points in the feasible region; in fact, $z = 6000$ will occur at any point in the feasible region that lies on the line $300x + 200y = 6000$. (See Figure 3–15b.) Furthermore, for any point (x, y) in the feasible region, one of the following will occur:

$$300x + 200y = 6000$$
$$300x + 200y < 6000$$

or

$$300x + 200y > 6000$$

We have already observed that equality holds for points on the line $300x + 200y = 6000$. The point $(0, 0)$ satisfies $300x + 200y < 6000$, so all points below the line do also. From graphing inequalities we then know that the points in the half plane *above* the line $300x + 200y = 6000$ must satisfy

$$300x + 200y > 6000$$

From Figure 3–15b, we observe that the point $(20, 20)$ lies in the feasible region and lies above $300x + 200y = 6000$, so that point will yield a larger value of the objective function. It yields $z = 300(20) + 200(20) = 10,000$. Again, points above the line $300x + 200y = 10,000$ exist in the feasible region, so there are points in the feasible region that make the objective function greater than 10,000 (See Figure 3–15c). When, if ever, will we reach the maximum value of the objective function $z = 300x + 200y$? As we move the line representing the objective function farther away from the origin, we obtain larger values of the objective function.

As shown in Figure 3–15e, we can move the line completely out of the feasible region, where points on the line cannot be considered. So, where do we stop moving the line away from the origin so that it still intersects the feasible region and gives the largest possible value of the objective function?

Look at Figures 3–15b and 3–15c. As we moved the line through $(10, 15)$ away from the origin to the point $(20, 20)$, the value of the objective function increased from 6000 to 10,000. If we continue to move it farther away, we will get larger values of the objective function. Observe in Figure 3–15d that the farthest we can move it and still intersect the feasible region is at the point $(30, 20)$. At $(30, 20)$, $z = 300(30) + 200(20) = 13,000$. Thus, the line $300x + 200y = 13,000$ divides the plane into half planes with all points on the line, giving

$$300x + 200y = 13,000$$

all points above the line, giving

$$300x + 200y > 13,000$$

and all points below the line, giving

$$300x + 200y < 13,000$$

Because all points above the line are outside the feasible region, we cannot consider them. The only point in the feasible region lying on the line is $(30, 20)$. Thus, $(30, 20)$ gives the value 13,000 to the objective function, and all other points in the feasible region give smaller values. The maximum value of the objective function indeed occurs at the corner point $(30, 20)$.

In this example, the feasible region is **bounded,** because it can be enclosed in a finite rectangle. For a bounded feasible region, the objective function will have both a maximum and a minimum.

In some cases the constraints lead to an inconsistency, so there are no points in the feasible region — it is **empty** (see Example 4 of Section 3.2).

THEOREM
Bounded Feasible Region

> When the feasible region is not empty and is bounded, the objective function has both a maximum and a minimum value, which must occur at corner points.

Example 1 and the discussion that follows show a setup of the problem, how it is solved, and why an optimal solution occurs at a corner. With an understanding of those ideas, we can solve a problem like Example 1 in a more concise manner, as shown in the next example.

EXAMPLE 2 ➤

Find the maximum value of the objective function

$$z = 10x + 15y$$

subject to the constraints

$$x + 4y \leq 360$$
$$2x + y \leq 300$$
$$x \geq 0, y \geq 0$$

SOLUTION

Graph the feasible region of the system of inequalities (Figure 3–16). The corner points of the feasible region are $(0, 90)$, $(0, 0)$, $(150, 0)$, and $(120, 60)$. The point $(120, 60)$ is found by solving the system

$$x + 4y = 360$$
$$2x + y = 300$$

Find the value of z at each corner point.

Corner	$z = 10x + 15y$
$(0, 90)$	1350
$(0, 0)$	0
$(150, 0)$	1500
$(120, 60)$	2100

The maximum value of z is 2100 and occurs at the corner $(120, 60)$.

■ **Now You Are Ready to Work Exercise 7**

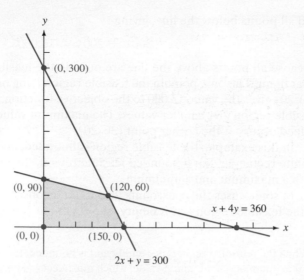

FIGURE 3–16 The maximum value of z occurs at the corner $(120, 60)$.

Unbounded Feasible Region

When we seek optimal values of an objective function, the value of the objective function tends to increase for points farther and farther from the origin and tends to decrease at points closer to the origin. We observe that the maximum value of the objective function occurs at a corner point farthest from the origin. In an unbounded feasible region similar to the one shown in Figure 3–17, you can find points in the feasible region that are infinitely far from the origin. Thus, the objective function keeps getting larger for points farther and farther away, so we say that no maximum value exists. The boundaries of the feasible region limit how close a point can be to the origin, so the objective function takes on its minimum value at a corner point.

THEOREM
Unbounded Feasible Region

When a feasible region with nonnegative conditions is unbounded, an objective function assumes a minimum at a corner point of the feasible region. However, the objective function can be arbitrarily large for points in the feasible region, so no optimal maximum solution exists.

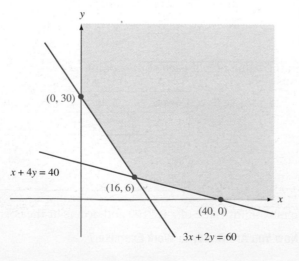

FIGURE 3–17 The minimum value of z occurs at the corner $(16, 6)$.

Here is an example of an unbounded feasible region with a minimum solution.

 **EXAMPLE 3** ➤

Find the minimum value of $z = 5x + 10y$ subject to the constraints

$$3x + 2y \geq 60$$
$$x + 4y \geq 40$$
$$x \geq 0, y \geq 0$$

SOLUTION

The corner points of the feasible region are $(0, 30)$, $(16, 6)$, and $(40, 0)$. (See Figure 3–17.) The value of z at each of these points is $z = 300$ at $(0, 30)$; $z = 140$ at $(16, 6)$; and $z = 200$ at $(40, 0)$. Thus, the minimum value of z is 140 and occurs at $(16, 6)$.

■ **Now You Are Ready to Work Exercise 11**

 **EXAMPLE 4** ➤

Find the maximum and minimum values of

$$z = 4x + 6y$$

subject to the constraints

$$5x + 3y \geq 15$$
$$x + 2y \leq 20$$
$$7x + 9y \leq 105$$
$$x \geq 0, y \geq 0$$

SOLUTION

The feasible region of this system and its corners are shown in Figure 3–18. Compute the value of z at each corner to determine the maximum and minimum values of z.

Corner	Value of z
$(0, 5)$	30
$(0, 10)$	60
$(6, 7)$	66
$(15, 0)$	60
$(3, 0)$	12

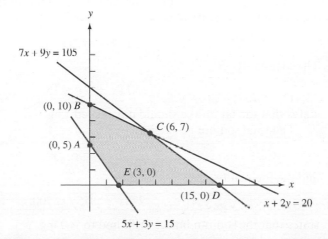

FIGURE 3–18   The maximum value of z occurs at the corner $(6, 7)$, and the minimum value of z occurs at the corner $(3, 0)$.

The maximum value of z is 66 and occurs at the corner $(6, 7)$. The minimum value of z is 12 and occurs at the corner $(3, 0)$.

■ **Now You Are Ready to Work Exercise 23**

We now look at two simple applications.

EXAMPLE 5 ➤

Lisa's grandfather, "Daddy Bill," survived triple bypass surgery and was placed on a strict diet. His wife, "Mamma Lou," monitored his food carefully. One Tuesday, she served him soup and a sandwich that contained the amounts of fat, sodium, and protein shown in the table.

	Amounts per Serving		
	Fat (g)	Sodium (mg)	Protein (g)
Sandwich	11	275	17
Soup	2	875	4

The lunch is to provide no more than 15 grams of fat and 700 milligrams of sodium. How many servings of soup and sandwich should "Daddy Bill" have for lunch in order to maximize his protein intake?

SOLUTION

Two restrictions apply to this problem: the amount of fat intake and the amount of sodium intake. At the same time, the amount of protein is to be maximized. The solution to the problem involves the following steps:

1. Identify the variables to be used.
2. State the restrictions and the quantity to be maximized with mathematical statements
3. Sketch the graph of given conditions.
4. Use the graph to find the solution.

The quantities that can be varied are the number of servings of soup and sandwiches.

$$\text{Let}\quad x = \text{number of servings of sandwiches}$$
$$y = \text{number of servings of soup}$$

The fat in x servings of sandwiches is $11x$ and the fat in y servings of soup is $2y$, so the total fat intake is

$$11x + 2y$$

The inequality

$$11x + 2y \leq 15$$

states that the fat intake should not exceed 15 g.
    The total sodium intake is

$$275x + 875y$$

and the inequality

$$275x + 875y \leq 700$$

states that the sodium intake is limited to 700 mg.

The total protein intake is

$$17x + 4y$$

We call this quantity to be maximized or minimized the objective function, in this case, $z = 17x + 4y$. Because x and y represent numbers of servings, they can never be negative, so $x \geq 0$ and $y \geq 0$.

We can now state this linear programming problem as:

Maximize $z = 17x + 4y$, subject to the constraints

$$\begin{aligned} 11x + \quad 2y &\leq 15 \\ 275x + 875y &\leq 700 \\ x \geq 0, \; y &\geq 0 \end{aligned}$$

We want to find the solutions of this system of inequalities that maximize $z = 17x + 4y$. A graph of the system and its feasible region is shown in Figure 3–19.

We find the corner points of the feasible region at the points where pairs of lines intersect. We find them from the following system.

Corner A	Corner B	Corner C	Corner D
$275x + 875y = 700$	$11x + 2y = 15$	$11x + 2y = 15$	$x = 0$
$x \qquad\quad = \quad 0$	$275x + 875y = 700$	$y = 0$	$y = 0$

These systems give the corners (shown to three decimals)

$$A: (0, 0.800) \qquad B: (1.292, 0.394) \qquad C: (1.364, 0) \qquad D: (0, 0)$$

The maximum value of $z = 17x + 4y$ is attained at one of these corners. Examine each one:

At $(0, 0.800)$: $z = 17(0) + 4(0.800) = 3.200$

At $(1.292, 0.394)$: $z = 17(1.292) + 4(0.394) = 23.540$

At $(1.364, 0)$: $z = 17(1.364) + 4(0) = 23.188$

At $(0, 0)$: $z = 0$

FIGURE 3–19

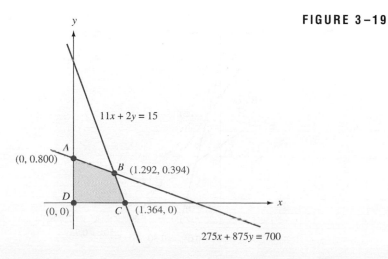

The maximum value of $z = 17x + 4y$ occurs when $x = 1.292$ and $y = 0.394$. This indicates that approximately 1.3 servings of sandwiches and 0.4 serving of soup will yield the greatest quantity of protein while keeping the fat and sodium intake at acceptable levels.

■ **Now You Are Ready to Work Exercise 39**

EXAMPLE 6 ➤ Mom's Old-Fashioned Casseroles produces a luncheon casserole that consists of 50% carbohydrates, 30% protein, and 20% fat. The dinner casserole consists of 75% carbohydrates, 20% protein, and 5% fat. The luncheon casserole costs $2.00 per pound and the dinner casserole costs $2.50 per pound. How much of each type of casserole should be used to provide at least 3 pounds of carbohydrates, 1.50 pounds of protein, and 0.50 pound of fat at a minimum cost?

SOLUTION

Let $x =$ the number of pounds of the luncheon casserole and $y =$ the number of pounds of dinner casserole. We wish to minimize the cost, $z = 2x + 2.50y$. The constraints are

$$
\begin{array}{ll}
0.50x + 0.75y \geq 3 & \text{(Total carbohydrates)} \\
0.30x + 0.20y \geq 1.50 & \text{(Total protein)} \\
0.20x + 0.05y \geq 0.50 & \text{(Total fat)} \\
x \geq 0, y \geq 0 &
\end{array}
$$

Figure 3–20 shows the feasible region with corners $(0, 10)$, $(1, 6)$, $(4.2, 1.2)$, and $(6, 0)$. The values of z at each of these corners are $z = 25$ at $(0, 10)$; $z = 17$ at $(1, 6)$; $z = 11.4$ at $(4.2, 1.2)$; and $z = 12$ at $(6, 0)$. Then the minimum cost is $11.40 when 4.2 pounds of the luncheon casserole and 1.2 pounds of the dinner casserole are used.

■ **Now You Are Ready to Work Exercise 49**

A common optimization problem is called the **transportation problem.** It seeks the minimum cost of shipping goods from several sources to several destinations. The next example shows a simple case.

FIGURE 3–20

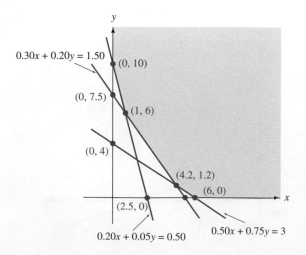

EXAMPLE 7 ➤

The Garden Center has two stores, one in Clifton and one in Marlin. They sell a popular garden tractor that is supplied to them from Supplier A and Supplier B. The shipping costs (in dollars) per tractor from suppliers A and B to the Clifton and Marlin stores are the following:

	Store	
Supplier	**Clifton**	**Marlin**
A	70	55
B	85	60

One spring, the Garden Center ordered 40 tractors for Clifton and 50 for Marlin. Supplier A had 45 tractors in stock, and Supplier B had 70 tractors in stock. Find the number of tractors that should be shipped from each supplier to each store to minimize shipping costs.

SOLUTION

Let x = number of tractors shipped from Supplier A to Clifton
       y = number of tractors shipped from Supplier A to Marlin
The balance of the tractors to each store must come from Supplier B, so

$40 - x$ = number of tractors shipped from Supplier B to Clifton
$50 - y$ = number of tractors shipped from Supplier B to Marlin

We now summarize the information with the following table:

	Store		
	Clifton	**Marlin**	**Available**
Supplier A			45
Cost (each)	$70	$55	
Number	x	y	
Supplier B			70
Cost (each)	$85	$60	
Number	$40 - x$	$50 - y$	
Number needed	40	50	

From this we find the total shipping cost,

$$C = 70x + 55y + 85(40 - x) + 60(50 - y)$$
$$= -15x - 5y + 6400$$

The constraints are

$$x +\quad y \le 45 \quad \text{(Available from A)}$$
$$40 - x + 50 - y \le 70 \quad \text{(Available from B)}$$
$$x \ge 0, y \ge 0, 40 - x \ge 0, 50 - y \ge 0$$

(No negative quantities are delivered.)
   These constraints simplify to

$$x + y \le 45$$
$$x + y \ge 20$$
$$x \ge 0, y \ge 0, x \le 40, y \le 50$$

FIGURE 3–21

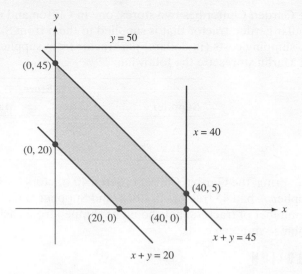

This problem is a linear programming problem that requires that we minimize

$$C = 6400 - 15x - 5y$$

subject to

$$x + y \leq 45$$
$$x + y \geq 20$$
$$x \geq 0, \, y \geq 0, \, x \leq 40, \, y \leq 50$$

Figure 3–21 shows the feasible region and the corner points.

We now find the value of $C = 6400 - 15x - 5y$ at each corner:

Corner	C
(20, 0)	6100
(0, 20)	6300
(0, 45)	6175
(40, 5)	5775
(40, 0)	5800

The minimum cost is $5775 and occurs when $x = 40$, $y = 5$. Thus, Supplier A should ship 40 tractors to Clifton and 5 to Marlin. Supplier B should ship none to Clifton and 45 to Marlin.

■ **Now You Are Ready to Work Exercise 61**

Unusual Linear Programming Situations

Multiple Optimal Solutions.   In the preceding example, just one corner point gave an optimal solution. It is possible for more than one optimal solution to exist. In other cases, there may be no solution at all. First, look at an example with multiple solutions.

EXAMPLE 8 ➤

Find the maximum value of

$$z = 12x + 9y$$

subject to the constraints

$$4x + 3y \leq 36$$
$$8x + 3y \leq 48$$
$$x \geq 0, y \geq 0$$

SOLUTION

Figure 3–22 shows the feasible region with corners $(0, 0)$, $(6, 0)$, $(3, 8)$, and $(0, 12)$. The value of z at each corner point is

At $(0, 0)$: $z = 0$
At $(6, 0)$: $z = 72$
At $(3, 8)$: $z = 108$
At $(0, 12)$: $z = 108$

In this case, the maximum value of $z = 12x + 9y$ occurs at two corners, $(3, 8)$ and $(0, 12)$.

Actually, the value of $z = 12x + 9y$ is 108 for every point on the line segment between the points $(3, 8)$ and $(0, 12)$. For example, the points $(1.5, 10)$ and $(0.75, 11)$ lie on the line and

For $(1.5, 10)$: $z = 12(1.5) + 9(10) = 108$
For $(0.75, 11)$: $z = 12(0.75) + 9(11) = 108$

Generally, when two different points yield a maximum for the objective function, we say the problem has **multiple solutions.** (See Figure 3–22.)

■ **Now You Are Ready to Work Exercise 27**

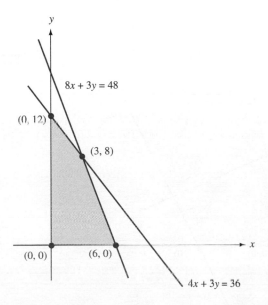

FIGURE 3–22   All points on the line segment between $(0, 12)$ and $(3, 8)$ give optimal solutions.

Multiple Solutions

> When two corners of a feasible region yield maximum solutions to a linear programming problem, then all points on the line segment joining those corners also yield maximum values. Note that the two corners are adjacent.

Multiple solutions allow a number of choices of x and y that yield the same optimal value for the objective functions. For example, if x and y represent production quantities, then management can achieve optimal production with a variety of production levels. If problems should occur with one production line, management might adjust production on the other line and still achieve optimal production. When just one optimal solution exists, management has no flexibility; there is one choice of x and y by which management can achieve its objective.

You might wonder how to tell whether there are multiple solutions. It turns out that this happens when a boundary line has the same slope as the objective function. *(See Exercise 64.)*

Unbounded Feasible Region.   The constraints of a linear programming problem might define an unbounded feasible region for which the objective function has no maximum value. In such a case, the problem has no solution. Here is an illustration.

EXAMPLE 9 ➤ Find the maximum value of the objective function $z = 2x + 5y$ subject to the constraints

$$2x - y \le 16$$
$$-3x + y \le 5$$
$$x \ge 0, y \ge 0$$

SOLUTION

Figure 3–23 shows the feasible region of this system. Observe that the feasible region extends upward indefinitely. Suppose someone claims that the maximum value of the objective function is 50. This is equivalent to stating that

$$2x + 5y = 50$$

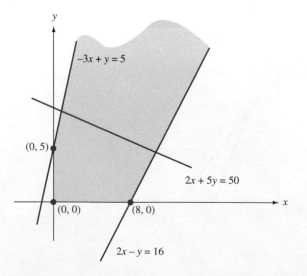

FIGURE 3–23

for points in the feasible region that lie on that line and that no other values of x and y in the feasible region give larger values. You recognize this as the equation of a straight line. Figure 3–23 shows that this line crosses the feasible region. Next, observe that the test point $(0, 0)$ does *not* satisfy the inequality $2x + 5y \geq 50$, so the points in the half plane above the line *must* satisfy it. Thus, every point in the feasible region that lies above the line $2x + 5y = 50$ will give a larger value of the objective function. If we substitute larger values — say, 100, 5000, and so on — instead of 50 in the equation, we essentially determine lines that are parallel to $2x + 5y = 50$ but are farther away from the origin. In each case, points in the feasible region that lie above the line give an even larger value of $2x + 5y$. Because the feasible region extends upward indefinitely, we can never find a largest value.

We point out that the objective function $z = 2x + 5y$ does have a *minimum* value at $(0, 0)$. So, an unbounded feasible region does not rule out an optimal solution. It depends on the region and the kind of optimal solution sought.

■ **Now You Are Ready to Work Exercise 33**

No Optimal Solution Because There Is No Feasible Region. The system of inequalities

$$5x + 7y \geq 35$$
$$3x + 4y \leq 12$$
$$x \geq 0, y \leq 0$$

is a system of inequalities that has no solution. (See Example 4 of Section 3.2.) Whenever the constraints of a linear programming problem do not define a feasible region, there can be no solution.

▦ 3.3 EXERCISES

LEVEL 1

1. *(See Example 1)* Set up the constraints and objective function for the following linear programming problem. A discount store is offered two styles of slightly damaged coffee tables. The store has storage space for 80 tables and 110 hours of labor for repairing the defects. Each table of style A requires 1 hour of labor to repair, and each table of style B requires 2 hours. Style A is priced at $50 each, and style B at $40 each. How many of each style should be ordered to maximize gross sales?

2. The winner of a writing contest is awarded a choice of books. Some of the books are worth $25, and the rest are worth $40. The winner may select as many as ten books, but their total value must not exceed $350. Describe the constraints on each type of book.

For each of the Exercises 3 through 6, find the maximum and minimum value of the objective function given.

3. $z = 6x + 15y$

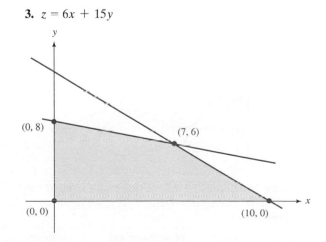

4. $z = 25x + 40y$

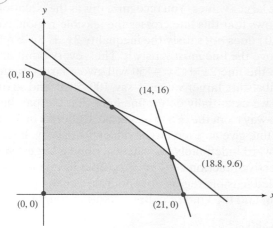

Points shown: (0, 18), (14, 16), (18.8, 9.6), (0, 0), (21, 0)

5. $z = 8x + 24y$

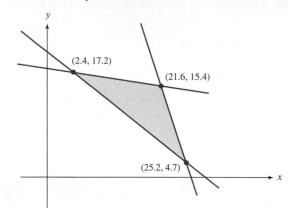

Points shown: (2.4, 17.2), (21.6, 15.4), (25.2, 4.7)

6. $z = 14x + 8y$

Points shown: (0, 10), (0, 4), (6, 3.5), (5, 0), (8, 0)

7. *(See Example 2)* Find the maximum value of the objective function $z = 20x + 12y$, subject to the constraints

$$3x + 2y \leq 18$$
$$3x + y \leq 15$$
$$x \geq 0, y \geq 0$$

8. Maximize $z = 2x + 3y$, subject to

$$3x + y \leq 24$$
$$3x + 2y \leq 2$$
$$x \geq 0, y \geq 0$$

9. Maximize $z = 9x + 2y$, subject to

$$5x + y \leq 35$$
$$3x + y \leq 27$$
$$x \geq 0, y \geq 0$$

10. Maximize $z = 21x + 10y$, subject to

$$x + 3y \leq 24$$
$$7x + 2y \leq 35$$
$$x \geq 0, y \leq 0$$

11. *(See Example 3)* Minimize $z = 2x + 3y$, subject to

$$4x + y \geq 40$$
$$4x + 3y \geq 64$$
$$x \geq 0, y \geq 0$$

12. Minimize $z = 5x + 4y$, subject to

$$x + y \geq 8$$
$$5x + 3y \geq 30$$
$$x \geq 0, y \geq 0$$

13. Maximize $z = 9x + 13y$, subject to

$$10x + 11y \leq 330$$
$$4x + 6y \leq 156$$
$$x \geq 0, y \geq 0$$

14. Minimize $z = 7x + 5y$, subject to

$$9x + 5y \geq 45$$
$$3x + 10y \geq 60$$
$$x \geq 0, y \geq 0$$

15. Maximize $z = 20x + 30y$, subject to

$$-x + 2y \leq 40$$
$$x + 4y \leq 54$$
$$3x + y \leq 63$$
$$x \geq 0, y \geq 0$$

16. Maximize $z = 10x + 15y$, subject to

$$x + 5y \leq 60$$
$$5x + y \leq 60$$
$$x + y \leq 16$$
$$x \geq 0, y \geq 0$$

17. Maximize $z = 320x + 140y$, subject to

$$x + 5y \leq 250$$
$$2x + 5y \leq 300$$
$$x \leq 75$$
$$x \geq 0, y \geq 0$$

18. (a) Find the values of x and y that maximize $z = 5x + 4y$, subject to the constraints

$$\begin{align}
x + 4y &\le 220 \\
x + 2y &\le 120 \\
x + y &\le 100 \\
x \ge 0, y &\ge 0
\end{align}$$

(b) Replace the constraint $x + 4y \le 220$ with $x + 4y \le 230$. (Additional resources became available.) Comment on the effect this has on the optimal solution.

(c) Replace the constraint $x + y \le 100$ with $x + y \le 105$. Comment on the effect this has on the optimal solution.

19. Minimize $z = 5x + 3y$, subject to

$$\begin{align}
6x + y &\ge 52 \\
2x + y &\ge 20 \\
x + 4y &\ge 24 \\
x \ge 0, y &\ge 0
\end{align}$$

20. Minimize $z = 6x + 2y$, subject to

$$\begin{align}
x + 5y &\ge 40 \\
x + y &\ge 16 \\
4x + 2y &\ge 48 \\
x \ge 0, y &\ge 0
\end{align}$$

21. Minimize $z = 5x + 3y$, subject to

$$\begin{align}
3x + y &\ge 30 \\
4x + 3y &\ge 60 \\
x + 2y &\ge 20 \\
x \ge 0, y &\ge 0
\end{align}$$

22. Minimize $z = 2x + 8y$, subject to

$$\begin{align}
x + 3y &\ge 29 \\
x + y &\ge 21 \\
2x + y &\ge 26 \\
x \ge 0, y &\ge 0
\end{align}$$

LEVEL 2

23. *(See Example 4)* Find the maximum and minimum values of $z = 20x + 30y$, subject to

$$\begin{align}
2x + 10y &\le 80 \\
6x + 2y &\le 72 \\
3x + 2y &\ge 6 \\
x \ge 0, y &\ge 0
\end{align}$$

24. Find the maximum and minimum values of $z = 5x + 12y$, subject to

$$\begin{align}
x + y &\le 20 \\
2x + 3y &\ge 30 \\
-x + 2y &\le 10 \\
x \ge 0, y &\ge 0
\end{align}$$

25. Find the maximum and minimum values of $z = 5x + 6y$, subject to

$$\begin{align}
6x + 8y &\le 300 \\
15x + 22y &\ge 330 \\
x \le 40, y &\le 21 \\
x \ge 0, y &\ge 0
\end{align}$$

26. Find the maximum and minimum values of $z = 10x + 8y$, subject to

$$\begin{align}
x + 2y &\le 54 \\
x + y &\ge 28 \\
4x - 3y &\le 84 \\
x \ge 0, y &\ge 0
\end{align}$$

27. *(See Example 8)* Maximize $z = 15x + 9y$, subject to

$$\begin{align}
5x + 3y &\le 30 \\
5x + y &\le 20 \\
x \ge 0, y &\ge 0
\end{align}$$

28. Maximize $z = 6x + 3y$, subject to

$$\begin{align}
2x + y &\le 11 \\
3x + 4y &\le 24 \\
x \ge 0, y &\ge 0
\end{align}$$

29. Minimize $z = 9x + 6y$, subject to

$$\begin{align}
3x + 2y &\ge 60 \\
10x + 3y &< 180 \\
y &\le 24 \\
x \ge 0, y &\ge 0
\end{align}$$

30. Maximize $z = 2x + 6y$, subject to

$$\begin{align}
x + y &\ge 10 \\
x + 3y &\le 72 \\
10x + 3y &\le 180 \\
x \ge 0, y &\ge 5
\end{align}$$

31. Minimize $z = 5x + 10y$, subject to

$$\begin{align}
3x + y &\ge 150 \\
x + 2y &\ge 100 \\
y &\ge 20 \\
x \ge 0, y &\ge 0
\end{align}$$

32. Maximize $z = 8x + 3y$, subject to

$$\begin{align}
2x - 5y &\le 10 \\
-2x + y &\le 2 \\
x \ge 0, y &\ge 0
\end{align}$$

33. *(See Example 9)* Maximize $z = 3x + y$, subject to

$$\begin{align}
2x - 3y &\ge 10 \\
x - 3y &\ge 8 \\
x \ge 0, y &\ge 0
\end{align}$$

34. Maximize $z = 4x + 5y$, subject to

$$-3x + y \le 3$$
$$x - 2y \le 4$$
$$x \ge 0, y \ge 0$$

35. Minimize $z = x + 2y$, subject to

$$-3x + y \le 4$$
$$-2x - y \ge 1$$
$$x \ge 0, y \ge 0$$

36. Maximize $z = 5x + 4y$, subject to

$$-5x + y \ge 4$$
$$-2x + y \le 3$$
$$x \ge 0, y \ge 0$$

37. Maximize $z = 9x + 13y$, subject to

$$3x - y \le 8$$
$$x - 2y \ge 5$$
$$x \ge 0, y \ge 0$$

Set up the objective function and the constraints for Exercises 38 through 44, but do not solve.

38. Jack has a casserole and salad dinner. Each serving of casserole contains 250 calories, 3 milligrams of vitamins, and 9 grams of protein. Each serving of salad contains 30 calories, 6 milligrams of vitamins, and 1 gram of protein. Jack wants to consume at least 23 milligrams of vitamins and 28 grams of protein but keep the calories at a minimum. How many servings of each food should he eat?

39. *(See Example 5)* Wilson Electronics produces a standard VCR and a deluxe VCR. The company has 2200 hours of labor and $18,000 in operating expenses available each week. It takes 8 hours to produce a standard VCR and 9 hours to produce a deluxe VCR. Each standard VCR costs $115, and each deluxe VCR costs $136. The company is required to produce at least 35 standard VCRs. The company makes a profit of $39 for each standard VCR and $26 for each deluxe VCR. How many of each type of VCR should be produced to maximize profit?

40. A company has two skill levels of production workers, I and II. A level I worker is paid $8.60 per hour and produces 15 items per hour. The level II worker is paid $12.25 per hour and produces 22 items per hour. The company is required to use at least 2600 employee hours per week, and it must produce at least 45,000 items per week. How many hours per week should the company use each skill level to minimize labor costs?

41. Home Furnishings has contracted to make at least 250 sofas per week, which are to be shipped to two distributors, A and B. Distributor A has a maximum capacity of 140 sofas, and distributor B has a maximum capacity of 165 sofas. It costs $13 to ship a sofa to A and $11 to ship to B. If A already has 30 sofas and B has 18, how many sofas should be produced and shipped to each distributor to minimize shipping costs?

42. A company makes a single product on two separate production lines, A and B. The company's labor force is equivalent to 1000 hours per week, and it has $3000 outlay weekly for operating costs. It takes 1 hour and 4 hours to produce a single item on lines A and B, respectively. The cost of producing a single item on A is $5 and on B is $4. How many items should be produced on each line to maximize total output?

43. The maximum production of a soft-drink bottling company is 5000 cartons per day. The company produces two kinds of soft drinks, regular and diet. It costs $1.00 to produce each carton of regular and $1.20 to produce each carton of diet. The daily operating budget is $5400. The profit is $0.15 per carton on regular and $0.17 per carton on diet drinks. How much of each type of drink is produced to obtain the maximum profit?

44. Chemical Products makes two insect repellents, Regular and Super. The chemical used for Regular is 15% DEET, and the chemical used for Super is 25% DEET. Each carton of repellent contains 24 ounces of the chemical. In order to justify starting production, the company must produce at least 12,000 cartons of insect repellent, and it must produce at least twice as many cartons of Regular as of Super. Labor costs are $8 per carton for Regular and $6 per carton for Super. How many cartons of each repellent should be produced to minimize labor costs if 59,400 ounces of DEET are available?

Solve Exercises 45 through 63.

45. Lamps Inc. makes desk lamps and floor lamps. The company has 1200 hours of labor and $4200 to purchase materials each week. It takes 0.8 hour of labor to make a desk lamp and 1.0 hour to make a floor lamp. The materials cost $4 for each desk lamp and $3 for each floor lamp. The company makes a profit of $2.65 on each desk lamp and $3.15 on each floor lamp. How many of each should be made each week to maximize profit?

46. A T-shirt company has three machines, I, II, and III, which can be used to produce two types of T-shirts, standard design and custom design. The following table shows the number of minutes required on each machine to produce the designs.

	Machine		
	I	II	III
Standard	1	1	1
Custom	1	4	5

For efficient use of equipment, the company uses machine I at least 240 minutes per day, machine II at least 660 minutes per day, and machine III at least 1000 minutes per day. Each standard T-shirt costs $3, and each custom T-shirt costs $4. Find the number of each type of T-shirt that should be produced to minimize costs.

LEVEL 3

47. In order to finish a term paper, Erin plans to work in her room all day on a Saturday and eat all her meals at a nearby deli. She plans to eat only steak sandwiches and cream of potato soup. The nutritional contents, per serving, are:

	Calories	Fat	Dietary Fiber
Steak Sandwich	400	12 g	9 g
Cream of Potato Soup	200	8.5 g	5 g

Erin wants to restrict her daily diet to no more than 2000 calories, and no more than 65 g of fat. How many servings of each food should she choose to maximize dietary fiber?

48. A sewing machine operator may sew coats or trousers. The trousers require 3 minutes of sewing time, and the operator receives $0.50 each. A coat requires 8 minutes of sewing time, and the operator receives $1.00 each. The operator must sew at least three coats per hour. How many coats and how many trousers should the operator sew each hour to maximize hourly income?

49. *(See Example 6)*   Two foods, I and II, contain the following percentages of carbohydrates, protein, and fat:

Food	Carbohydrates	Protein	Fat
I	40%	50%	6%
II	60%	20%	4%

Food I costs 3¢ per gram, and food II costs 4¢ per gram. Determine the amount of each that should be served to produce at least 10 grams of carbohydrates, 7.5 grams of protein, and 1.2 grams of fat, if cost is to be minimized.

50. The Nut Factory produces a mixture of peanuts and cashews. The company guarantees that at least 40%

of the total weight is cashews. It has a contract to produce 1000 pounds or more of the mixture. The peanuts cost $0.80 per pound, and the cashews cost $1.50 per pound. Find the amount of each kind of nut the company should use to minimize the cost:
(a) If 720 pounds of peanuts are available
(b) If 500 pounds of peanuts are available

51. Precision Machinists makes two grades of gears for industrial machinery, standard and heavy duty. The process requires two steps. Step 1 takes 8 minutes for the standard gear and 10 minutes for the heavy duty. Step 2 takes 3 minutes for the standard gear and 10 minutes for the heavy duty. The company's labor contract requires that it use at least 200 labor-hours per week on the step 1 equipment and 140 labor-hours per week on the step 2 equipment. The materials cost $15 for each standard gear and $22 for each heavy duty. How many of each gear should be made each week to minimize costs?

52. Beauty Products makes two styles of hair dryers, the Petite and the Deluxe. It requires 1 hour of labor to make the Petite and 2 hours of labor to make the Deluxe. The materials cost $4 for each Petite and $3 for each Deluxe. The profit is $5 for the Petite and $6 for the Deluxe. The company has 3950 labor-hours available each week and a materials budget of $9575 per week. How many of each dryer should be made each week to maximize profit?

53. A delivery service is considering the purchase of vans. The model SE costs $16,000 with an expected annual operating–maintenance cost of $2700. The model LE costs $20,000 with an expected annual operating–maintenance cost of $2400. The company needs at least nine vans and has $160,000 available for purchase. How many of each should it purchase to minimize the expected operating–maintenance costs?

54. Solve Exercise 42.

55. *(This exercise is based on Exercise 43)* The maximum production of a soft-drink bottling company is 5000 cartons per day. They produce two kinds of soft drinks, regular and diet. It costs $1.00 to produce each carton of regular and $1.20 for each carton of diet. The daily operating budget is $5400. The profit is $0.15 per carton on regular and $0.17 per carton on diet drinks.

(a) How much of each type of drink is produced to obtain the maximum profit?

(b) Changes in the profit per carton may or may not change the optimal solution. In the following, the constraints remain the same, but a profit changes. Comment on the change, if any, in the optimal solution in each case.

(i) The profit on the regular drinks increases to $0.20 per carton, and the profit on the diet drinks remains $0.17.

(ii) The profit on the diet drinks increases to $0.19, and the profit on the regular remains $0.15.

(iii) The profit on the diet drinks decreases to $0.16, and the profit on the regular drinks remains $0.15.

56. Solve Exercise 44.

57. The design of a building calls for at least 4000 square feet of exterior glass. Two types of glass are available. The ability of glass to transfer heat from the warm side to the cooler side is called *conductance*. It is measured in the number of BTUs transferred each hour per square foot and per degree of difference between inside and outside temperature. The glass that conducts the least is the most desirable. The contract allows up to $4500 to be spent on glass. The following holds for each type of glass:

	Type A	**Type B**
Conductance (BTU per sq ft)	1	0.25
Cost (per sq ft)	$0.80	$1.20

(a) Find the number of square feet of each type of glass that will minimize conductance.

(b) How many square feet of each type of glass should be used if the total conductance can be no more than 2200 BTU per hour and the cost is to be minimized?

58. Omega Saw and Tool uses two kinds of machines, A and B, in a manufacturing process. The company rates the machines by operating costs per hour and productivity in units produced per hour. It has space for at most 22 machines. The machines are rated as follows:

	A	**B**
Operating cost (per hour)	$16	$24
Productivity (units per hour)	8	10

(a) If operating costs cannot exceed $600 per hour, find the number of each that should be used to maximize hourly productivity.

(b) If productivity must be at least 200 units per hour, find the number of each machine that will minimize hourly costs.

59. Brent Publishing has two plants. The plant on Glen Echo Lane produces 200 paperback books and 300 hardback books per day, and the plant on Speegleville Road produces 300 paperback books and 200 hardback books per day. An order is received for 2400 paperbacks and 2100 hardbacks. Find the number of days each plant should operate so that the combined number of days in operation is a minimum.

60. A syrup manufacturer blends 14 gallons of maple syrup and 18 gallons of corn syrup to make two new syrups, Maple Flavored and Taste of Maple. Each gallon of Maple Flavored contains 0.3 gallon of corn syrup and 0.4 gallon of maple syrup. Each gallon of Taste of Maple contains 0.4 gallon of corn syrup and 0.2 gallon of maple syrup. The company makes a profit of $4.50 per gallon on Maple Flavored and $3.00 per gallon on Taste of Maple. How many gallons of each should be produced to maximize profits?

61. *(See Example 7)* Anticipating a successful response to a special sale, Computer Headquarters orders 165 of their most popular model for their Raleigh store and 190 for their Greensboro store from two suppliers, A and B. Supplier A has 200 of the model in stock and supplier B has 230. The shipping costs from the suppliers to the stores is the following:

Shipping Costs per Computer

Supplier	Store	
	Raleigh	Greensboro
A	$35	$30
B	$45	$50

Find the number that should be shipped from each supplier to each store in order to minimize shipping costs

62. The Motorcycle Shop has a shop in Edmond and one in Altus. They order bikes from two suppliers, A and B. The shipping costs from each supplier to each shop is as follows:

Shipping Costs per Bike

Supplier	Shop	
	Edmond	Altus
A	$40	$20
B	$30	$35

One month, Edmond orders 65 bikes and Altus orders 70. Supplier A has 75 bikes in stock and supplier B has 90 bikes. Find the number of bikes supplied from each supplier to each store that minimizes shipping costs.

63. The Furniture Mart has two suppliers, A and B, which supply their Emporia and Ardmore stores with lamps. One month, they ordered 25 lamps for the Emporia store and 20 for the Ardmore store. Supplier A had 10 in stock, and Supplier B had 40 in stock. The shipping charges from each supplier to each store are the following:

Shipping Costs per Lamp

Supplier	Store	
	Emporia	Ardmore
A	$5	$6
B	$7	$4

Find the number of lamps that should be shipped from each supplier to each store in order to minimize shipping costs.

EXPLORATIONS

64. The general form of the objective function is $z = Ax + By$. For a given value of z, the resulting line has a slope of $-A/B$. Show that each of the following linear programming problems has multiple optimal solutions. Verify in each case that the objective function has the same slope as one of the boundary lines. (The constraint $Cx + Dy \leq E$ has the boundary $Cx + Dy = E$, and its slope is $-C/D$.)

(a) Maximize $z = 10x + 4y$, subject to

$$5x + 2y \leq 50$$
$$x + 4y \leq 28$$
$$x \geq 0, y \geq 0$$

(b) Maximize $z = 5x + 6y$, subject to

$$5x + 12y \leq 300$$
$$10x + 12y \leq 360$$
$$10x + 6y \leq 300$$
$$x \geq 0, y \geq 0$$

(c) Minimize $z = 10x + 15y$, subject to

$$3x + 2y \geq 50$$
$$2x + 3y \geq 60$$

$$x + 4y \geq 40$$
$$x \geq 0, y \geq 0$$

(d) Maximize $z = 10x + 24y$, subject to

$$5x + 12y \leq 1200$$
$$5x + 4y \leq 600$$
$$x \geq 0, y \geq 0$$

65. The following figure shows the graph of a linear programming problem with two constraints that are to maximize profit. The profit line shown indicates the slope of the objective function. The corners do not occur at points with integer coordinates, so the optimal solution gives fractional values of x and y.

In some applications, x and y must have integer values. For example, in a problem where x represents the number of model A cars produced and y represents the number of model B cars produced, only integer values of x and y make sense. For the graph shown, find the integer values of x and y that yield maximum profit. Explain why you made that choice.

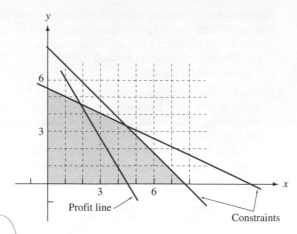

Profit line

Constraints

66. The following graph shows a feasible region and a line determined by the objective function. At which corner will the objective function be maximum?

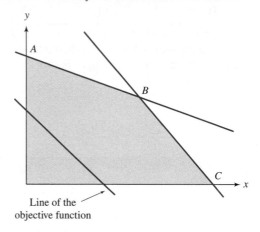

Line of the objective function

67. The following graph shows a feasible region and a line determined by the objective function. At which corner will the objective function be maximum?

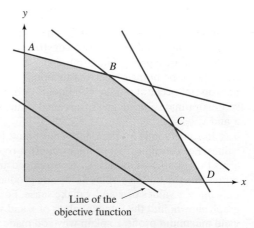

Line of the objective function

68. The following graph shows a feasible region and a line determined by the objective function.

At which corner will the objective function be maximum?

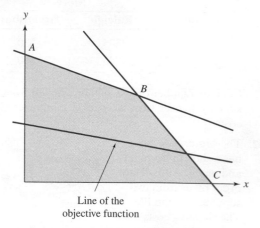

Line of the objective function

69. The following graph shows a feasible region and a line determined by the objective function. At which corner will the objective function be maximum?

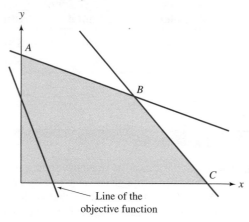

Line of the objective function

70. The following graph shows a feasible region and a line determined by the objective function. Which integer values of x and y give a maximum value of the objective function?

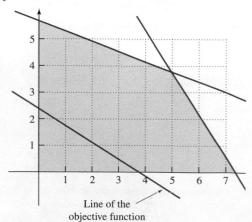

Line of the objective function

71. A maximization linear programming problem has three constraints L_1, L_2, and L_3, as shown in the following figure. Line M shows the direction of the objective function.

(a) At which corner point does the optimal solution occur?

(b) Additional resources become available such that the constraint L_1 is moved to the position L_1'. Explain why this does not change the optimal solution.

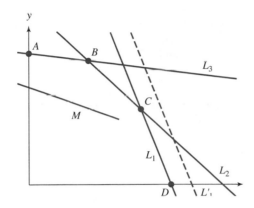

72. In the following figure, look at the feasible region defined by the constraints

$$-x + y \le 2$$
$$-x + 2y \ge 2$$
$$y \ge 3$$

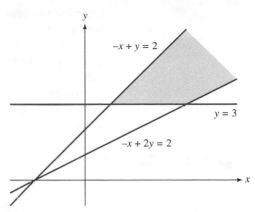

If they exist, find the locations and values of the maximum and minimum values of the following objective functions:

(a) $z = 2x - y$ (b) $z = -2x + y$
(c) $z = 2x + y$

Use a graphing calculator to solve (to two decimals) Exercises 73 through 75.

73. Maximize $z = 5.2x + 7.3y$, subject to

$$3.22x + y \le 12.45$$
$$1.15x + y \le 8.32$$
$$x \ge 0, y \le 0$$

74. Minimize $z = 6.3x + 4.4y$, subject to

$$0.7x + y \ge 8.6$$
$$2.3x + y \ge 12.4$$
$$x \ge 0, y \ge 0$$

75. Maximize $z = 2.2x + 4.8y$, subject to

$$2.2x + y \le 6.6$$
$$3.5x + y \le 11.5$$
$$x \ge 0, y \ge 0$$

3.4   Applications

The geometric method of solving linear programming problems works reasonably well when there are only two variables and a few constraints. In general, linear programming problems will likely require dozens of variables, not just two. In those cases, the geometric method is not practical. We address the method of solving problems with more than two variables in Chapter 4. The method applies to problems with a few or a very large number of variables and constraints. The procedure has the advantage of being a well-defined routine, so problems can be solved with pencil and paper or with computer programs.

Do not be misled by the fact that a solution can be obtained by a routine. A correct analysis and description of the problem must occur before applying the method. You may expect an erroneous constraint to yield an erroneous solution.

Defining the problem and identifying the constraints requires the work of a human mind.

In this section, you will concentrate on analyzing and setting up linear programming problems with more than two variables. When you have learned to correctly set up a problem, the methods of Chapter 4 can be used in a meaningful way.

To prepare you for solving larger systems, we now look at some examples of linear programming problems with more than two variables. We remind you that we do not yet have procedures to solve these problems.

EXAMPLE 1 ➤

Cox Department Store's
Huge 3-Day
Going-Out-of-Business
SALE
Everything must GO!
Open 24 hrs daily for
these 3 days only...
then we're gone for good!

Cox's Department Store plans a major advertising campaign for its going out of business sale. They plan to advertise in the newspaper and on radio and TV. To provide a balance among the three types of media, they will place no more than 10 ads in newspaper, at least 15% of the ads will be on TV, and no more than 40% will be on radio. The budget for this campaign is $30,000. These are the costs and expected audience exposure for the three types of media:

	Radio	TV	Newspaper
Cost per ad	$250	$2200	$700
Audience per ad	15,000	95,000	45,000

How many ads should be run in each type of media to maximize audience exposure? Find the constraints and objective function that describe this problem.

SOLUTION

The first task in setting up a problem is to identify the variables. The term *variable* suggests something that can vary. In this case, which items are fixed, and which can assume different values? The cost per ad is fixed, as well as the audience per each type of ad. The number of ads, and consequently the total audience exposure, can vary. In fact, the question "How many ads . . . ?" suggests that we want variables to represent the number of ads. Because total audience exposure can also vary, we would expect that to be a variable. However, the number of ads determines total audience exposure. Because exposure depends on ads, we call total audience exposure a **dependent variable.** We call the number of ads in this case **independent variables.**

The first step in setting up a problem is to determine the independent variables — in this case, three variables:

$$x_1 = \text{number of ads on the radio}$$
$$x_2 = \text{number of ads on TV}$$
$$x_3 = \text{number of ads in the newspaper}$$

The dependent variable, audience exposure, is to be maximized, so we let $z =$ total audience exposure and write it in terms of the number of ads to obtain the objective function:

$$z = 15,000x_1 + 95,000x_2 + 45,000x_3$$

The constraints are inequalities that describe the restrictions imposed. One restriction requires that total cost must not exceed $30,000, so we have the cost constraint

$$250x_1 + 2200x_2 + 700x_3 \le 30,000$$

Other constraints are

$x_3 \leq 10$ (No more than 10 ads are in the paper)
$x_2 \geq 0.15(x_1 + x_2 + x_3)$ (At least 15% of the ads are on TV)
$x_1 \leq 0.40 (x_1 + x_2 + x_3)$ (No more than 40% of the ads are on radio)

No variables are negative because we could not place a negative number of ads. The last two constraints can be written

$$0.15x_1 - 0.85x_2 + 0.15x_3 \leq 0$$
$$-0.60x_1 + 0.40x_2 + 0.40x_3 \geq 0$$

As a linear programming problem, we can state it in the following way:

Maximize $z = 15{,}000x_1 + 95{,}000x_2 + 45{,}000x_3$, subject to

$$0.15x_1 - 0.85x_2 + 0.15x_3 \leq 0$$
$$-0.60x_1 + 0.40x_2 + 0.40x_3 \geq 0$$
$$x_1 \geq 0, x_2 \geq 0, 10 \geq x_3 \geq 0$$

■ **Now You Are Ready to Work Exercise 3**

The next example illustrates how some scheduling problems can be modeled as a linear programming problem.

EXAMPLE 2 ➤   Adventure Time offers one-week summer vacations in the Rocky Mountains during the month of August. The package includes round-trip transportation and a week's accommodations at Rocky Mountain Lodge. The Lodge gives a discount to Adventure Time if they rent two- or three-week blocks of condos, giving a rent of $1000 per condo for a two-week period and $1300 per condo for a three-week period.
    Adventure Time expects to need the following number of condos:

	Number of Condos Needed
First week	30
Second week	42
Third week	21
Fourth week	32

How many condos should Adventure Time rent for two weeks, and how many should be rented for three weeks, to meet the needed number and to minimize Adventure Time's rental costs? Set this up as a linear programming problem.

SOLUTION

First, we need to determine the possible ways to schedule two- and three-week blocks in August.
    A glance at a "calendar" (Figure 3–24) shows three possible two-week periods and two possible three-week periods.
    We label them x_1, x_2, x_3, x_4, and x_5 as shown. The x_i's will also represent the number of condos rented during that period.

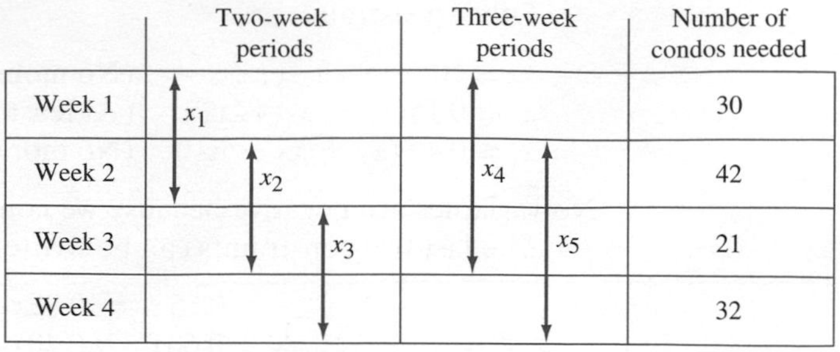

	Two-week periods	Three-week periods	Number of condos needed
Week 1	x_1		30
Week 2	x_2	x_4	42
Week 3	x_3	x_5	21
Week 4			32

FIGURE 3–24   Possible ways to schedule two-week and three-week blocks in August.

Notice that the 30 condos needed for the first week will come from the x_1 and x_4 group, and the 42 needed for the second week will come from the x_1, x_2, x_4, and x_5 groups. We want to minimize rent, that is,

$$\text{Minimize} \quad z = 1000x_1 + 1000x_2 + 1000x_3 + 1300x_4 + 1300x_5$$

We can summarize the relationship between the number of condos needed and the five time periods as follows.

$$
\begin{aligned}
x_1 + \qquad\qquad x_4 \qquad &\geq 30 \\
x_1 + x_2 + \qquad x_4 + x_5 &\geq 42 \\
x_2 + x_3 + x_4 + x_5 &\geq 21 \\
x_3 + \qquad x_5 &\geq 32 \\
\end{aligned}
$$
$$x_1 \geq 0, x_2 \geq 0, x_3 \geq 0, x_4 \geq 0, x_5 \geq 0$$

■ **Now You Are Ready to Work Exercise 6**

We now turn to an example of a transportation problem.

EXAMPLE 3 ➤ Spokes and Things has two bicycle stores, one in Calvert and one in Hico. They sell a popular model of a mountain bike that they obtain from two suppliers, A and B. In preparation for the upcoming biking season, they order a minimum of 80 bikes for the Calvert store and a minimum of 65 bikes for the Hico store. At that time, Supplier A has 62 bikes, and Supplier B has 90 bikes. The shipping cost from supplier to store is given in this chart:

Shipping Cost per Bike

	Store	
Supplier	**Calvert**	**Hico**
A	$ 8	$6
B	$10	$7

Find the number each supplier ships to each store to minimize shipping costs.

SOLUTION

We summarize the given information in the following table:

	Store		No. Available
	Calvert	**Hico**	
Supplier A			62
Cost (each)	$8	$6	
Number supplied	x_1	x_2	
Supplier B			90
Cost (each)	$10	$7	
Number	x_3	x_4	
Number needed	80 or more	65 or more	

Let x_1 = number shipped from Supplier A to Calvert

x_2 = number shipped from Supplier A to Hico

x_3 = number shipped from Supplier B to Calvert

x_4 = number shipped from Supplier B to Hico

Notice that we cannot use $80 - x_1$ for the number shipped from supplier B to Calvert (as we did in Example 7 of Section 3.3) because the exact number shipped to Calvert is not specified, it is 80 *or more*.

We now state the linear programming problem.

Minimize $z = 8x_1 + 6x_2 + 10x_3 + 7x_4$, subject to

$$x_1 + x_2 \leq 62 \text{ (Available from A)}$$
$$x_3 + x_4 \leq 90 \text{ (Available from B)}$$
$$x_1 + x_3 \geq 80 \text{ (Needed at Calvert)}$$
$$x_2 + x_4 \geq 65 \text{ (Needed at Hico)}$$
$$x_1 \geq 0, x_2 \geq 0, x_3 > 0, x_4 > 0 \quad \text{(No negative amounts are ordered)}$$

■ **Now You Are Ready to Work Exercise 9**

3.4 EXERCISES

Set up the linear programming problems in this exercise set. Do not attempt to solve them.

1. A vegetable farmer has 75 acres of land and grows onions, carrots, and lettuce. It costs $250 per acre to produce onions, and the profit is $65 per acre. It costs $300 per acre to grow carrots, and the profit is $70 per acre. It costs $325 per acre to produce lettuce, and the profit is $50 per acre. Production costs cannot exceed $225,000. Find the number of acres of each vegetable to maximize profit.

2. The three most popular models at the Computer Store are the M-140, M-180, and P-210. The M-140 costs the store $940 and yields a profit of $330. The M-180 costs $1120 and yields a profit of $290. The P-120 costs $1280 and yields a profit of $460. The Computer Store projects maximum monthly sales of the computers to be a total of 175. The store does not want the inventory, based on their costs, to exceed $195,000. Find the number of each model computer that should be purchased to maximize profit.

3. *(See Example 1)*   An import company obtains a license to import up to 200 pieces of hand-carved furniture. The importer may import chests, desks, and silverware boxes. The cost, volume, and profit are the following:

	Chest	Desk	Silverware Box
Cost	$270	$310	$90
Profit	$180	$300	$45
Volume (cubic feet)	7	18	1.5

The import company may purchase up to $5000 worth of furniture, and the furniture must all fit into a 1500-cubic-foot shipping container. Find the number of each type of furniture that should be ordered so that profit is maximum.

4. The Humidor blends regular coffee, High Mountain coffee, and chocolate to obtain four kinds of coffee: Early Riser, Coffee Time, After Dinner, and Deluxe. The blends and profit for each blend are given in the following chart:

	Blend			
	Early Riser	Coffee Time	After Dinner	Deluxe
Regular	80%	75%	75%	50%
High Mountain	20%	23%	20%	40%
Chocolate	0%	2%	5%	10%
Profit/pound	$1.00	$1.10	$1.15	$1.20

The shop has 260 pounds of regular coffee, 90 pounds of High Mountain coffee, and 20 pounds of chocolate. How many pounds of each blend should be produced to maximize profit?

5. The Health Fare Cereal Company makes four cereals using wheat, oats, raisins, and nuts. The portions and profit for each cereal are shown in the following chart:

	Portion of Each Pound of Cereal				
Cereal	Wheat	Oats	Raisins	Nuts	Profit/ Pound
Lite	0.75	0.20	0.05	0	$0.25
Trim	0.50	0.25	0.20	0.05	$0.25
Regular	0.80	0.20	0	0	$0.27
Health Fare	0.15	0.50	0.25	0.10	$0.32

The company has 2400 pounds of wheat, 1400 pounds of oats, 700 pounds of raisins, and 250 pounds of nuts available. How many pounds of each cereal should be produced to maximize profit?

6. *(See Example 2)*   The Modern Language Association plans a three-day meeting. The program committee requests 15 overhead projectors for the first day, 28 for the second day, and 19 for the third day. Projectors may be rented for a single day for $17, and they may be rented for two consecutive days for $30. How many should be rented for a single day and how many should be rented for two consecutive days to minimize rental costs?

7. The American Mathematics Teachers Association holds a four-day conference on technology in the classroom. The organizers plan to rent computers to be placed in the location of each session. They need the following number of computers:

	No. of Computers
First day	12
Second day	15
Third day	18
Fourth day	20

They can rent computers for two consecutive days for $80 each or for three consecutive days for $100 each. Find the number that should be rented for each two-day and for each three-day period so that rental cost is a minimum.

8. Citywide Ministries plans a five-day ski trip for children. The children may ski from one to five of the days, but the number scheduled to ski each day is the following:

Day	No. of Skiers
Monday	15
Tuesday	22
Wednesday	19
Thursday	24
Friday	31

Citywide Ministries can obtain two-day lift tickets for $36 each or three-day lift tickets for $47. Find the number of tickets they should purchase for each two-day period and each three-day period so that each child can ski on the day scheduled and that minimizes the cost of lift tickets.

9. *(See Example 3)*   The Hawaiian Pineapple Company has three pineapple plantations and two processing plants. The monthly production of pineapple generally exceeds 250, 275, and 310 tons at plantations P-1, P-2, and P-3. The cost of shipping pineapple, per ton, from each plantation to the processing plants is given by this chart:

Cost of Shipping per Ton

Plant	Plantation		
	P-1	**P-2**	**P-3**
A	$50	$65	$58
B	$40	$55	$69

It is desired that plant A process at least 540 tons per month and plant B process at least 450 tons per month. Find the number of tons of pineapple that should be shipped from each plantation to each processing plant to minimize shipping costs.

10. A freight company has two sizes of trucks for hauling freight. Type A has a capacity of 8 tons, and type B has a capacity for 11 tons. An order is received to haul 230 tons of freight to Columbus and 145 tons to Anderson. The cost of making the trip is the following:

Hauling Cost per Trip

Truck	Columbus	Anderson
A	$600	$450
B	$720	$525

Find the number of trips each type truck must make to each city to fill the order and to minimize the hauling costs.

11. Alabama Mining must produce at least 1800 tons of low-grade iron ore and 1350 tons of high-grade iron ore weekly. The company operates three mines. The daily production and daily operating costs are summarized in this chart:

Mine	Low Grade (Tons)	High Grade (Tons)	Cost
Red Mountain	105	90	$8,000
Cahaba	295	200	$14,000
Clear Creek	270	85	$12,000

How many days per week should each mine operate to meet the production goals at the minimum cost?

12. At State University, the finite mathematics classes are taught by graduate students, lecturers, and pro-fessors. Departmental policy limits the size of classes to 25 for graduate students, 35 for lecturers, and 40 for professors. The teaching cost per class is $4000 for graduate students, $5000 for lecturers, and $10,000 for professors. The department also requires that the lecturers teach at least 60% more classes than the graduate students and that the professors teach at least as many classes as the total for graduate students and lecturers. If the total finite mathematics enrollment is 1150 students, find the number of sections taught by graduate students, lecturers, and professors to minimize teaching costs.

13. The Spring Valley PTA plans a day at the amusement park for the elementary school. The rides at the park are classified as A rides and more expensive B rides. The PTA can obtain discounts by purchasing books of tickets. The MiniPacket contains eight A tickets and two B tickets and costs $19. The MidPacket contains seven A tickets and seven B tickets and costs $30. The MaxiPacket contains six A tickets and 14 B tickets and costs $45.

The PTA needs to purchase a total of 260 A tickets and 175 B tickets. Find the number of each packet that should be purchased to minimize costs.

14. George eats a meal of steak, potato with butter, salad with dressing, and bread. One ounce of each contains the indicated calories, protein, and fat.

	Calories	Protein (g)	Fat (g)
Salad	20	0.5	1.5
Potato	50	1.0	3.0
Bread	25	1.0	2.0
Steak	56	9.0	5.0

He is to consume no more than 1800 calories and 30 g of fat. How much of each should he eat to maximize the protein consumed?

15. Mike is training for a marathon, so he wants his diet to be heavy on carbohydrates. He wants to take in a maximum of 40 grams of fat per day and a maximum of 35 grams of protein each day. The number

of grams per serving in the foods he eats are the following:

	Milk	Vegetables	Fruit	Bread	Meat
Carbo-hydrates	12	7	11	15	0
Protein	8	2	0	2	7
Fat	10	0	0	1	5

Find the number of daily servings of each food that will maximize the carbohydrate intake.

16. Ralph and Jean are proud new grandparents. They immediately set up an investment fund to help provide for their grandchild's college education. They invest in stocks, treasury bonds, municipal bonds, and corporate bonds. To balance risks, they want no more than 40% of their investment to be in stocks, at least 15% in treasury bonds, no more than 30% in municipal bonds, and no more than 25% in corporate bonds. They have $10,000 to invest.

They expect stocks to return 10% on their investment; treasury bonds, 6%; municipal bonds, 7%; and corporate bonds, 8%. How much should they invest in each type of investment to maximize return on their investment?

17. A bond mutual fund invests only in bonds rated A, AA, or AAA. The fund's policy requires that no more than 65% of the total investment be in A and AA bonds and at least 50% of the total investment be in AA and AAA bonds. The expected return on the bonds is 7.2% for the A bonds, 6.8% for the AA bonds, and 6.5% for the AAA bonds. Find the percentage of the investment in each category that maximizes returns.

18. Moses Plumbing Supply stocks pipe in 32-foot lengths. From these, they cut pipes of lengths ordered by plumbers. A plumber orders 75 pipes, 17 feet long; 82 pipes, 12 feet long; and 47 pipes, 9 feet long. There are five cutting patterns that can be used to cut pipes of the specified lengths:

Pattern	Number of Pipes			
	17 ft.	12 ft.	9 ft.	Waste
1	1	1	0	3
2	0	1	2	2
3	0	2	0	8
4	0	0	3	5
5	1	0	1	6

Find the number of 32-foot pipes cut into each pattern so the order will be filled and waste is minimized.

19. A cabinet company makes cabinet doors from plywood sheets that measure 48 inches by 96 inches. The doors are 32 inches high and 15, 16, or 18 inches wide. For example, by cutting a piece 15 inches wide, the length of the sheet, three doors 15 inches wide are obtained. The remainder of the sheet can be used to cut doors 16 or 18 inches wide with a strip of waste. The cabinet company has seven cutting patterns that can be used to obtain doors of desired width:

Number of Doors Obtained per Sheet

Cutting Plan	15 in.	16 in.	18 in.	Waste
1	0	9	0	0
2	3	6	0	1 in.
3	9	0	0	3 in.
4	6	3	0	2 in.
5	0	0	6	12 in.
6	6	0	3	0
7	0	3	3	14 in.

The cabinet shop receives an order for 165 of the 15-inch doors, 200 of the 16-inch doors, and 85 of the 18-inch doors. Find the number of sheets of each cutting plan that are needed to fill the order and that will minimize the waste.

20. Cedric is on a Subway diet, and he is planning his breakfast and lunch menu. His plan has the following limits for the breakfast–lunch combination.

Calories: No more than 1200 calories.

Fat: No more than 45 grams.

Cholesterol: No more than 220 mg.

Dietary Fiber: At least 15 grams.

He limits his choices to the Classic Tuna, the Under 6 Roast Beef, the Under 6 Turkey, and the Western Egg sandwiches. The nutrition information per serving for those sandwiches is

	Calories	Fat	Choles-terol	Fiber	Carbo-hydrate
Tuna	450	22	40	5	4
Roast Beef	290	5	20	6	5
Turkey	280	4.5	20	4	6
Egg	300	12	180	3	36

Cedric wants to know the number of servings of each sandwich that satisfies the limits and maximizes carbohydrates. Set up the constraints and objective function.

IMPORTANT TERMS

3.1

Linear Inequality
Half Plane
Graph of a Linear Inequality

3.2

System of Linear Inequalities
Solutions to a System of Linear
    Inequalities
Graph of a System of Linear
    Inequalities

Feasible Solution
Feasible Region
Bounded Feasible Region
Unbounded Feasible Region
Boundary of a Feasible Region
Corners of a Feasible Region
No Feasible Solution

3.3

Constraints
Nonnegative Conditions

Objective Function
Maximize Objective Function
Minimize Objective Function
Optimal Solution
Multiple Optimal Solutions
Bounded Feasible Region
Unbounded Feasible Region

3.4

Dependent Variable
Independent Variable

REVIEW EXERCISES

1. Graph the solution to the following inequalities.
   (a) $5x + 7y < 70$ (b) $2x - 3y > 18$
   (c) $x + 9y \leq 21$ (d) $-2x + 12y \geq 26$
   (e) $y \geq -6$ (f) $x \leq 3$

2. Graph the following systems of inequalities.
   (a) $2x + y \leq 4$ (b) $x + y \leq 5$
       $x + 3y < 9$ $x - y > 3$
                           $x \geq 1, y \leq 3$

**Find the feasible region and corner points of the systems
in Exercises 3 through 7.**

3. $x - 3y \geq 6$
   $x - y \leq 4$
   $y \geq -5$

4. $5x + 2y \leq 50$
   $x + 4y \leq 28$
   $x \geq 0$

5. $-3x + 4y \leq 20$
   $x - y \geq -2$
   $8x + y \leq 40$
   $y \geq 0$

6. $3x + 10y \leq 150$
   $2x + y \leq 32$
   $x \leq 14$
   $x \geq 0, y \geq 0$

7. $x - 2y \leq 0$
   $-2x + y \leq 2$
   $x \leq 2, y \leq 2$

8. Maximize $z = x + 2y$, subject to
$$x + y \leq 8$$
$$x \leq 5$$
$$x \geq 0, y \geq 0$$

9. Maximize $z = 5x + 4y$, subject to
$$3x + 2y \leq 12$$
$$x + y \leq 5$$
$$x \geq 0, y \geq 0$$

10. Find the maximum and minimum values of
$z = 2x + 5y$, subject to
$$2x + y \geq 9$$
$$4x + 3y \geq 23$$
$$x \geq 0, y \geq 0$$

11. (a) Find the minimum value of $z = 5x + 4y$, sub-
    ject to
$$3x + 2y \geq 18$$
$$x + 2y \geq 10$$
$$5x + 6y \geq 46$$
$$x \geq 0, y \geq 0$$

    (b) Find the minimum value of
$$z = 10x + 12y$$
    subject to the constraints of part (a).

12. Maximize $z = x + 5y$, subject to
$$x + y \leq 10$$
$$2x + y \geq 10$$
$$x + 2y \geq 10$$
$$x > 0, y > 0$$

13. Maximize $z = 4x + 7y$, subject to
$$2x + y \leq 90$$
$$x + 2y \leq 80$$
$$x + y \leq 50$$
$$x \geq 0, y \geq 0$$

14. Minimize $z = 7x + 3y$, subject to
$$x + 2y \geq 16$$
$$3x + 2y \geq 32$$
$$5x + 2y \geq 40$$
$$x \geq 0, y \geq 0$$

15. An assembly plant has two production lines. Line A can produce 65 items per hour, and line B can produce 105 per hour. The loading dock can ship a maximum of 700 items per 8-hour day.
(a) Express the information with an inequality.
(b) Graph the inequality.

16. A building supplies truck has a load capacity of 25,000 pounds. A delivery requires at least 21 pallets of brick weighing 950 pounds each and at least 15 pallets of roofing material weighing 700 pounds each. Express these restrictions with a system of inequalities.

17. The Ivy Twin Theater has a seating capacity of 275 seats. Adult tickets sell for $4.50 each, and children's tickets sell for $3.00 each. The theater must have ticket sales of at least $1100 to break even for the night. Write these constraints as a system of inequalities.

18. A tailor makes suits and dresses. A suit requires 1 yard of polyester and 4 yards of wool. Each dress requires 2 yards of polyester and 2 yards of wool. The tailor has a supply of 80 yards of polyester and 150 yards of wool. What restrictions does this place on the number of suits and dresses she can make?

19. The Hoover Steel Mill produces two grades of stainless steel, which is sold in 100-pound bars. The standard grade is 90% steel and 10% chromium by weight, and the premium grade is 80% steel and 20% chromium. The company has 80,000 pounds of steel and 12,000 pounds of chromium on hand. If the price per bar is $90 for the standard grade and $100 for the premium grade, how much of each grade should it produce to maximize revenue?

20. The Nut Factory produces a mixture of peanuts and cashews. It guarantees that at least one third of the total weight is cashews. A retailer wants 1200 pounds or more of the mixture. The peanuts cost the Nut Factory $0.75 per pound, and the cashews cost $1.40 per pound. Find the amount of each kind of nut the company should use to minimize the cost:
(a) if 600 pounds of peanuts are available.
(b) if 900 pounds of peanuts are available.

21. A massive inoculation program is initiated in an area devastated by floods. Doctors and nurses form teams of the following sizes: A-teams are composed of one doctor and three nurses, B-teams are composed of one doctor and two nurses, C-teams are composed of one doctor and one nurse. An A-team can inoculate an estimated 175 people per hour, a B-team can inoculate an estimated 110 people per hour, and a C-team can inoculate an estimated 85 people per hour. There are 75 doctors and 200 nurses available. How many teams of each type should be formed to maximize the number of inoculations per hour? Set up the constraints and objective function. Do not solve.

22. A school cafeteria serves three foods for lunch: A, B, and C. There is pressure on the cafeteria director to reduce lunch costs. Help the director by finding the quantities of each food that will minimize costs and still maintain the desired nutritional level. The three foods have the following nutritional characteristics

| | Foods | | |
Per Unit	A	B	C
Protein (g)	15	10	23
Carbohydrates (g)	20	30	11
Calories	500	400	200
Cost ($)	1.40	1.65	1.95

A lunch must contain at least 80 grams of protein, 95 grams of carbohydrates, and 1500 calories. How many units of each food should be served to minimize cost? Set up the problem. Do not solve.

23. Southside Landscapes proposes to plant the flower beds for the historic Earle House gardens. The company uses four patterns. Pattern I uses 40 tulips, 25 daffodils, and 6 boxwood. Pattern II uses 25 tulips, 50 daffodils, and 4 boxwood. Pattern III uses 30 tulips, 40 daffodils, and 8 boxwood. Pattern IV uses 45 tulips, 45 daffodils, and 2 boxwood. The profit for each pattern is $48 for Pattern I, $45 for Pattern II, $55 for Pattern III, and $65 for Pattern IV. Southside Landscape has 1250 tulips, 1600 daffodils, and 195 boxwood available. How many of each pattern should be used to maximize profit? Set up the problem. Do not solve.

CHAPTER 4

Linear Programming: The Simplex Method

Chapter 3 introduced you to the basic ideas of linear programming. Perhaps you noticed that most of the examples and problems involved two variables. In practice, linear programming problems involve dozens of variables. The graphical method is not practical in problems having more than two variables. Fortunately, we have a procedure for solving linear programming problems involving several variables, thanks to the mathematician George B. Dantzig. Dantzig and his colleagues observed and defined the class of linear programming problems and originally had no efficient method for solving them. This changed in the mid-1940s when Dantzig invented a procedure that became the workhorse in solving linear programming problems. He called it the **simplex method,** a procedure for examining corners of a feasible region in an intelligent manner that speeds the process of finding the optimal solution.

Basically, the simplex method moves along a boundary from one vertex to another, seeking the vertex that gives the optimal solution. When the problem involves many variables, many boundaries may leave from the current vertex, and the best route may be difficult to ascertain. Many mathematicians, including Dantzig, thought there should be a better search technique. A great deal of research focused on developing a better algorithm. Then, in 1984, a breakthrough came when 28-year-old Narendra Karmarkar announced an improved method for large-scale problems (those involving thousands of variables). His method is considerably more complicated than the simplex method, so we study only the simplex method. ∎

4.1   Setting Up the Simplex Method

- Standard Maximum
- The Geometric Form of the Problem
- Slack Variables
- Simplex Tableau

We now study an algebraic technique that applies to any number of variables and enables us to solve larger linear programming problems. It has the added advantage of being well suited to a computer, thereby making it possible to avoid tedious pencil-and-paper solutions. This technique is called the **simplex method.**

In practice, a linear programming application of any consequence does not lend itself to a pencil-and-paper solution. Thus, you will use a computer if you are asked to solve a real-world problem. You may wonder why the exercises in this chapter are mainly pencil-and-paper exercises. The reason is that a correct solution to an application requires an understanding of the procedure used so that you can set up the analysis correctly. It helps if you understand the purpose and the restrictions of the procedure. Consequently, this chapter contains relatively simple linear programming exercises to give you experience in working through the details of the procedure, as well as exercises in which you will learn to analyze and set up linear programming problems.

Basically, the linear programming method involves modifying the constraints so that one has a system of linear equations and then finding selected solutions of the system. Remember how we solved a system of linear equations using an augmented matrix and reducing it with row operations? The simplex method follows a similar procedure. We introduce the simplex method in several steps, and we refer to the graphical method to illustrate the steps involved.

Standard Maximum

The following linear programming problem will be referred to several times as we develop the concepts.

ILLUSTRATIVE EXAMPLE ➤

Maximize the objective function
$$z = 4x_1 + 12x_2$$
subject to the constraints
$$3x_1 + x_2 \le 180$$
$$x_1 + 2x_2 \le 100$$
$$-2x_1 + 2x_2 \le 40 \tag{1}$$
and the nonnegative conditions
$$x_1 \ge 0, x_2 \ge 0$$

Notice that we now use the notation x_1 and x_2 for the variables instead of x and y. This notation allows us to use several variables without running out of letters for variables. ∎

In this section, we deal only with **standard maximum** linear programming problems. They are problems like the illustrative example that have the following properties.

Standard Maximum Problem

1. The objective function is to be maximized.
2. Each constraint is written using the $\leq$ inequality (excluding the nonnegative conditions).
3. The constants in the constraints to the right of $\leq$ are never negative (180, 100, and 40 in the example).
4. The variables are restricted to nonnegative values (nonnegative conditions).

The Geometric Form of the Problem

We now show the graph of the feasible region of the illustrative example to help us follow the steps of the simplex method. The problem is as follows:

Maximize $z = 4x_1 + 12x_2$, subject to

$$3x_1 + x_2 \leq 180$$
$$x_1 + 2x_2 \leq 100$$
$$-2x_1 + 2x_2 \leq 40$$
$$x_1 \geq 0, x_2 \geq 0$$

Recall that the boundary to the feasible region is formed by the lines

$$3x_1 + x_2 = 180$$
$$x_1 + 2x_2 = 100$$
$$-2x_1 + 2x_2 - 40$$
$$x_1 = 0, x_2 = 0$$

The feasible region is shown in Figure 4–1.

In the geometric approach, we look at the corners of the feasible region for optimal values. A corner point occurs at the intersection of a pair of boundary lines such as $(52, 24)$ and $(60, 0)$. Not all pairs of boundary lines intersect at a corner point. Notice the point $(40, 60)$ is not a corner point; it lies outside the feasible region.

The equation of the boundary lines can give us only points on the boundaries. To represent points in the interior of the feasible region, we must introduce some new variables called *slack variables*. Slack variables allow us to view the problem as a system of equations.

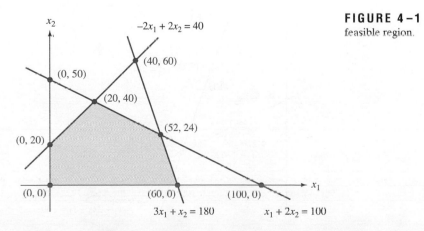

FIGURE 4–1 The feasible region.

Slack Variables

The first step in the simplex method converts the constraints to linear equations. To do this, we must introduce additional variables called **slack variables.** One slack variable is introduced for each constraint. The first constraint, $3x_1 + x_2 \leq 180$, is true for pairs of numbers such as $x_1 = 10$ and $x_2 = 20$ because $3(10) + 20 \leq 180$. Observe that $3x_1 + x_2 + 130 = 180$ when $x_1 = 10$ and $x_2 = 20$. When $x_1 = 30$ and $x_2 = 50$, the constraint $3x_1 + x_2 \leq 180$ is true, as is the equation $3x_1 + x_2 + 40 = 180$. In general, when a pair of values for x_1 and x_2 make the statement $3x_1 + x_2 \leq 180$ true, there will also be a value of s_1 (s_1 depends on the values of x_1 and x_2) that makes the statement $3x_1 + x_2 + s_1 = 180$ true. In each case, s_1 is not negative. We call s_1 a *slack variable* because it takes up the slack between $3x_1 + x_2$ and 180. Whenever $s_1 = 0$, then x_1 and x_2 must be coordinates of a point on the boundary line $3x_1 + x_2 = 180$.

The values $x_1 = 10$ and $x_2 = 20$ also make the constraint $x_1 + 2x_2 \leq 100$ true, and also make the equation $x_1 + 2x_2 + 50 = 100$ true. When we used these same values for x_1 and x_2 in the constraint $3x_1 + x_2 \leq 180$, we used the value 130 to take up the slack. We point this out to illustrate that whenever values of x_1 and x_2 make two constraints true, then *different* values of the slack variable may be required to take up the slack. Thus, one nonnegative slack variable is needed for each constraint.

Let's look at Figure 4–2 to illustrate the slack variables' relationship to the feasible region determined by

$$3x_1 + x_2 \leq 180$$
$$x_1 + 2x_2 \leq 100$$
$$x_1 \geq 0, x_2 \geq 0$$

that is,

$$3x_1 + x_2 + s_1 = 180$$
$$x_1 + 2x_2 + s_2 = 100$$
$$x_1 \geq 0, x_2 \geq 0$$

when the constraints are written with slack variables.

Notice that points that lie on the boundary $3x_1 + x_2 + s_1 = 180$ yield $s_1 = 0$ [like $(40, 60)$ and $(52, 24)$], and points that lie on $x_1 + 2x_2 + s_2 = 100$ yield

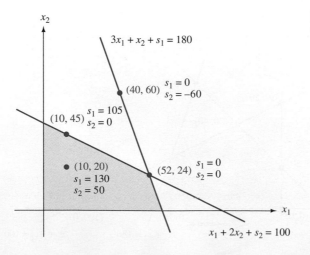

FIGURE 4–2   Values of slack variables at selected points.

$s_2 = 0$ [like $(10, 45)$ and $(52, 24)$]. A point that lies in the interior of the feasible region, such as $(10, 20)$, yields positive values for both s_1 and s_2. The point $(40, 60)$, which lies outside the feasible region, yields a negative value for s_2. In general, a point outside the feasible region will yield one or more negative slack variables.

| **Criteria for the Feasible Region** | A point $(x_1, x_2, \ldots, x_n)$ is in the feasible region of a linear programming problem if and only if all variables $x_1, x_2, \ldots, x_n$ and all slack variables $s_1, s_2, \ldots, s_k$ are *nonnegative*. |

EXAMPLE 1 ➤

With the addition of slack variables, the constraints (1) in the illustrative example become

$$\begin{aligned}
3x_1 + x_2 + s_1 \qquad\qquad &= 180 \\
x_1 + 2x_2 \qquad + s_2 \qquad &= 100 \\
-2x_1 + 2x_2 \qquad\qquad + s_3 &= 40
\end{aligned}$$

$$x_1 \geq 0,\, x_2 \geq 0,\, s_1 \geq 0,\, s_2 \geq 0,\, s_3 \geq 0 \qquad \textbf{(2)}$$

The last line gives the nonnegative conditions that apply to the variables. ∎

You might wonder how this system of equations will help. Here's how. When nonnegative values of x_1, x_2, and s_1 are chosen to satisfy $3x_1 + x_2 + s_1 = 180$, then the point obtained from the values of x_1 and x_2 will satisfy the constraint $3x_1 + x_2 \leq 180$. Furthermore, when x_1 and x_2 come from an interior point such as $(20, 30)$, for example, then the nonnegative values of the slack variables $(s_1 = 90, s_2 = 20,$ and $s_3 = 20,$ in this case$)$ satisfy the system of equations (2).

In general, the points in the feasible region are determined from the solutions to a system of equations like (2), provided that we use only those solutions in which *all* the variables $(x_1, x_2, s_1, s_2, s_3$ in this case$)$ are nonnegative.

EXAMPLE 2 ➤

Write the following constraints as a system of equations using slack variables:

$$\begin{aligned}
5x_1 + 3x_2 + 17x_3 &\leq 140 \\
7x_1 + 2x_2 + 4x_3 &\leq 256 \\
3x_1 + 9x_2 + 11x_3 &\leq 540 \\
2x_1 + 16x_2 + 8x_3 &\leq 99
\end{aligned}$$

SOLUTION

Introduce a slack variable for each equation:

$$\begin{aligned}
5x_1 + 3x_2 + 17x_3 + s_1 \qquad\qquad\qquad &= 140 \\
7x_1 + 2x_2 + 4x_3 \qquad + s_2 \qquad\qquad &= 256 \\
3x_1 + 9x_2 + 11x_3 \qquad\qquad + s_3 \qquad &= 540 \\
2x_1 + 16x_2 + 8x_3 \qquad\qquad\qquad + s_4 &= 99
\end{aligned}$$

■ **Now You Are Ready to Work Exercise 2**

The objective function needs to be included in the system of equations because we want to find the value of z that comes from the solution of the above system. Its form needs to be modified by writing all terms on the left-hand side. For example,

$$z = 3x_1 + 5x_2$$

is modified to

$$z - 3x_1 - 5x_2 = 0$$

which we write as

$$-3x_1 - 5x_2 + z = 0$$

because we want to write the x's first, as we do in the constraints.

The objective function

$$z = 6x_1 + 7x_2 + 15x_3 + 2x_4$$

is modified to

$$z - 6x_1 - 7x_2 - 15x_3 - 2x_4 = 0$$

and then to

$$-6x_1 - 7x_2 - 15x_3 - 2x_4 + z = 0$$

EXAMPLE 3 ➤ Include the objective function

$$z = 20x_1 + 35x_2 + 40x_3$$

with the constraints of Example 2 and write them as a system of equations.

SOLUTION

$$
\begin{aligned}
5x_1 + 3x_2 + 17x_3 + s_1 &= 140 \\
7x_1 + 2x_2 + 4x_3 + s_2 &= 256 \\
3x_1 + 9x_2 + 11x_3 + s_3 &= 540 \\
2x_1 + 16x_2 + 8x_3 + s_4 &= 99 \\
-20x_1 - 35x_2 - 40x_3 + z &= 0 \quad \blacksquare
\end{aligned}
$$

EXAMPLE 4 ➤ Write the following as a system of equations. Maximize $z = 4x_1 + 12x_2$, subject to

$$
\begin{aligned}
3x_1 + x_2 &\le 180 \\
x_1 + 2x_2 &\le 100 \\
-2x_1 + 2x_2 &\le 40 \\
x_1 \ge 0, x_2 &\ge 0
\end{aligned}
$$

SOLUTION

$$
\begin{aligned}
3x_1 + x_2 + s_1 \qquad\qquad\qquad &= 180 \\
x_1 + 2x_2 \qquad + s_2 \qquad\qquad &= 100 \\
-2x_1 + 2x_2 \qquad\qquad + s_3 \qquad &= 40 \\
-4x_1 - 12x_2 \qquad\qquad\qquad + z &= 0
\end{aligned}
$$

■ **Now You Are Ready to Work Exercise 5**

Let's pause to make some observations about the systems of equations that we obtain.

1. One slack variable is introduced for each constraint (four in Example 3 and three in Example 4).
2. The total number of variables in the system is the number of original variables (number of x's) plus the number of constraints (number of slack vari-

ables) plus one for z. (The total is $3 + 4 + 1 = 8$ in Example 3 and $2 + 3 + 1 = 6$ in Example 4.)

3. The introduction of slack variables always results in fewer equations than variables. Recall from the summary in Chapter 2 that this situation generally yields an infinite number of solutions. In fact, one variable for each equation can be written in terms of some other variables.

Look back at the system of equations in Example 4. One general form of the solution occurs when we solve for s_1, s_2, s_3, and z in terms of x_1 and x_2.

$$
\begin{aligned}
s_1 &= 180 - 3x_1 - x_2 \\
s_2 &= 100 - x_1 - 2x_2 \\
s_3 &= 40 + 2x_1 - 2x_2 \\
z &= 4x_1 + 12x_2
\end{aligned}
$$

You find a particular solution when you substitute values for x_1 and x_2. The first step of the simplex method always uses zero for that substitution because it gives corner points. So when $x_1 = 0$ and $x_2 = 0$, we obtain $s_1 = 180$, $s_2 = 100$, $s_3 = 40$, and $z = 0$.

The preceding discussion enables us to make the first step in solving a linear programming problem with the simplex method, setting up the simplex tableau.

Simplex Tableau

The simplex method uses matrices and row operations on matrices to determine an optimal solution. Let's use the problem from Example 4 to set up the matrix that is called the **simplex tableau.**

EXAMPLE 5 ➤

Set up the simplex tableau that represents the following problem.
Maximize $z = 4x_1 + 12x_2$, subject to

$$
\begin{aligned}
3x_1 + x_2 &\leq 180 \\
x_1 + 2x_2 &\leq 100 \\
-2x_1 + 2x_2 &\leq 40 \\
x_1 \geq 0, x_2 &\geq 0
\end{aligned}
$$

SOLUTION

First, write the problem as a system of equations using slack variables:

$$
\begin{aligned}
3x_1 + x_2 + s_1 &&&&&= 180 \\
x_1 + 2x_2 &+ s_2 &&&&= 100 \\
-2x_1 + 2x_2 &&+ s_3 &&&= 40 \\
-4x_1 - 12x_2 &&&+ z &&= 0
\end{aligned}
$$

Next, form the augmented matrix of this system:

$$
\begin{array}{cccccc}
x_1 & x_2 & s_1 & s_2 & s_3 & z \\
\end{array}
$$
$$
\left[
\begin{array}{cccccc|c}
3 & 1 & 1 & 0 & 0 & 0 & 180 \\
1 & 2 & 0 & 1 & 0 & 0 & 100 \\
-2 & 2 & 0 & 0 & 1 & 0 & 40 \\
\hline
-4 & -12 & 0 & 0 & 0 & 1 & 0
\end{array}
\right]
$$

This is the initial simplex tableau.

■ **Now You Are Ready to Work Exercise 9**

The line drawn above the bottom row emphasizes that the bottom row is the objective function.

EXAMPLE 6 ➤

Set up the initial simplex tableau for the following problem:

Garden Tools, Inc., manufactures three items: hoes, rakes, and shovels. It takes 3 minutes of labor to produce each hoe, 5 minutes of labor to produce each rake, and 4 minutes of labor to produce each shovel. Each hoe costs $2.50 to produce, each rake costs $3.15, and each shovel costs $3.35. The profit is $3.20 per hoe, $3.35 per rake, and $4.10 per shovel. If the company has 108,000 minutes of labor and $6800 in operating funds available per week, how many of each item should it produce to maximize profit?

SOLUTION

If you are puzzled about which quantities to call variables, ask yourself what quantities are mentioned, which quantities are given and therefore fixed, which quantities are mentioned and no values given, and, finally, how are the quantities related?

The problem gives information about three items: hoes, rakes, and shovels. The minutes of labor and the cost to produce each item are given. The profit for each item is also given. The available minutes of labor and operating funds are known. These last two quantities are related to the number of items produced, but those numbers are not given. In fact, you are asked to find the number of each item. This suggests that the variables represent the number of hoes, rakes, and shovels. Those are the quantities that can vary, so we let

$$x_1 = \text{number of hoes}$$
$$x_2 = \text{number of rakes}$$
$$x_3 = \text{number of shovels}$$

It often helps to summarize the given information in a chart before writing the constraints. In this problem, there are three variables and three types of information: labor time, cost, and profit. Look at the information and observe that it can be summarized as follows:

	x_1 (Hoes)	x_2 (Rakes)	x_3 (Shovels)	Maximum Available
Labor (minutes)	3	5	4	108,000
Cost	$2.50	$3.15	$3.35	$6800
Profit	$3.20	$3.35	$4.10	

The information on labor requirements, operating costs, and profit can be written as follows:

$$3x_1 + 5x_2 + 4x_3 \leq 108{,}000 \quad \text{(Labor)}$$
$$2.50x_1 + 3.15x_2 + 3.35x_3 \leq 6{,}800 \quad \text{(Operating cost)}$$
$$\text{Maximize:} \quad 3.20x_1 + 3.35x_2 + 4.10x_3 = z \quad \text{(Profit)}$$

We introduce slack variables into the constraints to form equations and rewrite the objective function (profit) as follows:

$$3x_1 + 5x_2 + 4x_3 + s_1 \qquad = 108{,}000$$
$$2.50x_1 + 3.15x_2 + 3.35x_3 \qquad + s_2 \qquad = \quad 6{,}800$$
$$-3.20x_1 - 3.35x_2 - 4.10x_3 \qquad\qquad + z = \qquad 0$$

The simplex tableau is

$$
\begin{array}{c}
\quad x_1 \qquad x_2 \qquad x_3 \quad s_1 \;\; s_2 \;\; z \\
\left[
\begin{array}{ccc|ccc|c}
3 & 5 & 4 & 1 & 0 & 0 & 108{,}000 \\
2.50 & 3.15 & 3.35 & 0 & 1 & 0 & 6{,}800 \\
\hline
-3.20 & -3.35 & -4.10 & 0 & 0 & 1 & 0
\end{array}
\right]
\end{array}
$$

■ **Now You Are Ready to Work Exercise 13**

4.1 EXERCISES

LEVEL 1

In Exercises 1 through 4, convert the systems of inequalities to systems of equations using slack variables.

1. $2x_1 + 3x_2 \le 9$
$\quad x_1 + 5x_2 \le 16$

2. *(See Example 2)*
$\quad 3x_1 - 4x_2 \le 24$
$\quad 9x_1 + 5x_2 \le 16$
$\quad -x_1 + x_2 \le 5$

3. $x_1 + 7x_2 - 4x_3 \le 150$
$\quad 5x_1 + 9x_2 + 2x_3 \le 435$
$\quad 8x_1 - 3x_2 + 16x_3 \le 345$

4. $x_1 + x_2 + x_3 + x_4 \le 78$
$\quad 3x_1 + 2x_2 + x_3 - x_4 \le 109$

In Exercises 5 through 8, express the problems as a system of equations using slack variables.

5. *(See Example 4)* Maximize $z = 3x_1 + 7x_2$, subject to

$$2x_1 + 6x_2 \le 9$$
$$x_1 - 5x_2 \le 14$$

$$-3x_1 + x_2 \le 8$$
$$x_1 \ge 0, x_2 \ge 0$$

6. Maximize $z = 150x_1 + 280x_2$, subject to

$$12x_1 + 15x_2 \le 50$$
$$8x_1 + 22x_2 \le 65$$
$$x_1 \ge 0, x_2 \ge 0$$

7. Maximize $z = 420x_1 + 260x_2 + 50x_3$, subject to

$$6x_1 + 7x_2 + 12x_3 \le 50$$
$$4x_1 + 18x_2 + 9x_3 \le 85$$
$$x_1 - 2x_2 + 14x_3 \le 66$$
$$x_1 \ge 0, x_2 \ge 0, x_3 \ge 0$$

8. Maximize $z = 3x_1 + 4x_2 + 7x_3 + 2x_4$, subject to

$$x_1 + 5x_2 + 7x_3 + x_4 \le 82$$
$$3x_1 + 6x_2 + 12x_3 \qquad \le 50$$
$$2x_1 \qquad + 15x_3 + 19x_4 \le 240$$
$$x_1 \ge 0, x_2 \ge 0, x_3 \ge 0, x_4 \ge 0$$

LEVEL 2

In Exercises 9 through 12, set up the simplex tableau. Do not solve.

9. *(See Example 5)* Maximize $z = 3x_1 + 17x_2$, subject to

$$4x_1 + 5x_2 \le 10$$
$$3x_1 + x_2 \le 25$$
$$x_1 \ge 0, x_2 \ge 0$$

10. Maximize $z = 140x_1 + 245x_2$, subject to

$$85x_1 + 64x_2 \le 560$$
$$75x_1 + 37x_2 \le 135$$
$$24x_1 + 12x_2 \le 94$$
$$x_1 \ge 0, x_2 \ge 0$$

11. Maximize $z = 20x_1 + 45x_2 + 40x_3$, subject to

$$16x_1 - 4x_2 + 9x_3 \le 128$$
$$8x_1 + 13x_2 + 22x_3 \le 144$$
$$5x_1 + 6x_2 - 15x_3 \le 225$$
$$x_1 \ge 0, x_2 \ge 0, x_3 \ge 0$$

12. Maximize $z = 18x_1 + 24x_2 + 95x_3 + 50x_4$, subject to

$$x_1 + 2x_2 + 5x_3 + 6x_4 \le 48$$
$$4x_1 + 8x_2 - 15x_3 + 9x_4 \le 65$$
$$3x_1 - 2x_2 + x_3 - 8x_4 \le 50$$
$$x_1 \ge 0, x_2 \ge 0, x_3 \ge 0, x_4 \ge 0$$

LEVEL 3

Set up the simplex tableau for each of Exercises 13 through 21. Do not solve.

13. *(See Evample 6)* Hardware Supplies manufactures three items: screwdrivers, chisels, and putty knives. It takes 3 hours of labor per carton to produce screwdrivers, 4 hours of labor per carton of chisels, and 5 hours of labor per carton of putty knives. Each carton of screwdrivers costs $15 to produce, each carton of chisels costs $12, and each carton of putty knives costs $11. The profit per carton is $5 for screwdrivers, $6 for chisels, and $5 for putty knives. If the company has 2200 hours of labor and $8500 in operating funds available per week, how many cartons of each item should it produce to maximize profit?

14. The maximum daily production of an oil refinery is 1900 barrels. The refinery can produce three types of fuel: gasoline, diesel, and heating oil. The production cost per barrel is $6 for gasoline, $5 for diesel, and $8 for heating oil. The daily production budget is $13,400. The profit is $7 per barrel on gasoline, $6 on diesel, and $9 for heating oil. How much of each should be produced to maximize profit?

15. Gina eats a meal of steak, baked potato with butter, and salad with dressing. One ounce of each contains the indicated calories, protein, and fat:

	Calories	Protein (mg)	Fat (g)
Salad (oz)	20	0.5	1.5
Potato (oz)	50	1.0	3.0
Steak (oz)	56	9.0	2.0

She is to consume no more than 1000 calories and 35 g of fat. How much of each should she eat to maximize the protein consumed?

16. The Book Fair sponsors book fairs at elementary schools and sells good books to the children, which are written to capture their interest. In preparation for a fair, they put together three different packages of paperback children's books. Each package contains short story books, science books, and history books. Pack I contains three short story, one science, and two history books. Pack II contains two short story, four science, and one history book. Pack III contains one short story, two science, and three history books. The Book Fair has 660 short story books, 740 science books, and 800 history books available to use in the packs. The profit on each pack is $1.25 for Pack I, $2.00 for Pack II, and $1.60 for Pack III. How many of each pack should be made to maximize profit?

17. The Health Fare Cereal Company makes three cereals using wheat, oats, and raisins. The portions and profit of each cereal are shown in the following chart:

Cereal	Portion of Each Pound of Cereal			Profit/Pound
	Wheat	Oats	Raisins	
Lite	0.75	0.25	0	$0.25
Trim	0.50	0.25	0.25	$0.25
Health Fare	0.15	0.60	0.25	$0.32

The company has 2320 pounds of wheat, 1380 pounds of oats, and 700 pounds of raisins available. How many pounds of each cereal should it produce to maximize profit?

18. The Humidor blends regular coffee, High Mountain coffee, and chocolate to obtain three kinds of coffee: Early Riser, After Dinner, and Deluxe. The blends and profit for each blend are given in the following chart:

	Blend		
	Early Riser	After Dinner	Deluxe
Regular	80%	75%	50%
High Mountain	20%	20%	40%
Chocolate	0%	5%	10%
Profit/pound	$1.00	$1.10	$1.20

The shop has 255 pounds of regular coffee, 80 pounds of High Mountain coffee, and 15 pounds of chocolate. How many pounds of each blend should be produced to maximize profit?

19. The Williams Trunk Company makes trunks for the military, for commercial use, and for decorative pieces. Each military trunk requires 4 hours for assembly, 1 hour for finishing, and 0.1 hour for packaging. Each commercial trunk requires 3 hours for assembly, 2 hours for finishing, and 0.2 hour for packaging. Each decorative trunk requires 2 hours for assembly, 4 hours for finishing, and 0.3 hour for packaging. The profit on each trunk is $6 for military, $7 for commercial, and $9 for decorative. If 4900 hours are available for assembly work, 2200 for finishing, and 210 for packaging, how many of each type of trunk should be made to maximize profit?

20. The Snack Shop makes three nut mixes from peanuts, cashews, and pecans in 1-kilogram packages. (1 kilogram = 1000 grams.) The composition of each mix is as follows:

   TV Mix: 600 grams of peanuts, 300 grams of cashews, and 100 grams of pecans

   Party Mix: 500 grams of peanuts, 300 grams of cashews, and 200 grams of pecans

   Dinner Mix: 400 grams of peanuts, 200 grams of cashews, and 400 grams of pecans

The shop has 39,500 grams of peanuts, 22,500 grams of cashews, and 18,000 grams of pecans. The selling price per package of each mix is $4.40 for TV Mix, $4.80 for Party Mix, and $5.20 for Dinner Mix. How many packages of each should be made to maximize revenue?

21. The Clock Works produces three clock kits for amateur woodworkers: the Majestic Grandfather Clock, the Traditional Clock, and the Wall Clock. The following chart shows the times required for cutting, sanding, and packing each kit and the price of each kit:

Process	Majestic	Traditional	Wall
Cutting	4 hours	2 hours	1 hour
Sanding	3 hours	2 hours	1 hour
Packing	1 hour	1 hour	0.5 hour
Price	$400	$250	$160

The cutting machines are available 120 hours per week, the sanding machines 80 hours, and the packing machines 40 hours. How many of each type of kit should be produced each week to maximize revenue?

22. Set up the constraints and the initial simplex tableau for this problem that would be used to find the number of each type of furniture that should be ordered so that profit is maximum. Do not solve.

An import company obtains a license to import up to 200 pieces of hand-carved furniture. The importer may import chests, desks, and silverware boxes. The cost, volume, and profit are the following:

	Chest	Desk	Silverware Box
Cost	$270	$310	$90
Profit	$180	$300	$45
Volume (cu ft)	7	18	1.5

The import company may purchase up to $5000 worth of furniture, and the furniture must all fit into a 1500-cubic-foot shipping container.

23. Set up the constraints and the initial simplex tableau for the following problem that would be used to find the number of pounds of each blend of coffee that should be produced to maximize profit. Do not solve.

The Humidor blends regular coffee, High Mountain coffee, and chocolate to obtain four kinds of coffee: Early Riser, Coffee Time, After Dinner, and Deluxe. The blends and profit for each blend are the following:

	Blend			
	Early Riser	Coffee Time	After Dinner	Deluxe
Regular	80%	75%	75%	50%
High Mountain	20%	23%	20%	40%
Chocolate	0%	2%	5%	10%
Profit/pound	$1.00	$1.10	$1.15	$1.20

The shop has 260 pounds of regular coffee, 90 pounds of High Mountain coffee, and 20 pounds of chocolate.

24. Set up the constraints and the initial simplex tableau for this problem that would be used to find the number of pounds of each cereal that should be produced in order to maximize profit. Do not solve.

The Health Fare Cereal Company makes four cereals using wheat, oats, raisins, and nuts.

The portions and profit for each cereal are the following:

Cereal	Portion of Each Pound of Cereal				Profit/ Pound
	Wheat	Oats	Raisins	Nuts	
Lite	0.75	0.20	0.05	0	$0.25
Trim	0.50	0.25	0.20	0.05	$0.25
Regular	0.80	0.20	0	0	$0.27
Health Fare	0.15	0.50	0.25	0.10	$0.32

The company has 2400 pounds of wheat, 1400 pounds of oats, 700 pounds of raisins, and 250 pounds of nuts available.

25. Set up the constraints and the initial simplex tableau for the following problem that would be used to find how much a couple should invest in each type of investment to maximize return on their investment. Do not solve.

Ralph and Jean are proud new grandparents. They immediately set up an investment fund to help provide for their grandchild's college education. They invest in stocks, treasury bonds, municipal bonds, and corporate bonds. To balance risks, they want no more than 40% of their investment to be in stocks, no more than 30% in municipal bonds, no

more than 25% in corporate bonds, and no more than 15% in treasury bonds. They have $10,000 to invest.

They expect stocks to return 10% on their investment, treasury bonds to return 6%, municipal bonds 7%, and corporate bonds 8%.

26. Set up the constraints and the initial simplex tableau for this problem that would be used to find how many of each pattern should be used to maximize profit. Do not solve.

Southside Landscapes proposes to plant the flower beds for the historic Earle House gardens. The company uses four patterns. Pattern I uses 40 tulips, 25 daffodils, and 6 boxwood. Pattern II uses 25 tulips, 50 daffodils, and 4 boxwood. Pattern III uses 30 tulips, 40 daffodils, and 8 boxwood. Pattern IV uses 45 tulips, 45 daffodils, and 2 boxwood. The profit for each pattern is $48 for Pattern I, $45 for Pattern II, $55 for Pattern III, and $65 for Pattern IV. Southside Landscape has 1250 tulips, 1600 daffodils, and 195 boxwood available.

EXPLORATIONS

27. The feasible region of a linear programming problem is determined by the constraints

$$12x + 10y \le 120$$
$$3x + 10y \le 60$$
$$7x + 12y \le 84$$

and the nonnegative conditions

$$x \ge 0, y \ge 0$$

(a) Convert the constraints to a system of equations using the slack variables s_1, s_2, and s_3, respectively.
(b) A point (x, y, s_1, s_2, s_3), with all coordinates nonnegative, satisfies the first equation in part (a). Comment on the relationship of the point to the feasible region.
(c) A point $(x, y, 0, s_2, s_3)$, x, y, s_2, s_3 nonnegative, satisfies the first equation in part (a). Comment on the relationship of this point to the feasible region.
(d) A point $(x, y, -3, s_2, s_3)$, x, y, s_2, and s_3 nonnegative, satisfies the first equation in part (a).

Comment on the relationship of this point to the feasible region.
(e) A point $(x, y, 0, s_2, 0)$, x and y and s_2 nonnegative, satisfies both the first and third equations in part (a). Comment on the relationship of this point to the feasible region.

28. The feasible region of a linear programming problem is determined by the constraints

$$3x_1 + 4x_2 + 7x_3 + x_4 \le 85$$
$$2x_1 + x_2 + 6x_3 + 5x_4 \le 110$$

and the nonnegative conditions

$$x_1 \ge 0, x_2 \ge 0, x_3 \ge 0, x_4 \ge 0$$

(a) Convert the constraints to a system of equations using slack variables s_1 and s_2.
(b) Suppose a point $(x_1, x_2, x_3, x_4, s_1, s_2)$, all coordinates nonnegative, satisfies the first equation in part (a). Comment on the relationship of the point (x_1, x_2, x_3, x_4) to the feasible region defined by the original constraints if s_1 is positive.

(c) Comment on the relationship of the point (x_1, x_2, x_3, x_4) to the feasible region determined by the original constraints if the point $(x_1, x_2, x_3, x_4, s_1, s_2)$, all coordinates nonnegative, satisfies both equations in part (a) and
  (i) $s_1 > 0$ and $s_2 > 0$.
  (ii) $s_1 = 0$ and $s_2 > 0$.
  (iii) $s_1 = 0$ and $s_2 = 0$.

29. Enter the initial simplex tableau into your graphing calculator or spreadsheet for each of the following linear programming problems.

(a) Maximize $z = 40x + 22y$, subject to

$$6x + 3y \le 18$$
$$5x + 2y \le 27$$
$$x \ge 0, y \ge 0$$

(b) Maximize $z = 134x + 109y$, subject to

$$10x + 14y \le 73$$
$$6x + 21y \le 67$$
$$15x + 8y \le 48$$
$$x \ge 0, y \ge 0$$

(c) Maximize $z = 15x_1 + 23x_2 + 34x_3$, subject to

$$x_1 + x_2 + x_3 \le 24$$
$$3x_1 + x_2 + 4x_3 \le 37$$
$$2x_1 + 5x_2 + 3x_3 \le 41$$
$$x_1 \ge 0, x_2 \ge 0, x_3 \ge 0$$

(d) Maximize $z = 24x_1 + 19x_2 + 15x_3 + 33x_4$, subject to

$$7x_1 + 4x_2 + x_3 + 2x_4 \le 435$$
$$5x_1 + 3x_2 + 6x_3 + x_4 \le 384$$
$$2x_1 + 8x_2 + 4x_3 + 5x_4 \le 562$$
$$x_1 \ge 0, x_2 \ge 0, x_3 \ge 0, x_4 \ge 0$$

(e) Maximize $z = 12.9x_1 + 11.27x_2 + 23.85x_3$, subject to

$$4.7x_1 + 3.2x_2 + 1.58x_3 \le 40.6$$
$$2.14x_1 + 1.82x_2 + 5.09x_3 \le 61.7$$
$$1.63x_1 + 3.44x_2 + 2.84x_3 \le 54.8$$
$$x_1 \ge 0, x_2 \ge 0, x_3 \ge 0$$

4.2 The Simplex Method

- System of Equations: Many Solutions
- Basic Solution
- Pivot Column, Row, and Element
- Final Tableau

System of Equations: Many Solutions

In Section 4.1, you learned how to convert a linear programming problem to a system of equations using slack variables and then putting the system into a matrix form called a simplex tableau. The simplex method uses the simplex tableau to find the optimal solution. By the nature of a linear programming problem, the simplex method involves a system of equations in which the number of variables exceeds the number of equations. Such a system generally has an infinite number of solutions where some of the variables can be chosen arbitrarily. We will see that the simplex method chooses the variables in a way that gives corner points of the feasible region. Recall that when the solutions of a system are represented by an augmented matrix such as

$$
\begin{array}{c}
\begin{array}{cccc} x_1 & x_2 & x_3 & x_4 \end{array} \\
\left[\begin{array}{ccc|cc}
1 & 0 & 0 & 2 & 3 \\
\hline
0 & 1 & 0 & -1 & 5 \\
\hline
0 & 0 & 1 & 4 & 7
\end{array}\right]
\end{array}
$$

then the solutions can be written as

$$x_1 = 3 - 2x_4$$
$$x_2 = 5 + x_4$$
$$x_3 = 7 - 4x_4$$

where any value may be assigned to x_4, thereby determining values of x_1, x_2, and x_3. We solve for x_1, x_2, and x_3 in terms of x_4 because x_1 occurs in only one row. (All entries in the x_1 column are zero except for one entry.) Similarly, x_2 and x_3 occur only once. The variable x_4 occurs in all rows, so it appears in each equation of the solution. Observe that the entries in the x_1, x_2, and x_3 columns form a column from an identity matrix (call these **unit columns**). The x_4 column is not like a column from an identity matrix. In general, we solve for variables corresponding to unit columns in terms of variables whose columns are not unit columns. For example, in the augmented matrix

$$\begin{array}{ccccc} x_1 & x_2 & x_3 & x_4 & x_5 \\ \left[\begin{array}{ccccc|c} 3 & 0 & -1 & 0 & 1 & 4 \\ 1 & 1 & 0 & 0 & 0 & -6 \\ 2 & 0 & 5 & 1 & 0 & 10 \end{array}\right] \end{array}$$

the unit columns are the x_2, x_4, and x_5 columns, so we solve for x_2, x_4, and x_5 in terms of x_1 and x_3, giving

$$\begin{array}{ll} x_2 = -6 - x_1 & \text{(From row 2)} \\ x_4 = 10 - 2x_1 - 5x_3 & \text{(From row 3)} \\ x_5 = 4 - 3x_1 + x_3 & \text{(From row 1)} \end{array}$$

Obtain specific solutions by assigning arbitrary values to x_1 and x_3.

We will see how the simplex method selects certain solutions to a system by assigning zeros to the arbitrary variables.

Basic Solution

The simplex method finds a sequence of selected solutions to a system of equations. We make the selections so that we find the optimal solution in a relatively small number of steps.

Let's look at the illustrative example in Section 4.1 again. It is the following: Maximize the objective function $z = 4x_1 + 12x_2$, subject to the constraints

$$\begin{array}{rcl} 3x_1 + x_2 & \leq & 180 \\ x_1 + 2x_2 & \leq & 100 \\ -2x_1 + 2x_2 & \leq & 40 \\ x_1 \geq 0, x_2 \geq 0 \end{array} \qquad \textbf{(1)}$$

We introduce slack variables to obtain the following. Maximize $z = 4x_1 + 12x_2$, subject to

$$\begin{array}{rcl} 3x_1 + x_2 + s_1 & = & 180 \\ x_1 + 2x_2 + s_2 & = & 100 \\ -2x_1 + 2x_2 + s_3 & = & 40 \end{array} \qquad \textbf{(2)}$$

where x_1, x_2, s_1, s_2, and s_3 are all nonnegative.

The simplex tableau is

$$\begin{array}{ccccccc} x_1 & x_2 & s_1 & s_2 & s_3 & z & \\ \left[\begin{array}{cccccc|c} 3 & 1 & 1 & 0 & 0 & 0 & 180 \\ 1 & 2 & 0 & 1 & 0 & 0 & 100 \\ -2 & 2 & 0 & 0 & 1 & 0 & 40 \\ \hline -4 & -12 & 0 & 0 & 0 & 1 & 0 \end{array}\right] \end{array}$$

Notice that unit columns occur in the s_1, s_2, and s_3 columns. Because the feasible region is determined by the constraints, for now we will ignore the bottom row (objective function) and the z column. Thus we can solve for s_1, s_2, and s_3 as

$$s_1 = 180 - 3x_1 - x_2$$
$$s_2 = 100 - x_1 - 2x_2$$
$$s_3 = 40 + 2x_1 - 2x_2$$

We can substitute any values whatever for x_1 and x_2 to obtain s_1, s_2, and s_3. Those five numbers will form a solution to the system of equations. However, only those nonnegative values of x_1 and x_2 that also yield nonnegative values for s_1, s_2, and s_3 give **feasible solutions.** The simplest choices are $x_1 = 0$ and $x_2 = 0$. This gives $s_1 = 180$, $s_2 = 100$, and $s_3 = 40$. Solutions like this, where the arbitrary variables are set to zero, are called **basic solutions.** The number of variables set to zero in this case was two, which happens to be the number of x's involved. In general, the number of x's determines the number of arbitrary variables set to zero.

We chose zero for the values of x_1 and x_2, not just for simplicity, but also because that gives a corner point $(0, 0)$ of the feasible region.

DEFINITION
Basic Solution

If a linear programming problem has k x's in the constraints, then a **basic solution** is obtained by setting k variables (except z) to zero and solving for the others.

Actually, we can find a basic solution by setting any two variables to zero and solving for the others.

If we set $s_1 = 0$ and $s_3 = 0$, we can obtain another basic solution by solving for the other variables from the simplex tableau. If $s_1 = 0$ and $s_3 = 0$, the system (2) reduces to

$$
\begin{aligned}
3x_1 + x_2 &= 180 \\
x_1 + 2x_2 + s_2 &= 100 \\
-2x_1 + 2x_2 &= 40 \\
-4x_1 - 12x_2 + z &= 0
\end{aligned}
$$

We have four equations in four unknowns, x_1, x_2, s_2, and z, to solve. This system has the solution (we omit the details)

$$x_1 = 40, \quad x_2 = 60, \quad s_2 = -60, \quad z = 880$$

Notice that s_2 is *negative.* This violates the nonnegative conditions on the x's and slack variables. Although the solution

$$x_1 = 40, \quad x_2 = 60, \quad s_1 = 0, \quad s_2 = -60, \quad s_3 = 0, \quad z = 880$$

is a basic solution, it is not feasible.

If you look at Figure 4–1, you will see that $(40, 60)$ is a point where two boundary lines intersect, but it lies outside the feasible region. Properly carried out, the simplex method finds only basic *feasible* solutions that are corner points of the feasible region.

DEFINITION
Basic Feasible Solution

Call the number of x variables in a linear programming problem k. A **basic feasible solution** of the system of equations is a solution with k variables (except z) set to zero and with none of the slack variables or x's negative.

The variables set to zero are called **nonbasic variables.** The others are called **basic variables.**

With this background, let's proceed to solve this example by finding the appropriate basic feasible solution.

Solve by the simplex method:

Maximize $z = 4x_1 + 12x_2$, subject to

$$3x_1 + x_2 \leq 180$$
$$x_1 + 2x_2 \leq 100$$
$$-2x_1 + 2x_2 \leq 40$$
$$x_1 \geq 0, x_2 \geq 0$$

Step 1. Begin with the *initial tableau:*

x_1	x_2	s_1	s_2	s_3	z		Basic variables
3	1	1	0	0	0	180	s_1
1	2	0	1	0	0	100	s_2
−2	2	0	0	1	0	40	s_3
−4	−12	0	0	0	1	0	

We find the first basic feasible solutions. Notice that the unit columns are for s_1, s_2, and s_3, so we can solve for them in terms of the x's. Now we can set the x's to zero to obtain $s_1 = 180$, $s_2 = 100$, and $s_3 = 40$. (You can read these from the tableau.) Because this basic feasible solution is the first found, we call it the **initial basic feasible solution.** It always occurs at the origin of the feasible region.

In this case, the unit columns give s_1, s_2, and s_3 as the basic variables. The nonbasic variables are x_1 and x_2.

■ **Now You Are Ready to Work Exercise 1**

Basic and Nonbasic Variables—Summary

A linear programming problem with n constraints using k variables is converted into a system of equations by including a slack variable for each constraint.

- A system of n constraints in k variables converts to a system of n equations in $n + k$ variables.
- The $n + k$ variables are classified as either *basic* or *nonbasic* variables.
- There is a basic variable for each equation, giving n basic variables. The other k variables are nonbasic.
- The values of the basic variables are found by setting all nonbasic variables to zero in each of the equations and then solving for the basic variables.

Pivot Column, Row, and Element

Step 2. Next we want to modify the tableau so that the new tableau has a basic feasible solution that increases the value of z. This step requires the selection of a **pivot element** from the tableau as follows:

(a) To select the column containing the pivot element, do the following. Select the *most negative* entry from the *bottom* row:

$$\begin{array}{ccccccc} x_1 & x_2 & s_1 & s_2 & s_3 & z & \\ \begin{bmatrix} 3 & 1 & 1 & 0 & 0 & 0 & 180 \\ 1 & 2 & 0 & 1 & 0 & 0 & 100 \\ -2 & 2 & 0 & 0 & 1 & 0 & 40 \\ \hline -4 & \boxed{-12} & 0 & 0 & 0 & 1 & 0 \end{bmatrix} \end{array}$$

Most negative entry gives pivot column.

This selects the **pivot column** containing the pivot element. The pivot element itself is an entry in this column *above* the line. We must now determine which row contains the pivot element.

(b) To select the row, called the **pivot row,** containing the pivot element, do the following. Divide each constant above the line in the last column by the corresponding entries in the pivot column. The ratios are written to the right of the tableau. In this example, all ratios are positive. However, negative or zero ratios can occur.

   (i) If a negative ratio occurs, do not use that row for the pivot row.

   (ii) If all ratios are positive (they will be in most cases), select the *smallest positive ratio* (20 in this case). The row containing this ratio is the pivot row.

Now look at the ratios on the right of the tableau.

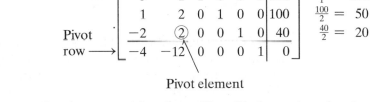

$$\begin{array}{ccccccc} & x_1 & x_2 & s_1 & s_2 & s_3 & z & \\ & \begin{bmatrix} 3 & 1 & 1 & 0 & 0 & 0 & 180 \\ 1 & 2 & 0 & 1 & 0 & 0 & 100 \\ -2 & \boxed{2} & 0 & 0 & 1 & 0 & 40 \\ \hline -4 & -12 & 0 & 0 & 0 & 1 & 0 \end{bmatrix} \end{array}$$

$\frac{180}{1} = 180$

$\frac{100}{2} = 50$

$\frac{40}{2} = 20$

Pivot row →

Pivot element

NOTE

The **pivot element** will always lie above the line drawn above the bottom row.

Because all ratios are positive, the smallest, 20, determines the pivot row. The entry 2 in the pivot row and pivot column is the **pivot element.**

■ **Now You Are Ready to Work Exercise 7**

Step 3. Move to the next basic feasible solution. We call this process **pivoting** on the pivot element. In this case we pivot on 2 in row 3, column 2.

To pivot on 2, use row operations to modify the tableau so that the pivot element becomes a 1 and the rest of the pivot column contains zeros. (You recognize that this is part of the Gauss-Jordan Method for solving systems.)

Multiply each entry in the third row (pivot row) by $\frac{1}{2}$ so that the pivot entry becomes 1. The third row becomes

$$[-1 \quad 1 \quad 0 \quad 0 \quad \tfrac{1}{2} \quad 0 \quad 20\,]]$$

giving the tableau

$$\begin{array}{cccccc} x_1 & x_2 & s_1 & s_2 & s_3 & z & \\ \begin{bmatrix} 3 & \boxed{1} & 1 & 0 & 0 & 0 & 180 \\ 1 & \boxed{2} & 0 & 1 & 0 & 0 & 100 \\ -1 & 1 & 0 & 0 & \tfrac{1}{2} & 0 & 20 \\ \hline -4 & \boxed{-12} & 0 & 0 & 0 & 1 & 0 \end{bmatrix} \end{array}$$

We now need zeros in the circled locations of the pivot column.

We replace each row where a zero is needed by multiplying row 3 by a constant and adding it to the row to be replaced. Each time, use the constant that gives a zero in the pivot column. This is accomplished as follows:

Replace row 1 with (row 1 − row 3):

$$= [4 \quad 0 \quad 1 \quad 0 \quad -\tfrac{1}{2} \quad 0 \quad 160]$$

Replace row 2 with (row 2 + (−2)row 3):

$$= [3 \quad 0 \quad 0 \quad 1 \quad -1 \quad 0 \quad 60]$$

Replace row 4 with (row 4 + (12)row 3):

$$= [-16 \quad 0 \quad 0 \quad 0 \quad 6 \quad 1 \quad 240]$$

This gives the tableau

$$
\begin{array}{c}
 \\
\begin{array}{cccccc}
x_1 & x_2 & s_1 & s_2 & s_3 & z
\end{array} \\
\left[
\begin{array}{cccccc|c}
4 & 0 & 1 & 0 & -\tfrac{1}{2} & 0 & 160 \\
3 & 0 & 0 & 1 & -1 & 0 & 60 \\
-1 & 1 & 0 & 0 & \tfrac{1}{2} & 0 & 20 \\
\hline
-16 & 0 & 0 & 0 & 6 & 1 & 240
\end{array}
\right]
\end{array}
\qquad
\begin{array}{l}
\text{Basic} \\
\text{variables} \\
s_1 \\
s_2 \\
x_2 \\

\end{array}
$$

■ **Now You Are Ready to Work Exercise 17**

To determine the basic feasible solution from this tableau, observe that the columns under x_2, s_1, and s_2 are unit columns. These variables are the *basic variables,* the ones we solve for. The other two variables, x_1 and s_3, are *nonbasic variables,* the ones we set to zero.

The basic feasible solution from this tableau is

$$x_1 = 0, \qquad x_2 = 20, \qquad s_1 = 160, \qquad s_2 = 60, \qquad s_3 = 0, \qquad z = 240$$

The initial solution gave $z = 0$, and this solution gives $z = 240$, so we do indeed have a larger value of the objective function.

We need to know when we have reached the optimal solution, the maximum value of z. We can tell when the maximum has been achieved from the simplex tableau.

Step 4. Is z maximum?

If the last row contains any negative coefficients, z is not maximum. Because −16 is a coefficient from the last row, 240 is not the maximum value of z, so we proceed to move to another basic feasible solution.

Final Tableau

Step 5. Find another basic feasible solution.

Proceed as in Steps 2 and 3 with the most recent tableau:

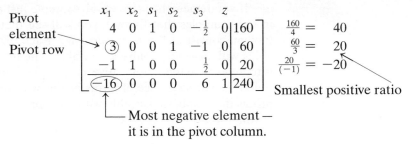

$$\begin{array}{c} \text{Pivot} \\ \text{element} \\ \text{Pivot row} \end{array}\quad \begin{array}{cccccc} x_1 & x_2 & s_1 & s_2 & s_3 & z \end{array}$$

Most negative element —
it is in the pivot column.

The smallest positive ratio determines the pivot row, row 2 in this case, and 3 is the pivot element.

We now use row operations to obtain a 1 in the pivot element position and zeros in the rest of the pivot column. Replace row 2 with $\left(\frac{1}{3}\right)$row $2 =$
$[\,1\ \ 0\ \ 0\ \ \frac{1}{3}\ \ -\frac{1}{3}\ \ 0\ \ 20\,]$ to obtain a 1 in the pivot position:

$$\begin{array}{c}\text{Need}\\ \text{zeros}\\ \text{here.}\end{array}\quad \begin{bmatrix} 4 & 0 & 1 & 0 & -\frac{1}{2} & 0 & 160 \\ 1 & 0 & 0 & \frac{1}{3} & -\frac{1}{3} & 0 & 20 \\ -1 & 1 & 0 & 0 & \frac{1}{2} & 0 & 20 \\ -16 & 0 & 0 & 0 & 6 & 1 & 240 \end{bmatrix}$$

We obtain the desired zeros by replacing the rows as follows:

Replace row 1 with $(\text{row }1 + (-4)\text{row }2)$:

$$= [\,0\ \ 0\ \ 1\ \ -\tfrac{4}{3}\ \ \tfrac{5}{6}\ \ 0\ \ 80\,]$$

Replace row 3 with $(\text{row }3 + \text{row }2)$:

$$= [\,0\ \ 1\ \ 0\ \ \tfrac{1}{3}\ \ \tfrac{1}{6}\ \ 0\ \ 40\,]$$

Replace row 4 with $(\text{row }4 + (16)\text{row }2)$:

$$= [\,0\ \ 0\ \ 0\ \ \tfrac{16}{3}\ \ \tfrac{2}{3}\ \ 1\ \ 560\,]$$

giving the tableau

$$\begin{array}{ccccccc} x_1 & x_2 & s_1 & s_2 & s_3 & z & \quad\begin{array}{c}\text{Basic}\\\text{variables}\end{array} \end{array}$$
$$\begin{bmatrix} 0 & 0 & 1 & -\frac{4}{3} & \frac{5}{6} & 0 & 80 \\ 1 & 0 & 0 & \frac{1}{3} & -\frac{1}{3} & 0 & 20 \\ 0 & 1 & 0 & \frac{1}{3} & \frac{1}{6} & 0 & 40 \\ 0 & 0 & 0 & \frac{16}{3} & \frac{2}{3} & 1 & 560 \end{bmatrix}\begin{array}{c} s_1 \\ x_1 \\ x_2 \\ \end{array}$$

The basic variables are x_1, x_2, and s_1, because these columns are unit columns. The basic feasible solution is obtained by setting s_2 and $s_3 = 0$ and solving for the others. The solution is

$$x_1 = 20, \quad x_2 = 40, \quad s_1 = 80, \quad s_2 = 0, \quad s_3 = 0, \quad z = 560$$

with x_1, x_2, and s_1 the basic variables and s_2 and s_3 nonbasic.

The value of z can be increased by going to another tableau only when there is a negative number in the bottom row. The last tableau, the **final tableau**, has no negative numbers in the last row, so $z = 560$ is the maximum value of z.

If you review the steps of the example just completed (and observe the steps in the examples that follow), you will see that the basic variables of the initial tableau consist of the slack variables. Each pivot replaces one of the basic variables with another, nonbasic, variable. The ultimate goal of the simplex method is to determine the nonnegative x's that yield the maximum value of the objective function. The sequence of pivots begins with only slack variables as basic, and it brings in a nonbasic variable with each pivot to replace a basic variable until an optimal solution is reached.

EXAMPLE 1 ➤

Use the simplex method to maximize $z = 2x_1 + 3x_2 + 2x_3$, subject to

$$2x_1 + x_2 + 2x_3 \leq 13$$
$$x_1 + x_2 - 3x_3 \leq 8$$
$$x_1 \geq 0, x_2 \geq 0, x_3 \geq 0$$

SOLUTION

We first write the problem as a system of equations:

$$2x_1 + x_2 + 2x_3 + s_1 \qquad\qquad = 13$$
$$x_1 + x_2 - 3x_3 \qquad + s_2 \qquad = 8$$
$$-2x_1 - 3x_2 - 2x_3 \qquad\qquad + z = 0$$

The initial simplex tableau is

	x_1	x_2	x_3	s_1	s_2	z		Basic variables
	2	1	2	1	0	0	13	s_1
	1	1	-3	0	1	0	8	s_2
	-2	-3	-2	0	0	1	0	

Because we have three x's, all basic solutions will have three variables set to zero. From the initial tableau, the initial basic feasible solution is

$$x_1 = 0, \qquad x_2 = 0, \qquad x_3 = 0, \qquad s_1 = 13, \qquad s_2 = 8, \qquad z = 0$$

As there are negative entries in the last row, the solution is not optimal.
Find the pivot element:

	x_1	x_2	x_3	s_1	s_2	z		
	2	1	2	1	0	0	13	$\frac{13}{1}$
	1	①	-3	0	1	0	8	$\frac{8}{1}$ Pivot row
	-2	-3	-2	0	0	1	0	

Pivot element ← →

Pivot column, most negative entry

Because all ratios are positive, the smallest, 8, determines the pivot row. Use row 2 to pivot on 1 in row 2, column 2 to find the next tableau. Use the following row operations:

Replace row 1 with (row 1 − row 2):

$$= [1 \quad 0 \quad 5 \quad 1 \quad -1 \quad 0 \quad 5]$$

Replace row 3 with (row 3 + (3)row 2):

$$= [1 \quad 0 \quad -11 \quad 0 \quad 3 \quad 1 \quad 24]$$

giving

$$\begin{array}{cccccc} x_1 & x_2 & x_3 & s_1 & s_2 & z \end{array} \qquad \begin{array}{c}\text{Basic}\\\text{variables}\end{array}$$

$$\begin{bmatrix} 1 & 0 & 5 & 1 & -1 & 0 & 5 \\ 1 & 1 & -3 & 0 & 1 & 0 & 8 \\ 1 & 0 & -11 & 0 & 3 & 1 & 24 \end{bmatrix} \begin{array}{c} s_1 \\ x_2 \\ \\ \end{array}$$

Again, the solution is not optimal because a negative entry, -11, occurs in the last row. Find the new pivot element:

$$\begin{array}{ccccccc} & x_1 & x_2 & x_3 & s_1 & s_2 & z \end{array}$$

Pivot
element

$$\begin{bmatrix} 1 & 0 & ⑤ & 1 & -1 & 0 & 5 \\ 1 & 1 & -3 & 0 & 1 & 0 & 8 \\ 1 & 0 & -11 & 0 & 3 & 1 & 24 \end{bmatrix} \qquad \begin{array}{l} \frac{5}{5} = 1 \text{ (smallest positive ratio)} \\ \frac{8}{-3} = -2.67 \end{array}$$

Pivot column

This tableau has no zero ratios, so choose the smallest positive, 1, which gives row 1 as the pivot row. Now use row 1 to pivot on 5 in row 1, column 3. First, divide row 1 by 5:

$$\begin{array}{cccccc} x_1 & x_2 & x_3 & s_1 & s_2 & z \end{array}$$

$$\begin{bmatrix} \frac{1}{5} & 0 & 1 & \frac{1}{5} & -\frac{1}{5} & 0 & 1 \\ 1 & 1 & -3 & 0 & 1 & 0 & 8 \\ 1 & 0 & -11 & 0 & 3 & 1 & 24 \end{bmatrix}$$

Next, use the following row operations:

Replace row 2 with (row 2 + (3)row 1):

$$= [\tfrac{8}{5} \quad 1 \quad 0 \quad \tfrac{3}{5} \quad \tfrac{2}{5} \quad 0 \quad 11]$$

Replace row 3 with (row 3 + (11)row 1):

$$= [\tfrac{16}{5} \quad 0 \quad 0 \quad \tfrac{11}{5} \quad \tfrac{4}{5} \quad 1 \quad 35]$$

giving the tableau

$$\begin{array}{cccccc} x_1 & x_2 & x_3 & s_1 & s_2 & z \end{array}$$

$$\begin{bmatrix} \frac{1}{5} & 0 & 1 & \frac{1}{5} & -\frac{1}{5} & 0 & 1 \\ \frac{8}{5} & 1 & 0 & \frac{3}{5} & \frac{2}{5} & 0 & 11 \\ \frac{16}{5} & 0 & 0 & \frac{11}{5} & \frac{4}{5} & 1 & 35 \end{bmatrix}$$

The basic feasible solution from the tableau is

$$x_1 = 0, \qquad x_2 = 11, \qquad x_3 = 1, \qquad s_1 = 0, \qquad s_2 = 0, \qquad z = 35$$

As there are no negative entries in the last row, $z = 35$ is a maximum.

■ **Now You Are Ready to Work Exercise 21**

Summary of the Simplex Method

Standard Maximization Problem

1. Convert the problem to a system of equations:
   (a) Convert each inequality to an equation by adding a slack variable.
   (b) Write the objective function

   $$z = ax_1 + bx_2 + \cdots + kx_n$$

   as

   $$-ax_1 - bx_2 - \cdots - kx_n + z = 0$$

2. Form the initial simplex tableau from the equations.
3. Locate the pivot element of the tableau:
   (a) Locate the most negative entry in the bottom row. It is in the pivot column. If there is a tie for most negative, choose either.
   (b) Divide each entry in the last column (above the line) by the corresponding entry in the pivot column. Choose the smallest positive ratio. It is in the pivot row. In case of a tie for pivot row, choose either.
   (c) The element where the pivot column and pivot row intersect is the pivot element.
4. Modify the simplex tableau by using row operations to obtain a new basic feasible solution.
   (a) Divide each entry in the pivot row by the pivot element to obtain a 1 in the pivot position.
   (b) Use the pivot row and row operations to obtain zeros in the other entries of the pivot column.
5. Determine whether z has reached its maximum.
   (a) If there is a negative entry in the last row of the tableau, z is not maximum. Repeat the process in steps 3 and 4.
   (b) If the bottom row contains no negative entries, z is maximum and the solution is available from the final tableau.
6. Determine the solution from the final tableau.
   (a) Set k variables to zero, where k is the number of x's used in the constraints. These are the nonbasic variables. They correspond to the columns that contain more than one nonzero entry.
   (b) Determine the values of the basic variables. These basic variables correspond to unit columns.

> **NOTE**
>
> There may be occasions when the ratio is zero. If the ratio giving zero has a *positive* divisor, then choose the row where this occurs as the pivot row. If all zero ratios have *negative* divisors, then choose the smallest *positive* ratio.

The next example illustrates a case where you have a tie for the choice of pivot column.

EXAMPLE 2 ➤

Maximize $z = 9x_1 + 5x_2 + 9x_3$, subject to

$$6x_1 + x_2 + 4x_3 \leq 72$$
$$3x_1 + 4x_2 + 2x_3 \leq 30$$
$$x_1 \geq 0, x_2 \geq 0, x_3 \geq 0$$

SOLUTION

Form the system of equations

$$6x_1 + x_2 + 4x_3 + s_1 \qquad\qquad = 72$$
$$3x_1 + 4x_2 + 2x_3 \qquad + s_2 \qquad = 30$$
$$-9x_1 - 5x_2 - 9x_3 \qquad\qquad + z = 0$$

From this system, we write the initial tableau:

$$
\begin{array}{cccccc|c}
x_1 & x_2 & x_3 & s_1 & s_2 & z & \\
6 & 1 & 4 & 1 & 0 & 0 & 72 \\
3 & 4 & 2 & 0 & 1 & 0 & 30 \\
-9 & -5 & -9 & 0 & 0 & 1 & 0
\end{array}
\quad
\begin{array}{l}
\text{Basic} \\
\text{variables} \\
s_1 \\
s_2 \\
\end{array}
$$

Because there is a tie in the last row for the most negative entry (-9), we have two choices for the pivot column. We use the first one.

Pivot element

$$
\begin{array}{cccccc|c}
x_1 & x_2 & x_3 & s_1 & s_2 & z & \\
6 & 1 & 4 & 1 & 0 & 0 & 72 \\
\boxed{3} & 4 & 2 & 0 & 1 & 0 & 30 \\
-9 & -5 & -9 & 0 & 0 & 1 & 0
\end{array}
\qquad
\begin{array}{l}
\frac{72}{6} = 12 \\
\frac{30}{3} = 10
\end{array}
$$

We give the sequence of tableaux to find the optimal solution but leave out some of the details. Be sure you follow each step.

$$
\begin{array}{cccccc|c}
6 & 1 & 4 & 1 & 0 & 0 & 72 \\
1 & \frac{4}{3} & \frac{2}{3} & 0 & \frac{1}{3} & 0 & 10 \\
-9 & -5 & -9 & 0 & 0 & 1 & 0
\end{array}
$$

$$
\begin{array}{cccccc|c}
0 & -7 & 0 & 1 & -2 & 0 & 12 \\
1 & \frac{4}{3} & \boxed{\frac{2}{3}} & 0 & \frac{1}{3} & 0 & 10 \\
0 & 7 & -3 & 0 & 3 & 1 & 90
\end{array}
\quad
\begin{array}{l}
\text{Basic} \\
\text{variables} \\
s_1 \\
x_1 \\
\end{array}
$$

Not a maximum yet

New pivot element

$$
\begin{array}{cccccc|c}
0 & -7 & 0 & 1 & -2 & 0 & 12 \\
\frac{3}{2} & 2 & 1 & 0 & \frac{1}{2} & 0 & 15 \\
0 & 7 & -3 & 0 & 3 & 1 & 90
\end{array}
$$

$$
\begin{array}{cccccc|c}
0 & -7 & 0 & 1 & -2 & 0 & 12 \\
\frac{3}{2} & 2 & 1 & 0 & \frac{1}{2} & 0 & 15 \\
\frac{9}{2} & 13 & 0 & 0 & \frac{9}{2} & 1 & 135
\end{array}
\quad
\begin{array}{l}
\text{Basic} \\
\text{variables} \\
s_1 \\
x_3 \\
\end{array}
$$

This is the final tableau with the solution

$$
x_1 = 0, \qquad x_2 = 0, \qquad x_3 = 15, \qquad s_1 = 12, \qquad s_2 = 0, \qquad z = 135
$$

We note that if we had chosen the 2 in row 2, column 3 as the pivot element, we have the sequence:

$$
\begin{array}{cccccc|c}
6 & 1 & 4 & 1 & 0 & 0 & 72 \\
3 & 4 & 2 & 0 & 1 & 0 & 30 \\
-9 & -5 & -9 & 0 & 0 & 1 & 0
\end{array}
$$

$$
\begin{array}{cccccc|c}
6 & 1 & 4 & 1 & 0 & 0 & 72 \\
\frac{3}{2} & 2 & 1 & 0 & \frac{1}{2} & 0 & 15 \\
-9 & -5 & -9 & 0 & 0 & 1 & 0
\end{array}
$$

$$\begin{bmatrix} 0 & -7 & 0 & 1 & -2 & 0 & | & 12 \\ \frac{3}{2} & 2 & 1 & 0 & \frac{1}{2} & 0 & | & 15 \\ \frac{9}{2} & 13 & 0 & 0 & \frac{9}{2} & 1 & | & 135 \end{bmatrix}$$

This gives the same final tableau in fewer steps, but you could not have predicted that this solution would be shorter.

■ **Now You Are Ready to Work Exercise 29**

In case of a tie for the most negative entry in the last row, each one will lead to the same answer. However, one might lead to the answer in fewer steps. You have no way of knowing which choice is shorter.

Now let's look at an example that has a tie for pivot row.

EXAMPLE 3 ➤

Maximize $z = 3x_1 + 8x_2$, subject to

$$\begin{aligned} x_1 + 2x_2 &\le 80 \\ 4x_1 + x_2 &\le 68 \\ 5x_1 + 3x_2 &\le 120 \\ x_1 \ge 0, x_2 &\ge 0 \end{aligned}$$

SOLUTION

The initial tableau is

$$\begin{bmatrix} 1 & 2 & 1 & 0 & 0 & 0 & | & 80 \\ 4 & 1 & 0 & 1 & 0 & 0 & | & 68 \\ 5 & 3 & 0 & 0 & 1 & 0 & | & 120 \\ -3 & -8 & 0 & 0 & 0 & 1 & | & 0 \end{bmatrix} \qquad \begin{matrix} \frac{80}{2} = 40 \\ \frac{68}{1} = 68 \\ \frac{120}{3} = 40 \end{matrix}$$

└─ Pivot column

As there are two ratios of 40 each, either row 1 or row 3 may be selected as the pivot row. If we select row 1, then 2 is the pivot element, and our sequence of tableaux is the following. You should work the row operations so that you see that each tableau is correct.

$$\begin{bmatrix} \frac{1}{2} & 1 & \frac{1}{2} & 0 & 0 & 0 & | & 40 \\ 4 & 1 & 0 & 1 & 0 & 0 & | & 68 \\ 5 & 3 & 0 & 0 & 1 & 0 & | & 120 \\ -3 & -8 & 0 & 0 & 0 & 1 & | & 0 \end{bmatrix}$$

Basic
variables

$$\begin{bmatrix} \frac{1}{2} & 1 & \frac{1}{2} & 0 & 0 & 0 & | & 40 \\ \frac{7}{2} & 0 & -\frac{1}{2} & 1 & 0 & 0 & | & 28 \\ \frac{7}{2} & 0 & -\frac{3}{2} & 0 & 1 & 0 & | & 0 \\ 1 & 0 & 4 & 0 & 0 & 1 & | & 320 \end{bmatrix} \qquad \begin{matrix} x_2 \\ s_2 \\ s_3 \\ \end{matrix}$$

This final tableau gives the optimal solution:

$$x_1 = 0, \quad x_2 = 40, \quad s_1 = 0, \quad s_2 = 28, \quad s_3 = 0, \quad z = 320$$

■ **Now You Are Ready to Work Exercise 33**

We now give an example that arises less frequently, but it is mentioned in the note to the summary of the simplex method. It is possible for a situation to arise where a ratio used to determine the pivot row is zero. Here is an example where that occurs.

EXAMPLE 4 ➤ Maximize $z = 6x + 5y$, subject to

$$3x + y \le 30$$
$$3x + 4y \le 48$$
$$4x + y \le 40$$
$$x \ge 0, y \ge 0$$

SOLUTION

The initial tableau is

$$\begin{bmatrix} 3 & 1 & 1 & 0 & 0 & 0 & 30 \\ 3 & 4 & 0 & 1 & 0 & 0 & 48 \\ 4 & 1 & 0 & 0 & 1 & 0 & 40 \\ \hline -6 & -5 & 0 & 0 & 0 & 1 & 0 \end{bmatrix} \qquad \begin{array}{l} \text{Ratio} \\ \frac{30}{3} = 10 \\ \frac{48}{3} = 16 \\ \frac{40}{4} = 10 \end{array}$$

The pivot column is column 1, and the pivot row is either row 1 or row 3 because the smallest ratio, 10, occurs twice. First, let's use row 1 as the pivot row and see what happens. We get the tableau

$$\begin{bmatrix} 1 & \frac{1}{3} & \frac{1}{3} & 0 & 0 & 0 & 10 \\ 0 & 3 & -1 & 1 & 0 & 0 & 18 \\ 0 & -\frac{1}{3} & -\frac{4}{3} & 0 & 1 & 0 & 0 \\ \hline 0 & -3 & 2 & 0 & 0 & 1 & 60 \end{bmatrix}$$

The solution is not optimal, so we use column 2 as the next pivot column. We check the ratios and obtain

$$\frac{10}{\frac{1}{3}} = 30, \qquad \frac{18}{3} = 6, \qquad \frac{0}{-\frac{1}{3}} = 0$$

The smallest of these is zero, and it came from a ratio with a negative divisor. According to the summary, we should use the smallest positive ratio, 6, to determine the pivot row. So we pivot on 3 in row 2, column 2 and obtain the tableau:

$$\begin{bmatrix} 1 & 0 & \frac{4}{9} & -\frac{1}{9} & 0 & 0 & 8 \\ 0 & 1 & -\frac{1}{3} & \frac{1}{3} & 0 & 0 & 6 \\ 0 & 0 & -\frac{13}{9} & \frac{1}{9} & 1 & 0 & 2 \\ \hline 0 & 0 & 1 & 1 & 0 & 1 & 78 \end{bmatrix}$$

This solution is optimal with maximum $z = 78$ at $(8, 6)$.

Now let's go back to the initial tableau and observe what happens if we choose the other of two possible pivot rows. The initial tableau is

$$\begin{bmatrix} 3 & 1 & 1 & 0 & 0 & 0 & 30 \\ 3 & 4 & 0 & 1 & 0 & 0 & 48 \\ 4 & 1 & 0 & 0 & 1 & 0 & 40 \\ \hline -6 & -5 & 0 & 0 & 0 & 1 & 0 \end{bmatrix}$$

The first time we used row 1 as the pivot row. Now we use row 3 as the pivot row. When we pivot on 4 in row 3, column 1, we obtain

$$\begin{bmatrix} 0 & \frac{1}{4} & 1 & 0 & -\frac{3}{4} & 0 & 0 \\ 0 & \frac{13}{4} & 0 & 1 & -\frac{3}{4} & 0 & 18 \\ 1 & \frac{1}{4} & 0 & 0 & \frac{1}{4} & 0 & 10 \\ 0 & -\frac{7}{2} & 0 & 0 & \frac{3}{2} & 1 & 60 \end{bmatrix}$$

For the next pivot, use column 2 as the pivot column. From it, we have the ratios

$$\frac{0}{\frac{1}{4}} = 0, \qquad \frac{18}{\frac{13}{4}} = 5.54, \qquad \frac{10}{\frac{1}{4}} = 40$$

In this case, the zero ratio has a positive divisor, $\frac{1}{4}$, and the summary instructions indicate that this determines the pivot row. When we pivot on $\frac{1}{4}$ in row 1, column 2, we obtain

$$\begin{bmatrix} 0 & 1 & 4 & 0 & -3 & 0 & 0 \\ 0 & 0 & -13 & 1 & 9 & 0 & 18 \\ 1 & 0 & -1 & 0 & 1 & 0 & 10 \\ 0 & 0 & 14 & 0 & -9 & 1 & 60 \end{bmatrix}$$

which is not optimal. The ratios

$$\frac{0}{-3} = 0, \qquad \frac{18}{9} = 2, \qquad \frac{10}{1} = 10$$

indicate that row 2 is the pivot row. Pivoting on 9 in row 2, column 5 we obtain

$$\begin{bmatrix} 0 & 1 & -\frac{1}{3} & \frac{1}{3} & 0 & 0 & 6 \\ 0 & 0 & -\frac{13}{9} & \frac{1}{9} & 1 & 0 & 2 \\ 1 & 0 & \frac{4}{9} & -\frac{1}{9} & 0 & 0 & 8 \\ 0 & 0 & 1 & 1 & 0 & 1 & 78 \end{bmatrix}$$

which is optimal with maximum $z = 78$ at $(8, 6)$, the same result as before.

■ **Now You Are Ready to Work Exercise 47**

To review: When we had two choices for a pivot row, either one led to the same solution. However, one sequence required more steps.

Now we want to show you what happens had we not used the row with a zero for the pivot row. Look back at the step where the tableau was

$$\begin{bmatrix} 0 & \frac{1}{4} & 1 & 0 & -\frac{3}{4} & 0 & 0 \\ 0 & \frac{13}{4} & 0 & 1 & -\frac{3}{4} & 0 & 18 \\ 1 & \frac{1}{4} & 0 & 0 & \frac{1}{4} & 0 & 10 \\ 0 & -\frac{7}{2} & 0 & 0 & \frac{3}{2} & 1 & 60 \end{bmatrix}$$

with ratios

$$\frac{0}{\frac{1}{4}} = 0, \qquad \frac{18}{\frac{13}{4}} = 5.54, \qquad \frac{10}{\frac{1}{4}} = 40$$

If we use the smallest *positive* ratio, 5.54, and pivot on $\frac{13}{4}$ in row 2 column 2, we would obtain the tableau

$$\begin{bmatrix} 0 & 0 & 1 & -\frac{1}{13} & -\frac{9}{13} & 0 & -\frac{18}{13} \\ 0 & 1 & 0 & \frac{4}{13} & -\frac{3}{13} & 0 & \frac{72}{13} \\ 1 & 0 & 0 & -\frac{1}{13} & \frac{4}{13} & 0 & \frac{112}{13} \\ 0 & 0 & 0 & \frac{14}{13} & \frac{9}{13} & 1 & \frac{1032}{13} \end{bmatrix}$$

The negative entry in the last column, $-\frac{18}{13}$, is not a valid entry; it indicates we are not in the feasible region. The pivot on $\frac{13}{4}$ took us outside the feasible region. Thus, if a zero ratio ever occurs, you must be careful to pivot correctly, or an invalid situation may arise.

4.2 EXERCISES

LEVEL 1

Write the basic feasible solution from the tableau given in Exercises 1 through 4. Indicate which variables are basic and which are nonbasic.

1. *(See Step 1)*

$$\begin{array}{ccccc} x_1 & x_2 & s_1 & s_2 & z \\ \end{array}$$
$$\left[\begin{array}{ccccc|c} 1 & 3 & 2 & 0 & 0 & 8 \\ 0 & -1 & 1 & 1 & 0 & 10 \\ \hline 0 & -4 & 2 & 0 & 1 & 14 \end{array}\right]$$

2.

$$\begin{array}{cccccc} x_1 & x_2 & s_1 & s_2 & s_3 & z \\ \end{array}$$
$$\left[\begin{array}{cccccc|c} 0 & 0 & 3 & \frac{1}{2} & 1 & 0 & 8 \\ 0 & 1 & 1 & -\frac{3}{2} & 0 & 0 & 15 \\ 1 & 0 & 4 & \frac{5}{2} & 0 & 0 & 2 \\ \hline 0 & 0 & -8 & \frac{1}{2} & 0 & 1 & 19 \end{array}\right]$$

3.

$$\begin{array}{ccccccc} x_1 & x_2 & x_3 & s_1 & s_2 & s_3 & z \\ \end{array}$$
$$\left[\begin{array}{ccccccc|c} 5 & 0 & -3 & 1 & 6 & 0 & 0 & 54 \\ 8 & 1 & 5 & 0 & 14 & 0 & 0 & 86 \\ -2 & 0 & 1 & 0 & -8 & 1 & 0 & 39 \\ \hline -4 & 0 & 3 & 0 & -2 & 0 & 1 & 148 \end{array}\right]$$

4.

$$\begin{array}{cccccc} x_1 & x_2 & x_3 & s_1 & s_2 & z \\ \end{array}$$
$$\left[\begin{array}{cccccc|c} 8 & 6 & -1 & 1 & 0 & 0 & 160 \\ 5 & 2 & 4 & 0 & 1 & 0 & 148 \\ \hline -6 & -10 & -5 & 0 & 0 & 1 & 0 \end{array}\right]$$

5. Each of the following describe a system of constraints for a linear programming problem. In each case find
   (i) The number of variables in the associated system of equations.
   (ii) The number of basic variables.
   (iii) The number of nonbasic variables.
   (a) Three constraints in four variables.

(b) Four constraints in three variables.
(c) Four constraints in five variables.

6. The constraints of a linear programming problem use six variables. The associated system of equations uses ten variables.
   (a) How many constraints are there?
   (b) How many basic variables?
   (c) How many nonbasic variables?

Determine the pivot element in each of the simplex tableaux in Exercises 7 through 16.

7. *(See Step 2)*

$$\left[\begin{array}{ccccccc|c} 5 & 4 & 3 & 1 & 0 & 0 & 0 & 8 \\ 2 & 7 & 1 & 0 & 1 & 0 & 0 & 15 \\ 6 & 8 & 5 & 0 & 0 & 1 & 0 & 24 \\ \hline -8 & -10 & -4 & 0 & 0 & 0 & 1 & 0 \end{array}\right]$$

8.

$$\left[\begin{array}{cccccc|c} 3 & 4 & 2 & 1 & 0 & 0 & 15 \\ 5 & 2 & 6 & 0 & 1 & 0 & 10 \\ \hline -8 & -3 & 10 & 0 & 0 & 1 & 0 \end{array}\right]$$

9.

$$\left[\begin{array}{ccccccc|c} 2 & 5 & 3 & 1 & 0 & 0 & 0 & 15 \\ 4 & 1 & 4 & 0 & 1 & 0 & 0 & 12 \\ 7 & 3 & -5 & 0 & 0 & 1 & 0 & 10 \\ \hline -25 & -30 & -50 & 0 & 0 & 0 & 1 & 0 \end{array}\right]$$

10.

$$\left[\begin{array}{ccccc|c} 6 & 8 & 1 & 0 & 0 & 75 \\ -2 & 5 & 0 & 1 & 0 & 20 \\ \hline -20 & -5 & 0 & 0 & 1 & 0 \end{array}\right]$$

11.

$$\left[\begin{array}{cccccc|c} 2 & 1 & 1 & 0 & 0 & 0 & 7 \\ 3 & 4 & 0 & 1 & 0 & 0 & 12 \\ 2 & 5 & 0 & 0 & 1 & 0 & 15 \\ \hline -5 & -8 & 0 & 0 & 0 & 1 & 0 \end{array}\right]$$

12.
$$\begin{bmatrix} 2 & 5 & 1 & 0 & 0 & 0 & 8 \\ 1 & 9 & 0 & 1 & 0 & 0 & 4 \\ 3 & 4 & 0 & 0 & 1 & 0 & 20 \\ \hline -9 & -4 & 0 & 0 & 0 & 1 & 0 \end{bmatrix}$$

13.
$$\begin{bmatrix} 3 & 5 & 6 & 1 & 0 & 0 & 0 & 9 \\ 2 & 8 & 2 & 0 & 1 & 0 & 0 & 6 \\ 5 & 4 & 3 & 0 & 0 & 1 & 0 & 15 \\ \hline -6 & -12 & -12 & 0 & 0 & 0 & 1 & 0 \end{bmatrix}$$

14.
$$\begin{bmatrix} 8 & 4 & 3 & 2 & 1 & 0 & 0 & 50 \\ 5 & 6 & 1 & 7 & 0 & 1 & 0 & 65 \\ \hline -6 & -20 & 5 & -20 & 0 & 0 & 1 & 0 \end{bmatrix}$$

15.
$$\begin{bmatrix} 6 & 2 & 1 & 0 & 0 & 0 & 3 \\ 4 & 3 & 0 & 1 & 0 & 0 & 0 \\ 3 & 5 & 0 & 0 & 1 & 0 & 8 \\ \hline -12 & -3 & 0 & 0 & 0 & 1 & 0 \end{bmatrix}$$

16.
$$\begin{bmatrix} 1 & 6 & 1 & 0 & 0 & 0 & 16 \\ 3 & -1 & 0 & 1 & 0 & 0 & 0 \\ 5 & 4 & 0 & 0 & 1 & 0 & 16 \\ \hline -3 & -8 & 0 & 0 & 0 & 1 & 0 \end{bmatrix}$$

In each tableau in Exercises 17 through 20, pivot on the circled entry.

17. *(See Step 3)*
$$\begin{bmatrix} 2 & 3 & 1 & 0 & 0 & 0 & 12 \\ ① & 2 & 0 & 1 & 0 & 0 & 6 \\ 2 & 5 & 0 & 0 & 1 & 0 & 20 \\ \hline -4 & -3 & 0 & 0 & 0 & 1 & 0 \end{bmatrix}$$

18.
$$\begin{bmatrix} \frac{1}{2} & \frac{1}{4} & 1 & \frac{1}{4} & 0 & 0 & 0 & 85 \\ -3 & \frac{5}{2} & 0 & -\frac{5}{2} & 1 & 0 & 0 & 50 \\ \frac{7}{2} & \frac{11}{4} & 0 & -\frac{1}{4} & 0 & 1 & 0 & 425 \\ \hline -\frac{1}{2} & -\frac{21}{4} & 0 & \frac{15}{4} & 0 & 0 & 1 & 1275 \end{bmatrix}$$

19.
$$\begin{bmatrix} 6 & 11 & 4 & 1 & 0 & 0 & 0 & 250 \\ -5 & -14 & -8 & 0 & 1 & 0 & 0 & -460 \\ -1 & -1 & -3 & 0 & 0 & 1 & 0 & -390 \\ \hline -10 & -50 & -30 & 0 & 0 & 0 & 1 & 0 \end{bmatrix}$$

20.
$$\begin{bmatrix} \frac{10}{3} & 0 & \frac{25}{3} & 1 & \frac{4}{3} & 0 & \frac{86}{3} \\ -\frac{1}{3} & 1 & -\frac{1}{3} & 0 & -\frac{1}{3} & 0 & \frac{10}{3} \\ \hline -\frac{16}{3} & 0 & -\frac{1}{3} & 0 & 0 & 1 & \frac{10}{3} \end{bmatrix}$$

LEVEL 2

Use the simplex method to solve Exercises 21 through 36.

21. *(See Example 1)* Maximize $z = 2x_1 + x_2$, subject to
$$3x_1 + x_2 \le 22$$
$$3x_1 + 4x_2 \le 34$$
$$x_1 \ge 0, x_2 \ge 0$$

22. Maximize $z = x_1 - 3x_2$, subject to
$$x_1 + x_2 \le 5$$
$$x_1 + 5x_2 \le 13$$
$$x_1 \ge 0, x_2 \ge 0$$

23. Maximize $z = 4x_1 + 5x_2$, subject to
$$x_1 + 4x_2 \le 9$$
$$4x_1 + x_2 \le 6$$
$$x_1 \ge 0, x_2 \ge 0$$

24. Maximize $z = 3x_1 + 2x_2$, subject to
$$-3x_1 + 2x_2 \le 8$$
$$5x_1 + 2x_2 \le 16$$
$$x_1 \ge 0, x_2 \ge 0$$

25. Maximize $z = 8x_1 + 4x_2$, subject to
$$x_1 + x_2 \le 240$$
$$4x_1 + 3x_2 \le 720$$
$$x_1 \ge 0, x_2 \ge 0$$

26. Maximize $z = 2x_1 + x_2 + x_3$, subject to
$$4x_1 + 12x_2 + x_3 \le 124$$
$$8x_1 + 5x_3 \le 152$$
$$x_1 \ge 0, x_2 \ge 0, x_3 \ge 0$$

27. Maximize $z = 100x_1 + 200x_2 + 50x_3$, subject to
$$5x_1 + 5x_2 + 10x_3 \le 1000$$
$$10x_1 + 8x_2 + 5x_3 \le 2000$$
$$10x_1 + 5x_2 \le 500$$
$$x_1 \ge 0, x_2 \ge 0, x_3 \ge 0$$

28. Maximize $z = 10x_1 + 24x_2 + 13x_3$, subject to
$$x_1 + 6x_2 + 3x_3 \le 36$$
$$3x_1 + 6x_2 + 6x_3 \le 45$$
$$5x_1 + 6x_2 + x_3 \le 46$$
$$x_1 \ge 0, x_2 \ge 0, x_3 \ge 0$$

29. *(See Example 2)* Maximize $z = 3x_1 + 5x_2 + 5x_3$, subject to
$$x_1 + x_2 + x_3 \le 100$$
$$3x_1 + 2x_2 + 4x_3 \le 210$$
$$x_1 + 2x_2 \le 150$$
$$x_1 \ge 0, x_2 \ge 0, x_3 \ge 0$$

30. Maximize $z = 8x_1 + 8x_2$, subject to
$$4x_1 + x_2 \le 32$$
$$4x_1 + 3x_2 \le 48$$
$$x_1 \ge 0, x_2 \ge 0$$

31. Maximize $z = 15x_1 + 9x_2 + 15x_3$, subject to

$$2x_1 + x_2 + 4x_3 \leq 360$$
$$2x_1 + 5x_2 + 10x_3 \leq 850$$
$$3x_1 + 3x_2 + x_3 \leq 510$$
$$x_1 \geq 0, x_2 \geq 0, x_3 \geq 0$$

32. Maximize $z = 8x_1 + 6x_2 + 8x_3$, subject to

$$x_1 - 3x_2 + 5x_3 \leq 50$$
$$2x_1 + 4x_2 \leq 40$$
$$x_1 \geq 0, x_2 \geq 0, x_3 \geq 0$$

33. *(See Example 3)* Maximize $z = 33x_1 + 9x_2$, subject to

$$x_1 + 8x_2 \leq 66$$
$$3x_1 + 9x_2 \leq 72$$
$$2x_1 + 6x_2 \leq 48$$
$$x_1 \geq 0, x_2 \geq 0$$

34. Maximize $z = 4x_1 + 3x_2$, subject to

$$2x_1 + 3x_2 \leq 12$$
$$x_1 + 2x_2 \leq 6$$
$$2x_1 + 5x_2 \leq 20$$
$$x_1 \geq 0, x_2 \geq 0$$

35. Maximize $z = 22x_1 + 20x_2 + 18x_3$, subject to

$$2x_1 + x_2 + 2x_3 \leq 100$$
$$x_1 + 2x_2 + 2x_3 \leq 100$$
$$2x_1 + 2x_2 + x_3 \leq 100$$
$$x_1 \geq 0, x_2 \geq 0, x_3 \geq 0$$

36. Maximize $z = x_1 + 2x_2 + 3x_3$, subject to

$$2x_1 + x_2 + 2x_3 \leq 330$$
$$x_1 + 2x_2 + 2x_3 \leq 330$$
$$-2x_1 - 2x_2 + x_3 \leq 132$$
$$x_1 \geq 0, x_2 \geq 0, x_3 \geq 0$$

LEVEL 3

37. A hardware manufacturing company makes three items: screwdrivers, chisels, and putty knives. It takes 3 hours of labor per carton to produce screwdrivers, 4 hours of labor per carton to produce chisels, and 5 hours of labor per carton to produce putty knives. Each carton of screwdrivers costs $15 to produce, each carton of chisels costs $12, and each carton of putty knives costs $11. The profit per carton is $5 for screwdrivers, $6 for chisels, and $5 for putty knives. If the company has 2200 hours of labor and $8500 in operating funds available per week, how many cartons of each item should it produce to maximize profit?

38. The maximum daily production of an oil refinery is 1900 barrels. The refinery can produce three types of fuel: gasoline, diesel, and heating oil. The production cost per barrel is $6 for gasoline, $5 for diesel, and $8 for heating oil. The daily production budget is $13,400. The profit is $8 per barrel on gasoline, $6 on diesel, and $9 for heating oil. How much of each should be produced to maximize profit?

39. The Book Fair sponsors book fairs at elementary schools and sells good books to children which are written to capture their interest. In preparation for a fair they put together three different packages of paperback children's books. Each package contains short story books, science books, and history books. Pack I contains three short story, one science, and two history books. Pack II contains two short story, four science, and one history book. Pack III contains one short story, two science, and three history books. The Book Fair has 660 short story books, 740 science books, and 853 history books available to use in the packs. The profit on each pack is $1.25 for Pack I, $2.00 for Pack II, and $1.60 for Pack III. How many of each pack should be made to maximize profit?

40. The Health Fare Cereal Company makes three cereals using wheat, oats, and raisins. The portions and profit of each cereal are the following:

| Cereal | Portion of Each Pound of Cereal | | | |
	Wheat	Oats	Raisins	Profit/Pound
Lite	0.75	0.25	0	$0.25
Trim	0.50	0.25	0.25	$0.25
Health Fare	0.25	0.50	0.25	$0.32

The company has 2320 pounds of wheat, 1380 pounds of oats, and 700 pounds of raisins available. How many pounds of each cereal should it produce to maximize profit?

41. The Humidor blends regular coffee, High Mountain coffee, and chocolate to obtain three kinds of coffee: Early Riser, After Dinner, and Deluxe.

The blends and profit for each blend are the following:

| | **Blend** | | |
	Early Riser	After Dinner	Deluxe
Regular	80%	75%	50%
High Mountain	20%	20%	40%
Chocolate	0%	5%	10%
Profit/pound	$1.00	$1.10	$1.20

The shop has 255 pounds of regular coffee, 80 pounds of High Mountain coffee, and 15 pounds of chocolate. How many pounds of each blend should be produced to maximize profit?

42. The Williams Trunk Company makes trunks for the military, for commercial use, and for decorative pieces. Each military trunk requires 4 hours for assembly, 1 hour for finishing, and 0.1 hour for packaging. Each commercial trunk requires 3 hours for assembly, 2 hours for finishing, and 0.2 hour for packaging. Each decorative trunk requires 2 hours for assembly, 4 hours for finishing, and 0.3 hour for packaging. The profit on each trunk is $6 for military, $7 for commercial, and $9 for decorative. If 4900 hours are available for assembly work, 2200 for finishing, and 210 for packaging, how many of each type of trunk should be made to maximize profit?

43. The Snack Shop makes three nut mixes from peanuts, cashews, and pecans in 1-kilogram packages. (1 kilogram = 1000 grams.) The composition of each mix is as follows:

   TV Mix: 600 grams of peanuts, 300 grams of cashews, and 100 grams of pecans

   Party Mix: 500 grams of peanuts, 300 grams of cashews, and 200 grams of pecans

   Dinner Mix: 400 grams of peanuts, 200 grams of cashews, and 400 grams of pecans

The shop has 39,500 grams of peanuts, 22,500 grams of cashews, and 18,000 grams of pecans. The selling price per package of each mix is $4.40 for TV Mix, $4.80 for Party Mix, and $5.20 for Dinner Mix. How many packages of each should be made to maximize revenue?

44. A craftsman makes two kinds of jewelry boxes for craft shows. The oval box requires 30 minutes of machine work and 20 minutes of finishing. The square box requires 20 minutes of machine work and 40 minutes of finishing. Machine work is limited to 600 minutes per day and finishing to 800 minutes. If there is $3 profit on the oval box and $4 profit on the square box, how many of each should be produced to maximize profit?

45. The Clock Works produces three clock kits for amateur woodworkers: the Majestic Grandfather Clock, the Traditional Clock, and the Wall Clock. The following chart shows the times required for cutting, sanding, and packing each kit and the price of each kit:

Process	Majestic	Traditional	Wall
Cutting	4 hours	2 hours	1 hour
Sanding	3 hours	1 hour	1 hour
Packing	1 hour	1 hour	0.5 hour
Price	$400	$250	$160

The cutting machines are available 124 hours per week; the sanding machines, 81 hours; and the packing machines, 46 hours. How many of each type of kit should be produced each week to maximize revenue?

EXPLORATIONS

46. What should be used for the pivot element in this simplex tableau?

$$\begin{bmatrix} 1 & 4 & 5 & 5 & 1 & 0 & 0 & 0 & 0 & 60 \\ 3 & 8 & 4 & 7 & 0 & 1 & 0 & 0 & 0 & 130 \\ 4 & 6 & 9 & 7.5 & 0 & 0 & 1 & 0 & 0 & 90 \\ 6 & 3 & 2 & 4 & 0 & 0 & 0 & 1 & 0 & 70 \\ \hline -10 & -50 & -30 & -50 & 0 & 0 & 0 & 0 & 1 & 0 \end{bmatrix}$$

In Exercises 47 through 49 a zero ratio was used to determine the pivot row. At some step, you will have two choices for the pivot row. Graph the feasible region. Work the problem using both choices of a feasible row and observe the sequence of corner points the tableau takes you through.

47. (See Example 4)   Maximize $z = 5x + 4y$, subject to

$$2x + y \le 80$$
$$2x + 3y \le 120$$
$$4x + y \le 160$$
$$x \ge 0, y \ge 0$$

48. Maximize $z = 15x + 20y$, subject to

$$4x + 3y \le 300$$
$$2x + 3y \le 180$$
$$x + 2y \le 120$$
$$x \ge 0, y \ge 0$$

49. Maximize $z = 7x + 8y$, subject to

$$x + 2y \le 100$$
$$x + y \le 60$$
$$4x + 5y \le 265$$
$$3x + 5y \le 255$$
$$x \ge 0, y \ge 0$$

50. The initial simplex tableau for a linear programming problem is

$$\begin{bmatrix} 5 & 5 & 10 & 1 & 0 & 0 & 0 & 1000 \\ 10 & 8 & 5 & 0 & 1 & 0 & 0 & 2000 \\ 10 & 5 & 0 & 0 & 0 & 1 & 0 & 500 \\ -100 & -200 & -50 & 0 & 0 & 0 & 1 & 0 \end{bmatrix}$$

The correct pivotal element is the 5 in row 3, column 2. What happens if you pivot on the 5 in row 1, column 2 instead?

51. The initial simplex tableau for a linear programming problem is

$$\begin{bmatrix} 60 & 50 & 40 & 1 & 0 & 0 & 0 & 395 \\ 30 & 30 & 20 & 0 & 1 & 0 & 0 & 230 \\ 10 & 20 & 40 & 0 & 0 & 1 & 0 & 200 \\ -4 & -5 & -6 & 0 & 0 & 0 & 1 & 0 \end{bmatrix}$$

The correct pivotal element is the 40 in row 3, column 3.

(a) What happens if you pivot on the 40 in row 1, column 3 instead?

(b) What happens if you pivot on 60 in row 1, column 1 instead?

Use a graphing calculator or spreadsheet to perform the indicated pivots in the following exercises.

52. Pivot on 2 in row 1, column 1.

$$\begin{bmatrix} 2 & -1 & 8 & 1 & 0 & 0 & 0 & 0 & 15 \\ -3 & 2 & -3 & 0 & 1 & 0 & 0 & 0 & 18 \\ 1 & 0 & 1 & 0 & 0 & 1 & 0 & 0 & 24 \\ 1 & -1 & 0 & 0 & 0 & 0 & 1 & 0 & 0 \\ -12 & -8 & -10 & 0 & 0 & 0 & 0 & 1 & 0 \end{bmatrix}$$

53. Pivot on 1.5 in the $(1, 3)$ position.

$$\begin{bmatrix} 1 & 0 & 1.5 & 1 & 0 & -0.5 & 0 & 50 \\ 0 & 0 & 2.5 & 1 & 1 & -1.5 & 0 & 75 \\ 0 & 1 & -1 & -1 & 0 & 1 & 0 & 25 \\ 0 & 0 & -5 & 2 & 0 & 9 & 1 & 1100 \end{bmatrix}$$

54. Pivot on the $(3, 1)$ entry.

$$\begin{bmatrix} 1 & 1 & 1 & -1 & 0 & 0 & 0 & 100 \\ 3 & 0 & -4 & 0 & 1 & 0 & 0 & 128 \\ 9 & 0 & 4 & 4 & 0 & 1 & 0 & 352 \\ -10 & 0 & -5 & 20 & 0 & 0 & 1 & 980 \end{bmatrix}$$

55. Pivot on the $(2, 3)$ entry.

$$\begin{bmatrix} 0.65 & 1 & 0.60 & 0 & 0.05 & 0 & 0 & 19.2 \\ -0.80 & 0 & 4.80 & 1 & 0.60 & 0 & 0 & 81.6 \\ 2.40 & 0 & 8.60 & 0 & 0.20 & 1 & 0 & 91.2 \\ 5.6 & 0 & -11.60 & 0 & -1.20 & 0 & 1 & 460.8 \end{bmatrix}$$

56. Use the SMPLX program on your calculator to solve the following linear programming problem:

Maximize $z = 52x_1 + 54x_2 + 48x_3 + x_4$, subject to

$$50x_1 + 60x_2 + 40x_3 + x_4 \le 660$$
$$5x_1 + 2x_2 + 4x_3 + 3x_4 \le 40$$
$$6x_1 + 4x_2 + 8x_3 + 5x_4 \le 60$$
$$x_1 \ge 0, x_2 \ge 0, x_3 \ge 0, x_4 \ge 0$$

57. Solve this problem.

Southside Landscapes proposes to plant the flower beds for the historic Earle House gardens. The company uses four patterns. Pattern I uses 40 tulips, 25 daffodils, and 6 boxwood. Pattern II uses 25 tulips, 50 daffodils, and 4 boxwood. Pattern III uses 30 tulips, 40 daffodils, and 8 boxwood. Pattern IV uses 45 tulips, 45 daffodils, and 2 boxwood. The profit for each pattern is $48 for Pattern I, $45 for Pattern II, $55 for Pattern III, and $65 for Pattern IV. Southside Landscape has 1250 tulips, 1600 daffodils, and 195 boxwood available. How many of each pattern should be used to maximize profit?

USING YOUR TI-83

A PROGRAM FOR THE SIMPLEX METHOD

A TI-83 program for the simplex method, **SMPLX,** allows you to enter the location of the pivot element, and it computes the next tableau. It also computes the ratios used to determine the next pivot row. The initial tableau is stored in [A]. Here is the **SMPLX** program.

: [A] → [B]
: dim ([A]) → L1
: Lbl 7
: Disp "PIVOT COL"
: Input J
: 1 → L
: Lbl 1
: If abs([B](L,J)) < 10 ^-7
: Goto 2
: [B](L,L1(2))/[B](L,J) → P
: round(P,2) → P
: Disp P
: Lbl 3
: L + 1→ L
: If L < L1(1)
: Goto 1
: Pause
: Goto 4
: Lbl 2
: Disp "ZERO DIV"

: Goto 3
: Lbl 4
: Disp "PIVOT ROW"
: Input I
: *Row(1/[B](I,J),[B],I) → [B]
: 1 → K
: Lbl 5
: If K = I
: Goto 6
: *Row+(-[B](K,J),[B],I,K) → [B]
: Lbl 6
: 1 + K → K
: If K ≤ L1(1)
: Goto 5
: round([B],2) → [C]
: Pause [C]
: Goto 7
: End

We use the **SMPLX** program to solve the following problem: Maximize $z = 5x_1 + 3x_2$, subject to

$$x_1 + x_2 \leq 240$$
$$2x_1 + x_2 \leq 300$$
$$x_1 \geq 0, x_2 \geq 0$$

First, enter the initial tableau in matrix [A]:

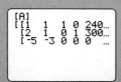

(*Note:* The z-column has been omitted throughout, because it does not change when pivoting and would only take up space.)

To initiate the program, select PRGM <SMPLX> ENTER , and you will see the screens

  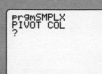

Enter the first pivot column, column 1 and ENTER , to obtain the screens

The ratios 240 and 150 indicate the second row is the pivot row. Press ENTER and enter 2.

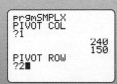

Press ENTER to obtain the next tableau.

USING EXCEL

We illustrate how to set up a sequence of tableaux for the simplex method. We use the problem:

Maximize $z = 3x + 5y$, subject to

$$3x + y \leq 180$$
$$x + 2y \leq 100$$
$$x + 5y \leq 220$$
$$x \geq 0, y \geq 0$$

The initial tableau is

$$
\begin{bmatrix}
3 & 1 & 1 & 0 & 0 & 0 & 180 \\
1 & 2 & 0 & 1 & 0 & 0 & 100 \\
1 & 5 & 0 & 0 & 1 & 1 & 220 \\
-3 & -5 & 0 & 0 & 0 & 0 & 0
\end{bmatrix}
$$

Note that the z-column has been omitted because it does not change when pivoting.

The following figure shows this matrix on a spreadsheet in cells A2:F5. The sequence of tableaux leading to a solution is also shown. Reference will be made to this figure throughout the discussion.

(continued)

	A	B	C	D	E	F	G	H	I
1	Initial	Tableau	in A2:F5					Ratios	
2	3	1	1	0	0	180		180	
3	1	2	0	1	0	100		50	
4	1	5	0	0	1	220		44	Pivot Row
5	-3	-5	0	0	0	0			
6		Pivot col.							
7									
8	Pivot on	Column 2, row 3		gives the	next tableau in A9:F12				
9	2.8	0	1	0	-0.2	136		48.5714	
10	0.6	0	0	1	-0.4	12		20	Pivot row
11	0.2	1	0	0	0.2	44		220	
12	-2	0	0	0	1	220			
13	Pivot col.								
14									
15	Pivot on	Column 1, row 2		gives the	next tableau in A16:F19				
16	0	0	1	-4.6667	1.66667	80		48	Pivot row
17	1	0	0	1.66667	-0.6667	20		-30	
18	0	1	0	-0.3333	0.33333	40		120	
19	0	0	0	3.33333	-0.3333	260			
20					Pivot col.				
21									
22	Pivot on	Column 5, row 1		gives the	final tableau in A23:F26				
23	0	0	0.6	-2.8	1	48			
24	1	0	0.4	-0.2	0	52			
25	0	1	-0.2	0.6	0	24			
26	0	0	0.2	2.4	0	276			
27									

The first pivot column is the second column. The ratios used to determine the pivot row are shown in H2:H4. The formula =F2/B2 is entered in H2 and dragged down to H4 to obtain the ratios.

For reference purposes, we will call a tableau obtained in the sequence the *current tableau* and the tableau we obtain by pivoting we call the *next tableau*. At any stage in the sequence, the formulas used to obtain the next tableau are constructed in a similar way.

- We need only to enter formulas in the first column of the next tableau (column A in the example). Drag each formula across the row to column F.

- Using the initial tableau as the current tableau, we observe the pivot element is in cell B4. The next tableau obtained by pivoting on B4 is located in A9:F12.

- The formulas in A9 through A12 are

$$\begin{aligned}
\text{A9:} \quad &=A2-\$B\$2*A11 \\
\text{A10:} \quad &=A3-\$B\$3*A11 \\
\text{A11:} \quad &=A4/\$B\$4 \\
\text{A12:} \quad &=A5-\$B\$5*A11
\end{aligned}$$

Now let us look at the pattern.

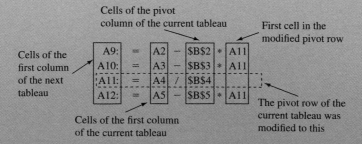

Note: The $ signs are used to create **absolute cell references.** Normally, when a formula is dragged from column A to column F, the cell addresses in the formula change so the formulas correctly refer to the data needed for each cell. In some cases, cell addresses should remain fixed. In the above, $B$2 means that cell B2 is the multiplier for all cells in the row. In pivoting, these multipliers always come from the pivot column, so they remain fixed across a row.

As you go through the sequence of tableaux in the simplex method, the next tableau becomes the current tableau, and a new next tableau is created. The formulas for the new next tableau follow the format above as long as you use the cells from the new current and next tableaux. The pivot row and column will change, so use the cells from the new pivot row and column.

Study the sequence of formulas above to make sure you understand their format.

EXERCISES

Use an EXCEL spreadsheet to solve the following exercises.

1. Maximize $z = 12x + 9y$, subject to

$$7x + 6y \leq 72$$
$$5x + 3y \leq 45$$
$$x \geq 0, y \geq 0$$

2. Maximize $z = 11x + 12y$, subject to

$$2x + 3y \leq 36$$
$$7x + 5y \leq 115$$
$$x \geq 0, y \geq 0$$

3. Maximize $z = 20x_1 + 8x_2 + 15x_3$, subject to

$$4x_1 + x_2 + 3x_3 \leq 132$$
$$3x_1 + x_2 + 2x_3 \leq 96$$
$$4x_1 + 2x_2 + 3x_3 \leq 144$$
$$x_1 \geq 0, x_2 \geq 0, x_3 \geq 0$$

4.3 | The Standard Minimum Problem: Duality

- Standard Minimum Problem
- Dual Problem
- Solve the Minimization Problem Using the Dual Problem

Standard Minimum Problem

We have used the simplex method to solve standard maximum problems. However, a variety of optimization problems that are not standard maximum need to be solved. One form that we consider in this section is the **standard minimum,** which can be solved by a procedure called the **dual method.**

The standard minimum problem can be solved with other techniques, to be discussed in Section 4.4. Those techniques also provide a means of solving other problems that are not standard. We include the dual method because it is a commonly used method.

The dual method converts a standard minimum problem to a standard maximum problem. Before we show you this method, we define a standard minimum problem.

A linear programming problem is **standard minimum** if

1. The objective function is to be minimized.
2. All the inequalities are $\geq$.
3. The constants to the right of the inequalities are nonnegative.
4. The variables are restricted to nonnegative values (nonnegative conditions).

Dual Problem

We can solve the standard minimum problem by converting it to a *dual* maximum problem. Let's look at an example to describe how to set up the **dual problem.**

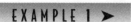

A doctor specifies that a patient's diet contain certain minimum amounts of iron and calcium, but calories are to be held to a minimum.

Two foods, A and B, are used in a meal, and the amounts of iron, calcium, and calories are given in the following table:

	Amount Provided by One Unit of		Amount Required
	A	**B**	
Iron	4	1	12 or more
Calcium	2	3	10 or more
Calories	90	120	

We convert this information into a linear programming form as follows:

$$\text{Let}\quad x_1 = \text{the number of units of A}$$
$$x_2 = \text{the number of units of B}$$

The iron requirement is

$$4x_1 + x_2 \geq 12$$

The calcium requirement is

$$2x_1 + 3x_2 \geq 10$$

and the calorie count is

$$90x_1 + 120x_2$$

where $x_1 \geq 0$ and $x_2 \geq 0$.

The problem is

Minimize $z = 90x_1 + 120x_2$, subject to

$$4x_1 + x_2 \geq 12$$
$$2x_1 + 3x_2 \geq 10$$
$$x_1 \geq 0, x_2 \geq 0$$

SOLUTION

To form the dual problem, we first write the minimum problem in a matrix form, using an augmented matrix of the constraints and objective function. For this ex-

CAUTION

This is *not* a simplex tableau, because it does not contain slack variables and the objective function has not been rewritten. We wrote the matrix in this form because it is used to obtain a maximization problem first.

ample, the matrix takes the following form. Notice its similarity to the augmented matrix of a system of equations.

$$\begin{bmatrix} 4 & 1 & 12 \\ 2 & 3 & 10 \\ 90 & 120 & 1 \end{bmatrix}$$

Next, we obtain a new matrix by taking each *row* of

$$A = \begin{bmatrix} 4 & 1 & 12 \\ 2 & 3 & 10 \\ 90 & 120 & 1 \end{bmatrix},$$

and making it the *column* of the new matrix. (The new matrix is called the **transpose of the matrix.**) The new matrix is

$$B = \begin{bmatrix} 4 & 2 & 90 \\ 1 & 3 & 120 \\ 12 & 10 & 1 \end{bmatrix}$$

From the new matrix, B, we set up a standard maximum problem. To do so, we introduce new variables, y_1 and y_2, because they play a different role from the original ones. We use the rows above the line to form constraints for the maximum problem. The row below the line forms the new objective function. Because we want this matrix to give a standard maximum problem, all inequalities are $\leq$. We write the new constraints and objective functions next to their rows:

$$\begin{array}{cc} y_1 & y_2 \\ \begin{bmatrix} 4 & 2 & 90 \\ 1 & 3 & 120 \\ 12 & 10 & 1 \end{bmatrix} \end{array} \begin{array}{l} \text{New constraint: } 4y_1 + 2y_2 \leq 90 \\ \text{New constraint: } y_1 + 3y_2 \leq 120 \\ \text{New objective function: } w = 12y_1 + 10y_2 \end{array}$$

This gives the dual problem:
    Maximize $w = 12y_1 + 10y_2$, subject to

$$\begin{aligned} 4y_1 + 2y_2 &\leq 90 \\ y_1 + 3y_2 &\leq 120 \\ y_1 \geq 0, y_2 &\geq 0 \quad \blacksquare \end{aligned}$$

EXAMPLE 2 ➤

Set up the dual problem to the following standard minimum problem.
    Minimize $z = 30x_1 + 40x_2 + 50x_3$, subject to

$$\begin{aligned} 10x_1 + 14x_2 + 5x_3 &\geq 220 \\ 5x_1 + 3x_2 + 9x_3 &\geq 340 \\ x_1 \geq 0, x_2 \geq 0, x_3 &\geq 0 \end{aligned}$$

SOLUTION

Form the augmented matrix of the problem with the objective function written in the last row:

$$\begin{array}{ccc} x_1 & x_2 & x_3 \\ A = \begin{bmatrix} 10 & 14 & 5 & 220 \\ 5 & 3 & 9 & 340 \\ 30 & 40 & 50 & 1 \end{bmatrix} \end{array}$$

Form the transpose of A:

$$B = \begin{bmatrix} 10 & 5 & | & 30 \\ 14 & 3 & | & 40 \\ 5 & 9 & | & 50 \\ \hline 220 & 340 & | & 1 \end{bmatrix} \begin{matrix} y_1 \quad\; y_2 \\ \\ \\ \\ \end{matrix}$$

with column headers y_1 and y_2 over the first two columns.

Set up the dual problem from this matrix using $\leq$ on all constraints:
Maximize $w = 220y_1 + 340y_2$, subject to

$$10y_1 + 5y_2 \leq 30$$
$$14y_1 + 3y_2 \leq 40$$
$$5y_1 + 9y_2 \leq 50$$
$$y_1 \geq 0,\ y_2 \geq 0$$

■ **Now You Are Ready to Work Exercise 5**

Set Up the Dual Problem of a Standard Minimum Problem

1. Start with a standard minimum problem.
2. Write the augmented matrix, A, of the minimum problem. Write the objective function in the last row.
3. Write the transpose of the matrix A to obtain matrix B. Each row of A becomes the corresponding column of B.
4. Form a constraint for the dual problem from each row of B (except the last) using the new variables and $\leq$.
5. Form the objective function of the dual problem from the last row of B. It is to be maximized.

Solve the Minimization Problem Using the Dual Problem

The theory relating a standard minimum problem to its dual problem is beyond the level of this course. The relationship between the solution of a minimum problem and its dual problem is a fundamental theorem, which we will use.

THEOREM
Fundamental Theorem of Duality

A standard minimum problem has a solution if and only if its dual problem has a solution. If a solution exists, the standard minimum problem and its dual problem *have the same* optimal value.

This theorem states that the maximum value of the dual problem objective function is the minimum value of the objective function for the minimum problem. To help see this, let's work through the diet example at the beginning of the section. The problem and its dual are as follows:

Standard Minimum Problem	Dual Problem
Minimize $z = 90x_1 + 120x_2$, subject to	Maximize $w = 12y_1 + 10y_2$, subject to
$4x_1 + x_2 \geq 12$	$4y_1 + 2y_2 \leq 90$
$2x_1 + 3x_2 \geq 10$	$y_1 + 3y_2 \leq 120$
$x_1 \geq 0,\ x_2 \geq 0$	$y_1 \geq 0,\ y_2 \geq 0$

The procedure is straightforward: Solve the dual problem by the simplex method. We first write the dual problem as a system of equations using slack variables and obtain

$$
\begin{aligned}
4y_1 + 2y_2 + x_1 \quad\quad\quad &= 90 \\
y_1 + 3y_2 \quad\quad + x_2 \quad\quad &= 120 \\
-12y_1 - 10y_2 \quad\quad\quad + w &= 0
\end{aligned}
$$

Notice that x_1 and x_2 are used for slack variables. These are intended to be the same as the variables in the original minimum problem, because it turns out that certain values of the slack variables of the dual problem give the desired values of the original variables in the minimum problem. Let's set up the simplex tableau and work through the solution.

The initial tableau is

$$
\begin{array}{ccccc}
y_1 & y_2 & x_1 & x_2 & w \\
\end{array}
$$
$$
\left[\begin{array}{ccccc|c}
4 & 2 & 1 & 0 & 0 & 90 \\
1 & 3 & 0 & 1 & 0 & 120 \\
\hline
-12 & -10 & 0 & 0 & 1 & 0
\end{array} \right]
$$

We now proceed to find the pivot element and perform row operations in the usual manner. You should fill in details that are omitted.

$$
\begin{array}{ccccc}
y_1 & y_2 & x_1 & x_2 & w \\
\end{array}
$$
$$
\left[\begin{array}{ccccc|c}
4 & 2 & 1 & 0 & 0 & 90 \\
1 & 3 & 0 & 1 & 0 & 120 \\
\hline
-12 & -10 & 0 & 0 & 1 & 0
\end{array} \right] \quad \tfrac{1}{4}R1 \rightarrow R1
$$

$$
\begin{array}{ccccc}
y_1 & y_2 & x_1 & x_2 & w \\
\end{array}
$$
$$
\left[\begin{array}{ccccc|c}
1 & \frac{1}{2} & \frac{1}{4} & 0 & 0 & \frac{90}{4} \\
1 & 3 & 0 & 1 & 0 & 120 \\
\hline
-12 & -10 & 0 & 0 & 1 & 0
\end{array} \right] \quad \begin{array}{l} -R1 + R2 \rightarrow R2 \\ 12R1 + R3 \rightarrow R3 \end{array}
$$

$$
\begin{array}{ccccc}
y_1 & y_2 & x_1 & x_2 & w \\
\end{array}
$$
$$
\left[\begin{array}{ccccc|c}
1 & \frac{1}{2} & \frac{1}{4} & 0 & 0 & \frac{90}{4} \\
0 & \frac{5}{2} & -\frac{1}{4} & 1 & 0 & \frac{390}{4} \\
\hline
0 & -4 & 3 & 0 & 1 & 270
\end{array} \right] \quad \tfrac{2}{5}R2 \rightarrow R2
$$

$$
\begin{array}{ccccc}
y_1 & y_2 & x_1 & x_2 & w \\
\end{array}
$$
$$
\left[\begin{array}{ccccc|c}
1 & \frac{1}{2} & \frac{1}{4} & 0 & 0 & \frac{90}{4} \\
0 & 1 & -\frac{1}{10} & \frac{2}{5} & 0 & 39 \\
\hline
0 & -4 & 3 & 0 & 1 & 270
\end{array} \right] \quad \begin{array}{l} -\tfrac{1}{2}R2 + R1 \rightarrow R1 \\ 4R2 + R3 \rightarrow R3 \end{array}
$$

$$
\begin{array}{ccccc}
y_1 & y_2 & x_1 & x_2 & w \\
\end{array}
$$
$$
\left[\begin{array}{ccccc|c}
1 & 0 & \frac{6}{20} & -\frac{1}{5} & 0 & 3 \\
0 & 1 & -\frac{1}{10} & \frac{2}{5} & 0 & 39 \\
\hline
0 & 0 & \frac{26}{10} & \frac{8}{5} & 1 & 426
\end{array} \right]
$$

As no entries of the last row are negative, the solution is optimal, and the maximum value is 426. The maximum value occurs when

$$
y_1 = 3 \quad\quad and \quad\quad y_2 = 39
$$

By the fundamental theory of duality, the *minimum* value of the original objective function, $z = 90x_1 + 120x_2$, is also 426. The values of x_1 and x_2 that yield this minimum value are found in the bottom row of the final tableau of the dual problem. That bottom row is

$$\begin{array}{ccccc} y_1 & y_2 & x_1 & x_2 & w \\ [0 & 0 & 2.6 & 1.6 & 1 & 426] \end{array}$$

The numbers under x_1 and x_2 are the values of x_1 and x_2 that give the optimal value of the original minimization problem. So the objective function $z = 90x_1 + 120x_2$ has the minimum value of 426 at $x_1 = 2.6$, $x_2 = 1.6$.

> **NOTE**
>
> To find the solution of a standard minimum problem, *look at the bottom row of the final tableau of the dual problem.*

EXAMPLE 3 ➤

Solve the following minimization problem by the dual problem method:
Minimize $z = 8x_1 + 15x_2$, subject to

$$4x_1 + 5x_2 \geq 80$$
$$2x_1 + 5x_2 \geq 60$$
$$x_1 \geq 0, x_2 \geq 0$$

SOLUTION

The augmented matrix of this problem is

$$A = \begin{bmatrix} 4 & 5 & | & 80 \\ 2 & 5 & | & 60 \\ \hline 8 & 15 & | & 1 \end{bmatrix}$$

The transpose of A is

$$B = \begin{bmatrix} 4 & 2 & | & 8 \\ 5 & 5 & | & 15 \\ \hline 80 & 60 & | & 1 \end{bmatrix}$$

B represents the following maximization problem:
Maximize $w = 80y_1 + 60y_2$, subject to

$$4y_1 + 2y_2 \leq 8$$
$$5y_1 + 5y_2 \leq 15$$
$$y_1 \geq 0, y_2 \geq 0$$

The initial simplex tableau of this problem is

$$\begin{array}{ccccc} y_1 & y_2 & x_1 & x_2 & w \\ \begin{bmatrix} 4 & 2 & 1 & 0 & 0 & | & 8 \\ 5 & 5 & 0 & 1 & 0 & | & 15 \\ \hline -80 & -60 & 0 & 0 & 1 & | & 0 \end{bmatrix} \end{array}$$

Now proceed with the pivot and row operations to obtain the sequence of tableaux:

$$\begin{array}{cccccc} y_1 & y_2 & x_1 & x_2 & w \end{array}$$

Pivot element → $\begin{bmatrix} \textcircled{4} & 2 & 1 & 0 & 0 & 8 \\ 5 & 5 & 0 & 1 & 0 & 15 \\ -80 & -60 & 0 & 0 & 1 & 0 \end{bmatrix}$ $\frac{1}{4}R1 \rightarrow R1$

$\begin{bmatrix} 1 & \frac{1}{2} & \frac{1}{4} & 0 & 0 & 2 \\ 5 & 5 & 0 & 1 & 0 & 15 \\ -80 & -60 & 0 & 0 & 1 & 0 \end{bmatrix}$ $\begin{array}{l} -5R1 + R2 \rightarrow R2 \\ \\ 80R1 + R3 \rightarrow R3 \end{array}$

Pivot element → $\begin{bmatrix} 1 & \frac{1}{2} & \frac{1}{4} & 0 & 0 & 2 \\ 0 & \textcircled{\frac{5}{2}} & -\frac{5}{4} & 1 & 0 & 5 \\ 0 & -20 & 20 & 0 & 1 & 160 \end{bmatrix}$ $\frac{2}{5}R2 \rightarrow R2$

$\begin{bmatrix} 1 & \frac{1}{2} & \frac{1}{4} & 0 & 0 & 2 \\ 0 & 1 & -\frac{1}{2} & \frac{2}{5} & 0 & 2 \\ 0 & -20 & 20 & 0 & 1 & 160 \end{bmatrix}$ $\begin{array}{l} -\frac{1}{2}R2 + R1 \rightarrow R1 \\ \\ 20\,R2 + R3 \rightarrow R3 \end{array}$

$\begin{bmatrix} 1 & 0 & \frac{1}{2} & -\frac{1}{5} & 0 & 1 \\ 0 & 1 & -\frac{1}{2} & \frac{2}{5} & 0 & 2 \\ 0 & 0 & 10 & 8 & 1 & 200 \end{bmatrix}$

This is the final tableau of the dual problem. The last row gives the solution to the minimum problem:

$$x_1 = 10, \qquad x_2 = 8, \qquad z = 200$$

■ **Now You Are Ready to Work Exercise 13**

4.3 EXERCISES

LEVEL 1

Write the transpose of the matrices in Exercises 1 through 4.

1. $\begin{bmatrix} 2 & 1 & 3 \\ 4 & 0 & 2 \end{bmatrix}$

2. $\begin{bmatrix} 5 & -1 & 6 \\ 2 & 1 & 2 \\ 3 & -3 & 5 \end{bmatrix}$

3. $\begin{bmatrix} 4 & 3 & 2 \\ 1 & 8 & -2 \\ 6 & -7 & 1 \\ 2 & 4 & 6 \end{bmatrix}$

4. $\begin{bmatrix} 2 & 1 & 5 & 4 & 3 \\ 10 & -2 & 7 & 9 & 14 \\ 8 & 15 & -3 & 6 & 1 \end{bmatrix}$

For each of the minimization problems in Exercises 5 through 8:

(a) set up the augmented matrix of the problem,
(b) find the transpose of the matrix in part (a), and
(c) set up the initial tableau for the dual problem.

5. *(See Example 2)*   Minimize $z = 25x_1 + 30x_2$, subject to

$$\begin{aligned} 6x_1 + 5x_2 &\geq 30 \\ 8x_1 + 3x_2 &\geq 42 \\ x_1 \geq 0, x_2 &\geq 0 \end{aligned}$$

6. Minimize $z = 14x_1 + 27x_2 + 9x_3$, subject to

$$\begin{aligned} 7x_1 + 9x_2 + 4x_3 &\geq 60 \\ 10x_1 + 3x_2 + 6x_3 &\geq 80 \\ 4x_1 + 2x_2 + x_3 &\geq 48 \\ x_1 \geq 0, x_2 \geq 0, x_3 &\geq 0 \end{aligned}$$

7. Minimize $z = 500x_1 + 700x_2$, subject to

$$\begin{aligned} 22x_1 + 30x_2 &\geq 110 \\ 15x_1 + 40x_2 &\geq 95 \\ 20x_1 + 35x_2 &\geq 68 \\ x_1 \geq 0, x_2 &\geq 0 \end{aligned}$$

8. Minimize $z = 40x_1 + 60x_2 + 50x_3 + 35x_4$, subject to

$$7x_1 + 6x_2 - 5x_3 + x_4 \geq 45$$
$$12x_1 + 18x_2 + 4x_3 + 6x_4 \geq 86$$
$$x_1 \geq 0, x_2 \geq 0, x_3 \geq 0, x_4 \geq 0$$

12.

y_1	y_2	y_3	x_1	x_2	w	
1	0	9	$\frac{5}{2}$	$-\frac{3}{2}$	0	15
0	1	-1	$-\frac{1}{2}$	$\frac{1}{2}$	0	12
0	0	900	750	150	1	1870

Exercises 9 through 12 give the final tableau of the dual problem of a standard minimum problem. From the tableau, determine the solution to the original problem.

9.

y_1	y_2	x_1	x_2	w	
0	1	1	-1	0	1
1	0	-1	2	0	2
0	0	6	4	1	40

10.

y_1	y_2	y_3	x_1	x_2	w	
0	16	1	$\frac{5}{2}$	$-\frac{3}{2}$	0	5
1	-1	0	$-\frac{1}{4}$	$\frac{1}{4}$	0	$\frac{5}{2}$
0	8	0	8	5	1	110

11.

y_1	y_2	y_3	x_1	x_2	x_3	w	
$\frac{1}{6}$	0	1	$\frac{1}{5}$	$-\frac{1}{8}$	0	0	$\frac{7}{6}$
$\frac{1}{3}$	1	0	$-\frac{1}{5}$	$\frac{1}{4}$	0	0	$\frac{4}{3}$
-1	0	0	$-\frac{2}{5}$	-1	1	0	1
45	0	0	12	10	0	1	510

Solve Exercises 13 through 18 by solving the dual problem.

13. *(See Example 3)*   Minimize $z = 4x_1 + 3x_2$, subject to

$$x_1 + x_2 \geq 8$$
$$2x_1 + x_2 \geq 14$$
$$x_1 \geq 0, x_2 \geq 0$$

14. Minimize $z = 42x_1 + 70x_2$, subject to

$$5x_1 + 4x_2 \geq 30$$
$$3x_1 + 4x_2 \geq 22$$
$$x_1 \geq 0, x_2 \geq 0$$

LEVEL 2

15. Minimize $z = 10x_1 + 16x_2 + 20x_3$, subject to

$$3x_1 + x_2 + 6x_3 \geq 9$$
$$x_1 + x_2 \geq 9$$
$$4x_2 + x_3 \geq 12$$
$$x_1 \geq 0, x_2 \geq 0, x_3 \geq 0$$

16. Minimize $z = 20x_1 + 30x_2$, subject to

$$6x_1 + 10x_2 \geq 60$$
$$10x_1 + 6x_2 \geq 60$$
$$x_1 + x_2 \geq 8$$
$$x_1 \geq 0, x_2 \geq 0$$

17. Minimize $z = 8x_1 + 5x_2 + 12x_3$, subject to

$$x_1 + x_2 + x_3 \geq 37$$
$$3x_1 + x_2 + 3x_3 \geq 81$$
$$3x_1 + 6x_2 + 8x_3 \geq 216$$
$$x_1 \geq 0, x_2 \geq 0, x_3 \geq 0$$

18. Minimize $z = 30x_1 + 15x_2 + 28x_3$, subject to

$$5x_1 + 3x_2 + 4x_3 \geq 45$$
$$5x_1 + 6x_2 + 8x_3 \geq 120$$
$$20x_1 + 6x_2 + 14x_3 \geq 300$$
$$x_1 \geq 0, x_2 \geq 0, x_3 \geq 0$$

LEVEL 3

19. A tire company has two plants, one in Dallas and one in New Orleans. The Dallas plant can make 800 radial and 280 standard tires per day. The New Orleans plant can make 500 radial and 150 standard tires per day. It costs $22,000 per day to operate the Dallas plant and $12,000 per day to operate the New Orleans plant. The company has a contract to make at least 28,000 radial and 9000

standard tires. How many days should each plant be scheduled to minimize operating costs?

20. A plant makes two models of an item. Each model A requires 3 hours of skilled labor and 6 hours of unskilled labor. Each model B requires 5 hours of skilled labor and 4 hours of unskilled labor. The plant's labor contract requires that it employ at least 3000 hours of skilled labor and at least 4200 hours of unskilled labor. Each model A costs $21, and each model B costs $25. How many of each model should be produced to minimize costs if the plant must produce a total of 900 items or more?

21. A tire company has plants in Chicago and Detroit. The Chicago plant can make 600 radial and 100 standard tires per day. The Detroit plant can make 300 radial and 100 standard tires per day. It costs $20,000 per day to operate the Chicago plant and $15,000 per day to operate the Detroit plant. The company has a contract to make at least 24,000 radial and 5000 standard tires. How many days should each plant be scheduled to minimize operating costs?

22. A bricklayers' union agrees to furnish at least 50 bricklayers to a shopping mall builder. The bricklayers are classified in three categories according to skill: low, medium, and high. The union requires that the total number of medium- and high-skilled bricklayers be at least four times the number of low skilled. The average number of bricks laid per

hour for each skill level is: low, 40 per hour; medium, 60 per hour; high, 75 per hour. The builder knows that the bricklayer crew must lay at least 3100 bricks per hour to stay on schedule. If the wages per hour are $10, $15, and $18 for low, medium, and high skills, respectively, how many of each type should be hired to minimize total hourly wages?

23. Elizabeth wants to choose her day's menu from cereal, cheeseburgers, and French fries. However she must choose so that she gets at least 210 grams of carbohydrates, 20 grams of dietary fiber, and 80% of minimum daily requirements of vitamin A. The foods contain the following:

	Calories	Carbohydrates	Fiber	Vitamin A
Cereal	250	30 g	8 g	20%
Cheeseburger	350	45 g	3 g	10%
Fries	500	60 g	4 g	0

How many of each food should she use in order to meet the restrictions and minimize calories?

4.4 Mixed Constraints

- Minimizing a Function
- Problems with ≥ or = Constraints
- Negative Constant in ≤ Constraints
- Examples and Applications

The simplex method has been used to solve standard maximum and standard minimum problems. Although these are important problems, other types of optimization problems arise.

In this section, we study more general problems. The constraints may be a mixture of ≤, ≥, or =, and we may wish to either maximize or minimize the objective function. Because such problems contain a mixture of ≤, ≥, or =, they are referred to as having **mixed constraints.**

Minimizing a Function

A minimization problem may or may not be standard. In either case, we can make an adjustment that converts a minimization problem to a maximization problem whose solution enables us to find the solution to the minimization problem.

The adjustment is simple. If z is the objective function to be minimized, then solve the maximization problem using $w = -z$ as the objective function. This works because if k is the maximum value of w, then $-k$ is the minimum value of z. For example, when you multiply a set of numbers by -1, you reverse the order. The set of numbers $\{1, 5, 7, 16\}$ has 1 as the smallest number and 16 as the largest number. The set made of the negatives of these numbers is $\{-1, -5, -7, -16\}$. It has -1 as the *largest* number and -16 as the *smallest.*

Here is an example that illustrates the simplex solution of a minimization problem.

EXAMPLE 1 ➤

Minimize $z = 2x_1 - 3x_2$, subject to

$$x_1 + 2x_2 \leq 10$$
$$2x_1 + x_2 \leq 11$$
$$x_1 \geq 0, x_2 \geq 0$$

SOLUTION

Convert the objective function to $w = -2x_1 + 3x_2$. We now seek to maximize $w = -2x_1 + 3x_2$, subject to the original constraints.

The tableaux for the solution are as follows:

$$
\begin{array}{ccccc}
x_1 & x_2 & s_1 & s_2 & w \\
\end{array}
$$

$$
\left[\begin{array}{ccccc|c}
1 & 2 & 1 & 0 & 0 & 10 \\
2 & 1 & 0 & 1 & 0 & 11 \\
\hline
2 & -3 & 0 & 0 & 1 & 0
\end{array}\right]
\qquad \tfrac{1}{2}\text{R1} \to \text{R1}
$$

$$
\left[\begin{array}{ccccc|c}
\tfrac{1}{2} & 1 & \tfrac{1}{2} & 0 & 0 & 5 \\
2 & 1 & 0 & 1 & 0 & 11 \\
\hline
2 & -3 & 0 & 0 & 1 & 0
\end{array}\right]
\qquad \begin{array}{l} -\text{R1} + \text{R2} \to \text{R2} \\ 3\text{R1} + \text{R3} \to \text{R3} \end{array}
$$

$$
\left[\begin{array}{ccccc|c}
\tfrac{1}{2} & 1 & \tfrac{1}{2} & 0 & 0 & 5 \\
\tfrac{3}{2} & 0 & -\tfrac{1}{2} & 1 & 0 & 6 \\
\hline
\tfrac{7}{2} & 0 & \tfrac{3}{2} & 0 & 1 & 15
\end{array}\right]
$$

The optimal solution (maximum) is $w = 15$ when $x_1 = 0$, $x_2 = 5$. The original problem then has as its optimal (minimum) solution $z = -15$ at $x_1 = 0$, $x_2 = 5$.

■ **Now You Are Ready to Work Exercise 1**

Problems with ≥ or = Constraints

The simplex method assumes all constraints are of the form

$$a_1 x_1 + a_2 x_2 + \cdots + a_n x_n \leq b$$

Realistically, some of the constraints can be of the form

$$a_1 x_1 + a_2 x_2 + \cdots + a_n x_n \geq b$$

or

$$a_1 x_1 + a_2 x_2 + \cdots + a_n x_n = b$$

We can use the simplex method by modifying any constraints of these types in the following way.

We modify a ≥ constraint by multiplying through by −1, which reverses the sign, giving a ≤ constraint. This may introduce a negative constant in the new constraint, but we will describe how to handle that situation later.

We modify an = constraint by replacing it with two constraints, a ≥ and a ≤ constraint. We can do this because the statement $c = d$ is equivalent to $c \geq d$ and $c \leq d$.

Modification of ≥ and = Constraints

For the simplex method:

Replace

$$a_1x_1 + a_2x_2 + \cdots + a_nx_n \geq b$$

with

$$-a_1x_1 - a_2x_2 - \cdots - a_nx_n \leq -b.$$

Replace

$$a_1x_1 + a_2x_2 + \cdots + a_nx_n = b$$

with

$$a_1x_1 + a_2x_2 + \cdots + a_nx_n \leq b$$

and

$$a_1x_1 + a_2x_2 + \cdots + a_nx_n \geq b$$

which in turn should be written

$$a_1x_1 + a_2x_2 + \cdots + a_nx_n \leq b$$

and

$$-a_1x_1 - a_2x_2 - \cdots - a_nx_n \leq -b.$$

Here are two examples that illustrate the modifications.

EXAMPLE 2 ➤

Modify the following problem and set up the initial simplex tableau:

Maximize $z = 8x_1 + 2x_2 + 6x_3$, subject to

$$6x_1 + 4x_2 + 5x_3 \leq 68$$
$$4x_1 + 3x_2 + x_3 \geq 32$$
$$2x_1 + 4x_2 + 3x_3 \geq 36$$
$$x_1 \geq 0, x_2 \geq 0, x_3 \geq 0$$

SOLUTION

Write all the constraints, other than the nonnegative conditions, as $\leq$ constraints. The problem then becomes

Maximize $z = 8x_1 + 2x_2 + 6x_3$, subject to

$$6x_1 + 4x_2 + 5x_3 \leq 68$$
$$-4x_1 - 3x_2 - x_3 \leq -32$$
$$-2x_1 - 4x_2 - 3x_3 \leq -36$$
$$x_1 \geq 0, x_2 \geq 0, x_3 \geq 0$$

The initial simplex tableau is

$$\begin{bmatrix} 6 & 4 & 5 & 1 & 0 & 0 & 0 & 68 \\ -4 & -3 & -1 & 0 & 1 & 0 & 0 & -32 \\ -2 & -4 & -3 & 0 & 0 & 1 & 0 & -36 \\ -8 & -2 & -6 & 0 & 0 & 0 & 1 & 0 \end{bmatrix}$$

■ **Now You Are Ready to Work Exercise 5**

EXAMPLE 3 ➤

Modify the following problem and set up the initial simplex tableau.

Minimize $z = 3x_1 + 5x_2 + 7x_3$, subject to

$$6x_1 + 8x_2 + 3x_3 \leq 72$$
$$3x_1 + 2x_2 + x_3 \geq 40$$
$$x_1 + 5x_2 + 3x_3 = 56$$
$$x_1 \geq 0, x_2 \geq 0, x_3 \geq 0$$

SOLUTION

Modify the problem to a maximization problem by writing the objective function as maximize $w = -3x_1 - 5x_2 - 7x_3$. Replace the $\geq$ and $=$ constraints to obtain the constraints

$$6x_1 + 8x_2 + 3x_3 \leq 72$$
$$-3x_1 - 2x_2 - x_3 \leq -40$$
$$x_1 + 5x_2 + 3x_3 \leq 56$$
$$-x_1 - 5x_2 - 3x_3 \leq -56$$
$$x_1 \geq 0, x_2 \geq 0, x_3 \geq 0$$

The initial tableau is

$$
\begin{bmatrix}
6 & 8 & 3 & 1 & 0 & 0 & 0 & 0 & 72 \\
-3 & -2 & -1 & 0 & 1 & 0 & 0 & 0 & -40 \\
1 & 5 & 3 & 0 & 0 & 1 & 0 & 0 & 56 \\
-1 & -5 & -3 & 0 & 0 & 0 & 1 & 0 & -56 \\
\hline
3 & 5 & 7 & 0 & 0 & 0 & 0 & 1 & 0
\end{bmatrix}
$$

■ **Now You Are Ready to Work Exercise 7**

In the last two examples, the last column contains negative constants. The standard simplex method assumes nonnegative entries in the last column, so we need to know how to proceed in such a situation.

Negative Constant in $\leq$ Constraints

Let's use a simple example to illustrate the adjustments that enable us to deal with a negative constant.

EXAMPLE 4 ▶ Maximize $z = 20x + 50y$, subject to

$$3x + 4y \leq 72$$
$$-5x + 2y \leq -16$$
$$x \geq 0, y \geq 0$$

SOLUTION

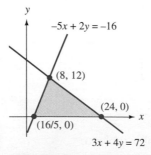

$-5x + 2y = -16$

$(8, 12)$

$(24, 0)$

$(16/5, 0)$

$3x + 4y = 72$

FIGURE 4–3 The feasible region of Example 4.

The graph of the feasible region is shown in Figure 4–3. Notice that the corners of the feasible region are $\left(\frac{16}{5}, 0\right)$, $(24, 0)$, and $(8, 12)$.

If we set up the initial tableau for this problem in the usual way, we obtain

$$
\begin{bmatrix}
3 & 4 & 1 & 0 & 0 & 72 \\
-5 & 2 & 0 & 1 & 0 & -16 \\
\hline
-20 & -50 & 0 & 0 & 1 & 0
\end{bmatrix}
$$

Generally, the initial basic solution in the simplex method is the origin. That is true in this case with $x_1 = 0$, $x_2 = 0$, $s_1 = 72$, and $s_2 = -16$. However, this solution is *not feasible* because s_2 is negative. Figure 4–3 shows that the origin is outside the feasible region. Because the basic solution is not feasible, we need to adjust the tableau so that the simplex procedure can be used. We do so by a sequence of pivots that move us from the origin to a corner point of the feasible region. Here's how.

In the initial tableau, select the row with a negative constant in the last column (row 2 because of -16). Next, select another negative entry in that row (-5 in the first column). This entry becomes a pivot element. Divide the second row by -5 to obtain the following tableau:

$$\begin{bmatrix} 3 & 4 & 1 & 0 & 0 & 72 \\ 1 & -\frac{2}{5} & 0 & -\frac{1}{5} & 0 & \frac{16}{5} \\ \hline -20 & -50 & 0 & 0 & 1 & 0 \end{bmatrix}$$

After pivoting on 1 in row 2, column 1, we have the following tableau:

$$\begin{bmatrix} 0 & \frac{26}{5} & 1 & \frac{3}{5} & 0 & \frac{312}{5} \\ 1 & -\frac{2}{5} & 0 & -\frac{1}{5} & 0 & \frac{16}{5} \\ \hline 0 & -58 & 0 & -4 & 1 & 64 \end{bmatrix}$$

This tableau yields the basic feasible solution $x = \frac{16}{5}$, $y = 0$, $s_1 = \frac{312}{5}$, and $s_2 = 0$. This solution takes us to the point $\left(\frac{16}{5}, 0\right)$, which is a corner point of the feasible region. Because the basic solution is feasible, we can apply the simplex method to this tableau in the usual manner. The next pivot element is $\frac{26}{5}$ in row 1. Pivoting gives

$$\begin{bmatrix} 0 & 1 & \frac{5}{26} & \frac{3}{26} & 0 & 12 \\ 1 & 0 & \frac{1}{13} & -\frac{2}{13} & 0 & 8 \\ \hline 0 & 0 & \frac{145}{13} & \frac{35}{13} & 1 & 760 \end{bmatrix}$$

This tableau gives the optimal solution of $z = 760$ at $x = 8$, $y = 12$.

■ **Now You Are Ready to Work Exercise 9**

Look back over the solution to the problem and observe two phases that generally apply to problems of this type.

Phase I. Set up the initial tableau. When a negative constant appears in the rightmost column of the tableau and the basic solution is not feasible, the tableau must be modified so that the basic solution is feasible. Do this by selecting a row with a negative entry in the last column and using another negative entry in that row as a pivot element. It might be necessary to pivot more than once in Phase I in order to remove all negative entries in the last column.

Phase II. When the modifications in Phase I produce a tableau with feasible basic solutions, then proceed with the usual simplex method.

Now look back at Figure 4–3 to see what happened in Phase I. The initial tableau gives the basic solution with $x = 0$, $y = 0$. In this case, this point $(0, 0)$ lies outside the feasible region. (That's why we got an infeasible solution.) When we pivot on -5, the next basic solution has $x = \frac{16}{5}$, $y = 0$ with both slack variables nonnegative. This solution is feasible and represents the corner $\left(\frac{16}{5}, 0\right)$ of the feasible region. Thus, Phase I moved from the origin to a corner of the feasible region.

When a feasible solution is reached in Phase I, we enter Phase II and follow the simplex procedure. In this example, the optimal solution is reached in one more pivot. Notice that the pivot takes us from the corner $\left(\frac{16}{5}, 0\right)$ to the corner $(8, 12)$, the optimal solution.

Look at another example to be sure that you understand the procedure followed in Phase I and Phase II.

EXAMPLE 5 ➤ Maximize $z = 3x_1 + 8x_2 + 4x_3$, subject to

$$
\begin{aligned}
x_1 + x_2 + x_3 &\leq 12 \\
2x_1 + 6x_2 + 3x_3 &\leq 42 \\
x_1 - 2x_2 &\geq 6 \\
x_1 \geq 0,\ x_2 \geq 0,\ x_3 &\geq 0
\end{aligned}
$$

SOLUTION

We need to replace the constraint

$$x_1 - 2x_2 \geq 6$$

with

$$-x_1 + 2x_2 \leq -6$$

Phase I. The initial simplex tableau is

$$
\left[
\begin{array}{ccccccc|c}
1 & 1 & 1 & 1 & 0 & 0 & 0 & 12 \\
2 & 6 & 3 & 0 & 1 & 0 & 0 & 42 \\
-1 & 2 & 0 & 0 & 0 & 1 & 0 & -6 \\
\hline
-3 & -8 & -4 & 0 & 0 & 0 & 1 & 0
\end{array}
\right]
$$

Pivot on -1 in row 3:

$$
\left[
\begin{array}{ccccccc|c}
0 & 3 & 1 & 1 & 0 & 1 & 0 & 6 \\
0 & 10 & 3 & 0 & 1 & 2 & 0 & 30 \\
1 & -2 & 0 & 0 & 0 & -1 & 0 & 6 \\
\hline
0 & -14 & -4 & 0 & 0 & -3 & 1 & 18
\end{array}
\right]
$$

This tableau gives a feasible basic solution, $x_1 = 6$, $x_2 = 0$, $x_3 = 0$, $s_1 = 6$, $s_2 = 30$, $s_3 = 0$, so proceed to Phase II.

Phase II. Pivot on 3 in row 1:

$$
\left[
\begin{array}{ccccccc|c}
0 & 1 & \frac{1}{3} & \frac{1}{3} & 0 & \frac{1}{3} & 0 & 2 \\
0 & 0 & -\frac{1}{3} & -\frac{10}{3} & 1 & -\frac{4}{3} & 0 & 10 \\
1 & 0 & \frac{2}{3} & \frac{2}{3} & 0 & -\frac{1}{3} & 0 & 10 \\
\hline
0 & 0 & \frac{2}{3} & \frac{14}{3} & 0 & \frac{5}{3} & 1 & 46
\end{array}
\right]
$$

The optimal solution is $z = 46$ at $(10, 2, 0)$.

■ **Now You Are Ready to Work Exercise 11**

Examples and Applications

We now give some examples to illustrate problems with mixed constraints.

EXAMPLE 6 ➤ Minimize $z = 8x_1 + 5x_2$, subject to

$$
\begin{aligned}
x_1 + x_2 &\leq 8 \\
5x_1 + 3x_2 &\geq 21 \\
x_1 + 3x_2 &\geq 9 \\
x_1 \geq 0,\ x_2 &\geq 0
\end{aligned}
$$

SOLUTION

We need to make the following modifications to obtain the simplex tableau.

Change minimize $z = 8x_1 + 5x_2$ to maximize $w = -8x_1 - 5x_2$, change $5x_1 + 3x_2 \geq 21$ to $-5x_1 - 3x_2 \leq -21$, and change $x_1 + 3x_2 \geq 9$ to $-x_1 - 3x_2 \leq -9$. This converts the problem to

Maximize $w = -8x_1 - 5x_2$, subject to

$$
\begin{array}{rcl}
x_1 + x_2 &\leq& 8 \\
-5x_1 - 3x_2 &\leq& -21 \\
-x_1 - 3x_2 &\leq& -9 \\
x_1 \geq 0, x_2 &\geq& 0
\end{array}
$$

(See Figure 4–4.)

The initial simplex tableau is

$$
\left[
\begin{array}{rrrrrr|r}
1 & 1 & 1 & 0 & 0 & 0 & 8 \\
-5 & -3 & 0 & 1 & 0 & 0 & -21 \\
-1 & -3 & 0 & 0 & 1 & 0 & -9 \\
\hline
8 & 5 & 0 & 0 & 0 & 1 & 0
\end{array}
\right]
$$

This has the basic solution

$$x_1 = 0, \quad x_2 = 0, \quad s_1 = 8, \quad s_2 = -21, \quad s_3 = -9$$

which is not feasible. [It gives the origin, $(0, 0)$, in Figure 4–4.] Because the solution is not feasible, we enter Phase I to modify the negative entries in the last column.

Phase I. Pivot on a negative entry in row 2 or row 3. We choose to pivot on -3 in row 2. This gives the next tableau

$$
\left[
\begin{array}{rrrrrr|r}
\frac{2}{3} & 0 & 1 & \frac{1}{3} & 0 & 0 & 1 \\
\frac{5}{3} & 1 & 0 & -\frac{1}{3} & 0 & 0 & 7 \\
4 & 0 & 0 & -1 & 1 & 0 & 12 \\
\hline
-\frac{1}{3} & 0 & 0 & \frac{5}{3} & 0 & 1 & 35
\end{array}
\right]
$$

FIGURE 4–4

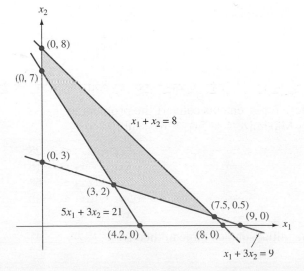

This has the basic solution

$$x_1 = 0, \qquad x_2 = 7, \qquad s_1 = 1, \qquad s_2 = 0, \qquad s_3 = 12$$

which is feasible (corner $(0, 7)$ in Figure 4–4), so we proceed with Phase II.

Phase II.   Column 1 is the pivot column and row 3 is the pivot row, so we pivot on 4 to obtain the tableau

$$\begin{bmatrix} 0 & 0 & 1 & \frac{1}{6} & \frac{1}{6} & 0 & 3 \\ 0 & 1 & 0 & \frac{1}{12} & -\frac{5}{12} & 0 & 2 \\ 1 & 0 & 0 & -\frac{1}{4} & \frac{1}{4} & 0 & 3 \\ \hline 0 & 0 & 0 & \frac{19}{12} & \frac{1}{12} & 1 & -34 \end{bmatrix}$$

This tableau yields the maximum solution $w = -34$ at $(3, 2)$, so the original problem has a minimum value of $z = 34$ at $(3, 2)$.

■ **Now You Are Ready to Work Exercise 17**

Next, we illustrate a problem with an equality constraint.

EXAMPLE 7 ➤   Maximize $z = 5x_1 + 6x_2$, subject to

$$\begin{aligned} x_1 + 2x_2 &\leq 28 \\ x_1 + x_2 &\geq 16 \\ 5x_1 + 2x_2 &= 68 \\ x_1 \geq 0, x_2 &\geq 0 \end{aligned}$$

SOLUTION

To obtain the initial simplex tableau we need to modify the constraints so all are $\leq$ constraints. First, replace

$$x_1 + x_2 \geq 16$$

with

$$-x_1 - x_2 \leq -16$$

and then replace

$$5x_1 + 2x_2 = 68$$

with

$$5x_1 + 2x_2 \leq 68 \qquad and \qquad -5x_1 - 2x_2 \leq -68$$

These replacements convert the problem to
    Maximize $z = 5x_1 + 6x_2$, subject to

$$\begin{aligned} x_1 + 2x_2 &\leq 28 \\ -x_1 - x_2 &\leq -16 \\ 5x_1 + 2x_2 &\leq 68 \\ -5x_1 - 2x_2 &\leq -68 \\ x_1 \geq 0, x_2 &> 0 \end{aligned}$$

The initial simplex tableau then is

$$\begin{bmatrix} 1 & 2 & 1 & 0 & 0 & 0 & 0 & 28 \\ -1 & -1 & 0 & 1 & 0 & 0 & 0 & -16 \\ 5 & 2 & 0 & 0 & 1 & 0 & 0 & 68 \\ -5 & -2 & 0 & 0 & 0 & 1 & 0 & -68 \\ -5 & -6 & 0 & 0 & 0 & 0 & 1 & 0 \end{bmatrix}$$

which has a basic solution

$$x_1 = 0, \quad x_2 = 0, \quad s_1 = 28, \quad s_2 = -16, \quad s_3 = 68, \quad s_4 = -68$$

which is not feasible. This occurs at the origin in Figure 4–5. A negative number in the last column gives the clue to a nonfeasible solution. As the solution is not feasible, we enter Phase I.

Phase I. Pivot on a -1 in row 2 or -5 or -2 in row 4. We choose to pivot on -1 in row 2, column 1. This gives the tableau

$$\begin{bmatrix} 0 & 1 & 1 & 1 & 0 & 0 & 0 & 12 \\ 1 & 1 & 0 & -1 & 0 & 0 & 0 & 16 \\ 0 & -3 & 0 & 5 & 1 & 0 & 0 & -12 \\ 0 & 3 & 0 & -5 & 0 & 1 & 0 & 12 \\ 0 & -1 & 0 & -5 & 0 & 0 & 1 & 80 \end{bmatrix}$$

Again, the -12 in the last column indicates the basic solution is not feasible. The basic solution is

$$x_1 = 16, \quad x_2 = 0, \quad s_1 = 12, \quad s_2 = 0, \quad s_3 = -12, \quad s_4 = 12$$

Notice that this occurs at the point $(16, 0)$ in Figure 4–5.

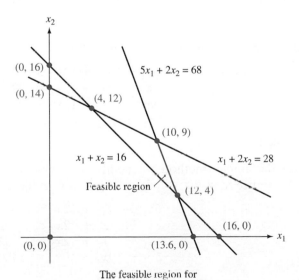

FIGURE 4–5 The feasible region is the line segment from $(10, 9)$ to $(12, 4)$.

The feasible region for
$$5x_1 + 2x_2 = 68$$
$$x_1 + 2x_2 \leq 28$$
$$x_1 + x_2 \geq 16$$
$$x_1 \geq 0, x_2 \geq 0$$
is the line segment from $(10, 9)$ to $(12, 4)$.

Now pivot on -3 in row 3 to obtain the following tableau:

$$\begin{bmatrix} 0 & 0 & 1 & \frac{8}{3} & \frac{1}{3} & 0 & 0 & 8 \\ 1 & 0 & 0 & \frac{2}{3} & \frac{1}{3} & 0 & 0 & 12 \\ 0 & 1 & 0 & -\frac{5}{3} & -\frac{1}{3} & 0 & 0 & 4 \\ 0 & 0 & 0 & 0 & 1 & 1 & 0 & 0 \\ \hline 0 & 0 & 0 & -\frac{20}{3} & -\frac{1}{3} & 0 & 1 & 84 \end{bmatrix}$$

This has the basic solution

$$x_1 = 12, \quad x_2 = 4, \quad s_1 = 8, \quad s_2 = 0, \quad s_3 = 0, \quad s_4 = 0$$

which is a feasible solution, so we enter Phase II.

Notice that this solution moves to the corner $(12, 4)$ in Figure 4–5, which is in the feasible region.

Phase II. Proceeding in the usual manner for the simplex method, we choose column 4 for the pivot column and row 1 for the pivot row. Pivoting on $\frac{8}{3}$ gives the following tableau:

$$\begin{bmatrix} 0 & 0 & \frac{3}{8} & 1 & \frac{1}{8} & 0 & 0 & 3 \\ 1 & 0 & -\frac{1}{4} & 0 & \frac{1}{4} & 0 & 0 & 10 \\ 0 & 1 & \frac{5}{8} & 0 & -\frac{1}{8} & 0 & 0 & 9 \\ 0 & 0 & 0 & 0 & 1 & 1 & 0 & 0 \\ \hline 0 & 0 & \frac{5}{2} & 0 & \frac{1}{2} & 0 & 1 & 104 \end{bmatrix}$$

This gives the optimal solution of $z = 104$ at $(10, 9)$.

■ **Now You Are Ready to Work Exercise 21**

The next example illustrates a minimization problem with mixed constraints.

EXAMPLE 8 ➤

Minimize $z = 3x_1 + 5x_2 + 2x_3$, subject to

$$\begin{aligned} 6x_1 + 9x_2 + 12x_3 &\le 672 \\ x_1 - x_2 + 2x_3 &= 92 \\ 5x_1 + 10x_2 + 10x_3 &\ge 480 \\ x_1 \ge 0, x_2 \ge 0, x_3 &\ge 0 \end{aligned}$$

SOLUTION

To use the simplex method, we modify this problem to the following:

Maximize $w = -z = -3x_1 - 5x_2 - 2x_3$, subject to

$$\begin{aligned} 6x_1 + 9x_2 + 12x_3 &\le 672 \\ x_1 - x_2 + 2x_3 &\le 92 \\ x_1 - x_2 + 2x_3 &\ge 92 \\ -5x_1 - 10x_2 - 10x_3 &\le -480 \\ x_1 \ge 0, x_2 \ge 0, x_3 &\ge 0 \end{aligned}$$

The following sequence of tableaux begins with the initial tableau and ends with the tableau having the optimal solution.

Phase I. The initial tableau is

$$
\begin{bmatrix}
6 & 9 & 12 & 1 & 0 & 0 & 0 & 0 & 672 \\
1 & -1 & 2 & 0 & 1 & 0 & 0 & 0 & 92 \\
-1 & 1 & -2 & 0 & 0 & 1 & 0 & 0 & -92 \\
-5 & -10 & -10 & 0 & 0 & 0 & 1 & 0 & -480 \\
3 & 5 & 2 & 0 & 0 & 0 & 0 & 1 & 0
\end{bmatrix}
$$

Pivot on -10 in row 4, column 2 to obtain

$$
\begin{bmatrix}
\frac{3}{2} & 0 & 3 & 1 & 0 & 0 & \frac{9}{10} & 0 & 240 \\
\frac{3}{2} & 0 & 3 & 0 & 1 & 0 & -\frac{1}{10} & 0 & 140 \\
-\frac{3}{2} & 0 & -3 & 0 & 0 & 1 & \frac{1}{10} & 0 & -140 \\
\frac{1}{2} & 1 & 1 & 0 & 0 & 0 & -\frac{1}{10} & 0 & 48 \\
\frac{1}{2} & 0 & -3 & 0 & 0 & 0 & \frac{1}{2} & 1 & -240
\end{bmatrix}
$$

Pivot on -3 in row 3 to obtain

$$
\begin{bmatrix}
0 & 0 & 0 & 1 & 0 & 1 & 1 & 0 & 100 \\
0 & 0 & 0 & 0 & 1 & 1 & 0 & 0 & 0 \\
\frac{1}{2} & 0 & 1 & 0 & 0 & -\frac{1}{3} & -\frac{1}{30} & 0 & \frac{140}{3} \\
0 & 1 & 0 & 0 & 0 & \frac{1}{3} & -\frac{1}{15} & 0 & \frac{4}{3} \\
2 & 0 & 0 & 0 & 0 & -1 & \frac{2}{5} & 1 & -100
\end{bmatrix}
$$

Phase II. Pivot on the 1 in row 2, column 6 to obtain

$$
\begin{bmatrix}
0 & 0 & 0 & 1 & -1 & 0 & 1 & 0 & 100 \\
0 & 0 & 0 & 0 & 1 & 1 & 0 & 0 & 0 \\
\frac{1}{2} & 0 & 1 & 0 & \frac{1}{3} & 0 & -\frac{1}{30} & 0 & \frac{140}{3} \\
0 & 1 & 0 & 0 & -\frac{1}{3} & 0 & -\frac{1}{15} & 0 & \frac{4}{3} \\
2 & 0 & 0 & 0 & 1 & 0 & \frac{2}{5} & 1 & -100
\end{bmatrix}
$$

The optimal solution is maximum $w = -100$ at $x_1 = 0$, $x_2 = \frac{4}{3}$, $x_3 = \frac{140}{3}$. Thus, the original problem has the optimal solution $z = -w = 100$ at $x_1 = 0$, $x_2 = \frac{4}{3}$, $x_3 = \frac{140}{3}$.

■ **Now You Are Ready to Work Exercise 27**

EXAMPLE 9 ➤

An investment firm offers three types of investments to its clients. To help a client make a better-informed decision, each investment is assigned a risk factor. The risk factor and expected return of each investment are the following:

Investment A: 12% return per year, risk factor $= 0.50$

Investment B: 15% return per year, risk factor $= 0.75$

Investment C: 9% return per year, risk factor $= 0.40$

A client wishes to invest up to $50,000. He wants an annual return of at least $6300 and at least $10,000 invested in type C investments. How much should be invested in each type to minimize his total risk? (*Note:* If $20,000 is invested in A, that risk totals $0.50 \times 20,000 = 10,000$.)

SOLUTION

Let x_1 = amount invested in A, x_2 = amount invested in B, and x_3 = amount invested in C. The total risk is to be minimized, so the objective function is

Minimize $z = 0.50x_1 + 0.75x_2 + 0.40x_3$.

The constraints are

$$
\begin{array}{rl}
x_1 + x_2 + x_3 \leq 50{,}000 & \text{(Total investment)} \\
0.12x_1 + 0.15x_2 + 0.09x_3 \geq 6{,}300 & \text{(Total annual return)} \\
x_3 \geq 10{,}000 & \text{(At least \$10,000 in C)} \\
x_1 \geq 0, x_2 \geq 0, x_3 \geq 0
\end{array}
$$

Modify the objective function to a maximum and the $\geq$ constraints to $\leq$ and obtain

Maximize $w = -0.50x_1 - 0.75x_2 - 0.40x_3$, subject to

$$
\begin{array}{rl}
x_1 + x_2 + x_3 \leq & 50{,}000 \\
-0.12x_1 - 0.15x_2 - 0.09x_3 \leq & -6{,}300 \\
-x_3 \leq & -10{,}000 \\
x_1 \geq 0, x_2 \geq 0, x_3 \geq 0 &
\end{array}
$$

The sequence of tableaux that lead to the optimal solution follows:

Phase I.

$$
\left[
\begin{array}{ccccccc|c}
1 & 1 & 1 & 1 & 0 & 0 & 0 & 50{,}000 \\
-0.12 & -0.15 & -0.09 & 0 & 1 & 0 & 0 & -6{,}300 \\
0 & 0 & -1 & 0 & 0 & 1 & 0 & -10{,}000 \\
\hline
0.50 & 0.75 & 0.40 & 0 & 0 & 0 & 1 & 0
\end{array}
\right]
$$

Pivot on -1 in row 3 to obtain

$$
\left[
\begin{array}{ccccccc|c}
1 & 1 & 0 & 1 & 0 & 1 & 0 & 40{,}000 \\
-0.12 & -0.15 & 0 & 0 & 1 & -0.09 & 0 & -5{,}400 \\
0 & 0 & 1 & 0 & 0 & -1 & 0 & 10{,}000 \\
\hline
0.5 & 0.75 & 0 & 0 & 0 & \frac{2}{5} & 1 & -4{,}000
\end{array}
\right]
$$

Pivot on -0.15 in row 2 to obtain

$$
\left[
\begin{array}{ccccccc|c}
\frac{1}{5} & 0 & 0 & 1 & \frac{20}{3} & \frac{2}{5} & 0 & 4{,}000 \\
\frac{4}{5} & 1 & 0 & 0 & -\frac{20}{3} & \frac{3}{5} & 0 & 36{,}000 \\
0 & 0 & 1 & 0 & 0 & -1 & 0 & 10{,}000 \\
\hline
-\frac{1}{10} & 0 & 0 & 0 & 5 & -\frac{1}{20} & 1 & -31{,}000
\end{array}
\right]
$$

Phase II.   Pivot on $\frac{1}{5}$ in row 1 to obtain

$$
\left[
\begin{array}{ccccccc|c}
1 & 0 & 0 & 5 & \frac{100}{3} & 2 & 0 & 20{,}000 \\
0 & 1 & 0 & -4 & -\frac{100}{3} & -1 & 0 & 20{,}000 \\
0 & 0 & 1 & 0 & 0 & -1 & 0 & 10{,}000 \\
\hline
0 & 0 & 0 & \frac{1}{2} & \frac{25}{3} & \frac{3}{20} & 1 & -29{,}000
\end{array}
\right]
$$

This tableau gives the optimal solution maximum $w = -29{,}000$ when $x_1 = 20{,}000$, $x_2 = 20{,}000$, $x_3 = 10{,}000$. Therefore, the original problem has the optimal solution minimum $z = 29,000$ when $x_1 = 20,000$, $x_2 = 20,000$, $x_3 = 10,000$. The minimum risk occurs when \$20,000 is invested in A, \$20,000 in B, and \$10,000 in C.

■ **Now You Are Ready to Work Exercise 31**

EXAMPLE 10 ➤

A convenience store has to order three items, A, B, and C. The following table summarizes information about the items.

Item	Cost	Selling Price	Storage Space Required	Weight
A	\$10	\$19	0.6 cu ft	2 lb
B	\$12	\$22	0.4 cu ft	3 lb
C	\$ 8	\$13	0.2 cu ft	4 lb

The purchasing agent must abide by the following guidelines:

The order must provide at least 3700 items.

The total cost of the order must not exceed \$36,000.

The total storage space available is 1420 cubic feet.

The total weight must not exceed 11,400 pounds.

How many of each item should be ordered to maximize profit?

SOLUTION

Let x_1 = number of items A, x_2 = number of items B, and x_3 = number of items C. The objective function and constraints described by the given information are the following:

Maximize $z = 9x_1 + 10x_2 + 5x_3$ (profit = selling price − cost), subject to

$$
\begin{aligned}
x_1 + x_2 + x_3 &\geq 3,700 & \text{(Total number of items)} \\
10x_1 + 12x_2 + 8x_3 &\leq 36,000 & \text{(Total cost)} \\
0.6x_1 + 0.4x_2 + 0.2x_3 &\leq 1,420 & \text{(Storage space)} \\
2x_1 + 3x_2 + 4x_3 &\leq 11,400 & \text{(Total weight)} \\
x_1 \geq 0, x_2 \geq 0, x_3 &\geq 0 & \text{(Nonnegative conditions)}
\end{aligned}
$$

The initial tableau and subsequent tableaux that lead to the optimal solution are as follows.

The initial tableau is

$$
\begin{bmatrix}
-1 & -1 & -1 & 1 & 0 & 0 & 0 & 0 & -3,700 \\
10 & 12 & 8 & 0 & 1 & 0 & 0 & 0 & 36,000 \\
0.6 & 0.4 & 0.2 & 0 & 0 & 1 & 0 & 0 & 1,420 \\
2 & 3 & 4 & 0 & 0 & 0 & 1 & 0 & 11,400 \\
\hline
-9 & -10 & -5 & 0 & 0 & 0 & 0 & 1 & 0
\end{bmatrix}
$$

Phase I. Because a negative number occurs in the last column, pivot on a −1 in that row (we use column 3, row 1) to obtain

$$
\begin{bmatrix}
1 & 1 & 1 & -1 & 0 & 0 & 0 & 0 & 3,700 \\
2 & 4 & 0 & 8 & 1 & 0 & 0 & 0 & 6,400 \\
0.4 & 0.2 & 0 & 0.2 & 0 & 1 & 0 & 0 & 680 \\
-2 & -1 & 0 & 4 & 0 & 0 & 1 & 0 & 3,400 \\
\hline
4 & 5 & 0 & -5 & 0 & 0 & 0 & 1 & 18,500
\end{bmatrix}
$$

Pivot on -2 in row 4 to obtain

$$\begin{bmatrix} 0 & \frac{1}{2} & 1 & 1 & 0 & 0 & \frac{1}{2} & 0 & 2{,}000 \\ 0 & 3 & 0 & 12 & 1 & 0 & 1 & 0 & 3{,}000 \\ 0 & 0 & 0 & 1 & 0 & 1 & 0.2 & 0 & 0 \\ 1 & \frac{1}{2} & 0 & -2 & 0 & 0 & -\frac{1}{2} & 0 & 1{,}700 \\ \hline 0 & -3 & 0 & -13 & 0 & 0 & -2 & 1 & 25{,}300 \end{bmatrix}$$

Phase II. Column 4 is the pivot column. The ratio $0/1$ determines the pivot row, so the 1 in row 3, column 4 is the pivot element:

$$\begin{bmatrix} 0 & \frac{1}{2} & 1 & 0 & 0 & -1 & \frac{3}{10} & 0 & 2{,}000 \\ 0 & 3 & 0 & 0 & 1 & -12 & -\frac{7}{5} & 0 & 3{,}000 \\ 0 & 0 & 0 & 1 & 0 & 1 & 0.2 & 0 & 0 \\ 1 & \frac{1}{2} & 0 & 0 & 0 & 2 & -\frac{1}{10} & 0 & 1{,}700 \\ \hline 0 & -3 & 0 & 0 & 0 & 13 & \frac{3}{5} & 1 & 25{,}300 \end{bmatrix}$$

The next pivot element is the 3 in column 2:

$$\begin{bmatrix} 0 & 0 & 1 & 0 & -\frac{1}{6} & 1 & \frac{8}{15} & 0 & 1{,}500 \\ 0 & 1 & 0 & 0 & \frac{1}{3} & -4 & -\frac{7}{15} & 0 & 1{,}000 \\ 0 & 0 & 0 & 1 & 0 & 1 & 0.2 & 0 & 0 \\ 1 & 0 & 0 & 0 & -\frac{1}{6} & 4 & \frac{2}{15} & 0 & 1{,}200 \\ \hline 0 & 0 & 0 & 0 & 1 & 1 & -\frac{4}{5} & 1 & 28{,}300 \end{bmatrix}$$

Finally, pivot on 0.2 in column 7 to obtain the optimal solution:

$$\begin{bmatrix} 0 & 0 & 1 & -\frac{8}{3} & -\frac{1}{6} & -\frac{5}{3} & 0 & 0 & 1{,}500 \\ 0 & 1 & 0 & \frac{7}{3} & \frac{1}{3} & -\frac{5}{3} & 0 & 0 & 1{,}000 \\ 0 & 0 & 0 & 5 & 0 & 5 & 1 & 0 & 0 \\ 1 & 0 & 0 & -\frac{2}{3} & -\frac{1}{6} & \frac{10}{3} & 0 & 0 & 1{,}200 \\ \hline 0 & 0 & 0 & 4 & 1 & 5 & 0 & 1 & 28{,}300 \end{bmatrix}$$

The optimal solution is maximum $z = 28{,}300$ when $x_1 = 1200$, $x_2 = 1000$, and $x_3 = 1500$. The purchasing agent should order 1200 of item A, 1000 of item B, and 1500 of item C to provide a maximum profit of \$28,300.

■ **Now You Are Ready to Work Exercise 33**

Summary of the Simplex Method for Problems with Mixed Constraints

1. For minimization problems, maximize $w = -z$.
2. ($\geq$ constraint) For each constraint of the form

$$a_1 x_1 + a_2 x_2 + \cdots + a_n x_n \geq b$$

multiply the inequality by -1 to obtain

$$-a_1 x_1 - a_2 x_2 - \cdots - a_n x_n \leq -b$$

3. ($=$ constraint) Replace each constraint of the form

$$a_1 x_1 + a_2 x_2 + \cdots + a_n x_n = b$$

with

$$a_1 x_1 + a_2 x_2 + \cdots + a_n x_n \leq b$$

and

$$a_1 x_1 + a_2 x_2 + \cdots + a_n x_n \geq b$$

The latter is written

$$-a_1 x_1 - a_2 x_2 - \cdots - a_n x_n \leq -b$$

4. Form the initial simplex tableau.
5. If no negative entry appears in the last column of the initial tableau, proceed to Phase II; otherwise, proceed to Phase I.
6. (Phase I) If there is a negative entry in the last column, change it to a positive entry by pivoting in the following manner. (Ignore a negative entry in the objective function [last row] for this step.)
   (a) The pivot row is the row containing the negative entry in the last column.
   (b) Select a negative entry in the pivot row that is to the left of the last column. The most negative entry is often a good choice. This entry is the pivot element.
   (c) Reduce the pivot element to 1 and the other entries of the pivot column to 0 using row operations.
7. Repeat the parts of step 6 as long as a negative entry occurs in the last column. When no negative entries remain in the last column (except possibly in the last row), proceed to Phase II.
8. (Phase II) The basic solution to the tableau is now feasible. Use the standard simplex procedure to obtain the optimal solution.

▦ 4.4 EXERCISES

LEVEL 1

1. *(See Example 1)* Minimize $z = 2x_1 - 5x_2$, subject to

$$\begin{aligned} 4x_1 + 3x_2 &\leq 120 \\ 2x_1 + x_2 &\leq 50 \\ x_1 > 0, x_2 &> 0 \end{aligned}$$

2. Minimize $z = 15x_1 - 40x_2$, subject to

$$\begin{aligned} 5x_1 + 3x_2 &\leq 210 \\ 2x_1 - x_2 &\leq 250 \\ x_1 \geq 0, x_2 &\geq 0 \end{aligned}$$

3. Minimize $z = 4x_1 + 5x_2 - 9x_3$, subject to

$$\begin{aligned} 3x_1 + 2x_2 - 12x_3 &\leq 120 \\ 2x_1 + 4x_2 + 6x_3 &\leq 120 \\ x_1 - 2x_2 + 3x_3 &\leq 52 \\ x_1 \geq 0, x_2 \geq 0, x_3 &\geq 0 \end{aligned}$$

4. Minimize $z = -15x_1 - 20x_2 + 5x_3$, subject to

$$\begin{aligned} 72x_1 - 48x_2 + 94x_3 &\leq 2360 \\ 5x_1 + 4x_2 - 2x_3 &\leq 30 \\ -2x_1 + 8x_2 + x_3 &\leq 40 \\ x_1 \geq 0, x_2 \geq 0, x_3 &\geq 0 \end{aligned}$$

5. *(See Example 2)* Set up the initial tableau for the following problem:

Maximize $z = 5x_1 + 3x_2 + 8x_3$, subject to

$$\begin{aligned} 9x_1 + 7x_2 + 10x_3 &\leq 154 \\ 3x_1 + 5x_2 + 8x_3 &\geq 106 \\ 6x_1 + 12x_2 + x_3 &\geq 98 \\ x_1 \geq 0, x_2 \geq 0, x_3 &\geq 0 \end{aligned}$$

6. Set up the initial tableau for the following problem:

Maximize $z = 15x_1 + 23x_2 + 7x_3$, subject to

$$\begin{aligned} 8x_1 + 9x_2 + 6x_3 &\leq 215 \\ 5x_1 + 4x_2 + 12x_3 &\geq 144 \\ 6x_1 + 8x_2 + 3x_3 &\geq 122 \\ x_1 \geq 0, x_2 \geq 0, x_3 &\geq 0 \end{aligned}$$

7. *(See Example 3)* Set up the initial tableau for the following problem:

Minimize $z = 7x_1 + 5x_2 + 8x_3$, subject to

$$\begin{aligned} 15x_1 + 23x_2 + 9x_3 &\leq 85 \\ 7x_1 + 9x_2 + 15x_3 &\geq 48 \\ x_1 + 3x_2 + 5x_3 &= 27 \\ x_1 \geq 0, x_2 \geq 0, x_3 &\geq 0 \end{aligned}$$

8. Set up the initial tableau for the following problem:

Maximize $z = 5x_1 + 8x_2 + 9x_3 + 4x_4$, subject to

$$2x_1 + 7x_2 + 3x_3 + 5x_4 \le 88$$
$$x_1 + 4x_2 + x_3 + 9x_4 \ge 67$$
$$2x_1 + 3x_2 + 2x_3 + 6x_4 = 56$$
$$x_1 \ge 0, x_2 \ge 0, x_3 \ge 0, x_4 \ge 0$$

9. *(See Example 4)*   Maximize $5x_1 + 2x_2$, subject to

$$3x_1 + 2x_2 \le 36$$
$$-2x_1 + x_2 \le -3$$
$$x_1 \ge 0, x_2 \ge 0$$

10. Maximize $11x_1 + 15x_2$, subject to

$$x_1 + 3x_2 \le 6$$
$$x_1 - 3x_2 \le -3$$
$$x_1 \ge 0, x_2 \ge 0$$

Solve Exercises 11 through 30.

11. *(See Example 5)*   Maximize $z = 15x_1 + 22x_2$, subject to

$$5x_1 + 11x_2 \le 350$$
$$15x_1 + 8x_2 \ge 300$$
$$x_1 \ge 0, x_2 \ge 0$$

12. Maximize $z = x_1 - 2x_2$, subject to

$$x_1 + x_2 \ge 10$$
$$2x_1 + 5x_2 \le 60$$
$$x_1 \ge 0, x_2 \ge 0$$

13. Maximize $z = 11x_1 + 20x_2$, subject to

$$5x_1 + 8x_2 \le 180$$
$$3x_1 + 6x_2 \ge 120$$
$$x_1 \ge 0, x_2 \ge 0$$

14. Maximize $z = 14x_1 + 24x_2 + 26x_3$, subject to

$$7x_1 + 12x_2 + 12x_3 \le 312$$
$$13x_1 + 20x_2 + 12x_3 \ge 384$$
$$5x_1 + 4x_2 + 12x_3 \ge 192$$
$$x_1 \ge 0, x_2 \ge 0, x_3 \ge 0$$

15. Maximize $z = 10x_1 + 50x_2 + 30x_3$, subject to

$$6x_1 + 12x_2 + 4x_3 \le 900$$
$$5x_1 + 16x_2 + 8x_3 \ge 120$$
$$3x_1 + x_2 + x_3 \ge 300$$
$$x_1 \ge 0, x_2 \ge 0, x_3 \ge 0$$

16. Maximize $z = 6x_1 + 6x_2 + 3x_3$, subject to

$$x_1 + 3x_2 + 3x_3 \le 120$$
$$5x_1 + 10x_2 + 5x_3 \ge 300$$
$$4x_1 + 2x_2 + x_3 \le 90$$
$$x_1 \ge 0, x_2 \ge 0, x_3 \ge 0$$

17. *(See Example 6)*   Minimize $z = 15x_1 + 8x_2$, subject to

$$x_1 + 2x_2 \le 20$$
$$3x_1 + 2x_2 \ge 36$$
$$x_1 \ge 0, x_2 \ge 0$$

18. Minimize $z = 30x_1 + 10x_2$, subject to

$$3x_1 + 8x_2 \le 120$$
$$2x_1 + x_2 \le 50$$
$$x_1 + x_2 \ge 20$$
$$x_1 \ge 0, x_2 \ge 0$$

19. Minimize $z = 4x_1 + 5x_2 + x_3$, subject to

$$10x_1 + 12x_2 + 5x_3 \ge 100$$
$$5x_1 + 7x_2 + 5x_3 \le 75$$
$$x_1 \ge 0, x_2 \ge 0, x_3 \ge 0$$

20. Minimize $z = 12x_1 + 6x_2 + 3x_3$, subject to

$$8x_1 + 2x_2 + 3x_3 \ge 144$$
$$6x_1 + x_2 + 3x_3 \le 120$$
$$x_1 \ge 0, x_2 \ge 0, x_3 \ge 0$$

21. *(See Example 7)*   Maximize $z = 8x_1 + 4x_2$, subject to

$$3x_1 + 2x_2 \le 48$$
$$2x_1 + 4x_2 \le 64$$
$$4x_1 + 6x_2 \ge 84$$
$$x_1 \ge 0, x_2 \ge 0$$

22. Maximize $z = 3x_1 + 4x_2$, subject to

$$5x_1 + 2x_2 \le 10$$
$$x_1 + 2x_2 = 6$$
$$x_1 \ge 0, x_2 \ge 0$$

23. Maximize $z = 6x_1 + 4x_2$, subject to

$$3x_1 + 2x_2 \le 60$$
$$2x_1 + 3x_2 \ge 24$$
$$x_1 + x_2 = 25$$
$$x_1 \ge 0, x_2 \ge 0$$

24. Maximize $z = 6x_1 + 5x_2 + 3x_3$, subject to

$$3x_1 + x_2 + 3x_3 \le 9$$
$$2x_1 + 3x_2 + x_3 \le 12$$
$$x_1 + x_2 - x_3 = 3$$
$$x_1 \ge 0, x_2 \ge 0, x_3 \ge 0$$

25. Maximize $z = 10x_1 + 24x_2 + 26x_3$, subject to

$$7x_1 + 12x_2 + 12x_3 \le 312$$
$$13x_1 + 20x_2 + 12x_3 \ge 384$$
$$5x_1 + 4x_2 + 12x_3 = 168$$
$$x_1 \ge 0, x_2 \ge 0, x_3 \ge 0$$

26. Maximize $z = 7x_1 + 7x_2 + 3x_3$, subject to

$$x_1 + 4x_2 + 3x_3 \le 134$$
$$2x_1 + 10x_2 + 5x_3 \ge 280$$
$$5x_1 + x_2 + 3x_3 = 100$$
$$x_1 \ge 0, x_2 \ge 0, x_3 \ge 0$$

27. (*See Example 8*) Minimize $z = 9x_1 + 5x_2$, subject to

$$-2x_1 + 5x_2 \leq 90$$
$$4x_1 + 3x_2 = 80$$
$$2x_1 - x_2 \geq 20$$
$$x_1 \geq 0, x_2 \geq 0$$

28. Minimize $z = 30x_1 + 10x_2$, subject to

$$3x_1 + 8x_2 \geq 120$$
$$2x_1 + x_2 \leq 50$$
$$x_1 + x_2 = 20$$
$$x_1 \geq 0, x_2 \geq 0$$

29. Minimize $z = 8x_1 + 10x_2 + 2x_3$, subject to

$$10x_1 + 12x_2 + 5x_3 \geq 100$$
$$5x_1 + 7x_2 + 5x_3 \leq 75$$
$$10x_1 + 2x_2 + 10x_3 = 120$$
$$x_1 \geq 0, x_2 \geq 0, x_3 \geq 0$$

30. Minimize $z = 30x_1 + 15x_2 + 16x_3$, subject to

$$5x_1 + 3x_2 + 4x_3 \geq 45$$
$$5x_1 + 6x_2 + 8x_3 \leq 120$$
$$20x_1 + 6x_2 + 14x_3 = 300$$
$$x_1 \geq 0, x_2 \geq 0, x_3 \geq 0$$

LEVEL 3

Set up the initial tableau for Exercises 31 through 38. Do not solve.

31. (*See Example 9*) A distributor offers a store a special on two models of night stands, the Custom and the Executive, if the store buys at least 100. The Custom costs $70 each, and the store will sell them for $90 each. The Executive costs $80 each and sells for $120 each. The store has 800 square feet of storage space available. Each Custom requires 4 square feet, and each Executive requires 5 square feet. The store manager wants gross sales of at least $10,800. How many of each type should be ordered so that the total cost will be minimized?

32. A bricklayers' union agrees to furnish 50 bricklayers to a shopping mall builder. The bricklayers are classified in three categories according to skill: low, medium, and high. The union requires that the total number of medium- and high-skilled bricklayers be at least four times the number of low skilled. The average number of bricks laid per hour for each skill level is: low, 40 per hour; medium, 60 per hour; high, 75 per hour. The builder knows that the bricklayer crew must lay at least 3100 bricks per hour to stay on schedule. If the wages per hour are $10, $15, and $18 for low, medium, and high skills, respectively, how many of each type should be hired to minimize total hourly wages?

33. (*See Example 10*) A store orders three items, A, B, and C. The following table summarizes information about the items:

Item	Cost	Selling Price	Storage Space Required	Weight
A	$20	$26	1 cu ft	8 lb
B	$25	$34	3 cu ft	10 lb
C	$15	$21	2 cu ft	15 lb

The purchasing agent has the following restrictions:

The order must provide at least 6600 items.

The total cost must not exceed $133,000.

The total storage space available is 13,600 cubic feet.

The total weight must not exceed 73,000 pounds.

How many of each item should be ordered to maximize profit?

34. Change the first restriction in Exercise 33 to: The order must provide at least 6800 items. Leave the other information as is and find how many of each item should be ordered to maximize profit.

35. Change the first and fourth restrictions in Exercise 33 to: The order must provide at least 6800 items. Total weight must not exceed 80,000 pounds. Add the restriction that at least 2000 of item B must be ordered. Leave the other information as is and find how many of each item should be ordered to maximize profit.

36. A college wishes to offer admission to exactly 1000 incoming freshmen. The college will give $3000 scholarships based on need and $2000 scholarships based on merit, but the total scholarships cannot exceed $900,000. The college will give at least 100 students scholarships based on need. Past experience indicates that the students who receive merit scholarships will have an average SAT score of 1200, those who receive need scholarships will have an average SAT score of 1000, and those receiving no scholarships will have an average SAT score of 900. How many need, merit, and other freshmen should the college admit to maximize the total SAT scores of all entering freshmen?

37. Roseanne jogs, plays handball, and swims at the athletic club. Jogging uses 13 calories per minute; hand-

ball, 11; and swimming, 7. She spends equal amounts of time jogging and swimming. She plays handball at least twice as long as she jogs.

(a) How long should she participate in each activity to use at least 660 calories in minimum time?

(b) She has a maximum of 90 minutes to exercise. How long should she participate in each activity to maximize the calories used?

38. Carl jogs, plays handball, and swims at the athletic club. Jogging uses 15 calories per minute; handball, 11; and swimming, 7. He swims 30 minutes and plays handball at least twice as long as he jogs.

(a) If he has 90 minutes to exercise, how long should he participate in each activity to maximize calories used?

(b) How long should he participate in each activity to use at least 715 calories in minimum time?

39. The Health Fare Cereal Company makes four cereals using wheat, oats, raisins, and nuts. The portions and profit for each cereal are shown in the following chart:

| | **Portion of Each Pound of Cereal** | | | | **Profit/** |
Cereal	**Wheat**	**Oats**	**Raisins**	**Nuts**	**Pound**
Lite	0.75	0.20	0.05	0	$0.25
Trim	0.50	0.25	0.20	0.05	$0.25
Regular	0.80	0.20	0	0	$0.27
Health Fare	0.15	0.50	0.25	0.10	$0.32

The company has 2400 pounds of wheat, 1400 pounds of oats, 700 pounds of raisins, and 250 pounds of nuts available. To use all nuts that are approaching the freshness expiration date, at least 200 pounds of nuts must be used. How many pounds of each cereal should be produced in order to maximize profit? Set up the initial simplex tableau. Do not solve.

40. The Humidor blends regular coffee, High Mountain coffee, and chocolate to obtain four kinds of coffee: Early Riser, Coffee Time, After Dinner, and Deluxe. The blends and profit for each blend are given in the following chart:

| | **Blend** | | | |
	Early Riser	**Coffee Time**	**After Dinner**	**Deluxe**
Regular	80%	75%	75%	50%
High Mountain	20%	23%	20%	40%
Chocolate	0%	2%	5%	10%
Profit/pound	$1.00	$1.10	$1.15	$1.20

The shop has 260 pounds of regular coffee, 90 pounds of High Mountain coffee, and 30 pounds of chocolate. To use chocolate while it is still fresh, at least 20 pounds must be used. How many pounds of each blend should be produced to maximize profit? Set up the initial simplex tableau. Do not solve.

41. A school cafeteria serves three foods for lunch, A, B, and C. There is pressure on the cafeteria director to reduce lunch costs. Help the director by finding the quantities of each food that will minimize costs and still maintain the desired nutritional level. The three foods have the following nutritional characteristics:

| | **Foods** | | |
Per Unit Info	**A**	**B**	**C**
Protein (g)	15	10	23
Carbohydrates (g)	20	30	11
Calories	500	400	200
Fat	8	3	6
Cost ($)	1.40	1.65	1.95

A lunch must contain at least 80 grams of protein, 95 grams of carbohydrates, and 1200 calories. It must contain no more than 35 grams of fat. How many units of each food should be served in order to minimize cost? Set up the initial tableau. Do not solve.

42. The Harpeth Heights School library has $40,000 with which to purchase history, literature, sociology, foreign language, chemistry, biology, and mathematics books. The average cost of books is history, $42; literature, $48; sociology, $35; foreign language, $37; chemistry, $65; biology, $58; and mathematics, $50.

To balance holdings, some restrictions are placed on the number ordered. The total number of books in history, literature, and sociology should be at least 300 and no more than 500. The number of foreign-language books should be no more than 12% of the total number of history, literature, and sociology books. The total number of chemistry, biology, and mathematics books should be at least 200. It is desired to find the number of books ordered in each subject to maximize the number of books purchased. Set up the initial simplex tableau. Do not solve.

43. Solve Exercise 31.

44. Solve Exercise 32.

45. Solve Exercise 33.

46. Solve Exercise 34.

47. Solve Exercise 35.

48. Solve Exercise 36.

49. Solve Exercise 37.

50. Solve Exercise 38.

EXPLORATIONS

51. For the problem

Maximize $z = 8x + 10y$, subject to

$$6x + 10y \geq 60$$
$$x + 6y \leq 72$$
$$10x + 3y \leq 150$$
$$x \geq 0, y \geq 0$$

sketch the feasible region and corner points.

Compute the sequence of simplex tableaux leading to an optional solution. Record the following information for each pivot:

Corner (x, y)	Value of z	Is the Corner in the Feasible Region?

52. For the problem

Maximize $z = 15x + 12y$, subject to

$$3x + 10y \leq 120$$
$$2x + 7y \geq 49$$
$$3x + 2y \leq 48$$
$$x \geq 0, y \geq 0$$

sketch the feasible region and its corner points.

Compute the sequence of simplex tableaux leading to an optional solution. Record the following information for each pivot.

Basic Solution (x, y)	Value of z	Is the Corner in the Feasible Portion?

53. (a) Sketch the feasible region and corners of the following linear programming problem:

Maximize $z = 10x + 15y$, subject to

$$-3x + 4y = 18$$
$$5x + 4y \leq 66$$
$$x + 4y \geq 26$$
$$x \geq 0, y \geq 0$$

(b) Describe the feasible region in part (a).

(c) Describe the feasible region for a linear programming problem like the following.

Maximize $z = cx + dy$, subject to

$$a_1 x + b_1 y = k_1$$
$$a_2 x + b_2 y \leq k_2$$
$$a_3 x + b_3 y \geq k_3$$
$$x \geq 0, y \geq 0$$

54. The problem

Minimize $z = 30x_1 + 10x_2$, subject to

$$3x_1 + 8x_2 \geq 120$$
$$2x_1 + x_2 \leq 50$$
$$x_1 + x_2 = 20$$
$$x_1 \geq 0, x_2 \geq 0$$

has an initial tableau with two rows that have a negative number in the right-hand column. For the first pivot there are four possible choices of pivot elements in the two rows. (They are $-3, -8, -1$, and -1.)

(a) Sketch the feasible region.

(b) Solve this problem four times by beginning with each possible choice of a pivot element and tracing it through the sequence of steps to the solution.

Compare the paths followed to get to the first feasible solution. Did the choice of pivot element affect the length of path?

55. Foster Corporation makes heating equipment for commercial buildings and has production operations in Cleveland, Ohio; St. Louis, Missouri; and Pittsburgh, Pennsylvania. They distribute the equipment through three distribution centers at Chicago, Dallas, and Atlanta.

The following chart shows the production capacity for the next three months for the most popular model:

Plant	Production Capacity (3 months)
Cleveland	4,200
St. Louis	4,800
Pittsburgh	3,700
Total	12,700

The following chart shows the forecast for the demand for the next three months at the distribution centers:

Distributor	Demand (3 months)
Chicago	5,400
Dallas	3,800
Atlanta	2,700
Total	11,900

The following chart shows the transportation costs of shipping a unit from the production plants to the distribution centers:

	Distribution Center		
Plant	Chicago	Dallas	Atlanta
Cleveland	35	64	60
St. Louis	37	59	51
Pittsburgh	49	68	57

Find the quantity to be shipped from each production plant to each distribution center that minimizes transportation costs. The number shipped to each distribution center should equal the demand.

(a) Set up the constraints and objective function.

(b) Set up the initial simplex tableau.

(c) Do not solve.

56. Suppose that in Exercise 55, extreme flooding prohibits shipping from St. Louis to Atlanta. Find the constraints and objective function that minimize shipping costs.

For Exercises 57 through 59, use the SMPLX program on a graphing calculator, or use pivoting on a spreadsheet.

57. Solve the following:

Maximize $z = 7x_1 + 7x_2 + 3x_3$, subject to

$$x_1 + 4x_2 + 3x_3 \leq 134$$
$$2x_1 + 10x_2 + 5x_3 \geq 280$$
$$5x_1 + x_2 + 3x_3 \leq 100$$
$$x_1 \geq 0, x_2 \geq 0, x_3 \geq 0$$

58. Solve the following:

Maximize $z = 45x_1 + 27x_2 + 18x_3 + 36x_4$, subject to

$$5x_1 + x_2 + x_3 + 8x_4 \leq 90$$
$$2x_1 + 4x_2 + 3x_3 + 2x_4 \leq 81$$
$$2x_1 + x_2 + x_3 + 4x_4 \geq 45$$
$$x_1 \geq 0, x_2 \geq 0, x_3 \geq 0, x_4 \geq 0$$

59. Solve the following:

Maximize $z = 16x_1 + 14x_2 + 22x_3 + 28x_4$, subject to

$$4x_1 + 3x_2 + 2x_3 + 5x_4 \leq 113$$
$$x_1 + 2x_2 + 6x_3 + 4x_4 \leq 92$$
$$8x_1 + 4x_2 + 4x_3 + 10x_4 \leq 212$$
$$2x_1 + 3x_2 + 4x_3 + 5x_4 \leq 107$$
$$x_1 \geq 0, x_2 \geq 0, x_3 \geq 0, x_4 \geq 0$$

4.5 Multiple Solutions, Unbounded Solutions, and No Solutions

- Multiple Solutions
- No Solutions
- Unbounded Solutions
- No Feasible Solution

Generally, a mathematics textbook introduces a new concept or method with examples and exercises that have nice, neat solutions.

Although well-behaved problems form a valid starting point, texts sometimes avoid problems that stray from these nice forms. In this chapter on the simplex method, we have studied the basic method and some variations, but the examples and problems generally have had unique solutions. That is not always the case. Reflect a moment on the fact that the simplex method converts a linear programming problem to a system of equations. We know that a system of equations can have a unique solution, no solution, or an infinity of solutions. Thus, with good reason, we should expect situations to arise when a linear program has no solutions or many solutions. Fortunately, the simplex method can signal when no solutions are possible or when multiple solutions exist.

Multiple Solutions

We look at a problem with multiple optimal solutions.

EXAMPLE 1 ➤

Maximize $z = 18x + 24y$, subject to

$$3x + 4y \le 48$$
$$x + 2y \le 22$$
$$3x + 2y \le 42$$
$$x \ge 0, y \ge 0$$

SOLUTION

Figure 4–6 shows the graph of the feasible region. Notice that the corners are $(0, 0)$, $(0, 11)$, $(4, 9)$, $(12, 3)$, and $(14, 0)$, so the maximum value occurs at one or more of these corners.

The initial tableau is

$$
\begin{array}{cccccc|c}
x & y & s_1 & s_2 & s_3 & z & \\
3 & 4 & 1 & 0 & 0 & 0 & 48 \\
1 & 2 & 0 & 1 & 0 & 0 & 22 \\
3 & 2 & 0 & 0 & 1 & 0 & 42 \\
\hline
-18 & -24 & 0 & 0 & 0 & 1 & 0
\end{array}
$$

The pivot element is 2 in row 2, column 2. Pivoting on this gives the tableau

$$
\begin{array}{cccccc|c}
x & y & s_1 & s_2 & s_3 & z & \\
1 & 0 & 1 & -2 & 0 & 0 & 4 \\
\frac{1}{2} & 1 & 0 & \frac{1}{2} & 0 & 0 & 11 \\
2 & 0 & 0 & -1 & 1 & 0 & 20 \\
\hline
-6 & 0 & 0 & 12 & 0 & 1 & 264
\end{array}
$$

This is not optimal, so we pivot on 1 in row 1, column 1 to obtain the following tableau:

$$
\begin{array}{cccccc|c}
x & y & s_1 & s_2 & s_3 & z & \\
1 & 0 & 1 & -2 & 0 & 0 & 4 \\
0 & 1 & -\frac{1}{2} & \frac{3}{2} & 0 & 0 & 9 \\
0 & 0 & -2 & 3 & 1 & 0 & 12 \\
\hline
0 & 0 & 6 & 0 & 0 & 1 & 288
\end{array}
$$

FIGURE 4–6

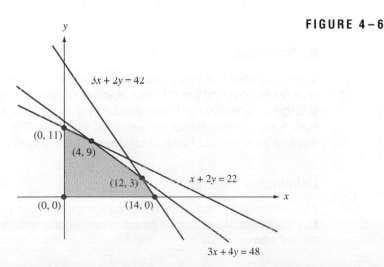

This tableau gives an optimal solution $z = 288$ when $x = 4$, $y = 9$. Recall that we say that x, y, and s_3 are basic variables and s_1 and s_2 are nonbasic variables. Notice that there is a zero in the bottom row of the s_2 column. This zero is the clue that there might be a different optimal solution.

To determine whether there is another optimal solution, use the s_2 column as the pivot column. The ratios of the entries in the last column are -2, 6, and 4. Because 4 is the smallest *nonnegative* ratio, row 3 is the pivot row, and 3 in row 3, column 4 is the pivot element. After pivoting, we obtain the tableau

$$
\begin{array}{cccccc}
x & y & s_1 & s_2 & s_3 & z \\
\end{array}
$$
$$
\left[
\begin{array}{cccccc|c}
1 & 0 & -\frac{1}{3} & 0 & \frac{2}{3} & 0 & 12 \\
0 & 1 & \frac{1}{2} & 0 & -\frac{1}{2} & 0 & 3 \\
0 & 0 & -\frac{2}{3} & 1 & \frac{1}{3} & 0 & 4 \\
\hline
0 & 0 & 6 & 0 & 0 & 1 & 288 \\
\end{array}
\right]
$$

This tableau gives an optimal solution $z = 288$ when $x = 12$, $y = 3$. Thus, the same maximum value of z, 288, occurs at another point. The optimal solution occurs at the corners $(4, 9)$ and $(12, 3)$ of the feasible region. Actually, *all* points on the line segment between $(4, 9)$ and $(12, 3)$ also yield the maximum value of $z = 288$.

■ **Now You Are Ready to Work Exercise 5**

Multiple Solutions

To determine whether a problem has more than one optimal solution:

1. Find an optimal solution by the usual simplex method.
2. Look at zeros in the bottom row of the final tableau. If a zero appears in the bottom row of a column for a *nonbasic* variable, there might be other optimal solutions.
3. To find another optimal solution, if any, use the column of a nonbasic variable with a zero at the bottom as the pivot column. Find the pivot row in the usual manner, and then pivot on the pivot element.
4. If this new tableau gives the same optimal value of z at another point, then multiple solutions exist.
5. Given the two optimal solutions, all points on the line segment joining them are also optimal solutions.

No Solutions

Two conditions give rise to no solution in a linear programming problem. In one case, an unbounded feasible region exists, so the objective function can be made arbitrarily large by selecting points farther away in the feasible region. In another case, some constraints are inconsistent, so no feasible region exists. We now find out how to recognize these situations from the simplex tableau.

Unbounded Solutions

The following example illustrates a problem with an unbounded feasible region. In such a case, the objective function has no maximum value because it can be arbitrarily large.

EXAMPLE 2 ➤ Maximize $z = x_1 + 4x_2$, subject to

$$x_1 - x_2 \leq 3$$
$$-4x_1 + x_2 \leq 4$$
$$x_1 \geq 0, x_2 \geq 0$$

SOLUTION

The graph of the feasible region is shown in Figure 4–7. This problem converts to the system

$$x_1 - x_2 + s_1 \quad\quad = 3$$
$$-4x_1 + x_2 \quad\quad + s_2 \quad\quad = 4$$
$$-x_1 - 4x_2 \quad\quad\quad + z = 0$$

and to the initial simplex tableau

$$
\begin{array}{ccccc}
x_1 & x_2 & s_1 & s_2 & z \\
\end{array}
$$
$$
\left[
\begin{array}{ccccc|c}
1 & -1 & 1 & 0 & 0 & 3 \\
-4 & \textcircled{1} & 0 & 1 & 0 & 4 \\
\hline
-1 & -4 & 0 & 0 & 1 & 0 \\
\end{array}
\right]
$$

Pivot element

Except for the pivot element, convert all entries in the pivot column to 0 to obtain the next tableau:

$$
\left[
\begin{array}{ccccc|c}
-3 & 0 & 1 & 1 & 0 & 7 \\
-4 & 1 & 0 & 1 & 0 & 4 \\
\hline
-17 & 0 & 0 & 4 & 1 & 16 \\
\end{array}
\right]
$$

Because of the -17 in the last row, we know that z is not maximal. When we check for the pivot row, we get the ratios $-7/3$ and -1. We cannot proceed with the simplex method because *all* ratios are negative. When this occurs, there is no maximum. Because the feasible region is unbounded (Figure 4–7), a maximum value of the objective function does not exist.

■ **Now You Are Ready to Work Exercise 9**

Unbounded Solutions

When you arrive at a simplex tableau that has no positive entries in the pivot column, the feasible region is unbounded, and the objective function is unbounded. There is no maximum value.

No Feasible Solution

The following example illustrates what happens in a simplex tableau when the problem has no feasible solution.

EXAMPLE 3 ➤ Maximize $z = 8x_1 + 24x_2$, subject to

$$x_1 + x_2 \leq 10$$
$$2x_1 + 3x_2 \geq 60$$
$$x_1 \geq 0, x_2 \geq 0$$

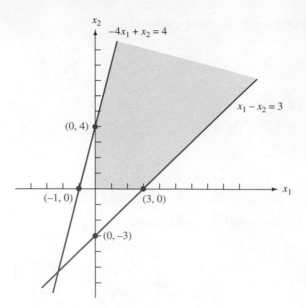

FIGURE 4–7   Unbounded feasible region.

SOLUTION

The graph of the constraints (Figure 4–8) shows that the half plane below $x_1 + x_2 = 10$ and the half plane above $2x_1 + 3x_2 = 60$ do not intersect in the first quadrant, so no feasible solution exists. Let's attempt to solve the problem with the simplex method and see what happens.

Here is the initial tableau:

$$
\begin{array}{ccccc|c}
x_1 & x_2 & s_1 & s_2 & z & \\
\begin{bmatrix}
1 & 1 & 1 & 0 & 0 & 10 \\
-2 & -3 & 0 & 1 & 0 & -60 \\
\hline
-8 & -24 & 0 & 0 & 1 & 0
\end{bmatrix}
\end{array}
$$

To remove the negative entry in the last column, we pivot on -3 in row 2, column 2 and obtain the next tableau:

$$
\begin{array}{ccccc|c}
x_1 & x_2 & s_1 & s_2 & z & \\
\begin{bmatrix}
\frac{1}{3} & 0 & 1 & \frac{1}{3} & 0 & -10 \\
\frac{2}{3} & 1 & 0 & -\frac{1}{3} & 0 & 20 \\
\hline
8 & 0 & 0 & -8 & 1 & 480
\end{bmatrix}
\end{array}
$$

The first row indicates no feasible solution. Here's why. The first row represents the equation $\frac{1}{3}x_1 + s_1 + \frac{1}{3}s_2 = -10$. Because x_1, s_1, and s_2 cannot be negative, there are no values that can be used on the left-hand side that will give a negative number. Therefore, there is no feasible solution.

■ **Now You Are Ready to Work Exercise 13**

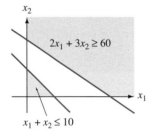

FIGURE 4–8   No feasible region.

No Feasible Solution	When a simplex tableau has a negative entry in the last column and no other entries in that row are negative, then there is no feasible solution to the problem.

EXAMPLE 4 ➤

A machine shop makes standard and heavy-duty gears. The process requires two steps. Step 1 takes 8 minutes for the standard gear and 10 minutes for the heavy-duty gear. Step 2 takes 6 minutes for the standard gear and 10 minutes for the heavy-duty gear. The company's labor contract requires that it use at least 200 labor-hours (12,000 minutes) per week on the step 1 equipment. The maintenance required on the step 2 machine restricts it to 140 hours per week or less (8400 minutes). The materials cost $15 for each standard gear and $22 for each heavy-duty gear. How many of each type of gear should be made each week to minimize material costs?

Show that this problem has no solution.

SOLUTION

Let x = number of standard gears and y = number of heavy-duty gears. The problem is

Minimize $z = 15x + 22y$, subject to

$$8x + 10y \geq 12,000$$
$$6x + 10y \leq 8,400$$
$$x \geq 0, y \geq 0$$

To use the simplex method, we modify the problem to:

Maximize $w = -15x - 22y$, subject to

$$-8x - 10y \leq -12,000$$
$$6x + 10y \leq 8,400$$
$$x \geq 0, y \geq 0$$

The initial tableau is

$$\left[\begin{array}{ccccc|c} -8 & -10 & 1 & 0 & 0 & -12,000 \\ 6 & 10 & 0 & 1 & 0 & 8,400 \\ \hline 15 & 22 & 0 & 0 & 1 & 0 \end{array}\right]$$

Because the basic solution is not feasible, we must apply Phase I and pivot on -10 in row 1:

$$\left[\begin{array}{ccccc|c} \frac{8}{10} & 1 & -\frac{1}{10} & 0 & 0 & 1,200 \\ 6 & 10 & 0 & 1 & 0 & 8,400 \\ \hline 15 & 22 & 0 & 0 & 1 & 0 \end{array}\right]$$

$$\left[\begin{array}{ccccc|c} \frac{8}{10} & 1 & -\frac{1}{10} & 0 & 0 & 1,200 \\ -2 & 0 & 1 & 1 & 0 & -3,600 \\ \hline -\frac{13}{5} & 0 & \frac{11}{5} & 0 & 1 & -26,400 \end{array}\right]$$

Now pivot on -2 in row 2:

$$\left[\begin{array}{ccccc|c} 0 & 1 & \frac{3}{10} & \frac{2}{5} & 0 & -240 \\ 1 & 0 & -\frac{1}{2} & -\frac{1}{2} & 0 & 1,800 \\ \hline 0 & 0 & \frac{9}{10} & -\frac{13}{10} & 1 & -21,720 \end{array}\right]$$

Row 1 has a negative number in the constant column, and all coefficients to the left of the line are nonnegative. This indicates that there is no feasible region and therefore no solution.

■ **Now You Are Ready to Work Exercise 28**

■■ 4.5 EXERCISES

From the tableaux in Exercises 1 through 4, determine whether the linear programming problem has no feasible solution, an unbounded solution, or multiple solutions.

1.
$$\begin{bmatrix} 0 & \frac{16}{5} & 1 & \frac{1}{5} & 0 & -20 \\ 1 & \frac{4}{5} & 0 & -\frac{1}{5} & 0 & 40 \\ 0 & -4 & 0 & -1 & 1 & 200 \end{bmatrix}$$

2.
$$\begin{bmatrix} 0 & 0 & 30 & 15 & -5 & 1 & 0 & 750 \\ 1 & 0 & \frac{5}{8} & \frac{1}{2} & -\frac{1}{8} & 0 & 0 & 30 \\ 0 & 1 & \frac{1}{24} & -\frac{1}{6} & \frac{1}{8} & 0 & 0 & 10 \\ 0 & 0 & 0 & 1 & 1 & 0 & 1 & 360 \end{bmatrix}$$

3.
$$\begin{bmatrix} 0 & 0 & 1 & -\frac{1}{5} & \frac{1}{5} & 0 & 0 & 3 \\ 0 & 1 & 0 & -\frac{1}{6} & \frac{1}{6} & \frac{1}{6} & 0 & \frac{5}{2} \\ 1 & 0 & 0 & -\frac{7}{20} & -\frac{13}{20} & -\frac{1}{4} & 0 & \frac{31}{4} \\ 0 & 0 & 0 & -\frac{7}{5} & \frac{3}{5} & \frac{3}{4} & 1 & -60 \end{bmatrix}$$

4.
$$\begin{bmatrix} 1 & \frac{3}{5} & 0 & -\frac{1}{5} & 0 & 0 & 360 \\ 0 & \frac{1}{5} & 1 & -\frac{1}{5} & 0 & 0 & 90 \\ 0 & \frac{1}{10} & 0 & \frac{2}{5} & 1 & 0 & -40 \\ 0 & \frac{2}{25} & 0 & \frac{1}{25} & 0 & 1 & -300 \end{bmatrix}$$

Find the multiple solutions in Exercises 5 through 8.

5. *(See Example 1)*   Maximize $z = 15x_1 + 15x_2$, subject to

$$3x_1 + 5x_2 \le 60$$
$$x_1 + x_2 \le 14$$
$$2x_1 + x_2 \le 24$$
$$x_1 \ge 0, x_2 \ge 0$$

6. Maximize $z = 15x_1 + 10x_2$, subject to

$$2x_1 + 9x_2 \le 144$$
$$2x_1 + 3x_2 \le 60$$
$$3x_1 + 2x_2 \le 75$$
$$x_1 \ge 0, x_2 \ge 0$$

7. Maximize $z = 3x_1 + 3x_2 + 4x_3$, subject to

$$2x_1 + x_2 + 2x_3 \le 20$$
$$x_1 + 2x_2 + 2x_3 \le 20$$
$$x_1 + x_2 + 4x_3 \le 20$$
$$x_1 \ge 0, x_2 \ge 0, x_3 \ge 0$$

8. Maximize $z = 30x_1 + 70x_2 + 20x_3$, subject to

$$32x_1 + 40x_2 + 63x_3 \le 1600$$
$$2x_1 + 2x_2 + x_3 \le 40$$
$$4x_1 + 12x_2 + 3x_3 \le 120$$
$$x_1 \ge 0, x_2 \ge 0, x_3 \ge 0$$

Show that the feasible regions in Exercises 9 through 12 are unbounded.

9. *(See Example 2)*   Maximize $z = 8x_1 + 3x_2$, subject to

$$2x_1 - 5x_2 \le 10$$
$$-2x_1 + x_2 \ge 2$$
$$x_1 \ge 0, x_2 \ge 0$$

10. Maximize $z = 5x_1 + 4x_2$, subject to

$$-5x_1 + x_2 \le 3$$
$$-2x_1 + x_2 \ge 3$$
$$x_1 \ge 0, x_2 \ge 0$$

11. Maximize $z = 8x_1 + 6x_2 + 2x_3$, subject to

$$x_1 - 3x_2 + 2x_3 \le 50$$
$$-2x_1 + 4x_2 + 5x_3 \le 40$$
$$x_1 \ge 0, x_2 \ge 0, x_3 \ge 0$$

12. Maximize $z = 2x_1 + 4x_2 + x_3$, subject to

$$-x_1 + 2x_2 + 3x_3 \le 6$$
$$-x_1 + 4x_2 + 5x_3 \le 5$$
$$-x_1 + 5x_2 + 7x_3 \le 7$$
$$x_1 \ge 0, x_2 \ge 0, x_3 \ge 0$$

Show that Exercises 13 through 16 have no feasible solutions.

13. *(See Example 3)*   Maximize $z = 12x_1 + 20x_2$, subject to

$$-x_1 + x_2 \ge 13$$
$$2x_1 + 9x_2 \le 72$$
$$x_1 \ge 0, x_2 \ge 0$$

14. Minimize $z = 15x_1 + 9x_2$, subject to

$$3x_1 + 20x_2 \le 60$$
$$-x_1 + 2x_2 \ge 30$$
$$x_1 \ge 0, x_2 \ge 0$$

15. Maximize $z = 20x_1 + 30x_2 + 15x_3$, subject to

$$6x_1 + 4x_2 + 3x_3 \le 60$$
$$3x_1 + 6x_2 + 4x_3 \le 48$$
$$x_1 + x_2 - 2x_3 \le -60$$
$$x_1 \ge 0, x_2 \ge 0, x_3 \ge 0$$

16. Minimize $z = 8x_1 + 5x_2 + 10x_3$, subject to

$$4x_1 + 6x_2 + 3x_3 \le 36$$
$$14x_1 + 21x_2 + 12x_3 \ge 168$$
$$2x_1 - 3x_2 + x_3 \le 12$$
$$x_1 \ge 0, x_2 \ge 0, x_3 \ge 0$$

LEVEL 2

Determine whether each of Exercises 17 through 27 has multiple solutions, unbounded solutions, or no feasible solutions.

17. Maximize $z = x_1 + 4x_2$, subject to

$$-4x_1 + x_2 \leq 2$$
$$2x_1 - x_2 \leq 1$$
$$x_1 \geq 0, x_2 \geq 0$$

18. Maximize $z = 15x_1 + 9x_2$, subject to

$$5x_1 + 3x_2 \leq 30$$
$$5x_1 + x_2 \leq 20$$
$$x_1 \geq 0, x_2 \geq 0$$

19. Maximize $z = 18x_1 + 15x_2 + 8x_3$, subject to

$$-9x_1 + 4x_2 + 6x_3 \geq 36$$
$$4x_1 + 5x_2 + 8x_3 \leq 40$$
$$6x_1 - 2x_2 + x_3 \leq 18$$
$$x_1 \geq 0, x_2 \geq 0, x_3 \geq 0$$

20. Minimize $z = 12x_1 + 3x_2 + 18x_3$, subject to

$$3x_1 + 4x_2 \leq 12$$
$$2x_2 + 5x_3 \leq 10$$
$$-10x_2 + 30x_3 \geq 75$$
$$x_1 \geq 0, x_2 \geq 0, x_3 \geq 0$$

21. Maximize $z = 15x_1 + 9x_2$, subject to

$$5x_1 + 3x_2 \leq 30$$
$$2x_1 + x_2 \geq 20$$
$$x_1 \geq 0, x_2 \geq 0$$

22. Maximize $z = 2x_1 + 3x_2$, subject to

$$-5x_1 + x_2 \leq 5$$
$$x_1 - 4x_2 \leq 8$$
$$x_1 \geq 0, x_2 \geq 0$$

23. Maximize $z = 4x_1 + 12x_2$, subject to

$$10x_1 + 15x_2 \leq 150$$
$$6x_1 + 3x_2 \geq 180$$
$$x_1 \geq 0, x_2 \geq 0$$

24. Maximize $z = 10x_1 + 18x_2 + 7x_3$, subject to

$$2x_1 + 3x_2 - 9x_3 \leq 72$$
$$2x_1 + 5x_2 + 10x_3 \geq 100$$
$$x_1 \geq 0, x_2 \geq 0, x_3 \geq 0$$

LEVEL 3

25. Minimize $z = 2x_1 + 3x_2 + x_3$, subject to

$$2x_1 + 3x_2 + 4x_3 \geq 60$$
$$2x_1 + 3x_2 - 6x_3 \geq 30$$
$$6x_2 - 5x_3 \leq 30$$
$$x_1 \geq 0, x_2 \geq 0, x_3 \geq 0$$

26. Minimize $z = 6x_1 + 14x_2 + 4x_3$, subject to

$$10x_1 + 12x_2 + 39x_3 \geq 480$$
$$2x_1 + 2x_2 + x_3 \geq 40$$
$$4x_1 + 12x_2 + 3x_3 \geq 120$$
$$x_1 \geq 0, x_2 \geq 0, x_3 \geq 0$$

27. Maximize $z = 8x_1 + 5x_2 + 8x_3$, subject to

$$3x_1 - 10x_2 + 6x_3 \leq 60$$
$$-15x_1 + 4x_2 + 6x_3 < 60$$
$$4x_1 + 5x_2 - 20x_3 \leq 100$$
$$x_1 \geq 0, x_2 \geq 0, x_3 \geq 0$$

28. *(See Example 4)* A plant makes Gadgets and Widgets. The materials to make Gadgets cost $0.20 each, and those to make Widgets cost $0.25 each.

The plant sells Gadgets for $2.50 each and Widgets for $3.00 each. The operating costs are $0.10 per Gadget and $0.20 per Widget. The plant must make at least a total of 4000 items, must gross at least $12,000, and must keep the materials cost at $900 or less. How many of each should the plant make to minimize operating costs? Show that this problem has no solution.

29. Show that the following problem has no solution. Beauty Products makes two styles of hair dryers, the Petite and the Deluxe. It requires 1 hour of labor to make the Petite and 2 hours to make the Deluxe. The materials cost $4 for each Petite and $3 for each Deluxe. The profit is $5 for each Petite and $6 for each Deluxe. The company has 3950 labor-hours available each week

and a materials budget of $9575 per week and must make at least 2000 Deluxe hair dryers. Find the number of each type that must be made per week to maximize profit.

30. The Crystal Cola bottling company produces two kinds of soft drinks, regular and diet. Management has ordered that at least 5000 cartons per day be produced within an operating budget of $5400 per day. It costs $1.10 per carton to produce the regular cola and $1.30 per carton to produce the diet cola. The company makes a profit of $0.15 per carton on the regular cola and $0.17 per carton on the diet cola. How much of each type of drink produces the maximum profit?

31. An author plans to invest at least $10,000 per year from book royalties in stocks and bonds. In anticipation of long-term growth, the amount invested in stocks will be at least $5000 more than that invested in bonds. If the expected return is 10% on stocks and 7% on bonds, how much should be invested in each to maximize returns?

32. Alex, Blake, and Cindy earn money for their club by packing books for the School Book Fair. They package fiction and nonfiction books. Their assignments are the following:

    Alex packs 6 fiction and 9 nonfiction books in each package. Blake packs 8 fiction and 9 nonfiction books in each package. Cindy packs 9 fiction and 12 nonfiction books in each package. Cindy is to

pack at least 72 more fiction books than Alex and Blake combined. Alex and Blake are to pack at least 90 more nonfiction books than Cindy. Blake is to pack at most 10 more packages than Alex.

If Alex is paid $2.00 per package, Blake is paid $2.50 per package, and Cindy is paid $3.00 per package, how many packages does each one pack in order to maximize their total earnings?

33. A school cafeteria serves three foods for lunch, A, B, and C. There is pressure on the cafeteria director to reduce lunch costs. Help the director by finding the quantities of each food that will minimize costs and still maintain the desired nutritional level. The three foods have the following nutritional characteristics:

Per Unit	Foods		
	A	**B**	**C**
Protein (g)	15	10	23
Carbohydrates (g)	20	30	11
Calories	200	160	160
Fat (g)	10	5	6
Cost ($)	1.40	1.00	1.50

A lunch must contain at least 80 grams of protein, 95 grams of carbohydrates, and 1400 calories. It must contain no more than 40 grams of fat. How many units of each food should be served to minimize cost?

EXPLORATIONS

34. Andrea is ordering supplies for summer camp and notices the supplier has a special offer on four items on her list: caps, T-shirts, socks, and shoes. The supplier packaged these items for some special orders, and a number were returned that the supplier would like to move. There are three types of packages, I, II, III.

    I contains 5 caps, 10 pairs of socks, 2 T-shirts, and 4 pairs of shoes.

    II contains 5 caps, 6 pairs of socks, 3 T-shirts, and 3 pairs of shoes.

    III contains 4 caps, 5 pairs of socks, 6 T-shirts, and no shoes.

    Andrea estimates they will need up to 200 caps, up to 300 pairs of socks, up to 210 T-shirts, and up to 60 pairs of shoes.

Based on the special price and their published camp fees, the camp can make a profit of $6 on each package I, $5 on each package II, and $2 on each package III. How many of each type of package should Andrea order to achieve maximum profit?

35. Construct a linear programming problem with two variables that has an unbounded feasible region.

36. Construct a linear programming problem with two variables that has no feasible region.

37. Construct a linear programming problem with two variables that has multiple solutions.

4.6   What's Happening in the Simplex Method? (Optional)

The simplex method can be performed in a rather mechanical manner, thereby making it a procedure that can run on a computer. This makes it possible to solve linear programming problems with hundreds of constraints and variables. As much as we appreciate a machine handling the routine, tedious computation, the human mind sometimes becomes curious about what lies behind the computations. In this section, we look at the simplex process to better understand why the steps are performed. We outline the steps of the simplex method and explain why we perform these steps. We use Example 5 from Section 4.1 again.

Maximize $z = 4x_1 + 12x_2$, subject to

$$3x_1 + x_2 \leq 180$$
$$x_1 + 2x_2 \leq 100$$
$$-2x_1 + 2x_2 \leq 40$$
$$x_1 \geq 0, x_2 \geq 0$$

The graph of the feasible region and the lines forming its boundary is shown in Figure 4–9.

1. We convert the problem to a system of equations by adding a nonnegative slack variable to each inequality

$$3x_1 + x_2 + s_1 \qquad\qquad = 180$$
$$x_1 + 2x_2 \qquad + s_2 \qquad\qquad - 100$$
$$-2x_1 + 2x_2 \qquad\qquad + s_3 \qquad = 40$$
$$-4x_1 - 12x_2 \qquad\qquad\qquad + z = 0$$

where $x_1, x_2, s_1, s_2,$ and s_3 are all nonnegative.

2. The simplex method searches for solutions to this system of equations. Each simplex tableau gives a basic feasible solution. Recall that we set a variable to zero for each x in the system to obtain a basic solution. The points so obtained are points where two of the boundary lines of the feasible region intersect.

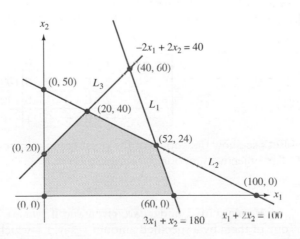

FIGURE 4–9

Look at Figure 4–9, which illustrates this. Setting a slack variable to zero gives a boundary line. The boundary lines are

$$L_1: \quad 3x_1 + x_2 = 180 \quad \text{(Where } s_1 = 0\text{)}$$
$$L_2: \quad x_1 + 2x_2 = 100 \quad \text{(Where } s_2 = 0\text{)}$$
$$L_3: \quad -2x_1 + 2x_2 = 40 \quad \text{(Where } s_3 = 0\text{)}$$
$$x_1\text{-axis:} \quad x_2 = 0$$
$$x_2\text{-axis:} \quad x_1 = 0$$

The corner points of the feasible region occur at intersections of boundary lines and are $(0, 0)$, $(0, 20)$, $(20, 40)$, $(52, 24)$, and $(60, 0)$. Some boundary lines intersect outside the feasible region, such as $(40, 60)$, and give no corner point (see Figure 4–9).

We have already seen that all slack variables are nonnegative for points in the feasible region and that the optimal solution occurs at a corner point.

The simplex method finds the corner at which the objective function is maximum in the following manner:

1. Begin at the origin, $(0, 0)$.
2. Select a pivot column that tends to increase the value of z the most.
3. At each step, move along a boundary line to an adjacent corner of the feasible region.
4. When z can no longer be increased, the procedure stops.

Now let's see how these relate to the simplex tableaux. We refer to the tableaux in Section 4.2 that were obtained in finding the optimal solution.

1. The initial tableau is

$$\begin{bmatrix} 3 & 1 & 1 & 0 & 0 & 0 & 180 \\ 1 & 2 & 0 & 1 & 0 & 0 & 100 \\ -2 & 2 & 0 & 0 & 1 & 0 & 40 \\ \hline -4 & -12 & 0 & 0 & 0 & 1 & 0 \end{bmatrix}$$

which has the basic solution $x_1 = 0$, $x_2 = 0$, $s_1 = 180$, $s_2 = 100$, $s_3 = 40$, $z = 0$. This gives the point $(0, 0)$, the origin.

2. The choice of the pivot element determines the solution in the next step, and it increases z as much as possible.

Let's look at the initial tableau to illustrate this:

$$\begin{array}{cccccc} x_1 & x_2 & s_1 & s_2 & s_3 & z \\ \begin{bmatrix} 3 & 1 & 1 & 0 & 0 & 0 & 180 \\ 1 & 2 & 0 & 1 & 0 & 0 & 100 \\ -2 & 2 & 0 & 0 & 1 & 0 & 40 \\ \hline -4 & -12 & 0 & 0 & 0 & 1 & 0 \end{bmatrix} \end{array}$$

Let's see how the choice of the pivot column increases z as much as possible. The objective function in this problem is

$$z = 4x_1 + 12x_2$$

If values of x_1 and x_2 are given and if you are allowed to increase either one of them by a specified amount — say, 1 — which one would you change to increase z the most? The coefficients of x_1 and x_2 hold the key to your response. If x_1 is increased by 1, then the coefficient 4 causes z to increase by 4. Similarly, an increase of 1 in x_2 causes z to increase by 12. The greatest increase in z is gained by increasing the variable with the largest positive coefficient, 12 in this case. In the tableau, $z = 4x_1 + 12x_2$ is written as $-4x_1 - 12x_2 + z = 0$. In this form, the choice of the *most negative* coefficient is equivalent to choosing the variable that will increase z the most. So in the simplex method, the pivot column is chosen by the most negative entry in the bottom row because this gives the variable that will tend to increase z the most.

3.   Let's see how the choice of the pivot row restricts basic solutions to corner points.

In the initial tableau, the basic solution assumes that both x_1 and x_2 are zero. Then the x_2 column becomes the pivot column to obtain the next tableau because that column contains the most negative entry of the last row. Because this means that we want to increase x_2, x_1 remains zero. Using $x_1 = 0$, let's write each row of the initial tableau in equation form:

$$[3 \quad 1 \quad 1 \quad 0 \quad 0 \quad 0 \quad 180] \qquad \text{becomes}$$
$$x_2 + s_1 = 180$$

$$[1 \quad 2 \quad 0 \quad 1 \quad 0 \quad 0 \quad 100] \qquad \text{becomes}$$
$$2x_2 + s_2 = 100$$

and

$$[-2 \quad 2 \quad 0 \quad 0 \quad 1 \quad 0 \quad 40] \qquad \text{becomes}$$
$$2x_2 + s_3 = 40$$

(The first number in each row doesn't appear in the equation because it is the coefficient of x_1, which we are using as 0.)

We can write these three equations in the following form:

$$s_1 = 180 - x_2$$
$$s_2 = 100 - 2x_2$$
$$s_3 = 40 - 2x_2$$

Keep in mind that we want to increase x_2 to achieve the largest increase in z. The larger the increase in x_2, the more z increases, but be careful: We must remain in the feasible region. Because s_1, s_2, and s_3 must not be negative, x_2 must be chosen to avoid making any one of them negative.

The equations

$$s_1 = 180 - x_2$$
$$s_2 = 100 - 2x_2$$
$$s_3 = 40 - 2x_2$$

and the nonnegative condition $s_1 \geq 0$, $s_2 \geq 0$, and $s_3 \geq 0$ indicate that

$$180 - x_2 \geq 0$$
$$100 - 2x_2 \geq 0$$
$$40 - 2x_2 \geq 0$$

must all be true. Solving each of these inequalities for x_2 gives

$$\frac{180}{1} \geq x_2$$

$$\frac{100}{2} \geq x_2$$

$$\frac{40}{2} \geq x_2$$

In order for all three of the ratios to be larger than x_2, the smallest ratio must be larger. Thus, for *all three* of s_1, s_2, and s_3 to be nonnegative, the smallest value of x_2, 20, must be used. The ratios

$$\frac{180}{1}, \quad \frac{100}{2}, \quad \frac{40}{2}$$

are exactly the ratios that are used in the simplex method to determine the pivot row. These ratios are also the x_2-coordinates where the boundary lines cross the x_2-axis. By the nature of the feasible region, the lowest point is the one in the feasible region. The selection of the smallest nonnegative ratio makes a basic solution a feasible basic solution; that is, a corner point is chosen.

4. Recall that the maximum value of z occurs when the last row of the simplex tableau contains no negative entries. The final tableau of this problem was

$$
\begin{array}{ccccccc}
x_1 & x_2 & s_1 & s_2 & s_3 & z & \\
\left[\begin{array}{cccccc|c}
0 & 0 & 1 & -\frac{4}{3} & \frac{5}{6} & 0 & 80 \\
1 & 0 & 0 & \frac{1}{3} & -\frac{1}{3} & 0 & 20 \\
0 & 1 & 0 & \frac{1}{3} & \frac{1}{6} & 0 & 40 \\
\hline
0 & 0 & 0 & \frac{16}{3} & \frac{2}{3} & 1 & 560
\end{array}\right]
\end{array}
$$

This tableau tells us that s_2 and s_3 are set to zero (their columns are not unit columns) in the optimal solution; and as the last row contains no negative entries, we know that we cannot increase z further. Here's why: Write the last row of this tableau in equation form. It is

$$\frac{16}{3} s_2 + \frac{2}{3} s_3 + z = 560$$

which can be written as

$$z = 560 - \frac{16}{3} s_2 - \frac{2}{3} s_3$$

This form tells us that if we use any positive number for s_2 or s_3, we will *subtract* something from 560, thereby making z smaller. So we stop because another tableau will move us to another corner point, and s_2 or s_3 will become positive and therefore reduce z.

Let's compare this situation with the next to last tableau. Its last row was

$$\begin{bmatrix} -16 & 0 & 0 & 0 & 6 & 1 & 240 \end{bmatrix}$$

This row represents the equation

$$-16x_1 + 6s_3 + z = 240$$

which may be written

$$z = 240 + 16x_1 - 6s_3$$

If x_1 is increased from 0 to a positive number, then a positive quantity will be added to z, thereby increasing it. If you look back at this step in the solution, you will find that x_1 was increased from 0 to 10.

The value of z can be increased as long as there is a negative entry in the last row. It can be increased no further when no entry is negative.

5. Reviewing the steps of this example, we observe that the sequence of simplex tableaux and the corner points determined are the following:

Initial tableau:

$$\begin{bmatrix} 3 & 1 & 1 & 0 & 0 & 0 & 180 \\ 1 & 2 & 0 & 1 & 0 & 0 & 100 \\ -2 & 2 & 0 & 0 & 1 & 0 & 40 \\ -4 & -12 & 0 & 0 & 0 & 1 & 0 \end{bmatrix}$$

Corner $(0, 0)$ $z = 0$, slack $(180, 100, 40)$.

Tableau 1:

$$\begin{bmatrix} 4 & 0 & 1 & 0 & -\frac{1}{2} & 0 & 160 \\ 3 & 0 & 0 & 1 & -1 & 0 & 60 \\ -1 & 1 & 0 & 0 & \frac{1}{2} & 0 & 20 \\ -16 & 0 & 0 & 0 & 6 & 1 & 240 \end{bmatrix}$$

Corner $(0, 20)$ $z = 240$, slack $(160, 60, 0)$.

Tableau 2:

$$\begin{bmatrix} 0 & 0 & 1 & -\frac{4}{3} & \frac{5}{6} & 0 & 80 \\ 1 & 0 & 0 & \frac{1}{3} & -\frac{1}{3} & 0 & 20 \\ 0 & 1 & 0 & \frac{1}{3} & \frac{1}{6} & 0 & 40 \\ 0 & 0 & 0 & \frac{16}{3} & \frac{2}{3} & 1 & 560 \end{bmatrix}$$

Corner $(20, 40)$ $z = 560$, slack $(80, 0, 0)$.

Notice that the initial corner point is the origin, $(0, 0)$. The first pivot moves to $(0, 20)$, which is a corner adjacent to $(0, 0)$. The second pivot moves to $(20, 40)$, which is a corner adjacent to $(0, 20)$. A pivot in the simplex method moves from the current corner to an adjacent corner (see Figure 4–10).

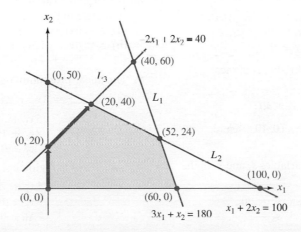

FIGURE 4–10 Sequence of corner points generated by the simplex tableaux.

On the basis of the preceding discussion, the simplex method can be summarized this way:

The simplex method maximizes the objective function by computing the objective function at selected corner points of the feasible region until the optimal solution is reached. The method begins at the origin and moves at each stage to an adjacent corner point determined by the variable that yields the largest increase in z.

■ 4.6 EXERCISES

LEVEL 1

1. The constraint $4x_1 + 3x_2 \leq 17$ is written as

$$4x_1 + 3x_2 + s_1 = 17$$

using a slack variable.
(a) Find the value of s_1 for each of the points $(0, 0)$, $(2, 2)$, $(5, 10)$, $(2, 1)$, and $(2, 3)$.
(b) Which points are in the feasible region?
(c) Which of the above points lie on the line $4x_1 + 3x_2 = 17$?

2. The constraint $6x_1 + 5x_2 \leq 25$ is written as

$$6x_1 + 5x_2 + s_1 = 25$$

using a slack variable. Find the value of s_1 for each of the points $(0, 5)$, $(2, 2)$, $(1.5, 3)$, and $(2.5, 2)$.

3. Find the corner corresponding to $x_2 = 0$, $s_2 = 0$, in the feasible region determined by the constraints

$$\begin{aligned} 3x_1 + 4x_2 + s_1 &= 24 \\ 5x_1 + 2x_2 + s_2 &= 30 \\ x_1 \geq 0, x_2 &\geq 0 \end{aligned}$$

4. Find the corner corresponding to $s_1 = 0$, $s_2 = 0$, in the feasible region determined by the constraints

$$\begin{aligned} x + 2y + s_1 &= 28 \\ 2x + y + s_2 &= 26 \\ x \geq 0, y &\geq 0 \end{aligned}$$

5. Find the values of the slack variable s_1 in the constraint

$$3x_1 + 4x_2 + x_3 \leq 40$$

for the points $(0, 0, 0)$, $(1, 2, 3)$, $(0, 10, 0)$, and $(4, 2, 7)$.

6. Find the values of the slack variables s_1 and s_2 in the constraints

$$\begin{aligned} 2x_1 + x_2 + 5x_3 + s_1 &= 30 \\ x_1 + 6x_2 + 4x_3 + s_2 &= 28 \end{aligned}$$

for the points $(1, 1, 1)$, $(2, 1, 5)$, $(0, 4, 1)$, and $(5, 10, 2)$.

7. Using slack variables, the constraints of a linear programming problem are

$$\begin{aligned} x_1 + 5x_2 + s_1 &= 70 \\ 6x_1 + x_2 + s_2 &= 72 \\ 7x_1 + 6x_2 + s_3 &= 113 \end{aligned}$$

For each point listed in the following table, find the missing entries.

Point	s_1	s_2	s_3	Is Point on Boundary?	Is Point in Feasible Region?
$(5, 10)$					
$(8, 10)$					
$(5, 13)$					
$(11, 13)$					
$(10, 12)$					
$(15, 11)$					

8. Which single variable contributes the most to increasing z in the following objective functions?
(a) $z = 6x_1 + 5x_2 + 14x_3$
(b) $z = 8x_1 - 12x_2 + 3x_3$
(c) $z = 8x_1 + 4x_2 - 7x_3 + 5x_4$

9. The linear programming problem:

Maximize $z = 3x_1 + 2x_2$, subject to

$$\begin{aligned} 5x_1 + 2x_2 &\leq 900 \\ 8x_1 + 10x_2 &\leq 2800 \\ x_1 \geq 0, x_2 &\geq 0 \end{aligned}$$

has a feasible region bounded by

$$\begin{aligned} 5x_1 + 2x_2 &= 900 \\ 8x_1 + 10x_2 &= 2800 \\ x_1 = 0, x_2 &= 0 \end{aligned}$$

The simplex method solution to this problem includes the following tableaux:

(i)
$$\begin{array}{ccccc|c} x_1 & x_2 & s_1 & s_2 & z & \\ 5 & 2 & 1 & 0 & 0 & 900 \\ 8 & 10 & 0 & 1 & 0 & 2800 \\ \hline -3 & -2 & 0 & 0 & 1 & 0 \end{array}$$

(ii)
$$\begin{bmatrix} 1 & \frac{2}{5} & \frac{1}{5} & 0 & 0 & 180 \\ 0 & \frac{34}{5} & -\frac{8}{5} & 1 & 0 & 1360 \\ 0 & -\frac{4}{5} & \frac{3}{5} & 0 & 1 & 540 \end{bmatrix}$$

(iii)
$$\begin{bmatrix} 1 & 0 & \frac{25}{85} & -\frac{1}{17} & 0 & 100 \\ 0 & 1 & -\frac{4}{17} & \frac{5}{34} & 0 & 200 \\ 0 & 0 & \frac{7}{17} & \frac{2}{17} & 1 & 700 \end{bmatrix}$$

(a) Write the basic solution from each tableau.
(b) Determine the two boundary lines whose intersection gives the basic solution.

LEVEL 2

10. Given the constraint

$$x_1 + 5x_2 + s_1 = 48$$

(a) When $x_1 = 0$, what is the largest value x_2 can have so that s_1 meets the nonnegative condition?
(b) When $x_1 = 6$, what is the largest value x_2 can have so that s_1 meets the nonnegative condition?
(c) When $x_2 = 0$, what is the largest possible value of x_1 so that the point is in the feasible region?

11. One constraint to a linear programming problem is

$$6x_1 + 4x_2 + s_1 = 24.$$

This constraint determines a boundary line to the feasible region.
(a) Is the point $(3, 5)$ on the boundary line?
(b) Can $(3, 5)$ be in the feasible region?

12. One constraint to a linear programming problem is

$$8x_1 + 5x_2 + s_2 = 59.$$

This constraint determines a boundary line to the feasible region.
(a) Is the point $(3, 7)$ on the boundary line?
(b) Can $(2, 4)$ be in the feasible region?

LEVEL 3

13. Given the constraints

$$\begin{aligned} 6x_1 + 7x_2 + s_1 \quad\quad &= 36 \\ 2x_1 + 5x_2 \quad\quad + s_2 &= 32 \end{aligned}$$

(a) When $x_1 = 0$, what is the largest possible value of x_2 so that the point is in the feasible region?
(b) When $x_2 = 0$, what is the largest possible value

of x_1 so that the point is in the feasible region?

14. Given the constraint

$$4x_1 + 5x_2 + x_3 + s_1 = 45$$

When $x_1 = 0$, what is the largest possible value of x_2 so that s_1 is nonnegative?

EXPLORATIONS

For Exercises 15 through 19, trace the sequence of corner points in the simplex procedure. At each pivot, record the basic solution and the value of z and fill in the following table.

Corner (x, y)	Value of z	Increase in z	Slack Variables (s_1, s_2, s_3)

15. Maximize $z = 17x + 20y$, subject to

$$\begin{aligned} -7x + 10y &\le 50 \\ 3x + 5y &\le 90 \\ 4x + 5y &\le 105 \\ x \le 20, \; x \ge 0, \; y &\ge 0 \end{aligned}$$

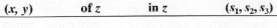
16. Maximize $z = 11x + 20y$, subject to

$$\begin{aligned} -7x + 10y &\le 50 \\ 3x + 5y &\le 90 \\ 4x + 5y &\le 105 \\ x \le 20, \; x \ge 0, \; y &\ge 0 \end{aligned}$$

17. Maximize $z = 10x_1 + 24x_2 + 13x_3$, subject to

$$x_1 + 6x_2 + 3x_3 \leq 36$$
$$3x_1 + 6x_2 + 6x_3 \leq 45$$
$$5x_1 + 6x_2 + x_3 \leq 46$$
$$x_1 \geq 0, x_2 \geq 0, x_3 \geq 0$$

18. Maximize $z = 22x_1 + 20x_2 + 18x_3$, subject to

$$2x_1 + x_2 + 2x_3 \leq 100$$
$$x_1 + 2x_2 + 2x_3 \leq 100$$
$$2x_1 + 2x_2 + x_3 \leq 100$$
$$x_1 \geq 0, x_2 \geq 0, x_3 \geq 0$$

19. Maximize $z = x_1 + 2x_2 + 3x_3$, subject to

$$2x_1 + x_2 + 2x_3 \leq 330$$
$$x_1 + 2x_2 + 2x_3 \leq 330$$
$$-2x_1 - 2x_2 + x_3 \leq 132$$
$$x_1 \geq 0, x_2 \geq 0, x_3 \geq 0$$

20. Exercise 15 dealt with the problem

Maximize $z = 17x + 20y$, subject to

$$-7x + 10y \leq 50$$
$$3x + 5y \leq 90$$
$$4x + 5y \leq 105$$
$$x \leq 20, x \geq 0, y \geq 0$$

The standard simplex routine gave the following sequence of corners and values of z:

Corner	z
$(0, 0)$	0
$(0, 5)$	100
$(10, 12)$	410
$(15, 9)$	435
$(20, 5)$	440

The first pivot column was column 2.

Now solve the same problem, but use column 1 as the first pivot column. You should obtain the following sequence of corners and values of z:

Corner	z
$(0, 0)$	0
$(20, 0)$	340
$(20, 5)$	440

Notice that this nonstandard procedure increased z by 340 the first pivot and found the optimal solution with the second pivot, whereas the standard procedure increased z by 100 the first pivot and required four pivots to find the optimal solution.

Recall that the standard simplex procedure is designed to give the maximum increase in z at each pivot. Explain why that is not the case in this example.

21. In Exercises 47 through 49 of Section 4.2, you solved problems that had two choices for the pivot row at some step. These gave a zero ratio for the pivot row. Review those exercises with attention to the feasible region and the sequence of corner points followed. Recall that where a slack variable is zero, it indicates that corner point lies on the line that corresponds to that slack variable. Recall that a pivot of a tableau moves along a boundary line from one corner of the feasible region to another corner.

Analyze Exercises 47 through 49 and determine what happens when a pivot is made on a tableau with a zero ratio.

4.7 Sensitivity Analysis

- Changes in the Objective Function
- Resource Changes

In the linear programming examples and exercises we have used throughout this chapter, we have tacitly assumed the quantities, costs, and other data are accurately known. In practice, those quantities and costs are often estimates or subject to change. Thus, this question arises: "If some of the data change, how will that affect the optimal solution?" A study of the consequences of changes in data is called **sensitivity analysis.** Such an analysis can become complicated, but we look at two simpler situations: (a) when changes occur in the objective function, and (b) when changes occur in the resources available.

We use the following elementary example to illustrate the analysis.

EXAMPLE 1 ➤

Electronic Design makes two electronic control devices for an equipment manufacturer, the Primary Control and the Auxiliary Control. These controls require two types of circuit boards, the C-8 board and the H-2 board. The Primary Control requires 5 C-8 boards and 1 H-2 board. The Auxiliary Control requires 4 C-8 boards and 2 H-2 boards. The company expects to make a $16 profit on the Primary Control devices and a $24 profit on the Auxiliary Control devices. The company has 88 C-8 and 32 H-2 boards. How many of each control boards should be made to maximize profit?

SOLUTION

Let's set up the constraints and objective function.

$$\text{Let } x = \text{number of Primary Control devices}$$
$$y = \text{number of Auxiliary Control devices}$$

Maximize $z = 16x + 24y$, subject to

$$5x + 4y \leq 88 \quad \text{(Number of C-8 boards)}$$
$$x + 2y \leq 32 \quad \text{(Number of H-2 boards)}$$
$$x \geq 0, y \geq 0$$

Figure 4–11 shows the corners of the feasible region.
The corners and the corresponding values of the objective function are the following:

Corner	$z = 16x + 24y$
$(0, 0)$	0
$(0, 16)$	384
$(8, 12)$	416
$(17.6, 0)$	281.6

The maximum profit is $416, with 8 Primary Control and 12 Auxiliary Control devices made.

At its maximum, the objective function gives $16x + 24y = 416$, a line through $(8, 12)$ as shown in Figure 4–12.

The fact that the line $16x + 24y = 416$ lies between the lines $5x + 4y = 88$ and $x + 2y = 32$ is significant. In fact, for the objective funtion to attain its maximum value at $(8, 12)$, it *must* lie between the boundary lines that intersect at $(8, 12)$. ∎

FIGURE 4–11

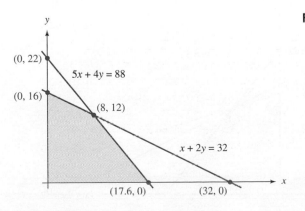

FIGURE 4-12

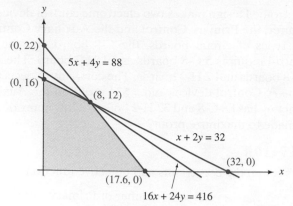

Changes in the Objective Function

In Figure 4–13, we show three other lines, L_1, L_2, and L_3, that pass through $(8, 12)$ but have different slopes from the objective function.

Because the maximum value of the objective function must occur at a corner point, for our example, it must be at $(8, 12)$, $(0, 16)$, or $(17.6, 0)$.

In the case where the profit on a device changes, we could get a line with a slope similar to L_1, L_2, or L_3. Recall that the objective function increases in value as it moves away from the origin. It reaches its maximum value at the last corner point it touches as it moves out. Notice that the point $(8, 12)$ is the last corner L_1 touches as it moves away from the origin, so $(8, 12)$ is the point of maximum value for *any* line that lies between $5x + 4y = 88$ and $x + 2y = 32$. Also observe that L_2 can move farther away from the origin and still cross the feasible region. The corner $(0, 16)$ is the last corner touched by L_2, so the maximum value of the L_2 objective function occurs at $(0, 16)$.

Likewise, L_3 can move farther out until it touches the corner $(17.6, 0)$, which is the point that makes the L_3 objective function maximum.

The discussion of Figure 4–13 can help to answer questions such as "What profit values can we have for each device so the point $(8, 12)$ remains the point of maximum profit?" and "What profit values for each device will cause the point of maximum profit to move to $(0, 16)$ or to $(17.6, 0)$?" The next three examples show how to answer these questions.

FIGURE 4-13

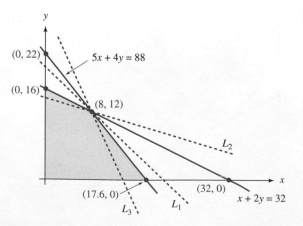

EXAMPLE 2 ➤ The problem

Maximize $z = 16x + 24y$, subject to

$$5x + 4y \leq 88 \quad \text{(Number of C-8 boards)}$$
$$x + 2y \leq 32 \quad \text{(Number of H-2 boards)}$$
$$x \geq 0, y \geq 0$$

has the solution

Maximum $z = 416$ at $(8, 12)$.

How much change can occur in a device profit, such that the point $(8, 12)$ still yields the maximum profit?

SOLUTION

We use the slopes of the boundary lines $5x + 4y = 88$ and $x + 2y = 32$ and the slope of an arbitrary objective function, $z = Ax + By$, to find the answer. Because the objective function with slope $-A/B$ through $(8, 12)$ must lie between the lines $5x + 4y = 88$ (slope $= -\frac{5}{4}$) and $x + 2y = 32$ (slope $= -\frac{1}{2}$), the slope of the objective function must lie between the slopes of the boundary lines; that is,

$$-\frac{5}{4} \leq -\frac{A}{B} \leq -\frac{1}{2}$$

which we can write as

$$\frac{1}{2} \leq \frac{A}{B} \leq \frac{5}{4}$$

Thus, the profits of the devices whose ratio lies between 0.5 and 1.25 will yield a maximum total profit at $(8, 12)$. For example, device profits of $A = 18$ and $B = 24$ give $\frac{A}{B} = \frac{18}{24} = 0.75$, giving a total maximum profit at $(8, 12)$.

■ **Now You Are Ready to Work Exercise 1**

EXAMPLE 3 ➤ For the problem of Example 2, determine the device profits that will cause the maximum total profit to move to $(0, 16)$.

SOLUTION

For the objective function $z = Ax + By$ to be maximum at the point $(0, 16)$, it must pass through $(0, 16)$ and lie above $x + 2y = 32$, like the dotted line labeled L_1 in Figure 4-14.

FIGURE 4–14

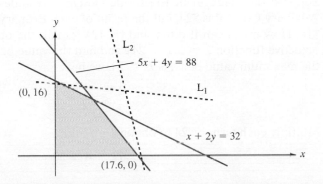

Because the slopes are negative and line L_1 is more nearly horizontal than $x + 2y = 32$, the slope of line L_1 is nearer to zero than the slope $-\frac{1}{2}$. We indicate this by

$$-\frac{1}{2} \le -\frac{A}{B} \le 0$$

and conclude that $z = Ax + By$ has its maximum value at $(0, 16)$ when

$$0 \le \frac{A}{B} \le \frac{1}{2} \quad \text{which is equivalent to } A \le \frac{1}{2}B$$

One such example is $z = 10x + 24y$.

■ **Now You Are Ready to Work Exercise 5**

EXAMPLE 4 ➤ For the problem of Example 2, determine the device profits that will cause the maximum total profit to move to $(17.6, 0)$.

SOLUTION

For the objective function $z = Ax + By$ to be maximum at the point $(17.6, 0)$, it must pass through $(17.6, 0)$ and lie above $5x + 4y = 88$, like the dotted line labeled L_2 in Figure 4–14. Because the slopes are negative and line L_2 is more nearly vertical than $5x + 4y = 88$, the slope of line L_2 is farther from zero than the slope $\frac{-5}{4}$. We indicate this by

$$-\frac{A}{B} \le -\frac{5}{4} \le 0$$

and conclude that $z = Ax + By$ has its maximum value at $(17.6, 0)$ when

$$0 \le \frac{5}{4} \le \frac{A}{B} \quad \text{or} \quad \frac{5}{4} \le \frac{A}{B}$$

which is equivalent to $A \ge 1.25B$. One such example is $z = 25x + 14y$.

■ **Now You Are Ready to Work Exercise 7**

Let's continue the analysis of Example 2 to see the effect a change in the objective function has on the point of maximum profit. Recall that the profit on each Primary Control is \$16 and the profit on each Auxiliary Control is \$24, so the objective function is $z = 16x + 24y$, and it attains its maximum value at $(8, 12)$.

EXAMPLE 5 ➤ Suppose the manager of Electronic Design is confident that the profit of each Auxiliary Control is \$24, but the profit of each Primary Control may differ from \$16. How much can it differ and $(8, 12)$ remain the optimal point? We use the objective function $z = Ax + 24y$ and find the value of A for which $(8, 12)$ gives the maximum value of z. As $B = 24$, we have

$$\frac{1}{2} \le \frac{A}{24} \le \frac{5}{4}$$

Multiply each term by 24 to obtain

$$12 \le A \le 30$$

The profit on each Primary Control device can vary from $12 to $30, and (8, 12) remains the optimal point.

■ **Now You Are Ready to Work Exercise 11**

EXAMPLE 6 ➤ Let's look at the situation when the profit for each Primary Control, $16, is considered accurate, but the profit for each Auxiliary Control might differ from $24. As $A = 16$, we have

$$\frac{1}{2} \leq \frac{16}{B} \leq \frac{5}{4}$$

We can take the reciprocal of each term and reverse the inequality signs to obtain

$$2 \geq \frac{B}{16} \geq \frac{4}{5}$$

Multiply each term in

$$2 \geq \frac{B}{16} \geq \frac{4}{5}$$

by 16 to obtain

$$32 \geq B \geq 12.8$$

In this case, the profit per Auxiliary Control can vary from $12.80 to $32 and (8, 12) remains the optimal point.

■ **Now You Are Ready to Work Exercise 13**

<table>
<tr><td>

NOTE

If you are in doubt about this, consider the numbers

$$\frac{1}{2} < 3 < 5$$

and observe that

$$2 > \frac{1}{3} > \frac{1}{5}$$

</td></tr>
</table>

Resource Changes

A manager may determine the number of devices produced that yield maximum profit for the number of circuit boards available. That, however, can change. Additional circuit boards might become available, or a shipment may be delayed so that fewer circuit boards are available. We should not be surprised if the optimal solution changes as resources change, because corner points change when a boundary line is moved. Let's look at Example 1 and change the number of H-2 boards available from 32 to 38, with other data unchanged. The problem then becomes the next example.

EXAMPLE 7 ➤ Maximize $z = 16x + 24y$, subject to

$$5x + 4y \leq 88$$
$$x + 2y \leq 38$$
$$x \geq 0, y \geq 0$$

SOLUTION

Figure 4–15 shows the feasible region and corner points resulting from the change in available H-2 boards.

FIGURE 4–15

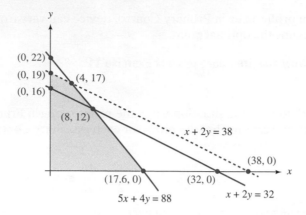

The corners are $(0, 0)$, $(0, 19)$, $(4, 17)$, and $(17.6, 0)$. The corresponding values of z are the following:

Corner	$z = 16x + 24y$
$(0, 0)$	0
$(0, 19)$	456
$(4, 17)$	472
$(17.6, 0)$	281.6

The maximum profit is \$472 and occurs when 4 Primary Control and 17 Auxiliary Control devices are made. When 6 H-2 boards were added, profits increased by \$56. We observe that the maximum value of z still occurs at the corner formed by the intersection of the two constraints. This will not always happen. If the number of available H-2 boards increases enough — say, to 48 — we have the situation shown by Figure 4–16. The constraint lines intersect outside the first quadrant, so the constraint $x + 2y = 48$ can be ignored because the constraint $5x + 4y = 88$ completely determines the feasible region.

■ **Now You Are Ready to Work Exercise 15**

So we ask, "How does a change in the available number of H-2 boards affect maximum profit when the constraint lines still intersect in the first quadrant?"

We answer the question by looking at a more general form of the H-2 constraint in the next example.

FIGURE 4–16

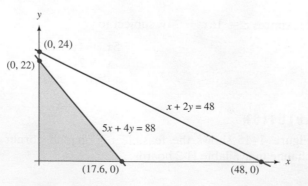

EXAMPLE 8 ➤

In the problem
    Maximize $z = 16x + 24y$, subject to

$$5x + 4y \le 88 \quad \text{(Number of C-8 boards)}$$
$$x + 2y \le 32 \quad \text{(Number of H-2 boards)}$$
$$x \ge 0, y \ge 0$$

How does a change in the available number of H-2 boards affect maximum profit when the constraint lines still intersect in the first quadrant?

SOLUTION

Let D = the change in the number of H-2 boards, where D can be positive or negative. The problem becomes
    Maximize $z = 16x + 24y$, subject to

$$5x + 4y \le 88$$
$$x + 2y \le 32 + D$$
$$x \ge 0, y \ge 0$$

Figure 4–17 shows the feasible region, where the y-intercept of $x + 2y = 32 + D$ is

$$\left(0, 16 + \frac{D}{2}\right)$$

and the x-intercept is $(32 + D, 0)$.
    Solving the system

$$5x + 4y = 88$$
$$x + 2y = 32 + D$$

we find at the intersection of the two lines

$$x = 8 - \frac{2D}{3}$$

$$y = 12 + \frac{5D}{6}$$

and the value of z at that point is

$$16\left(8 - \frac{2D}{3}\right) + 24\left(12 + \frac{5D}{6}\right) = 128 - \frac{32}{3}D + 288 + 20D$$

$$= 416 + \frac{28}{3}D$$

FIGURE 4–17

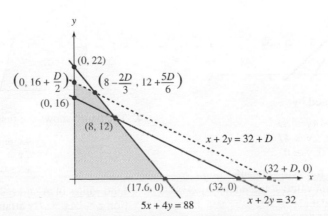

This tells us that for each unit change in the number of H-2 boards available, the number of Primary devices decreases by $\frac{2}{3}$, the number of Auxiliary devices increases by $\frac{5}{6}$, and profit increases by $\frac{28}{3} = 9.33$. However, we must be careful. For the constraint lines to intersect in the first quadrant, neither x nor y can be negative. We must have

$$x = 8 - \frac{2D}{3} \geq 0$$

which gives $12 \geq D$ and

$$y = 12 + \frac{5D}{6} \geq 0$$

which gives $D \geq -\frac{72}{5} = -14.4$. This indicates that the corner

$$\left(8 - \frac{2D}{3}, 12 + \frac{5D}{6} \right)$$

is the point where z is maximum when D is restricted to the interval $-14.4 \leq D \leq 12$.

■ **Now You Are Ready to Work Exercise 16**

■■ 4.7 EXERCISES

1. *(See Example 2)* A feasible region is defined by

$$3x + 2y \leq 30$$
$$5x + 2y \leq 40$$
$$x \geq 0, y \geq 0$$

Determine which of the following objective functions have their maximum value at the intersection of $3x + 2y = 30$ and $5x + 2y = 40$.
(a) $z = 7x + 7y$ **(b)** $z = 8x + 4y$
(c) $z = 34x + 20y$

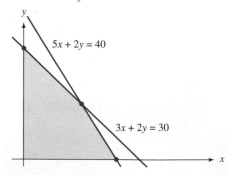

2. A feasible region is defined by

$$3x + 4y \leq 39$$
$$7x + 2y \leq 47$$
$$x \geq 0, y \geq 0$$

Determine which of the following objective functions have their maximum value at the intersection of $3x + 4y = 39$ and $7x + 2y = 47$.

(a) $z = 9x + 3y$
(b) $z = 2x + 8y$
(c) $z = 10x + 9y$

3. A linear programming problem seeks to maximize $z = Ax + By$, subject to

$$10x + 15y \leq 255$$
$$18x + 5y \leq 261$$
$$x \geq 0, y \geq 0$$

Find the range of slopes of the objective function so that the maximum z occurs at $(12, 9)$.

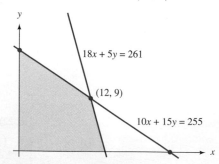

4. The figure shows the feasible region determined by

$$4x + 5y \leq 182$$
$$7x + 2y \leq 170$$
$$x \geq 0, y \geq 0$$

Find values of A and B so that the objective function $z = Ax + By$ attains its maximum at $(18, 22)$.

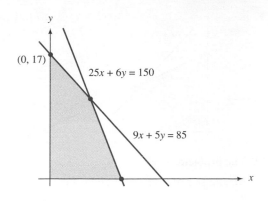

5. *(See Example 3)* A feasible region is defined by

$$5x + 3y \le 150$$
$$7x + 10y \le 280$$
$$x \ge 0, y \ge 0$$

Determine which of the following objective functions have their maximum value at $(0, 28)$.
(a) $z = 5x + 5y$ **(b)** $z = 9x + 20y$
(c) $z = 8x + 3y$

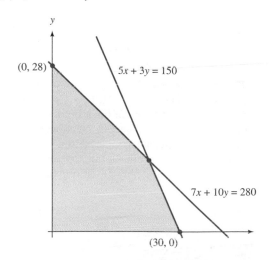

6. A linear programming problem seeks to maximize $z = Ax + By$, subject to

$$9x + 5y \le 85$$
$$25x + 6y \le 150$$
$$x \ge 0, y \ge 0$$

Find the range of slopes of the object
that the maximum z occurs at $(0, 1$

7. *(See Example 4)* Which of the following objective functions have their maximum value at the corner $(30, 0)$ of the feasible region in Exercise 5?
(a) $z = 4x + 7y$
(b) $z = 5x + 2y$
(c) $z = 30x + 15y$

8. A linear programming problem seeks to maximize $z = Ax + By$, subject to

$$5x + 2y \le 240$$
$$2x + 3y \le 140$$
$$x \ge 0, y \ge 0$$

Find the range of slopes of the objective function $z = Ax + By$ so that the maximum occurs at $(48, 0)$.

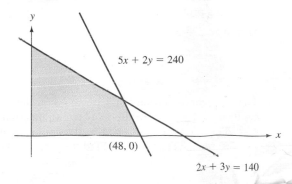

9. A linear programming ~~maximize~~ ≤ 126
$z = Ax + B \ge 0, y \ge 0$ $8y \le 96$

~~...nge of slopes of the objective function so~~
~~...e maximum z occurs at (21, 0).~~

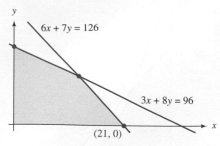

10. The problem

Maximize $z = 10x + 8y$, subject to

$$4x + 5y \le 107$$
$$7x + 3y \le 101$$
$$x \ge 0, y \ge 0$$

has the solution maximum $z = 200$ at $(8, 15)$. Find the range of values that can be used for the coefficient of x in the objective function $z = Ax + 8y$ so that the optimal point $(8, 15)$ remains the same.

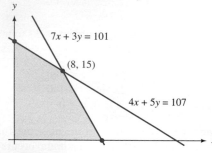

11. *(See Example 2b)*
Tatum and Associates used the following linear programming problem to maximize profit.

Maximize $z = 30x + 20y$, subject to

$$9x + 8y \le 684$$
$$4x + 10y \le 536$$
$$x \ge 0, y \ge 0$$

The manager asks what range of values for the coefficient of x (now 30) in the objective function $z = Ax + 20y$ is possible so that the optimal point $(44, 36)$ remains unchanged. Find the range for her.

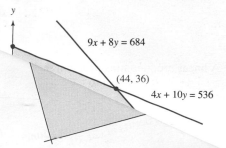

12. The manager of Nick's Inc. solved the following linear programming problem:

Maximize $z = 18x + 20y$, subject to

$$5x + 3y \le 121$$
$$4x + 10y \le 264$$
$$x \ge 0, y \ge 0$$

He found the maximum $z = 638$ occurred at $(11, 22)$. He was unsure of the coefficient of y, 20, in the objective function. Find the values of B in $z = 18x + By$ that can be used so that the optimal point $(11, 22)$ remains unchanged.

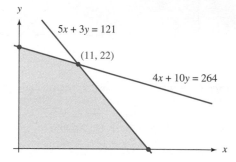

13. *(See Example 6)*
Ortin Company used the following linear programming problem to maximize production.

Maximize $z = 6x + 4y$, subject to

$$4x + 5y \le 174$$
$$8x + 4y \le 240$$
$$x \ge 0, y \ge 0$$

The maximum value of the given objective function is 198 and occurs at $(21, 18)$. Although the manager was confident of the accuracy of the x-coefficient, 6, in the objective function, she realized that the y-coefficient, 4, might differ. Find the range of values of the y coefficient in $z = 6x + By$ for which the optimal point $(21, 18)$ remains unchanged.

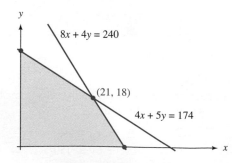

14. Lauren, the production manager of Michna's Manufacturing, solves the following linear programming problem:

Maximize $z = 18x + 12y$, subject to
(a) $8x + 5y \leq 200$
(b) $4x + 3y \leq 108$
    $x \geq 0, y \geq 0$

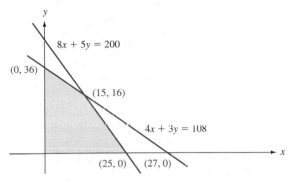

Lauren found that the maximum $z = 162$ occurs at the point $(15, 16)$. Later she finds additional resources become available and constraint (a) becomes $8x + 5y \leq 208$. With no change in constraint (b), find the resulting optimal solution.

15. *(See Example 7)*   A supervisor asks Anna to solve the linear programming problem

Maximize $z = 52x + 25y$, subject to

$$8x + 6y \leq 480$$
$$3x + y \leq 150$$
$$x \geq 0, y \geq 0$$

Anna finds the feasible region as shown with maximum $z = 2784$ at $(42, 24)$.

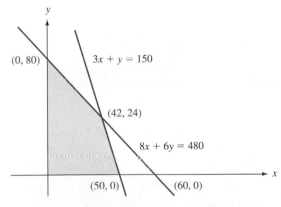

When Anna reports her findings to the supervisor, she is greeted with "Oh, we received more resources," so the constraints are now

$$8x + 6y \leq 480$$
$$3x + y \leq 186$$

Help Anna by finding the new optimal solution.

16. *(See Example 8)*
The linear programming problem

Maximize $z = 12x + 10y$, subject to

$$8x + 10y \leq 320$$
$$12x + 5y \leq 280$$
$$x \geq 0, y \geq 0$$

has the solution maximum $z = 380$ at $(15, 20)$.

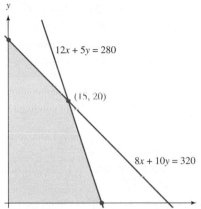

Suppose the resources for the first constraint change from 320 to $320 + D$, changing the problem to

Maximize $z = 12x + 10y$, subject to

$$8x + 10y \leq 320 + D$$
$$12x + 5y \leq 280$$
$$x \geq 0, y \geq 0$$

(a) Find the corner point determined by the intersection of $8x + 10y = 320 + D$ and $12x + 5y = 280$.
(b) Find $z = 12x + 10y$ for the point of intersection found in part (a).
(c) How does a change in D affect the maximum value of z?
(d) What restrictions apply to D?

17. The linear programming problem

Maximize $z = 10x + 4y$, subject to

$$4x + 5y \leq 140$$
$$6x + 2y \leq 144$$
$$x \geq 0, y \geq 0$$

has the solution maximum $z = 248$ at $(20, 12)$.

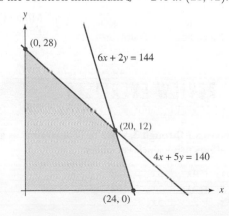

Suppose the resources for the second constraint change from 144 to 144 + D, changing the problem to

Maximize $z = 10x + 4y$, subject to

$$4x + 5y \le 140$$
$$6x + 2y \le 144 + D$$
$$x \ge 0, y \ge 0$$

(a) Find the corner point determined by the intersection of $4x + 5y = 140$ and $6x + 2y = 144 + D$.
(b) Find $z = 10x + 4y$ for the point of intersection found in part (a).
(c) How does a change in D affect the maximum value of z?
(d) What restrictions apply to D?

18. The following problem has as its solution, maximum $z = 152$ at $(5, 12)$.

Maximize $z = 4x + 11y$, subject to

$$6x + 11y \le 162$$
$$3x + y \le 54$$
$$x + 5y \le 65$$
$$x \ge 0, y \ge 0$$

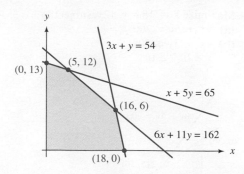

What changes can be made to the coefficient of x in the objective function so that the maximum value occurs at the corner $(16, 6)$? That is, find the values of A such that $z = Ax + 11y$ has its maximum value at $(16, 6)$.

IMPORTANT TERMS

4.1

Simplex Method
Standard Maximum Problem
Slack Variable
Simplex Tableau

4.2

Unit Column
Feasible Solution
Basic Solution
Basic Feasible Solution
Nonbasic Variable
Basic Variable
Initial Basic Feasible Solution

Pivot Element
Pivot Column
Pivot Row
Pivoting
Final Tableau

4.3

Standard Minimum Problem
Dual Problem
Transpose of a Matrix

4.4

Mixed Constraints
Phase I

Phase II
Equality Constraint

4.5

Multiple Solutions
No Solutions
Unbounded Solution
No Feasible Solution

4.7

Sensitivity Analysis

REVIEW EXERCISES

In Exercises 1 through 3, write the constraints as a system of equations using slack variables.

1. $6x_1 + 4x_2 + 3x_3 \le 220$
$x_1 + 5x_2 + x_3 \le 162$
$7x_1 + 2x_2 + 5x_3 \le 139$

2. $5x_1 + 3x_2 \le 40$
$7x_1 + 2x_2 \le 19$
$6x_1 + 5x_2 \le 23$

3. $6x_1 + 5x_2 + 3x_3 + 3x_4 \le 89$
$7x_1 + 4x_2 + 6x_3 + 2x_4 \le 72$

In Exercises 4 through 6, write the constraints and objective function as a system of equations.

4. Objective function: $z = 3x_1 + 7x_2$
Constraints:
$$7x_1 + 5x_2 \leq 14$$
$$3x_1 + 6x_2 \leq 25$$
$$4x_1 + 3x_2 \leq 29$$

5. Objective function: $z = 20x_1 + 36x_2 + 19x_3$
Constraints:
$$10x_1 + 12x_2 + 8x_3 \leq 24$$
$$7x_1 + 13x_2 + 5x_3 \leq 35$$

6. Objective function: $z = 5x_1 + 12x_2 + 8x_3 + 2x_4$
Constraints:
$$9x_1 + 7x_2 + x_3 + x_4 \leq 84$$
$$x_1 + 3x_2 + 5x_3 + x_4 \leq 76$$
$$2x_1 + x_2 + 6x_3 + 3x_4 \leq 59$$

Write Exercises 7 through 10 as systems of equations.

7. Maximize $z = 9x_1 + 2x_2$, subject to
$$3x_1 + 7x_2 \leq 14$$
$$9x_1 + 5x_2 \leq 18$$
$$x_1 - x_2 \leq 21$$
$$x_1 \geq 0, x_2 \geq 0$$

8. Maximize $z = x_1 + 5x_2 + 4x_3$, subject to
$$x_1 + x_2 + x_3 < 20$$
$$4x_1 + 5x_2 + x_3 \leq 48$$
$$2x_1 - 6x_2 + 5x_3 \leq 38$$
$$x_1 \geq 0, x_2 \geq 0, x_3 \geq 0$$

9. Maximize $z = 6x_1 + 8x_2 + 4x_3$, subject to
$$x_1 + x_2 + x_3 \leq 15$$
$$2x_1 + 4x_2 + x_3 \leq 44$$
$$x_1 \geq 0, x_2 \geq 0, x_3 \geq 0$$

10. Maximize $z = 5x_1 + 5x_2$, subject to
$$5x_1 + 3x_2 \leq 15$$
$$2x_1 + 3x_2 \leq 12$$
$$x_1 \geq 0, x_2 \geq 0$$

11. Find the pivot element in each of the following tableaux.

(a)
$$\begin{bmatrix} 5 & 3 & 2 & 1 & 0 & 0 & 0 & 660 \\ 4 & 6 & 1 & 0 & 1 & 0 & 0 & 900 \\ 1 & 2 & 3 & 0 & 0 & 1 & 0 & 800 \\ \hline -5 & -8 & -4 & 0 & 0 & 0 & 1 & 0 \end{bmatrix}$$

(b)
$$\begin{bmatrix} 1 & 4 & 0 & 3 & 0 & -2 & 0 & 60 \\ 0 & 6 & 1 & 5 & 0 & 4 & 0 & 60 \\ 0 & -3 & 0 & 1 & 1 & 2 & 0 & 60 \\ \hline 0 & -1 & 0 & -2 & 0 & 3 & 1 & 48 \end{bmatrix}$$

12. Find the pivot element in each of the following tableaux.

(a)
$$\begin{bmatrix} 3 & 0 & 5 & 1 & 0 & 0 & 20 \\ 2 & 0 & -1 & 0 & 1 & 0 & 6 \\ 1 & 1 & 4 & 0 & 0 & 0 & 0 \\ \hline 4 & 0 & -8 & 0 & 0 & 1 & 145 \end{bmatrix}$$

(b)
$$\begin{bmatrix} 6 & 1 & 0 & -5 & 0 & 2 & 0 & 0 \\ 4 & 0 & 1 & 2 & 0 & -4 & 0 & 10 \\ -2 & 0 & 0 & 7 & 1 & 3 & 0 & 21 \\ \hline -1 & 0 & 0 & -9 & 0 & 6 & 1 & 572 \end{bmatrix}$$

13. Write the basic feasible solution for each of the following tableaux.

(a)
$$\begin{bmatrix} 6 & 0 & 10 & 1 & 0 & 42 \\ 5 & 1 & 8 & 0 & 0 & 80 \\ \hline -2 & 0 & 5 & 0 & 1 & 98 \end{bmatrix}$$

(b)
$$\begin{bmatrix} 0 & 1 & 0 & 8 & 6 & 4 & 0 & 42 \\ 1 & 0 & 0 & -2 & 3 & 3 & 0 & 73 \\ 0 & 0 & 1 & 5 & -1 & 6 & 0 & 15 \\ \hline 0 & 0 & 0 & 4 & 2 & 5 & 1 & 138 \end{bmatrix}$$

14. Write the initial simplex tableau for each of the following problems. Do not solve.
(a) Minimize $z = 4x_1 - 5x_2 + 3x_3$, subject to
$$9x_1 + 7x_2 + x_3 \leq 45$$
$$3x_1 + 2x_2 + 4x_3 \leq 39$$
$$x_1 + 5x_2 + 12x_3 \leq 50$$
$$x_1 \geq 0, x_2 \geq 0, x_3 \geq 0$$

(b) Maximize $z = 8x_1 + 13x_2$, subject to
$$9x_1 + 5x_2 \leq 45$$
$$6x_1 + 8x_2 \geq 48$$
$$x_1 \geq 0, x_2 \geq 0$$

15. Write the initial simplex tableau for each of the following problems. Do not solve.
(a) Maximize $z = 3x_1 + 5x_2 + 4x_3$, subject to
$$11x_1 + 5x_2 + 3x_3 \leq 142$$
$$3x_1 + 4x_2 + 7x_3 \geq 95$$
$$2x_1 + 15x_2 + x_3 \leq 124$$
$$x_1 \geq 0, x_2 \geq 0, x_3 \geq 0$$

(b) Minimize $z = 14x_1 + 22x_2$, subject to
$$7x_1 + 4x_2 \leq 28$$
$$x_1 + 3x_2 \geq 6$$
$$x_1 \geq 0, x_2 \geq 0$$

16. Write the initial simplex tableau for each of the following problems. Do not solve.
(a) Maximize $z = 24x_1 + 36x_2$, subject to
$$2x_1 + 9x_2 \leq 18$$
$$5x_1 + 7x_2 \leq 35$$
$$x_1 + 8x_2 = 8$$
$$x_1 \geq 0, x_2 \geq 0$$

(b) Minimize $z = 6x_1 + 11x_2$, subject to

$$5x_1 + 4x_2 \geq 20$$
$$3x_1 + 8x_2 \geq 24$$
$$x_1 + x_2 = 22$$
$$x_1 \geq 0, x_2 \geq 0$$

17. Write the initial simplex tableau for each of the following problems. Do not solve.

(a) Maximize $z = 5x_1 + 12x_2$, subject to

$$15x_1 + 8x_2 \geq 120$$
$$10x_1 + 12x_2 \leq 120$$
$$15x_1 + 5x_2 = 75$$
$$x_1 \geq 0, x_2 \geq 0$$

(b) Minimize $z = 3x_1 + 2x_2$, subject to

$$14x_1 + 9x_2 \leq 126$$
$$10x_1 + 11x_2 \geq 110$$
$$-5x_1 + x_2 = 9$$
$$x_1 \geq 0, x_2 \geq 0$$

Solve Exercises 18 through 23.

18. Maximize $z = 4x_1 + 5x_2$, subject to

$$x_1 + 3x_2 \leq 12$$
$$2x_1 + 4x_2 \leq 16$$
$$x_1 \geq 0, x_2 \geq 0$$

19. Maximize $z = 3x_1 + 5x_2 + 2x_3$, subject to

$$2x_1 + 4x_2 + 2x_3 \leq 34$$
$$3x_1 + 6x_2 + 4x_3 \leq 57$$
$$2x_1 + 5x_2 + x_3 \leq 30$$
$$x_1 \geq 0, x_2 \geq 0, x_3 \geq 0$$

20. Maximize $z = 3x_1 + 4x_2$, subject to

$$x_1 - 3x_2 \leq 6$$
$$x_1 + x_2 \leq 8$$
$$x_1 \geq 0, x_2 \geq 0$$

21. Maximize $z = 10x_1 + 15x_2$, subject to

$$-4x_1 + x_2 \leq 3$$
$$x_1 - 2x_2 \leq 12$$
$$x_1 \geq 0, x_2 \geq 0$$

22. Maximize $z = 3x_1 + 6x_2 + x_3$, subject to

$$4x_1 + 4x_2 + 8x_3 \leq 800$$
$$8x_1 + 6x_2 + 4x_3 \leq 1800$$
$$8x_1 + 4x_2 \leq 400$$
$$x_1 \geq 0, x_2 \geq 0, x_3 \geq 0$$

23. Maximize $z = x_1 + 3x_2 + x_3$, subject to

$$4x_1 + x_2 + x_3 \leq 372$$
$$x_1 + 8x_2 + 6x_3 \leq 1116$$
$$x_1 \geq 0, x_2 \geq 0, x_3 \geq 0$$

24. Find the values of the slack variables in the constraints

$$4x_1 + 3x_2 + 6x_3 + s_1 = 68$$
$$x_1 + 2x_2 + 5x_3 + s_2 = 90$$

for the points $(3, 2, 1)$ and $(5, 10, 3)$.

25. Which variable contributes the most to increasing z in the following objective functions of a maximization problem?

(a) $z = 7x_1 + 2x_2 + 9x_3$
(b) $z = x_1 + 8x_2 + x_3 + 7x_4$

26. Given the constraint

$$7x_1 + 4x_2 + 17x_3 + s_1 = 56$$

when $x_2 = 0$, what is the largest possible value of x_3 so that s_1 is nonnegative?

27. Write the transpose of the matrices

$$\begin{bmatrix} 3 & 1 & -2 \\ 4 & 0 & 6 \\ 5 & 7 & 8 \end{bmatrix} \quad \text{and} \quad \begin{bmatrix} 4 & 3 & 2 & 1 \\ -5 & 0 & 12 & 9 \end{bmatrix}$$

Solve Exercises 28 through 36.

28. Minimize $z = 3x_1 + 5x_2 + 4x_3$, subject to

$$3x_1 - 3x_2 + x_3 \leq 54$$
$$x_1 + x_2 + x_3 \geq 24$$
$$-3x_1 + 2x_3 \leq 15$$
$$x_1 \geq 0, x_2 \geq 0, x_3 \geq 0$$

29. Minimize $z = 6x_1 + 8x_2 + 16x_3$, subject to

$$2x_1 + x_2 \geq 6$$
$$x_2 + 2x_3 \geq 8$$
$$x_1 \geq 0, x_2 \geq 0, x_3 \geq 0$$

30. Minimize $z = 10x_1 + 20x_2 + 15x_3$, subject to

$$x_1 + x_2 + x_3 \geq 100$$
$$9x_1 - 4x_3 \leq 128$$
$$x_1 + 4x_2 = 48$$
$$x_1 \geq 0, x_2 \geq 0, x_3 \geq 0$$

31. Maximize $z = 6x_1 + 11x_2 + 8x_3$, subject to

$$2x_1 + 5x_2 + 4x_3 \leq 40$$
$$40x_1 + 45x_2 + 30x_3 \leq 430$$
$$6x_1 + 3x_2 + 4x_3 \geq 48$$
$$x_1 \geq 0, x_2 \geq 0, x_3 \geq 0$$

32. Maximize $z = 3x_1 + 4x_2 + x_3$, subject to

$$x_1 + x_2 + x_3 \leq 28$$
$$3x_1 + 4x_2 + 6x_3 \geq 60$$
$$2x_1 + x_2 = 30$$
$$x_1 \geq 0, x_2 \geq 0, x_3 \geq 0$$

33. Maximize $z = 2x_1 + 5x_2 + 3x_3$, subject to

$$x_1 + x_2 + x_3 \geq 6$$
$$2x_1 + x_2 + 3x_3 \leq 10$$
$$2x_2 - x_3 \leq 5$$
$$x_1 \geq 0, x_2 \geq 0, x_3 \geq 0$$

34. Minimize $z = 18x_1 + 24x_2$, subject to

$$3x_1 + 4x_2 \geq 48$$
$$x_1 + 2x_2 \leq 22$$
$$3x_1 + 2x_2 \leq 42$$
$$x_1 \geq 0, x_2 \geq 0$$

35. Maximize $z = 20x_1 + 32x_2$, subject to

$$x_1 - 3x_2 \leq 24$$
$$-5x_1 + 4x_2 \leq 20$$
$$x_1 \geq 0, x_2 \geq 0$$

36. Minimize $z = 8x_1 + 10x_2 + 25x_3$, subject to

$$x_1 + x_3 \geq 30$$
$$2x_1 + 4x_2 + 5x_3 \geq 70$$
$$2x_2 + x_3 \geq 27$$
$$x_1 \geq 0, x_2 \geq 0, x_3 \geq 0$$

Solve Exercises 37 through 41.

37. Minimize $z = 18x_1 + 36x_2$, subject to

$$3x_1 + 2x_2 \geq 24$$
$$5x_1 + 4x_2 \geq 46$$
$$4x_1 + 9x_2 \geq 60$$
$$x_1 \geq 0, x_2 \geq 0$$

38. Maximize $z = 5x_1 + 15x_2$, subject to

$$4x_1 + x_2 \leq 200$$
$$x_1 + 3x_2 \geq 120$$
$$x_1 \geq 0, x_2 \geq 0$$

39. Maximize $z = 5x_1 + 15x_2$, subject to

$$4x_1 + x_2 \leq 180$$
$$x_1 + 3x_2 \geq 120$$
$$-x_1 + 3x_2 = 150$$
$$x_1 \geq 0, x_2 \geq 0$$

40. Minimize $z = 3x_1 - 2x_2$, subject to

$$x_1 + 3x_2 \leq 30$$
$$3x_1 + x_2 \leq 21$$
$$x_1 \geq 0, x_2 \geq 0$$

41. Maximize $z = 2x_1 + x_2$, subject to

$$x_1 + 3x_2 \leq 9$$
$$x_1 - x_2 \leq -2$$
$$x_1 \geq 0, x_2 \geq 0$$

42. For each of the following minimization problems, set up the augmented matrix and the initial tableau for the dual problem. Do not solve.

(a) Minimize $z = 30x_1 + 17x_2$, subject to

$$4x_1 + 5x_2 \geq 52$$
$$7x_1 + 14x_2 \geq 39$$
$$x_1 \geq 0, x_2 \geq 0$$

(b) Minimize $z = 100x_1 + 225x_2 + 145x_3$, subject to

$$20x_1 + 35x_2 + 15x_3 \geq 130$$
$$40x_1 + 10x_2 + 6x_3 \geq 220$$
$$35x_1 + 22x_2 + 18x_3 \geq 176$$
$$x_1 \geq 0, x_2 \geq 0, x_3 \geq 0$$

Set up the initial simplex tableau for Exercises 43 and 44. Do not solve.

43. A company manufactures three items: hunting jackets, all-weather jackets, and ski jackets. It takes 3 hours of labor per dozen to produce hunting jackets, 2.5 hours per dozen for all-weather jackets, and 3.5 hours per dozen for ski jackets. The cost per dozen is $26 for hunting jackets, $20 for all-weather jackets, and $22 for ski jackets. The profit per dozen is $7.50 for hunting jackets, $9 for all-weather jackets, and $11 for ski jackets. The company has 3200 hours of labor and $18,000 in operating funds available. How many of each jacket should it produce to maximize profits?

44. A fertilizer company produces two kinds of fertilizers, Lawn and Tree. It has orders on hand that call for the production of at least 20,000 bags of Lawn fertilizer and 5000 bags of Tree fertilizer. Plant A can produce 1500 bags of Lawn and 300 bags of Tree fertilizer per day. Plant B can produce 750 bags of Lawn and 250 bags of Tree fertilizer per day. It costs $18,000 per day to operate plant A and $12,000 per day to operate plant B. How many days should the company operate each plant to minimize operating costs?

45. The constraints to a linear programming problem are

$$5x + 2y \leq 155$$
$$3x + 4y \leq 135$$
$$x \geq 0, y \geq 0$$

Find the possible values of A so that the objective function $z = Ax + 8y$ has its maximum value at $(25, 15)$.

46. The constraints to a linear programming problem are

$$4x + 5y \leq 205$$
$$16x + 5y \leq 336$$
$$x \geq 0, y \geq 0$$

Find the possible values of B so that the objective function $z = 10x + By$ has its maximum value at $(0, 41)$.

CHAPTER 5

Mathematics of Finance

Our modern economy depends on borrowed money. If you have a credit card or a student loan, you have firsthand experience with a loan. Borrowed money enables students to obtain an education or to own an automobile. Few families can own a home without borrowing money. Business depends on borrowed money for day-to-day operations and major expansions. Governments at all levels, schools, churches, and other institutions borrow money. Banks depend on loans for a major source of their income. Our economy would collapse if financial institutions quit making loans.

Here are some trivia questions for you: "What is the current national debt?" "When was the last time the federal government balanced its budget?" "When was the last time there was no national debt?"

"Rented money" describes "borrowed money" because a fee is paid for the use of money for a period of time. Just as you pay a rental fee for the use of an apartment for a semester, you pay a rental fee, or **interest** as it is called, for the use of money.

Even those few who may not borrow money may place money in a savings account or a certificate of deposit. In doing so, that person loans money to the bank, and the bank pays interest. In turn, the bank loans the money to an individual or business (at a higher rate, of course).

Some day in the distant future, you hope to retire with adequate funds to enjoy a reasonable quality of life. To do so, you, your employer, or both need to invest in a retirement plan that will pay you an **annuity** in your retirement years. An analysis of investments and annuities shows that the earlier you invest for retirement, the more likely you will have adequate retirement income.

We now look at some methods used to determine the fee charged for the use of money. ■

5.1   Simple Interest

- Simple Interest
- Future Value
- Treasury Bills—Simple Discount

Simple Interest

Simple interest is most often used for loans of shorter duration, in situations like a construction company that may be expecting payment in a week for work done, but it is payday and the workers are due their checks so the company borrows money for a week, or a store may obtain a 30-day loan to purchase an inventory for a sale.

How does a simple interest loan work? First, let's mention some standard terms used when discussing simple interest. The money borrowed in a loan is called the **principal.** The number of dollars received by the borrower is the **present value.** In a simple interest loan, the principal and the present value are the same. The fee for a simple interest loan is usually expressed as a percentage of the principal and is called the **interest rate.** If the interest rate is 10% per year, then each year the borrower pays 10% of the principal, the amount borrowed. If the interest rate is 1.5% per month, then the borrower pays 1.5% of the principal for each month the money is borrowed. You should be aware that the formulas in this chapter require that the interest rate be written in decimal form, not in percent. For example, 7.5% must be written as 0.075.

We denote the amount of the principal by P. If the interest rate is 10% per year, then the interest paid is $0.10P$ for each year of the loan. Thus, if the money is borrowed for three years, the total interest paid is $3(0.10P)$.

This suggests the general formula that gives the total fee, interest, which is paid for a simple interest loan.

Simple Interest

$$I = Prt$$

where

P = principal (amount borrowed)
r = interest rate per year (expressed in decimal form)
t = time in years

An interest rate may be stated as 10% per year, 1% per month, or in terms of other time units. Sometimes no time units are specified. In such cases it is understood that the time unit is years. A statement that the interest rate is 12% should be interpreted as 12% *per year.* The time period will be stated when it is not annual. Simple interest is paid on the principal borrowed and is not paid on interest already earned.

EXAMPLE 1 ➤

Compute the interest paid on a loan of $1400 at a 9% interest rate for 18 months.

SOLUTION

$$P = \$1400$$
$$r = 9\% = 0.09 \text{ in decimal form}$$
$$t = 18 \text{ months} = \frac{18}{12} \text{ years} = 1.5 \text{ years}$$
$$I = 1400 \times 0.09 \times 1.5 = 189$$

So, the interest paid is $189.

 **CAUTION**

The time units for r and t must be consistent, so months were converted to years.

■ **Now You Are Ready to Work Exercise 1**

EXAMPLE 2 ➤

An individual borrows $300 for 6 months at 1% simple interest per month. How much interest is paid?

SOLUTION

Notice that the interest rate is given as 1% *per month.* We can still use the $I = Prt$ formula, provided that r and t are consistent in time units — in this case, months.

$$I = 300 \times 0.01 \times 6 = \$18$$

■ **Now You Are Ready to Work Exercise 11**

EXAMPLE 3 ➤

Jose borrows money at 8% for 2 years. He paid $124 interest. How much did he borrow?

SOLUTION

In this case, $I = 124, r = 0.08,$ and $t = 2,$ so

$$124 = P(0.08)(2) = 0.16P$$
$$P = \frac{124}{0.16} = 775$$

The loan was $775.

■ **Now You Are Ready to Work Exercise 14**

EXAMPLE 4 ➤

Jane borrowed $950 for 15 months. The interest was $83.13. Find the interest rate.

SOLUTION

We are given $P = 950, t = \frac{15}{12} = 1.25$ years, and $I = 83.13,$ so

$$83.13 = 950r(1.25)$$
$$= 1187.5r$$
$$r = \frac{83.13}{1187.5} = 0.07$$

So, the interest rate was 7% per year.

■ **Now You Are Ready to Work Exercise 18**

Future Value

A loan made at simple interest requires that the borrower pay back the sum borrowed (principal) plus the interest. This total, $P + I$, is called the **future value,** or **amount,** of the loan. Because $I = Prt$, we can write the future value as $A = P + Prt$.

Amount, or Future Value, of a Loan

$$A = P + I$$
$$= P + Prt$$
$$= P(1 + rt)$$

where

$P = $ principal, or present value
$r = $ annual interest rate
$t = $ time in years
$A = $ amount, or future value

EXAMPLE 5 ➤

Find the amount (future value) of a $2400 loan for 9 months at 11% interest rate.

SOLUTION

We want to find A in $A = P + I = P + Prt$. We know that $P = 2400, r = 0.11$, and $t = \frac{9}{12} = 0.75$ years, so

$$I = 2400(0.11)(0.75) = 198$$

and

$$A = 2400 + 198 = 2598$$

We can also use the formula $A = P(1 + rt)$ and compute A as

$$A = 2400(1 + 0.11(0.75))$$
$$= 2400(1 + 0.0825)$$
$$= 2400(1.0825)$$
$$= 2598$$

The total of principal and interest is $2598.

■ **Now You Are Ready to Work Exercise 22**

Up to this point, the examples have dealt with the cost (interest) of a loan. In order for a person to obtain a loan, another party must provide the money to be borrowed. That party invests money and is interested in the income received from the loan. The same formulas apply to the investor as to the borrower.

EXAMPLE 6 ➤

How much should you invest at 12% for 21 months to have $3000 at the end of the 21 months?

SOLUTION

In this example, you are given the future value, $3000, and are asked to find the present value, P. We know that $r = 0.12, t = \frac{21}{12} = 1.75$ years, and $A = 3000$. Then,

$$3000 = P(1 + 0.12(1.75))$$
$$= P(1 + 0.21)$$
$$= 1.21P$$

So, $P = \frac{3000}{1.21} = \$2479.34$ (rounded to nearest cent).

■ **Now You Are Ready to Work Exercise 26**

EXAMPLE 7 ➤ Your friend loaned some money. The debtor is scheduled to pay him $550 in 4 months. (The future value is $550.) Your friend needs the money now, so you agree to pay him $525 for the note, and the debtor will pay you $550 in 4 months. What annual interest rate will you earn?

SOLUTION

For the purposes of this problem, $A = 550$, $P = 525$, and $t = 4$ months $= \frac{1}{3}$ year. Using the formula $A = P + Prt$, we have

$$550 = 525 + 525r\left(\frac{1}{3}\right)$$
$$= 525 + 175r$$
$$25 = 175r$$
$$r = \frac{25}{175} = 0.1429$$

You will earn about 14.3% annual interest.

■ **Now You Are Ready to Work Exercise 47**

Treasury Bills—Simple Discount

Treasury bills are short-term securities issued by the federal government. The bills do not specify a rate of interest. They are sold at weekly public auctions with financial institutions making competitive bids. For example, a bank may bid $978,300 for a 90-day $1 million treasury bill. At the end of 90 days, the bank receives $1 million, which covers the cost of the bill and interest earned on the bill. This is basically what we call a **simple discount** transaction.

The simple discount loan differs from the simple interest loan in that the interest is *deducted* from the principal and the borrower receives less than the principal. For example, if a person borrows $1000 at 6% for 1 year, the interest is $60, which is deducted from the $1000, and the borrower receives $940. When the loan is repaid, the borrower pays $1000. We use the terminology **simple discount note** for this type of loan. The interest deducted is the **discount,** the amount received by the borrower is the **proceeds,** the **discount rate** is the percentage used, and the amount repaid is the **maturity value.** Here's how they are related.

Simple Discount

$$D = Mdt$$
$$PR = M - D$$
$$= M - Mdt$$
$$= M(1 - dt)$$

where

$M =$ maturity value (principal)
$d =$ annual discount rate, written in decimal form
$t =$ time in years
$D =$ discount
$PR =$ proceeds, the amount the borrower receives

EXAMPLE 8 ➤

Find the discount and the amount a borrower receives (proceeds) on a $1500 simple discount loan at 8% discount rate for 1.5 years.

SOLUTION

In this case, $M = 1500$, $d = 0.08$, and $t = 1.5$ years. Then,

$$D = 1500(0.08)(1.5)$$
$$= 180$$
$$PR = 1500 - 180$$
$$= 1320$$

So, the bank keeps the discount, $180, and the borrower receives $1320.

■ **Now You Are Ready to Work Exercise 29**

EXAMPLE 9 ➤

A bank wants to earn 7.5% simple discount interest on a 90-day $1 million treasury bill. How much should it bid?

SOLUTION

We are given the maturity value, $M = 1,000,000$, the discount rate, $d = 0.075$, and the time, $t = \frac{90}{360}$. (Banks often use 360 days per year when computing daily interest.) We want to find the proceeds, PR.

Substituting these in the simple discount formula, we get

$$PR = 1,000,000\left[1 - (0.075)\frac{90}{360}\right]$$
$$= 1,000,000(1 - 0.01875)$$
$$= 1,000,000(0.98125)$$
$$= 981,250$$

So, the bank should bid $981,250.

■ **Now You Are Ready to Work Exercise 59**

EXAMPLE 10 ➤

A bank paid $983,000 for a 90-day $1 million treasury bill. Find the simple discount rate.

SOLUTION

We are given $M = 1,000,000$, $PR = 983,000$, and $t = \frac{90}{360}$. We want to find d.

$$983,000 = 1,000,000\left[1 - d\left(\frac{90}{360}\right)\right]$$
$$= 1,000,000 - 250,000d$$
$$-17,000 = -250,000d$$
$$d = \frac{17,000}{250,000} = 0.068$$

So, the annual discount rate was 6.8%.

■ **Now You Are Ready to Work Exercise 61**

▣ 5.1 EXERCISES

LEVEL 1

1. *(See Example 1)*   Compute the interest on a loan of $1100 at 8% interest rate for 9 months.

Compute I for Exercises 2 through 5.

2. $P = \$2300, r = 9\%, t = 1.25$ years

3. $P = \$600, r = 10\%, t = 18$ months

4. $P = \$4200, r = 8\%, t = 16$ months

5. $P = \$745, r = 8.5\%, t = 6$ months

6. Compute the simple interest on $500 for 1 year at 7%.

7. Compute the simple interest and the amount for $300 at 6% for 1 year.

8. Compute the simple interest on $400 at 6% for 3 months.

9. Jones borrows $500 for 3 years at 6% simple interest (annual rate). How much interest will be paid?

10. Smith borrows $250 at 7% (annual rate) for 3 months. How much interest will be paid?

11. *(See Example 2)*   Mrs. Witkowski borrows $700 for 5 months at 1.5% simple interest per month. How much interest will she pay?

12. $1450 is borrowed for 3 months at a 1.5% monthly rate. How much interest is paid?

13. How much interest is paid on a $950 loan for 7 months at a 1.75% monthly rate?

How much was borrowed for the loans in Exercises 14 through 17?

14. *(See Example 3)* $I = \$24.19, r = 7.5\%, t = 9$ months

15. $I = \$21.45, r = 6.5\%, t = 6$ months

16. $I = \$38.00, r = 5\%, t = 10$ months

17. Matt paid $115.50 interest on a loan at 7% simple interest for 1.5 years. How much did he borrow?

Find the interest rate in Exercises 18 through 21.

18. *(See Example 4)* $P = \$650, I = \$91.00, t = 2$ years

19. $P = \$1140, I = \$49.40, t = 8$ months

20. $P = \$75, I = \$1.50, t = 2$ months

21. $P = \$5800, I = \$616.25, t = 1.25$ years

Find the amount (future value) of the loans in Exercises 22 through 25.

22. *(See Example 5)* $P = \$1800, r = 5.5\%, t = 7$ months

23. $P = \$2700, r = 4\%, t = 1.5$ years

24. $P = \$240, r = 6.3\%, t = 10$ months

25. $P = \$6500, r = 3.6\%, t = 1.75$ years

26. *(See Example 6)*   How much should you invest at 4% for 16 months to have $3000 at the end of that period?

27. How much should be invested at 6% to have $1800 at the end of 1.5 years?

28. Sandi wants to buy a $400 television set in 6 months. How much should she invest at 7% simple interest now to have the money then?

Find the discount and the amount the borrower receives (proceeds) for the discount loans in Exercises 29 through 32.

29. *(See Example 8)* $M = \$1850, d = 4.5\%, t = 1$ year

30. $M = \$960, d = 5.25\%, t = 10$ months

31. $M = \$485, d = 3.8\%, t = 1.5$ years

32. $M = \$9540, d = 6.5\%, t = 8$ months

LEVEL 2

33. Andy borrowed $3500 for $2\frac{1}{2}$ years at a simple interest rate of 4.6%. Find the total interest paid over this period.

34. Unique Antiques borrowed $8500 for 2 years at a simple interest rate of 6.35%. Find the total interest paid.

35. The Fort House Tea Room borrowed $17,500 for renovations at a simple interest rate of 6.85% for 4 years. The interest is to be paid annually. Find the amount of each annual interest payment.

36. Austin Avenue Accessories borrowed $14,000 for 3 years at a simple interest rate of 6.6%. The interest is to be paid semiannually.
   (a) Find the amount of the semiannual interest payment.
   (b) Find the total interest paid.

37. Hinton Inc. borrowed $85,000 for 5 years at a simple interest rate of 5.1%. The interest was to be paid monthly.
   (a) Find the monthly interest payment.
   (b) Find the total amount of interest paid.

38. Joe borrowed $150 from a loan company. At the end of 1 month he paid off the loan with $152.25. What annual interest rate did he pay?

39. Kathy borrowed $800 at 6% interest. The amount of interest paid was $144. What was the length of the loan?

40. For how long did Amy borrow $1200 if she paid $112 in interest at a 7% interest rate?

41. How long will it take $800 to earn $18 at 6% simple interest?

42. A student borrows $200 at 6% simple interest (annual rate). When the loan is repaid, the amount of the principal and interest is $218. What was the length of the loan?

43. A boy borrowed some money from the bank at 7.5% simple interest for 1 year. He paid $48.75 in interest. How much money was loaned?

44. The Yeager family borrowed some money for 15 months (1.25 years). The interest rate was 9%, and they paid $9.90 in interest. How much did they borrow?

45. A family borrows $20,000 to buy a home. The interest rate is 9%, and each monthly payment is $179.95. How much of the first payment goes for interest and how much for principal?

46. Mr. O'Neill borrows $750 for 4 months at 8% interest. How much interest does he pay?

47. *(See Example 7)* You pay $860 for a loan that has a future value of $900 in 6 months. What annual interest rate will you earn?

48. How much should you pay for a loan that has a future value of $320 in 3 months to earn 9% simple interest?

49. A man borrowed $950. Six months later, he repaid the loan (principal and interest) with $1000. What simple interest rate did he pay?

50. Find the future value of a $1300 loan for 14 months at 9.5% simple interest.

LEVEL 3

51. A jeweler wants to obtain $3000 for 4 months. She has the choice of a simple interest loan at 10.3% or a discount loan at 10.1%. Which one will result in the lower fee for use of the money?

52. The low point in gross public debt was $38,000 in 1836. If the government paid 6% simple interest, how much were the annual interest payments?

53. The total government debt (federal, state, and local) in 1970 was $450 billion. How much interest was paid for 1 year if the interest rate averaged 11%?

54. In 2003, the U.S. national debt reached $7 trillion dollars. The U.S. population was approximately 285 million people.
   (a) Find the per capita debt in 2003.
   (b) Assuming an interest rate of 4.5%, find the annual interest paid on the national debt.
   (c) Find the per capita interest paid on the national debt in 2003.
   (d) How long would it take to eliminate the 2003 national debt if no interest is charged and if the debt is reduced $100 billion each year?

55. A city has raised $500,000 for the purpose of putting in a new sewer system. This money was raised by issuing bonds that mature after 5 years. The interest on the bonds is 6.5% per annum. Determine the total interest paid.

56. The city of Hewitt sold $365,000 in bonds to build a new city park. The interest rate is 7.2% and is to be paid annually for 4 years. Find the annual interest and the total interest.

57. Monty's Manufacturing issued $625,000 in 8-year bonds. The interest rate was 5.5% simple interest and the interest was to be paid semiannually.
   (a) Find the semiannual interest that will be paid the bondholders.
   (b) Find the total interest paid.

58. An investment company bought $50,000 of 3-year bonds at 7% in simple interest payable quarterly. Find the quarterly interest payment and the total interest paid over the 3-year period.

59. *(See Example 9)*   A bank wants to earn 7% simple discount interest on a 90-day $1 million treasury bill. How much should it pay?

60. How much should a bank pay for a 90-day $1 million treasury bill to earn 6.5% simple discount?

61. *(See Example 10)*   A bank paid $982,000 for a 90-day $1 million treasury bill. What was the simple discount rate?

62. A bank paid $987,410 for a 90-day $1 million treasury bill. What was the simple discount rate?

63. How much should a bank bid on a 30-day $2 million treasury bill to receive 5.125% interest?

64. The Midway School District sold $1.5 million in bonds to build an addition to the high school. The interest on the bonds is to be paid annually at 4.5% simple interest over a period of 4 years. Find the annual interest payments and the total interest over the 4 years.

65. The McGregor Water Company sold $350,000 in bonds to finance a water system in a new addition to the city. The bonds were issued for 6 years at 4.2% simple interest. The interest is payable quarterly. Find the quarterly payments that will be paid to the bondholders and the total interest that will be paid.

EXPLORATIONS

66. Report on interest rates for a $5000 certificate of deposit for 1 year and for 5 years at a credit union, a bank, and a savings and loan institution.

67. Report on the interest rates charged by a local bank on
   (a) House loans.   **(b)** Automobile loans.
   (c) Personal loans.   **(d)** Commercial loans.

68. Report on the interest rates and the collateral required by a nonbank finance company.

69. Report on the interest rates and the collateral required by a pawn shop.

70. Check a newspaper and report on interest rates charged by automobile dealers.

71. Check with a savings and loan bank to determine the interest paid on savings accounts and the in-

terest charged on loans. Which interest rate is higher? Why?

72. Graph the simple interest function $I = Prt$ as $y = Prx$ where $P = \$100$ and $r = 7\%$. Observe the growth of simple interest for x in the interval 0 to 5 years.

73. Graph the simple interest function $y = Pxt$ where $P = \$100$ and $t = 1$ year for values of x ranging from 1% to 10% and observe how the interest paid changes.

74. On one screen, graph three simple interest functions $y = Prx$ where $P = \$100$ and x ranges from 0 to 5 years using $r = 2.3\%$, 6%, and 9.5%.

5.2 Compound Interest

- Compound Interest
- Present Value
- Doubling an Investment
- Effective Rate

Compound Interest

When you deposit money into a savings account, the bank will pay for the use of your money. Normally, the bank pays interest at specified periods of time, such as every three months. Unless instructed otherwise, the bank credits your account with the interest, and for the next time period, the bank pays interest on the new total. This is called **compound interest.** Let's look at a simple example.

EXAMPLE 1 ➤ You put $1000 into an account that pays 8% annual interest. The bank will compute interest and add it to your account at the end of each year. This is called **compounding interest annually.** Here's how your account builds up. We use the formula $A = P(1 + rt)$, where $r = 0.08$ and $t = 1$.

End of Year	Balance in Your Account
0 (start)	$1000.00
1	$1000(1.08) = 1080.00$
2	$1080(1.08) = 1166.40 = 1000(1.08)^2$
3	$1166.40(1.08) = 1259.71 = 1000(1.08)^3$
4	$1259.71(1.08) = 1360.49 = 1000(1.08)^4$
5	$1360.49(1.08) = 1469.33 = 1000(1.08)^5$

So, at the end of 5 years, the account has grown to $1469.33.

■ **Now You Are Ready to Work Exercise 8**

Notice how the pattern of growth involves *powers* of 1.08. The amount at the end of 1 year is just 1.08 times the amount at the end of the preceding year. Thus, the amount at the end of 4 years is 1.08 times the amount at the end of 3 years, the amount at the end of 5 years is 1.08 times the amount at the end of 4 years; or for longer periods of time, the amount at the end of 17 years is 1.08 times the amount at the end of 16 years, and so on. A more helpful form gives the amount after, say, 5 years in terms of the *original* investment and 1.08. Notice the amount after 5 years is $1000(1.08)^5$.

This follows because multiplying 1000 by 1.08 to obtain the amount for the first year, multiplying the first year total by 1.08 to obtain the amount for the second year, and so on for 5 years is equivalent to multiplying the original investment by 1.08 five times. You may expect, quite correctly, that the amount after 10 years is equal to $1000(1.08)^{10}$.

In general, we have the following formula for interest compounded annually.

Amount of Annual Compound Interest

When P dollars are invested at an annual interest rate r and the interest is compounded annually, the amount A at the end of t years is

$$A = P(1 + r)^t$$

EXAMPLE 2 ➤

$800 is invested at 6%, and it is compounded annually. What is the amount in the account at the end of 4 years?

SOLUTION

Here, $P = 800$, $r = 0.06$, and $t = 4$, so

$$A = 800(1.06)^4 = 800(1.26248) = 1009.98$$

■ **Now You Are Ready to Work Exercise 10**

In the preceding examples, interest was compounded annually. Interest is often calculated and added to the principal at other regular intervals. The most common intervals are semiannually, quarterly, monthly, and daily.

What do we mean by compounding semiannually, quarterly, and so on? It means that at the end of a fixed time period, interest is calculated and added to the account. Here is a summary of common compound interest intervals.

Interest Period	Length of Period	Frequency of Interest Payments
Annually	1 year	Once a year
Semiannually	6 months	2 times a year
Quarterly	3 months	4 times a year
Monthly	1 month	12 times a year
Daily	1 day	365 times a year

We use the quarterly interest rate when interest is compounded quarterly, the monthly interest rate for monthly compounding, and so on.

Because interest rates are usually stated as an annual rate, we need to convert to the appropriate **periodic interest rate.** We do this by dividing the annual rate by the number of interest periods in a year. An 8% annual rate becomes $8\%/4 = 2\%$ quarterly rate and a 6% annual rate becomes $6\%/12 = 0.5\%$ monthly rate.

Generally, we will use the letter r to represent the annual rate and the letter i to represent the periodic rate.

Periodic Interest Rate

Given the annual interest rate, r, with interest compounded m times a year, the periodic interest rate, i, is

$$i = \frac{r}{m}$$

Let's repeat Example 1 but compound the interest quarterly.

EXAMPLE 3 ➤

$1000 is invested at 8% annual rate. The interest is compounded quarterly. Find the amount in the account at the end of 5 years.

SOLUTION

Because interest is compounded quarterly, it must be computed every three months and added to the principal. The quarterly interest rate is 2%, so the computations are the following:

End of Quarter	Amount in Account
0	1000.00
1	$1000(1.02) = 1020.00$
2	$1020(1.02) = 1040.40 = 1000(1.02)^2$
3	$1040.40(1.02) = 1061.21 = 1000(1.02)^3$
4	$1061.21(1.02) = 1082.43 = 1000(1.02)^4$

At this stage we have the amount at the end of 1 year. We will not continue for another 16 computations, but observe the pattern. At the end of ten quarters, we correctly expect the amount to be $A = 1000(1.02)^{10}$. At the end of 5 years, 20 quarters, the amount is

$$A = 1000(1.02)^{20} = 1000(1.48595) = 1485.95$$

Notice that compounding quarterly gives an amount of $1485.95, a larger amount than the amount $1469.33 obtained from compounding annually. ■

The general formula for finding the **amount** after a specified number of compound periods is the following.

Compound Interest—Amount (Future Value)

$$A = P(1 + i)^n$$

where

r = annual interest rate
m = number of times compounded per year
$i = r/m$ = the interest rate per period
n = the number of periods, $n = mt$ where t is the number of years
A = amount (**future value**) at the end of n compound periods
P = principal (present value)

 NOTE

Two people may obtain different results of calculations in the examples and exercises. Calculators may use a different number of digits in computations. A person who rounds the calculations at each step will likely obtain different results from the person who uses all digits in a calculator at each step.

In this chapter, the examples and exercises are worked using all digits generated by the calculator. Some of the intermediate steps are shown with fewer than the 12 digits used by the calculator. When computing dollar amounts, the final answer is rounded to the nearest penny, although the intermediate steps are not rounded.

An example of the effect of rounding intermediate steps follows: If $1000 is invested at a 7% annual rate compounded monthly, the amount after 2 years is given by

$$1000\left(1 + \frac{.07}{12}\right)^{24} \quad \text{where} \quad \frac{0.07}{12} = 0.00583333\ldots$$

with as many 3's as the calculator will contain. The person who rounds to 0.00583 will obtain $A = 1000(1.00583)^{24} = 1149.71$ (to the nearest penny). The person who uses all digits in the calculator will obtain $A = 1000(1.0058333\ldots)^{24} = 1149.81$ (rounded to the nearest penny). The answers differ by $0.10.

EXAMPLE 4 ➤

$800 is invested at 12% for 2 years. Find the amount at the end of 2 years if the interest is compounded (a) annually, (b) semiannually, and (c) quarterly.

SOLUTION

In this example, $P = 800$, $r = 0.12$, and $t = 2$ years.

(a) $m = 1$, so $i = 0.12$, $n = 2$:

$$A = 800(1 + 0.12)^2 = 800(1.12)^2 = 800(1.2544) = 1003.52$$

(b) $m = 2$, so $i = \dfrac{0.12}{2} = 0.06$, $n = 4$:

$$A = 800(1 + 0.06)^4 = 800(1.06)^4 = 800(1.26248) = 1009.98$$

(c) $m = 4$, so $i = \dfrac{0.12}{4} = 0.03$, $n = 8$:

$$A = 800(1 + 0.03)^8 = 800(1.03)^8 = 800(1.26677) = 1013.42$$

■ **Now You Are Ready to Work Exercise 12**

EXAMPLE 5 ➤

$3000 is invested at 7.2%, compounded quarterly. Find the amount at the end of 5 years.

SOLUTION

$P = 3000$, $m = 4$, $r = 0.072$, $i = \dfrac{0.072}{4} = 0.018$, and $n = 4 \times 5 = 20$.

$$A = 3000(1.018)^{20} = 3000(1.428748) = 4286.24$$

The amount at the end of 5 years is $4286.24. Table 5–1 and Figure 5–1 show the growth by each quarter.

TABLE 5–1

Quarter	Interest	Amount	Quarter	Interest	Amount
0	0	3000.00	11	64.54	3650.45
1	54.00	3054.00	12	65.71	3716.16
2	54.97	3108.97	13	66.89	3783.05
3	55.96	3164.93	14	68.10	3851.15
4	56.97	3221.90	15	69.32	3920.47
5	58.00	3279.90	16	70.57	3991.04
6	59.03	3338.93	17	71.84	4062.88
7	60.11	3399.04	18	73.12	4136.01
8	61.18	3460.22	19	74.46	4210.46
9	62.28	3522.50	20	75.78	4286.24
10	63.41	3585.91			

Notice that the interest earned each quarter grows steadily, from $54.00 the first quarter to $75.78 the 20th quarter.

■ **Now You Are Ready to Work Exercise 14**

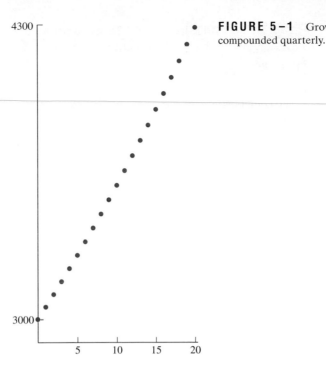

FIGURE 5−1 Growth of $3000 at 7.2% compounded quarterly.

EXAMPLE 6 ➤

A department store charges 1% per month on the unpaid balance of a charge account. This means that 1% of the bill is added to the account each month it is unpaid. This makes it compound interest. A customer owes $135.00, and the bill is unpaid for 4 months. What is the amount of the bill at the end of 4 months?

SOLUTION

The interest rate is 1% per month and is compounded monthly. The value of i is then 0.01, and the number of periods is 4. The amount of the bill is

$$A = 135(1.01)^4$$
$$= 135(1.04060)$$
$$= 140.48$$

Notice that i and n were given in this example, so they can be substituted directly into the formula.

■ **Now You Are Ready to Work Exercise 19**

Money invested at compound interest grows faster for shorter compound periods. Table 5–2 shows how $1000 grows over 10 years at 7.2% interest with different compounding periods. Notice that the future value increases with more frequent compounding.

Although the future value increases with more frequent compounding, it increases more slowly. In fact, even with more frequent compounding such as each hour, minute, or second, the future value will never exceed $2054.44.

Present Value

We use the term **present value** to designate the principal that must be invested now to accumulate an amount at a specified time in the future.

TABLE 5–2 $1000 invested at 7.2% for 10 years

Compound Frequency	Number of Periods	Future Value
Simple interest (no compounding)		$1000[1 + 0.072(10)] = 1720$
Annually	10	$1000(1.072)^{10} = 2004.23$
Quarterly	40	$1000\left(1 + \dfrac{0.072}{4}\right)^{40} = 2041.32$
Monthly	120	$1000\left(1 + \dfrac{0.072}{12}\right)^{120} = 2050.02$
Weekly	520	$1000\left(1 + \dfrac{0.072}{52}\right)^{520} = 2053.41$
Daily	3650	$1000\left(1 + \dfrac{0.072}{365}\right)^{3650} = 2054.29$

EXAMPLE 7 ➤   How much should Josh invest at 8%, compounded quarterly, so that he will have $5000 at the end of 7 years?

SOLUTION

In this case, we are given $A = 5000$, $r = 0.08$, $m = 4$, and $t = 7$. Then, $i = \dfrac{0.08}{4} = 0.02$ and $n = 4(7) = 28$. Using the compound interest formula, we have

$$5000 = P(1.02)^{28}$$
$$= 1.74102P$$

Solve the equation for P:

$$P = \frac{5000}{1.74102} = 2871.87$$

Josh should invest $2871.87 to have $5000 in 7 years. In other words, the present value of $5000 due in 7 years is $2871.87.

■ **Now You Are Ready to Work Exercise 23**

Doubling an Investment

An investor wishes to know how fast an investment will grow. One measure is the time it takes for an investment to double in value.

EXAMPLE 8 ➤   An investor has visions of doubling her money in 6 years. What interest rate is required for her to do so if the investment draws interest compounded quarterly?

SOLUTION

P dollars are invested in order to have $2P$ dollars in 6 years. The future value formula for interest compounded quarterly becomes

$$2P = P(1 + i)^{24}$$

where we wish to find i. We do not need to know the value of P because we can divide both sides by P and have $2 = (1 + i)^{24}$.

You can solve this by taking the 24th root of both sides of the equation

$$2 = (1 + i)^{24}$$

to get

$$2^{1/24} = 1 + i$$

to four decimal places, $2^{1/24} = 1.0293$, or $i = 2.93\%$ per quarter or 11.72% annual rate.

■ **Now You Are Ready to Work Exercise 33**

Effective Rate

For a given annual rate, a more frequent compounding of interest gives a larger value of an investment at the end of the year. A 10% annual rate, compounded monthly, gives a larger amount at the end of the year than does 10%, compounded semiannually. However, a lower rate, compounded more frequently, may or may not give a larger return. For example, which yields the better return, 9% compounded semiannually or 8.8% compounded quarterly? To answer this, let's assume that $1000 is invested in each case and compute the amount at the end of 1 year.

1. $r = 9\%$ is compounded semiannually. Let $P = 1000, t = 1$ year, $m = 2$, and
$$i = \frac{0.09}{2} = 0.045. \text{ Then,}$$
$$A = 1000(1.045)^2 = 1000(1.09203)$$
$$= 1092.03$$

2. $r = 8.8\%$ is compounded quarterly. Let $P = 1000, t = 1$ year, $m = 4$, and
$$i = \frac{0.088}{4} = 0.022. \text{ Then,}$$
$$A = 1000(1.022)^4$$
$$= 1000(1.090947) = 1090.95$$

Thus, 9% compounded semiannually is a slightly better investment.

Another way to put different rates and frequency of compounding on a comparable basis is to find the **effective rate.** The effective rate is the percentage increase of an investment in 1 year — that is, the simple interest rate that gives the same annual increase as the compound rate. If an investment increases by 5.9%, for example, the effective rate is 5.9%.

DEFINITION
Effective Rate

The **effective rate** of an annual interest rate r compounded m times per year is the simple interest rate that produces the same total value of investment per year as the compound interest.

Here's how we find the effective rate.

EXAMPLE 9 ➤

Find the effective rate of 8% compounded quarterly.

SOLUTION

If we invest P dollars, the amount of the investment at the end of the year is

$$A = P(1.02)^4$$

If we invest the same amount, P, at a simple interest rate, r, the amount of the investment at the end of the year is $A = P + Pr(1)$.

Now r is the unknown simple interest rate that gives the same amount A as does compound interest, so

$$P + Pr = P(1.02)^4$$

We can divide throughout by P to get

$$1 + r = (1.02)^4$$

and so

$$\begin{aligned} r &= (1.02)^4 - 1 \\ &= 1.08243 - 1 \quad \text{(Rounded)} \\ &= 0.08243 \end{aligned}$$

In percentage form, $r = 8.243\%$ is the effective rate of 8% compounded quarterly.

■ **Now You Are Ready to Work Exercise 26**

This method works generally, so we can make the following statement:

Effective Rate

If money is invested at an annual rate r and compounded m times per year, the effective rate, x, in decimal form is

$$x = (1 + i)^m - 1 \qquad \text{where} \qquad i = \frac{r}{m}$$

Now let's look at an example with a given effective rate and find the annual interest rate, compounded periodically.

EXAMPLE 10 ➤

The Mattson Brothers Investment Firm advertises certificates of deposit paying a 7.2% effective rate. Find the annual interest rate, compounded quarterly, that gives the effective rate.

SOLUTION

We let $i =$ quarterly rate. Then,

$$\begin{aligned} 0.072 &= (1 + i)^4 - 1 \\ 1.072 &= (1 + i)^4 \\ \sqrt[4]{1.072} &= 1 + i \\ 1.017533 &= 1 + i \\ i &= 0.017533 \end{aligned}$$

The annual rate $= 4(0.017533) = 0.070133 = 7.013\%$ (rounded). The annual rate just found is also called the **nominal rate.**

■ **Now You Are Ready to Work Exercise 29**

5.2 EXERCISES

LEVEL 1

1. The annual interest rate is 6.6%. Find
   (a) The semiannual interest rate.
   (b) The quarterly interest rate.
   (c) The monthly interest rate.

2. The annual interest rate is 5.4%. Find
   (a) The semiannual interest rate.
   (b) The quarterly interest rate.
   (c) The monthly interest rate.

In Exercises 3 through 7, find (a) the final amounts and (b) the total interest earned on the original investment.

3. $4500 is invested at 7.5% compounded annually for 6 years.

4. $14,500 is invested at 8% compounded quarterly for 3 years.

5. $31,000 is invested at 7.5% compounded annually for 8 years.

6. $3200 is invested at 7.2% compounded quarterly for 5 years.

7. $5000 is invested at 7% compounded annually for 4 years.

8. *(See Example 1)* A family deposits $1500 into an account that pays 6% compounded quarterly. Find the amount in the family's account at the end of each year for 4 years.

9. $800 is invested at 6% compounded quarterly. Find the amount at the end of the first four quarters.

10. *(See Example 2)* $2600 is invested at 5% compounded annually. What is the amount of the account at the end of 3 years?

11. $1800 is invested at 9% compounded semiannually. How much is in the account at the end of 1 year?

12. *(See Example 4)* $4500 is invested at 8% annual rate. Find the amount at the end of 2 years if the interest is compounded
    (a) annually   (b) semiannually   (c) quarterly

13. $12,000 is invested at 10% interest. Find the amount at the end of 3 years if the interest is compounded
    (a) annually   (b) semiannually   (c) quarterly

Find the amount at the end of the specified time in Exercises 14 through 17.

14. *(See Example 5)* $P = \$7000, r = 3.2\%, t = 4$ years, $m = 4$ (compounded quarterly)

15. $P = \$10,000, r = 6\%, t = 5$ years, $m = 4$

16. $P = \$550, r = 5\%, t = 4$ years, $m = 2$ (compounded semiannually)

17. $P = \$460, r = 5.4\%, t = 6$ months, $m = 12$ (compounded monthly)

LEVEL 2

18. $3500 is invested at 6.2% compounded quarterly. Find the amount at the end of 6 years.

19. *(See Example 6)* A loan shark charges 2% per month on the unpaid balance of a loan. A student's loan was for $640. He was unable to pay for 6 months. What was his loan balance at the end of 6 months?

20. A store has an interest rate of 1.5% per month on the unpaid balance of charge accounts. (Interest is compounded monthly.) A customer charges $60 but allows it to become 4 months overdue. What is the bill at that time?

21. Your bookstore charges 1% per month on your unpaid balance. You charge $232.75 for books. You do not pay the bill until 4 months have passed after the first billing. What is the bill that you owe after 4 months of accumulated interest?

22. The Brunners fail to pay a bill for items they charged. If the original bill was $140 and the store charges 1.5% interest per month on the unpaid balance, what does the bill total when it is 8 months overdue? (Interest is compounded monthly.)

23. *(See Example 7)* A woman buys an investment that pays 6% compounded semiannually. She wants $25,000 when she retires in 15 years. How much should she invest? (Find the present value.)

24. Hank and Gretel want to put money in a savings account now so that they will have $1800 in 5 years. The savings bank pays 6% interest compounded quarterly. How much should Hank and Gretel invest?

25. Alex expects to graduate in 3.5 years and hopes to buy a new car then. He will need a 20% down payment that will be about $3600 for the car he wants.

How much should he save now to have $3600 when he graduates if he can invest it at 6% compounded monthly?

26. *(See Example 9)*   A savings company pays 8% interest compounded semiannually. What is the effective rate of interest?

27. What is the effective interest rate of 5.4% interest compounded quarterly?

28. What is the effective interest rate of 5.1% interest compounded monthly?

Find the annual nominal rate for the effective rates given in Exercises 29 through 32.

29. *(See Example 10)*   Effective rate = 5.302%. Annual rate is compounded quarterly.

30. Effective rate = 6.765%. Annual rate is compounded quarterly.

31. Effective rate = 7.422%. Annual rate is compounded monthly.

32. Effective rate = 7.123%. Annual rate is compounded semiannually.

33. *(See Example 8)*   An investor wants to double her money in 7 years. What interest rate, compounded quarterly, will enable her to do so?

34. $2400 is invested at 4.5% compounded quarterly. Find the amount at the end of 2.5 years.

35. A store charges 1.6% per month on the unpaid balance of a charge account. A bill of $260 is unpaid for 5 months. Find the amount of the bill at the end of 5 months.

36. How much should be invested at 7.6% compounded quarterly in order to have $12,000 at the end of 5 years?

37. Ken invested $1000 at 5.2% compounded quarterly, and Barb invested $1000 at 5.13% compounded monthly. Which one had the largest amount at the end of 5 years?

38. Jody invests $1500 at 6.5% interest and plans to leave it for 10 years. Find the value after 10 years if
(a) The interest paid is simple interest.
(b) The interest paid is compounded quarterly.

39. Carmen invests $5400 in a retirement fund that pays 4.2% interest. Compare the value of the fund after 25 years if
(a) The interest is simple interest.
(b) The interest is compounded quarterly.

LEVEL 3

40. What interest rate enables an investor to obtain a 60% increase in an investment in 3 years if the interest is compounded quarterly?

In Exercises 41 through 43, determine which is the better investment.

41. 6.8% compounded semiannually or 6.6% compounded quarterly?

42. 5.2% compounded semiannually or 4.8% compounded monthly?

43. 6.4% compounded annually or 6.2% compounded quarterly?

44. A credit union pays interest at 4% compounded quarterly on accounts totaling $275,000. It is considering compounding monthly. How much would this increase the interest the credit union will pay in the period of one year?

45. Which is the better investment, 5.1% compounded annually or 5.0% compounded quarterly?

46. A loan company charges 24% a year compounded monthly for small loans. What will be the amount of $100 after 6 months?

47. On her 58th birthday, a woman invests $15,000 in an account that pays 5.6% compounded quarterly. How much will be in her account when she retires on her 65th birthday?

48. An investment company advertises that an investment with them will yield a 12% annual rate compounded monthly. Another firm advertises that its investments will yield 12.5% compounded semiannually. Which of these gives the better yield?

49. Daniel received a $1000 gift that he deposited in a savings bank that compounded interest quarterly. After 5 years of accumulating interest, the account had grown to $1485.95. What was the annual interest rate of the bank?

50. The cost of an average family house increased from $32,500 in 1975 to $52,950 in 1985. Determine approximately the average annual inflation in house prices over this 10-year period.

51. Jerri placed $500 in a credit union that compounded interest semiannually. Her account had grown to $633.39 after 4 years. What was the annual interest rate of the credit union?

52. The Martinelli family sells some property obtained from an inheritance. How much should they invest at 5% compounded annually in order to have $28,000 in 6 years?

53. Which should you choose: a savings account that starts with $6000 and earns interest at 10% compounded quarterly for 10 years or a lump sum of $16,000 at the end of 10 years?

54. On graduating from college, a student gets a job that pays $25,000 a year. If inflation is averaging 6%, what will he have to be making 5 years after graduation to keep pace with inflation? What will he have to make 5 years after graduation to increase his standard of living by 4%?

55. In 14 years, an investment of $8000 increases to $18,000 in an account that pays interest compounded semiannually. Find the annual interest rate.

56. A company acquires an asset and signs an agreement to pay $10,000 for it 3 years later. The interest rate is to be 6% compounded semiannually. What is the current market value of the asset to the company?

57. A company will need $240,000 cash to modernize machinery in 5 years time. A financial institution will invest the company's money in a fund at 8% interest compounded quarterly. Determine the cash that must be deposited at present to meet this need.

58. Jim McAtee bought a 1928 Model A Ford for $8250 and sold it 2 years later for $9300. Find the annual compound rate of return earned on his investment.

59. A zero coupon bond pays no interest during its lifetime. Upon maturity, the investor receives face value. The purchase price determines the rate of return.
   (a) Henri paid $10,250 for a $20,000 face value zero coupon bond that matures in 10 years. Find the annual compound rate of interest received.
   (b) Sandi plans to purchase a zero coupon bond with a face value of $10,000 and matures in 6 years. She wants to earn 6.2% compounded annually. Find the price she should pay.

60. The rental cost of office space increased 3% annually for the last three years. If the current cost is $25 per square foot,
   (a) Find the cost three years ago.
   (b) At the same rate of increase, find the cost three years in the future.

61. Beth bought an old trunk at an estate sale. Among its contents she found a document, dated 1845, stating that George Woolfolk loaned Dora Grau $500 at 7% compounded annually. Apparently, the loan was never repaid. Find the amount of the loan, with accumulated interest, that would be required to repay the loan in 2002.

62. When Anna's grandfather retired, he received Social Security payments of $1258 per month. Each year Social Security makes a cost of living adjustment due to inflation. If the adjustments average 1.4% each year, find the grandfather's monthly Social Security payments (to the nearest dollar) in
   (a) 5 years
   (b) 12 years.

63. An annual interest rate, r, is compounded monthly, and its effective rate is 0.0585. Find the annual rate r. (Round to a tenth of a percent).

EXPLORATIONS

64. Report on the interest rates paid on investments by a savings and loan company and the compounding period for
   (a) Passbook savings.
   (b) 6-month money market.
   (c) 12-month money market.
   (d) 5-year money market.

65. One rule of thumb states that the number of years it takes for an investment to double in value using compound interest is 72 divided by the annual percentage interest rate; for example, it takes $\frac{72}{8} = 9$ years to double an investment at 8% compounded annually. For 6% compound interest, it takes about $\frac{72}{6} = 12$ years. Fill in the table.

Interest Rate Compounded Annually	Years, n, to Double Value Using Rule of Thumb	Actual Value of $1 After n Years
3%		
5%		
6%		
7%		
8%		
9%		
10%		

Comment on the accuracy of the rule.

66. Many people "invest" in the lottery each week in hopes of becoming instant millionaires. Realistically, it is most unlikely that a ticket will be a winner. However, it is considerably more likely that a person can become a millionaire if he or she is willing to start young enough and is willing to reach that status slowly. Here are some exercises that will help clarify what it takes to become a millionaire.

First, let's look at fixed-income investments that pay a fixed rate each year. We use a 7% annual rate compounded quarterly with an investment of $5000.

(a) **(i)** At age 25, a person has saved $5000 to invest. Find the value of the investment at age 65. At age 70.

(ii) Every 5 years, at ages 25, 30, 35, . . . 65, this person invests $5000. How many times must these investments be made in order to have $1 million at age 70?

(iii) How much should be invested at age 25 to have $1 million at age 70?

(b) The results of part (a) suggest it is not easy to accumulate $1 million with fixed-income investments. Over a long period of time, an investment in stocks will generally yield greater returns than a fixed-income investment even though the value of stocks will fluctuate. Let's look at investments of $5000 into a stock mutual fund that averages 10% growth per year.

(i) At age 25, a person invests $5000 into a mutual fund that averages 10% growth compounded annually. Find the estimated value at age 65. At age 70.

(ii) Every 5 years, at ages 25, 30, 35, . . . 65, this person invests $5000. How many times must these investments be made to have $1 million at age 70?

(iii) Beginning at age 25, this person invests $5000 each year into a mutual fund that averages 10% growth compounded annually. How many such investments must be made to accumulate $1 million by age 70?

(iv) How much should be invested at age 25 at 10% compounded annually to have $1 million at age 70?

(c) In part (b), the growth was compounded annually. Let's compare the results of part (b) with the case in which the 10% annual growth is compounded quarterly.

(i) At age 25, a person invests $5000 into a mutual fund that averages 10% annual growth compounded quarterly. Find the estimated value at age 65. At age 70.

(ii) Every 5 years, at ages 25, 30, 35, . . . 60, this person invests $5000. How many times must these investments be made to accumulate $1 million by age 70?

(iii) Beginning at age 25, this person invests $5000 each year into a mutual fund that averages 10% annual growth compounded quarterly. How many such investments must be made to accumulate $1 million by age 70?

(iv) How much should be invested at age 25 to have $1 million by age 70?

(d) Based on this exercise, write a paragraph on the effect of increased interest rates, early investments, and more frequent compounding on the growth of an investment.

67. Graph $y = (1 + i)^x$ to visualize the growth of $1 invested at compound interest. Use [0, 120] scale for x and [0, 10] scale for y.
(a) $i = 0.020$ **(b)** $i = 0.025$ **(c)** $i = 0.015$

68. Graph $y = (1.02)^x$ to determine the value of x that corresponds to $y = 2, 3, 4$, and 8.
(a) What is the meaning of the values of x found?
(b) What relationship exists between the values of x corresponding to $y = 2, 4$, and 8?

69. On one screen, graph the amount of $100 with interest compounded annually using 6%, 7%, and 8% for 10 years. Compare the amounts of each rate.

70. Graph the difference of the function $y = (1.085)^x$ and $y = (1.02)^{(4x)}$ using [0, 25] for the range of x and [−0.5, 0.5] for the range of y.

Based on the graph, which is the better investment, 8.5% compounded annually or 8% compounded quarterly? Do the amounts accumulated grow closer together or further apart over a period of years?

71. In 1790, Benjamin Franklin left $4600 to the city of Boston with the stipulation that it be invested at compound interest and allowed to accumulate for at least 100 years. Graph the growth of this investment and evaluate the function at $x = 50, 100, 150$, and 200 years, assuming an interest rate of 5% compounded annually. Compare the amounts after 200 years if the interest rate was 4.5% instead of 5%. Do the same for an interest rate of 5.5%.

72. Graph $y = (1 + x)^{40}$ using $[0, 0.1]$ for the x-scale and $[0, 40]$ for the y-scale. Interpret the meaning of the graph.

73. Graph the effective rate of 8% using

$$y = \left(1 + \frac{0.08}{x}\right)^x - 1$$

and use $[0, 30]$ for the x-scale and $[0.07, 0.09]$ for the y-scale. Interpret the meaning of the graph.

74. Graph $y = (1 + x)^n$ and trace the graph to find the interest rate compounded quarterly that will double an investor's money in
- **(a)** 10 years ($n = 40$).
- **(b)** 8 years ($n = 32$).
- **(c)** 15 years ($n = 60$).

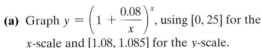

75. $y = \left(1 + \dfrac{i}{x}\right)^x$ gives the growth factor of compound interest for one year where

i = annual interest rate
x = number of times per year the interest is compounded

(a) Graph $y = \left(1 + \dfrac{0.08}{x}\right)^x$, using $[0, 25]$ for the x-scale and $[1.08, 1.085]$ for the y-scale.
(b) On the same screen, graph $y = e^{0.08}$.
(c) Describe how the two graphs are related.
(d) Change the x-scale to $[0, 100]$ and repeat the graphs. How does the graph of $y = \left(1 + \dfrac{0.08}{x}\right)^x$ behave when x is in the range of 25 to 100?
(e) As larger values of x are used, the interest is compounded more frequently, x times per year. As x gets quite large, the interest is compounded *continuously* and $A = e^{rt}$ gives the accumulated amount after t years where r is the annual rate.

Graph, on the same screen, $y = e^{0.06x}$ (for continuous compounding) and

$$y = \left(1 + \frac{0.06}{12}\right)^{(12x)} \text{ (for monthly compounding)}$$

and compare the graphs. Are they just about the same?

Repeat for other interest rates and compounding periods such as

5% compounded quarterly and continuously.

8% compounded weekly and continuously.

7% compounded annually and continuously.

76. If you invest $1000 at 8% compounded quarterly, how long will it take to double your investment? This question is equivalent to: Find n where

$$2000 = 1000(1.02)^n$$

This can be simplified by dividing through by 1000 to obtain

$$2 = (1.02)^n$$

Two methods can be used to solve this kind of problem.
(a) Find n by graphing $y = (1.02)^x$ and use TRACE to locate where $y = 2$ on the graph. The corresponding value of x gives the number of periods required to double the investment.

The doubling time can also be found by finding the intersection of $y = (1.02)^x$ and $y = 2$. Note that the time required to double an investment is independent of the amount invested because the equation

$$2P = P(1 + i)^n$$

reduces to

$$2 = (1 + i)^n$$

(i) How long does it take for an investment to double if the interest is 10% compounded quarterly?
(ii) How long does it take for an investment to double if the interest is 12% compounded monthly?
(iii) How long does it take for an investment to double if the interest is 8% compounded semiannually?
(iv) How long does it take for an investment to double if the interest is 6% compounded annually?
(v) How long does it take for an investment to triple if the interest is 10% compounded quarterly?
(vi) How long does it take for an investment to increase 50% if the interest is 1.5% per month compounded monthly?

(b) Use the log key on your calculator. The number of compounding periods, n, it takes to double an investment at the interest rate of i per period is

$$n = \frac{\log 2}{\log(1 + i)}$$

Example: 8% compounded quarterly doubles in

$$n = \frac{\log 2}{\log(1.02)} \text{ quarters}$$

Use this method to find the doubling time for the following:
(i) 5% compounded quarterly
(ii) 8% compounded monthly
(iii) 10% compounded quarterly

The log key method can also be used to determine the time it takes to triple, increase by 50%, and so on. Replace the numerator, log 2, with log 3, log 1.5, and so on.

77. Population growth and inflation behave in the same manner as compound interest, with population and inflation compounding annually.

  (a) If the inflation rate is 5% per year, how much will a $100 item cost in 10 years?

  (b) Panama has a population of 2,500,000 and an annual population growth rate of 2.0%. If this growth rate continues, what will be the population in 50 years?

  (c) The annual growth rates of the following countries are

Spain 0.1%	$(i = 0.001)$
United States 0.8%	$(i = 0.008)$
Kenya 3.7%	$(i = 0.037)$

  On one screen, graph $y = (1 + i)^x$ for each value of i above to obtain the growth of each country for the next 200 years. Use $[0, 200]$ for the x-scale and $[0, 20]$ for the y-scale.

  Discuss the prospects of long-term population problems, if any, of these countries.

78. Compare the growth of $1 from simple interest and interest compounded quarterly. Use a 6% annual rate. Graph on the same screen

$$y = 1 + .06x$$
$$y = (1.015)^{(4x)}$$

  (a) Use $[0, 10]$ for the range of x.

  (b) Use $[0, 25]$ for the range of x.

  Comment on the relative values of the growth of the investments.

79. Lauren and Anna each invest $1000 at 5.2% interest. Lauren's investment earns interest compounded annually and Anna's investment earns simple interest. Make a table showing the amount of each one at the end of each year for 10 years.

80. $1000 is invested at 4.8% compounded quarterly for 5 years. Make a chart showing the amount of the investment and the interest earned each quarter.

USING YOUR TI-83

MAKING A TABLE

In Section 1.1, we illustrated how to make a table of values of a function for several values of x. We can make tables of a period-by-period amounts of an investment in a similar way. A built-in table has a column for a list of values of the variable x, and columns for $y_1, y_2, \ldots$ that contain the values of y calculated from the corresponding formulas in the $\boxed{Y=}$ menu.

We illustrate by making a year-by-year table for two investments, both investing $1000 at 5% interest for 10 years. For the first investment, the interest is compounded annually, and the second draws simple interest. Briefly, here's how.

- Press $\boxed{\text{TblSet}}$ to display the **TABLE SETUP** screen.
- Enter 0 for **TblStart** and press $\boxed{\text{ENTER}}$.
- Enter 1 for **ΔTbl** and press $\boxed{\text{ENTER}}$.
- Select the **Auto** option for **Indpnt** and press $\boxed{\text{ENTER}}$.
  (This sequence will enter 0, 1, 2, ... in x. To enter 0, 2, 4, ... use 2 for **ΔTbl**)
- Select **Auto** for **Depend** and press $\boxed{\text{ENTER}}$.

 NOTE

This will give a list of x values from zero to the end of the table, so you pick out the ones you want.

To calculate the values of compound interest in y_1 and simple interest in y_2:

- Select $\boxed{Y=}$ and enter 1000(1.05)^x as the y_1 function,
- To calculate the simple interest amounts, enter 1000(1+.05x) for the y_2 function.

- Press ⬚Table⬚ to view the tables.
- You may also graph the two functions by setting the Window to 0–20 for the x-range and 0–2000 for the y-range and press ⬚GRAPH⬚.

USING INTERSECT

We can use the **intersect** command (See Section 1.3) to determine, for example, when an investment of $700 doubles in value if it is invested at 4.4% compounded quarterly. Do so by plotting the two functions $y_1 = 700(1.011)^x$ and $y_2 = 1400$, and then selecting the **intersect** command.

EXERCISES

1. Make a table of the year-by-year amounts of a $200 investment at 5.1% compounded annually for 10 years.
2. Make a table, for eight years, of the year-by-year amounts of two $100 investments, one at 4.5% compounded annually and the second at 5.5% compounded annually.
3. Find how long it takes for $5,000 to increase to $12,000 when invested at 5.5% compounded quarterly.

USING EXCEL

We will show how to graph functions using EXCEL. Let's illustrate by graphing the amount of compound interest and the amount of simple interest. We use an initial investment of $1000, interest of 7%, for a period of 20 years. The compound interest investment is compounded annually. Thus, we graph the equations

Compound interest: $A = 1000(1.07)^x$ where x ranges from 0 to 20.

Simple interest: $A = 1000(1 + 0.07x)$ where x ranges from 0 to 20.

Excel makes a graph by plotting points on the graph and connecting them with a smooth curve. This means we compute a list of points to be used by Excel. Here's how:

- In column A, list values of x to be used. We use $0, 2, 4, \ldots, 20$, listed in A2:A12.
- *Note:* We could use $0, 1, 2 \ldots, 20$ or some other spacing for x.
- In column B, calculate the amount of compound interest corresponding to the number of periods, x. We use =1000*(1.07)^A2 in B2 and drag the formula down to B12.
- In column C, we calculate the amounts of simple interest using =1000*(1+.07*A2) in C2 and dragging the formula down to C12.

	A	B	C
	x	A=1000(1.07)^x	A=1000(1+0.07x)
1			
2	0	1000.00	1000
3	2	1144.90	1140
4	4	1310.80	1280
5	6	1500.73	1420
6	8	1718.19	1560
7	10	1967.15	1700
8	12	2252.19	1840
9	14	2578.53	1980
10	16	2952.16	2120
11	18	3379.93	2260
12	20	3869.68	2400

(continued)

We now use these points to graph both functions on the same graph.

- Select the **Wizard** icon from the Menu Bar.
- Select **XY(Scatter)** for **Chart type:**
- Click on the first option of the second column (Two curves) under **Chart sub-type:**
- Click **Next**
- Select the cells A2:C12 for **Data range.**
- Click **Finish**

In case you want to adjust the x-scale or y-scale, double-click on the scale to be adjusted, make the changes in the dialog box that appears, and click on **OK**. The Series 1 and Series 2 legends that appear with the graph refer to the function defined in column B (for Series 1), and to the function defined in Column C (for Series 2).

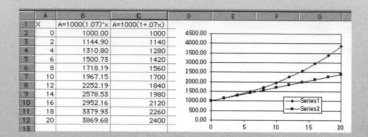

We can use **Goal Seek** to determine when an investment reaches a certain value, doubles in value, increases 60%, and so forth. Let's illustrate with the following. $1500 is invested at 4.8% compounded quarterly. How long will it take for it to reach $2000 in value? We let $x =$ the required number of quarters, $i = 0.012$, and $A = 1500(1.012)^x$. We enter zero for x in A2, and =1500*(1.012)^A2, in B2.

- Select **Goal Seek** under **Tools** in the menu bar.
- In the **Goal Seek** dialog box, enter B2 for **Set Cell**, 2000 for **To Value**, and A2 for **By changing cell.**
- Click **OK,** and A2 shows 24.11. Thus, the investment reaches $2000 in the 24th quarter.

EXERCISES

1. Make a table of year-by-year amounts of $500 invested at 6% compounded annually for six years.

2. Make a table of quarter-by-quarter amounts of $800 invested at 5.6% compounded quarterly for four years.

3. Make a table of year-by-year amounts of two investments of $2000 each. The first is invested at 6% compounded annually and the second is invested at 5.5% compounded annually. Both are invested for five years.

4. Graph the two functions in Exercise 3 on one graph.

5. Find how long it will take for $2500 invested at 5.2% compounded quarterly to reach $4000.

6. Find the interest rate, compounded quarterly, that will enable an $800 investment to double in eight years.

5.3 | Annuities and Sinking Funds

- Ordinary Annuity
- Sinking Funds

Ordinary Annuity

Most of us have encountered or will encounter debt several times. You may now have a student loan or owe some credit card charges. You may be looking forward to life after graduation when you can begin a good job or upgrade your present job. Early in your career, you may experience a struggle to pay off a student loan, to make payments on a car, and to keep up the monthly payments on your dream house. Then there might be a time when you decide to invest money periodically into a fund to provide a college education for your children. Later in your career, you will focus more on your retirement and the income you might expect during that time.

These financial activities have a common characteristic, that of periodic payments. Car and house loans have periodic payments, usually monthly, of a specified amount that is to be paid until the loan is paid off. A college fund receives periodic payments to accumulate an amount for use at a future date. A retiree receives monthly payments from a retirement fund.

A series of equal periodic payments like these is an example of what we call an annuity. Whether the payments are used to pay off a loan, build a college fund, or provide retirement income, equal payments paid at equal time intervals form an **annuity.**

We use some standard terminology when discussing annuities. We call the time between successive payments the **payment period** and the amount of each payment the **periodic payment.** The interest on an annuity is compound interest. The payments may be made annually, semiannually, quarterly, or at any specified time interval. Monthly payments are common. The annuity may take on different forms, but we will study just one form, the **ordinary annuity.** Furthermore, we assume the payment period and the interest compounding period are the same.

Ordinary Annuity

> An **ordinary annuity** is an annuity with periodic payments made at the *end* of each payment period.

We first look at annuities in which payments are invested to build up a fund for future use.

EXAMPLE 1 ➤   A family enters a savings plan whereby they will invest $1000 at the end of each year for 5 years. The annuity will pay 7% interest, compounded annually. Find the value of the annuity at the end of the 5 years.

SOLUTION

Because $1000 is deposited each year, the first deposit will draw interest longer than subsequent deposits. The value of each deposit at the end of 5 years is the original $1000 plus the compound interest for the time it draws interest. Use the formula for the amount at compound interest. Here is a summary of how payments of $1000 deposited at the end of each year grow over a 5-year period. The interest rate, $r = 0.07$, is compounded annually.

Year Deposited	Length of Time Deposit Draws Interest	Value of Deposit at End of 5 Years
1	4 years	$1000(1.07)^4 = 1310.80$
2	3 years	$1000(1.07)^3 = 1225.04$
3	2 years	$1000(1.07)^2 = 1144.90$
4	1 year	$1000(1.07) = 1070.00$
5	0 year	$1000(1) = 1000.00$
		Final value: 5750.74

NOTE

Payments could be made at times other than the end of a payment period. This would result in a different analysis and a different formula. If a person wishes to make five periodic payments beginning now, the formulas for ordinary annuities hold by declaring that the end of the first year has arrived when the first payment is made, and the last payment is then at the end of the fifth year.

You obtain the final value by adding the five payments and the interest accumulated on each one. The final value is called the **amount,** or **future value,** of the annuity.

■ **Now You Are Ready to Work Exercise 1**

Let's use this simple example to observe some general relationships. In the example, five payments are made, and the accumulated value is obtained by adding five amounts, the first of which is $1000(1.07)^4$ and the last is 1000, the annual payment. If $1000 were deposited for 10 years, a similar pattern holds with ten amounts, the first being $1000(1.07)^9$ and the last 1000. In general, the first amount has an exponent on 1.07 that is one less than the number of payments. The total accumulated value of five payments is

$$1000 + 1000(1.07) + 1000(1.07)^2 + 1000(1.07)^3 + 1000(1.07)^4$$

In a like manner, the total of ten payments is

$$1000 + 1000(1.07) + 1000(1.07)^2 + \cdots + 1000(1.07)^9$$

where the terms are written in reverse order. In general, the sum of the values of n payments is

$$A = 1000 + 1000(1.07) + 1000(0.07)^2 + \cdots + 1000(1.07)^{n-1}$$

NOTE

For other interest rates and amounts of periodic payments, use that interest rate instead of 0.07 and the amount invested instead of 1000. You will obtain a similar expression for the total, A.

Clearly, computing the total value (future value) of an annuity can be tedious for a number of payments. Let's examine a way to avoid a period-by-period computation by using the 5-year future value in Example 1. The sum of the five terms is

$$A = 1000 + 1000(1.07) + 1000(1.07)^2 + 1000(1.07)^3 + 1000(1.07)^4$$

Now multiply both sides of this equation by 1.07 to obtain

$$A(1.07) = 1000(1.07) + 1000(1.07)^2 + 1000(1.07)^3 + 1000(1.07)^4 + 1000(1.07)^5$$

We subtract the first equation from this to obtain

$$\begin{aligned} A(1.07) - A = {} & 1000(1.07) + 1000(1.07)^2 + 1000(1.07)^3 + 1000(1.07)^4 \\ & + 1000(1.07)^5 - 1000 - 1000(1.07) - 1000(1.07)^2 \\ & - 1000(1.07)^3 - 1000(1.07)^4 \end{aligned}$$

and obtain the form

$$A(1.07) - A = -1000 + 1000(1.07)^5$$

From this we get

$$A(1.07 - 1) = 1000(1.07^5 - 1)$$
$$A = \frac{1000(1.07^5 - 1)}{1.07 - 1} = \frac{1000(1.07^5 - 1)}{0.07}$$

Check this computation to verify that it gives the same total as in Example 1, $5750.74. Notice in the final result that 1000 represents the periodic payment, 1.07 is 1 plus the periodic interest rate, the exponent 5 is the number of time periods, and the denominator 0.07 is the periodic interest rate.

We would like a formula that computes the future value for any amount for the periodic payment — say, R — for any periodic interest rate — say, i — and any number of periods — say, n. If we follow the steps above using R, i, and n instead of 1000, 0.7, and 5, we will end up with a similar formula giving the value of the amount accumulated, or future value. The general form of the formula is the following.

Future Value (Amount) of an Ordinary Annuity

(Payments are made at the end of each period)

$$A = R\left[\frac{(1 + i)^n - 1}{i}\right]$$

where

$$i = \text{interest rate per period}$$
$$n = \text{number of periods}$$
$$R = \text{amount of each periodic payment}$$
$$A = \text{future value or amount}$$

The next example gives you an idea how an annuity grows over time.

EXAMPLE 2 ➤ Tiffany invests $1000 at the end of each year into an ordinary annuity that pays 6% compounded annually. Figure 5–2 and Table 5–3 show how the annuity grows year by year for 15 years. ∎

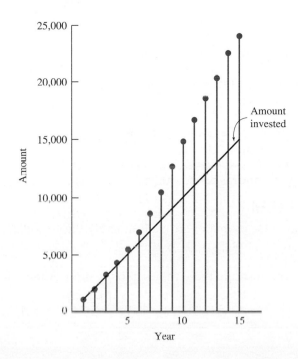

FIGURE 5–2   Growth of an annuity with $1000 invested annually at 6%.

TABLE 5-3			
Year	Payment	Interest	Total
1	1000	0	1,000.00
2	1000	60.00	2,060.00
3	1000	123.60	3,183.60
4	1000	191.02	4,374.62
5	1000	262.47	5,637.09
6	1000	338.23	6,975.32
7	1000	418.52	8,393.84
8	1000	503.63	9,897.47
9	1000	593.85	11,491.32
10	1000	689.47	13,180.79
11	1000	790.85	14,971.64
12	1000	898.30	16,869.94
13	1000	1012.20	18,882.14
14	1000	1132.93	21,015.07
15	1000	1260.90	23,275.97

Using Table 5–3, notice the growth over 5-year periods. In the first 5 years, a total of $5000 was invested and grew to a value of $5637, a gain of $637 over the amount invested. During the sixth through tenth year, a total of $5000 was invested, and the value of the investment grew to $13,181 — a gain of $7544, of which $2544 was interest earned. The last 5 years a total of $5000 was invested, and the value of the investment increased to $23,276 — a gain of $10,095, of which $5095 was interest earned. This acceleration in the increase in value occurs because more money becomes available to earn interest. The compounding effect of interest contributes greatly to the growth. This illustrates that regular investment over a long period of time is a smart way to invest.

Figure 5–2 shows how the total value of the investment increases with time relative to the amount invested.

We need not compute a period-by-period table like Table 5–3 to find the value at some future date. We use the future value formula as illustrated in the next example.

EXAMPLE 3 ➤

How much money will you have when you retire if you save $20 each month from graduation until retirement? We need some definite numbers to answer this question, so let's assume you start saving at age 22 until age 65, 43 years, and the interest rate averages 6.6% annual rate compounded monthly.

SOLUTION

Because the payments are made monthly, the periodic rate = $0.066/12 = 0.0055$, the number of periods is $n = 12(43) = 516$, and the periodic payments are $R = 20$.

We substitute these values into the future value formula

$$A = R\left[\frac{(1 + i)^n - 1}{i}\right]$$

to obtain

$$A = 20 \left[\frac{1.0055^{516} - 1}{0.0055} \right] = 57,997.30$$

You will accumulate $57,997.30 in 43 years.

■ **Now You Are Ready to Work Exercise 3**

Sinking Funds

There are times when a company or an individual expects to need a specified amount at a time in the future. For example, a company may expect to replace a machine costing $150,000 in 6 years, a family expects to need a new car in 4 years, or new parents wish to accumulate $150,000 for their child's college expenses in 18 years. In cases like this when an amount of money will be needed at some future date, the company or person can systematically accumulate a fund that will build to the desired amount at the time needed. We call such a fund a **sinking fund.**

EXAMPLE 4 ➤

Susie wants to deposit her savings at the end of every three months so that she will have $7500 available in four years. The account will pay 8% interest per annum, compounded quarterly. How much should she deposit every quarter?

SOLUTION

Susie is accumulating a sinking fund with a future value of $7500, periodic rate $i = 0.08/4 = 0.02$, and $n = 16$ periods. Use the formula for the future value of an annuity,

$$A = R \left[\frac{(1 + i)^n - 1}{i} \right]$$

to find the periodic payments, R:

$$7500 = R \left[\frac{(1.02)^{16} - 1}{0.02} \right]$$
$$7500 = R(18.63929)$$
$$R = \frac{7500}{18.63929} = 402.38$$

Susie should deposit $402.38 every quarter to accumulate the desired $7500.

■ **Now You Are Ready to Work Exercise 12**

The formula for periodic payments into a sinking fund is obtained from the future value formula for an annuity by solving for R.

Formula for Periodic Payments of a Sinking Fund

$$R = \frac{Ai}{(1 + i)^n - 1}$$

where

A = value of the annuity after n payments
n = number of payments
i = periodic interest rate
R = amount of each periodic payment

EXAMPLE 5 ▶

Darden Publishing Company plans to replace a piece of equipment at an expected cost of $65,000 in 10 years. The company establishes a sinking fund with annual payments. The fund draws 7% interest, compounded annually. What are the periodic payments?

SOLUTION

$A = 65{,}000$, $i = 0.07$, and $n = 10$, so

$$R = \frac{65000(0.07)}{(1.07)^{10} - 1} = \frac{4550}{0.9671513} = 4704.54$$

The annual payment is $4704.54.

■ **Now You Are Ready to Work Exercise 18**

A sinking fund problem is a variation of an annuity problem. The growth of an annuity can generally be described as "I can save $25 a month. How much will I have after 5 years?" The sinking fund problem is like "I will need $15,000 in five years to buy a new car. How much should I save each month to have the amount I need?" In the growth of an annuity, the amount of the periodic payment is known, and the future value is sought. In the sinking fund problem, the future value is known, and the periodic payments that will accrue to that value are sought.

5.3 EXERCISES

LEVEL 1

Reminder: The compounding periods and the deposit periods are the same for an ordinary annuity.

1. *(See Example 1)* $600 is deposited into an account at the end of each year for 4 years. The money earns 5% compounded annually. Determine the value of each deposit at the end of 4 years and the total in this account at that time.

2. A boy deposits $500 into an account at the end of each year for 3 years. The money earns 6% compounded annually. Determine the value of each deposit at the end of 3 years and the total amount in the account.

In Exercises 3 through 11, determine the amount of the ordinary annuities at the end of the given periods.

3. *(See Example 3)* $16,000 deposited annually at 7% for 15 years

4. $500 deposited annually at 5.5% for 10 years

5. $250 deposited quarterly at 4% for 5 years

6. $300 deposited quarterly at 6.5% for 6 years

7. $200 deposited monthly at 1% per month for 20 months

8. $50 deposited per month at 1.5% per month for 15 months

9. $4000 deposited semiannually at 5.8% for 10 years

10. $800 deposited semiannually at 6.2% for 8 years

11. $750 deposited quarterly at 6.8% for 4 years

In Exercises 12 through 17, the amount (future value) of an ordinary annuity is given. Find the periodic payments.

12. *(See Example 4)* $A = \$8000$, and the annuity earns 7% compounded annually for 15 years.

13. $A = \$2500$, and the annuity earns 6.5% compounded annually for 4 years.

14. $A = \$25{,}000$, and the annuity earns 5.6% compounded quarterly for 4 years.

15. $A = \$14{,}500$, and the annuity earns 6% compounded semiannually for 10 years.

16. $A = \$50,000$, and the annuity earns 8% compounded semiannually for 8 years.

17. $A = \$10,000$, and the annuity earns 6% compounded quarterly for 3 years.

In Exercises 18 through 21, the amount desired in a sinking fund is given. Find the periodic payments required to obtain the desired amount.

18. *(See Example 5)* $A = \$12,000$, the interest rate is 8%, and payments are made semiannually for 5 years.

19. $A = \$75,000$, the interest rate is 7.2%, and payments are made quarterly for 4 years.

20. $A = \$40,000$, the interest rate is 6%, and payments are made annually for 8 years.

21. $A = \$15,000$, the interest rate is 6.3%, and payments are made monthly for 18 months.

LEVEL 2

22. An executive prepares for retirement by depositing $2500 into an annuity each year for 10 years. The annuity earns 7% per year. Find the future value of the annuity at the end of 10 years.

23. A young couple saves for a down payment on a house by depositing $100 each month into an annuity that pays a 4.8% annual rate. Find the amount in the annuity at the end of 2 years.

24. The Cooper Foundation contributes $25,000 per year into an annuity fund for building a new zoo. The fund earns 9% interest. Find the amount in the fund at the end of 15 years.

25. A couple pays $400 at the end of each 6 months for 5 years into an ordinary annuity paying 8% compounded semiannually. What is the future value at the end of 5 years?

26. Delores invests $2000 a year in a tax-exempt bond for 20 years. By reinvesting interest the fund in-

creases 6% per year. What will be the value of her shares when she makes the 20th payment?

27. Sam wants to invest the same amount at the end of every 3 months so that he will have $4000 in 3 years. The account will pay 6% compounded quarterly. How much should he deposit each quarter?

28. How much should a family deposit at the end of every 6 months in order to have $8000 at the end of 5 years? The account pays 6% interest compounded semiannually.

29. A 13-year-old child received an inheritance of $5000 per year. This was to be invested and allowed to accumulate until the child reached 21 years of age. The first payment was made on the child's 13th birthday and the last on the 21st birthday. If the money was invested at 7% compounded annually, what did the child receive at age 21?

30. A condominium association decided to set up a sinking fund to accumulate $50,000 by the end of 4 years to build a new sauna and swimming pool. What quarterly deposits are required if the annual interest rate is 8% and it is compounded quarterly?

31. A company projects that it will need to expand its plant in 6 years. It expects the expansion to cost $150,000. How much should it put into a sinking fund each year at 7% compounded annually?

LEVEL 3

32. The Citizen's Bank sets up a scholarship fund by making deposits of $1000 every 6 months into an annuity earning 8% interest. Find the amount of the annuity at the end of 11 years.

33. A city has issued bonds to finance a new convention center. The bonds have a total face value of $750,000 and are payable in 8 years. A sinking fund has been opened to meet this obligation. If the interest rate on the fund is 6.3% compounded quarterly, what will be the quarterly payments?

34. A couple plan to start a business of their own in 6 years. They plan to have $10,000 cash available at the time for this purpose. To raise the $10,000, a fund has been started that earns interest at 8% compounded quarterly. What would the quarterly payments into this fund have to be to raise the $10,000?

35. Electronic Instruments plans to establish a debt retirement fund. The company wants at least $23,800 in 5 years. Deposits of $4000 are to be made to a trustee each year. Can they meet these requirements if the interest rate falls to 6% compounded annually?

36. A company requires each of two subsidiaries to make deposits into debt retirement funds over the next 3 years. One subsidiary is to contribute $2000 quarterly, and the other is to contribute $6000 semi-annually. Interest at 8% compounded quarterly and semiannually will be paid on the respective funds. Determine the combined amounts of the funds at the end of 3 years.

37. A savings institution advertised "Invest $1000 a year for 10 years and we will pay you $1000 a year forever." The institution pays an 8% interest rate compounded annually.
   (a) Find the value of the investment after 10 years.
   (b) The investor ceases payment at the end of 10 years, and the investment firm pays $1000 at the end of each year. Will the interest on the investment (still 8%) provide for the $1000 annual payment?

38. (a) Stephanie invested $2000 per year in an IRA each year for 10 years. The interest rate was 7% compounded annually. At the end of 10 years, she ceased the IRA payments but left the total of her investments at 7% compounded annually for the next 25 years.
   (i) What was the value of her IRA investments at the end of 10 years?
   (ii) What was the value of her investment at the end of the next 25 years?
   (b) Stephanie's friend, Roy, started his IRA investments the tenth year and invested $2000 each year for the next 25 years at 7% compounded annually. What was the value of his investment at the end of the 25 years?
   (c) Which of the investments had the greater value at the end of the 35 years?
   (d) What does this exercise suggest to you as the best time to save for retirement or for some amount needed some time in the future?

39. $100 is deposited monthly into an ordinary annuity that earns 1.6% per month compounded monthly. Find the amount at the end of 30 months.

40. $500 is deposited quarterly into an annuity that earns 8.6% compounded quarterly. Find the amount of the annuity at the end of 10 years.

41. Mr. Nakamura wants to make monthly deposits into an annuity for his grandchild so that $15,000 will be available in 5 years. If the interest rate is 9% compounded monthly, find the monthly deposit.

42. The Otwell Company puts $14,000 per year into a sinking fund to purchase major equipment. The fund earns 7.8% interest compounded annually. Find the amount in the fund after 8 years.

43. The Hargis Company makes annual deposits into a sinking fund to expand facilities. If the fund pays 8.1% compounded annually, find the annual payments so that $100,000 is accumulated in 6 years.

44. Joey starts a paper route at age 15 and saves $15 each month in an annuity that pays 6.9%. He continues this habit until he retires at age 68, 53 years later.
   (a) What will be the value of the annuity at his retirement?
   (b) Find the total amount invested during this time.

45. Evelyn starts a retirement fund 10 years before retirement. She pays $100 per month into the annuity for 10 years. Her total investment is $12,000. Esther starts a retirement fund 20 years before retirement. She pays $50 per month into the annuity for 20 years. Her total investment is $12,000. Lois starts a retirement fund 30 years before retirement. She pays $25 per month into the annuity for 30 years. Her total investment is $9,000. In each case the annuities pay 6% interest compounded monthly.
   (a) Find the value of each annuity at the time of retirement.
   (b) What lesson should be learned from the answers in part (a)?

46. Lakisha stops in the Common Ground Coffee Shop 20 times a month to drink a $2.75 latte. After studying annuities, she decides to forgo the latte and, at the end of each month, invest the amount she had been spending on latte in an annuity paying 5.7% compounded monthly. If she continues this for 40 years, how much will the annuity be worth?

47. Ricardo is a 30-year-old pack-and-a-half per day smoker who pays $3.40 for a pack of cigarettes. Observing an uncle suffering from emphysema, he decides to stop smoking. To help motivate and reward himself, he decides to put the cost of cigarettes into an annuity until he is 65. Assuming an annuity with an interest rate of 5.1% and investing the cost of 45 packs each month, find the future value when Ricardo is 65.

EXPLORATIONS

48. This exercise compares the future value of investing an amount of money in different time frames. A total of $20,000 is invested into funds paying 6% interest over a 20-year period. Compute the future value at the end of 20 years for each of the following:
  (a) $1000 is invested in an annuity each year for 20 years.
  (b) $2000 is invested in an annuity each year for 10 years. After the first 10 years, the money remains in the fund, drawing 6% interest compounded annually.
  (c) The entire $20,000 is invested at the beginning and remains in the fund, drawing 6% interest compounded annually.
  (d) Comment on the best strategy to accumulate wealth over the long term.

49. Beginning at age 40, Yoshi paid $200 per month into an ordinary annuity that paid 6.9% interest. As retirement age approached, Yoshi wanted to know the amount of the annuity at several different ages to help plan when to retire. Help Yoshi by finding the amount of the annuity when Yoshi is
  (a) 65 years old.   **(b)** 66 years old.
  (c) 67 years old.   **(d)** 68 years old.
  (e) 69 years old.   **(f)** 70 years old.

50. This exercise compares the effect of the starting age in building a retirement account. How much should be invested annually to have a retirement fund of $500,000 at age 65? Assume 5% interest. Each of the following ages is the age at which the annuity is established. Find the annual investment required in each case.
  (a) 25 years   **(b)** 30 years
  (c) 40 years   **(d)** 45 years
  (e) 50 years

51. Pablo likes to plan ahead, so he investigates the effect of investing $2000 annually into an IRA account and starting the investment at different ages. Find the amount accumulated at age 65 for each of the following. Assume a 6% interest rate compounded annually with $2000 invested at the end of each year.
  (a) Start the investments at age 25.
  (b) Start the investments at age 30.
  (c) Start the investments at age 40.
  (d) Start the investments at age 50.

52. Compare the growth of an annuity with quarterly deposits of $1 for 20 years at various interest rates by graphing the future value formula
$$y = \frac{(1 + x)^{80} - 1}{x}.$$ Evaluate this formula at the quarterly rates of

(a) $x = 0.0125$ **(b)** $x = 0.015$
(c) $x = 0.0175$ **(d)** $x = 0.020$
(e) $x = 0.025$
Use the **value** function or a spreadsheet

53. Graph the growth of an ordinary annuity with $1 invested periodically with 2% periodic interest rate for x time periods.
$$y = \frac{(1.02)^x - 1}{0.02}$$
Use [0, 50] for the x-scale and [0, 80] for the y-scale. Interpret the meaning of this graph.

54. Graph the growth of an annuity given by
$$y = \frac{(1 + x)^{40} - 1}{x}$$ using the x-scale [0, 0.05] for periodic interest rate and the y-scale [20, 100] for accumulated total. Interpret this graph.

55. Walt deposits $120 each month in an ordinary annuity. He wants to have $8700 at the end of 5 years. Graph the future value function
$$y = R\left[\frac{(1 + x)^{60} - 1}{x}\right],$$ using $R = 120$ and determine the monthly interest rate that will be required for Walt to meet his goal.

56 Baldemar invested a graduation gift of $3000 in a fund that paid 6.2% compounded quarterly. At the same time, Sami obtained a job and began investing $50 per month in an annuity that paid 5.7% compounded monthly. When, if ever, will Sami's investment catch up with Baldemar's investment assuming Baldemar invests no additional principal?

57. Andrew makes a single investment of $5000 at 8% compounded quarterly, and Cutter invests $150 each quarter into an annuity.
Let
$$x = \text{number of quarters and}$$
$$y = \text{value of the investment}$$
Graph the values of the investments and determine when, if ever, Cutter's investment catches up with Andrew's if
  (a) Cutter's annuity pays 8% compounded quarterly.
  (b) Cutter's annuity pays 6.8% compounded quarterly.
  (c) Cutter's annuity pays 6% compounded quarterly. Assume that Andrew invests no additional principal.

58. Oscar contributes $100 each month to an annuity that pays a 6.9% annual rate (.00575 monthly rate). Use the ANNGRO program to compute the value of the annuity for each of the first 18 months, or make a table using a spreadsheet.

We have used the formula $A = R\left[\dfrac{(1 + i)^n - 1}{i}\right]$ to find the future value of an *ordinary annuity* where payments are made at the end of each period. Another option allows the payments to be made at the *beginning* of each period. Such an annuity is called an **annuity due**. The future value of an annuity due at the end of n periods with periodic payments, R, at the beginning of each period is given by

$$A = R\left[\frac{(1 + i)^{n+1} - 1}{i}\right]$$

Notice this just increases the exponent on $1 + i$ by 1.

59. Find the future value of an annuity due with monthly payments of $50 for 60 months. The annual interest rate is 4.5%.

60. **(a)** Find the future value of an ordinary annuity at the end of 48 months with monthly payments of $75 at an interest rate of 4.2% compounded monthly.

(b) Find the future value of an annuity due at the end of 48 months with monthly payments of $75 at an interest rate of 4.2% compounded monthly.

(c) How much more is the amount of an annuity due than an ordinary annuity?

(d) Compare $75(1.0035)^{48}$ to the answer in part (c).

USING YOUR TI-83

GROWTH OF ANNUITIES

We have formulas to compute the amount of an annuity after a length of time. To see how it grows month by month when a fixed amount is invested on a regular monthly basis requires tedious computing. A TI-83 program, **ANNGRO,** will do the tedious calculating. The program requires that you enter the amount invested periodically (R), the periodic interest rate (I), and the number of periods (N). The program computes the amount of the annuity period by period. Here are the program instructions.

```
: Lbl 1
: Prompt R,I,N
: 0 → J
: R → A
: Disp "PERIOD AMOUNT"
: Lbl 2
: J + 1 → J
: round(A,2) → B
: Disp J,B
: Pause
: A(1 + I) + R → A
: If J < N
: Goto 2
: Goto 1
: End
```

After you have entered the program into your calculator, use it as follows. We will illustrate with monthly payments of $100, 9% interest (.0075 monthly) for 12 months.

Press ⌐PRGM⌐<ANNGRO> ⌐ENTER⌐, and R=? will appear on the screen, asking you to enter the periodic payment. Then press ⌐ENTER⌐ and continue entering I and N

```
prgmANNGRO
R=?100
I=?.0075
N=?12
```

The calculator will pause after showing each period number and amount. Press ENTER to continue to the next period.

When you reach the end, the program will start over, requesting new values of R, I, and N.

You can exit from the program by pressing ON, and this screen will appear:

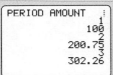

```
PERIOD AMOUNT    :
            1
          100
            2
       200.75
            3
       302.26
```

```
        927.48
          10
      1034.43
          11
      1142.19
          12
      1250.76
R=?
```

```
ERR:BREAK
1■Quit
2:Goto
```

Press ENTER to quit.

EXERCISES

1. Find the month-by-month amounts of an annuity with monthly payments of $200 and annual interest rate of 9%, for 15 months.

2. Find the quarterly amounts of an annuity with quarterly payments of $500 and 8% annual interest rate, for 5 years.

3. Find the annual amounts of an annuity with annual payments of $1000 and annual interest rate of 7.2%, for 10 years.

USING EXCEL

We illustrate how to show the period-by-period amounts of an annuity. We use a monthly investment of $100, 6.6% interest (0.0055 per month) for 12 months. Column A will list the months, and column B will list the corresponding amounts of the annuity.

- Enter zero in A2
- Enter =A2+1 in A3 and drag through A14.
- Enter=100*(1.0055^A2-1)/.0055 in B3 and drag through B14 and you get

	A	B
1	Month	Amount
2	0	0.00
3	1	100.00
4	2	200.55
5	3	301.65
6	4	403.31
7	5	505.53
8	6	608.31
9	7	711.66
10	8	815.57
11	9	920.06
12	10	1025.12
13	11	1130.75
14	12	1236.97

(continued)

EXERCISES

1. Find the annual amounts of an annuity with annual payments of $1500 and an annual interest rate of 5.7% for ten years.
2. Find the quarterly amounts of an annuity with quarterly payments of $250 and a 5.6% annual interest, for two years.
3. Find the monthly amounts of an annuity with monthly payments of $75 and an annual interest rate of 5.4% for 12 months.

5.4   Present Value of an Annuity and Amortization

- Present Value
- Amortization

Present Value

Let's look at some variations of the ordinary annuity. We have studied the result of accumulating equal payments made at regular intervals with the object to provide a fund at a future date. Now we want to look at two variations of this problem.

1. How much should be put into a savings account in *one lump sum* at compound interest so that the amount accumulated at the end of 5 years equals the amount accumulated by an annuity, say investing $25 each month for 5 years?

   The lump sum payment that yields the same total amount as that obtained through equal periodic payments made over the same period of time is called the **present value of the annuity.**

2. An example of a second variation is the problem of finding the monthly payments that will pay off a 4-year car loan.

   A similar problem seeks to find the amount that grandparents should place in a college fund to provide, say $5000 each year, to help with a grandchild's college expenses. We comment that $20,000 is not required to provide $5000 each year for 4 years because the fund will draw interest on the balance of the fund that is available as part of the $5000 annual payment.

Solve the first problem, finding the present value of an annuity, as follows. Recall that the amount of an annuity is

$$A = R\left[\frac{(1 + i)^n - 1}{i}\right]$$

and the amount of compound interest is

$$A = P(1 + i)^n$$

where i is the periodic interest rate, n is the number of periods, R is the periodic payment for the annuity, and P is the lump sum invested at compound interest. We want to find P so that the amount of compound interest equals the amount of the annuity; that is

$$P(1 + i)^n = R\left[\frac{(1 + i)^n - 1}{i}\right]$$

Solving for P gives

$$P = R\left[\frac{(1 + i)^n - 1}{i(1 + i)^n}\right]$$

When we divide the numerator and denominator by $(1 + i)^n$, we obtain the equivalent form

$$P = R\left[\frac{1 - (1 + i)^{-n}}{i}\right]$$

We have obtained the formula for present value of an annuity.

Present Value of an Annuity

$$P = R\left[\frac{(1 + i)^n - 1}{i(1 + i)^n}\right] = R\left[\frac{1 - (1 + i)^{-n}}{i}\right]$$

where

i = periodic rate
n = number of periods
R = periodic payments
P = present value of the annuity

You might ask which form should be used to find present value

$$P = R\left[\frac{(1 + i)^n - 1}{i(1 + i)^n}\right] \quad \text{or}$$

$$P = R\left[\frac{1 - (1 + i)^{-n}}{i}\right]$$

Either form is acceptable. The second form has the advantage that it uses fewer steps to evaluate it, but be sure you use a negative n for the exponent. You will see both forms used in examples.

EXAMPLE 1 ➤

Find the present value of an annuity with periodic payments of $2000, semiannually, for a period of 10 years at an interest rate of 6% compounded semiannually.

SOLUTION

Here $R = 2000$, $i = \frac{0.06}{2} = 0.03$, and $n = 20$. We use these values in the formula for present value

$$P = R\left[\frac{(1 + i)^n - 1}{i(1 + i)^n}\right]$$

and obtain

$$P = 2000\left[\frac{(1.03)^{20} - 1}{0.03(1.03)^{20}}\right]$$

or the form

$$P = 2000\left[\frac{1 - (1.03)^{-20}}{0.03}\right]$$

can be used.

$$P = 2000(14.8774749)$$
$$= 29754.95$$

The present value of the annuity is $29,754.95. This lump sum will accumulate the same amount in 10 years as investing $2000 semiannually for 10 years.

■ **Now You Are Ready to Work Exercise 1**

Another form of periodic equal payments occurs when an established fund pays out money periodically. For example, a college student's grandparents established a fund to pay $5000 each semester to help pay the student's education expenses. This "annuity in reverse" does not require a fund that is 8 times $5000 for eight semesters of payments, because the money in the fund draws interest. Let's see how much money is needed to make equal periodic payments for a specified number of times with no money left when the last payment is made.

To analyze this kind of problem, we use a simpler example from which we draw a general formula: What amount of money is required to make payments of $100 each month for 4 months? The fund draws 12% annual interest and is compounded monthly.

Let's think of the money needed as divided into four parts that we label A_1 for the investment that will grow to the amount needed to make the first payment, A_2 for the investment that will grow to the amount needed for the second payment, and A_3 and A_4 likewise for the third and fourth payments. These investments are not equal because A_4, for example, draws interest longer than the other investments, so it can be smaller. The total $A_1 + A_2 + A_3 + A_4$ is the amount needed initially.

At the beginning of the plan, the amount A_1 is needed to enable a payment of $100 at the end of one month. Because the monthly interest rate is 1%, A_1 will draw interest for one month and will be worth $1.01A_1$ when the payment is made. Thus,

$$1.01A_1 = 100 \quad \text{and} \quad A_1 = \frac{100}{1.01}$$

The amount A_2 in the original investment will draw interest for two months before it makes the second payment, so

$$(1.01)^2 A_2 = 100 \quad \text{and} \quad A_2 = \frac{100}{(1.01)^2}$$

Likewise,

$$(1.01)^3 A_3 = 100 \quad \text{and} \quad A_3 = \frac{100}{(1.01)^3}$$

and

$$(1.01)^4 A_4 = 100 \quad \text{and} \quad A_4 = \frac{100}{(1.01)^4}$$

The investment needed in the fund at the beginning of the fund is

$$S = A_1 + A_2 + A_3 + A_4 = \frac{100}{1.01} + \frac{100}{(1.01)^2} + \frac{100}{(1.01)^3} + \frac{100}{(1.01)^4}$$

We find S by first computing

$$1.01S - S = 100 + \frac{100}{1.01} + \frac{100}{(1.01)^2} + \frac{100}{(1.01)^3}$$
$$- \frac{100}{1.01} - \frac{100}{(1.01)^2} - \frac{100}{(1.01)^3} - \frac{100}{(1.01)^4}$$

which reduces to

$$(1.01 - 1)S = 100 - \frac{100}{(1.01)^4}$$

$$0.01S = 100\left[1 - \frac{1}{(1.01)^4}\right]$$

$$= 100\left[\frac{(1.01)^4 - 1}{(1.01)^4}\right]$$

$$S = 100\left[\frac{(1.01)^4 - 1}{0.01(1.01)^4}\right]$$

By dividing numerator and denominator by $(1.01)^4$, this can be reduced to an alternate form

$$S = 100\left[\frac{1 - (1.01)^{-4}}{0.01}\right]$$

We can now do the arithmetic computations to obtain

$$100\left[\frac{1 - 1.01^{-4}}{0.01}\right] = 100(3.9019656)$$
$$= 390.20 \qquad \text{(Rounded)}$$

A total of $390.20 will provide $100 a month for 4 months.

If we let P = amount needed at the beginning, we have

$$P = 100\left[\frac{1 - 1.01^{-4}}{0.01}\right]$$

Now observe the following:

100 is the amount of a monthly payment (in general, periodic payment).

4 is the number of payments.

0.01 is the monthly interest rate (in general, the periodic rate).

This is an example of the general form.

Amount Needed to Provide Equal Periodic Payments

$$P = R\left[\frac{(1 + i)^n - 1}{i(1 + i)^n}\right]$$

or equivalently,

$$P = R\left[\frac{1 - (1 + i)^{-n}}{i}\right]$$

where

P = amount needed in the fund
R = amount of periodic payments
i = periodic interest rate
n = number of payments

Notice that this is just the present value formula for an annuity.

Thus, this type of problem does not require a new formula. We can call the amount needed in the fund at the beginning the *present value of an annuity*.

EXAMPLE 2 ➤ Find the present value of an annuity (lump sum investment) that will pay $1000 per quarter for 4 years. The annual interest rate is 10%, compounded quarterly.

SOLUTION

$R = 1000$, $i = \frac{0.10}{4} = 0.025$, and $n = 16$ quarters.

$$P = 1000\left[\frac{1 - (1.025)^{-16}}{0.025}\right]$$

$$P = 1000(13.0550027) = 13,055$$

A lump sum investment of $13,055 will provide $1000 per quarter for 4 years.

■ **Now You Are Ready to Work Exercise 34**

Amortization

We now analyze the problem of paying a debt with equal periodic payments. A standard method of paying off a car or house loan requires that the borrower pay equal monthly payments until the debt is paid. This is called **amortization.** Each monthly payment pays all of the interest charged for that month and repays a part of the loan. Because the loan is reduced each month, the interest decreases a little each month, and the amount repaid increases by the same amount.

To illustrate the process, we use a simplified example of a $10,000 car loan with equal payments at the end of each year for 4 years. The interest rate is 8%. We let R represent the annual payments.

To find the annual payments, we will find the amount still owed, the **balance,** at the end of each year. The following observations are helpful.

Observation 1. The original amount of the loan is $10,000.

Observation 2. As the end of the first year approaches, the amount owed increases from $10,000 to $10,800, because 8% interest, 0.08($10,000) = $800, is now to be paid.

Observation 3. The balance owed, $10,800, is reduced by the payment R. Clearly, R must be greater than $800 so that all of the interest and some of the loan are paid.

Observation 4.

First year: We can summarize the first year balance as

$$\text{Balance} = 10,000 + 800 - R$$

which can also be written as

$$\text{Balance} = 10,000 + 0.08(10,000) - R$$

and as

$$\text{Balance} = 10,000(1.08) - R$$

Observation 5.

Second year: The balance at the end of the second year is

First year's balance $+$ interest on first year's balance $- R$

This can be written as

$$\begin{aligned}
\text{Balance} &= 10{,}000(1.08) - R + 0.08\,[\,10{,}000(1.08) - R\,] - R \\
&= [\,10{,}000(1.08) - R\,](1.08) - R \\
&= 10{,}000(1.08)^2 - 1.08R - R
\end{aligned}$$

Observation 6.   The balances for the third and fourth year are

Third year:

$$\begin{aligned}
\text{Balance} &= \text{second year's balance} + \text{interest} - R \\
&= 10{,}000(1.08)^2 - 1.08R - R + [10{,}000(1.08)^2 - 1.08R - R]\,(0.08) - R \\
&= [10{,}000(1.08)^2 - 1.08R - R](1.08) - R \\
&= 10{,}000(1.08)^3 - (1.08)^2R - 1.08R - R
\end{aligned}$$

Fourth year:

$$\begin{aligned}
\text{Balance} &= [10{,}000(1.08)^3 - (1.08)^2R - 1.08R - R](1.08) - R \\
&= 10{,}000(1.08)^4 - (1.08)^3R - (1.08)^2R - 1.08R - R \\
&= 10{,}000(1.08)^4 - R[(1.08)^3 + (1.08)^2 + 1.08 + 1\,]
\end{aligned}$$

We now find the sum in the second term above:

$$S = (1.08)^3 + (1.08)^2 + (1.08) + 1$$

It will help in more general cases if we find this sum in the following way. First, write $1.08S$ as follows:

$$1.08S = (1.08)^4 + (1.08)^3 + (1.08)^2 + 1.08$$

Then,

$$\begin{aligned}
1.08S - S &= (1.08)^4 + (1.08)^3 + (1.08)^2 + 1.08 \\
&\quad - (1.08)^3 - (1.08)^2 - 1.08 - 1 \\
&= (1.08)^4 - 1 \\
1.08S - S - (1.08 - 1)S &= (1.08)^4 - 1 \\
S = \frac{(1.08)^4 - 1}{1.08 - 1} &= \frac{(1.08)^4 - 1}{0.08}
\end{aligned}$$

We can now write the fourth year's balance as

$$\text{Balance} = 10{,}000(1.08)^4 - R\left[\frac{(1.08)^4 - 1}{0.08}\right]$$

This is an example of the general formula for the balance of a loan. The general form is

$$\text{Balance} = P(1 + i)^n - R\left[\frac{(1 + i)^n - 1}{i}\right]$$

where P = the amount of the loan, i = periodic interest rate, n = number of periods, and R = periodic payments.

Now we are ready to find R, the annual payment. Because the loan is to be repaid in 4 years, the fourth year's balance must be zero. We then have

$$10{,}000(1.08)^4 - R\left[\frac{(1.08)^4 - 1}{0.08}\right] = 0$$

and

$$10{,}000(1.08)^4 = R\left[\frac{(1.08)^4 - 1}{0.08}\right]$$

We now calculate $\dfrac{(1.08)^4 - 1}{0.08} = 4.506112$

and

$$10{,}000(1.08)^4 = R\left[\frac{(1.08)^4 - 1}{0.08}\right] = R[4.506112]$$

so

$$R = \frac{10{,}000(1.08)^4}{4.506112} = 3019.21$$

The annual payments are \$3019.21.

Let's observe that the result

$$10{,}000(1.08)^4 = R\left[\frac{(1.08)^4 - 1}{0.08}\right]$$

when written as

$$10{,}000 = R\left[\frac{(1.08)^4 - 1}{0.08(1.08)^4}\right]$$

is exactly the form of present value of an annuity. In this case,

$$P = 10{,}000, \text{ the amount of the loan}$$
$$i = 0.08, \text{ the periodic interest rate}$$
$$R = \text{ periodic payments}$$
$$n = 4, \text{ the number of time periods}$$

This relationship between the amount of a loan and the periodic payments holds when payments are made monthly, quarterly, or any other time period provided that the interest rate and number of periods are also monthly rates and number of months, and so on.

We have gone to some effort to show the following:

> The amortization of a debt (repayment of a debt) requires no new formula because *the amount borrowed is just the present value of an annuity.*

This method usually applies to car payments and house payments. We can use the present value formula of an annuity to find the periodic payments.

Debt Payments

> **Amortization of a Loan**
>
> The amount borrowed, P, is related to the periodic payments, R, by the formula
>
> $$P = R\left[\frac{(1 + i)^n - 1}{i(1 + i)^n}\right] \quad \text{or} \quad P = R\left[\frac{1 - (1 + i)^{-n}}{i}\right]$$
>
> where
>
> i = periodic interest rate   and n = number of payments
>
> (*Note:* This is the present value formula for an ordinary annuity.)

EXAMPLE 3 ➤

An employee borrows $8000 from the company credit union to purchase a car. The interest rate is 12%, compounded monthly, with payments every month. The employee wants to pay off the loan in 3 years. (The loan is amortized over 3 years.) How much are the monthly payments?

SOLUTION

Here we have $P = \$8000$, $i = \frac{0.12}{12} = 0.01$, $n = 36$ months. Substituting into the present value formula, we have

$$8000 = R\left[\frac{1 - (1.01)^{-36}}{0.01}\right]$$

$$8000 = R(30.107505)$$

Solving for R, we have

$$R = \frac{8000}{30.107505} = 265.71$$

The monthly payments are $265.71 each.

■ **Now You Are Ready to Work Exercise 13**

In general, when we solve for R in the present value formula, we have the amortization payment formula.

You should be aware that the amount of the periodic payment may involve a fraction of a cent. In that case the bank rounds up to the next cent. Consequently, the final payment may be a little less than the other payments.

EXAMPLE 4 ➤

A student obtained a 24-month loan on a car. The monthly payments are $395.42 and are based on a 12% interest rate. What was the amount borrowed?

SOLUTION

The amount borrowed is just the present value of the annuity. We then have

$$R = 395.42$$

$$i = \frac{0.12}{12} = 0.01 \quad \text{(Monthly rate)}$$

$$n = 24 \quad\quad\quad \text{(Number of months)}$$

So,

$$P = 395.42\left[\frac{1 - (1.01)^{-24}}{0.01}\right]$$

$$= 395.42(21.2433873) = 8400.06$$

It is reasonable to round this to $8400, the amount borrowed.

■ **Now You Are Ready to Work Exercise 9**

EXAMPLE 5 ➤ Habitat for Humanity helps low-income families build affordable homes. In one southwest area, they can build a house for $32,000. The family makes monthly payments of $130, with no interest, until the loan is paid.

(a) How long will it take a family to pay off the loan?
(b) What monthly payments would it take to pay off the loan if they were charged 6% interest and the length of the loan was the same as in part (a)?

SOLUTION

(a) It will take $\frac{32,000}{130} = 246$ months (rounded) to complete the payments.
(b) If $P = 32,000$, $i = \frac{0.06}{12} = 0.005$, and $n = 246$, then

$$32,000 = R\left[\frac{1 - 1.005^{-246}}{0.005}\right]$$

$$R = \frac{32,000}{141.3620467} = 226.37$$

It would require monthly payments of $226.37 to pay the loan in 246 months if 6% interest were charged.

■ **Now You Are Ready to Work Exercise 22**

We can gain other information related to amortization as illustrated in following examples.

EXAMPLE 6 ➤ A family borrowed $60,000 to buy a house. The loan was for 30 years at 12% interest rate. The monthly payments were $617.17.

(a) How much of the first month's payment was interest and how much was principal?
(b) What total amount did the family pay over the 30 years?

SOLUTION

(a) The monthly interest rate was $1\% = 0.01$, so the first month's interest was $60,000(0.01) = 600.00$. The family paid $600 interest the first month. The rest of the payment, $17.17, went to repay part of the principal.
(b) The family paid $617.17 each month for 360 months, so the total amount paid was $617.17(360) = \$222,181.20$. You may be surprised at this figure, but it is true. Notice that the total amount paid for interest was

$$\$222,181.20 - 60,000 = \$162,181.20$$

■ **Now You Are Ready to Work Exercise 27**

When a family makes monthly payments on a house mortgage, some of each month's payment goes to reduce the loan. In Example 6, the first payment reduced the loan by $17.17. To find the balance of the loan after a period of time — say, 5 years — you can find the amount repaid each month for 5 years and deduct these from the loan. For example, here is an amortization schedule for the first 12 months of the loan:

NOTE

In order to qualify for a Habitat House, a family must put in 300 hours of "sweat equity" in building their house and houses of other families

$60,000 Loan for 30 Years at 12%

Month	Monthly Payment	Interest Paid	Principal Paid	Balance
0				$60,000.00
1	$617.17	$600.00	$17.17	59,982.83
2	617.17	599.83	17.34	59,965.49
3	617.17	599.65	17.52	59,947.97
4	617.17	599.48	17.69	59,930.28
5	617.17	599.30	17.87	59,912.41
6	617.17	599.12	18.05	59,894.36
7	617.17	598.94	18.23	59,876.13
8	617.17	598.76	18.41	59,857.72
9	617.17	598.58	18.59	59,839.13
10	617.17	598.39	18.78	59,820.35
11	617.17	598.20	18.97	59,801.38
12	617.17	598.01	19.16	59,782.22

We see that the balance declines very slowly during the first year of the loan. However, it gradually declines faster, and by the end of the 29th year the balance is $6938.72. The amortization schedule for the last 12 months is the following:

Month from End	Monthly Payment	Interest Paid	Principal Paid	Balance
12	$617.17	$69.39	$547.78	$6390.94
11	617.17	63.91	553.26	5837.68
10	617.17	58.38	558.79	5278.89
9	617.17	52.79	564.38	4714.51
8	617.17	47.15	570.02	4144.49
7	617.17	41.44	575.73	3568.76
6	617.17	35.69	581.48	2987.28
5	617.17	29.87	587.30	2399.98
4	617.17	24.00	593.17	1806.81
3	617.17	18.07	599.10	1207.71
2	617.17	12.08	605.09	602.62
1	608.65	6.03	602.62	—

Notice that the last payment totals less than $617.17 because payments were rounded to the nearest penny.

A chart like this helps to see how fast a debt is reduced (not very fast in the early stages). This approach is too tedious in finding the balance after a number of payments. Actually, we found a formula for the balance early in the discussion of amortization. It is

$$\text{Balance} = P(1 + i)^n - R\left(\frac{(1 + i)^n - 1}{i}\right)$$

Notice that the balance after n periods is

(amount of compound interest) − (amount of an annuity)

The Balance of an Amortization

$$\text{Balance} = P(1 + i)^n - R\left[\frac{(1 + i)^n - 1}{i}\right]$$

where

P = the amount borrowed
i = periodic interest rate
n = number of time periods elapsed
R = monthly payments

EXAMPLE 7 ➤

What is the balance of the loan in Example 6 after two years?

SOLUTION

In Example 6,

$P = 60{,}000$
$i = 1\% = 0.01$ per month
$R = 617.17$
$n = 24$ months   (The balance after 24 months is desired)

To find the balance after two years, substitute these values in the Balance of an Amortization formula above.

$$\begin{aligned}
\text{Balance} &= 60{,}000(1.01)^{24} - 617.17\left[\frac{(1.01)^{24} - 1}{0.01}\right] \\
&= 60{,}000(1.01)^{24} - 617.17(26.9734649) \\
&= 60{,}000(1.26973) - 617.17(26.9734649) \\
&= 76{,}184.08 - 16{,}647.21 \\
&= 59{,}536.87
\end{aligned}$$

So the balance owed after two years is \$59,536.87. The part of the loan repaid is the **equity:**

$$\text{Equity} = \text{loan} - \text{balance}$$

In this case, the equity after two years is

$$\text{Equity} = 60{,}000 - 59{,}536.87 = \$463.13$$

The results of these computations might seem wrong. Hardly any principal is repaid each month. In two years, the 24 payments total \$14,812.08, but only \$463.41 of the \$60,000 loan has been repaid. However, the amount repaid increases a little each month (about 18¢; see the amortization table for this loan on p. 387). Although an 18¢ per month increase hardly seems worthwhile, it eventually becomes a significant increase, and the loan is paid off. Notice from Example 6 that the total interest on the loan of the example is more than \$160,000.

■ **Now You Are Ready to Work Exercise 28**

Now let's look at a situation that combines the future value and present value of an annuity.

EXAMPLE 8 ➤

The parents of a baby want to provide for the child's college education. How much should be deposited on each of the child's first 17 birthdays to be able to withdraw \$10,000 on each of the next four birthdays? Assume an interest rate of 8%.

SOLUTION

First, compute the amount that must be in the account on the child's 17th birthday to withdraw $10,000 per year for 4 years. This is the present value of an annuity, where

$$P \text{ is to be found}$$
$$i = 0.08$$
$$n = 4$$
$$R = 10,000$$

So, substituting in the present value formula

$$P = R\left[\frac{1 - (1 + i)^{-n}}{i}\right]$$

we have

$$P = 10,000\left[\frac{1 - (1.08)^{-4}}{0.08}\right]$$
$$= 10,000(3.3121268) = \$33,121.27$$

the total necessary on the 17th birthday.

Next, find the annual payments that will yield a future value of $33,121.27 in 17 years at 8% using the formula for the future value of an annuity

$$A = R\left[\frac{(1 + i)^n - 1}{i}\right]$$

Given this formula, we have:

$$A = 33,121.27$$
$$i = 0.08$$
$$n = 17$$
$$33,121.27 = R\left[\frac{(1.08)^{17} - 1}{0.08}\right]$$
$$33,121.27 = 33.7502256R$$
$$R = \frac{33,121.27}{33.7502257}$$
$$= 981.36$$

So, $981.36 should be deposited every birthday for 17 years to provide $10,000 per year for 4 years.

■ **Now You Are Ready to Work Exercise 39**

▦ 5.4 EXERCISES

LEVEL 1

Reminder: The compounding periods and the payment periods are the same for an annuity and for an amortization. In Exercises 1 through 8, determine the present values of the annuities that will pay the given periodic payments.

1. *(See Example 1)* Periodic payments of $4000 annually for 8 years. The interest is 7% compounded annually.

2. Periodic payments of $2300 annually for 15 years. The interest is 5% compounded annually.

3. Monthly payments of $508.80 for 3 years at 5.7% interest.

4. Monthly payments of $425 for 5 years with a monthly interest rate of 0.8%.

5. Monthly payments of $240 for 10 years at an annual interest rate of 6.6%.

6. Semimonthly payments of $25 for 18 months at a monthly rate of 1.2%.

7. Periodic payments of $300 per quarter for 5.5 years. The interest is 6.8% compounded quarterly.

8. Periodic payments of $75 per month for 2 years. The interest is 1% compounded monthly.

In Exercises 9 through 12, find the amount borrowed for each loan described.

9. *(See Example 4)* $R = $218.66, the interest rate is 5.6%, and the payments are made quarterly for 3 years.

10. $R = $948.70, the interest rate is 5.8%, and the payments are made semiannually for 6 years.

11. $R = $381.04, the annual interest rate is 5.4%, and payments are made monthly for 24 months.

12. $R = $239.92, the annual interest rate is 5.7%, and payments are made monthly for 5 years.

In Exercises 13 through 19, the present value of an annuity is given. Find the periodic payments.

13. *(See Example 3)* Present value = $9500, and the interest rate is 6.9% compounded annually for 5 years.

14. Present value = $12,000, and the interest rate is 5% compounded semiannually for 10 years.

15. Present value = $7500, and the interest rate is 7.2% compounded quarterly for 15 years.

16. Present value = $14,000, and the interest rate is 7.5% compounded monthly for 7 years.

17. Present value = $32,000, and the interest rate is 7.2% compounded monthly for 8 years.

18. Present value = $6200, and the interest rate is 6.4% compounded quarterly for 5 years.

19. Present value = $96,000, and the interest rate is 5.7% compounded monthly for 15 years.

In Exercises 20 through 23, find the monthly payments for the loans indicated.

20. Amount of loan = $14,500 at an annual interest rate of 5.4% for 3 years.

21. Amount of loan = $12,750 at an annual interest rate of 6.9% for 4 years.

22. *(See Example 5)* To purchase a home, a family borrowed $85,000 at an annual interest rate of 5.7% for 20 years. Find their monthly payments.

23. Amount of loan = $68,000 at an annual interest rate of 6.6% for 30 years.

24. The monthly payments on a 20-year loan of $25,000 at 6.3% interest are $183.46.
   (a) What is the total amount paid over the 20 years?
   (b) What is the total amount of interest paid?

LEVEL 2

25. A family obtains a $75,000 house loan for 30 years at 8% interest. The monthly payments are $550.32 each.
   (a) What is the total amount paid over the 30 years?
   (b) What is the total amount of interest paid? Comment on the relative sizes of the loan, the total amount of interest paid, and the total paid over the 30 years.

26. Curtis and Sue Mathis obtained an $85,000 house loan at 6% interest for 30 years. For each of the first three monthly payments, find the amount that is paid for interest and the amount that is paid toward principal. Write a brief statement about the rate at which the loan is being reduced.

27. *(See Example 6)* A family borrowed $68,000 to buy a house. The loan was at 6.9% and for 25 years. The monthly payments were $476.28 each.

   (a) How much of the first month's payment was interest, and how much was principal?
   (b) What was the total amount paid over the 25 years?

28. *(See Example 7)* The Hiles family has obtained a house loan of $48,000 for 25 years at 7.5% interest. The monthly payments are $354.72. What is the balance of their loan after
   (a) 1 year?   **(b)** 2 years?
   (c) What is their equity after 1 and 2 years? Compare the total payments made over 1 and 2 years with the equity after 1 and 2 years.

29. A company amortizes a $75,000 loan, at 5.8% interest, with quarterly payments of $2484.21 for 10 years. What is the balance of the loan after 5 years?

30. Find the present value of the annuity that will pay $1500 every 6 months for 9 years from an account paying interest at a rate of 8% compounded semiannually.

31. Ms. Greenberg obtained a 24-month loan on a car. The monthly payments are $400.03 based on a 10.8% interest rate. How much did Ms. Greenberg borrow?

32. A druggist borrows $4500 from a bank to stock her drugstore. The interest rate is 8% compounded semiannually with payments due every 6 months. She wants to repay the loan in 18 months. How much are the semiannual payments?

33. A mechanic borrows $7500 to expand his garage. The interest rate is 10% compounded quarterly with payments due every quarter.

What are the quarterly payments if the loan is to be paid off in 4 years?

34. *(See Example 2)* An executive wants to invest a lump sum that will provide $7500 per year for 15 years for his wife. If the investment earns 8% compounded annually, how much should he invest?

35. A donor wants to establish a fund that will pay the Marlin Public Library $10,000 per year for 20 years. If the fund earns 5% interest compounded annually, how much should be put into the fund?

LEVEL 3

36. Raul wins $1 million in a lottery. He is paid the winnings at $50,000 for 20 years. The lottery establishes an annuity that makes the annual payments. How much should the lottery place in the annuity if it earns 7% compounded annually?

37. Alice borrows $12,000 to buy a car. She pays 6.3% interest compounded monthly and the loan is for 2 years. Find the monthly payments.

38. Rachel can amortize a $1400 music center with monthly payments at 1% per month for 2 years. Find the monthly payments.

39. *(See Example 8)* How much should parents invest on each of their child's first 18 birthdays to provide $15,000 per year for the next 4 years if the investment pays 8% interest?

40. How much should be invested each year for 10 years to provide you with $5000 per year for the next 25 years? Assume a 5.6% interest rate.

41. Find the present value of an annuity that will pay $200 each month for 5 years from an account that pays a 6% annual rate compounded monthly.

42. Brett borrowed $13,000 at 6% interest to buy a car. It is a 4-year loan. Find the monthly payments.

43. Shelley borrowed $9700 to buy a car at 6.3% interest. If the loan is for 5 years, find the monthly payments.

44. How much should be placed in an annuity that earns 7.5% compounded annually to provide college expenses of $10,000 per year for 4 years?

45. The Big Red Million Dollar Shootout is an annual hole-in-one contest benefiting the Waco Baptist Academy. Golfers compete in a preliminary round to qualify for the semifinal and final round. The first

round is an "inside the circle" contest, with golfers paying $1 to attempt to shoot a ball within a three-foot circle from 120 yards. A shot within the circle qualifies the golfer for the semifinals, and a hole-in-one in the first round qualifies a person for the finals.

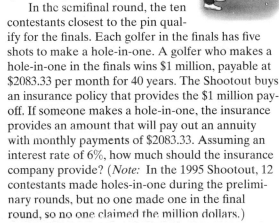

In the semifinal round, the ten contestants closest to the pin qualify for the finals. Each golfer in the finals has five shots to make a hole-in-one. A golfer who makes a hole-in-one in the finals wins $1 million, payable at $2083.33 per month for 40 years. The Shootout buys an insurance policy that provides the $1 million payoff. If someone makes a hole-in-one, the insurance provides an amount that will pay out an annuity with monthly payments of $2083.33. Assuming an interest rate of 6%, how much should the insurance company provide? (*Note:* In the 1995 Shootout, 12 contestants made holes-in-one during the preliminary rounds, but no one made one in the final round, so no one claimed the million dollars.)

46. Carissa's parents were unable to pay for her last year of college, so she obtained a student loan of $9,500. The conditions of the loan were: She would make no payments while in college, but the interest would accumulate at 3.3% compounded monthly. Upon graduation she would begin equal monthly payments that would repay the loan in five years.
   (a) What was the amount of the loan when she graduated one year later?
   (b) What monthly payments will repay the loan in five years?

47. Beth obtained a student loan of $75,000 to finish her last two years of medical school. She would make no payments until she finished, but the loan would accumulate interest at 2.7% compounded monthly. When she finished school, Beth was to begin monthly payments that would repay the loan in eight years at 2.7% interest.
(a) Find the amount of the loan when Beth finished.
(b) Find the monthly payments.

EXPLORATIONS

48. Scott and Beth Walker were Georgians who moved to the heart of Texas. They were delighted to find a southern-style house for sale and proceeded with plans to buy it. They needed to borrow $120,000 and decided to obtain a 20-year loan at a 9.6% interest rate. However, the monthly payments were higher than they could afford, so they decided on a 30-year loan to obtain lower monthly payments.
(a) Find the monthly payments for the 20-year and the 30-year loans.
(b) Find the total amount paid over the life of each loan.
(c) Find the savings per month by going to the 30-year loan.
(d) Find the increased total cost by going to the 30-year loan.

49. Jim and Ann bought a house with a down payment of $10,000 and an $80,000 loan. The loan was for 25 years at a 9% interest rate. Closing costs amounted to an additional 1.5%. Two years later they were transferred and sold the house for what they paid for it, $90,000. The real estate agent charged a 6% fee for selling the house. Find the average monthly cost of the house taking into consideration the monthly payments, the costs of buying and selling, and the equity built up over 2 years.

50. Beginning at age 30, Ms. Trinh invests $2000 each year into an IRA account until she retires. When she retires she plans to withdraw equal amounts each year that will deplete the account when she is 80. Find the annual amounts she will receive for each of the following retirement ages. Assume the account pays 6% compounded annually.
(a) Retires at age 60.   (b) Retires at age 65.
(c) Retires at age 70.

51. Show that the balance after n months of P dollars borrowed at $i\%$ monthly interest rate with monthly payments for 30 years (360 months) is

$$\text{Bal} = P\left[(1+i)^n - \left[\frac{(1+i)^{360}}{(1+i)^{360}-1}\right]\left((1+i)^n - 1\right)\right]$$

52. The size of the national debt concerns many people, but it continues to grow. Let's look at three possible ways to respond.

   1. Pay interest only each year with the principal unchanged.

   2. Pay neither principal nor interest so the interest compounds annually. (This is basically the situation in many recent years.)

   3. Pay off the principal and interest on a regular basis.

Assume a debt of $1 billion, an interest rate of 6%, and a 30-year period. Look at the total obligation resulting from each of these three strategies. By total obligation, we mean the amount paid plus the amount owed at the end of 30 years.

   Which strategy results in the smallest total obligation?

53. Suppose you buy a house for $90,000 and pay $15,000 as a down payment, so you obtain a loan of $75,000 at 6% for 30 years. Determine the following:
(a) Closing costs (ask a real estate agent).
(b) Monthly payments.
(c) Monthly average of taxes.
(d) Monthly average of insurance.
(e) The balance of the loan after 3 years.
(f) The amount you have invested in the home after 3 years: down payment, monthly payments, insurance, and taxes.
(g) You sell the house after 3 years for what you paid for it, 6% of the selling price goes to the real estate agent, and the balance of the loan goes to the bank. How much of the sale price do you receive? Find the average monthly cost of the house.
(h) Repeat part (g), assuming you sell the house for $100,000.
(i) Repeat part (g), assuming you sell the house for $85,000.
(j) Assume you had rented an apartment for $750 per month for the 3 years instead of buying a house. Compare the cost of renting with the cost of buying in the three cases above.
(k) You can receive a deduction on income taxes of 20% of the interest and tax costs. So, you may reduce the cost of buying by that amount. Find the average monthly reduction of purchasing costs due to this tax deduction.

54. I once heard that houses in Japan are so expensive that many house loans are made for 100 years. (When you inherit a house, you inherit the loan.) Assume a $100,000 loan at 9% annual interest rate.

(a) Find the monthly payments for a 20-, 30-, 50-, 75-, and 100-year loan.

(b) Find the total of the monthly payments over the life of the loan in each case in part (a).

(c) Compare the monthly payments and the total of monthly payments for 30, 50, and 100 years and discuss if a 100-year loan is practical compared to a 30- or 50-year loan.

55. For a 30-year house mortgage of $75,000 at 5.7% interest, find

(a) The amount of the first monthly payment that goes to repay principal.

(h) The amount of the 181st month's payment (after 15 years) that goes toward payment of principal.

56. The Outlet Mall Auto Store runs a TV ad advertising autos for $299 down and payments of $199 per month for 3 years. The fine print disappears before it can be read carefully, but it indicates a sum of money will still be owed at the end of 3 years. The owner is expected to pay that balance (called a balloon payment) at that time.

(a) For a car that sells for $14,990, find the balance remaining after 3 years if the interest rate is 6.9%.

(b) What monthly payments would be required to completely pay off the original loan in 3 years?

57. The Dietze family purchased a home and obtained a 30-year loan at 10.5%; they borrowed $125,000. After they made payments for 10 years, the interest rate fell to 8.1%, so they considered refinancing the balance of the loan for the remaining 20 years.

(a) Find the balance remaining.

(b) Find the monthly payments if the balance is refinanced at 8.1%.

(c) How much will total payments over the 20 years be reduced if the loan is refinanced?

58. Andrea found it necessary to buy a car 8 months before graduating from college. Her finances were insufficient to buy a car or even make monthly payments. Thus, she was delighted when a banker friend agreed to loan her $6000 at 6.9% interest compounded monthly; furthermore, she would pay neither interest nor principal during the 8 months before graduation. However, the interest would accumulate monthly and be added to the loan. She agreed to begin monthly payments when she graduated, and the payments were to repay the loan in 3 years. Find Andrea's monthly payments.

59. Upon retirement, a retiree usually has several options on retirement benefits received. Report on options available.

60. Report on the IRA annuity plan.

61. Report on a local savings and loan company's house loan policy. Assume the purchase of a $100,000

house and include the following information for two cases, a 20-year mortgage and a 30-year mortgage.

(a) Down payment required

(b) Interest rate

(c) Insurance required

(d) Closing costs

(e) Approximate tax rates

(f) Monthly payments

(g) Total paid for the home

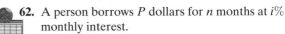

62. A person borrows P dollars for n months at i% monthly interest.

(a) Show that the total paid to the bank over the total life of the loan, n months, is

$$\frac{nPi(1 + i)^n}{(1 + i)^n - 1}$$

(b) Graph the total paid for a $10,000 loan at 0.75% per month interest for x months where $60 \le x \le 360$ (5 to 30 years).

(c) Graph the total paid for a $10,000 loan for 30 years at i% per month where i varies from 0.5% to 2% per month.

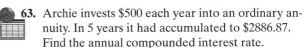

63. Archie invests $500 each year into an ordinary annuity. In 5 years it had accumulated to $2886.87. Find the annual compounded interest rate.

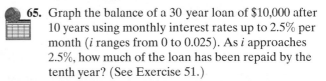

64. Graph the balance function for month x for a $75,000 loan at 9% annual interest with monthly payments of $603.47 for 30 years:

$$\text{Bal} = 75,000(1.0075)^x - 603.47\left[\frac{(1.0075)^x - 1}{0.0075}\right]$$

with the x-scale [0, 360] and the y-scale [0, 75,000]. Use the graph to determine

(a) The amount repaid after 1, 2, 3, 4, and 5 years.

(b) How long it takes to repay 20% of the loan.

(c) How long it takes to repay 50% of the loan.

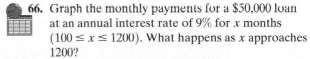

65. Graph the balance of a 30 year loan of $10,000 after 10 years using monthly interest rates up to 2.5% per month (i ranges from 0 to 0.025). As i approaches 2.5%, how much of the loan has been repaid by the tenth year? (See Exercise 51.)

66. Graph the monthly payments for a $50,000 loan at an annual interest rate of 9% for x months ($100 \le x \le 1200$). What happens as x approaches 1200?

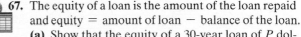

67. The equity of a loan is the amount of the loan repaid and equity = amount of loan − balance of the loan.

(a) Show that the equity of a 30-year loan of P dollars after n months, where the monthly interest rate is i%, can be expressed as

$$\text{Equity} = P\left[1 - (1 + i)^n + \left[\frac{(1 + i)^{360}((1 + i)^n - 1)}{(1 + i)^{360} - 1}\right]\right]$$

(b) Graph the equity of a 30-year, $10,000 loan after x months where the monthly interest rate is 0.75% ($i = 0.0075$).

(c) On the same graph, draw the graph of the equity of a 20-year, $10,000 loan after x months where the monthly interest rate is 0.75%. Use the formula in part (a) but replace the exponent 360 with 240.

 68. The Martinez family obtained a 30-year house loan of $80,000 at 9.3% interest.

(a) Graph the equation of the balance of the loan after x months. Estimate the number of months it takes for the balance to be reduced to half of the loan.

(b) Assume the loan is for 25 years instead of 30 years. Graph the equation of the balance of the loan after x months and estimate the number of months it takes for the balance to be reduced to half the loan.

(c) Assume the loan is for 20 years. Again, graph the equation of the balance and estimate when the balance is reduced to half the loan.

(d) In each of parts (a), (b), and (c), find the ratio of the number of months until the balance is half the loan to the total number of months of the loan. Does this ratio change with the length of the loan? If so, which loan length enables the Martinez family to most quickly reduce the loan by a half?

 69. The examples and exercises on the future value of savings and annuities have not taken into consideration the effect of inflation on the value of the investment. We now look at the present value required of an annuity to maintain constant purchasing power. Rather than ask for the fund amount that will provide for an annual income of some fixed amount, say $50,000, we want the fund amount that will provide an increasing annual income that will provide an income adjusted for inflation that has the same purchasing power as $50,000 at the beginning of retirement. The formula used is

$$P = R\left(\frac{1 + I}{r - I}\right)\left[1 - \left(\frac{1 + I}{1 + r}\right)^n\right]$$

where R = the first annual payment. Subsequent payments are to have the same purchasing power as R does at the beginning.

I = annual inflation rate

r = interest rate of the investment

P = amount necessary to provide for annual payments; present value.

n = number of payments

(*Note:* For year k the annual payment is $R(1 + I)^k$.)

(a) Find the present value required to provide the equivalent of $50,000 income each year for 15 years where the inflation rate is 3% and the interest rate is 7%.

(b) Find the present value required to provide a fixed amount of $50,000 per year for 15 years where the interest rate is 7%. Note the difference in the amount required in parts (a) and (b).

(c) Find the present value required to provide the equivalent of $35,000 income each year for 10 years if the inflation rate is 5% and the interest rate is 8%.

(d) Find the present value required to provide the equivalent of $35,000 income each year for 10 years if the inflation rate is 10% and the interest rate is 12%.

(e) Find the present value required to provide the equivalent of $35,000 each year for 10 years if the inflation rate is 1% and the interest rate is 4%.

(f) Find the present value required to provide the equivalent of $35,000 each year for 10 years if the inflation rate is 10% and the interest rate is 13%.

(g) In parts (c), (e), and (f), the interest rate is 3% more than the inflation rate. Based on your answers to those exercises, which is better for a retired person, low inflation and low interest rates or high inflation and high interest rates?

(h) Now find the present value required to provide a fixed income of $35,000 per year for 10 years with no inflation and a 3% interest rate. How does this amount compare to the present values in parts (c), (e), and (f)?

 70. Graph the present value function

$$y = x\left[\frac{1 - 1.015^{-40}}{0.015}\right]$$

using the x-scale [25, 100] for periodic payments and the y-scale [500, 4000] for the accumulated total. Interpret this graph.

 71. Graph the present value function

$$y = \frac{1 - 1.015^{-x}}{0.015}$$

using [0, 250] for the x-scale and [0, 75] for the y-scale. Interpret the graph.

 72. Does the size of a loan affect the rate at which the balance of the loan decreases? Answer this by comparing the results of the two loans described below.

For a $100,000 loan and a $60,000 loan, each with an interest rate of 8% and a loan length of 30 years, find the following:

(a) The monthly payments for each loan.

(b) Graph the balance of each loan after x months.

(c) Use **intersect** or **Goal Seek** to determine when the balance is 75%, 50%, and 25% of the loan.

(d) For the two loans, do the balances reach 75% at different times? Do the balances reach 50% at different times? Do the balances reach 25% at different times? If so, how does the size of the loan affect the rate at which the balance decreases?

73. Does the interest rate affect the rate at which the balance of a loan decreases? Answer this by comparing the results of three loans: For a 30-year loan of $100,000, find the following, using interest rates of 6%, 8.1%, and 10.2%.
  (a) Find the monthly payments for each interest rate.
  (b) Graph the balance of the loan after x months for each interest rate.
  (c) For each interest rate, find when the balance reaches 75% of the loan, 50% of the loan, and 25% of the loan.
  (d) Describe the effect, if any, a difference in interest rate has on the balance of a loan.

74. Orvilla borrowed $8400 to buy a car. She has a 2-year loan at 7.2% (0.006 per month). Use the **AMLN** program or a spreadsheet to find the monthly payment, the amount each month for interest paid, principal repaid, and the balance of the loan.

The following kind of problem cannot be solved directly with algebraic methods, but it can be solved by finding the intersection of graphs.

Doris borrows $3000 at a 7.2% interest rate and agrees to make monthly payments of $30 per month. How long will it take to repay the loan?
  This is an amortization loan with $P = 3000$, $R = 30$, and $i = \frac{0.072}{12} = 0.006$, so

$$3000 = 30\left[\frac{1 - 1.006^{-n}}{0.006}\right]$$

We can find n by finding the intersection of the graphs

$$y = 30\left[\frac{1 - 1.006^{-x}}{0.006}\right] \quad \text{and}$$
$$y = 3000$$

This occurs at $x = 153.17$. It will take about 153 months, or 12.75 years, to pay off the loan.

Use this technique to solve Exercises 75 through 78.

75. Harold borrows $7600 at 7.5% interest rate and agrees to repay the loan at $60 per month. How long will it take to repay the loan?

76. Amber borrowed $3500 at 6.9% and agreed to repay the loan at $100 per month. How long will it take her to repay the loan?

77. When Mike received his credit card bill, his first reaction was "I've got to pay this off." The bill was $1426, the annual interest rate 19.5%, and the minimum monthly payment was $29. How long will it take Mike to pay the bill if he makes no additional charges and pays the $29 payment each month?

78. Brenda considers buying a car. The interest rate on a car loan is 8.7%, and she can afford car payments of $225 per month. Graph

$$P = 225\left[\frac{1 - 1.00725^{-x}}{.00725}\right]$$

which relates the amount she can borrow and the length of the loan.
  (a) Use the graph to estimate the size of the loan that can be paid off in 3 years. In 5 years.
  (b) Use the graph to estimate how long it will take to pay off a loan of $8000. To pay off a loan of $15,000.

Annual Percentage Rate, APR

We have discussed the amortization of a loan where the monthly interest is based on the unpaid balance of the loan. A lender might give a simple interest loan where the interest is calculated on the original principal and spread evenly over the life of the loan, as illustrated in the following example.

Harlan borrows $2000 for 2 years and the lender gives him a "good deal," 4% interest. Because 4% simple interest for 2 years is $160, Harlan is to repay a total of $2160 in 24 equal payments, which he repays at $90 per month.

Harlan pays 4% on the $2000 for the entire 2 years, though he repays some of the principal each month. So he pays the same dollar amount for interest the second year as he does the first year, but he owes less for the second year. Thus, he really pays more than 4% on what he owes.

What interest rate does Harlan pay if it is computed as interest on the unpaid balance?

Here's how we solve the problem. We want to find the interest rate that will amortize a loan of $2000 in 24 monthly payments of $90. We use the amortization formula

$$P = R\left[\frac{1 - (1 + i)^{-n}}{i}\right]$$

with $P = 2000$, $R = 90$, and $n = 24$:

$$2000 = 90\left[\frac{1 - (1 + i)^{-24}}{i}\right]$$

We cannot solve for i directly using algebra, but we can solve by using the intersection of two graphs.
  Find the intersection of the graphs

$$y = 90\left[\frac{1 - (1 + x)^{-24}}{x}\right] \quad \text{and} \quad y = 2000$$

This occurs at $x = 0.0062507$ for the monthly rate. The annual interest rate is $12(0.0062507) = 0.0750084$, which we can round to 7.5%.

Federal law now requires that lenders quote **APR** (annual percentage rate) **interest rates**—the interest rate used to amortize a loan, with interest paid on the unpaid balance. In the example above, the APR rate is 7.5%.

Use the method in the above example to solve Exercises 79 through 85.

79. Vicki borrows $2500 for 18 months and repays the loan with $150 monthly payments. Find the APR.

80. Elsie borrows $1200 for 1 year and repays the loan with monthly payments of $103. Find the APR.

81. The Maxey family purchases dining room furniture for $1800. The store allows them to pay for it in 12 monthly payments of $161.25, for a total cost of $1935. The clerk stated they were paying 7.5% interest. Mrs. Maxey disagreed, stating that would be correct if they made a single payment of $1935 at the end of the 12

months. Estimate the correct interest rate based on the given monthly payments. (Find the APR.)

Find the APR for Exercises 82 through 85.

82. A loan of $3000 for 2 years, with monthly payments of $135

83. A loan of $1800 for 18 months, with monthly payments of $115

84. A loan of $2400 for 12 months, with monthly payments of $210

85. A loan of $1600 for 15 months, with monthly payments of $112

USING YOUR TI-83

MONTHLY LOAN PAYMENTS

To see how much of a monthly loan payment goes to interest and how much goes to reduce the loan, we have a TI-83 program that shows this and the balance of the loan as well. It requires that you enter the amount of the loan, the annual interest rate, and the number of years. It assumes monthly payments. The program is named **AMLN.**

An Amortization Schedule of a Loan

: Lbl 1
: Disp "AMOUNT BORROWED"
: Input P
: Disp "ANNUAL RATE"
: Input K
: Disp "NO. YEARS"
: Input M
: K/12 → I
: 12*M → N
: (P*I)/(1 - (1 + I) ^ (-N)) → R
: Disp "MONTHLY PAY"
: Round(R,2) → R
: Pause R
: 1 → J
: Lbl 2
: Disp "MO I PR BAL"

: I*P → A
: Round(A,2) → A
: R - A → B
: P - B → P
: Disp J,A,B,P
: Pause
: J + 1 → J
: If J ≤ N
: Goto 2
: Pause
: Goto 1
: End

Enter the program into your calculator. We illustrate its use with a loan of $1500 at 7.5% annual interest for 1 year.

Press ⌷PRGM⌷ <AMLN> ⌷ENTER⌷, and AMT BORROWED will appear on the screen, asking you to enter the amount of the loan. Then press ⌷ENTER⌷ and continue entering *I* and number of years.

```
PrgmAMLN
AMT BORROWED
?1500
ANNUAL RATE
?.075
NO. YEARS
?1
```

Each time the calculator pauses showing a calculation, press ENTER to continue to the next step. The screen shows (a) the amount of each monthly payment, $130.14. It shows for month 1 the amount paid for interest, $9.38; the amount of principal repaid, $120.76; and the balance of the loan, $1379.24. Screen (b) shows the interest, principal repaid, and balance for month 6.

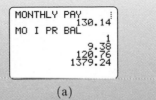

```
MONTHLY PAY
         130.14
MO I PR BAL
              1
          9.38
        120.76
       1379.24
```
(a)

```
            123.81
            888.59
MO I PR BAL
                 6
             5.55
           124.59
              764
```
(b)

EXERCISES

1. Find the monthly interest and principal payments and the monthly balance for a loan of $2000 at 9% annual interest for 1 year.

2. Find the monthly interest and principal payments and the monthly balance for a loan of $4500 at 8.5% annual interest for 2 years.

3. Find the monthly interest and principal payments and the monthly balance for a loan of $85,000 at 7.8% annual interest for 15 years. This loan schedule is 180 months long, so don't step through the entire schedule. Run the steps for only the first year.

USING EXCEL

We illustrate how much of a monthly payment goes for interest and how much goes to repay the loan. We also show how to find the balance for each month.

We illustrate with a loan of $2500 at 6.6% annual interest for one year. We use cell B2 for the amount borrowed, C2 for the annual interest rate, D2 for the monthly interest rate, E2 for the number of years, F2 for the number of months, and G2 for the monthly payments.

We show the completed table and then indicate the entries made in various cells.

	A	B	C	D	E	F	G
1		Borrowed	Annual %	Monthly %	Years	Months	Month. Pay
2		2500	0.066	0.0055	1	12	215.86
3	Month	Interest	Repaid	Balance			
4	0			2500			
5	1	13.75	202.11	2297.89			
6	2	12.64	203.22	2094.68			
7	3	11.52	204.34	1890.34			
8	4	10.40	205.46	1684.88			
9	5	9.27	206.59	1478.29			
10	6	8.13	207.73	1270.57			
11	7	6.99	208.87	1061.70			
12	8	5.84	210.02	851.68			
13	9	4.68	211.17	640.51			
14	10	3.52	212.33	428.18			
15	11	2.35	213.50	214.68			
16	12	1.18	214.68	0.00			

CELL ENTRIES

Numbers entered are: 2500 in B2, 0.066 in C2, 1 in E2, and 0 in A4. As you change these numbers for other problems, the spreadsheet will automatically compute the appropriate new values.

Formulas entered: In D2, =C2/12; in F2, =E2*12; in G2, the formula $\dfrac{Pi(1+i)^n}{(1+i)^n - 1}$, i.e., =B2*D2*(1+D2)^F2/((1+D2)^F2-1).

Formulas entered and dragged through row 16: In A5, =A4+1; in B5, =$D$2*D4; in C5, =$G$2-B5; in D5, =D4-C5.

IMPORTANT TERMS

5.1

Simple Interest
Principal
Present Value
Interest Rate
Future Value Amount
Simple Discount
Proceeds
Discount Rate
Maturity Value

5.2

Compound Interest
Periodic Interest Rate Amount

Future Value of Compound Interest
Nominal Rate
Present Value
Effective Rate

5.3

Annuity
Payment Period
Periodic Payment
Ordinary Annuity Amount
Future Value
Sinking Fund
Annuity Due

5.4

Present Value of an Annuity
Amortization
Balance
Equity
APR

SUMMARY OF FORMULAS

Simple interest

Interest $\qquad I = Prt$

Future Value $\qquad A = P + I = P(I + rt)$

Simple Discount

Discount $\qquad D = Mdt$

Proceeds $\qquad PR = M - D = M(1 - dt)$

Compound Interest

Future Value $\qquad A = P(1 + i)^n$

Effective Rate $\qquad x = (1 + i)^m - 1$

Annuity

Future Value $\qquad A = R\left[\dfrac{(1 + i)^n - 1}{i}\right]$

Present Value $\qquad P = R\left[\dfrac{(1 + i)^n - 1}{i(1 + i)^n}\right] = R\left[\dfrac{1 - (1 + i)^{-n}}{i}\right]$

Amortization

Periodic Payments $\qquad R = \dfrac{P}{\left[\dfrac{(1 + i)^n - 1}{i(1 + i)^n}\right]} = \dfrac{P}{\left[\dfrac{1 - (1 + i)^{-n}}{i}\right]}$

$\qquad\qquad\qquad\qquad = \dfrac{P(i)(1 + i)^n}{(1 + i)^n - 1} = \dfrac{Pi}{1 - (1 + i)^{-n}}$

Balance of a Loan $\qquad \text{Bal} = P(1 + i)^n - R\left[\dfrac{(1 + i)^n - 1}{i}\right]$

▥ REVIEW EXERCISES

1. A loan is made for $500 at 9% simple interest for 2 years. How much interest is paid?

2. How much simple interest was paid on an $1100 loan at 6% for 10 months?

3. A loan, principal and interest, was paid with $1190.40. The loan was made at 8% simple interest for 3 years. How much was borrowed?

4. The interest on a loan was $94.50, the simple interest rate was 7.5%, and the loan was for 1.5 years. What was the amount borrowed?

5. A sum of $3000 is borrowed for a period of 5 years at simple interest of 9% per annum. Compute the total interest paid over this period.

6. Find the future value of a $6500 loan at 7.5% simple interest for 18 months.

7. A simple discount note at 9% discount rate for 2 years has a maturity value of $8500. What are the discount and the proceeds?

8. How much should be borrowed on a discount note with 6.1% discount rate for 1.5 years so that the borrower obtains $900?

9. Compute the interest earned on $5000 in 3 years if the interest is 7% compounded annually.

10. Compute the interest and the amount of $500 after 2 years if the interest is 10% compounded semi-annually.

11. Missy invested $1000 at 6.8% compounded quarterly. Can she expect it to double in value after ten years?

12. A sum of $1000 is deposited in a savings account that pays interest of 5%, compounded annually. Determine the amount in the account after 4 years.

13. Find the effective rate of 6% compounded semi-annually.

14. Find the effective rate of 5% compounded semi-annually.

15. Find the effective rate of 7.2% compounded monthly.

16. Which is the better investment, one that pays 8% compounded quarterly or one that pays 8.3% compounded annually?

17. Which account gives the better interest, one that gives 5.6% per annum compounded quarterly or one that gives 5.7% compounded annually?

18. The price of an automobile is now $5000. What would be the anticipated price of that automobile in 2 years' time if prices are expected to increase at an annual rate of 8%?

Find (a) future values and (b) the total interest earned on the amounts in Exercises 19 through 21.

19. $8000 at 6% compounded annually for 4 years

20. $3000 at 5% compounded annually for 2 years

21. $5000 at 5.8% compounded quarterly for 6 years

22. A company sets aside $120,000 cash in a special building fund to be used at the end of 5 years to construct a new building. The fund will earn 6% compounded semiannually. How much money will be in the fund at the end of the period?

23. On January 1, a Chicago firm purchased a new machine to be used in the plant. The list price of the machine was $15,000, payable at the end of 2 years with interest of 5.5% compounded annually. How much was due on the machine on January 1 two years later?

24. Anticipating college tuition for their child in 10 years, the Heggens want to deposit a lump sum of money into an account that will provide $96,000 at the end of that 10-year period. The account selected pays 6.2%, compounded semiannually. How much should they deposit?

25. An executive thinks that it is a good time to sell a certain stock. She wants to take a leave of absence in 5 years. How much stock should she sell and invest at 8% compounded quarterly so that $50,000 will be available in 5 years?

26. Will a $1000 investment increase to $3000 in 15 years if the interest rate is 7.5% compounded quarterly?

27. Jonel is advised that her investment of $2000 will grow to $3500 in six years if it is invested at 6.6% compounded quarterly. Determine if this advice is correct.

Find the future value of each of the ordinary annuities in Exercises 28 through 30.

28. $R = \$500, i = 0.05, n = 8$

29. $R = \$1000, i = 0.06, n = 5$

30. $R = \$300, i = 0.025, n = 20$

31. Find the future value of $600 paid into an annuity at the end of every 6 months for 5 years. The annual interest rate is 6.4%.

32. Find the future value of $1000 paid into an annuity at the end of every year for 6 years. The interest rate is 5.9%.

33. $250 is invested in an annuity at the end of every 3 months at 6.4%. Find the value of the annuity at the end of 6 years.

34. An annuity consists of payments of $1000 at the end of each year for a period of 5 years. Interest is paid at 6% per year compounded annually. Determine the amount of the annuity at the end of 5 years.

35. Midway School District sold $2 million of bonds to construct an elementary school. The bonds must be paid in 10 years. The district establishes a sinking fund that pays 4.8% interest. Find the annual payments into the sinking fund that will provide the necessary funds at the end of 10 years.

36. A couple wants to start a fund with annual payments that will give them $20,000 cash on retirement in 10 years. The proposed fund will give interest of 7% compounded annually. What will be the annual payments?

In Exercises 37 through 40, determine the present value of the given amounts using the given interest rate and length of time.

37. $5000 over 5 years at 6% compounded annually

38. $4750 over 4 years at 5% compounded semiannually

39. $6000 over 5 years at 6.2% compounded quarterly

40. $1000 over 8 years at 5.8% compounded quarterly

41. A company wants to deposit a certain sum at the present time into an account that pays compound interest at 8% compounded quarterly to meet an expected expense of $50,000 in 5 years' time. How much should it deposit?

42. A corporation is planning plant expansion as soon as adequate funds can be accumulated. The corporation has estimated that the additions will cost approximately $105,000. At the present time it has $70,935 cash on hand that will not be needed in the near future. A local savings institution will pay 8% compounded semiannually. Can $70,935 accumulate to approximately $105,000 in 5 years?

43. A medical supplier is making plans to issue $200,000 in bonds to finance plant modernization. The interest rate on the bonds will be 8% compounded quarterly. The company estimates that it can pay up to a maximum of $97,000 in total interest on the bonds. The company wants to stretch out the period of payment as long as possible. Is it possible for the supplier to take longer than 5 years to pay off the bonds?

44. A corporation owed a $40,000 debt. Its creditor agreed to let the corporation pay the debt in five equal annual payments at 6.5% interest. Compute the annual payments.

45. A student borrowed $3000 from a credit union toward purchasing a car. The interest rate on such loans is 6.4% compounded quarterly with payments due every quarter. The student wants to pay off the loan in 3 years. Find the quarterly payments.

46. The interest on a house mortgage of $53,000 for 30 years is 6.9%, compounded monthly, with payments made monthly. Compute the monthly payments.

47. Andrew borrows $4800 at 7.4% interest. It is to be paid in five annual payments. Find the annual payments.

48. Holt's Clothing Store borrowed $125,000 to be repaid in semiannual payments. Find the semiannual payments if the loan is for 4 years and the interest rate is 7.4% compounded semiannually.

49. The city of McGregor borrowed $7.8 million to build a zoo. The interest rate is 4% compounded annually, and the loan is to be paid in 20 years. Find the annual payments.

50. The Thaxton family purchased a franchise with a loan of $25,000 at 8% compounded quarterly for 6 years. Find the quarterly payments.

51. The Bar X Ranch purchased additional land with a loan of $98,000 at 9% compounded annually for 8 years. Find the annual payments.

52. Fashion Floors borrowed $65,000 at 8.7% compounded semiannually for 3 years. Find the semiannual payments.

53. An investor invests $1000 at 6.2% compounded quarterly for 5 years. At the end of the 5 years, the total amount in the account is reinvested at 6.4% compounded quarterly for another 5 years. How much is in the account at the end of 10 years?

54. A family borrowed $85,000 at 8.7% for 30 years to buy a house. Their payments are $665.66 per month. How much of the first month's payment is interest? How much of the first month's payment is principal? What is the total amount they will pay over the 30 years?

55. If $1700 is invested at 5.8% compounded quarterly, find the amount at the end of 10 years.

56. Find the amount of $20,000 invested at 5.2% compounded quarterly for 50 years.

57. How much should be invested at 6.4% compounded quarterly in order to have $500,000 at the end of 40 years?

58. If $100 per month is deposited in an annuity earning 5.7% interest, find the amount at the end of 10 years.

59. How much should be deposited monthly into an annuity paying 6% in order to have $100,000 at the end of 6 years?

Let's look at some of the traditional terminology used in speaking about sets. The objects that form a set may be varied. We can have a set of people, a set of numbers, a set of ideas, or a set of raindrops. In mathematics, the general term for an object in a set is **element.** A set may have people, numbers, ideas, or raindrops as elements. So that the contents of a set will be clearly understood, we need to be rather precise in describing a set of elements. One way to describe a set is to explicitly list all its elements. We usually enclose the list with braces. When we write

$$A = \{\text{Tom, Dick, Harry}\}$$

we are stating that we have named the set "A," and its elements are Tom, Dick, and Harry. Customarily, capital letters designate sets, and lowercase letters represent elements in a set. The notation

$$x \in A$$

is read "x is an element of A." We may also say that "x is a member of the set A." The statement "x is not an element of the set A" is written

$$x \notin A$$

For the set $B = \{2, 4, 6, 8\}$, we may write $4 \in B$ to specify that 4 is one of the elements of the set B, and $5 \notin B$ to specify that 5 is not one of the elements of B. Here are some examples of other sets.

EXAMPLE 1 ➤   The set of positive integers less than 7 is $\{1, 2, 3, 4, 5, 6\}$. ■

EXAMPLE 2 ➤   The vowels of the English alphabet form the set $\{a, e, i, o, u\}$. ■

EXAMPLE 3 ➤   The countries of North America form the set $\{\text{Canada, USA, Mexico}\}$. ■

EXAMPLE 4 ➤   The set of all natural numbers is $\{1, 2, 3, 4, \ldots\}$. Notice that we use three dots to indicate that the natural numbers continue, without ending, beyond 4. This is an **infinite set.** The three dots are commonly used to indicate that the preceding pattern continues. This notation is the mathematical equivalent of *et cetera*. The context in which the three dots are used determines which elements are missing from the list. The notation $\{20, 22, 24, \ldots, 32\}$ indicates that 26, 28, and 30 are missing from the list. This set is finite because it contains a finite number of elements. When "$\ldots$" is used in representing a finite set, the last element of the set is usually given. An infinite set gives no indication of a last element, as in the set of positive even numbers $\{2, 4, 6, \ldots\}$. ■

Set-Builder Notation

Another way of describing a set uses the **implicit** or **set-builder notation.** The set of the vowels of the English alphabet may be described by

$$\{x \mid x \text{ is a vowel of the English alphabet}\}$$

Read this as "The set of all x such that x is a vowel of the English alphabet." Note that the vertical line | is read "such that." The symbol preceding the vertical line, x in this instance, designates a typical element of the set. The statement to the right of the vertical line *describes* a typical element of the set; it tells how to find a specific instance of x.

EXAMPLE 5 ➤ (a) The elements of the set $A = \{x \mid x$ is an odd integer between 10 and 20$\}$ are 11, 13, 15, 17, and 19.

(b) The set $\{x \mid x$ is a male student at Community College and x is over 6 ft tall$\}$ is the set of all male students at Community College who are over 6 ft tall.

(c) The set $\{x \mid x$ is a member of the United States Senate$\}$ is the set of all senators of the United States.

(d) The set $\{x \mid x = 3n - 2$ where n is a positive integer$\}$ contains $1, 4, 7, 10, \ldots$.

■ **Now You Are Ready to Work Exercise 2**

Empty Set

Sometimes a set is described but it has no elements. For example, the set of golfers who made nine consecutive holes-in-one at the Richland Golf Course has no elements. Such a set, consisting of no elements, plays an important role in set theory. It is called the **empty set** and is denoted by $\varnothing$. A set is called *nonempty* if it contains one or more elements. The set of people who are 10 feet tall is empty. The set of blue-eyed people is not empty. The set of numbers that belong to both $\{1, 3, 5\}$ and $\{2, 4, 6\}$ is empty.

Venn Diagrams

It usually helps to use a diagram to represent an idea. For sets, it is customary to use a rectangular area to represent the **universe** to which a set belongs. We have not used the term *universe* before. The universe is not some all-inclusive set that contains everything. When I talk about a set of students, I am generally thinking of college students, so the elements of my set of students are restricted to college students. In that context, the universe is the set of college students. When a student at Midway High speaks of a set of students, she restricts the elements of her set to students at her school, so the set of all students at Midway High School forms her universe. In a given context, the sets under discussion have elements that come from some restricted set. We call the set from which the elements come the **universe** or the **universal set.** We usually represent the universe by a rectangle. A circular area within the rectangle represents a set of elements in the universe. The diagram in Figure 6–1 represents a set A in a universe U. It is called a **Venn diagram.**

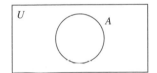

FIGURE 6–1 A Venn diagram.

Equal Sets

Two sets A and B are said to be **equal** if they consist of exactly the same elements; this is denoted by $A = B$. The sets $A = \{2, 4, 6, 8\}$ and $B = \{6, 6, 2, 8, 4\}$ are equal because they consist of the same elements. *Listing the elements in a different order or with repetition does not create a different set.*

We use the symbol $\neq$ to indicate that two sets are not equal:

$$\{1, 2, 3, 4\} \neq \{2, 5, 6\}$$

Subset

The set B is said to be a **subset** of A if every element of B is also an element of A. This is written $B \subseteq A$.

EXAMPLE 6 ➤

$\{2, 5, 9\} \subseteq \{1, 2, 4, 5, 7, 9\}$ because every element of $\{2, 5, 9\}$ is in the set $\{1, 2, 4, 5, 7, 9\}$. The set $\{3, 5, 9\}$ is not a subset of $\{1, 2, 4, 5, 7, 9\}$ because 3 is in the first set but not in the second. Figure 6–2 shows the Venn diagram.

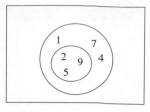

FIGURE 6–2
$\{2, 5, 9\} \subseteq \{1, 2, 4, 5, 7, 9\}$.

■ **Now You Are Ready to Work Exercise 13**

When B is not a subset of A, we write $B \nsubseteq A$:

$$\{3, 5, 9\} \nsubseteq \{1, 2, 4, 5, 7, 9\}$$

EXAMPLE 7 ➤

The subsets of $\{1, 2, 3\}$ are

$$\varnothing, \{1\}, \{2\}, \{3\}, \{1, 2\}, \{1, 3\}, \{2, 3\}, \{1, 2, 3\}$$

Notice that a set is a subset of itself; $\{1, 2, 3\}$ is a subset of $\{1, 2, 3\}$. Because $\varnothing$ contains no elements, it is true that every element of $\varnothing$ is in $\{1, 2, 3\}$, so $\varnothing \subseteq \{1, 2, 3\}$. In the same sense, $\varnothing$ is a subset of every set. The set $\{1, 2\}$ is a **proper subset** of $\{1, 2, 3\}$ because it is a subset of and not equal to $\{1, 2, 3\}$. We point out that if $A \subseteq B$ and $B \subseteq A$, then $A = B$.

> **NOTE**
>
> The collection of all subsets of $\{1, 2, 3\}$ is called the **power set** of $\{1, 2, 3\}$.

■ **Now You Are Ready to Work Exercise 21**

We use $\subset$ to denote a proper subset:

$$\{1, 2, 3\} \subset \{1, 2, 3, 4\}.$$

The Union of Two Sets

Sometimes, we wish to construct a set using elements from two given sets. One way to obtain a new set is to combine the two given sets into one.

DEFINITION
Union of Sets

The **union** of two sets, A and B, is the set whose elements are from A or from B, or from both. Denote this set by $A \cup B$.
In set-builder notation,

$$A \cup B = \{x \mid x \in A \quad \text{or} \quad x \in B \quad \text{or} \quad x \text{ is in both}\}$$

In ordinary use, the word *or* is ambiguous, and its meaning may be confusing. The statement "To be admitted to an R-rated movie, a person must be at least 17 years old or accompanied by a parent" allows a 20-year-old to attend an R-rated movie alone. It also allows a 20-year-old to attend with a parent. This is an example of the **inclusive or;** it allows one or both options to occur. We interpret the statement "I will take history or language at 10:00 MWF" to allow only one course at 10:00 MWF. This is called the **exclusive or;** it does not allow both options. The word *or* in mathematics is interpreted as the inclusive or.

In standard mathematical terminology, the phrase "$x \in A$ or $x \in B$" includes the possibility of x existing in both. Therefore, the phrase "or x is in both" is mathematically redundant.

Figure 6–3 shows a Venn diagram that represents the union of two sets. The shaded area represents $A \cup B$.

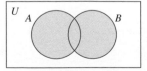

FIGURE 6–3 The union of sets A and B, $A \cup B$.

EXAMPLE 8 ➤

Given: $A = \{2, 4, 6, 8, 10\}$ and $B = \{6, 7, 8, 9, 10\}$. To determine $A \cup B$, list all elements of A and add those from B that are not already listed to obtain

$$A \cup B = \{2, 4, 6, 8, 10, 7, 9\}$$

■ **Now You Are Ready to Work Exercise 33**

EXAMPLE 9 ➤

(a) Given the sets

$$A = \{\text{Tom, Danny, Harry}\}$$
$$B = \{\text{Sue, Ann, Jo, Carmen}\}$$

then,

$$A \cup B = \{\text{Tom, Danny, Harry, Sue, Ann, Jo, Carmen}\}$$

(b) Given the sets

$$A = \{x \mid x \text{ is a letter of the word } radio\}$$
$$B = \{x \mid x \text{ is one of the first six letters of the English alphabet}\}$$

then,

$$A \cup B = \{\text{a, b, c, d, e, f, r, i, o}\} \quad ■$$

> **NOTE**
>
> The order in which the elements in a set are listed is not important. It might seem more natural to list the elements as $\{2, 4, 6, 7, 8, 9, 10\}$. Notice that an element that appears in both A and B is listed only once in $A \cup B$.

The Intersection of Two Sets

Another way to construct a set from two sets is by performing the operation called the **intersection** of two sets.

DEFINITION
Intersection of Sets

The **intersection** of two sets, A and B, is the set of all elements contained in both sets A and B — that is, those elements that A and B have in common. The intersection of A and B is denoted by

$$A \cap B = \{x \mid x \in A \quad \text{and} \quad x \in B\}$$

EXAMPLE 10 ➤

Figure 6–4 shows the Venn diagram of the intersection of two sets. The shaded area is $A \cap B$. ■

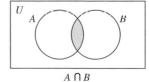

$A \cap B$

FIGURE 6–4 The intersection of sets A and B, $A \cap B$.

EXAMPLE 11 ➤

(a) If $A = \{2, 4, 5, 8, 10\}$ and $B = \{3, 4, 5, 6, 7\}$, then

$$A \cap B = \{4, 5\}$$

(b) If $A = \{a, b, c, d, e, f, g, h\}$ and $B = \{b, d, e\}$, then

$$A \cap B = \{b, d, e\}$$

■ **Now You Are Ready to Work Exercise 37**

EXAMPLE 12 ➤

Given the sets

$$A = \text{set of all positive even integers}$$
$$B = \text{set of all positive multiples of 3}$$

describe and list the elements of $A \cap B$ and $A \cup B$.

SOLUTION

Because an element of $A \cap B$ must be even (a multiple of 2) and a multiple of 3, each such element must be a multiple of 6. Thus $A \cap B$ is the set of all positive multiples of 6, that is, $\{6, 12, 18, 24, \dots\}$. As this is an infinite set, only the pattern of numbers can be listed. $A \cup B$ is the set of positive integers that are multiples of 2 or multiples of 3.

■ **Now You Are Ready to Work Exercise 61**

Note that the sets A and B are related to $A \cup B$ and $A \cap B$ in the following way.

$$A \subseteq A \cup B$$
$$B \subseteq A \cup B$$
$$A \cap B \subseteq A$$
$$A \cap B \subseteq B$$

In a finite set, you can count the elements and finish at a definite number. You can never count all the elements in an infinite set. No matter how high you count, there will be some left uncounted. The set of whole numbers between 10 and 20 is finite; there are 9 of them. The set of fractions between 10 and 20 is infinite; there is no end to them.

EXAMPLE 13 ➤

Let

$$U = \text{set of letters of the English alphabet}$$
$$A = \text{set of vowels}$$
$$B = \text{set of the first nine letters of the alphabet}$$

Place each of the letters a, c, d, e, i, u, x, y in the appropriate region in a Venn diagram.

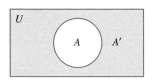

FIGURE 6–5

SOLUTION

The letters a, e, i, and u are elements of A; a, c, d, e, and i are elements of B; and x and y are in neither set. Notice that a, e, and i are in both A and B, so they are in their intersection. We may indicate this as shown in Figure 6–5. ■

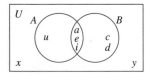

FIGURE 6–6 The shaded area, A', is the complement of A.

Complement

The elements in the universe U that lie outside A form a set called the **complement** of A, denoted by A'. (See Figure 6–6.)

EXAMPLE 14 ➤

Let $U = \{1, 2, 3, 4, 5, 6, 7, 8\}$ and $A = \{1, 2, 8\}$. Then $A' = \{3, 4, 5, 6, 7\}$.

From the definition of complement, it follows that for any set A in a universe U,

$$A \cap A' = \varnothing \quad \text{and} \quad A \cup A' = U$$

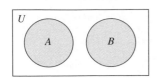

FIGURE 6−7   Because
A and *B* are disjoint, *A* ∩ *B*
has no elements; it is the
empty set: *A* ∩ *B* = ∅.
The shaded area is *A* ∪ *B*.

A set and its complement have no elements in common. The union of a set and its complement is the universal set.

■ **Now You Are Ready to Work Exercise 39**

Disjoint Sets

If *A* and *B* do not have any elements in common, they are said to be **disjoint sets,** and $A \cap B = \emptyset$. The sets $A = \{2, 4, 6, 8\}$ and $B = \{1, 3, 5, 7\}$ are disjoint.

Figure 6−7 shows two disjoint sets. Although their intersection is empty, their union is not, and it is shown by the shaded area.

6.1 EXERCISES

LEVEL 1

1. Establish whether each of the following statements is true or false for $A = \{1, 2, 3, 4\}$, $B = \{3, 4, 5, 6\}$, $C = \{5, 6, 7, 8\}$, and $D = \{1, 3, 5, 7, \dots\}$.
   (a) $2 \in A$
   (b) $5 \in A$
   (c) $8 \in B$
   (d) $5 \notin C$
   (e) $6 \notin A$
   (f) $5 \in D$
   (g) $20 \in D$
   (h) $49 \in D$
   (i) $200 \in D$

List the elements of each of the sets in Exercises 2 through 6.

2. *(See Example 5)*
   $A = \{x \mid x$ is a positive odd integer less than 10$\}$

3. $B = \{x \mid x$ is a letter in the word Mississippi$\}$

4. $A = \{x \mid x$ is an integer larger than 13 and less than 20$\}$

5. $C = \{x \mid x$ is an even integer larger than 15$\}$

6. $A = \{x \mid x$ is a prime less than 20$\}$

Which of the following pairs of sets in Exercises 7 through 12 are equal?

7. $A = \{1, 2, 3, 4\}$, $B = \{3, 1, 2, 5\}$

8. $A = \{1, -1, 0, 4\}$, $B = \{-1, 0, 1, 4\}$

9. $A = \{a, e, i, o, u\}$,
   $B = \{x \mid x$ is a vowel of the English alphabet$\}$

10. $A = \{x \mid x$ is a prime integer less than 20$\}$
    $B = \{2, 3, 5, 7, 11, 13, 17, 19\}$

11. $A = \{5, 10, 15, 20\}$, $B = \{x \mid x$ is a multiple of 5$\}$

12. $A = \{x \mid x$ is a letter in the word REPORTER$\}$
    $B = \{x \mid x$ is a letter in the word POTTER$\}$

In which of Exercises 13 through 20 is *A* a subset of *B*?

13. *(See Example 6)*
    $A = \{2, 4, 6, 8\}$, $B = \{1, 2, 3, 4, 5, 6, 7, 8, 9, 10\}$

14. $A = \{1, 2, 3, 5\}$, $B = \{1, 2, 3, 4, 5\}$

15. $A = \{a, b, c\}$, $B = \{a, e, i, o, u\}$

16. $A = \{1, 3, 7, 9, 11, 13\}$, $B = \{1, 9, 13\}$

17. $A = \{3, 8, 2, 5\}$, $B = \{2, 3, 5, 8\}$

18. $A = $ set of male college students,
    $B = $ set of college students

19. $A = $ set of even integers, $B = $ set of integers

20. $A = \{x \mid x$ is a prime number less than 15$\}$,
    $B = \{x \mid x$ is an odd integer less than 15$\}$

21. *(See Example 7)*   Determine all the subsets of *A*, *B*, and *C*.
    (a) $A = \{-1, 2, 4\}$ (b) $B = \{4\}$
    (c) $C = \{3, 5, 6, 8\}$

Which of the sets in Exercises 22 through 32 are empty?

22. Female presidents of the United States

23. A flock of extinct birds

24. The integers larger than 10 and less than 5

25. English words that begin with the letter *k* and end with the letter *e*

26. Families with five children

27. Odd integers divisible by 2
28. Integers larger than 0 and less than 37
29. Integers larger than 5 and less than 6
30. Fractions larger than 5 and less than 6
31. $\{2, 4, 6\} \cap \{1, 8, 13\}$
32. $\{5, 9, 27\} \cap \{8, 9, 10\}$

List the elements of the sets in Exercises 33 through 38.

33. *(See Example 8)* $\{2, 1, 7\} \cup \{4, 6, 7\}$
34. $\{h, i, s, t, o, r, y\} \cup \{m, u, s, i, c\}$
35. $\{a, b, c, x, y, z\} \cup \{b, a, d\}$
36. $\{5, 2, 9, 4\} \cup \{3, 0, 8\}$
37. *(See Example 11)*
    $\{1, 3, 5, 9, 12\} \cap \{6, 7, 8, 9, 10, 11, 12\}$
38. $\{h, i, s, t, o, r, y\} \cap \{e, n, g, l, i, s, h\}$
39. *(See Example 14)*   If
    $U = \{15, 16, 17, 18, 19, 20, 21\}$ and
    $A = \{15, 16, 20, 21\}$, find A'.
40. If $U = \{-1, 0, 1\}$ and $A = \{-1, 1\}$, find A'.
41. If $U = \{1, 2, 3, 11, 12, 13\}$ and $A = \{1, 2, 3\}$,
    find A'.

42. If $U = \{a, b, c, d, e, f\}$, $A = \{a, b\}$, and
    $B = \{b, c, d\}$, find
    (a) A' (b) B'
    (c) $(A \cup B)'$ (d) $(A \cap B)'$
43. If $U = \{-1, 0, 1, 11, 12, 13\}$, $A = \{1, 11, 13\}$, and
    $B = \{-1, 0, 11\}$, find
    (a) A' (b) B'
    (c) $(A \cup B)'$ (d) $(A \cap B)'$
44. If $U = \{a, b, c, d, e, f, g\}$, $A = \{a, c, f\}$, and
    $B = \{b, c, d\}$, find
    (a) A' (b) B' (c) $A' \cup B'$ (d) $A' \cap B'$

To determine the sets indicated in Exercises 45 through 53, use $A = \{1, 2, 3\}$, $B = \{1, 2, 3, 6, 9\}$, and $C = \{-3, -1, 0, 2, 3, 6, 7\}$.

45. $A \cap B$ 46. $A \cap C$
47. $A \cup B$ 48. $A \cup C$
49. $A \cap B \cap C$ 50. $A \cup B \cup C$
51. $A \cap \varnothing$ 52. $A \cup (B \cap C)$
53. $(A \cap C) \cup B$

LEVEL 2

List the first three elements of each of the sets in Exercises 54 through 56.

54. $\{x \mid x = 3n, n \text{ a positive integer}\}$
55. $\{x \mid x = n^2 - 3, n \text{ a positive integer greater than 3}\}$
56. $\{x \mid x = n^3, n \text{ a nonnegative integer}\}$

List all the elements of the sets in Exercises 57 and 58.

57. $\{x \mid x \text{ is a positive integer less than 5}\}$
58. $\{x \mid (x + 2)(x - 5)(x - 7) = 0\}$
59. Replace the asterisk in each of the following statements with $=, \subseteq,$ or $\not\subseteq$, whichever is most appropriate.
    (a) $\{2, 4, 6\} * \{1, 2, 3, 4, 5, 6\}$
    (b) $\{2, 4, 8\} * \{1, 3, 5, 9\}$
    (c) $\{2, 5, 17\} * \{17, 5, 2\}$
    (d) $\{8, 10, 12\} * \{8, 10\}$
    (e) $\{2, 4, 6, 8\} * \{x \mid x \text{ is an even integer}\}$
60. Find $A \cap B$ where

    $A = \{x \mid x \text{ is an integer larger than 12}\}$
    $B = \{x \mid x \text{ is an integer less than 21}\}$

61. *(See Example 12)*   Find $A \cap B$ where

    $A = \{x \mid x \text{ is an integer that is a multiple of 5}\}$
    $B = \{x \mid x \text{ is an integer that is a multiple of 7}\}$

62. Find $A \cup B$ where

    $A = \{x \mid x \text{ is an even integer greater than 3}\}$
    $B = \{x \mid x \text{ is an odd integer greater than 10}\}$

63. Find $A \cup B$ where

    $A = \{x \mid x \text{ is a positive integer that is a multiple of 5 and less than 30}\}$
    $B = \{x \mid x \text{ is a positive integer that is a multiple of 6 and less than 35}\}$

Determine which of the pairs of sets in Exercises 64 through 66 are disjoint.

64. $A = \{1, 2, 3, 4, 5, 6\}$, $B = \{2, 4, 6, 8, 10\}$
65. $A = \{3, 6, 9, 12\}$, $B = \{5, 10, 15, 20\}$
66. $A = $ all multiples of 4
    $B = $ all multiples of 3

LEVEL 3

67. Let A be the set of students at Miami Bay University who are taking finite mathematics. Let B be the set of students at Miami Bay University who are taking American history. Describe $A \cap B$.

68. Let A be the set of students in the Collegiate Choir and let B be the set of students in the University Orchestra.
(a) Describe $A \cup B$.   **(b)** Describe $A \cap B$.

69. A is the set of students at Winfield College who had a 4.0 GPA for the Fall, 1997 semester. B is the set of students at Winfield College who had a 4.0 GPA for the Spring, 1998 semester.
(a) Describe $A \cap B$.   **(b)** Is A a subset of B?
(c) Describe $A \cup B$.

EXPLORATIONS

70. Describe a set that is empty.

71. Give an example of two sets A and B for which A is not a subset of B and B is not a subset of A.

72. Give an example of two sets A and B for which A is a subset of B and B is not a subset of A.

73. Give an example of two sets A and B for which A is a subset of B and B is a subset of A.

74. Give an example of two sets that are disjoint.

75. Give an argument that for any two sets A and B, $A \cap B$ is a subset of both A and B. Begin by considering an element of $A \cap B$ and determine how it is related to A.

76. A, B, and C are sets for which A is a subset of B and B is a subset of C. What conclusion can be drawn about the relationship between A and C?

77. If A and B are sets such that $A \cap B = A \cup B$, how are A and B related?

6.2   Counting Elements in a Subset Using a Venn Diagram

- Number of Elements in a Subset
- Venn Diagrams Using Three Sets

Number of Elements in a Subset

We now turn our attention to counting the elements in a set. Sometimes we want to be able to use information about a set and determine the number of elements in it but avoid actually counting the elements one by one.

We use the notation $n(A)$, read **"n of A,"** to indicate the number of elements in set A. If A contains 23 elements, we write $n(A) = 23$.

Suppose you count 10 people in a group that like brand X cola and 15 that like brand Y. We denote this by $n(\text{brand X}) = 10$ and $n(\text{brand Y}) = 15$. How many people are involved in the count? The answer depends on the number who like both brands. In set terminology, the people who like brand X form one set, and those who like brand Y form another set. The totality of all people involved in the count form the union of the two sets, and those who like both brands form the intersection of the two sets. If we attempt to determine the total count by adding the number who like brand X and the number who like brand Y, then we count those who like both X and Y twice. We need to subtract the number who like both from the sum of those who like X and those who like Y to obtain the total number involved. If there are 4 people who like both brands, then the total count is $10 + 15 - 4 = 21$.

In general, the number of elements in the union of two sets is given by the following theorem called the **Inclusion–Exclusion Principle.**

THEOREM
**Inclusion–Exclusion
Principle $n(A \cup B)$**

$$n(A \cup B) = n(A) + n(B) - n(A \cap B)$$

where $n(A)$ represents the number of elements in set A, $n(B)$ represents the number of elements in set B, and $n(A \cap B)$ represents the number of elements of $A \cap B$.

EXAMPLE 1 ➤ $A = \{a, b, c, d, e, f\}, B = \{a, e, i, o, u, w, y\}$. Compute $n(A), n(B), n(A \cap B)$, and $n(A \cup B)$.

SOLUTION

$n(A) = 6$ and $n(B) = 7$. In this case, $A \cap B = \{a, e\}$, so $n(A \cap B) = 2$. $A \cup B = \{a, b, c, d, e, f, i, o, u, w, y\}$, so $n(A \cup B) = 11$. This checks with the formula, $n(A \cup B) = 6 + 7 - 2 = 11$.

■ **Now You Are Ready to Work Exercise 1**

EXAMPLE 2 ➤   Set A is the 9 o'clock English class of 15 students, so A contains 15 elements. Set B is the 11 o'clock history class of 20 students, so B contains 20 elements. $A \cap B$ is the set of students in both classes (there are seven), so $A \cap B$ contains seven elements. The number of elements in $A \cup B$ (a joint meeting of the classes) is

$$n(A \cup B) = n(A) + n(B) - n(A \cap B)$$

so

$$n(A \cup B) = 15 + 20 - 7 = 28$$

The sets A and B divide their union into three regions. The number in each region is shown in the Venn diagram in Figure 6–8.

7 elements in the intersection

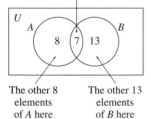

The other 8
elements
of A here

The other 13
elements
of B here

FIGURE 6–8

■ **Now You Are Ready to Work Exercise 3**

EXAMPLE 3 ➤   The union of two sets, $A \cup B$, has 48 elements. Set A contains 27 elements, and set B contains 30 elements. How many elements are in $A \cap B$?

SOLUTION

Using the relationship of Theorem 1, we have

$$48 = 27 + 30 - n(A \cap B)$$
$$48 - 27 - 30 = -n(A \cap B)$$
$$-9 = -n(A \cap B)$$

So, $n(A \cap B) = 9$.

■ **Now You Are Ready to Work Exercise 5**

EXAMPLE 4 ➤   One hundred students were asked whether they were taking psychology (P) or biology (B). The responses showed that

61 were taking psychology; that is, $n(P) = 61$.
18 were taking both; that is, $n(P \cap B) = 18$.
12 were taking neither.

Because $n(P) = 61$,
$61 - 18 = 43$ goes here.

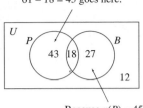

Because $n(B) = 45$,
27 goes here.

FIGURE 6–9

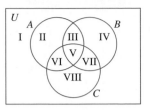

FIGURE 6–10   Three sets may divide the universe into eight regions.

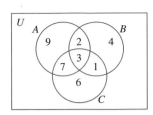

FIGURE 6–11

(a) How many were taking biology? [Find $n(B)$.]
(b) How many were taking psychology but not biology? [Find $n(P \cap B')$.]
(c) How many were not taking biology? [Find $n(B')$.]
(d) Find $n[(B \cap P)']$

SOLUTION

(a) Because 12 were taking neither, the rest, 88, were taking at least one of the courses, so $n(P \cup B) = 88$. We can find $n(B)$ from

$$n(P \cup B) = n(P) + n(B) - n(P \cap B)$$
$$88 = 61 + n(B) - 18$$
$$n(B) = 45$$

(b) Because 18 students were taking both psychology and biology, the remainder of the 61 psychology students were taking only psychology. So, $61 - 18 = 43$ students were taking psychology but not biology.

(c) The students not taking biology were those 12 taking neither and the 43 taking only psychology, a total of 55 not taking biology. (See Figure 6–9.)

(d) Figure 6–9 shows that the number outside $B \cap P$; that is, $n(B \cap P)'$, is $43 + 27 + 12 = 82$.

■ **Now You Are Ready to Work Exercise 19**

Venn Diagrams Using Three Sets

A general Venn diagram of three sets divides a universe into as many as eight nonoverlapping regions. (See Figure 6–10.) We can use information about the number of elements in some of the regions (subsets) to obtain the number of elements in other subsets.

EXAMPLE 5 ➤

The sets A, B, and C intersect as shown in Figure 6–11. The numbers in each region indicate the number of elements in that subset.

The number of elements in other subsets may be obtained from this diagram. For example,

$$n(A) = 9 + 2 + 3 + 7 = 21$$
$$n(B) = 2 + 3 + 1 + 4 = 10$$
$$n(A \cap B) = 2 + 3 = 5$$
$$n(A \cap B \cap C) = 3$$
$$n(A \cup B) = 9 + 2 + 3 + 7 + 4 + 1 = 26 \quad ■$$

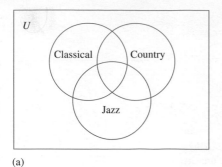
(a)

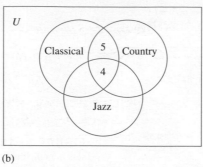
(b)

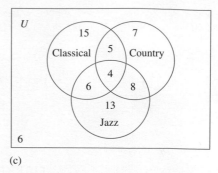
(c)

FIGURE 6–12 (a) A Venn diagram representing students grouped by musical preference. (b) Four students like all three types of music. A total of nine students like both classical and country. (c) The number in each category of musical preference.

EXAMPLE 6 ➤

A survey yields the following information about the musical preferences of students:

> 30 like classical.
>
> 24 like country.
>
> 31 like jazz.
>
> 9 like country and classical.
>
> 12 like country and jazz.
>
> 10 like classical and jazz.
>
> 4 like all three.
>
> 6 like none of the three.

Draw a diagram that shows this breakdown of musical tastes. Determine the total number of students interviewed.

SOLUTION

Begin by drawing a Venn diagram as shown in Figure 6–12a.

The universe is the set of college students interviewed. We want to determine the number of students in each region of the diagram. Because some students may like more than one kind of music, these sets may overlap. We begin where the three sets intersect, and because we know that 4 students like all three types of music, we place a 4 in the region where all three sets intersect. Of the 9 students who like both country and classical, we have already recorded 4 (those who like all three). The other 5 are in the intersection of classical and country that lies outside jazz (Figure 6–12b). In a similar fashion, the number who like both jazz and country breaks down into the 4 who like all three and the 8 who like jazz and country but are outside the region of all three. Because these three regions account for 17 of those who like country, the other 7 who like country are in the region where country does not intersect the jazz and classical. Fill in the rest of the regions; the results are shown in the diagram in Figure 6–12c.

Obtain the total number of students interviewed by adding the number in each region of the Venn diagram. The total is 64.

■ **Now You Are Ready to Work Exercise 23**

6.2 EXERCISES

LEVEL 1

1. *(See Example 1)* $A = \{a, b, c, d, e, f, g, x, y, z\}$, $B = \{c, r, a, z, y\}$. Find
   (a) $n(A)$
   (b) $n(B)$
   (c) $n(A \cap B)$
   (d) $n(A \cup B)$

2. $A = \{1, 3, 5, 7, 9, 11\}$, $B = \{3, 6, 9, 12\}$. Find
   (a) $n(A)$
   (b) $n(B)$
   (c) $n(A \cap B)$
   (d) $n(A \cup B)$

3. *(See Example 2)*   Given $n(A) = 120$, $n(B) = 100$, and $n(A \cap B) = 40$. Find $n(A \cup B)$.

4. Given $n(A) = 26$, $n(B) = 14$, and $n(A \cap B) = 6$, determine $n(A \cup B)$.

5. *(See Example 3)*   Given $n(A) = 15$, $n(B) = 22$, and $n(A \cup B) = 30$, determine $n(A \cap B)$.

6. Given $n(A) = 21$, $n(A \cup B) = 33$, and $n(A \cap B) = 5$, determine $n(B)$.

7. For two sets A and B, $n(A) = 14$, $n(A \cup B) = 28$, and $n(A \cap B) = 5$. Find $n(B)$.

8. If $n(A \cup B) = 249$, $n(A \cap B) = 36$, and $n(B) = 98$, find $n(A)$.

LEVEL 2

9. Two sets are formed using 100 elements. There are 60 elements in one set and 75 in the other. How many elements are in the intersection of the two sets?

10. Use Figure 6–13 to find the following.
    (a) $n(A')$
    (b) $n(B')$
    (c) $n(A' \cup B')$
    (d) $n(A' \cap B')$

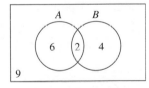

FIGURE 6–13

11. Look at Figure 6–14 to find the following.
    (a) $n(A)$
    (b) $n(B)$
    (c) $n(A \cup B)$
    (d) $n(A')$
    (e) $n(B')$
    (f) $n(A \cap B)'$
    (g) $n(A \cup B)'$
    (h) $n(A' \cap B')$
    (i) $n(A' \cup B')$

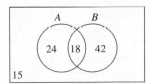

FIGURE 6–14

12. Use the Venn diagram in Figure 6–15. The number of elements in each subset is given. Compute
    (a) $n(A \cap B)$
    (b) $n(A \cap B \cap C)$
    (c) $n(A \cup B)$
    (d) $n(A \cup B \cup C)$
    (e) $n(A')$

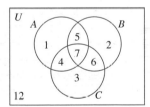

FIGURE 6–15

13. Use the Venn diagram in Figure 6–16. The number of elements in each subset is given. Compute
    (a) $n(A \cup B)$
    (b) $n(A \cup B)'$
    (c) $n(A \cap B)$
    (d) $n(A \cap B)'$
    (e) $n(A' \cup B')$
    (f) $n(B \cap C')$

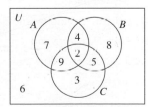

FIGURE 6–16

14. The following information gives the number of elements in some subsets of A, B, and C. Represent this information with a Venn diagram.

$n(A \cap B \cap C) = 2, n(A \cap B) = 6, n(A \cap C) = 5,$
$n(B \cap C) = 3, n(A) = 15, n(B) = 12, n(C) = 15.$

15. The following information gives the number of elements in some subsets of A, B, and C. Represent this information with a Venn diagram.

$n(A \cup B \cup C) = 33, n(A) = 14, n(B) = 20,$
$n(C) = 23, n(A \cap B) = 7, n(A \cap C) = 11,$
$n(A \cap B \cap C) = 6.$

16. A marketing class polled 150 people at a shopping center to determine how many read the *Daily News* and how many read the *Weekly Gazette*. They found the following:

126 read the *Daily News*.

31 read both.

10 read neither.

How many read the *Weekly Gazette?*

17. If a universal set contains 500 elements, $n(A) = 240$, $n(A \cup B) = 460, n(A \cap B) = 55$, find $n(B')$.

18. If a universal set contains 250 elements, $n(A) = 85$, $n(B) = 130$, and $n(A \cap B) = 35$, find $n(A \cup B)'$.

19. *(See Example 4)*   Twenty people belong to the Alpha Club, 30 people belong to the Beta Club, and 6 belong to both clubs.
   (a) How many belong only to the Alpha Club?
   (b) How many belong only to the Beta Club?
   (c) How many belong to just one club?
   (d) How many belong to one or both clubs?

20. Of the 19 fast-food businesses on Valley Mills Drive, the number that have a drive-up window, outside seating, or a pay telephone is summarized as follows:

15 have a drive-up window.

14 have outside seating.

5 have pay telephones.

10 have a drive-up window and outside seating.

4 have outside seating and a pay telephone.

3 have a drive-up window and a pay telephone.

2 have all three.

Find the number of businesses that have

(a) A drive-up window and no outside seating
(b) A drive-up window and outside seating, but no pay telephone
(c) No pay telephone
(d) Only a drive-up window
(e) Outside seating or a pay telephone

21. Of 50 students surveyed, 27 owned a microcomputer, 39 owned a graphic calculator, and 25 owned both.
   (a) How many students did not own a calculator?
   (b) How many students owned a calculator but did not own a microcomputer?
   (c) How many owned neither?
   (d) How many owned one or the other but not both?

22. The SWR Group advertised for applicants for secretarial, clerical, and typist positions. Respondents could apply for one or more of the positions. The responses were as follows:

12 applied for secretary.

10 applied for clerk.

14 applied for typist.

7 applied for secretary and typist.

5 applied for secretary and clerk.

3 applied for clerk and typist.

2 applied for all three.

3 applied for none of the positions.

Draw a Venn diagram and use it to find the following.
(a) What was the total number of respondents?
(b) How many applied for the typist position only?
(c) How many applied for the secretary position but not the clerk position?

23. *(See Example 6)*   A survey of 100 students at New England College showed the following:

48 take English.

49 take history.

38 take language.

17 take English and history.

15 take English and language.

18 take history and language.

7 take all three.

How many students
(a) Take history but neither of the other two?
(b) Take English and history but not language?
(c) Take none of the three?
(d) Take just one of the three?
(e) Take exactly two of the three?
(f) Do not take language?

LEVEL 3

24. Forty students in a music appreciation class could attend a piano recital or a voice recital for extra points. Twenty students attended the piano recital, 23 attended the voice recital, and 6 attended neither. How many attended both?

25. Professor Hickey assigned two homework problems to a class of 35 students. At the next meeting, 17 indicated that they had worked the first problem, 19 had worked the second problem, and 8 had worked both. Determine the number who
(a) worked at least one of the problems.
(b) worked only the first problem.
(c) worked exactly one of the problems.
(d) worked neither of the problems.

26. A home economics class surveyed 1100 students about which meals they ate in their dormitory cafeteria. The class found the following information:

425 ate breakfast.

680 ate lunch.

855 ate dinner.

275 ate breakfast and lunch.

505 ate lunch and dinner.

375 ate breakfast and dinner.

240 ate all three meals.

(a) How many ate only breakfast in the cafeteria?
(b) How many ate at least two meals?
(c) How many ate only one meal?
(d) How many ate exactly two meals?
(e) How many did not eat in the cafeteria?

27. Mr. X teaches freshman English, and Mrs. X teaches freshman history. They hosted a party at their home for members of their classes and their dates. Twenty-eight of the students were in Mr. X's class, 33 were in Mrs. X's class, 7 were in both Mr. and Mrs. X's classes, and 15 other students were in neither class. How many students were at the party?

28. A group of 63 music students were comparing notes on their high school activities. They found that

40 played in the band.

31 sang in a choral group.

9 neither played in the band nor sang in a choral group.

How many played in the band and sang in a choral group?

29. What is inconsistent about the following?

$$n(A) = 25, \quad n(B) = 10, \quad n(A \cup B) = 23$$

30. In the Rivercrest Apartments, the tenants owned answering machines (A), VCRs, or microwave ovens (M), as follows:

Item	Number
A	42
VCR	39
M	48
A and VCR	27
A and M	21
VCR and M	19
All three	12
None	8

(a) How many owned only answering machines?
(b) How many did not own a VCR?
(c) How many owned a VCR or answering machine but not a microwave oven?

31. A marketing class polled 85 faculty members to determine whether they had American Express or Visa cards. The class found that 42 had American Express and 35 had Visa. Of the Visa card holders, 12 also had American Express cards.
(a) How many American Express card holders did not have Visa?
(b) How many had neither?

EXPLORATIONS

32. There are 29 elements in sets A, B, and C. Is the following information consistent?

$$n(A) = 15$$
$$n(B) = 12$$
$$n(C) = 15$$

$$n(A \cap B) = 5$$
$$n(A \cap C) = 6$$
$$n(B \cap C) = 4$$
$$n(A \cap B \cap C) = 2$$

33. Charles Tally gave the following summary of his interviews with 135 students for a sociology project:

65 said they like to go to movies.

77 said they like to go to football games.

61 said they like to go to the theater.

28 said they like to attend movies and football games.

25 said they like to attend movies and the theater.

29 said they like to attend football games and the theater.

8 said they like to attend all three.

4 said they do not like to attend any of these.

The professor refused to accept Charles's paper because the information was inconsistent. Was the professor justified in claiming that the information was inconsistent?

34. A staff member of the news service of Great Lakes University gave out the following information in a news release:

Of a class of 500 students at Great Lakes, 281 are taking English, 196 are taking English and math, 87 are taking math and a foreign language, 143 are taking English and a foreign language, and 36 are taking all three.

The staff member was told that the information must not be released. Why?

35. The Richland Mall advertising manager polled 56 people to obtain their opinion on two holiday displays. Twenty-two people liked the traditional display, and 23 people liked the avant-garde display but did not like the traditional display. The advertising manager reported that 15 people liked neither display. The mall executive thought the information was inconsistent. Determine if the information is consistent or not.

36. The Groesbeck Industrial Development Board attempts to attract industry that will
(a) Hire employees locally.
(b) Buy a significant portion of their supplies and materials locally.
(c) Run an environmentally clean operation.
The chair of the board reported contacts with 32 companies for which

19 would hire locally.

18 would buy materials locally.

16 were environmentally clean.

9 would buy materials locally and were environmentally clean.

10 would hire locally and were environmentally clean.

11 would hire and buy materials locally.

8 would meet all three requirements.

3 would meet none of the three requirements.

A board member requested that the data be reviewed because there appeared to be an error. Determine if that is the case.

37. Given two sets A and B where $n(A) = 5$, $n(B) = 10$, and $n(A \cup B) = 15$:
(a) Find $n(A \cap B)$. **(b)** What is $A \cap B$?

38. Use a Venn diagram to establish that if $A \subseteq B$, then $A \cap B = A$.

39. Use a Venn diagram to simplify $A \cup B$ when $A \subseteq B$.

40. Draw the Venn diagram of $A \cup B$ when $A \cap B = \varnothing$.

For the 3 sets shown in the following figure, $n(A) + n(B) + n(C)$ counts some of the elements more than once. The elements in the regions labeled "1" are counted once, the elements in the regions labeled "2" are counted twice, and the elements in the regions labeled "3" are counted three times. Consequently, the Inclusion–Exclusion Principle for three sets is

$$n(A \cup B \cup C) = n(A) + n(B) + n(C) - n(A \cap B) \\ - n(A \cap C) - n(B \cap C) + n(A \cap B \cap C)$$

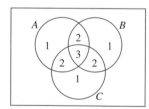

Use the Inclusion–Exclusion Principle for three sets in the following exercises:

41. Find $n(A \cup B \cup C)$ for the sets in Exercise 22.

42. Find $n(A \cup B \cup C)$ for the sets in Exercise 23.

43. Find $n(A \cup B \cup C)$ for the sets in Exercise 26.

6.3   Basic Counting Principles

- Tree Diagrams
- Multiplication Rule
- Addition Rule for Counting Elements in Disjoint Sets

Elizabeth Barrett Browning answered her own question "How do I love thee? Let me count the ways" in a poem that seems to be immortal. Few "how many?" questions are answered poetically, but such questions often need answers.

For example:

- The telephone company, for planning purposes, may ask "How many telephone exchanges will Memphis need in ten years?"
- The state highway department may ask "How many vehicles can I-70 carry safely?"
- A person interested in his or her chances of winning might ask "How many combinations of numbers are possible in a lottery drawing?"
- On most college campuses, someone asks "How many parking places are needed on campus?"
- A professor making out quizzes might ask "How many ways can I select five problems from a review sheet?"

In many "How many?" questions, it may not be possible, or practical, to list all possible ways and count them one by one. If possible, we would like to have a way to arrive at a total count without a tedious one-by-one listing.

Methods do exist; for example, we learned how to count the number of elements in the union of two sets (Section 6.2). We learn about some other methods in this section.

We illustrate one method by first making a list in a systematic manner and drawing conclusions from the list.

EXAMPLE 1 ➤

A teenager asked how many different outfits she could form if she had two skirts and three blouses. Let's list the different ways.

1. First skirt with first blouse
2. First skirt with second blouse
3. First skirt with third blouse
4. Second skirt with first blouse
5. Second skirt with second blouse
6. Second skirt with third blouse

Notice the pattern of the list. For each skirt, she can obtain three outfits by selecting each of the different blouses, so the total number of outfits is simply three times the number of skirts. ■

If you are to select one book from a list of five books and a second book from a list of seven books, you can determine the number of possible selections of the two books by writing the first book from list one with each of the seven books from list two. You will list the seven books five times, once with each of the books from list one, so you will have a list that is $5 \times 7 = 35$ long.

We can list all possible selections in another way, using what are called **tree diagrams.**

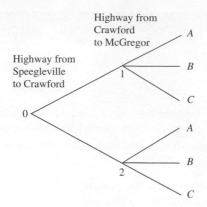

FIGURE 6-17

Tree Diagrams

We illustrate the construction and use of tree diagrams in the next two examples.

EXAMPLE 2 ➤

Let's look at another problem. Suppose there are two highways from Speegleville to Crawford and three highways from Crawford to McGregor. How many different routes can we choose to go from Speegleville to McGregor through Crawford? A tree diagram provides a visual means to list all possible routes (Figure 6–17).

Reading from left to right, starting at 0, draw two branches representing the two highways from Speegleville to Crawford (use 1 and 2 to designate the highways). At the end of each of these two branches, draw three branches representing the three highways from Crawford to McGregor. (Use *A*, *B*, and *C* to designate the highways.) A choice of a first-level branch and a second-level branch determines a route from Speegleville to McGregor. Notice that the total number of possible routes (branches that end at McGregor) is six because each first-level branch is followed by three second-level branches. ■

EXAMPLE 3 ➤

Cox's Department Store has two positions to fill, those of a department manager and an assistant manager. Three people are eligible for the manager position, and four people are eligible for the assistant manager position. Use a tree diagram to show the different ways in which the two positions can be filled.

SOLUTION

Label the candidates for department manager as *A*, *B*, and *C*. Label the candidates for assistant manager as *D*, *E*, *F*, and *G*.

The tree diagram in Figure 6–18 illustrates the 12 possible ways in which the positions can be filled. Reading from left to right, starting at 0, you find three possible "branches" (managers). A branch for each possible assistant manager is attached to the end of each branch representing a manager. In all, 12 paths begin at 0 and go to the end of a branch. For example, the path 0*BD* represents the selection of *B* as the manager and *D* as the assistant manager.

■ **Now You Are Ready to Work Exercise 1**

A tree diagram shows all possible ways to make a sequence of selections, and it shows the number of different ways the selections can be made.

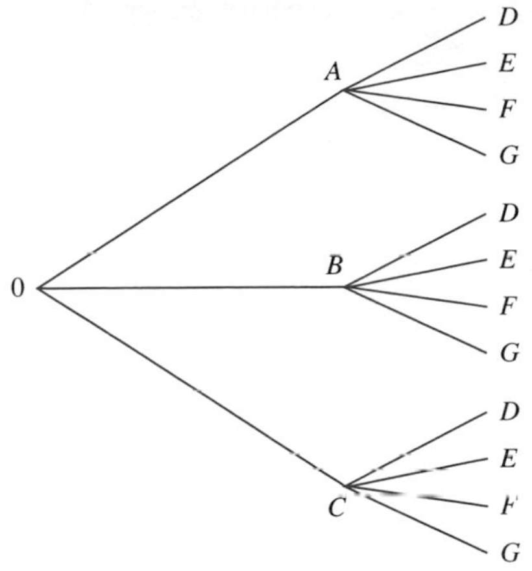

FIGURE 6-18

The total number of selections is the number of end points of branches and is the product of the number of first-level branches and the number of second-level branches; that is, the number of ways in which the first selection can be made times the number of ways in which the second selection can be made.

These three examples are fundamentally the same kind of problem. Let's make a general statement that includes each one.

Multiplication Rule

Suppose that there are two activities, A_1 and A_2. Each activity can be carried out in several ways. We want to determine in how many different ways the first activity followed by the second activity can be performed.

In the three preceding examples, A_1 and A_2 are the following:

Selections of an outfit: A_1 is the activity of selecting a skirt, and A_2 is the activity of selecting a blouse.

Routes from Speegleville to McGregor: A_1 is the selection of a highway from Speegleville to Crawford, and A_2 is the selection of a highway from Crawford to McGregor.

Filling two positions: A_1 is the selection of a manager, and A_2 is the selection of an assistant manager.

In many cases, we do not need a list of all possible selections, but we need their number. In such a case, the problem reduces to a question of the number of ways in which we can carry out the activity A_1 followed by the activity A_2. The solution is simple: Multiply the number of ways in which activity A_1 can be performed by the number of ways in which activity A_2 can be performed. This is often called the **Multiplication Rule.**

THEOREM
Multiplication Rule

Two activities A_1 and A_2 can be performed in n_1 and n_2 different ways, respectively. The total number of ways in which A_1 followed by A_2 can be performed is

$$n_1 \times n_2$$

Now let's apply the Multiplication Rule to some examples.

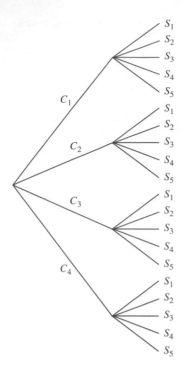

FIGURE 6–19

EXAMPLE 4 ➤

A taxpayers' association is to elect a chairman and a secretary. There are four candidates for chairman and five candidates for secretary. In how many different ways can a slate of officers be elected?

SOLUTION

This problem can be analyzed using a tree diagram. (See Figure 6–19.) The first activity, selecting a chairman, is represented by the four first-stage branches, C_1, C_2, C_3, and C_4. To each of these branches we attach the five possible selections of the second activity, selecting a secretary. This results in $4 \times 5 = 20$ branches representing the different slates possible.

■ Now You Are Ready to Work Exercise 5

The analysis of a problem using a tree diagram can quickly become unwieldy. You quickly see that you *don't* want to use a tree diagram to analyze the next example, even though it is theoretically possible to do so.

NOTE

Two slates of officers differ if one or more officers is replaced by another person or if two or more officers exchange positions.

EXAMPLE 5 ➤

Moody Library wishes to display two rare books, one from the history collection and one from the literature collection. If the library has 50 history and 125 literature books to select from, how many different ways can the books be selected for display?

SOLUTION

A tree diagram of this problem would have 50 branches at the first stage representing the 50 history books. Attached to the end of each of the history branches are 125 branches representing the literature books. The number of branches finally is the number of choices of the first-stage branch times the number of choices of the second-stage branch. In this case, there are $50 \times 125 = 6250$ possible displays.

It is not practical to draw a tree with 6250 branches, so use the Multiplication Rule to analyze the problem. The first activity is a selection of a history book, and the second activity is the selection of a literature book. This sequence can be performed in $50 \times 125 = 6250$ ways.

■ **Now You Are Ready to Work Exercise 8**

EXAMPLE 6 ➤

Jane selects one card from a deck of 52 different cards. The first card is *not* replaced before Joe selects the second one. In how many different ways can they select the two cards?

SOLUTION

Jane selects from a set of 52 cards, so $n_1 = 52$. Joe selects from the remaining cards, so $n_2 = 51$. Two cards can be drawn in $52 \times 51 = 2652$ ways.

■ **Now You Are Ready to Work Exercise 15**

Some problems ask for the number of ways a sequence of more than two activities can be performed. The Multiplication Rule can be applied to sequences of more than two activities.

Corollary

Activities $A_1, A_2, \ldots, A_k$ can be performed in $n_1, n_2, \ldots, n_k$ different ways, respectively. The number of ways in which one can perform A_1 followed by $A_2 \ldots$ followed by A_k is

$$n_1 \times n_2 \times \cdots \times n_k$$

EXAMPLE 7 ➤

A quiz consists of four multiple-choice questions with five possible responses to each question. In how many different ways can the quiz be answered?

SOLUTION

In this case, there are four activities — that is, answering each of four questions. Each activity (answering a question) can be performed (choosing a response) in five different ways. The answers can be given in

$$5 \times 5 \times 5 \times 5 = 625$$

different ways.

■ **Now You Are Ready to Work Exercise 21**

EXAMPLE 8 ➤

Three couples attend a movie and are seated in a row of six seats. How many different seating arrangements are possible if couples are seated together?

SOLUTION

Think of this as a sequence of six activities, that of assigning a person to sit in each of seats 1 through 6. The Multiplication Rule states that the number of ways in which this can be done is the product of the numbers of ways in which each selection can be made.

First seat: Any one of the six people may be chosen, so there are six choices.

Second seat: Only the partner of the person in the first seat may be chosen, so there is one choice.

Third seat: One couple is seated, so any one of the remaining four people may be chosen. Thus there are four choices.

Fourth seat: There is one choice only because it must be the partner of the person in the third seat.

Fifth seat: There are two choices because either of the two remaining people may be seated.

Sixth seat: The one remaining person is the only choice.

By the multiplication rule, the number of possible arrangements is the product of the number of choices for each seat:

$$6 \times 1 \times 4 \times 1 \times 2 \times 1 = 48$$

■ **Now You Are Ready to Work Exercise 41**

EXAMPLE 9 ➤   The Cameron Art Gallery has several paintings by each of five artists. A wall has space to hang four paintings in a row. How many different arrangements by artists are possible if

(a) The paintings are by different artists?
(b) More than one painting by an artist may be displayed, but they may not be hung next to each other?

SOLUTION

These problems may be viewed as a sequence of four activities, choosing an artist for each of the four spaces. The total number of ways is the product of the number of choices for each space.

(a) Because all artists must be different, there are five choices for the artist in the first space, four for the second, three for the third, and two for the fourth. The number of arrangements is $5 \times 4 \times 3 \times 2 = 120$.
(b) The first painting to be hung can be selected from any one of the five artists. A painting by that artist may not be used in the second space, but a painting by any of the remaining four artists may be used. The artist chosen for the second space may not be chosen for the third space, but the artist chosen for the first space may be used. (The first and third paintings are not hung next to each other.) Thus, there are four choices, for the third space. Likewise, there are four choices for the fourth space. The total number of arrangements is

$$5 \times 4 \times 4 \times 4 = 320$$

■ **Now You Are Ready to Work Exercise 45**

Addition Rule for Counting Elements in Disjoint Sets

In Section 6.2, we have the Inclusion–Exclusion Principle,

$$n(A \cup B) = n(A) + n(B) - n(A \cap B)$$

which gives the number of elements in the union of two sets. When the sets are disjoint, $n(A \cap B) = 0$, so the rule reduces to

$$n(A \cup B) = n(A) + n(B)$$

We can apply this counting principle to a situation in which you wish to determine the number of ways to perform one activity or another, but not both. For example, there are six movies and four talk shows on cable TV at the same time. You plan to watch either a movie or a talk show. How many choices do you have? Think of listing your choices. List the six movies and then the talk shows (or vice versa) and you have a list of $6 + 4 = 10$ choices.

This illustrates a basic principle, the **Addition Rule.**

Addition Rule

> If an activity A_1 can be performed in n_1 ways and activity A_2 can be performed in n_2 ways, then either A_1 or A_2, but not both simultaneously, can be performed in $n_1 + n_2$ ways.

EXAMPLE 10

Rhonda plans to buy a Saturn car. The Temple dealer has 15 Saturns in stock, and the Austin dealer has 23 in stock. How many choices does Rhonda have if she buys a Saturn from one of these dealers?

NOTE

We emphasize that the Addition Rule applies only to cases where there is no overlap of the sets A and B. When there is ovelap, the Inclusion–Exclusion Principle applies.

SOLUTION

Because Rhonda will buy from one dealer or the other (and not both), the Addition Rule applies. She has $15 + 23 = 38$ choices.

■ **Now You Are Ready to Work Exercise 19**

The next example illustrates a problem where both the Addition Rule and the Multiplication Rule apply.

EXAMPLE 11

In how many different ways can three science and two history books be arranged on a shelf if books on each subject are kept together?

SOLUTION

First, observe that the subjects can be arranged in two ways, science on the left and history on the right or vice versa.

Next, count the number of ways in which the books can be arranged within the subject arrangements:

> *Science-history:* There are five positions to be filled, science first and history second. The number of possible arrangements is $3 \times 2 \times 1 \times 2 \times 1 = 12$.

> *History-science:* Again there are five positions to be filled, but history books are placed first and science books second. The number of possible arrangements is $2 \times 1 \times 3 \times 2 \times 1 = 12$.

Because the arrangements can be either science-history or history-science, the Addition Rule states there are $12 + 12 = 24$ arrangements.

■ **Now You Are Ready to Work Exercise 47**

We point out that this can also be solved using the Multiplication Rule exclusively with six arrangements of the science books within the science group, two arrangements of the history books within the history group, and two arrangements of the subject groups, giving $6 \times 2 \times 2 = 24$ arrangements.

Some of the exercises of this section, and the sections that follow, refer to a bridge deck or a deck of 52 cards. The following figure shows the composition of such a deck of cards.

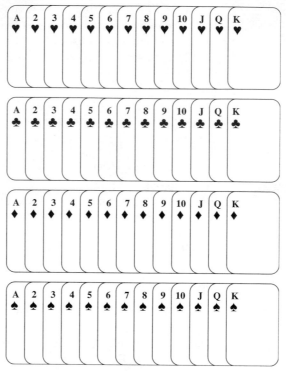

A Standard Deck or Bridge Deck has 52 cards with 13 cards in each suit, Hearts (♥), Clubs (♣), Diamonds (♦), and Spades (♠). Each suit contains an ace (A), 2 through 10, Jack (J), Queen (Q), and King (K). The hearts and diamonds are red cards. The clubs and spades are black cards. The Jack, Queen, and King are called face cards.

6.3 EXERCISES

LEVEL 1

1. *(See Example 3)* Andrew has a red tie and a green tie. He has a white shirt, a blue shirt, and a yellow shirt. Draw a tree diagram that shows all possible ways in which a tie and shirt can be selected.

2. Draw a tree diagram to show the different ways in which a boy and then a girl can be selected from the following set of children:

{Anita, Bobby, Carl, Debbie, Enrique, Flo}

3. Draw a tree diagram showing all sequences of heads and tails in two tosses of a coin.

4. For her breakfast, Susan always selects one item from each of the following:

   Orange juice or tomato juice

   Cereal or eggs

   Toast or muffins

   Draw a tree diagram to show all her possible menus.

5. *(See Example 4)*   Each day, Mr. Ling has a boy and a girl summarize the assignment for the day. There are 3 boys and 4 girls in his group. In how many different ways can the teacher select the two students?

6. Heidi has 6 necklaces and 8 pairs of earrings. In how many different ways can she select a necklace and a pair of earrings to wear?

7. A grocery store has 5 brands of crackers and 9 different varieties of cheese. How many ways can Josh select one brand of cheese and one brand of crackers?

8. *(See Example 5)*   A reading list for American history contains two groups of books. There are 16 books in the first group and 21 in the second. A student is to read one book from each group. In how many ways can the choice of books be made?

9. Sam has 6 suits and 7 shirts. How many different outfits can he form?

10. A house has 4 doors and 18 windows. In how many ways can a burglar pass through the house by entering through a window and leaving through a door?

11. The Hickory Stick has a selection of 4 meats and 7 vegetables. How many different selections of one meat and one vegetable are possible?

12. A traveling salesperson may take one of five different routes from Brent to Centreville and three different routes from Centreville to Moundville. How many different routes are possible from Brent to Moundville through Centreville?

13. **(a)** In how many different ways can a player select a diamond and a club from a deck of 52 bridge cards?

    (b) In how many different ways can a player select one card of each suit from a bridge deck?

14. A manufacturer of disks for microcomputers packs them 10 to a box. If a box contains 4 defective disks, in how many different ways can one defective and one good disk be selected from the box?

15. *(See Example 6)*   The Bookstore has a sale table of 22 different books. How many ways can Benni select two of the books?

16. Prof. Okamoto selects three students to administer a student evaluation of the professor. If there are 28 students in the class, in how many ways can Professor Okamoto make the selection?

17. A car manufacturer provides 6 exterior colors, 5 interior colors, and 3 different trims. How many different color–trim schemes are available?

18. Alfred is opening a new office and needs to decorate and furnish it. He has 3 sources for wallpaper, 5 sources for carpet, 4 sources for drapes, 8 sources for furniture, and 2 sources for pictures. How many ways can he select one of each?

19. *(See Example 10)*   Reshma stops at a newsstand to buy either a newspaper or news magazine to read on a business trip. The newsstand has 13 news magazines and 9 newspapers. How many choices does Reshma have?

20. Kirk's parents gave him a blazer for his birthday. He can choose one from Shellenbarger's or from Salesman's Sample Outlet. Shellenbarger's has 11 blazers of his size, and Salesman's Sample Outlet has 15. How many choices does Kirk have?

LEVEL 2

21. *(See Example 7)*   A quiz consists of 6 multiple-choice questions with 4 possible responses to each one. In how many different ways can the quiz be answered?

22. A coin is flipped three times. How many different sequences of heads and tails are possible?

23. A telephone number consists of 7 digits. How many phone numbers are possible if the first digit is neither 1 nor 0?

24. How many different radio call letters beginning with K and consisting of four letters can be assigned to radio stations?

25. How many four-letter passwords can be formed from the letters {A, B, J, K, X, Z}
    (a) If a letter can be repeated?
    (b) If a letter cannot be repeated?

26. How many different three-digit numbers can be formed from the digits {0, 1, 2, 3, 4, 5}
   (a) If a digit cannot be repeated?
   (b) If a digit can be repeated?

27. The identification code on a campus parking permit consists of a letter of the alphabet followed by 3 digits. How many different identification codes will this provide?

28. Will an identification code of one letter followed by 2 digits provide for 3000 parking permits?

29. A college basketball conference has 8 teams. Each team plays each of the other 7 teams twice, once at home and once away. How many conference games are played?

30. Dan's Bagel Shop has 6 choices of meat and 11 choices of bagels for a bagel sandwich. How many bagel–meat sandwich choices are possible?

31. Anna goes to the Frame Shop to get a frame and mat for a picture she made in art class. The clerk suggests she choose from a selection of 6 frames and 3 mats. Anna wants to look at all possible pairings of a mat and a frame to help her decide which one to use. How many ways can she pair a mat and a frame?

32. Draw a tree diagram showing all sequences of heads and tails that are possible in three tosses of a coin.

33. Use a tree diagram to show the different ways in which first, second, and third prizes can be awarded to three different contestants, Jones, Allen, and Cooper.

LEVEL 3

34. Professor Mueller gives a sociology exam consisting of 3 questions, one from each of 3 topics. Because of crowded conditions in the classroom, she likes to give different exams to students. She has a collection of questions, 34 on the first topic, 27 on the second topic, and 16 on the third topic. How many ways can she select 3 questions (i.e., How many exams can be formed)?

35. A portion of a computerized language test requires that a student conjugate a regular verb and an irregular verb. The regular verb is selected from a set of 32 verbs, and the irregular verb is selected from a set of 18 verbs. How many ways can a regular and an irregular verb be selected?

36. The Lahr family is redecorating their home, and the interior designer recommends they consider 4 paint colors for the dining room walls and 3 paint colors for the trim. Leslie Lahr asks that a painter make samples of a wall and a trim color to help her decide. If the painter makes a sample for each wall and trim combination, how many samples will be made?

37. Jessica and some friends are in Baltimore for a mathematics conference. They have an afternoon to see the sights. They agree to visit a museum and an art exhibit. Here is a list of their choices.

> Museums: Babe Ruth Museum, Cryptologic Museum, Edgar Allen Poe House and Museum, Ft. Henry National Monument, and the *USS Constellation* Museum.

> Art Exhibits: Baltimore Museum of Art, Contemporary Museum of Art, Maryland Art Place, and Walters Art Museum.

How many ways can they select one museum and one art exhibit?

38. A three-digit number is formed using digits from {1, 2, 3, 4, 5}.
   (a) How many numbers can be formed using three different digits?
   (b) How many numbers can be formed if digits may be repeated?
   (c) How many even numbers can be formed using three different digits?
   (d) How many even numbers can be formed if digits may be repeated?
   (e) How many multiples of 5 can be formed if digits may be repeated?

39. A child forms three-letter "words" using three different letters from HISTORY. A three-letter "word" is any arrangement of three letters, whether it is in the dictionary or not.
   (a) How many three-letter "words" are possible?
   (b) How many three-letter "words" beginning with H are possible?
   (c) How many three-letter "words" beginning with a vowel are possible?
   (d) How many three-letter "words" with a vowel for the middle letter are possible?
   (e) How many three-letter "words" with exactly one vowel are possible?

40. A child forms three-letter "words" using letters from INVOKED. A letter may be repeated within a "word."
   (a) How many different three-letter "words" are possible?
   (b) How many different three-letter "words" beginning with K are possible?
   (c) How many different three-letter "words" beginning with a vowel are possible?
   (d) How many different three-letter "words" with a vowel for a middle letter are possible?

(e) How many different three-letter "words" containing exactly one vowel are possible?

41. *(See Example 8)*   Four couples attend a play and are seated in a row of 8 chairs. How many different arrangements are possible if couples are seated together?

42. In how many different arrangements can 3 men and 3 women be seated in a row if no one sits next to a member of the same gender?

43. In how many different arrangements can 4 men and 3 women be seated in a row if no one sits next to a member of the same gender?

44. In how many different arrangements can 5 people be seated in a row if 2 certain people must be seated together?

45. *(See Example 9)*   An art gallery has several paintings by each of 6 artists. A wall has space to hang 4 paintings in a row. How many different arrangements by artists are possible if:
(a) The paintings are by different artists?
(b) More than one painting by an artist may be displayed but they may not be hung next to each other?

46. Jones starts a chain letter by sending letters to 10 people the first week. The second week, each person who received a letter sends 10 letters to other people. This continues each week. Assume that no one receives two letters.
(a) How many people receive letters the second week?
(b) How many people receive letters the third week?
(c) How many people receive letters the fourth week?

47. *(See Example 11)*   In how many different ways can 5 art and 3 music books be arranged on a shelf if books on the same subject are kept together?

48. A portion of a test contains 6 multiple-choice questions. For each question, a student may leave the answer blank or select 1 of the 4 answers given. How many different ways can the student respond to the questions?

49. A customer who orders a hamburger at the Burger Place has the choice of "with or without" for each of the following: lettuce, tomato, onion, catsup, mustard, mayonnaise, and cheese. In how many different ways can a hamburger be ordered?

50. In a group of 750 people, at least 2 have the same first and last initials. Explain why this is true.

51. The Panthers and Lions softball teams play a championship series. The first team to win two games (best two out of three) is the winner. Use a tree diagram to show the possible outcomes of the series.

52. A Boy Scout troop hikes from the mall to their camping area at Hilltop Farm. They have 3 possible routes from the mall to the farm: the highway access road, a gravel road, or through a subdivision. At the farm they can continue to the camping area by way of a trail through the woods, a lane across the pasture, or along a stream. Show all possible routes from the mall to their camping area with a tree diagram.

EXPLORATIONS

53. An art professor wants to index a collection of 22,500 slides with three initials. Is it possible to assign different initials to each slide? If not, how many initials are necessary?

54. The telephone area codes were established so we could direct dial any phone in the United States and Canada. The area codes were set up so that

1.   An area code consists of three digits.
2.   The first digit cannot be 0 or 1 (These digits would give an operator or long distance).
3.   The second digit could only be a 0 or 1.

With the rapid increase in phones, caused largely by the popularity of cellular phones, additional area codes were needed. (There were about 140 area codes in use in 1995.) Explain why the phone companies now use area codes with middle digits other than 0 or 1.

55. Currently a Social Security number is composed of nine digits. Any of the digits 0–9 may be used in any position.
(a) Find the maximum number of Social Security numbers possible under the system.
(b) John Morton proposes that life would be simpler if we used a Social Security number composed of six alphabet characters. Would this scheme provide as many possible social security numbers as the current system?

(c) Andre Lenz agreed with the six-character idea, but he thought we should use alphanumeric characters so there would be enough numbers to last for a long time. (An alphanumeric character can be either a letter from the alphabet or a numeric digit.) Compare the possible number of Social Security numbers if all 26 letters of the alphabet and all ten digits are used.

Compare the maximum number of Social Security numbers possible if all 26 letters of the alphabet and the digits 2–9 are used. (0 and 1 are not used because they might be confused with O and I).

(d) The Social Security Administration decides it needs at least 10 billion different numbers. Give two methods whereby this can be achieved.

56. Baylor University purchased a financial systems software package and converted all budget identification numbers to a 15-character identification code. Each character in the identification code can be any one of the 26 letters of the alphabet or any one of the digits 0 through 9.

(a) How many different budget numbers are possible in this system?

(b) If the Budget Office assigns 1 million budget numbers each second, 24 hours a day for every day in the year, how long will it take to use up all the budget numbers?

57. A computer password is an arrangement of six, seven, or eight letters of the alphabet, with repetitions allowed. How many different passwords are possible?

58. Two students sat next to each other on a multiple-choice test. Each question had four choices for an answer. Both students missed the same 14 questions, and the two students had exactly the same wrong answer on the 14 questions.

The instructor accused the students of cheating. Was the accusation justified? Give reasons for your answer.

6.4   Permutations

- Permutations
- Notations for Number of Permutations
- Permutation of Objects with Some Alike (Optional)

Permutations

The Multiplication Rule counts the number of ways a sequence of activities can be performed. A common application counts the ways a selection can be made from two or more sets such as the number of ways a person can select a salad, an entree, and a dessert from a menu. The Multiplication Rule can also apply to counting the number of ways a sequence of objects can be selected from *one* set. We next focus on such an application that occurs frequently enough to be given a name, a *permutation*.

The next example provides an illustration from which we will form a general conclusion.

EXAMPLE 1 ▶   The Art Department received a gift of seven paintings from an alumnus artist. An art major is to select three of the paintings and arrange them in three given locations in the foyer of the art building. How many different arrangements are possible?

SOLUTION

This problem can be viewed as a Multiplication Rule problem with three activities.

Activity 1: Select a painting from the set of seven for the first location. This can be done in seven ways because any of the seven paintings can be selected.

Activity 2: Select one of the remaining paintings for the second location. This can be done in six ways because any of the six remaining paintings can be selected.

Activity 3: This can be done in five ways because any of the five remaining paintings can be selected.

Therefore, there are $7 \times 6 \times 5 = 210$ possible arrangements.

■ **Now You Are Ready to Work Exercise 17**

Notice that this is not the same problem as selecting three paintings to send to a gallery for exhibition when just the collection, not the order of arrangement, form the activities. This example has properties that we identify with a permutation.

A permutation problem can be viewed in two ways:

1. Count the number of sequences in which elements can be selected from a set, removing an element at each step and making the next selection from the elements that remain. This fits the Multiplication Rule directly, with the number of choices decreasing by one at each step.
2. Select all subsets of elements with the designated number of elements from a set and count the number of ways the elements can be arranged.

Many problems are presented from the second viewpoint.

In either case, the number of sequences or arrangements of, say, four selections from a set of *n* elements is

$$n(n - 1)(n - 2)(n - 3) \text{ ways}$$

DEFINITION

> A subset of distinct elements selected from a given set and arranged in a specific order is called a **permutation.**

Because a problem may not be identified as a permutation problem, you need to be able to recognize it as such. Here are three keys that can help in recognizing a permutation.

Keys to Recognizing a Permutation

> 1. A permutation is an arrangement or sequence of selections of elements from a single set.
> 2. Repetitions are not allowed, which means that once an element is selected it is not available for a subsequent selection. The same element may not appear more than once in a particular arrangement.
> 3. The order in which the elements are selected or arranged is significant.

Students sometimes ask how to tell when order is significant. A different order occurs when two items exchange positions or the order of two selections is exchanged. If such an exchange makes a difference in the selection, then order is significant. If the exchange makes no difference, then order is not significant.

EXAMPLE 2 ➤

Ten students each submit one essay for competition. In how many ways can first, second, and third prizes be awarded?

SOLUTION

This is a permutation problem because

1. Each essay selected is from the same set.
2. No essay can be submitted more than once; that is, an essay cannot be awarded two prizes (no repetition).
3. The order (prize given) of the essays is important.

Any of the ten essays may be chosen for first prize. Then any of the remaining nine may be chosen for the second prize, and any of the other eight may be chosen for third prize. According to the Multiplication Rule, the three prizes may be awarded in

$$10 \times 9 \times 8 = 720 \text{ different ways}$$

■ **Now You Are Ready to Work Exercise 21**

EXAMPLE 3 ➤ In how many different ways can a penny, a nickel, a dime, and a quarter be given to four children if one coin is given to each child?

SOLUTION

Each child may be considered a "position" that receives a coin. The number of ways a coin may be given to each child is

First child: four possibilities of a coin

Second child: three possibilities of a coin

Third child: two possibilities of a coin

Fourth child: one possibility of a coin

Therefore, the coins may be distributed in

$$4 \times 3 \times 2 \times 1 = 24 \text{ different ways}$$

■ **Now You Are Ready to Work Exercise 25**

EXAMPLE 4 ➤ At the Cumberland River Festival, four young women (called "belles" in the South) are stationed at historic Fort House; one stands at the entrance, one in the living room, one in the dining room, and one on the back veranda. If there are ten belles, in how many different ways can four be selected for the stations?

SOLUTION

This is a permutation, because a woman can be selected for, at most, one station, and the order (place stationed) is significant. The permutation can be made in

$$10 \times 9 \times 8 \times 7 = 5040 \text{ different ways}$$

■ **Now You Are Ready to Work Exercise 29**

If the problem in Example 4 had been to select four belles to be present in the living room with no particular station for each one, then Example 4 would not have been a permutation problem, because the belles would not be arranged in any particular order. (The number of selections in this case uses a technique that we discuss later.)

Notations for Number of Permutations

The notation commonly used to represent the number of permutations for a set is written $P(8, 3)$, which is read "permutation of eight things taken three at a time." This notation represents the number of permutations of three elements from a set of eight elements. $P(10, 4)$ represents the number of permutations of four elements selected from a set of ten elements. [$P(7, 3)$ is the answer to Example 1, and $P(10, 4)$ is the answer to Example 4.]

We want you to understand the pattern for calculating numbers like $P(10, 4)$ so that you can do it routinely. Let's look at some examples.

$$P(10, 4) = 10 \times 9 \times 8 \times 7 \text{ permutations of four elements}$$
taken from a set of ten elements.

$$P(5, 3) = 5 \times 4 \times 3$$

$$P(7, 2) = 7 \times 6 \text{ permutations of two elements}$$
selected from a set of seven elements.

$$P(21, 3) = 21 \times 20 \times 19$$

In each case, the calculation begins with the first number in the parentheses, 21 in $P(21, 3)$ and 7 in $P(7, 2)$. The second number in $P(21, 3)$, $P(7, 2)$, and so on determines the number of terms in the product. Because the terms decrease by one to the next term, you need to know only the first term and how many are needed to calculate the answer.

This reasoning lets us know that $P(30, 5)$ is a product of five terms beginning with 30, each term thereafter decreasing by one, so

$$P(30, 5) = 30 \times 29 \times 28 \times 27 \times 26$$
$$P(105, 4) = 105 \times 104 \times 103 \times 102$$

and

$$P(4, 4) = 4 \times 3 \times 2 \times 1$$

Can you give the last term in $P(52, 14)$ without writing out all the terms? The preceding examples give a pattern that helps. To give a general description of the last term in the calculation of $P(n, r)$, we write the last terms of the examples in a way that seems unnecessarily complicated. However, it will help us obtain the general form.

Example	Last term	May be written as
$P(5, 3) = 5 \times 4 \times 3$	$3 = 5 - 2$	$5 - 3 + 1$
$P(10, 4) = 10 \times 9 \times 8 \times 7$	$7 = 10 - 3$	$10 - 4 + 1$
$P(105, 4) = 105 \times 104 \times 103 \times 102$	$102 = 105 - 3$	$105 - 4 + 1$

so we expect the last term of

$$P(52, 14) \quad \text{to be} \quad 52 - 13 = 39 \quad \text{(Also written } 52 - 14 + 1\text{)}$$

In general, $P(n, k)$ indicates the number of arrangements that can be formed by selecting k elements from a set of n elements. Following the observed pattern, it may be written

$$P(n, k) = n(n - 1)(n - 2)\cdots(n - k + 1)$$

Number of
Permutations, $P(n, k)$

$$P(n, k) = n(n-1)(n-2) \cdots (n - k + 1)$$

There is a special notation for the case when a permutation uses all elements of a set. Notice that $P(4, 4)$ is just the product of the integers 4 through 1, that is, $4 \times 3 \times 2 \times 1$. In general, $P(n, n)$ is the product of the integers n through 1. The following notation is used.

DEFINITION
$n!$

The product of the integers n through 1 is denoted by $n!$ (called **n factorial**).

$$1! = 1$$
$$n! = n(n-1)(n-2) \times \cdots \times 2 \times 1 \qquad \text{(for } n > 1\text{)}$$
$$0! = 1$$

Notice that $n!$ is not defined for negative values of n.

EXAMPLE 5 ➤

$$7! = 7 \times 6 \times 5 \times 4 \times 3 \times 2 \times 1 = 5040$$
$$2! = 2 \times 1 = 2$$
$$6! = 6 \times 5 \times 4 \times 3 \times 2 \times 1 = 720$$

Notice that $6! = 6 \times 5!$ and $4! = 4 \times 3!$, and so on.

$$1! = 1$$
$$0! \text{ is defined to be } 1$$

Arithmetic involving factorials can be carried out easily if you are careful to use the factorial as defined.

$$\frac{5!}{3!} = \frac{5 \times 4 \times \cancel{3} \times \cancel{2} \times \cancel{1}}{\cancel{3} \times \cancel{2} \times \cancel{1}}$$
$$= 5 \times 4$$
$$= 20$$
$$3! \, 4! = 3 \times 2 \times 1 \times 4 \times 3 \times 2 \times 1 = 144$$

■ **Now You Are Ready to Work Exercise 1**

EXAMPLE 6 ➤

How many different ways can six people be seated in a row of six chairs?

SOLUTION

This is a permutation because

1. The people are all selected from the same set.
2. Repetitions are not allowed, a person may not occupy two different seats at the same time.
3. Order is significant, because a different seating arrangement occurs when two people exchange seats.

Because six positions are to be filled from a set of six people, the number of arrangements is

$$P(6, 6) = 6! = 6 \times 5 \times 4 \times 3 \times 2 \times 1 = 720$$

■ **Now You Are Ready to Work Exercise 33**

Factorials allow us to write the expression for the number of permutations in another form that is sometimes useful. For example,

$$P(8, 3) = 8 \times 7 \times 6$$
$$= \frac{8 \times 7 \times 6 \times 5!}{5!}$$

Because $8 \times 7 \times 6 \times 5! = 8 \times 7 \times 6 \times 5 \times 4 \times 3 \times 2 \times 1 = 8!$, we can write

$$P(8, 3) = \frac{8!}{5!}$$

Be sure you understand that

$$P(6, 4) = \frac{6!}{2!} \qquad (2! \text{ came from } (6 - 4)!)$$

In general, we can write

$$P(n, k) = \frac{n!}{(n - k)!}$$

EXAMPLE 7 ➤

Many auto license plates have three letters followed by three digits. How many different license plates are possible if

(a) Letters and digits are not repeated on a license plate?
(b) Repetitions of letters and digits are allowed?

SOLUTION

(a) First of all, this may be viewed as a Multiplication Rule problem with two activities. The first activity is the selection and arrangement of letters; the second activity is the selection and arrangement of digits. The number of license plates is found by multiplying the number of selections of letters and the number of selections of digits. The selection of letters is a permutation, $P(26, 3)$ in number, and the selection of digits is a permutation, $P(10, 3)$ in number. The number of license plates is then $P(26, 3) \times P(10, 3) = 15,600 \times 720 = 11,232,000$.

(b) This is an ordered arrangement that is not a permutation, because a letter or digit may appear more than once on a license plate. This is a Multiplication Rule problem with six activities, the selection of three letters followed by the selection of three digits. This can be done in $26 \times 26 \times 26 \times 10 \times 10 \times 10 = 17,576,000$ ways.

■ **Now You Are Ready to Work Exercise 37**

Example 7(a) could also be worked as a Multiplication Rule with six activities, giving $26 \times 25 \times 24 \times 10 \times 9 \times 8 = 11,232,000$ ways.

Permutations of Objects with Some Alike *(Optional)*

So far the permutation problems have involved objects that are all different. Sometimes we arrange objects when some are alike. For example, we may ask for all arrangements of the letters of the word AGREE. Generally, we have said that we can arrange five objects in $P(5, 5) = 5! = 120$ ways. However, when we interchange the two E's in a word, we obtain the same word. Each time we "spell" a word, the E's are placed in certain positions. We can arrange the E's in those positions in 2! ways and still have the same "word." Therefore, the number of different "words" (arrangements) is 120/2!.

For example, in the word EAGER we can think of the two E's as E_1 and E_2 to distinguish them momentarily. One spelling is $E_1 AGE_2 R$, and another is $E_2 AGE_1 R$. The number of "different" spellings that give the same words depends on the number of arrangements of the identical letters. In this case the two E's can be arranged in 2! ways. In general, k identical objects can be arranged in $k!$ ways that leave the overall arrangement unchanged.

The number of distinguishable arrangements is the total number of arrangements (distinguishable and undistinguishable) divided by $k!$.

EXAMPLE 8 ➤ How many different words can be formed using all the letters of DEEPEN?

SOLUTION

Because three of the six letters (E's) are identical, the number of permutations is $\dfrac{6!}{3!} = 120$.

■ **Now You Are Ready to Work Exercise 41**

THEOREM
Permutation of Identical Objects

(a) The number of permutations of n objects with r of the objects identical is $\dfrac{n!}{r!}$.

(b) If a set of n objects contains k subsets of objects in which the objects in each subset are identical and objects in different subsets are not identical, the number of different permutations of all n objects is

$$\frac{n!}{r_1! r_2! \cdots r_k!}$$

where r_1 is the number of identical objects in the first subset, r_2 is the number of identical objects in the second subset, and so on.

Part (b) of the theorem tells how to compute the number of permutations when there are two or more categories of identical objects.

EXAMPLE 9 ➤ In how many ways can the letters of REARRANGE be permuted?

SOLUTION

There are nine letters, with three R's, two A's, and two E's. The number of permutations is

$$\frac{9!}{3!2!2!} = \frac{9 \times 8 \times 7 \times 6 \times 5 \times 4 \times 3 \times 2 \times 1}{3 \times 2 \times 1 \times 2 \times 1 \times 2 \times 1} = 15{,}120$$

■ **Now You Are Ready to Work Exercise 45**

EXAMPLE 10 ➤ Basketball teams X and Y are in a playoff. The team that wins three out of a possible five games is the winner. Denote the sequence of winners by a sequence of letters such as XXYYY. This indicates that X won the first two games and Y won the last three. How many different sequences are possible if X wins the playoff?

SOLUTION

In cases where a team wins the playoff in fewer than five games, such as XXYX, only one Y appears, so we can let Y "win" the unplayed game and have 3 X's and 2 Y's, XXYXY. This way we can represent all possible sequences of playoff games with 3 X's and 2 Y's. The number of games is

$$\frac{5!}{3!\,2!} = 10 \text{ different ways}$$

Similarly, there are ten different sequences possible for Y to win, giving $10 + 10 = 20$ possible sequences for the playoff to occur.

■ **Now You Are Ready to Work Exercise 49**

6.4 EXERCISES

LEVEL 1

Perform the computations in Exercises 1 through 15.

1. *(See Example 5)* $3!$

2. $7!$

3. $5!$

4. $3!\,2!$

5. $5!\,3!$

6. $\dfrac{5!}{6!}$

7. $\dfrac{7!}{3!}$

8. $\dfrac{10!}{4!\,6!}$

9. $\dfrac{12!}{7!}$

10. $P(12, 3)$

11. $P(6, 4)$

12. $P(6, 2)$

13. $P(100, 3)$

14. $P(5, 5)$

15. $P(7, 4)$

16. The number of permutations of eight objects taken three at a time.

LEVEL 2

17. *(See Example 1)*   The artist R. Locklen Jones selects 3 paintings from a collection of 6 to display in a row. How many different arrangements of the display are possible?

18. How many arrangements of 3 people seated along one side of a table are possible if there are 8 people to select from?

19. The program committee of a Blue Grass festival must arrange five numbers for an evening performance. Seven numbers are available. How many different arrangements of the evening performance are possible?

20. Four people are to be seated in a row. How many different arrangements of all 4 people are possible?

21. *(See Example 2)*   Eight fellows are candidates for Mr. Ugly. In how many different ways can first, second, and third places be awarded?

22. In how many ways can five essays be ranked in a contest?

23. Seven paintings are exhibited by art students.
   (a) An art appreciation class is asked to rank the paintings 1 through 7. How many different rankings are possible?
   (b) If the students are asked to rank only the top three, how many rankings are possible?

24. Six runners are competing in a 100-meter race. In how many different ways can runners finish in first, second, and third place?

25. *(See Example 3)*   Mr. Ling bought three different gifts. In how many different ways can he give one to each of his three children?

26. The Oxford Place bookstore gives 4 different books, one to each of 4 students. In how many different ways can this be done?

27. Three different door prizes are given at a club meeting at which 22 people are present. A person may receive at most one prize. In how many different ways can the prizes be awarded?

28. Three women are selected from the audience of a style show to receive a purse, a pair of gloves, and a scarf. If 30 women are present, in how many different ways may the gifts be given?

29. *(See Example 4)* There are four Pizza Places in Lorena, and the management must assign a manager to each place.
 (a) If four people are available, in how many ways can they be assigned?
 (b) If seven people are available, in how many ways can a manager be assigned?

30. Pi Mu Epsilon is to select a president and a vice-president from a group of five people. Find the total number of selections possible.

31. Seven school children compete for 3 different prizes. How many ways can the prizes be awarded if no child gets more than one prize?

32. In how many ways can a president, a vice-president, and a secretary be selected from an organization of 32 members?

33. *(See Example 6)* In how many ways can five people be seated in a row of five chairs?

34. Six horses are running at the Raton Racetrack. In how many different orders can the horses finish?

35. How many different "words" can be made using all the letters of MATH? A "word" is any arrangement of letters, not just arrangements that give words in a dictionary.

36. In how many ways can 6 children line up in a row to have their picture taken?

37. *(See Example 7)* How many three-digit numbers can be formed using the digits from {2, 4, 6, 8} if
 (a) Repetitions are not allowed?
 (b) Repetitions are allowed?

38. A music teacher selects 4 boys and 5 girls to sing at a PTA meeting. The children stand in a row with the boys on the left and the girls on the right. How many such arrangements of children are possible if the teacher can select from seven boys and nine girls?

39. A voice teacher selects nine students from the class to sing for the trustees. The class has 11 men and 12 women. In how many ways can 9 students be selected and arranged in a row with 5 men in the middle and 2 women on each end?

40. A club has 25 members and 12 pledges. A president, vice-president, and secretary are selected from the members, and a pledge chairman and pledge vice-chairman are selected from the pledges. In how many ways can these five officers be selected?

Exercises 41 through 49 are optional.

41. *(See Example 8)* How many different "words" are possible using all the letters of POSSIBLE?

42. How many different "words" are possible using all the letters of POPPER?

43. How many different color arrangements are possible by placing 3 green balls, 1 red ball, 1 yellow ball, and 1 tan ball in a row?

44. How many different "words" can be formed using all the letters of FALLEN?

45. *(See Example 9)* How many different "words" are possible using all the letters of:
 (a) MISSOURI   **(b)** MISSISSIPPI
 (c) BOOKKEEPER   **(d)** REARRANGED

46. How many color arrangements are possible by placing 4 red balls, 3 green balls, and 2 yellow balls in a row?

47. A coin is tossed seven times:
 (a) How many sequences of 4 heads and 3 tails are possible?
 (b) How many sequences of 5 heads and 2 tails are possible?

48. Codes to identify entries in a computer file are formed by using sequences of 4 zeros and 6 ones. How many such sequences are possible?

49. *(See Example 10)* Teams A and B play in the World Series. The team that wins four of seven games is the winner.
 (a) How many sequences of games are possible if A wins the series?
 (b) How many sequences of games are possible if B wins the series?
 (c) How many different sequences of games are possible?

50. A student did not study for a test, so he has no idea of the correct answers to 10 true/false questions. He decides to answer 5 true and 5 false. How many sequences of 5 true and 5 false answers are possible?

51. Ten students apply for the position of grader in mathematics. One grader is to be assigned to each of four teachers. In how many different ways can the teachers be assigned a grader if no student grades for two teachers?

52. Three children go to the ice cream store that serves 31 flavors. In how many different ways can the children be served one dip of ice cream each if
(a) Each child receives a different flavor?

(b) The same flavor can be served to more than one child?

53. A recreation center has 28 video games. Mia has time to play 3. In how many different orders can she play three different games?

54. How many four-digit numbers can be formed using the digits 1 through 9 if no digit can appear twice in a number?

LEVEL 3

55. Write out the following as expressions in n:
(a) $P(n, 2)$ (b) $P(n, 3)$
(c) $P(n, 1)$ (d) $P(n, 5)$

56. If $P(7, k) = 210$, what is k?

57. A password to a computer consists of five characters: a letter, a digit, a letter, a digit, and a letter in that order, where the numbers from 1 through 9 are allowed for digits. How many different passwords are possible?

58. Three pairs of shoes are displayed in a row.
(a) In how many ways can the six shoes be arranged with no restriction on their positions?
(b) In how many ways can the shoes be arranged if mates must be kept together but mates can be placed in either order?
(c) In how many ways can the shoes be arranged if mates must be kept together and left and right mates are placed in order, left shoe then right?

59. Fifteen students are competing for 4 Rotary scholarships, 1 in England, 1 in Germany, 1 in Brazil, and 1 in Japan. In how many ways can the scholarships be awarded to different students?

60. The Student Foundation has 120 members, of whom 80 are seniors, 25 are juniors, and 15 are sophomores. They select a Publicity Chair, a Financial Chair, a Solicitation Chair, and a Coordinator. The Coordinator and Solicitation Chair must be seniors, and the Publicity Chair and Financial Chair must be juniors. In how many different ways can the officers be selected?

61. Use these grids and, in each case, find the number of paths from A to B, traveling along grid lines. At any given intersection, you may proceed only to the right (R) or down (D). (*Hint:* A path is a sequence of R's and D's.)

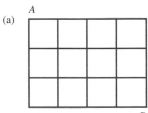

(a)

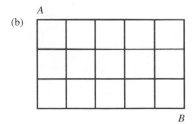

(b)

EXPLORATIONS

62. What is the meaning of $P(6, 8)$, if any? Explain.

63. What is the meaning of $P(7, -3)$ if any? Explain.

64. Find the largest value of n for which your calculator will compute $n!$

65. (a) For $n = 2, 3, 4$, and 5 show that $n! = n(n - 1)!$
(b) Give an argument that $n! = n(n - 1)!$ for $n > 1$.

66. Find the largest value of k so that your calculator or spreadsheet gives an answer for $P(15000, k)$.

67. Find the largest value of n such that your calculator or spreadsheet gives an answer for $P(n, 50)$.

68. Alex is a bridge fanatic who doesn't shuffle cards very well. He thinks it would be neat to have bridge decks shuffled in all possible orders, one deck for each way a bridge deck (52 cards) can be arranged in order.

(a) How many decks would Alex need?

(b) How big a box would Alex need to store his decks? Assume the box is in the shape of a cube (length = width = height) and find its dimensions. For a deck of cards use length = 3 inches, width = 2 inches, and thickness = 0.5 inches. (This exercise created by F. Eugene Tidmore.)

 69. At the Rotary Club meeting the week before the Triple Crown races (the Kentucky Derby, the Preakness Stakes, and the Belmont Stakes) Spencer claimed to predict the first-, second-, and third-place horses in the race by writing the winners on a paper that was sealed in an envelope and opened by Spencer after the race. How many different ways could the first three positions occur

(a) In the Kentucky Derby with 19 horses entered?

(b) In the Preakness Stakes with 12 horses entered?

(c) In the Belmont Stakes with 13 horses entered?

(d) After each race the envelope was opened and, sure enough, the top three winners were listed correctly. Do you think this happened by chance, Spencer had unusual prophetic powers, or there was some trickery involved?

USING YOUR TI-83

PERMUTATIONS

Using the **nPr** command, the number of permutations — say, $P(7, 4)$ — can be calculated on the TI-83 as follows.

First, enter 7, then select **nPr**, using [MATH] <PRB> <2:nPr>, which gives the screen:

```
MATH NUM CPX PRB
1:rand
2:nPr
3:nCr
4:!
5:randInt(
6:randNorm(
7:randBin(
```

Press [ENTER], then enter 4:

```
7 nPr 4■
```

Press [ENTER], which gives 840:

```
7 nPr 4
             840
■
```

EXERCISES

Calculate the following:

1. $P(12, 5)$

2. $P(22, 13)$

3. $P(8, 4)P(6, 3)$

4. $\dfrac{P(15, 7)}{P(7, 5)}$

USING EXCEL

PERMUTATIONS

The number of permutations, say $P(7, 4)$, can be calculated in EXCEL using the PERMUT function, written as PERMUT(7,4). The formula =PERMUT(A2,A3) will calculate the number of permutations using the numbers in A2 and A3.

EXERCISES

1. Find $P(6, 3)$.

2. Find $P(12, 7)$.

3. Find $P(50, 18)$.

6.5 Combinations

- Combinations
- Special Cases
- Problems Involving More Than One Counting Technique
- Binomial Theorem

Combinations

When you pay your bill at the Pizza Place, the cashier is interested in the collection of coins and bills you give her, not the order in which you present them. When you are asked to answer six out of eight test questions, the collection of questions is important, not the arrangement. Therefore, if the professor wishes to compute the number of different ways in which students can choose six questions from eight, she is not dealing with permutations. She wants the number of ways in which a subset of six elements can be obtained.

DEFINITION
Combination

A subset of elements chosen from a given set without regard to their arrangement is called a **combination.**

The notation $C(n, k)$, read "combinations of n things taken k at a time," represents the number of subsets consisting of k elements taken from a set of n elements.

$C(8, 3)$ denotes the number of ways in which three elements can be selected from a set of eight. $C(52, 6)$ denotes the number of ways in which six elements can be selected from a set of 52.

The keys to recognizing a combination are the following.

Combinations

1. A combination selects elements from a single set.
2. Repetitions are not allowed.
3. The order in which the elements are arranged is *not* significant.

Notice that a combination differs from a permutation only in that order is not significant in a combination, whereas it is important in a permutation.

EXAMPLE 1 ➤ Given the set $A = \{a, b, c, d, e, f\}$, the subset $\{b, d, f\}$ is a combination of three elements taken from a set of six elements.

Because the elements of the subset $\{b, d, f\}$ can be arranged in several ways, we expect there to be several permutations for each subset. This indicates that you should expect more permutations than combinations in a given set. ■

EXAMPLE 2 ➤ List all combinations of two elements taken from the set $\{a, b, c\}$.

SOLUTION

Because of the small number of elements involved, it is rather easy to list all subsets consisting of two elements. They are $\{a, b\}$, $\{a, c\}$, and $\{b, c\}$. Therefore, $C(3, 2) = 3$.

■ Now You Are Ready to Work Exercise 9

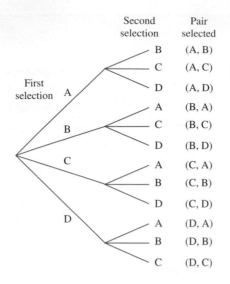

Second selection / Pair selected

Second selection	Pair selected
B	(A, B)
C	(A, C)
D	(A, D)
A	(B, A)
C	(B, C)
D	(B, D)
A	(C, A)
B	(C, B)
D	(C, D)
A	(D, A)
B	(D, B)
C	(D, C)

First selection A

B

C

D

FIGURE 6–20 Pairs selected from the set {A, B, C, D}.

If we are interested in lists like all 5-letter subsets of our 26-letter alphabet, the listing rapidly increases in difficulty. If we want only the number of such subsets, not the list, then the problem becomes easier. Let's look at two examples that illustrate how to determine the number of combinations.

We will see there is a definite relationship between the number of permutations and the number of combinations.

EXAMPLE 3 ➤ Select all two-letter combinations of letters from the set of four letters {A, B, C, D}.

SOLUTION

We use a tree diagram to show all the ways we can select a pair of letters. (See Figure 6–20.)

Because we are interested only in the pair of objects, not their order of selection, notice that (A, B) and (B, A) both appear and are equal sets, so (A, B) occurs twice. The list of all possible pairs then is

$$(A, B), (A, C), (A, D), (B, C), (B, D), (C, D)$$

each appearing twice in the tree diagram. It is no accident that the number of pairs, 6, is

$$6 = \frac{12}{2} = \frac{P(4, 2)}{2!}$$

where 12 is the number of permutations, $P(4, 2)$, and 2! is the number of arrangements of each pair. Thus,

$$C(4, 2) = \frac{P(4, 2)}{2!}$$

This shows 6 ways a subset of two letters can be chosen; that is, there are 6 combinations of two letters selected from a set of four. ∎

EXAMPLE 4 ➤

Let's look at the relationship between permutations and combinations again by making a list of combinations and corresponding permutations using subsets with more than two elements.

We use a set of four letters, $\{a, b, c, d\}$, and form all three-element subsets. There are four, each of which may be obtained by removing one letter from the set of four letters. Here is the list of combinations and all permutations that can be formed from the letters in the combination.

Subset	Arrangements	Subset	Arrangements
Combinations	Permutations	Combinations	Permutations
$\{a, b, c\}$	abc	$\{a, c, d\}$	acd
	acb		adc
	bac		cad
	bca		cda
	cab		dac
	cba		dca
$\{a, b, d\}$	abd	$\{b, c, d\}$	bcd
	adb		bdc
	bad		cbd
	bda		cdb
	dab		dbc
	dba		dcb

Notice there are six permutations for each combination. This is expected because three letters can be arranged in $3! = 6$ ways. We can get the total number of permutations with

(Number of combinations) $\times$ (Number of permutations per combination)

In the above that is

$$C(4, 3) \times 6 = P(4, 3)$$

which we write as

$$C(4, 3) \times 3! = P(4, 3)$$

Dividing by 3! we have

$$C(4, 3) = \frac{P(4, 3)}{3!}$$

This gives us what we need to compute $C(4, 3)$:

$$C(4, 3) = \frac{P(4, 3)}{3!} = \frac{4 \times 3 \times 2}{3 \times 2 \times 1} = \frac{24}{6} = 4 \quad\blacksquare$$

In general, the same type of result applies to other sizes of the set and the other numbers in the subset, so

$$C(11, 5) = \frac{P(11, 5)}{5!}, \qquad C(9, 2) = \frac{P(9, 2)}{2!}, \qquad C(101, 14) = \frac{P(101, 14)}{14!}$$

and generally we have the following theorem:

THEOREM

$$C(n, r) = \frac{P(n, r)}{r!}$$

or

$$P(n, r) = r!\, C(n, r)$$

Because $P(n, r)$ can be written as

$$\frac{n!}{(n - r)!}$$

$C(n, r)$ can also be written as

$$\frac{n!}{r!(n - r)!}$$

We now have a convenient way of calculating the number of combinations.

EXAMPLE 5 ➤

$$C(5, 2) = \frac{P(5, 2)}{2!} = \frac{5 \times 4}{2 \times 1} = 10$$

$$C(5, 3) = \frac{P(5, 3)}{3!} = \frac{5 \times 4 \times 3}{3 \times 2 \times 1} = 10$$

$$C(10, 4) = \frac{P(10, 4)}{4!} = \frac{10 \times 9 \times 8 \times 7}{4 \times 3 \times 2 \times 1} = 210$$

$$C(8, 6) = \frac{8!}{6!\,2!} = 28$$

$$C(15, 3) = \frac{15!}{3!\,12!} = 455$$

■ **Now You Are Ready to Work Exercise 1**

EXAMPLE 6 ➤ A student has seven books on his desk. In how many different ways can he select a set of three?

SOLUTION

Because the order is not important, this is a combination problem:

$$C(7, 3) = \frac{P(7, 3)}{3!} = \frac{7 \times 6 \times 5}{3 \times 2 \times 1} = 35$$

■ **Now You Are Ready to Work Exercise 19**

EXAMPLE 7 ➤ **(a)** In how many ways can a committee of four be selected from a group of ten people?
(b) In how many ways can a slate of officers consisting of a president, vice-president, and secretary be selected from a group of ten people?

SOLUTION

(a) The order of selection is not important in the selection of a committee, so this is a combination problem of taking four elements from a set of ten:

$$C(10, 4) = \frac{P(10, 4)}{4!} = 210$$

(b) In selecting a slate of officers, President Jones, Vice-President Smith, and Secretary Allen is a different slate than President Allen, Vice-President Smith, and Secretary Jones. Each office is a position to be filled, so order is significant. The number of slates is $P(10, 3) = 720$.

■ **Now You Are Ready to Work Exercise 23**

Notice the pattern used in computing combinations. To compute $C(10, 4)$, begin with 10 and write four integers in decreasing order. Then divide by 4!. This is true in general. To compute $C(15, 5)$, form the numerator using the five integers beginning with 15 and decreasing by 1. The denominator is 5!. In general, we can write $C(n, r)$ by forming the numerator from the product of r integers that begin with n and decrease by 1. The denominator is $r!$.

Special Cases

The form for the number of combinations,

$$C(n, r) = \frac{n!}{r!(n - r)!}$$

is a useful form. Let's use it to look at some special cases.

1. In how many ways can one element be selected from a set? $C(6, 1)$ is the number of ways one element can be selected from a set of six. It is

$$C(6, 1) = \frac{6!}{1!5!} = \frac{6 \times 5!}{1!5!} = 6$$

In general,

$$C(n, 1) = \frac{n!}{1!(n - 1)!}$$
$$= \frac{n(n - 1)!}{1!(n - 1)!} = n$$

So one item can be selected from a set of n items in n ways.

2. In how many ways can zero items be selected from a set? We write $C(6, 0)$ to represent the number of ways in which no elements can be selected from a set of six. The formula gives

$$C(6, 0) = \frac{6!}{0!6!}$$

Because $0! = 1$, this reduces to $C(6, 0) = 1$. In general,

$$C(n, 0) = \frac{n!}{0!n!} = 1$$

Does your intuition tell you that there is just one way to select zero elements from a set? The one way is to take none.

3. In how many ways can all the elements be selected from a set? Our intuition tells us there is just one way, namely, take all of them. The formula agrees.

$$C(6, 6) = \frac{6!}{6!0!} = 1$$

and

$$C(n, n) = \frac{n!}{n!0!} = 1$$

4. For positive integers, n, $P(n, 1) = n$, so when one element is selected from a set, the number of permutations equals the number of combinations, n.

Problems Involving More Than One Counting Technique

The solution to a problem may involve more than one counting technique. Often the first level is the Multiplication Rule with two or more activities involved. To count the number of ways in which each of these activities can occur may require permutations, combinations, or the Multiplication Rule again. The examples that follow involve more than one counting technique.

EXAMPLE 8 ➤

A cafeteria offers a selection of four meats, six vegetables, and five desserts. In how many ways can you select a meal consisting of two different meats, three different vegetables, and two different desserts?

SOLUTION

Basically, this is a problem whose solution first uses the Multiplication Rule. We obtain the possible number of meals by multiplying the number of ways in which you can select two meats, the number of ways in which you can select three vegetables, and the number of ways in which you can select two desserts.

Each of the numbers of ways in which you can select meats, vegetables, and desserts forms a combination problem. Therefore, we obtain the number of meals as

(Number of meat selections) $\times$ (number of vegetable selections)
$\times$ (number of dessert selections) $= C(4, 2) \times C(6, 3) \times C(5, 2)$

$$= \frac{4 \times 3}{2 \times 1} \times \frac{6 \times 5 \times 4}{3 \times 2 \times 1} \times \frac{5 \times 4}{2 \times 1}$$

$$= 6 \times 20 \times 10 = 1200$$

■ **Now You Are Ready to Work Exercise 25**

EXAMPLE 9 ➤

The Beta Club has 14 male and 16 female members. A committee composed of 3 men and 3 women is formed. In how many ways can this be done?

SOLUTION

Because we are trying to determine the number of ways a first event (selecting 3 males) *and* a second event (selecting 3 females) can occur, we need to compute the number of ways each can occur and then multiply these values.

The male members can be chosen in

$$C(14, 3) = \frac{14 \times 13 \times 12}{3 \times 2 \times 1} = 364 \text{ different ways}$$

The female members can be chosen in

$$C(16, 3) = \frac{16 \times 15 \times 14}{3 \times 2 \times 1} = 560 \text{ different ways}$$

By the Multiplication Rule, the committee can be chosen in $364 \times 560 = 203{,}840$ ways.

■ **Now You Are Ready to Work Exercise 27**

Instead of counting the number of outcomes for a sequence of activities, some counting problems seek the number of possible outcomes when the outcome selected is from one activity *or* another. We can also state this as a selection from one of two disjoint sets. This type of problem requires that we find the number of ways each activity can occur and *add* them.

EXAMPLE 10 ➤

How many different committees can be selected from eight men and ten women if a committee is composed of three men *or* three women?

SOLUTION

For a moment, think of listing all possible selections of a committee. The list has two parts, a list of committees composed of three women and a list of committees composed of three men. The total number of possible committees can be obtained by adding the number of all-female to the number of all-male committees. We get each of these by the following:

$$\begin{aligned}
\text{Number of all-female committees} &= C(10, 3) = 120 \\
\text{Number of all-male committees} &= C(8, 3) \ \ = \ \ \underline{56} \\
\text{Total number of committees} & \qquad\qquad\quad = 176
\end{aligned}$$

Do not confuse this problem with the number of ways in which a committee of three men *and* a committee of three women can be chosen. That calls for the selection of a *pair* of committees, one from *each* of two disjoint sets.

This example calls for the selection of *one* committee from *one* of two disjoint sets. So, this example uses the Addition Rule.

■ **Now You Are Ready to Work Exercise 29**

EXAMPLE 11 ➤

One freshman, three sophomores, four juniors, and six seniors apply for five positions on an Honor Council. If the council must have at least two seniors, in how many different ways can the council be selected?

SOLUTION

The council has at least two seniors when it has two, three, four, or five seniors. Because this situation asks for the number of ways one event or another event can occur, we need to compute the number of ways each event can occur and then *add* them. We must compute the number of councils possible with two seniors, with three seniors, and so on, and add:

2 seniors and 3 others: $C(6, 2) \times C(8, 3) = 15 \times 56 = 840$
3 seniors and 2 others: $C(6, 3) \times C(8, 2) = 20 \times 28 = 560$
4 seniors and 1 other: $C(6, 4) \times C(8, 1) = 15 \times 8 = 120$
5 seniors: $C(6, 5) = 6$

The total is $840 + 560 + 120 + 6 = 1526$.

■ **Now You Are Ready to Work Exercise 37**

EXAMPLE 12 ➤

The Huck Manufacturing firm forms a six-person advisory committee. The committee is composed of a chair, vice-chair, and secretary from the administrative staff and three members from the plant workers. Seven members from the administrative staff and eight plant workers are eligible for the committee positions. In how many different ways can the committee be formed?

SOLUTION

At the first level, this is a Multiplication Rule problem because two activities are involved: selecting officers from the administrative staff and selecting committee members from the plant workers. We compute the number of ways in which each can occur and then we multiply. The selection of officers is a permutation because repetitions are not allowed (a person may not hold two offices) and the different offices impose an order. The number of slates of officers is $P(7, 3)$. The selection of committee members from the plant workers is a combination because no distinction is made between those positions, and repetitions are not allowed. The number of selections is $C(8, 3)$.

The total number of ways in which the administrative committee can be selected is $P(7, 3) \times C(8, 3) = 11{,}760$.

■ **Now You Are Ready to Work Exercise 41**

Binomial Theorem

Perhaps you remember from your high school algebra that

$$(x + y)^2 = x^2 + 2xy + y^2$$

You are less likely to remember that

$$(x + y)^3 = x^3 + 3x^2y + 3xy^2 + y^3$$

and few remember that

$$(x + y)^4 = x^4 + 4x^3y + 6x^2y^2 + 4xy^3 + y^4$$

At this point, you may be asking why we discuss the expansion of $(x + y)^n$ in a section on counting combinations. We do so because we can use combinations to find the expansion of, say, $(x + y)^5$ without memorizing the coefficients and without multiplying

$$(x + y)(x + y)(x + y)(x + y)(x + y)$$

To see how this is done, let's make some helpful observations. In any term of $(x + y)^4$, the exponents on x and y add to 4. In any term of $(x + y)^3$, the exponents on x and y add to 3. In general, in any term of $(x + y)^n$, the exponents on x and y add to n.

In the expansion $(x + y)^4 = x^4 + 4x^3y + 6x^2y^2 + 4xy^3 + y^4$, the coefficients 1, 4, 6, 4, and 1 are equal, respectively, to

$$1 = C(4, 0)$$
$$4 = C(4, 1)$$
$$6 = C(4, 2)$$
$$4 = C(4, 3)$$
$$1 = C(4, 4)$$

In another example, $(x + y)^3 = x^3 + 3x^2y + 3xy^2 + y^3$, the coefficient of x^2y is $C(3, 1)$, the coefficient of x^8y^5 in $(x + y)^{13}$ is $C(13, 5)$, and in general, the coefficient of $x^{n-4}y^4$ in $(x + y)^n$ is $C(n, 4)$. Even more general, the coefficient of $x^{n-k}y^k$ in $(x + y)^n$ is $C(n, k)$. This enables us to write $(x + y)^n$ in the following way using what is called the **Binomial Theorem.**

Binomial Theorem

Expansion of $(x + y)^n$:

$$(x + y)^n = C(n, 0)x^n + C(n, 1)x^{n-1}y + C(n, 2)x^{n-2}y^2 + \cdots$$
$$+ C(n, k)x^{n-k}y^k + \cdots + C(n, n-1)xy^{n-1} + C(n, n)y^n$$

EXAMPLE 13 ➤

$$(x + y)^6 = C(6, 0)x^6 + C(6, 1)x^5y + C(6, 2)x^4y^2 + C(6, 3)x^3y^3$$
$$+ C(6, 4)x^2y^4 + C(6, 5)xy^5 + C(6, 6)y^6$$
$$= x^6 + 6x^5y + 15x^4y^2 + 20x^3y^3 + 15x^2y^4 + 6xy^5 + y^6 \ ∎$$

NOTE

We refer to this as the Binomial Theorem because an expression of two terms, $x + y$, is called a binomial.

Application to Subsets   We can apply the Binomial Theorem to the counting of all subsets of a set. First, we observe that we can write $2^n = (1 + 1)^n$, so we then can write, for example,

$$2^4 = (1 + 1)^4 = C(4, 0)(1)^4 + C(4, 1)(1)^3(1)$$
$$+ C(4, 2)(1)^2(1)^2 + C(4, 3)(1)(1)^3 + C(4, 4)(1)^4$$

Because $1 = 1^2 = 1^3 = \cdots$,

$$2^4 = (1 + 1)^4 = C(4, 0) + C(4, 1) + C(4, 2) + C(4, 3) + C(4, 4)$$

We recognize the terms in the sum above as

A set of four elements has $C(4, 0)$ subsets with 0 elements.

A set of four elements has $C(4, 1)$ subsets with 1 element.

A set of four elements has $C(4, 2)$ subsets with 2 elements.

A set of four elements has $C(4, 3)$ subsets with 3 elements.

A set of four elements has $C(4, 4)$ subsets with 4 elements.

Thus, the sum of $(1 + 1)^4$ gives the number of all subsets of a four-element set, namely, $2^4 = 16$.

The example of the number of subsets of a four-element set illustrates a more general principle.

Number of Subsets of a Set

The number of all subsets of a set with n elements is 2^n.

EXAMPLE 14 ➤

In how many ways can Suzie invite two or more of her five friends to her birthday party?

SOLUTION

This problem essentially asks for the number of subsets with two or more elements that can be formed from a five-element set. The number of subsets with two or more elements is the *number of all subsets minus the number of subsets with zero or one element*, which is:

$$\text{Number with two or more elements} = 2^5 - C(5, 0) - C(5, 1)$$
$$= 2^5 - 1 - 5 = 32 - 6 = 26$$

Two or more friends can be selected in 26 ways.

■ **Now You Are Ready to Work Exercise 33**

We conclude this section with an explanation of why, for example, $C(6, 4)$ is the correct coefficient of $x^2 y^4$ in the expansion of $(x + y)^6$. First, we write $(x + y)^6$ as

$$(x + y)(x + y)(x + y)(x + y)(x + y)(x + y)$$

Now, for the moment, think of each factor, $x + y$, as a pair of two elements, (x, y). We have six such pairs from the product.

When you obtain $(x + y)^6$ by multiplying the factors of

$$(x + y)(x + y)(x + y)(x + y)(x + y)(x + y)$$

different people may vary some in the way they find the product, but all will essentially take either an x or a y from each of the six factors and multiply. This is done in all possible ways and the terms added. If a y is taken from four of the factors and an x from the other two, then $x^2 y^4$ is obtained. In the process, this term is obtained several times. We want to determine exactly what is meant by "several times." We have six pairs (x, y), and we select four of them from which to take a y; then an x is automatically taken from the other two. We can select the four pairs from the six pairs in $C(6, 4)$ ways. Thus, when we add all the $x^2 y^4$ terms, we will have $C(6, 4)x^2 y^4$. If we use an integer n instead of 6 and an integer k instead of 4, a similar argument implies that the coefficient of $x^{n-k} y^k$ in $(x + y)^n$ is $C(n, k)$.

▦ 6.5 EXERCISES

LEVEL 1

Perform the computations in Exercises 1 through 7.

1. *(See Example 5)*
   $C(6, 2)$

2. $C(4, 3)$

3. $C(13, 3)$

4. $C(9, 4)$

5. $C(9, 5)$

6. $C(20, 3)$

7. $C(4, 4)$

8. Verify the following.
   (a) $C(7, 3) = C(7, 4)$ (b) $C(7, 2) = C(7, 5)$
   (c) $C(6, 4) = C(6, 2)$ (d) $C(9, 6) = C(9, 3)$
   (e) $C(8, 3) = C(8, 5)$
   These are examples of a general fact that
   $C(n, k) = C(n, n - k)$.

9. *(See Example 2)*   List all combinations of two different elements taken from $\{a, b, c, d\}$.

10. List all combinations of three different elements taken from {Tom, Dick, Harriet, Jane}.

11. List all combinations of four different elements taken from $\{a, b, c, d, e\}$.

12. List all combinations of two elements taken from {penny, nickel, dime, quarter}.

Use the Binomial Theorem to find the expansions in Exercises 13 through 16.

13. $(x + y)^5$

14. $(x + y)^7$

15. $(x + y)^{10}$

16. $(x + y)^{12}$

17. Find the number of subsets of a five-element set.

18. Find the number of subsets of an eight-element set.

LEVEL 2

19. *(See Example 6)* The Pizza Place must hire 2 employees from 6 applicants. In how many ways can this be done?

20. Students are to answer 4 out of 5 exam questions. In how many different ways can the questions be selected?

21. A Campfire Girls troop has 15 members. In how many different ways can the leader appoint three members to clean up camp?

22. Blackhawk Tech gives four presidential scholarships. If there are 50 nominees, in how many ways can the scholarships be awarded?

23. *(See Example 7)*
    (a) In how many ways can a committee of 4 be selected from a group of 7 people?
    (b) In how many ways can a slate of officers consisting of a chair, vice-chair, secretary, and treasurer be selected from a group of seven people?

24. An executive hires 3 office workers from 6 applicants.
    (a) In how many ways can the selection be made?
    (b) In how many ways can the selection be made if one worker is to be a receptionist, one a secretary, and one a technical typist?

25. *(See Example 8)* Collins Cafeteria offers a selection of 5 meats, 6 vegetables, and 8 desserts. In how many ways can you select a meal of 2 different meats, 3 different vegetables, and 2 different desserts?

26. From a set of 7 math books, 9 science books, and 5 literature books, in how many ways can a student select 2 from each set?

27. *(See Example 9)* Mid-State Trucking must hire three truck drivers and four clerks. There are 6 applicants for truck driver and 10 for clerk. In how many ways can the 7 employees be chosen?

28. A candy manufacturer makes 5 kinds of brown chocolate candies and 3 kinds of white chocolate candies. A sample package contains 2 kinds of brown and 2 kinds of white chocolate candies. How many different sample packages can be prepared?

29. *(See Example 10)* How many different committees can be formed from a group of 9 women and 11 men if a committee is composed of 3 women or 3 men?

30. A child has 3 pennies, 5 nickels, and 4 dimes. In how many ways can two coins of the same denomination be selected?

31. The Schmidts are considering the purchase of three paintings for their home. Their decorator advises that the selection be made from a collection of 7 landscapes or from a collection of 6 historical paintings. In how many different ways can they select three paintings from one collection or the other?

32. A desk holder contains 8 pens and 5 pencils. In how many ways can 4 pens or 4 pencils be selected?

33. *(See Example 14)* In how many ways can Joe select 1 or more of 6 friends to a dinner party?

34. The English Club received eight essays from members. They will publish at least three of the essays in their literary publication. In how many ways can they select at least three of the essays?

35. The KOT club has seven pledges. They will send at least 2 and no more than 6 pledges to work at Goodwill on Helping-Out Day. In how many ways can this be done?

36. Belmont Records sends a disk jockey ten new CD releases for possible use. In how many ways can the disk jockey select no more than eight CDs?

LEVEL 3

37. *(See Example 11)* A committee of 5 is selected from 5 men and 6 women. How many committees are possible if there must be at least 3 men on the committee?

38. Five freshmen, 4 sophomores, and 2 juniors are present at a meeting of students.
    (a) In how many ways can a six-member committee of three freshmen, two sophomores, and one junior be formed?

    (b) In how many ways can a six-member committee be selected with no more than two freshmen members?

39. An English reading list has 8 American novels and 6 English novels. A student must read 4 from the list, and at least 2 must be American novels. In how many different ways can the four books be selected?

40. Prof. Shanks assigns five students to work on a project. The group is to have no more than one freshman. In how many ways can the group be selected if there are 5 freshmen and 10 sophomores in the class?

41. *(See Example 12)* A club has 40 members and 10 pledges. The club selects a five-member executive committee composed of a chair and vice-chair, who must be members, and an advisory committee of three pledges. In how many ways can the executive committee be formed?

42. Scout Troop 456 has 22 members. The scoutmaster appoints a five-member campout crew composed of a leader, a cook, and a cleanup crew (three scouts). How many different campout crews are possible?

43. A Presidential Task Force is selected from 8 executives and 12 staff positions. The task force has a chair and secretary chosen from the executives and 4 members chosen from the staff positions. In how many different ways can the task force be selected?

44. The Temple Chamber of Commerce forms a five-person committee to attract industry to the city. The committee consists of an arrangements chair and a spokesperson from the Chamber office and three people from the business community. If 4 people from the Chamber office and 7 people from business volunteer, in how many ways can the committee be formed?

45. A business major needs 3 humanities courses to graduate. They can be selected from 28 courses offered. How many ways can the 3 courses be selected?

46. In how many ways can three aces be selected from a bridge deck? (Order is not significant.)

47. Ms. Wood has a collection of 20 questions. How many different tests of 5 questions each can be made from the set of questions?

48. In how many ways can one select 3 different letters from the word HISTORY and 2 different letters from the word ENGLISH?

49. A set contains four people. Denote it by
$A = \{$Alice, Bianca, Cal, Dewayne$\}$.
   (a) List all subsets of one person each.
   (b) List all subsets of two people.
   (c) List all subsets of three people.
   (d) List all subsets of four people.
   (e) List all subsets containing no people.
   (f) Find the total number of subsets.

(*Note:* You should obtain a total of 2^4. In general. a set of n elements has 2^n subsets.)

50. Each Thursday, Alan takes a break from studies and eats dinner with his grandparents. On one occasion, his choice for dessert is peppermint or vanilla ice cream with chocolate, strawberry, caramel, or pineapple topping.
   (a) How many ways can Alan choose one flavor of ice cream and one topping?
   (b) How many ways can Alan choose one flavor of ice cream and two toppings?

51. The Coffee Shop has a lunch buffet that offers 2 meats, 4 vegetables, 3 salads, and 4 desserts.
   (a) How many ways can Cindy select a lunch consisting of 1 meat, 1 vegetable, and 1 salad?
   (b) How many ways can George select a lunch consisting of 1 meat, 2 vegetables, 2 salads, and 1 dessert?

52. A quiz team of 5 children is to be selected from a class of 25 children. There are 15 girls and 10 boys in the class.
   (a) How many teams made up of 3 girls and 2 boys can be selected?
   (b) How many teams can be selected with at least 3 girls?

53. A test consists of 10 true–false and 8 multiple-choice questions.
   (a) In how many ways can a student select 6 true–false and 5 multiple-choice questions to answer?
   (b) In how many ways can a student select 10 questions, at least 6 of which are multiple-choice?

54. The Rodeo Club has 9 seniors, 16 juniors, 14 sophomores, and 10 freshmen. In how many ways can two representatives be selected if the selection must be composed of a senior and a freshman or a junior and a sophomore?

55. Each day after class, Prof. Phrind invites three of his class members to join him at coffee at the Student Center. He has 32 students in class.
   (a) How many days will it take for him to have coffee with all combinations of three students?
   (b) How many days will it take for him to have coffee with all the students if no student is invited more than once?

56. The Valley College basketball team has a serious losing streak. The coach vows to try every combination of players to find a winning combination. How many games will it take to try all possible combinations if there are 11 players and one combination is used per game? Can this be accomplished in one season?

57. Explain in your own words what the properties *permutations* and *combinations* have in common and how they differ.

58. Professor Puckett gives a ten-problem mathematics test. To encourage students to do their homework, Professor Puckett includes 3 problems taken directly from homework (80 problems). The other 7 problems are taken from a test file of 50 problems. How many ways can the 10 test problems be selected? Leave your answer in symbolic form.

59. For parts (a) through (e), make a table for $C(n, x)$, $x = 0, 1, 2, \ldots, n$ and observe the value(s) of x that yield the largest value for $C(n, x)$.
   (a) $C(6, x), x = 0, 1, 2, \ldots, 6$
   (b) $C(7, x), x = 0, 1, 2, \ldots, 7$
   (c) $C(8, x), x = 0, 1, 2, \ldots, 8$
   (d) $C(9, x), x = 0, 1, 2, \ldots, 9$
   (e) $C(10, x), x = 0, 1, 2, \ldots, 10$
   (f) Based on parts (a) through (e), what value of x makes $C(50, x)$ largest? Makes $C(200, x)$ largest?
   (g) Based on parts (a) through (e), what value of x makes $C(55, x)$ largest? Makes $C(301, x)$ largest?

EXPLORATIONS

60. What meaning, if any, do you attach to $C(4, 6)$? Why?

61. What meaning, if any, do you attach to $C(15, -6)$? Why?

62. Make an argument that $C(n, 1) = P(n, 1)$.

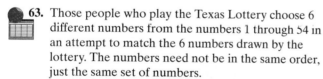

63. Those people who play the Texas Lottery choose 6 different numbers from the numbers 1 through 54 in an attempt to match the 6 numbers drawn by the lottery. The numbers need not be in the same order, just the same set of numbers.
   (a) How many different combinations of the six numbers are possible?
   (b) Suppose you obtain pennies, equal in number to the answer to part (a), and make a single stack of all the pennies. How tall is the stack?
   (c) Suppose one penny in the stack is secretly marked and placed in a random place in the stack. You are allowed to select one penny from the stack. If you select the marked penny, you win a million dollars. You have the same chance of winning a million dollars this way as you have in winning the Texas lottery.

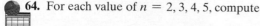

64. For each value of $n = 2, 3, 4, 5$, compute

$$C(n, 0) + C(n, 1) + C(n, 2) + \cdots + C(n, n)$$

What general conclusion do you draw from these computations?

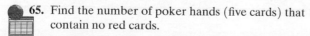

65. Find the number of poker hands (five cards) that contain no red cards.

Your calculator has a limit on the size of numbers that can be calculated. For example, some calculators allow only numbers less than 10^{100}. Thus, you cannot calculate $C(n, k)$ for indefinitely large values of n and k. For Exercises 66 through 68, experiment with your calculator to determine the largest value of n or k for which the requested $C(n, k)$ will be calculated.

66. Find the largest value of k for which your calculator will compute $C(1500, k)$.

67. Find the largest value of n for which your calculator will compute $C(n, 50)$.

68. Find the largest value of n for which your calculator will compute $C(n, 500)$.

69. Calculate the number of bridge hands (13 cards) that are possible (from a deck of 52 cards).

70. Calculate the number of poker hands (5 cards) that are possible (from a deck of 52 cards).

71. A person plays a lottery by selecting 6 numbers from the numbers 1 through 45. Find how many different selections are possible:
   (a) If the six numbers are different and must be in the same order as drawn by the lottery.
   (b) If the six numbers are different and the order is irrelevant.

72. Poker is a card game that is sometimes played for fun and is sometimes played for thousands of dollars by professional gamblers. Each player is dealt five cards (called a hand) and certain combinations of cards rank higher than others.
   Here are combinations in the order of rank with the first listed being the hand of highest rank.

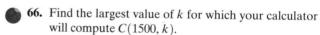

Straight Flush: Five cards in sequence all in the same suit.

> *Example:* 3 through 7 in clubs;
> Ace through 5 in diamonds;
> 10, Jack, Queen, King, Ace in spades

> Notice that the Ace can be used as either the lowest or highest card of the 13.

Four of a kind: Four cards of the same kind and one other card.

> *Example:* All four 6's and a 10.

Full house: Three cards of the same kind and two cards of another kind.

> *Example:* Three 8's and two Queens.

Flush: Five cards of the same suit but not in sequence.

> *Example:* 2, 3, 6, 9, and King of hearts.

Straight: Five cards in sequence but not all in the same suit.

> *Example:* 2 of hearts, 3 of clubs, 4 of clubs, 5 of spades, and 6 of hearts.

Three of a kind: Three cards of the same kind and two cards of two other kinds.

> *Example:* Three 7's, a 5, and a Jack.

Two Pairs: Two cards of the same kind, two more cards of a different kind, and one card of a third kind.

> *Example:* Two 5's, two 9's, and an Ace.

One Pair: Two cards of the same kind and three cards of three other kinds.

> *Example:* Two 8's, a Jack, a 4, and a 9.

Compute the number of ways each of the above five-card hands can be dealt.

USING YOUR TI-83

COMBINATIONS

Using the **nCr** command, the number of combinations — say, $C(7, 4)$ — can be calculated on the TI-83 as follows. First, enter 7; then select **nCr**, using

$$\boxed{\text{MATH}}\ \text{<PRB> <3:nCr>}$$

which gives the screen

Press $\boxed{\text{ENTER}}$, then enter 4:

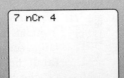

Press $\boxed{\text{ENTER}}$, which gives 35:

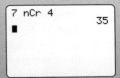

EXERCISES

Calculate the following:

1. $C(14, 6)$ **2.** $C(33, 20)$ **3.** $C(16, 9)C(10, 5)$ **4.** $C(54, 5)$ **5.** $\dfrac{C(23, 18)}{C(11, 5)}$

COMBINATIONS

The number of combinations, say $C(7, 4)$ can be calculated in EXCEL using the COMBIN function, written as COMBIN(7,4).

The formula $=$COMBIN(A2,A3) will calculate the number of combinations using the numbers in A2 and A3.

EXERCISES

1. Find $C(6, 3)$.　　　　　　**2.** Find $C(12, 7)$.　　　　　　**3.** Find $C(50, 18)$.

6.6　A Mixture of Counting Problems

We have studied four kinds of counting techniques: the Multiplication Rule, the Addition Rule, permutations (a special case of the Multiplication Rule), and combinations. One of the more difficult steps of counting problems is the determination of the appropriate counting technique to be used.

Usually, you expect to solve the exercises in the permutation section using the permutation counting technique. Because we want you to be able to analyze a problem and determine the counting technique, or techniques, to be used, this section has a mixture of problems for which you determine the technique to be used.

We offer a sequence of questions that should help you determine the appropriate counting technique. The basic questions to ask are

"How many sets are selections made from?"

"Is repetition of elements allowed?"

"Is the order of selection or arrangement significant?"

"Are elements selected from each set or is the selection made from one of several sets?"

Figure 6–21 shows how the answers to these questions lead to the appropriate counting technique. Let's apply it to some examples.

In some problems you will need to apply the procedure more than once, as in Examples 3 through 6.

EXAMPLE 1 ➤ Billy's mother allows him to select three Dr. Seuss books from eight such books in the Children's Library. How many ways can Billy make the selection?

SOLUTION

Here are the questions to ask, and the answers that lead us through the diagram:

1. How many sets are involved? One set of eight books.
2. Are repetitions allowed? No, we expect Billy to select three different books.
3. Is order significant? No, the order of selection is irrelevant.

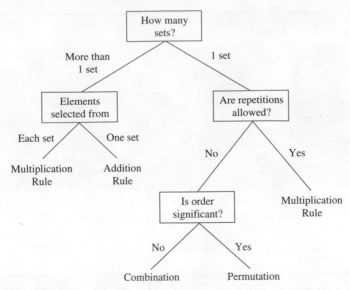

FIGURE 6–21　Procedure to determine the appropriate counting technique.

The tree diagram indicates this is a combination problem with a selection of three objects from a set of eight. The number of possible selections is

$$C(8, 3) = \frac{8 \times 7 \times 6}{3 \times 2 \times 1} = 56 \blacksquare$$

EXAMPLE 2 ➤

Billy is allowed to select one book from eight Dr. Seuss books and one book from six "What If . . . ?" books. How many ways can he make the selection?

SOLUTION

We ask the questions:

"How many sets are involved?" Two, the Dr. Seuss books and the "What If . . . ?" books.

"Is a selection made from each set?" Yes, so this is a Multiplication Rule problem.

Billy can make the selection in $8 \times 6 = 48$ ways. ■

Some problems are multilevel problems, where a determination of the appropriate counting technique must be made at each level, so you may need to use the tree diagram in Figure 6–21 more than once.

EXAMPLE 3 ➤

The Gamma Club has a membership of 10 freshmen, 15 sophomores, 22 juniors, and 12 seniors. The club has an executive group consisting of a chair, vice-chair, and secretary, all of whom must be seniors, and a three-member committee composed of juniors. How many different ways can the executive group be formed?

SOLUTION

The answer to the question "How many sets are involved?" is two because, for the purposes of this problem, one kind of selection is made from the group of juniors and another kind from the seniors so the set of juniors and the set of seniors are the only relevant sets.

Because selections are made from each of the two sets, this is a Multiplication Rule problem with

Number of ways the executive group can be selected
$$= \text{(number of ways the officers can be selected)}$$
$$\times \text{(number of ways the committee of juniors can be selected)}$$

This is a multilevel problem, because we still must determine the number of ways the committee and the officers can be selected.

For the officers there is one set, the seniors. There are no repetitions, because one person does not hold two offices, and order is significant because when two people exchange offices, a different slate is formed. The number of slates equals $P(12, 3) = 1320$.

For the committee, the selections are from one set, the juniors. There are no repetitions because one person cannot hold two committee positions, and order is not significant because no distinction is made in committee positions. This is a combination problem with the number of possible committee formations equaling $C(22, 3) = 1540$.

Finally, the executive group can be formed in $1320 \times 1540 = 2,032,800$ different ways. ∎

EXAMPLE 4 ➤

The Mathematics Department is hosting a high school mathematics contest. Professors Tidmore, Johns, and Cannon have volunteered students in their classes to help with the contest. The coordinator of the contest asks Prof. Tidmore to enlist four students, two female and two male, to serve as a welcoming committee. Prof. Johns is asked to enlist five students to host the lounge where visiting teachers and parents will wait, and Prof. Cannon is asked to provide three students who will help grade: one to grade algebra, one to grade geometry, and one to grade calculator problems. The composition of the classes is

Tidmore's class: 13 males and 15 females

Johns's class: 7 males and 5 females

Cannon's class: 9 females and 9 males

In how many ways can the requested students be selected?

SOLUTION

This is a multilevel selection process, so you may use the decision tree in Figure 6–21 at each level.

Students are selected from each of three sets — Tidmore's class, Johns's class, and Cannon's class — so at the first level, this is a Multiplication Rule problem.

We now need to determine the number of ways each selection can be made.

Tidmore's class: Two males are selected from the 13 males, and 2 females are selected from the 15 females, so the Multiplication Rule applies. The selection of 2 males is a combination, $C(13, 2) = 78$. The selection of 2 females is a combination, $C(15, 2) = 105$. The 4 students can be selected in $78 \times 105 = 8190$ ways.

Johns's class: The 5 students are selected from the 12 students in the class, and there is no assignment of position or order, so this is a combination, $C(12, 5) = 792$ ways.

Cannon's class: The 3 students are selected from the 18 students in the class, and they are assigned a position, or activity, so this is a permutation, $P(18, 3) = 4896$ ways.

Finally, the 12 students can be selected in

$$8190 \times 792 \times 4896 = 31,757,806,080 \text{ ways } \blacksquare$$

Some problems can be correctly worked in more than one way. Sometimes we are interested in just some of all possible outcomes. The outcomes can be classified as those we want and those we don't want. Sometimes it is easier to determine the number we don't want. We use that number to determine the number we do want, as illustrated in the following example.

EXAMPLE 5 ➤

The PIN number used at an ATM machine is a sequence of six digits, using the digits 0 through 9 with repetitions allowed.

(a) How many PIN numbers are possible?
(b) How many PIN numbers are possible with no repetition of digits?
(c) How many PIN numbers are possible with at least one digit repeated?

SOLUTION

(a) Each digit can be selected in ten ways, so the number of PIN numbers is
$10 \times 10 \times 10 \times 10 \times 10 \times 10 = 1,000,000$.
(b) Since digits are not repeated and the sequence of digits is significant, this is a permutation with $P(10, 6) = 151,200$.
(c) The PIN numbers with at least one digit repeated are all possible PIN numbers *except* those with *no* repetition. Thus, we subtract the number with no repetition from the number of all possible to obtain the number with at least one repeated digit:

$$\text{Number with at least one repeat} = 1,000,000 - 151,200$$
$$= 848,800 \ \blacksquare$$

Recall the Addition Rule that states when making selections from one of two or more disjoint sets, the number of possible selections is the sum of the number of selections from each set.

EXAMPLE 6 ➤

The PTA program committee decides to ask for 4 students to participate in the next PTA program. They want the students from either Mr. Dudley's class of 14 students, from Miss DeWeese's class of 17 students, or from Mrs. Bowden's class of 15 students. How many selections are possible?

SOLUTION

One group of 4 students is selected from one of the three classes, so the Addition Rule applies. The number of selections is $C(14, 4) + C(17, 4) + C(15, 4) = 1001 + 2380 + 1365 = 4746. \ \blacksquare$

6.6 EXERCISES

1. Mrs. Brown has 21 children in her first-grade class, and Mr. Lopez has 23 children in his second-grade class. The principal has been asked to select 1 student from one of the classes to appear at a PTA meeting. How many ways can the selection be made?

2. Ms. Ling has 14 boys and 17 girls in her third-grade class. How many ways can she select 2 boys or 2 girls to make a Thanksgiving poster?

3. Eight students are on standby to fly back to school at the end of spring break. The airline finds it has space for three of them. How many ways can the three be selected ignoring who were first to arrive at the terminal?

4. Slimp's Bookstore receives a box of books from each of five paperback publishers. Each box contains 24 different titles. The store manager decides there is room to display 4 books from each of 3 of the publishers. How many ways can the display be chosen?

5. Tony receives a gift certificate to buy a tie, a shirt, and a jacket at Sandy's Clothing. The store has 56 different ties, 8 styles of shirts, and 15 styles of jackets. How many ways can he select his tie, shirt, and jacket?

6. Duane goes to Golden's Used Books during their semiannual sale. There are 75 books on their 50% off table and 96 books on the 30% off table. How many ways can he select 4 books from 50% off and 3 books from 30% off?

7. How many ways can a class president, vice-president, and secretary be selected from each of the seventh grade and eighth grade to form a junior high student council? The seventh grade has 110 members, and the eighth grade has 90 members.

8. In an effort to publicize the importance of wearing seat belts, the local police will monitor vehicles leaving Robinson High school. They will issue three citations to three students leaving the east exit who are not wearing seat belts. One student will be required to write a 200-word paper on the dangers of not wearing a seat belt, one will be fined $25, and the third will make a three-minute speech in assembly on why one should wear a seat belt. Four of the students leaving at the south exit who are not wearing seat belts will be issued warnings. If 48 students leaving the east exit are not wearing seat belts and 28 leaving the south exit are not wearing seat belts, in how many ways can the seven citations/warnings be issued?

9. Courtney's mother took her to Story Time at the public library. While there, Courtney could check out two Dr. Seuss books or one Sesame Street video. The library had 11 Dr Seuss books and 23 Sesame Street videos. How many ways can Courtney make her choice?

10. The Saturday Night Jazz Band asked the music department of the community college to furnish a group of 3 trumpet players or a group of 4 saxophone players for prom weekend. The music department has 8 trumpet players and 11 saxophone players available. How many ways can the request be filled?

11. Mr. Albee has ten saxophone players in his class. He receives a request to provide a group of three saxophone players for a jazz band. How many ways can he select the group?

12. Billy has 43 baseball cards, and Scottie has 36. How many ways can Billy trade 2 of his cards for 2 of Scottie's cards?

13. Because of her work schedule, Molly Kay plans to take classes at 8:00, 10:00, and 11:00. Five courses that fit her program are offered at 8:00, eight at 10:00, and three at 11:00. How many different schedules are possible for Molly Kay?

14. A survey of 500 students found that during the last week 220 had eaten at Burger King, 165 had eaten at Whataburger, and 32 had eaten at both places.
    (a) How many had eaten at one or both places?
    (b) How many had eaten at neither of the two places?

15. A problem in Mrs. Schrader's geometry class has eight steps in the proof. She writes the steps, one per card, and gives them to a student to put in proper sequence. How many different sequences of steps are possible?

16. An ATM keypad has ten keys. A PIN code is entered by pressing a sequence of four keys.

(a) How many PIN codes are possible if no key is pressed twice in a code?

(b) How many PIN codes are possible if a key may be pressed more than once in a code?

(c) A bank changes the PIN codes to a sequence of six keys. Find the number of PIN codes possible if a key may be pressed more than once in a sequence. Discuss the increase in security by going to a six-key code.

17. Prof. Sendon and Prof. Barr were comparing their class rolls. They observed that Prof. Sendon had 28 students, Prof. Barr had 21 students, and 4 students were enrolled in both classes. If they hold a joint meeting of their classes and all students attend, how many students are present?

18. Student Congress has eight members who are seniors and six members who are juniors. At Spring Premiere for prospective students, one of the juniors greets prospective students, one works at the registration table, and a third gives a welcome to the campus greeting. One of the seniors introduces the dean of students, another introduces the academic vice-president, and the third discusses how to apply for financial aid. How many ways can these assignments be made?

19. Fab-Knit has 15 sales representatives. Atlanta and San Diego host trade shows the same week. The company plans to send 3 sales representatives to Atlanta and 4 to San Diego. How many ways can the selections be made?

20. The Mathematics Club is planning a picnic. They will select 3 freshmen for the cleanup crew, a menu committee composed of 1 sophomore and 1 junior, and a program committee of 3 seniors: one of whom plans entertainment, one presents the club awards, and one who tells about the project for the semester. The club has 6 freshmen members, 4 sophomore members, 8 junior members, and 12 senior members. In how many ways can the students be selected for the above assignments?

21. A telephone number has seven digits with repetition of digits allowed, but the first digit cannot be a 0 or a 1.
    (a) How many telephone numbers are possible?
    (b) How many telephone numbers have no repetition of digits?
    (c) How many telephone numbers have at least one digit repeated?

22. A speaker has a collection of 65 jokes. He uses three jokes in each speech.
    (a) In how many ways can he select three jokes for a speech?
    (b) How many speeches can he give without using a joke twice?

23. The Homeowner's Network has a "great deal." If you join their Book of the Month Club today, you will receive a free gift. Your gift choices are 2 books from the 7 books in their HandyMan Series, 3 books from the 10 books in their Gardening Series, or 1 book from the 5 books in their Interior Decorating Series. Mr. Hull decides to join. How many ways can he select his free gift?

24. The Hewitt Elementary School PTA plans to use students in their next program. They want 5 children from Mrs. Perry's PE class to give a PE-activity demonstration; 2 children from Mrs. Brown's class, a boy and a girl, to hand out programs; and 3 children from Ms. Gilchrest's class: 1 to give the welcome, 1 to lead in the pledge to the flag, and 1 to tell about the science fair. Mrs. Perry has 15 children in her PE class, Mrs. Brown has 9 boys and 13 girls in her class, and Ms. Gilchrest has 17 children in her class. In how many ways can the 10 children be selected?

25. A president, vice-president, and secretary are selected from a group of 6 men and 5 women. How many ways can the selection be made so the slate of officers contains at most two men?

26. Anita plans to give her husband 3 shirts for his birthday. She narrows the search to 3 shirts from a selection of 18 shirts at Dillon's Store or 3 from a selection of 23 shirts at The Men's Shop. In how many ways can she select the shirts?

27. Four service clubs are each asked to enlist two of their members to help with the Fall Review for high school students. The clubs are Alpha Club with 22 members; Beta Club, 19 members; Gamma Club, 25 members; and Zeta Club, 14 members. In how many ways can the helpers be selected?

28. A Social Security number has nine digits, using the digits 0 through 9 with repetition of digits allowed. How many Social Security numbers have at least one repeated digit?

29. The Student Life Committee asks that two members from a service club attend the next committee meeting. The two students will be selected from one of the following service clubs: Alpha Club with 22 members; Beta Club, 19 members; Gamma Club, 25 members; and Zeta Club, 14 members. In how many ways can the two students be selected?

30. A tray contains 12 chocolate chip cookies and 16 oatmeal raisin cookies. How many ways can a child select 4 cookies that include at least 1 chocolate chip cookie?

31. A committee of 5 people is selected from a group of 6 men and 7 women. In how many ways can the committee be selected so it contains at least one woman?

32. The reading list for a literature class has six fiction and eight nonfiction books. A student is to write a report on five books from the list during the semester. The reports must include at least 1 and at most 4 reports of fiction books. In how many ways can the selection of books be made?

6.7   Partitions (Optional)

- Ordered Partitions
- Number of Ordered Partitions
- Special Case: Partition into Two Subsets

- Unordered Partitions
- Number of Unordered Partitions

In this section we discuss an idea called the **partitioning** of a set. We want to determine the number of ways a set can be partitioned. We will look at two kinds of partitions, **ordered** and **unordered** partitions. Lets begin with an example to lead us into the ideas.

EXAMPLE 1 ➤

A group of 15 students is to be divided into three groups to be transported to a game. The three vehicles will carry four, five, and six students, respectively. In how many different ways can the three groups be formed?

SOLUTION

Select the four students that ride in the first vehicle. This can be done in

$$C(15, 4) = \frac{15!}{4!11!}$$

different ways. (Notice the form we use for $C(15, 4)$. It is more useful in this case.) After this selection, five students may be selected for the second vehicle in

$$C(11, 5) = \frac{11!}{5!6!}$$

different ways. (There are 11 students left after the first vehicle is filled.) There are six students left for the last vehicle, and they can be chosen in

$$C(6, 6) - \frac{6!}{6!0!}$$

different ways.

By the Multiplication Rule, the total number of different ways is

$$C(15, 4) \times C(11, 5) \times C(6, 6) = \frac{15!}{4!11!} \times \frac{11!}{5!6!} \times \frac{6!}{6!0!}$$

$$= \frac{15!}{4!5!6!} = 630,630$$

■ **Now You Are Ready to Work Exercise 9**

This partition problem has the following properties that make it a partition.

1. The set is divided into disjoint subsets (no two subsets intersect.)
2. Each member of the set is in one of the subsets.

The following is a more formal definition of a partition.

DEFINITION
Partition

> A set S is **partitioned** into k nonempty subsets $A_1, A_2, \dots, A_k$ if:
>
> 1. Every pair of subsets is disjoint: that is, $A_i \cap A_j = \varnothing$ when $i \neq j$.
> 2. $A_1 \cup A_2 \cup \cdots \cup A_k = S$.

Ordered Partitions

We first discuss ordered partitions.

DEFINITION
Ordered Partition

> A partition is **ordered** if different subsets of the partition have characteristics that distinguishes one from the other.

The characteristics that distinguish subsets may vary widely. For example, one subset may be males, another females; one subset may be A students, another C students; one subset is awarded a million dollar contract, another a $1000 contract; one subset is the first team, another the second team, another the third team; one subset contains ten elements, another eight elements.

A partition is ordered if an exchange of two subsets gives a different partition. For example, let 15 basketball players be divided into three teams of five players each, and let the teams be designated as first, second, and third teams. If the division of players is left unchanged, but the first team is now designated as the second team and the second team becomes the first team, then a different partition is obtained. Thus, we have an ordered partition.

Number of Ordered Partitions

We now determine the number of ways in which a set can be partitioned.

From Example 1, we see that the number of ways in which a set of 15 elements can be partitioned into subsets of four, five, and six elements may be expressed as

$$\frac{15!}{4!5!6!}$$

A commonly used notation for this quantity is

$$\binom{15}{4,\,5,\,6}$$

This is generalized in the following theorem.

THEOREM
Ordered Partitions

> A set with n elements can be partitioned into k ordered subsets of $r_1, r_2, \dots, r_k$ elements $(r_1 + r_2 + \cdots + r_k = n)$ in the following number of ways:
>
> $$\binom{n}{r_1, r_2, \dots, r_k} = \frac{n!}{r_1! r_2! \dots r_k!}$$

EXAMPLE 2 ➤ A set of 12 people ($n = 12$) can be divided into three groups of three, four, and five ($r_1, r_2,$ and r_3) in

$$\binom{12}{3, 4, 5} = \frac{12!}{3!4!5!} = 27,720$$

different ways.

■ **Now You Are Ready to Work Exercise 11**

EXAMPLE 3 ➤ The United Way Allocations Committee has 14 members. In how many ways can they be divided into the following subcommittees so that no member serves on two subcommittees?

   Scouting subcommittee: two members

   Salvation Army subcommittee: four members

   Health Services subcommittee: five members

   Summer Recreational Program subcommittee: three members.

SOLUTION

The subcommittees form a partition, because no one is on two subcommittees and all 14 members are used. The partitions are ordered for two reasons: The sub-committees are of different sizes, and they have different functions. The number of partitions is

$$\binom{14}{2, 4, 5, 3} = \frac{14!}{2!4!5!3!} = 2,522,520$$

■ **Now You Are Ready to Work Exercise 15**

EXAMPLE 4 ➤ A college basketball squad has 15 players. In how many ways can the coach form a first, second, and third team of five players each?

SOLUTION

This is an ordered partition because there is a distinction between teams. The number of partitions is

$$\binom{15}{5, 5, 5} = \frac{15!}{5!5!5!} = 756,756$$

■ **Now You Are Ready to Work Exercise 19**

Special Case: Partition Into Two Subsets

Let's look at a special case of partitions. Suppose a set of 8 objects is partitioned into two subsets of three and five objects. The formula for partitions gives

$$\binom{8}{3, 5} = \frac{8!}{3!5!}$$

Notice that the formulas for $C(8, 3)$ and $C(8, 5)$ both give

$$C(8, 3) = \frac{8!}{5!3!} = C(8, 5)$$

so the number of partitions into two subsets is just the number of ways in which a subset of one size can be selected. This result occurs because when one subset of three objects is selected, the remaining five objects automatically form the other subset in the partition.

In general the following is true:

> The number of partitions of a set into two ordered subsets is the number of ways in which one of the subsets can be formed.

Unordered Partitions

We now look at partitions that are not ordered.

DEFINITION
Unordered Partition

> A partition is **unordered** when no distinction is made between subsets.

For a partition to be unordered, all subsets must be the same size, otherwise, the different sizes would distinguish between subsets. When a teacher partitions a class into four equal groups, all groups working on the same problem, an **unordered partition** has been formed. If the four equal groups work on different problems, the partition is **ordered**. If eight members of a traveling squad are paired to room together on the trip, an unordered partition is formed. If the pairs are assigned to rooms 516, 517, 518, and 519, an ordered partition is formed.

Number of Unordered Partitions

A basketball squad of 15 members can be divided into first, second, and third teams of five players each in $\dfrac{15!}{5!5!5!}$ ways. Because a distinction is made between teams, this is an ordered partition. We ask in how many ways an unordered partition can be made; that is, no distinction is made between teams. We can find the number by relating the number of ordered and unordered partitions.

First, divide the 15 players into three teams of five each, with no distinction made between teams. Call these teams A, B, and C. These teams can be ordered into first, second, and third teams in six ways: ABC, ACB, BAC, BCA, CAB, and CBA. You recognize this as the 3! permutations of the three groups. In general, the ordered partitions can be obtained by forming three groups (an unordered partition) and then arranging them in 3! ways. If we let N be the number of unordered partitions, then

$$3!N = \text{number of ordered partitions} = \binom{15}{5, 5, 5}$$

This gives

$$N = \frac{1}{3!}\binom{15}{5, 5, 5} = \frac{15!}{3!5!5!5!}$$

This generalizes to the following theorem:

THEOREM

A set of n elements can be partitioned into k **unordered subsets** of r elements each $(kr = n)$ in the following number of ways:

$$\frac{1}{k!}\binom{n}{r, r, \ldots, r} = \frac{n!}{k!r!r! \ldots r!} = \frac{n!}{k!(r!)^k}$$

EXAMPLE 5 ➤

A set of 12 elements can be partitioned into three unordered subsets of four each in

$$\frac{12!}{3!4!4!4!} = 5775 \text{ ways}$$

■ **Now You Are Ready to Work Exercise 23**

Here is an example of partitioning a set when no distinction is made between some subsets and a distinction is made between others.

EXAMPLE 6 ➤

Find the number of partitions of a set of 12 elements into subsets of three, three, four, and two elements. No distinction is to be made between subsets except for their size.

SOLUTION

Because the two subsets of three elements are the same size, no distinction is made between them. Because they are of different sizes, a distinction is made between subsets of size 2 and 4 (or 2 and 3). The number of ordered partitions is $\dfrac{12!}{3!3!4!2!}$.

The number of unordered partitions is found by dividing by 2! because two sets (of size 3) are indistinct. Thus there are $\dfrac{12!}{2!3!3!4!2!}$ unordered partitions.

■ **Now You Are Ready to Work Exercise 27**

EXAMPLE 7 ➤

Find the number of unordered partitions of a set of 23 elements that is partitioned into two subsets of four elements and three subsets of five elements.

SOLUTION

Because there are two indistinct subsets of four elements and three indistinct subsets of five elements, we divide the number of *ordered* subsets by 2! and 3! to obtain $\dfrac{23!}{2!3!4!4!5!5!5!}$.

■ **Now You Are Ready to Work Exercise 31**

In general the number of unordered partitions is given by the following theorem:

THEOREM

A set of n elements is partitioned into unordered subsets with k subsets of r elements each and j subsets of t each $(kr + jt = n)$. The number of such partitions is

$$\frac{\text{number of ordered partitions}}{k!j!} = \frac{n!}{k!j!(r!)^k(t!)^j}$$

▙ 6.7 EXERCISES

LEVEL 1

Perform the computations in Exercises 1 through 8.

1. $\binom{12}{3,3,3,3}$

2. $\binom{8}{2,2,4}$

3. $\binom{7}{3,4}$

4. $\binom{10}{4,6}$

5. $\binom{9}{2,3,4}$

6. $\binom{8}{4,4}$

7. $\binom{6}{2,4}$

8. $\binom{7}{2,2,2,1}$

LEVEL 2

9. *(See Example 1)* In how many ways can a lab instructor assign nine students so that three perform experiment A, three perform experiment B, and three perform experiment C?

10. In how many ways can a set of nine objects be divided into subsets of two, three, and four objects?

11. *(See Example 2)* In how many ways can 14 people be divided into three groups of three, five, and six?

12. In how many ways can 16 different books be divided into stacks of four, five, and seven books?

13. An accounting instructor separates her 18 students into three groups of six each. Each group is assigned a different problem. In how many ways can the class be divided into these groups?

14. A store has 12 items to be displayed in three display windows. In how many ways can they be displayed if six are placed in one window, four in the second window and 2 in the third window?

15. *(See Example 3)* In how many different ways can a 15-person committee be subdivided into subcommittees having six, four, and five members?

16. The State University football team plays 11 games. In how many ways can they complete the season with four wins, six losses, and one tie?

17. A scholarship committee will award four $5000 scholarships, four $8000 scholarships, and two $10000 scholarships. Ten students are selected to receive scholarships. In how many ways can the scholarships be awarded?

18. A high school dance committee of 17 members is divided into the following subcommittees:

Decorations: four members

Music: two members

Refreshments: three members

Publicity: five members

Ticket sales: three members.

In how many different ways can the subcommittees be assigned?

19. *(See Example 4)* In how many ways can ten players be divided into first and second teams of five members each?

20. In how many different ways can 12 directors of a corporation be divided into four equal groups to discuss new products, sales forecasts, implications of recent legislation, and benefit programs, respectively?

21. A drama class of 18 students is divided into three equal groups to work on costumes, lighting, and backdrops, respectively. In how many ways can this be done?

22. A magazine has 20 articles that need to be evaluated. Four assistant editors are given five articles each to evaluate. In how many ways can this distribution of articles be made?

23. *(See Example 5)* In how many ways can a set of nine elements be partitioned into three unordered subsets of three elements each?

24. In how many ways can a set of eight elements be partitioned into four unordered subsets of two elements each?

25. In how many ways can a set of 12 elements be partitioned into three subsets of equal size?

26. **(a)** A teacher assigns six students to three groups of two each to work on three different homework problems. Each group has a different problem. In how many ways can this be done?

(b) A teacher assigns six students to three groups of two each to work on a homework problem. In how many different ways can this be done?

27. *(See Example 6)* Find the number of partitions of a set of 15 elements into subsets of five, five, three, and two elements. No distinction made between subsets except for their size.

28. Find the number of partitions of a set of 17 elements into subsets of five, five, five, and two elements.

29. An antique shop has 16 dining chairs; no two are alike. They plan to offer them at a reduced price in groups of six, six, and four chairs. In how many ways can the chairs be divided that way?

30. A group of 21 executives is divided into four groups of five, five, five, and six each for a brainstorming session. In how many ways can the division be made? Leave your answer in factorial form.

31. *(See Example 7)* Find the number of unordered partitions of a set of 22 objects that is partitioned into three subsets of two each and four subsets of four each. Leave your answer in factorial form.

32. A bookstore displays 16 different books in groups of three, three, four, two, two, and two books. In how many ways can this be done?

33. A university scholarship committee wishes to award three $5000 scholarships, four $3500 scholarships, and two $8000 scholarships. It selects nine students to receive scholarships. In how many different ways can they be awarded?

LEVEL 3

34. A party of 24 students goes to a restaurant for a study break. They are seated a tables seating two, two, two, four, four, four, and six people. In how many different ways can they be seated? Leave your answer in factorial form.

35. During a mixer for 50 students, the students are divided into ten groups of five students each. In how many different ways can this be done? Leave your answer in factorial form.

36. In the game of bridge, the deck of cards contains 52 cards. Each of the four players receives 13 cards. In how many different ways can this be done? Leave your answer in factorial form.

■■ IMPORTANT TERMS

6.1

Set
Element of a Set
Set-Builder Notation
Equal Sets
Empty Set
Universe (Universal Set)
Venn Diagram
Subset
Proper Subset
Union
Intersection

Complement
Disjoint Sets

6.2

$n(A)$
Inclusion–Exclusion Principle

6.3

Tree Diagram
Multiplication Rule
Addition Rule

6.4

Permutation
Factorial
Permutation with Identical Objects

6.5

Combination

6.7

Partition
Ordered Partition
Unordered Partition

■■ REVIEW EXERCISES

1. Let $A = \{6, 10, 15, 21, 30\}$, $B = \{6, 12, 24, 48\}$, $C = \{x \mid x$ is an integer divisible by 3$\}$. Identify the following as true or false.
   (a) $21 \in A$
   (b) $21 \in B$
   (c) $25 \in C$
   (d) $30 \notin A$
   (e) $16 \notin B$
   (f) $24 \notin C$
   (g) $6 \in A \cap B \cap C$
   (h) $12 \in A \cap B$
   (i) $10 \in A \cup B$
   (j) $A \subseteq B$
   (k) $B \subseteq C$
   (l) $C \subseteq A$
   (m) $\varnothing \subseteq B$
   (n) $A \subseteq C$
   (o) A and B are disjoint

2. Let the universe set $U = \{-2, -1, 0, 1, 2, 3, 4\}$, $A = \{-2, 0, 2, 4\}$, $B = \{-2, -1, 1, 2\}$. Find the following.
   (a) A'
   (b) B'
   (c) $(A \cap B)'$
   (d) $A' \cap B'$
   (e) $A' \cup B'$
   (f) $A \cup A'$

3. Which of the following pairs of sets are equal?
   (a) $A = \{x \mid x$ is a digit in the number 25102351$\}$
       $B = \{x \mid x$ is a digit in the number 5111023$\}$
   (b) $A = \{x \mid x$ is a letter in the word PATTERN$\}$
       $B = \{x \mid x$ is a letter in the word REPEAT$\}$

   (c) $A = \{2, 4, 9, 8\} \cap \{6, 7, 20, 22, 23\}$
   $B = \{x \mid x$ is a letter in both words STRESS
   and HAPPY$\}$

4. $n(A) = 27$, $n(B) = 30$, and $n(A \cap B) = 8$. Find
   $n(A \cup B)$.

5. $n(A \cup B) = 58$, $n(A) = 32$, and $n(B) = 40$. Find
   $n(A \cap B)$.

6. A and B are sets in a universe U with $n(U) = 42$,
   $n(A) = 15$, $n(B) = 24$, and $n(A \cup B)' = 8$. Find
   $n(A \cup B)$ and $n(A \cap B)$.

7. Draw a tree diagram showing the ways in which you
   can select a meat and then a vegetable from roast,
   fish, chicken, peas, beans, and squash.

8. The freshman class traditionally guards the school
   mascot the night before homecoming. There are five
   key locations where a freshman is posted. Nine
   freshmen volunteer for the 2:00 A.M. assignment. In
   how many different ways can they be assigned?

9. How many different license plates can be made using four digits followed by two letters
   (a) If repetitions of digits and letters are allowed?
   (b) If repetitions are not allowed?

10. Strecker Museum has a display case with four dis-
    play compartments. Eight antique vases are avail-
    able for display. How many ways can the display be
    arranged with one vase in each compartment?

11. A medical research team selects five patients at ran-
    dom from a group of 15 patients for special treat-
    ment. In how many different ways can the patients
    be selected?

12. In how many ways can Andrew invite one or more
    of his four friends to come to his house to play?

13. One student representative is selected from each of
    four clubs. In how many different ways can the four
    students be selected, given the following number of
    members in each club: Rodeo Club, 40 members;
    Kite Club, 27 members; Frisbee Club, 85 members;
    and Canoeing Club, 34 members.

14. In the finale of the University Sing, there are ten
    people in the first row. Club A has three members on
    the left end, club B has four members in the center,
    and Club C has three members on the right end. In
    how many different ways can the line be arranged?

15. A program consists of four musical numbers and
    three speeches. In how many ways can the program
    be arranged so that it begins and ends with a musi-
    cal number?

16. Students take four exams in Sociology 101. On each
    exam the possible grades are A, B, C, D, and F. How
    many sequences of grades can a student receive?

17. An advertising agency designs 11 full-page ads for
    Uncle Dan's Barbecue. In how many ways can one
    ad be selected for each of three different magazines
    (a) If the three ads are different?
    (b) If the ads need not be different?

18. A computer password is composed of six alpha-
    betic characters. How many different passwords
    are possible?

19. In how many different ways can a chairman, a secre-
    tary, and four other committee members be formed
    from a group of ten people?

20. The KOT club has 12 pledges. On a club workday,
    four pledges are assigned to the Red Cross, six are
    assigned to the Salvation Army, and two are not
    assigned. In how many ways can the groups be
    selected?

21. A survey of 60 people gave the following information:

    25 jog regularly.

    26 ride a bicycle regularly.

    26 swim regularly.

    10 both jog and swim.

    6 both swim and ride a bicycle.

    7 both jog and ride a bicycle.

    1 does all three.

    3 do none of the three.

    Show that there is an error in this information.

22. The Spirit Shop had a sale on records, books, and
    T-shirts. A cashier observed the purchases of 38
    people and found that

    16 bought records.

    15 bought books.

    19 bought T-shirts.

    5 bought books and records.

    7 bought books and T-shirts.

    6 bought records and T-shirts.

    3 bought all three.

    (a) How many bought records and T-shirts but no
    books?
    (b) How many bought records but no books?
    (c) How many bought T-shirts but no books and no
    records?
    (d) How many bought none of the three?

23. A poll was conducted among a group of teenagers
    to see how many have televisions, radios, and micro-
    computers. The results were as follows: T denotes
    television, R denotes radio, and M denotes micro-
    computer.

Item	Number of Teenagers Having This Item
T	39
R	73
M	10
T and R	22
M and R	3
T and M	4
T and R and M	2

Determine the following.
(a) How many had a radio and TV but no micro-computer?
(b) How many had a microcomputer and had no TV?
(c) How many had exactly two of the three items?

24. During the summer, 110 students toured Europe. Their language skills were as follows: 46 spoke German, 56 spoke French, 8 spoke Italian, 16 spoke French and German, 3 spoke French and Italian, 2 spoke German and Italian, and 1 spoke all three.
(a) How many spoke only French?
(b) How many spoke French or German?
(c) How many spoke French or Italian but not both?
(d) How many spoke none of the languages?

25. Mrs. Bass has five bracelets, eight necklaces, and seven sets of earrings. In how many ways can she select one of each to wear?

26. The Labor Day Raft Race has 110 entries. In how many ways is it possible to award prizes for the fastest raft, the slowest raft, and the most original raft?

27. From a group of five people, two are to be selected to be delegates to a conference. How many selections are possible?

28. In how many different ways can a group of 15 people select a president, vice-president, and secretary?

29. Twenty people attend a meeting at which three different door prizes are awarded by drawing names.
(a) If a name is drawn and replaced for the next drawing, in how many ways can the door prizes be awarded?
(b) If a name is drawn and not replaced, in how many ways can the door prizes be awarded?

30. A bag contains six white balls, four red balls, and three green balls. In how many ways can a person draw out two white balls, three red balls, and two green balls?

31. An Honor Council consists of 4 seniors, 4 juniors, 3 sophomores, and 1 freshman. Fifteen seniors, 20 juniors, 25 sophomores, and 11 freshmen apply. In how many ways can the Honor Council be selected? Leave your answer in symbolic form.

32. An art gallery has eight oil paintings and four water-colors. A display of five oil paintings and two water-colors arranged in a row is planned. How many different displays are possible with a watercolor at each end and the oils in the center?

33. A club agrees to provide five students to work at the school carnival. One sells balloons, one sells pop-corn, one sells cotton candy, one sells candied apples, and one sells soft drinks. Nine students agree to help. In how many ways can the assignments be made?

34. Prof. Goode gives a reading list of six books. A student is to read three. In how many ways can the selection be made?

35. Five students are to be chosen from a high school government class of 22 students to meet the governor when he visits the school. In how many ways can this be done?

36. How many different five-card hands can be obtained from a deck of 52 cards?

37. Compute:
(a) $P(8, 4)$ **(b)** $C(9, 5)$ **(c)** $P(7, 7)$ **(d)** $C(5, 5)$
(e) $4!$ **(f)** $\dfrac{7!}{3!4!}$ **(g)** $\dfrac{8!}{4!}$
(h) $\begin{pmatrix} 15 \\ 4, 5, 6 \end{pmatrix}$ **(i)** $\begin{pmatrix} 9 \\ 3, 3, 3 \end{pmatrix}$ (unordered)

38. One day a machine produced 50 good circuit boards and eight defective ones.
(a) In how many ways can two defective circuit boards be selected?
(b) In how many ways can three good circuit boards be selected?
(c) In how many ways can two defective and three good circuit boards be selected?

39. A club has 80 members of whom 20 are seniors, 15 are juniors, 25 are sophomores, and 20 are freshmen. A chair, vice-chair, secretary, and treasurer are to be selected. The chair and vice-chair must be seniors, the treasurer must be a junior, and the secretary must be a sophomore. How many different slates of officers can be formed?

40. The Campus Deli offers caffeinated and decaffeinated regular coffee, and caffeinated and decaffeinated French Roast coffee. Coffee creamer is available in plain, Irish Creme, and Hazelnut flavors. Sugar and two brands of sweetener are available.
(a) How many ways can a student select a coffee and a creamer?
(b) How many ways can a student select a decaffeinated coffee and a creamer, including the choice of no creamer?
(c) How many ways can a student select a coffee, one creamer, and sugar or a sweetener?

41. A social organization and a service club held a joint meeting. Of the 83 people present, 46 belonged to the social organization, and 51 belonged to the service club. How many belonged to both?

42. The digits {2, 3, 4, 5, 6, 7} are used to form three-digit numbers.
   (a) How many can be formed if repetitions are allowed?
   (b) How many can be formed if repetitions are not allowed?
   (c) How many larger than 500 can be formed with repetitions allowed?

43. The Sports Mart store has ten sportswear outfits for display purposes. In how many ways can a group of four outfits be selected for display?

44. Nye Printing has 16 female employees and 14 male employees. How many different advisory committees consisting of 2 males and 2 females are possible?

45. List all the subsets of {red, white, blue}.

46. A panel of four is selected from eight businessmen and seated in a row behind a table. In how many different orders can they be seated?

47. The Physics club has ten freshmen and eight sophomore members. At a club picnic, the cook and entertainment leader are freshmen, and the cleanup crew consists of three sophomores. In how many different ways can these five be selected?

48. Draw a tree diagram showing the ways a girl and then a boy can be selected from the children Carlos, Betty, Darla, Gary, and Natasha.

49. A student is allowed to check out four books from the reserve room. All the books must come from one collection of six books or from another collection of eight books. In how many different ways can the selection be made?

50. Three married couples are seated in a row. How many different seating arrangements are possible:
   (a) if there is no restriction of seating order?
   (b) if the men sit together and the women sit together?
   (c) if a husband and wife sit together?

51. How many different words are possible using all the letters of
   (a) RELAX?   **(b)** PUPPY?   **(c)** OFFICIAL?

Probability

When a weather forecaster predicts the weather, when a coach evaluates the team's chances of winning, or when a businessperson projects the success of the big clearance sale, an element of uncertainty exists. The weather forecaster knows he is often wrong, the coach knows there is no such thing as a sure win, and the businessperson knows the best advertised sale sometimes flops. Often in our daily lives we would like to measure the likelihood of an outcome of an event or activity. We ask such questions as "Which topics are more likely to be on the test?" "What are my chances of finding a parking place in time to get to my 9 o'clock class?" "Will we have a quiz today?" ■

7.1   Introduction to Probability

- Terminology
- Empirical Probability
- Properties of Probability
- Probability Assignments
- A Visual Model of Probability

An area of mathematics known as **probability theory** provides a measure of the likelihood of the outcome of phenomena and events. The government uses it to determine fiscal and economic policies; theoretical physicists use it to understand the nature of atomic-sized systems in quantum mechanics; and public-opinion polls, such as the Harris Poll, have their theoretical acceptability based on probability theory.

The theory of probability is said to have originated from the following gambling question: Two gamblers are playing for a stake that goes to the player who first wins a specified number of points. The game is interrupted before either player has won enough points to win the stake. (We don't know whether the game was raided.) The question is to determine a fair division of the stakes based on the number of points won by each player at the time of the interruption.

This problem was mentioned from time to time in the mathematical literature for a period of 150 years but received no widespread attention until it was proposed to the French mathematician Blaise Pascal about 1654. Pascal communicated the problem to another French mathematician, Pierre Fermat, who also became quite interested in it. The two mathematicians arrived at the same answer by different methods. Their discussions aroused quite a bit of interest in that type of problem. Pascal and Fermat are generally considered the founders of probability theory.

Probability turned out to have applications far beyond interrupted gambling games. J. C. Maxwell used probability theory to derive his famous gas laws in 1860. Edmund Halley, the first astronomer to predict the return of a comet, applied probability to actuarial science in 1693. Today, insurance companies depend on probability to determine competitive and profitable rates for their policies. Quality control in manufacturing and product development decisions are based on probability. The military uses it in the theory of search for enemy submarines. We will learn some elementary applications in this chapter.

Terminology

We use the terms *probability, experiment, outcome,* and *trial* in our discussions. When we ask the likelihood that a tossed coin will turn up heads, we call the activity of tossing the coin an **experiment,** the result of tossing the coin an **outcome.** For this experiment, two outcomes are possible, "heads" and "tails." Each time the coin is tossed we have a **trial.** We measure the likelihood of "heads" with a number, which we call the **probability** of heads, and we use the notation $P(\text{heads})$ to denote that number.

DEFINITION
Experiment, Outcome, Trial

An activity or phenomenon that is under consideration is called an **experiment.** The experiment can produce a variety of observable results, called **outcomes.** We study activities that can be repeated or phenomena that can be observed a number of times. We call each observation or repetition of the experiment a **trial.**

We can apply the terms *experiment, outcome,* and *trial* in a wide variety of ways, as illustrated in the next example.

EXAMPLE 1 ➤

(a) Drawing a number out of a hat is an experiment with the number drawn as an outcome. Each draw of a number is a trial.

(b) A test to determine the germination of flower seeds is an experiment with "germinated" and "not germinated" as possible outcomes. Each test conducted is a trial.

(c) A drawing of lottery numbers is an experiment with the numbers drawn as an outcome. Each draw is a trial. ■

In general, experiments involve chance or **random** results. This means that the outcomes do not occur in a set pattern but vary depending on impartial chance, and the outcome cannot be determined in advance. The order in which leaves fall off a tree, the number of cars that pass a checkpoint on the freeway, and the selection of a card from a well-shuffled deck are examples of experiments that have random outcomes.

Probability deals with random outcomes that have some long-term pattern for which it is not possible to predict what happens next, but a prediction can be made of what will happen in the long term.

Life insurance companies know quite accurately how many people will die by a certain age. They cannot tell when a certain individual will die, but they can predict general, long-term numbers of deaths.

The determination of probabilities can be a difficult and expensive process, but we can't escape the desire to measure the likelihood, or probability, that a certain outcome occurs. What is the probability of getting a ticket if I exceed the speed limit? What are the chances of my book being stolen if I leave it on a shelf in the cafeteria? What is the probability of a walk in history class today?

The losing basketball team may intentionally foul a member of the winning team in the closing minutes of a game in hopes that the player will miss the free throw. This gives the losing team a chance of getting the ball. The winning team wants Mike to have the ball, because he has made 80% of his free throws for the season. The losing team wants to foul Art, because he has made 55% of his free throws. Based on this past history, each team assumes that Mike is more likely to make the free throw than Art.

In order to compare the likelihood of two different outcomes, a number is assigned to each outcome. We consider the outcome that is assigned the larger number to be more likely. In the basketball game, it seems natural to use the numbers 80% and 55% as a measure of the probability of making free throws.

Let's look at one way to measure probability, called **empirical probability.**

Empirical Probability

Probabilities may be assigned by observing a number of trials and using the frequency of outcomes to estimate probability. For example, the operator of a concession stand at a park keeps a record of the kinds of drinks children buy. Her records show the following:

Drink	Frequency
Cola	150
Lemonade	275
Fruit juice	75
Total	500

To estimate the probability that a child will buy a certain kind of drink, we compute the **relative frequency** of each drink. We do this by dividing the frequency of each drink by the total number of drinks.

Drink	Frequency	Relative Frequency
Cola	150	$\dfrac{150}{500} = 0.30$
Lemonade	275	$\dfrac{275}{500} = 0.55$
Fruit juice	$\dfrac{75}{500}$	$\dfrac{75}{500} = 0.15$

This experiment has three outcomes, $S = \{\text{cola, lemonade, fruit juice}\}$. We use relative frequency to estimate probability. The probability a child will buy lemonade is 0.55, which we write symbolically as $P(\text{lemonade}) = 0.55$; also, $P(\text{cola}) = 0.30$ and $P(\text{fruit juice}) = 0.15$.

The determination of empirical probability requires gathering data from which relative frequencies are obtained. In practice, this may be a highly sophisticated operation such as a national political poll.

EXAMPLE 2 ➤   A college has an enrollment of 1210 students. The number in each class is as shown in the following table:

Class	Number of Students
Freshman	420
Sophomore	315
Junior	260
Senior	215
Total	1210

A student is selected at random. Determine the empirical probability that the student is

(a) a freshman.
(c) a junior.
(e) a freshman or sophomore.

(b) a sophomore.
(d) a senior.

(A **random selection** means that each student has a chance of being selected and any two students have equal chances of being selected. To select people in a random manner usually requires careful planning and methodology, often a difficult process.)

SOLUTION

Estimate the probability of each as the relative frequency.

Class	Number of Students	Relative Frequency
Freshman	420	$\dfrac{420}{1210} = 0.35$
Sophomore	315	$\dfrac{315}{1210} = 0.26$
Junior	260	$\dfrac{260}{1210} = 0.21$
Senior	215	$\dfrac{215}{1210} = 0.18$

This gives $P(\text{freshman}) = 0.35$, $P(\text{sophomore}) = 0.26$, $P(\text{junior}) = 0.21$, $P(\text{senior}) = 0.18$, and $P(\text{freshman or sophomore}) =$

$$\frac{420 + 315}{1210} = \frac{420}{1210} + \frac{315}{1210} = 0.35 + 0.26 = 0.61$$

■ Now You Are Ready to Work Exercise 19

Our intuition suggests that a tossed coin will turn up heads one half of the time. However, if you toss a coin 10 times, you should not expect 5 heads and 5 tails. (Try it.) Should you repeat the experiment several times, you should expect more than one value for the relative frequency. Thus, observations of past outcomes can only provide *estimates* of the probability. We can get better estimates by repeating the trials more times. As the number of trials increases, we expect to obtain better estimates of the probability. This idea is known as the **Law of Large Numbers.**

THEOREM
Law of Large Numbers

As an experiment is repeated more and more times, the relative frequency obtained approaches the actual probability.

The Law of Large Numbers has been tested for the coin tossing problem. The Comte de Buffon (1701–1788), Karl Pearson (1857–1936), and John Kerrich, a prisoner of war during World War II, each tossed a coin many times. The results of their efforts were,

	Tosses	Heads	Relative frequency
Buffon	4,040	2,048	0.5069
Pearson	12,000	6,019	0.5016
Pearson	24,000	12,012	0.5005
Kerrich	10,000	5,067	0.5067

Although these results fail to exactly give our intuitive value of 0.5000 for the probability of heads, they do suggest that 0.5000 is a reasonable probability.

Properties of Probability

Let's observe the properties of relative frequency that correspond to properties needed for probability assignments in general. First, let's discuss the terminology we use.

An experiment need not classify outcomes in a unique way. It depends on how the results are interpreted. When a multiple-choice test of 100 questions is given, the instructor wants to know the number of correct answers given by each student. For this purpose, an outcome can be any of the numbers 0 through 100. When the tests are returned to the students, they tend to ask, "What is an A?" They are interested in the outcomes A, B, C, D, and F. Then there might be the student who only asks, "What is passing?" To that student there are just two outcomes of interest, pass and fail.

When asked "What is today?" a person may respond in several ways, such as "It is April 1," "It is Friday," "It is payday," or any one of numerous responses. Depending on the focus of the individual, the set of possible responses may be all the days in a year, all the days of the week, or the two outcomes payday and not payday.

Because the outcomes of an experiment can be classified in a variety of ways, it is important that the appropriate set of outcomes be selected and that everyone understand which set of outcomes is used. We call the set of outcomes used a **sample space.**

DEFINITION
Sample Space

> A **sample space** is the set of all possible outcomes of an experiment. Each element of the sample space is called a **sample point** or **simple outcome.**

The next example illustrates some sample spaces. Notice that the sample space can be defined in more than one way for the same experiment.

EXAMPLE 3 ➤

(a) If the experiment is tossing a coin, the sample space is {heads, tails}.
(b) If the experiment is drawing a card from a bridge deck, one sample space is the set of 52 cards.
(c) If the experiment is drawing a number from the numbers 1 through 10, the sample space can be {1, 2, 3, 4, 5, 6, 7, 8, 9, 10}. Sometimes people are assigned a number and those assigned an odd number are placed in one group and those assigned an even number are placed in another group. In this case, the sample space of interest is {even, odd}.
(d) If the experiment is tossing a coin twice, a sample space is {HH, HT, TH, TT}.

■ **Now You Are Ready to Work Exercise 1**

We do not insist on just one correct sample space for an experiment because the situation dictates how to interpret the results. However, we do insist that a sample space conform to two properties.

Properties of a Sample Space

Let S be the sample space of an experiment.

1. Each element in the set S is an outcome of the experiment.
2. Each outcome of the experiment corresponds to exactly one element in S.

If a student is selected from a group of university students and the class standing of the student is the outcome of interest, then {freshman, sophomore, junior, senior} is a valid sample space. If the gender of the student is the outcome of interest, then {male, female} is a valid sample space. You can form other sample spaces using age, GPA, and so on as the outcomes of interest.

In defining the outcomes of an experiment, care must be taken that the properties of a sample space hold. In an experiment involving the GPA of students, the second property of a sample space is violated if we define the outcomes of interest as

Outcome	GPA
Unacceptable	0.0–0.9
Marginal	1.0–1.9
Acceptable	2.0–2.9
Superior	3.0–4.0

This definition provides no outcome for a GPA such as 1.95 or 2.97, so property 2 does not hold.

Property 2 is violated if we define the outcomes so that a GPA belongs to two outcomes:

Outcome	GPA
Unacceptable	0.0–1.0
Marginal	1.0–2.0
Acceptable	2.0–3.0
Superior	3.0–4.0

In this case, the GPA scores of 1.0, 2.0, and 3.0 are indicated to be in two different outcomes, so property 2 does not hold.

Property 1 would be violated if we defined the unacceptable outcome as -1.0 through -0.9, because there are no negative values of GPA.

In some instances our interest lies in a *collection* of outcomes in the sample space, not just one outcome. If I toss a coin twice, I may be interested in the likelihood that the coin will land with the same face up both times. I am interested in the subset of outcomes {HH, TT}, not just one of the possible outcomes. We call such a collection of simple outcomes an **event.** In Example 2, we might be interested in the event that a freshman or sophomore is selected.

DEFINITION
Event

An **event** is a subset of a sample space.
An event can be a subset consisting of a single outcome. Such an event is called a **simple event.** An event also can be as much as the entire sample space.

We focus primarily on the probability of events that includes the probability of a simple event. An event often can be formed in more than one way from a set of simple outcomes. The next example shows some variations in events.

EXAMPLE 4 ➤

(a) In the experiment of drawing a number from the numbers 1 through 10, the sample space is

$$S = \{1, 2, 3, 4, 5, 6, 7, 8, 9, 10\}$$

The event of drawing an odd number is the subset $\{1, 3, 5, 7, 9\}$. The event of drawing an even number is the subset $\{2, 4, 6, 8, 10\}$. The event of drawing a prime number is the subset $\{2, 3, 5, 7\}$.

(b) A teacher selects one student from a group of six students. The sample space is {Scott, Jane, Mary, Kaye, Ray, Randy}. The event of selecting a student with first initial R is {Ray, Randy}. The event of selecting a student with first initial J is {Jane}. The event of selecting a student with first initial A is the empty set. ∎

An Event Occurs

> We say that an **event occurs** if the trial yields an outcome that is in the event set.

Probability Assignments

To help understand the properties of a probability assignment, we look again at Examples 1 and 2 to observe some properties that hold for probability assignments generally.

We want to accomplish two goals when a probability assignment is made:

1. Assign a probability, a number, to each simple outcome in the sample space that will indicate the relative frequency with which the simple outcome occurs.
2. Assign a probability to each set of outcomes, that is, to each event, that also indicates the relative frequency with which the event occurs.

In Examples 1 and 2, notice the following:

- Each relative frequency can be as small as 0 or as large as 1. In general, a probability is in the interval 0 through 1.
- For a sample space, the relative frequencies add to 1. In general, the probabilities of all simple outcomes add to 1.
- The relative frequency of an event is the sum of the relative frequencies of the simple outcomes making up the event.

Although a general probability assignment need not be formed using relative frequencies, the assignment cannot be arbitrary. It must satisfy some standard conditions similar to those seen in relative frequency.

Properties of Probability

> Let $S = \{e_1, e_2, \ldots, e_n\}$ be a sample space with simple outcomes $e_1, e_2, \ldots e_n$.
>
> 1. Each simple outcome in a sample space, e_i, is assigned a probability denoted by $P(e_i)$.
> 2. The probability of an event E is determined by the simple outcomes making up E. $P(E)$ is the sum of the probabilities of all simple outcomes making up E. For example, if $E = \{e_1, e_2, e_3\}$, then $P(E) = P(e_1) + P(e_2) + P(e_3)$.

> **3.** Each probability is a number that is not negative and is no larger than 1; for simple events e_i, $0 \leq P(e_i) \leq 1$ and for each event E, $0 \leq P(E) \leq 1$.
> **4.** $P(S) = 1$, that is, $P(S) =$ the sum of probabilities of all simple events in a sample space
>
> $$P(S) = P(e_1) + P(e_2) + \cdots + P(e_n) = 1$$

The next three examples help to understand the properties of probability.

EXAMPLE 5 ➤

As part of a class assignment, Annette, Ben, and Casie are sent to different classroom buildings to poll students on the number of siblings they have. They are to classify them into five categories: 0, 1, 2, 3, and 4 or more. They are to report their findings as empirical probability. Here are their reports.

Annette reported:

$$P(0) = 0.20, \quad P(1) = 0.10, \quad P(2) = 0.15$$
$$P(3) - 0.30, \quad P(4 \text{ or more}) = 0.25$$

Is this a valid probability assignment? It is, because

1. Each outcome is assigned a probability.
2. Each probability is nonnegative and not larger than 1.
3. The sum of the probabilities of all simple events is 1.

Possibly, Annette made errors in counting so her report may not be an *accurate* empirical probability, but it is *valid* because it satisfies the properties of probability.

Ben reported:

$$P(0) = 0.2, \quad P(1) - 0.2, \quad P(2) = 0.3$$
$$P(3) - 0.25, \quad P(4 \text{ or more}) = 0.4$$

Is this a valid probability assignment? It is not valid, because the sum of all probabilities is 1.35.

Casie reported:

$$P(0) = 0.3, \quad P(1) - -0.1, \quad P(2) = 0.2$$
$$P(3) = 0.3, \quad P(4 \text{ or more}) = 0.3$$

Is this a valid probability assignment? The sum of all probabilities is 1.00, but the assignment is not valid because one of the probabilities, $P(1)$, is negative.

■ **Now You Are Ready to Work Exercise 5**

EXAMPLE 6 ➤

The sample space of an experiment is $\{A, B, C, D\}$, where $P(A) = 0.35, P(B) = 0.15, P(C) = 0.22, P(D) = 0.28$. The properties of probability enable us to find the probabilities of the following events:

$$P(\{A, B\}) = 0.35 + 0.15 = 0.50$$
$$P(\{A, C\}) = 0.35 + 0.22 = 0.57$$
$$P(\{B, C, D\}) = 0.15 + 0.22 + 0.28 = 0.65$$
$$P(\{A, B, C, D\}) = 0.35 + 0.15 + 0.22 + 0.28 = 1.00$$

■ **Now You Are Ready to Work Exercise 7**

EXAMPLE 7 ➤ An ice chest at a junior high picnic contains three brands of soft drinks, Pepsi, Coke, and Dr Pepper, and there are some regular and some diet drinks of each brand. The mathematics teacher had filled the chest and counted the number of each kind of drink, so she was able to tell the students "If you get a drink from the chest without looking, then for each kind of drink, here is the probability of its being selected."

	Pepsi	**Coke**	**Dr Pepper**
Regular	0.05	0.15	0.23
Diet	0.10	0.17	0.30

"For example, notice the probability of selecting a regular Dr Pepper is 0.23. Using our probability notation of $P(E)$, we write it as $P(\text{regular Dr Pepper})$. Now before you get your drinks, verify that the following are correct."

(a) What is the probability of drawing a Coke? In this example, the event is the subset {regular Coke, diet Coke}. We ask for the probability that the drink selected is in that subset. According to the second condition in the properties of probability, the probability of selecting a Coke is $0.15 + 0.17 = 0.32$, the sum of the probabilities of the simple outcomes making up the event.

In a similar manner, we obtain the probability of the following event:

(b) The probability of drawing a diet drink is $0.10 + 0.17 + 0.30 = 0.57$ $[P(\text{diet drink}) = 0.57]$.

(c) The probability of selecting a regular Pepsi or a diet Dr Pepper is $0.05 + 0.30 = 0.35$.

(d) The probability of selecting a drink that is not a Dr Pepper is $0.05 + 0.15 + 0.10 + 0.17 = 0.47$ $[P(\text{not Dr Pepper}) = 0.47]$.

■ **Now You Are Ready to Work Exercise 11**

EXAMPLE 8 ➤ A new student asks about the chances of finding a parking place on campus. His roommate, who is taking finite mathematics, responds with the following statement: "The probability of finding a parking place at remote parking is twice the probability of finding a place in the parking garage, and the probability of finding a place near your classroom building is one half the probability of finding a place in the parking garage." Assuming that the roommate was correct, find the probability of each outcome: nearby, parking garage, or remote parking.

SOLUTION

Let $P(G)$ = the probability of finding a place in the parking garage. Then, for remote parking, $P(R) = 2P(G)$, and for nearby, $P(N) = 0.5P(G)$. Because these probabilities must add to 1, we have

$$2P(G) + P(G) + 0.5P(G) = 1$$
$$3.5P(G) = 1$$
$$P(G) = \frac{1}{3.5} = 0.286 \quad \text{(rounded)}$$

From this we have $P(R) = 0.572$, $P(G) = 0.286$, and $P(N) = 0.143$. (Because we rounded to three decimal places, these add to 1.001.)

■ **Now You Are Ready to Work Exercise 15**

Recall that a probability is a number in the interval from 0 through 1, with 0 representing the probability that the event *cannot* occur and 1 representing the probability that the event *must* occur. If a selection is made from a sample space, it cannot come from the empty set because the empty set has no elements. Thus, the probability of the empty set is zero. The probability of the sample space, $P(S)$, is the probability that the outcome will come from the sample space. Because the selection is made from the sample space, the outcome must be in the sample space, so that makes $P(S) = 1$. We highlight these special cases.

Reminder: Two Important Special Cases

1. If an event is the empty set, the probability that an outcome is in the event is zero; that is, $P(\varnothing) = 0$.
2. If an event is the entire sample space, then $E = S$ and $P(E) = P(S) = 1$.

A Visual Model of Probability

NOTE

When we represent an event with a circle, we do not attempt to draw it to scale because the figure is intended only to show relationships and is not intended to be used to estimate probabilities.

We can use area in a Venn diagram to visualize the probability of an event. We let a rectangle, the universe, represent the sample space and a circle represent an event (Figure 7–1). We let the probability of an event be represented by the area of the figure representing it. Thus, we let the rectangle representing the sample space have area 1. If the probability of an event is 0.20, then we think of the area of circle E as 20% of the sample space.

Intuitively, a Venn diagram suggests that when an event A is a subset of an event B (Figure 7–2), then the probability of A cannot be larger than the probability of B.

Likewise, the probability that an outcome is in the intersection of events cannot be larger than the probability of either of the events (Figure 7–3) and the probability of an event A and the probability of an event B must not exceed the probability of their union (Figure 7–4).

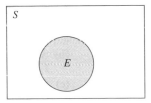

FIGURE 7–1   An event in sample space S.

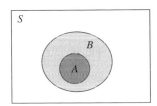

FIGURE 7–2
$P(A) \le P(B)$

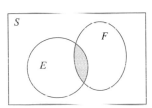

FIGURE 7–3
$P(E \cap F) \le P(E)$
$P(E \cap F) \le P(F)$

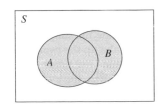

FIGURE 7–4
$P(A) \le P(A \cup B)$
$P(B) \le P(A \cup B)$

36. An inspection of desks made at a furniture plant reveals that 2% of the desks have structural defects, 3% have finish defects, and 1% have both. A desk is selected at random. Find the probability that the desk has
(a) at least one kind of defect.
(b) neither kind of defect.

37. A jeweler's bag contains six rings. One is plain gold, one is plain silver, one is gold with emeralds and rubies, one is silver with diamonds, one is gold with diamonds and emeralds, and one is silver with rubies. If a ring is selected at random, find the probability that it
(a) is a gold ring.
(b) is a ring with diamonds.
(c) is a ring with diamonds or emeralds.
(d) does not have rubies.

Determine whether or not the events given in Exercises 38 through 41 are mutually exclusive.

38. A person is selected from the customers at Fast Foods. F is the event that a female is selected, and G is the event that a child is selected.

39. Consider the experiment of selecting a student from a college. F is the event of an international student attending on a student visa, and G is the event of a U.S. student.

40. A child is selected from a group of children who were born prematurely. F is the event that the child suffers some loss of hearing, and G is the event that the child suffers some loss of vision.

41. A coin is tossed three times. F is the event that it lands heads up on the second toss, and G is the event that it lands tails up on the second toss.

42. A single card is drawn from a deck of 52 bridge cards. Find the probability that it is
(a) a 5.
(b) a club.
(c) a red spade.
(d) an even number.
(e) a 9 or a 10.
(f) a heart, diamond, or black card.

43. On a single toss of one die, find the probability of tossing
(a) a number less than 6.
(b) the number 4.
(c) 2, 4, or 5.
(d) an odd number less than 5.
(e) 1 or 3.

44. A card is drawn at random from a deck of 52 playing cards. Find the probability of drawing
(a) an ace or a 7.
(b) a 6, a 7, or an 8.

45. The Select-Three lottery has three containers each containing ten Ping-Pong balls numbered 1 through 10. A fan blows the balls upward until one comes out of the opening at the top of the container. Find the probability that the ball that comes out of the first container is
(a) number 10.
(b) number 1.
(c) an even number.
(d) an odd number.
(e) a number greater than 5.
(f) an even number or a number greater than 4.
(g) an even number or a number less than 8.

46. Two dice are thrown. What is the probability that the same number appears on both dice?

47. A card is drawn from a deck of 52 playing cards. Find the probability that it is an ace or a spade.

LEVEL 3

48. A person is selected at random from a pool of ten people consisting of six men and four women for a psychology study. Find the probability that the person selected is a
(a) man.
(b) woman.

49. For a proposed piece of legislation the probability of its passing the House is 0.76, the probability of its passing the Senate is 0.62, and the probability of its passing at least one is 0.89. Find the probability of its passing both.

50. A person's birthday is known to be in April. What is the probability that it is
(a) April 15?
(b) in the first seven days of April?
(c) in the first half of April?

51. In a corporation, 65% of the employees are female, executives, or both. Furthermore, 55% of the employees are female, and 5% are female executives. Find the percentage of employees who are male executives.

52. The professor reveals only the following information about a class: There are 21 seniors, 14 English majors, 8 students who are neither seniors nor English majors, and a total of 34 students.
(a) Determine the number who are both seniors and English majors.
(b) Determine the number who are English majors but are not seniors.
(c) If a student is selected at random, find the probability that the student is a senior.

53. A die is rolled. What is the probability that the number is odd or is a 2?

54. Lauren attended her 10th high school reunion. Of the 65 people present, she observed that 31 were married and in their first marriage, 22 were married and in their second marriage, and 12 were currently single.
  (a) If one person is selected at random from the group, find the probability that it is a single person.
  (b) If two people are selected at random, find the probability that both are in their second marriage.
  (c) If three people are selected at random, find the probability that two are in their first marriage and one is single.

55. There are 115 passengers waiting to board flight 622 at the Denver airport. The number of passengers who have two carry-on articles is 64, the number who have one carry-on is 43, and the number who have no carry-on is 8. The security personnel randomly selects 3 passengers to search their carry-on articles. Find the probability that
  (a) all 3 have one carry-on.
  (b) 1 has one carry-on and 2 have two carry-ons.
  (c) 1 has one carry-on, 1 has two carry-ons, and 1 has no carry-on.

56. Three numbers are drawn from $\{1, 2, 3, 4, 5, 6, 7\}$ and placed in a row in the order drawn to form a three-digit number. What is the probability that
  (a) the number is 456?
  (b) the first digit is 2, 4, or 6?
  (c) it is larger than 600?
  (d) it is less than 326?

57. A plumber needs a certain part to complete a repair job. The probability that one part supplier has the part is 0.85, the probability that the other part supplier has the part is 0.93, and the probability that both have the part is 0.81. Find the probability that the plumber will be able to complete the repair.

EXPLORATIONS

58. The English Department advertises for three faculty positions. They receive 87 applications, 54 women and 33 men. The department selection committee announces that the applicants are equally qualified and they select three by a random method. All three are women. A male applicant claims discrimination. Justify why you think his claim is or is not reasonable.

59. Alice has a single die, and Francine has a pair of dice. Which is more likely?
  (a) Alice obtains at least one 6 in 4 rolls of the die.
  (b) Francine obtains two 6's (on the same roll) at least once in 26 rolls of the dice.
  Justify your answer.

60. Here is an experiment for a group of people. Each person tosses a coin until heads appears twice in succession. Record the number of the toss on which the heads appeared the second time (of the two in a row). Repeat 15 times.

Each person computes the probability the two heads in a row will occur on the second toss, the third toss, and so on.

Find the group totals and use them to calculate two heads in a row on the second toss, the third toss, and so on.

61. To promote attendance at basketball games, County College conducts the following activity at halftime. A fan is selected who is given 20 green tennis balls and 20 white tennis balls. The fan is to divide the 40 balls between two boxes. The balls may be divided in any way the fan chooses. The fan is blindfolded, the boxes shuffled, and then the fan selects one of the boxes and a tennis ball from the box chosen. If the ball selected is white, the fan wins $1000.

How many balls of each color would you put in each box in order to maximize your chances of winning the $1000?

7.4   Conditional Probability

- Conditional Probability
- Multiplication Rule

We have computed the probability of events involving equally likely outcomes by determining the number of outcomes in the event and in the sample space. We have learned to compute the probability when events are used to form a compound event. We now turn our attention to some instances when these computations are affected by a related event or by additional conditions imposed. These may modify the sample space and thereby change the probability.

Conditional Probability

A simple example in which the probability is adjusted based on additional information is the following. Suppose you are taking a test with multiple-choice questions. A question has four possible answers listed, and you have no idea of the correct answer. If you make a wild guess, the probability of selecting the correct answer is $\frac{1}{4}$. However, if you know one of the answers cannot be correct, then your chance of guessing the correct answer improves because the sample space has been reduced to three elements. You now choose from three answers, increasing the probability of guessing correctly to $\frac{1}{3}$.

We denote this situation by **conditional probability.** In general terms we describe conditional probability as follows. We seek the probability of an event E. A related event F occurs, giving reason to change the sample space and thereby potentially changing the probability of E.

We say that we want to determine the probability of E given that F occurred and use the notation $P(E \mid F)$, which we read "the probability of E given F."

We can state the multiple-choice question example as "The probability of guessing the correct answer given that one answer is known to be incorrect is $\frac{1}{3}$." Or, we can write $P(\text{correct} \mid \text{one answer known incorrect}) = \frac{1}{3}$.

If a student guesses wildly at the correct answer from the four given ones, the sample space consists of the four possible answers. When one answer is ruled out, the sample space reduces to three possible answers. Sometimes it helps to look at a conditional probability problem as one in which the sample space changes when certain conditions exist or related information is given. Let's look at the following from that viewpoint.

EXAMPLE 1 ▶   A student has a job testing microcomputer chips. The chips are produced by two machines, I and II. It is known that 5% of the chips produced by machine I are defective and 15% of the chips produced by machine II are defective. The student has a batch of chips that she assumes is a mixture from both machines. If she selects one at random, what is the probability that it is defective? You cannot give a precise answer to this question unless you know the proportion of chips from each machine. It does seem reasonable to say that the probability lies in the interval from 0.05 through 0.15.

Now suppose that the student obtains more information: The chips are all from machine II. This certainly changes her estimate of the probability of a defective chip; she knows that the probability is 0.15. The sample space changes from a set of chips from both machines to a set of chips from machine II. This illustrates the point that when you gain information about the state of the experiment, you may need to change the probabilities assigned to the outcomes. ∎

Here's how we write some of the information from the preceding examples.

EXAMPLE 2 ➤

(a) Machines I and II produce microchips, with 5% of those from machine I being defective and 15% of those from machine II being defective. This can be stated as

$$P(\text{defective chip} \mid \text{machine I}) = 0.05$$
$$P(\text{defective chip} \mid \text{machine II}) = 0.15$$

(b) There are four possible answers to a multiple-choice question, one of which is correct. The probability of guessing the right answer is $\frac{1}{4}$. However, if one incorrect answer can be eliminated, the sample space is reduced from four to three answers, and the probability of guessing correctly becomes $\frac{1}{3}$. This is stated as

$$P(\text{guessing correct answer}) = \frac{1}{4}$$

$$P(\text{guessing correct answer} \mid \text{one incorrect answer eliminated}) = \frac{1}{3}$$

■ **Now You Are Ready to Work Exercise 1**

Let's look at an example of how to compute $P(E \mid F)$.

EXAMPLE 3 ➤

Professor Baird teaches two sections of philosophy. The regular section has 35 students, and the honors section has 25 students. The professor gives both sections the same test, and 14 students make an A, 5 in the regular section and 9 in the honors section.

(a) If a test paper is selected at random from all papers, what is the probability that it is an A paper?
(b) A test paper is selected at random. If it is known that the paper is from the honors section, what is the probability that it is an A paper?

SOLUTION

We will refer to Venn diagrams as we work this problem to visualize the solution and to illustrate some general principles of conditional probability.

In this example, the sample space is the collection of all 60 papers, the event E is the set of all A papers, and F is the set of papers from the honors class (Figure 7–15a).

(a) $P(E)$ = the probability a paper selected from the 60 papers is an A paper (Figure 7–15b).

$$P(E) = \frac{14}{60}$$

(b) The knowledge that the paper is from the honors section restricts the outcomes to F (Figure 7–15c), so F becomes the **reduced sample space.** If the paper selected is an A paper, it then must come from the A papers in the hon-

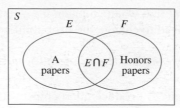

S = All of the exams, 60 total
E = All A exam papers, 14 total
F = Papers from Honors section, 25 total
$E \cap F$ = A papers from Honors section, 9 total

(a)

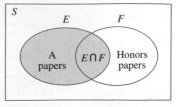

$P(E)$ can be represented by the area of E

(b)

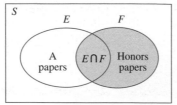

The paper is an Honors paper, so the paper must come from F. Thus, F becomes the reduced sample space.

(c)

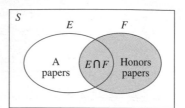

The event $E|F$ focuses on the outcomes in $E \cap F$
$n(E \cap F) = 9$, $n(F) = 25$

(d)

FIGURE 7–15

ors section, from $E \cap F$. Because the honors section contains 9 A papers and 25 papers total,

$$P(E \mid F) = \frac{9}{25}$$

We can express this as

$$P(E \mid F) = \frac{9}{25} = \frac{n(E \cap F)}{n(F)}$$

Let's take this a step farther and divide the numerator and denominator of the last fraction by $n(S)$, giving

$$P(E \mid F) = \frac{\dfrac{n(E \cap F)}{n(S)}}{\dfrac{n(F)}{n(S)}} = \frac{P(E \cap F)}{P(F)}$$

■ **Now You Are Ready to Work Exercise 3**

In terms of using area as a visual probability model, the probability $P(E \mid F)$ is the ratio

$$\frac{\text{area of } E \cap F}{\text{area of } F}$$

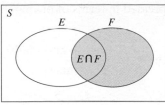

For $P(E|F)$
F = reduced sample sample space
$E \cap F$ = the event
$$P(E|F) = \frac{\text{Area of } E \cap F}{\text{Area of } F} = \frac{P(E \cap F)}{P(F)}$$

FIGURE 7–16

Be sure you understand that, in essence, F *becomes* the sample space when we compute $P(E \mid F)$, and we focus our attention on the contents of F. We sometimes call F the **reduced sample space.** Then, $E \cap F$ becomes the event of successful outcomes. (See Figure 7–16.)

DEFINITION	
Conditional Probability $P(E \mid F)$	E and F are events in a sample space S, with $P(F) \neq 0$. The conditional probability of E given F, denoted by $P(E \mid F)$, is **(a)** $P(E \mid F) = \dfrac{P(E \cap F)}{P(F)}$ This holds whether or not the outcomes of S are equally likely. **(b)** If the outcomes of S are equally likely, $P(E \mid F)$ may be written as $P(E \mid F) = \dfrac{n(E \cap F)}{n(F)}$

EXAMPLE 4 ➤

In a group of 200 students, 40 are taking English, 50 are taking mathematics, and 12 are taking both.

(a) If a student is selected at random, what is the probability that the student is taking English?
(b) A student is selected at random from those taking mathematics. What is the probability that the student is taking English?
(c) A student is selected at random from those taking English. What is the probability that the student is taking mathematics?
(d) A student is selected at random from those taking English. What is the probability that the student is not taking mathematics?

SOLUTION

Figure 7–17 represents the given information with the Venn diagrams.

(a) $P(\text{English}) = \dfrac{40}{200} = \dfrac{1}{5}.$

(b) This problem is that of finding $P(\text{English} \mid \text{math})$, so

$$P(\text{English} \mid \text{math}) = \frac{n(\text{English and math})}{n(\text{math})} = \frac{12}{50} = \frac{6}{25} = 0.24$$

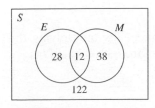

FIGURE 7–17

This may also be expressed in terms of probability:

$$P(\text{English}\,|\,\text{math}) = \frac{P(\text{English and math})}{P(\text{math})} = \frac{\dfrac{12}{200}}{\dfrac{50}{200}} = \frac{0.06}{0.25} = 0.24$$

(c) This asks for

$$P(\text{math}\,|\,\text{English}) = \frac{n(\text{math and English})}{n(\text{English})} = \frac{12}{40} = \frac{3}{10}$$

Parts (b) and (c) illustrate that $P(E\,|\,F)$ and $P(F\,|\,E)$ might not be equal.

(d) This asks for

$$P(\text{not math}\,|\,\text{English}) = \frac{n(\text{not math and English})}{n(\text{English})}$$

$$= \frac{28}{40} = \frac{7}{10}$$

Notice this is also

$$P(\text{not math}\,|\,\text{English}) = 1 - P(\text{math}\,|\,\text{English})$$

$$= 1 - \frac{3}{10}$$

Because {not math | English} is the complement of {math | English}.

■ **Now You Are Ready to Work Exercise 9**

In Section 7.3 we studied some useful properties of probability. These properties hold for conditional probability if the conditions are applied consistently. The following are properties that follow from the three theorems of Section 7.3.

Properties of $P(E\,|\,F)$

$$P(E\,|\,F) + P(E'\,|\,F) = 1$$
$$P([A \cup B]\,|\,F) = P(A\,|\,F) + P(B\,|\,F) - P((A \cap B)\,|\,F)$$

If A and B are mutually exclusive events, then

$$P((A \cup B)\,|\,F) = P(A\,|\,F) + P(B\,|\,F)$$

We now can find how to compute the probability of E and F, using the **Multiplication Rule.**

Multiplication Rule

We obtain a useful formula for the probability of E and F by multiplying the equation (a) in the definition of conditional probability throughout by $P(F)$:

$$P(E\,|\,F) = \frac{P(E \cap F)}{P(F)}$$

Multiply by $P(F)$ to obtain

$$P(F)P(E\,|\,F) = P(E \cap F)$$

Because

$$P(F\,|\,E) = \frac{P(F \cap E)}{P(E)} = \frac{P(E \cap F)}{P(E)}$$

we also have

$$P(E)P(F\,|\,E) = P(E \cap F)$$

Thus, we have two forms by which to compute $P(E \cap F)$.

THEOREM

Multiplication Rule for Conditional Probability

E and F are events in a sample space S.

$$P(E \text{ and } F) = P(E \cap F) = P(F)P(E\,|\,F)$$

or

$$P(E \text{ and } F) = P(E \cap F) = P(E)P(F\,|\,E)$$

This theorem states that we can find the probability of E and F by multiplying the probability of E by the conditional probability of F given E. Let's use the area model of probability again to visualize the Multiplication Rule.

First, look at $P(E \cap F)$ as the fraction of S occupied by $E \cap F$. We would like to find that fraction from information given about E and F.

If we know $P(F)$, we know the fraction of S occupied by F. (If $P(F) = 0.40$, then F occupies 40% of S in Figure 7–18a). If we know the fraction of F occupied by $E \cap F$, then we know $P(E\,|\,F)$. (If $E \cap F$ occupies 35% of F, $P(E\,|\,F) = 0.35$ in Figure 7–18b). Then, F occupies 40% of S and $E \cap F$ occupies 35% of F (35% of the 40% occupied by F). So, $E \cap F$ occupies $(0.35)(0.40) = 0.14$ of S, or $E \cap F$ occupies 14% of S.

Looking back over these computations, we have computed

$$P(E \cap F) = (0.35)(0.40) = P(E\,|\,F)P(F)$$

which is our Multiplication Rule.

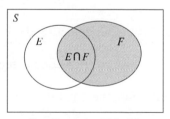

Say, F occupies 40% of S

(a)

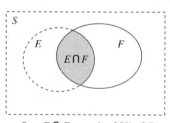

Say, $E \cap F$ occupies 35% of F

(b)

Then, $E \cap F$ occupies $(0.35)(0.40)$ of S

(c)

FIGURE 7–18

EXAMPLE 5 ➤ Two cards are drawn from a bridge deck, without replacement. What is the probability that the first is an ace and the second is a king?

SOLUTION

According to the Multiplication Rule for conditional probability, we want to find

$$P(\text{ace first and king second}) = P(\text{ace first}) \times P(\text{king second} \mid \text{ace first})$$

Because the first card is drawn from the full deck of 52 cards,

$$P(\text{ace first}) = \frac{4}{52}$$

This first card is not replaced, so the sample space for the second card is reduced to 51 cards. As we are assuming that the first card was an ace, there are still four kings in the deck. Then,

$$P(\text{king second} \mid \text{ace first}) = \frac{4}{51}$$

It then follows that

$$P(\text{ace first and king second}) = \left(\frac{4}{52}\right) \times \left(\frac{4}{51}\right) = \frac{4}{663}$$

■ **Now You Are Ready to Work Exercise 19**

EXAMPLE 6 ➤ A box contains 12 light bulbs, 3 of which are defective. If 3 bulbs are selected at random without replacement, what is the probability that all 3 are defective?

SOLUTION

$$P(\text{first defective}) = \frac{3}{12}$$

Given that the first bulb is defective, there are 2 defective bulbs left, so

$$P(\text{second defective}) = \frac{2}{11}$$

For the third selection, there is 1 defective bulb left, so

$$P(\text{third defective}) = \frac{1}{10}$$

According to the Multiplication Rule, applied twice,

$$P(\text{first defective and second defective and third defective})$$

$$= \left(\frac{3}{12}\right)\left(\frac{2}{11}\right)\left(\frac{1}{10}\right) = \frac{1}{220}$$

■ **Now You Are Ready to Work Exercise 23**

Here are two examples in which selections are made *with replacement.*

EXAMPLE 7 ➤

Two cards are drawn from a bridge deck. The first card is drawn, the outcome is observed, and the card is replaced and the deck shuffled before the second card is drawn. Find the probability that the first is an ace and the second is a king.

SOLUTION

The fact that the first card is replaced before the second card is drawn makes this example different from Example 5. We still use the Multiplication Rule, but notice the difference when we compute

$$P(\text{ace first and king second})$$
$$= P(\text{ace first}) \times P(\text{king second} \mid \text{ace first})$$

When we compute $P(\text{ace first})$, we get $\frac{4}{52}$, just as we did in Example 5.

When the second card is drawn, the deck still contains 52 cards, because the first card drawn was replaced. This gives

$$P(\text{king second} \mid \text{ace first}) = \frac{4}{52}$$

We now have

$$P(\text{ace first and king second})$$
$$= P(\text{ace first}) \times P(\text{king second} \mid \text{ace first})$$
$$= \frac{4}{52} \times \frac{4}{52} = \frac{1}{169}$$

Be sure you understand how this example differs from Example 5. They both use the basic property that

$$P(E \cup F) = P(E)P(F \mid E)$$

but replacing the first card in this example makes the sample space for the second draw different from the sample space in Example 5. In Example 5, the sample space for the second draw contains 51 cards, an ace having been removed. In this example, the sample space for the second draw contains 52 cards because the ace drawn first was replaced before the second draw occurred.

■ **Now You Are Ready to Work Exercise 27**

These two examples illustrate that you need to be sure you understand what effect the first action has on the second when you compute conditional probability.

The next two examples use the principles introduced in this section.

EXAMPLE 8 ➤

A box contains five red balls, six green balls, and two white balls. Three balls are drawn, but each one is replaced before the next one is drawn. Find the probability the first is red, the second is green, and the third is white.

SOLUTION

$$P(\text{red first and green second and white third})$$
$$= P(\text{red first}) \times P(\text{green second} \mid \text{red first})$$
$$\times P(\text{white third} \mid \text{red first and green second})$$
$$= \frac{5}{13} \times \frac{6}{13} \times \frac{2}{13} = \frac{60}{2197}$$

Notice that the sample space has 13 elements for each draw because the balls are replaced. If the balls are not replaced after each draw, the probability is

$$\frac{5}{13} \times \frac{6}{12} \times \frac{2}{11} = \frac{60}{1716}$$

■ **Now You Are Ready to Work Exercise 31**

EXAMPLE 9 ➤

A large corporation has 1500 male and 1200 female employees. In a fitness survey, it was found that 40% of the men and 30% of the women are overweight. An employee is selected at random. Find the probability that

(a) the person is male.
(b) the person is overweight, given the person is a male.
(c) the person is overweight, if the person is selected from the females.
(d) the person is overweight.
(e) the person is male if the person is selected from the overweight persons.

SOLUTION

Let O represent the event of an overweight person, let F represent the event of a female, and let M represent the event of a male.

(a) We want $P(M)$, which is $\dfrac{1500}{2700} = \dfrac{5}{9}$.

(b) $P(O|M)$ is the statement that 40% of the males are overweight, so $P(O|M) = 0.40$.

(c) This asks to find $P(\text{overweight}\,|\,\text{female})$ and it is given that 30% of the females are overweight, so $P(O|F) = 0.30$.

(d) The information given tells us that 600 males are overweight (40% of 1500) and 360 females are overweight (30% of 1200), giving a total of 960 overweight persons. Then,

$$P(O) = \frac{960}{2700} = \frac{32}{90}$$

(e) We want to find $P(M|O)$:

$$P(M \cap O) = \frac{600}{2700} = \frac{2}{9} \quad \text{and} \quad P(O) = \frac{32}{90}$$

So,

$$P(M|O) = \frac{P(M \cap O)}{P(O)} = \frac{\frac{2}{9}}{\frac{32}{90}} = \frac{10}{16} = 0.625$$

■ **Now You Are Ready to Work Exercise 35**

We now summarize the main ideas in this section.

Conditional Probability

If $P(F) \neq 0$,

$$P(E|F) = \frac{P(E \cap F)}{P(F)}$$

If the outcomes are equally likely, this may be written as

$$P(E|F) = \frac{n(E \cap F)}{n(F)}$$

Properties of $P(E|F)$

$$P(E|F) + P(E'|F) = 1$$
$$P(A \cup B|F) = P(A|F) + P(B|F) - P(A \cap B|F)$$

If A and B are mutually exclusive events, then

$$P(A \cup B|F) = P(A|F) + P(B|F)$$

Multiplication Rule

$$P(E \cap F) = P(F)P(E|F) = P(E)P(F|E)$$

■■ 7.4 EXERCISES

LEVEL 1

1. *(See Example 2)* The probability that Jane will solve a problem is $\frac{3}{4}$, and the probability that Jill will solve the problem is $\frac{1}{2}$.
   (a) What is $P(\text{problem will be solved} | \text{Jane})$?
   (b) What is $P(\text{problem will be solved} | \text{Jill})$?

2. TV Consolidated manufactures TV sets on two assembly lines, A and B. Quality control studies found 1% of TV sets assembled on line A are defective and 2.5% of those on line B are defective. State this information in terms of conditional probability.

3. *(See Example 3)* A French professor gives Sections 1 and 2 of a French course the same passage to translate. Section 1 has 24 students, and 7 translate the passage correctly. Section 2 has 21 students, and 9 translate the passage correctly.
   (a) If one of the papers is selected at random, what is the probability that it is a correct translation?
   (b) One of the papers is selected and is known to be from Section 1. What is the probability that it is a correct translation?
   (c) A paper is selected and is known to be a correct translation. What is the probability that it is from Section 2?

4. A letter is drawn at random from the 26 letters of the alphabet. Find the probability that it is an
   (a) a, given that it is a vowel.
   (b) x, given that it is not a vowel.
   (c) x, y, or z, given that it is not a vowel.

5. A wallet contains seven \$1 bills, three \$5 bills, and five \$10 bills. A bill is selected at random from the wallet. Find the probability that the bill is
   (a) a \$5 bill, given that it is not a \$1 bill.
   (b) a \$1 bill, given that it is smaller than \$10.

6. A creative writing class has 32 students, of whom 16 are seniors, 12 are juniors, and 4 are sophomores. Nine of the seniors, 5 of the juniors, and 2 of the sophomores are journalism majors. A student is selected at random. Find the probability that the student is
   (a) a journalism major, knowing the student is a junior.
   (b) not a senior, given that the student is a journalism major.

7. A class has 15 boys and 10 girls. One student is selected. F is the event of selecting a girl, and K is the event of selecting Kate, one of the girls in the class. Determine $P(K|F)$ and $P(F|K)$.

8. A card is drawn from a pack of 52 playing cards. Find the probability that the card will be a king, given that it is a face card. (The face cards are the jacks, queens, and kings.)

9. *(See Example 4)* In a group of 60 children, 28 are enrolled in a summer swimming program, 20 signed up for soccer, and 6 are in both.
   (a) If a child is selected at random, find the probability that the child is enrolled in swimming.

(b) If a child is selected from those signed up for soccer, what is the probability that the child is enrolled in swimming?

(c) If a child is selected from those enrolled in swimming, what is the probability that the child is signed up for soccer?

10. Northwest Furniture Factory makes desks. Inspectors find that the probability that a desk has a structural defect is 0.04, the probability that it has a defect in finish is 0.09, and the probability of both kinds of defect is 0.02. A desk is randomly selected. Find the probability that it has a

(a) structural defect, given that it has a defect in finish.

(b) defect in finish, given that it has a structural defect.

Compute $P(E \mid F)$ and $P(F \mid E)$ in Exercises 11 through 14.

11. $P(E) = 0.6, \quad P(F) = 0.7, \quad P(E \cap F) = 0.3$

12. $P(E) = \dfrac{4}{5}, \quad P(F) = \dfrac{3}{5}, \quad P(E \cap F) = \dfrac{1}{5}$

13. $P(E) = 0.60, \quad P(F) = 0.40, \quad P(E \cap F) = 0.24$

14. $P(E) = \dfrac{3}{7}, \quad P(F) = \dfrac{2}{7}, \quad P(E \cap F) = \dfrac{1}{7}$

15. A mathematics professor assigns two problems for homework and knows that the probability of a student solving the first problem is 0.75, the probability of solving the second is 0.45, and the probability of solving both is 0.20.

(a) Jed has solved the second problem. What is the probability he also solves the first problem?

(b) Edna has solved the first problem. What is the probability she also solves the second problem?

16. The following table summarizes the graduating class of Old Main University:

	B.A.	B.S.	B.B.A.	Total
Male	180	60	240	480
Female	159	23	194	376
Total	339	83	434	856

A student is selected at random from the graduation class. Find the probability that the student is

(a) male.

(b) receiving a B.A. degree.

(c) a female receiving a B.B.A. degree.

(d) a female, given that the student is receiving a B.S. degree.

(e) receiving a B.A. degree, given that the student is a female.

(f) a male, knowing that the student will be selected from those receiving a B.A. or B.S. degree.

(g) a male student who is receiving a B.A. or B.S. degree.

17. A university cafeteria surveyed the students who ate breakfast there for their coffee preferences. The findings are summarized as follows:

	Do Not Drink Coffee	Prefer Regular Coffee	Prefer Decaffeinated Coffee	Total
Female	23	145	69	237
Male	18	196	46	260
Total	41	341	115	497

A student is selected at random from this group. Find the probability that the student

(a) does not drink coffee.

(b) is male.

(c) is a female who prefers regular coffee.

(d) prefers decaffeinated coffee, the student being selected from the male students.

(e) is male, given that the student prefers decaffeinated coffee.

(f) is female, given that the student prefers regular coffee or does not drink coffee.

18. A standardized reading test was given to fourth- and fifth-grade classes at an elementary school. A summary of the results is the following:

	Scoring Below Grade Level	Scoring at Grade Level	Scoring Above Grade Level	Total
4th grade	120	342	216	678
5th grade	105	324	98	527
Total	225	666	314	1205

A student is selected at random. Find the probability the student is a

(a) fourth-grade student.

(b) fourth-grade student scoring above grade level.

(c) fifth-grade student scoring at or above grade level.

(d) fourth-grade student when the student is selected from those scoring below grade level.

(e) student scoring below grade level, given the student is in the fourth grade.

19. *(See Example 5)* Two cards are drawn from a bridge deck without replacement. Find the probability that the first is a 4 and the second is a 5.

20. Two people are selected at random from eight men and six women. Find the probability that the first is a woman and the second is a man.

21. A card is drawn at random from a deck of playing cards. It is replaced, and another card is drawn. Find the probability that
  (a) the first card is an 8 and the second is a 10.
  (b) both are clubs.
  (c) both are red cards.

22. The letters of the word BETTER appear on a child's blocks, one letter per block. The child arbitrarily selects two of the blocks. Find the probability that the child selects a T followed by R.

23. *(See Example 6)* A person draws three balls in succession from a box containing four red balls, two white balls, and six blue balls. Find the probability that the balls drawn are red, white, and blue, in that order.

24. Three people are selected from a group of seven men and five women. Find the probability that
  (a) all three are men.
  (b) the first two are women and the third is a man.

25. A company motor pool contains six Dodge and eight Ford cars. Each of two salespeople is randomly assigned a car. Find the probability that both are assigned Dodges.

26. A random selection of four books is made from a shelf containing six novels, five chemistry books, and three history books. Find the probability that all four are novels.

27. *(See Example 7)* The name of each person attending a club meeting is written on a card and placed in a box for a drawing of two door prizes. A name is drawn for one prize, and the name is replaced before the second one is drawn. If 32 people are present, find the probability that both prizes go to the Rosen family. Three members of the Rosen family are present.

28. In a second semester English class, 30% of the students are sophomores and the rest freshmen; 25% of the sophomores are repeating the course and 15% of the freshmen gained admission to the course by advanced placement. The professor randomly selects a student to comment on the assignment.
  (a) Find the probability that the professor selects a sophomore who is repeating the course.
  (b) Find the probability that the professor selects a freshman who gained advanced placement admission to the class.

29. Fifteen cards are numbered 1 through 15. Two cards are drawn with replacement.
  (a) Find the probability that both are less than 5.
  (b) Find the probability that the first is less than 5 and the second is greater than 12.

30. A pen holder contains 12 identical pens, 4 of which do not write. A child randomly selects a pen, replaces it, and selects again. Find the probability that both pens do not write.

31. *(See Example 8)* A box contains four black balls, six red balls, and five green balls. Three balls are drawn, and each one is replaced before the next one is drawn.
  (a) Find the probability that the balls drawn are red, black, and red in that order.
  (b) Find the probability that all three balls are green.

32. On a trip, the Shih family keeps the children entertained by having them guess the last digit of the license of each car they pass. Find the probability that the last digit of the license of the next three cars are 2, 6, and 9, in that order.

33. Two dice are tossed. What is the probability that their sum will be 6, given that exactly one face shows 2?

34. Two students are selected from a class of 30 students containing 10 freshmen and 8 sophomores. Find the probability that the first is a freshman and the second is a sophomore.

LEVEL 2

35. *(See Example 9)* In a group of college students, 60 males and 75 females, 35% of the males and 40% of the females are from out of state. A student is randomly selected. Find the probability that the person is
  (a) a female.
  (b) from out of state, given that the person is a male.
  (c) from out of state when the person is selected from the females.
  (d) from out of state.
  (e) female when the selection is made from the out-of-state persons.

36. In a group of 80 professional people, 25 are accountants, 15 are engineers, 30 are teachers, and

10 are nurses. The number of females within each profession is 11 accountants, 6 engineers, 24 teachers, and 9 nurses. A person is selected at random from the group. Find the probability that the selection is

(a) a female.
(b) an engineer.
(c) a female engineer.
(d) a female, given the person is an engineer.
(e) an engineer, when the selection is made from the females.
(f) a female, if the selection is made from those who are accountants or engineers.
(g) a teacher or nurse, if the selection is made from females.

37. Ms. Speegle's fourth-grade class has 11 boys and 9 girls. She randomly selects two students to return books to the library. Find the probability that the students selected are
(a) 2 girls.
(b) a boy and a girl, in that order.

(c) a boy and a girl, order being unimportant.

38. At a college, 32% of the freshmen score above 550 on the SAT mathematics test. Of those scoring above 550, 87% make a grade of C or higher in their mathematics course. A freshman is selected at random. What is the probability that the student selected at random scored above 550 and makes a C or above in mathematics?

39. As an incentive to do homework, Prof. Rouse randomly draws three students' names, and they are expected to explain a homework problem of the day. On a particular day, 5 students have not done homework, and 17 have.
Find the probability that
(a) the first two drawn have not done homework and the third has.
(b) all three drawn have not done homework.
(c) all three drawn have done homework.
(d) at least one of the three drawn has done homework.

LEVEL 3

40. The following experiment is used to check a person who claims to have powers of mental telepathy. Five cards are numbered 1 through 5. Seat the person being tested on one side of a screen and a person who will select a card on the other side. The person with the cards shuffles them, selects one at random, and turns it up. The person being tested "receives vibes" and tells which number he thinks is turned up. The card is replaced, and the process is repeated for a total of five times.
Based on random chance alone,
(a) find the probability that the person being tested identifies all five numbers correctly.
(b) find the probability that the person being tested identifies all five numbers incorrectly.
(c) find the probability that the person being tested identifies at least one of the five numbers correctly.
(d) if this sequence is repeated three times and each time the person being tested identifies at least one of the sequence of five numbers correctly, would you say this supports powers of mental telepathy?

41. If one check in 10,000 is forged, 5% of all checks are postdated, and 80% of forged checks are postdated, find the probability that a postdated check is forged.

42. In a corporation, 30% of the employees hold college degrees, and 85% of those holding college degrees earn over $30,000. If an employee is selected at random, what is the probability that the employee has a college degree and earns over $30,000?

43. Thirty percent of a freshman class are awarded scholarships, and the probability that a scholarship will continue to the sophomore year is 0.90. If a freshman is selected at random, what is the probability that it is a student who holds a scholarship that will be continued the next year?

44. The winning ticket of a lottery is a sequence of three numbers chosen from the numbers 1 through 100.
(a) If a number may repeat in the sequence, find the probability that a player guesses the winning sequence.
(b) If a number may not repeat in the winning sequence, find the probability that a player guesses the winning sequence.

45. The Pick Three lottery picks one number from each of three containers that each contain the numbers 1 through 9. Find the probability that
(a) a 1 is picked from the first container, a 2 from the second, and a 3 from the third.
(b) the three numbers add to 5.

46. Of a group of children, 0.4 are boys and 0.6 are girls. Of the boys, 0.5 have brown eyes; of the girls, 0.3 have brown eyes. A child is selected at random from the group.
(a) Find the probability that the child is a girl.
(b) Find $P(\text{brown eyes} \mid \text{boy})$.

(c) Find the probability that the child is a boy with brown eyes.

(d) Find the probability that the child is a girl with brown eyes.

(e) Find the probability that the child has brown eyes.

(f) Find the probability that the child is a girl, given that the child has brown eyes.

47. Prof. Martinez stimulates interest in homework by requiring two students each to draw a problem at random and solve the problem drawn. On a certain day, ten problems were assigned. Anita is to draw first and Al second. They had solved nine problems but had not solved one. Find the probability that

(a) Anita draws the unsolved problem.

(b) neither Anita nor Al draws the unsolved problem.

(c) Al does not draw the unsolved problem.

48. A national corporation schedules a sales meeting in Detroit and reserves hotel rooms, some of which allow smoking in the room and the rest of which are nonsmoking rooms. Access to the exercise room is available for an added fee. Past records show that 30% of sales persons request smoking rooms and 70% request nonsmoking. Of those requesting smoking rooms, 40% want access to exercise equipment. Of those requesting nonsmoking rooms, 65% want access to exercise equipment. A person requests a room reservation but fails to specify smoking or nonsmoking and does not indicate if access to exercise equipment is wanted. Find the probability that

(a) the person prefers a smoking room and wants access to exercise equipment.

(b) The person wants access to exercise equipment.

49. The High Plains Construction Company gives all applicants an aptitude test and a drug test. A total of 96 people applied when the company won a contract to build a convention center. Of those, 54 passed the aptitude test, 73 passed the drug test, and 42 passed both. If an applicant is selected at random, find the probability that the applicant passed

(a) both tests.

(b) the aptitude test.

(c) the drug test if it is known the applicant passed the aptitude test.

50. Michael Johnson, the world record holder of the 200-meter dash, preferred to run in lane 3. In a race with six runners, each runner takes turns drawing a number from 1 through 6 to determine their lane.

(a) If Michael Johnson draws first, what is the probability that he draws lane 3?

(b) If Michael Johnson draws second, what is the probability that he draws lane 3?

(c) If Michael Johnson draws third, what is the probability that he draws lane 3?

(d) If Michael Johnson draws fourth, what is the probability that he draws lane 3?

(e) If Michael Johnson draws last, what is the probability that he draws lane 3?

(f) Based on the results in parts (a) through (e), when should he draw to have the best chance of drawing lane 3?

51. A hotel located near a medical center has special rates for families receiving treatment at the medical center. Their records show that 75% of room reservations receive the medical rate and 5% of the guests receiving the medical rate request a room designed for the handicapped.

(a) A call is received requesting a room reservation. Find the probability that the person is eligible for the medical rate and requests a handicap room.

(b) For two consecutive calls, find the probability that the first is not eligible for the medical rate and the second is eligible but does not request a handicap room.

EXPLORATIONS

52. Use the definition of conditional probability to determine the following. Draw a supporting Venn diagram.

(a) $P(E|E)$.

(b) $P(E|F)$, if F is a subset of E.

(c) $P(F|E)$, if F is a subset of E.

53. Use the definition of conditional probability to verify that

$$P(E|F) + P(E'|F) = 1$$

Draw a supporting Venn diagram.

54. Use the definition of conditional probability to verify that

$$P((A \cup B)|F) = P(A|F) + P(B|F) - P((A \cap B)|F)$$

Draw a supporting Venn diagram.

55. A deck of cards contains cards of several colors with five cards of each color numbered 1 through 5. The deck is used for some friendly gambling. Determine which of the following are good wagers.

(a) First, discuss what is meant by a good wager.

(b) The deck contains two colors (10 cards). A player draws 2 cards without replacement.
   (i) The bet is won if the cards are different colors.
   (ii) The bet is won if the cards are different numbers.
(c) The deck contains three colors (15 cards). Three cards are drawn without replacement.
   (i) The bet is won if the cards are different colors.
   (ii) The bet is won if the cards are different numbers.
(d) The deck contains four colors (20 cards). Four cards are drawn without replacement.
   (i) The bet is won if the cards are different colors.
   (ii) The bet is won if the cards are different numbers.

56. The registrar's report at State University includes the following information:

Total enrollment — 47,348

Enrollment in the Arts and Science College — 27,837

Number of males enrolled in Arts and Sciences — 12,722

The alumni magazine wants to interview a randomly selected student about the student's attitude toward university athletics. The registrar's office makes a random selection and observes the student is enrolled in Arts and Sciences. Find the probability the student is a male.

57. In the city of Stevenville, 84,349 people are registered to vote in city elections. All are registered as either Democrats or Republicans. In an election for mayor, the candidates were Arnold and Betros.

The results of the election included the following information:

15,183 Democrats voted.

16,026 Republicans voted.

16,827 people voted for Arnold.

14,382 people voted for Betros.

TV reporters interviewed voters at the polls and reported that 39% of the Democrats voted for Arnold and 32% of the Republicans voted for Betros.

The next week a newspaper reporter selected a person at random from the voter registration rolls to interview about the election. Find the following. (Assume the TV reporter's poll was accurate.)

(a) Find the probability that the person selected did not vote.

(b) Find the probability that the person selected was a Democrat who voted.
(c) Find the probability that the person voted for Betros.
(d) It was observed that the person selected voted in the election. Find the probability that the person voted for Arnold.
(e) The reporter remembered that the person selected had been interviewed on TV at the polls and had stated that she voted for Betros. Find the probability that the person is a Republican.

58. Poker is a card game that is sometimes played for fun and is sometimes played for thousands of dollars by professional gamblers. Each player is dealt five cards (called a hand) and certain combinations of cards rank higher than others. Here are combinations in the order of rank with the first listed being the hand of highest rank.

Straight flush: Five cards in sequence all in the same suit.

   Example: 3 through 7 in clubs
   ace through 5 in diamonds
   10, jack, queen, king, ace in spades

Notice that the ace can be used as either the lowest or highest card of the 13.

Four of a kind: Four cards of the same kind and one other card.

   Example: All four 6's and a 10

Full house: Three cards of the same kind and two cards of another kind.

   Example: Three 8's and two queens

Flush: Five cards of the same suit but not in sequence.

   Example: 2, 3, 6, 9, and king of hearts

Straight: Five cards in sequence but not all in the same suit.

   Example: 2 of hearts, 3 of clubs, 4 of clubs, 5 of spades, and 6 of hearts

Three of a kind: Three cards of the same kind and two cards of two other kinds.

   Example: Three 7's, a 5, and a jack

Two pairs: Two cards of the same kind, two more cards of a second kind, and one card of a third kind.

   Example: Two 5's, two 9's, and an ace

One pair: Two cards of the same kind and three cards of three other kinds.

(a) Compute the probability of each of the above five-card hands dealt from a well-shuffled deck.

(b) If the above hands are ranked properly, they should be in order of increasing probability, with the straight flush the least probable and one pair the most probable. Is that the case?

7.5 Independent Events

- Independent Events
- Multiplication Rule for Independent Events
- Mutually Exclusive and Independent Events

Independent Events

One evening, Andy tosses a coin to determine which subject he will study. At the same time across campus, Bill tosses a coin to determine which movie he will see. Does the outcome of Andy's toss have any influence on how Bill's coin turns up? We have no difficulty in stating that whether Andy's coin comes up heads or tails is irrelevant to how Bill's coin turns up. We say that Andy's toss and Bill's toss are **independent,** because the outcome of one toss has no effect on the outcome of the other. In a similar situation when Andy tosses a coin twice, the fact that it turns up heads on the first toss has no effect on whether the second toss turns up heads. The outcome of the first toss is independent of the outcome of the second toss.

Generally speaking, sometimes the occurrence of an event E affects whether or not an event F will occur. In other cases, the occurrence of an event E has no effect whatever on the occurrence of event F. If one person selects a card from one deck and another person selects one from another deck, then an ace (or any other card) drawn from the first deck has no effect on whether or not a king is drawn from the second deck. If two people select one card each from the *same* deck, then the outcome of the first selection affects the outcome of the second person's selection because the first card drawn is excluded as a possible outcome of the second selection. In this case, the outcome of the second person's draw is not independent of the first draw, so we call the two draws **dependent.**

People often confuse *independent* with *mutually exclusive.* They are not the same. Mutually exclusive indicates that two events have no outcomes in common. Events are independent when their probabilities relate in a definite way. In fact, mutually exclusive events with nonzero probabilities are always dependent, whereas events having outcomes in common may or may not be independent.

Two events are independent when knowledge of one event provides no information on the other event. More precisely, we determine the independence of two events by how their probabilities relate.

Situations arise in which our intuition may not be able to decide if two events are independent or not. We have a means to help in those situations. Again, we appeal to our intuition to make a general definition.

If two events E and F have no effect on the occurrence of each other, we intuitively expect that the occurrence of F has no effect on the probability that E occurs, and vice versa. In terms of probability, this means that $P(E \mid F) = P(E)$. Similarly, $P(F \mid E) = P(F)$. This leads to the following definition of independence.

The events E and F are **independent** if

$$P(E \mid F) = P(E) \qquad \text{or} \qquad P(F \mid E) = P(F)$$

Otherwise, E and F are dependent. (*Note:* If $P(E \mid F) = P(E)$, then $P(F \mid E) = P(F)$ also holds, and vice versa.)

In the next example, we check $P(E \mid F)$ and $P(E)$ for two situations, one in which intuition suggests that the events are independent and one in which intuition suggests that the events are dependent.

EXAMPLE 1 ➤

Two children play with three toys, a red, a green, and a blue one. We denote the toys by the set $\{R, G, B\}$. We consider two situations.

(a) Each child selects a toy with the first toy replaced before the second one is selected.

(b) Each child selects a toy with the first toy *not* replaced before the second one is selected.

In each situation, determine if the events "first toy is red" and "second toy is green" are independent or not. Do so by comparing $P(\text{second is green})$ with $P(\text{second is green} \mid \text{first is red})$.

SOLUTION

We will use tree diagrams to help analyze this problem. The first stage of the tree diagram represents the possible selections by the first child and the second stage represents the possible selections by the second child.

(a) In this case, three outcomes are possible for the first selection, and for each of these selections, three possible outcomes exist for the second selection (Figure 7–19a). This gives a total of nine possible sequences of selection. To check the independence of selecting R first and G second, compare $P(G$ second) and $P(G$ second $\mid R$ first).

From the tree diagram, we see that three of the nine branches have G as the second selection. Thus,

$$P(G \text{ second}) = \frac{3}{9} = \frac{1}{3}$$

To find P(G second | R first), we focus only on the part of the tree where the first branch is red (Figure 7–19b). We see that one of the three branches has green as the second selection, so

$$P(G \text{ second} \mid R \text{ first}) = \frac{1}{3}$$

Intuitively, we expected that "R first" and "G second" would be independent. We have further verified our intuition by finding $P(G$ second) $= P(G$ second $\mid R$ first). We point out that we can also verify independence by comparing $P(R$ first) and $P(R$ first $\mid G$ second).

(b) Intuitively, we expect that "R first" and "G second" will be dependent. Again, we compare $P(G$ second) with $P(G$ second $\mid R$ first).

Figure 7–20(a) shows the possible sequence of selections, six in all. Observe that two of the six branches have G as the second selection, so

$$P(G \text{ second}) = \frac{2}{6} = \frac{1}{3}$$

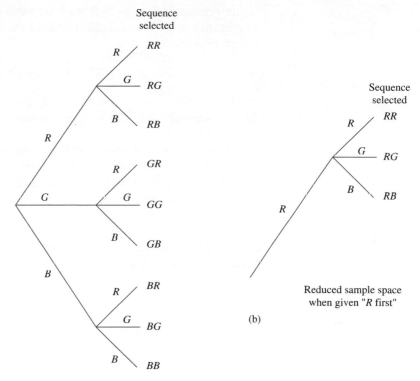

Sequence selected

RR
RG
RB
GR
GG
GB
BR
BG
BB

Sample space of two selections with the first
replaced before the second is selected

(a)

Sequence
selected

RR
RG
RB

Reduced sample space
when given "R first"

(b)

FIGURE 7-19

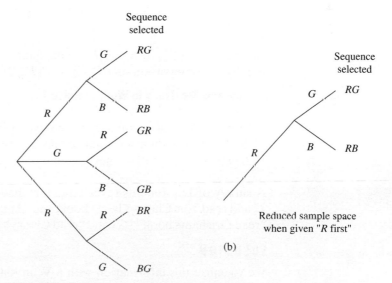

Sequence
selected

RG
RB
GR
GB
BR
BG

Sample space of two selections with the first
not replaced before the second is selected

(a)

Sequence
selected

RG
RB

Reduced sample space
when given "R first"

(b)

FIGURE 7-20

To determine $P(G \text{ second} \mid R \text{ first})$, we focus on the part of the tree where the first branch is R (Figure 7–20b). One of the two possible sequences has G as the second selection, so

$$P(G \text{ second} \mid R \text{ first}) = \frac{1}{2}$$

$P(G \text{ second}) = \frac{1}{3}$ is not equal to $P(G \text{ second} \mid R \text{ first}) = \frac{1}{2}$, so we conclude that "$G$ second" and "R first" are dependent. ∎

In Example 1, our intuition and the probabilities behaved in a way that was consistent with the definition of independent events. Now we go to an extreme and look at an example in which our intuition gives no indication of whether or not two events are independent.

EXAMPLE 2 ➤

A survey of 100 psychology majors revealed that 25 had taken an economics course (E), 35 had taken a French course (F), and 5 had taken both economics and French ($E \cap F$). Are the events "had taken economics" and "had taken French" independent or dependent?

SOLUTION

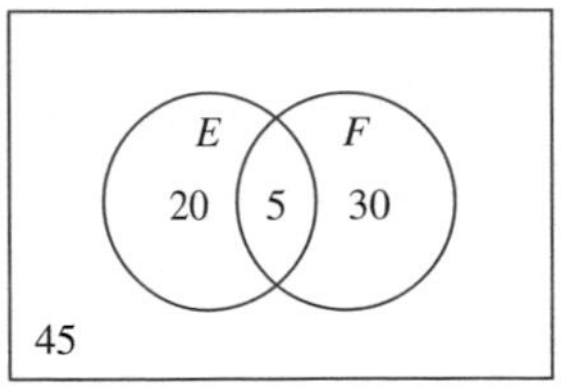

It is helpful to draw a Venn diagram of this information (see Figure 7–21). The sample space contains 100 elements, E contains 25, F contains 35, and $E \cap F$ contains 5. From the Venn diagram, we get the following probabilities:

$$P(E) = \frac{25}{100} = 0.25$$

$$P(F) = \frac{35}{100} = 0.35$$

$$P(E \mid F) = \frac{5}{35} = 0.1429$$

$$P(F \mid E) = \frac{5}{25} = 0.20$$

As $P(E) \neq P(E \mid F)$, E and F are dependent. We could reach the same conclusion by the observation that $P(F) \neq P(F \mid E)$.

FIGURE 7–21

■ **Now You Are Ready to Work Exercise 1**

Next we look at an example in which two events are independent, but our intuition provides no help in making that determination. We must find $P(E)$ and $P(E \mid F)$ and see whether they are equal or not.

EXAMPLE 3 ➤

A survey of 165 students revealed that 60 had read John Grisham's latest book, 22 had read Tom Clancy's latest book, and 8 had read both. Determine if the events "read Grisham's book" (G) and "read Clancy's book" (C) are independent or not.

SOLUTION

We visualize this information with a Venn diagram (Figure 7–22) and compare $P(G)$ with $P(G \mid C)$. We could use $P(C)$ and $P(C \mid G)$ just as well. As an area, $P(G)$ is the fraction of S that is occupied by G. Because $n(G) = 60$ and $n(S) = 165$,

$$P(G) = \frac{60}{165} = \frac{4}{11}$$

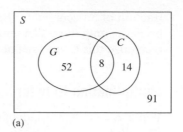

(a)

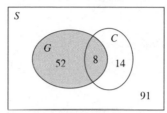

$P(G)$ = fraction of S occupied by G

$P(G) = \dfrac{60}{165}$

(b)

For $P(G|C)$, C is the sample space

$P(G|C) = \dfrac{8}{22}$

(c)

FIGURE 7–22

(Figure 7–22b).

For $P(G|C)$, C becomes the sample space and $G \cap C$ the event, so

$$P(G|C) = \frac{8}{22} = \frac{4}{11}$$

(Figure 7–22c).

Because $P(G) = P(G|C) = \frac{4}{11}$, the events "read Grisham" and "read Clancy" are independent. ∎

These examples illustrate that sometimes our intuition correctly tells us whether or not two events are independent, and sometimes it does not tell us. The best way to tell is to compute $P(E)$ and $P(E|F)$ and see whether they are equal or not.

Now, let's use the area model of probability to make a general statement of what we did in Examples 2 and 3.

To determine the independence of events E and F, we compare $P(E)$ and $P(E|F)$. We use the area of E in Figure 7–23a to determine $P(E)$. $P(E)$ is the

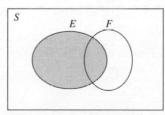

$P(E)$ = fraction of S occupied by E

(a)

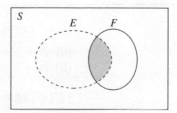

$P(E|F)$ = fraction of F occupied by $E \cap F$

(b)

FIGURE 7–23

fraction of S occupied by E. For $P(E\,|\,F)$, we determine the area of F occupied by E. As only those elements of E that are also in F can occupy space in F, you recognize this as the area of F occupied by $E \cap F$ (Figure 7–23b).

Multiplication Rule for Independent Events

When E and F are independent events, the Multiplication Rule $P(E \cap F) = P(E)P(F\,|\,E)$ simplifies to the following.

THEOREM
Multiplication Rule for Independent Events

> If E and F are **independent** events, then
> $$P(E \text{ and } F) = P(E \cap F) = P(E)P(F)$$

We can also use this theorem to determine whether two events are independent. For the events in Example 2,

$$P(E \cap F) = \frac{5}{100} = 0.05$$
$$P(E)P(F) = (0.25)(0.35) = 0.0875$$

As $P(E \cap F) \neq P(E)P(F)$, E and F are not independent — the same conclusion we reached by using the definition of independent events.

Although it may be intuitively clear that two tosses of a coin are independent, in other situations independence may be determined only by using the definition of independent events or the Multiplication Rule for independent events (as in Example 2). Actually, *any* one of the three conditions provides a test for independence.

Test for Independence of Events

> If any one of the following holds, then events E and F are independent.
> 1. $P(E\,|\,F) = P(E)$
> 2. $P(F\,|\,E) = P(F)$
> 3. $P(E \cap F) = P(E)P(F)$

NOTE

If any one of the three holds, then they all hold.

Now let's look at the relation between independent events and mutually exclusive events.

EXAMPLE 4 ➤

A card is selected at random from the following four cards: 10 of diamonds, 10 of spades, 8 of hearts, and 6 of clubs. Let E be the event "select a red card" and let F be the event "select a 10."

(a) Are E and F mutually exclusive?
(b) Are E and F independent?

SOLUTION

(a) E and F are not mutually exclusive because selecting a red card and selecting a 10 can occur at the same time — for example, selecting the 10 of diamonds.
(b) From the information given,

$$P(E) = P(\text{red card}) = \frac{1}{2}$$

$$P(F) = P(10) = \frac{1}{2}$$

$$P(E \cap F) = P(\text{red } 10) = \frac{1}{4}$$

As $P(E \cap F) = P(E)P(F) = \frac{1}{4}$, E and F are independent.

■ **Now You Are Ready to Work Exercise 11**

This example illustrates two points:

---

1.   **Two independent events need not be mutually exclusive.** In fact, mutually exclusive events with nonzero probabilities are always dependent. *(See Exercise 46.)*
2.   Sometimes the only way to determine whether events are independent is to perform the computation in the tests for independence.

We can now compute the compound probability, $P(E \cap F)$, by using the properties of $P(E \mid F)$ for independent events and the Multiplication Rule for independent events.

EXAMPLE 5 ➤ Ms. Bowden brought two bags of cookies to her second-grade class. The first bag contained 10 chocolate chip cookies and 14 oatmeal cookies. The second bag contained 8 peanut butter cookies and 12 sugar cookies. A student took one cookie from the first bag, and another student took one cookie from the second bag. Find the probability that the first was a chocolate chip cookie and the second was a sugar cookie.

SOLUTION

Because the cookies were taken from different bags, the events are independent, so

$$P(\text{first chocolate chip and second sugar}) = \left(\frac{10}{24}\right) \times \left(\frac{12}{20}\right) = \frac{1}{4}$$

■ **Now You Are Ready to Work Exercise 15**

EXAMPLE 6 ➤ Jack and Jill work on a problem independently. The probability that Jack solves it is $\frac{2}{3}$, and the probability that Jill solves it is $\frac{4}{5}$.

(a) What is the probability that both solve it?
(b) What is the probability that neither solves it?
(c) What is the probability that exactly one of them solves it?
(d) For each possible value of X, construct a table summarizing the probability that X people solve the problem.

SOLUTION

(a) Because their work is independent, the probability that both solve the problem is

$$P(\text{Jack solves it and Jill solves it}) = \left(\frac{2}{3}\right) \times \left(\frac{4}{5}\right) = \frac{8}{15}$$

(b) The probability that Jack does not solve the problem is $1 - \frac{2}{3} = \frac{1}{3}$, and the probability that Jill does not is $1 - \frac{4}{5} = \frac{1}{5}$. Then,

$$P(\text{Jack doesn't and Jill doesn't}) = \left(\frac{1}{3}\right) \times \left(\frac{1}{5}\right) = \frac{1}{15}$$

(c) There are two ways in which one of them will solve the problem, namely, Jack does and Jill doesn't or Jack doesn't and Jill does. These two ways are mutually exclusive events, so we need to compute the probability of each of these outcomes and then add.

$$P(\text{Jack does and Jill doesn't}) = \left(\frac{2}{3}\right) \times \left(\frac{1}{5}\right) = \frac{2}{15}$$

$$P(\text{Jack doesn't and Jill does}) = \left(\frac{1}{3}\right) \times \left(\frac{4}{5}\right) = \frac{4}{15}$$

Then, $P(\text{one of them solves the problem}) = \dfrac{2}{15} + \dfrac{4}{15} = \dfrac{6}{15} = \dfrac{2}{5}$

(d) Because two people are involved, two, one, or no people solve the problem ($X = 2, 1,$ or 0).

In part (a), we found $P(X = 2) = \dfrac{8}{15}$.

In part (b), we found $P(X = 0) = \dfrac{1}{15}$.

In part (c), we found $P(X = 1) = \dfrac{6}{15}$.

X	$P(X)$
0	1/15
1	6/15
2	8/15

Note that the probabilities in the table add to 1, because all possible values of X are in the table.

■ **Now You Are Ready to Work Exercise 19**

Recall that $P(E \cup F) = P(E) + P(F) - P(E \cap F)$. When E and F are independent, $P(E \cap F) = P(E)P(F)$, so we can substitute for $P(E \cap F)$ to obtain the following theorem:

THEOREM
Union of Independent Events

When E and F are independent,

$$P(E \cup F) = P(E) + P(F) - P(E)P(F)$$

In some problems, you have the choice of more than one approach to solving the problem. We illustrate with the following example.

EXAMPLE 7 ➤

Jack and Jill work on a problem independently. The probability that Jack solves it is $\frac{2}{3}$, and the probability that Jill solves it is $\frac{4}{5}$. What is the probability that at least one of them solves it?

SOLUTION

We will show you three ways to work this problem:

> *Case I:* Using the Union of Independent Events
>
> *Case II:* Using the Complement Theorem (Section 7.3)
>
> *Case III:* Using mutually exclusive events

Case I: If we let $A =$ the event that Jack solves the problem and $B =$ the event that Jill solves the problem, then $A \cup B$ is one or both (at least one) solves the problem. From the Union of Independent Events Theorem in this section,

$$P(A \cup B) = P(A) + P(B) - P(A)P(B)$$
$$= \frac{2}{3} + \frac{4}{5} - \frac{2}{3} \times \frac{4}{5}$$
$$= \frac{10 + 12 - 8}{15} = \frac{14}{15}$$

Case II: The complement of "at least one solves the problem" is "neither solves the problem," so

$$P(\text{at least one solves}) = 1 - P(\text{neither solves})$$

by the Complement Theorem (Section 7.3). This is

$$1 - \frac{1}{3} \times \frac{1}{5} = 1 - \frac{1}{15} = \frac{14}{15}$$

Case III: At least one of them solving the problem is equivalent to exactly one of them solving the problem (E) or both of them solving the problem (F). These events, E and F, are mutually exclusive, so

$$P(E \cup F) = P(E) + P(F)$$
$$P(E) = \frac{2}{5} \qquad \text{(From Example 6c)}$$
$$P(F) = \frac{8}{15} \qquad \text{(From Example 6a)}$$

so $P(\text{at least one}) = P(E \cup F) = \frac{2}{5} + \frac{8}{15} = \frac{14}{15}$.

■ **Now You Are Ready to Work Exercise 25**

Let's illustrate some of these concepts using a tree diagram.

EXAMPLE 8 ➤

The city council has money for one public service project: a recreation–sports complex, a performing arts center, or a branch library. The council polled 200 citizens for their preference, 120 men and 80 women. The men responded as follows: 45% preferred the recreation–sports complex, 20% the performing arts center, and 35% the branch library. The women responded as follows: 15% favored the recreation–sports complex, 40% the performing arts center, and 45% the branch library.

(a) Represent this information on a tree diagram.
(b) What is the probability that a person selected at random prefers the performing arts center?
(c) What is the probability that a person selected at random is a woman who prefers the performing arts center or the branch library?
(d) Are the events "male" and "prefers the recreation–sports complex" independent?

SOLUTION

Use the following abbreviations: M for male and F for female; RS, AR, and BL for recreation–sports complex, performing arts center, and branch library, respectively.

The information provided gives the following probabilities:

$$P(M) = 0.6, \qquad P(F) = 0.4$$
$$P(RS\,|\,M) = 0.45, \qquad P(AR\,|\,M) = 0.20, \qquad P(BL\,|\,M) = 0.35$$
$$P(RS\,|\,F) = 0.15, \qquad P(AR\,|\,F) = 0.40, \qquad P(BL\,|\,F) = 0.45$$

(a) Figure 7–24 shows the tree diagram with this information.
(b) The two branches that terminate at AR are $M \cap AR$ and $F \cap AR$, so

$$P(AR) = 0.6 \times 0.20 + 0.4 \times 0.40 = 0.12 + 0.16 = 0.28$$

(c) The branches $F \cap AR$ and $F \cap BL$ are the two outcomes in this event, so

$$P(F \text{ who prefers } AR \text{ or } BL) = 0.16 + 0.18 = 0.34$$

(d) If the events "male" and "prefers the recreation–sports complex" are independent, then $P(RS\,|\,M)$ must equal $P(RS)$. As $P(RS\,|\,M) = 0.45$ and $P(RS) = 0.27 + 0.06 = 0.33$, the events are not independent.

■ **Now You Are Ready to Work Exercise 33**

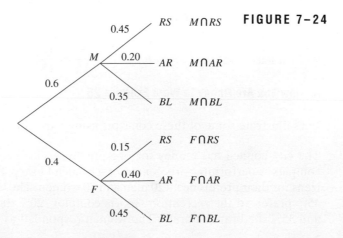

FIGURE 7–24

Mutually Exclusive and Independent Events

Sometimes students find it difficult to understand the concepts of independent, dependent, and mutually exclusive events. If you determine two events to be dependent, you still might not know whether they are mutually exclusive or not because two dependent events may or may not be mutually exclusive. Let's look at some examples to get a feeling of the concepts.

EXAMPLE 9 ➤ Nikki and Alice are two students at the university who do not know each other and have no contact with each other. In such a case, it seems reasonable that Nikki's decision to take art history or not has no influence on whether or not Alice takes art history. We say the events "Nikki takes art history" and "Alice takes art history" are independent because the probability that one of them takes the course has no influence on the probability that the other takes it. On the other hand, both could take the course, so the events are *not* mutually exclusive. To be mutually exclusive, the occurrence of one event *must exclude* the possibility of the other occurring.

Under different circumstances, as when Nikki and Alice are roommates, it seems reasonable to think that whether Nikki takes art history or not might influence whether or not Alice takes it; that is, the events "Nikki takes art history" and "Alice takes art history" might be dependent. Even if they are roommates, their decisions might still be independent, so we need information on the probability of each person's decision before we can say.

Even if their decisions are dependent, the events are not mutually exclusive because both could take art history. ■

Keep in mind that the test for mutually exclusive events is to determine whether or not the two events can occur at the same time. The test for the independence or dependence of two events depends on how their probabilities are related.

EXAMPLE 10 ➤ Some people's eyes become irritated and water when the pollen count rises above a certain threshold value. A similar reaction occurs when the ozone pollution rises above a threshold value. The city's Air Quality Board monitors the pollution levels and knows that 30% of the days the ozone will be above threshold value, 20% of the days pollen will be above threshold, and 6% of the days both will be above threshold.

(a) Are the events "ozone above threshold" and "pollen above threshold" mutually exclusive?

(b) Are the events "ozone above threshold" and "pollen above threshold" independent?

SOLUTION

(a) Both events can occur at the same time. In fact, they occur at the same time 6% of the time, so the occurrence of one does not exclude the possibility of the other. They are not mutually exclusive.

(b) To determine independence or dependence, we need to check the probabilities of the events. One test given earlier in this section is to determine whether

$$P(\text{ozone pollution}) \times P(\text{pollen pollution})$$

is equal to $P(\text{both ozone and pollen pollution})$.

From the given information, we have

$$P(\text{ozone pollution}) \times P(\text{pollen pollution}) = 0.30 \times 0.20 = 0.06$$

and

$$P(\text{both ozone and pollen pollution}) = 0.06$$

As these computations are equal, the events are independent. ∎

EXAMPLE 11 ➤ Look at the three situations in Figure 7–25 and determine whether the events are mutually exclusive and whether they are independent or dependent.

(a) Because E and F overlap, they have outcomes in common, which tells us that they are not mutually exclusive.

To check for independence, we must compute the probabilities

$$P(E) \times P(F) = \frac{48}{132} \times \frac{22}{132} = \frac{8}{132}$$

and

$$P(E \cap F) = \frac{8}{132}$$

E and F are independent because $P(E) \times P(F) = P(E \cap F)$.

(b) The events E and F are not mutually exclusive, because they have events in common.

Now compute

$$P(E) \times P(F) = \frac{45}{320} \times \frac{65}{320} = \frac{117}{4096}$$

and

$$P(E \cap F) = \frac{5}{320} = \frac{1}{64}$$

E and F are dependent because $P(E) \times P(F) \neq P(E \cap F)$.

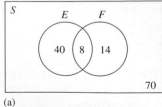

(a)

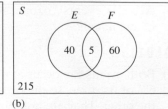

(b)

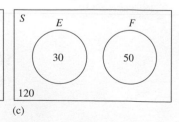

(c)

FIGURE 7–25

(c) The events E and F are mutually exclusive because they have no outcomes in common. E and F are dependent because $P(E) \times P(F) \neq P(E \cap F)$ as the following computations show:

$$P(E) \times P(F) = \frac{30}{200} \times \frac{50}{200} = \frac{3}{80}$$

and

$$P(E \cap F) = \frac{0}{200} = 0 \quad \blacksquare$$

7.5 EXERCISES

LEVEL 1

1. *(See Example 2)* Each of the Venn diagrams in Figure 7–26 shows the number of elements in each region determined by the two events E and F and the sample space S. In each case, determine (i) whether the events are mutually exclusive and (ii) whether the events are independent.

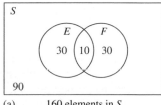

(a)   160 elements in S

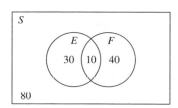

(b)   160 elements in S

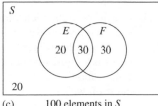

(c)   100 elements in S

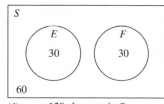

(d)   120 elements in S

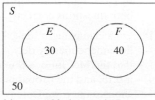

(e)   120 elements in S

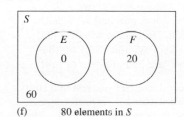

(f)   80 elements in S

FIGURE 7–26

Determine whether or not E and F are independent or dependent in Exercises 2 through 5.

2. $P(E) = 0.6$, $P(F) = 0.4$, $P(E \cap F) = 0.10$

3. $P(E) = 0.3$, $P(F) = 0.5$, $P(E \cap F) = 0.15$

4. $P(E) = 0.4$, $P(F) = 0.4$, $P(E \cap F) = 0.16$

5. $P(E) = 0.3$, $P(F) = 0.7$, $P(E \cap F) = 0.20$

6. Let E and F be events such that $P(E) = 0.4$, $P(F) = 0.6$, and $P(E \cap F) = 0.3$.
(a) Find $P(E \mid F)$ and $P(E \cup F)$.
(b) Are E and F independent?

7. E and F are events such that $P(E) = 0.35$, $P(F) = 0.60$, and $P(E \cap F) = 0.21$.
(a) Find $P(F \mid E)$ and $P(E \cup F)$
(b) Are E and F independent?

8. E and F are independent events such that $P(E) = 0.28$ and $P(F) = 0.45$. Find $P(E \cup F)$.

9. E and F are independent events such that $P(E) = 0.3$ and $P(F) = 0.8$. Find $P(E \cup F)$.

10. E and F are independent events such that $P(E) = 0.15$ and $P(F) = 0.40$. Find $P(E \cup F)$.

LEVEL 2

11. *(See Example 4)* A jewel box contains four rings:

One has a diamond and an emerald.

One has a diamond and a ruby.

One has a ruby and an emerald.

One has pearls.

A ring is selected at random from the box. Let the event E be "a ring has a diamond" and let the event F be "a ring has a ruby."
(a) Are the events E and F mutually exclusive?
(b) Are the events E and F independent?

12. A kennel raises purebred dogs. Several litters from one dog produced 16 puppies with the following markings:

6 had a white mark on the head only.

2 had a white mark on the forelegs only.

2 had a white mark on both head and forelegs.

6 had neither mark.

Determine whether the events "white mark on the head" and "white mark on the forelegs" are independent or not.

13. A study of 100 students who took a certain mathematics course revealed that 15 received a grade of A, 20 had SAT mathematics scores above 550, and 10 received both a grade of A and had SAT mathematics scores above 550. Determine whether the events "received a grade of A" and "SAT mathematics score above 550" are independent or not.

14. An advertising class conducted a taste test in a shopping mall. Passersby were asked their preference of brands X and Y coffee and brands A and B doughnuts. Of 500 people, 320 preferred brand X coffee, 240 preferred brand A doughnuts, and 110 preferred both brand X coffee and brand A doughnuts. Are the events "prefer brand X coffee" and "prefer Brand A doughnuts" independent?

15. *(See Example 5)* Mrs. Brown brought two bags of goodies to her first-grade class. The first bag contained 6 granola bars and 8 packages of cheese crackers. The second bag contained 9 peanut butter crackers and 12 boxes of raisins.

Lauren takes one item from the first bag and Anna takes one from the second bag. Find the probability that Lauren gets a granola bar and Anna gets a box of raisins.

16. Box A contains eight $1 bills and two $10 bills. Box B contains five $1 bills, four $5 bills, and one $10 bill. If a bill is drawn at random from each box, find the probability that
(a) no $10 bills are drawn.
(b) exactly one $10 bill is drawn.
(c) at least one $10 bill is drawn.

17. In a programmed learning module, if a student answers a question correctly, she proceeds to the next question; otherwise, more study is required. Question 1 is a multiple-choice question with four possible responses, and question 2 is a multiple-choice question with five possible responses. There is one correct response to each question. A student has not studied but tries to proceed through the questions by guessing. Find the probability that she answers both questions correctly.

18. On a Friday night the McClintock and Dietze families independently decide to eat out. The probability that the McClintock family chooses the Elite Cafe is 0.15, and the probability that the Dietze family chooses the Elite Cafe is 0.22. Find the probability that
(a) only the McClintock family chooses the Elite Cafe.

(b) both families choose the Elite Cafe.

(c) neither family chooses the Elite Cafe.

19. *(See Example 6)* The probabilities that two students will not show up for class on a beautiful spring day are 0.2 and 0.3, respectively. It is a beautiful spring day. Find the probability that

(a) neither will show up for class.

(b) both will show up for class.

(c) exactly one will show up for class.

20. The probabilities that Jack and Jill can solve a homework problem are $\frac{3}{5}$ and $\frac{2}{3}$, respectively. Find the probability that

(a) both will solve the problem.

(b) neither will solve the problem.

(c) Jack will and Jill won't solve the problem.

(d) Jack won't and Jill will solve the problem.

(e) exactly one of them will solve the problem.

21. Three people are shooting at a target. The probabilities that they hit the target are 0.5, 0.6, and 0.8. Find the probability that all three

(a) hit the target.

(b) miss the target.

22. The probabilities that three students will not show up for class on a cold winter day are 0.1, 0.2, and 0.4. On a cold winter day, what is the probability that

(a) none of the three will show up for class?

(b) all three will show up for class?

(c) exactly one of the three will show up for class?

23. In each of the following, E and F are independent. Find $P(E \cup F)$. *(See the Theorem on the Union of Independent Events.)*

(a) $P(E) = 0.3$, $P(F) = 0.5$

(b) $P(E) = 0.2$, $P(F) = 0.6$

(c) $P(E) = 0.4$, $P(F) = 0.6$

24. A sample space has 56 equally likely outcomes. E and F are independent events in the sample space. E contains 24 outcomes, and F contains 21 outcomes. Find $P(E \cup F)$.

Assume that events are independent in Exercises 25 through 28.

25. *(See Example 7)* The probabilities that two students will show up for class are 0.6 and 0.8. Find the probability that at least one

(a) shows up for class.

(b) does not show up for class.

26. The probabilities that Zack and Zelda can solve a homework problem are 0.4 and 0.7, respectively. Find the probability that at least one solves the problem.

27. Two people are shooting at a target. The probabilities that they hit the target are 0.3 and 0.9. Find the probability that at least one hits the target.

28. A private plane has two engines. The probability that the left one fails in flight is 0.01, and the probability that the right one fails is 0.03. Find the probability that at least one fails in flight.

29. A group of students consists of 12 females and 18 males. Four of the females and six of the males are mathematics majors. Are the events "male" and "mathematics majors" independent?

30. The owner of the Tiger Hut observed that the customers come in no particular order by gender, but overall 40% are consistently females and 60% are males. At a random time, find the probability that the next five customers are female.

31. Yolinda, a star basketball player, makes 80% of her free throws. In practice the coach tells her to take five. Find the probability that she makes the first three and misses the last two.

LEVEL 3

32. Streams near industrial plants may suffer chemical pollution or thermal pollution from wastewater released in the stream. An environmental task force estimates that 6% of the streams suffer both types of pollution, 40% suffer chemical pollution, and 30% suffer thermal pollution. Determine whether chemical pollution and thermal pollution are independent.

33. *(See Example 8)* The mall polled 300 customers — 180 women and 120 men — about their preference of the use of an open area. The choices were a driftwood display, a fountain, and an abstract sculpture. The men responded as follows: 44% preferred the driftwood display, 36% preferred the fountain, and 20% preferred the abstract sculpture. The women responded as follows: 32% preferred the driftwood display, 52% preferred the fountain, and 16% preferred the abstract sculpture.

(a) Represent this information with a tree diagram.

(b) What is the probability that a person selected at random prefers the fountain?

(c) What is the probability that a person selected at random is a male who prefers the driftwood display?

(d) Are the events "female" and "prefers the fountain" independent?

34. A record shop surveyed 250 customers — 90 high school students, 80 college students, and 80 college graduates — about their preference of the musical groups The Swingers and The Top Brass. They responded as follows:

High school students: 65% preferred The Swingers, 35% preferred The Top Brass.

College students: 54% preferred The Swingers, 46% preferred The Top Brass.

College graduates: 48% preferred The Swingers, 52% preferred The Top Brass.

(a) Represent this information with a tree diagram.

(b) If a person is selected at random, what is the probability that the person is a college student who prefers The Top Brass?

(c) What is the probability that a person selected at random prefers The Swingers?

(d) Are the events "high school student" and "prefers The Top Brass" independent?

35. A card is picked from a pack of 52 playing cards. Let F be the event of selecting a red card and let G be that of selecting a face card. Are F and G independent events?

36. A summary of the number of graduating seniors who received job offers is given in the following table:

| | **Job Offer** | | |
	Yes	**No**	**Total**
Male	112	72	184
Female	168	108	276
Totals	280	180	460

A graduating senior is selected at random.

(a) Find the probability that the student is a male.

(b) Find the probability that the student has a job offer.

(c) Find the probability that the student is a male with a job offer.

(d) Show that the events "male" and "has a job offer" are independent.

37. A bowler makes a strike if all ten pins are knocked down with the first ball of a frame. A perfect game consists of 12 consecutive strikes.

(a) Whenever Hoffman rolls a first ball of a frame, the probability of making a strike is 0.5. Find the probability that Hoffman rolls a perfect game.

(b) Whenever McKay rolls a first ball of a frame, the probability of a strike is 0.90. Find the probability that McKay rolls a perfect game.

38. Let E and F be events such that $P(E) = 0.6$, $P(F) = 0.5$, and $P(E \cup F) = 0.8$.

(a) Find $P(E \cap F)$.

(b) Find $P(E \mid F)$.

(c) Are E and F independent?

39. Tim has two irregular coins. The nickel comes up heads three fifths of the time, and the quarter comes up heads two thirds of the time. The nickel is tossed.

If the nickel comes up heads, then the quarter is tossed twice.

If the nickel comes up tails, the quarter is tossed once. Find the probability that exactly two tails come up.

40. Profs. Harris and Bryant have unusually good mathematics classes with no grades below C. Their grade distributions are as follows:

	Harris	**Bryant**	**Total**
A	10	7	17
B	6	4	10
C	8	5	13
Total	24	16	40

A student is selected at random from the 40 students.

(a) Find the probability that the student is in Bryant's class.

(b) Find the probability that the student is in Harris' class.

(c) Find the probability that the student is an A student.

(d) Find the probability that the student is an A student from Bryant's class.

(e) Find the probability that the student is an A student, given that the student is from Bryant's class.

(f) Find the probability that the student is a B student, given that the student is from Bryant's class.

(g) Show that the events "A student" and "student from Bryant's class" are dependent.

(h) Show that the events "B student" and "student form Bryant's class" are independent.

41. The entrance requirements for Alliance College include at least 620 on the verbal SAT test and at least 530 on the mathematics SAT test. They received 3200 applications and 2850 met the verbal requirement, 2965 met the mathematics requirement, and 2795 met both requirements. Based on these data, determine if meeting the verbal requirement and meeting the mathematics requirement are independent.

42. At Valley View College, the entrance requirements include at least 500 on the SAT verbal and at least 500 on the

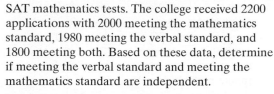

SAT mathematics tests. The college received 2200 applications with 2000 meeting the mathematics standard, 1980 meeting the verbal standard, and 1800 meeting both. Based on these data, determine if meeting the verbal standard and meeting the mathematics standard are independent.

43. The following table shows the number of mathematics, science, and humanities majors by gender at Foster College. Based on this information, determine if the events "the student is female" and "the student is a science major" are independent or not.

	Math	**Science**	**Humanities**	**Total**
Female	28	40	67	135
Male	24	48	90	162
Total	52	88	157	297

44. A coin is tossed three times. Show that the events "at most one head turns up" and "at least one head and one tail turn up" are independent.

45. A coin is tossed four times. Show that the events "at most one head turns up" and "at least one head and one tail turns up" are dependent.

EXPLORATIONS

46. Show that if E and F are mutually exclusive events, neither with probability zero, then they are dependent. *(Hint:* Consider $P(E \cap F)$ and $P(E)P(F)$.)

47. If E and F are mutually exclusive events, is it possible for E and F to be independent? If so, when? Give examples to support your conclusion.

48. If E and F are mutually exclusive events, are E and F dependent events? If so, when? Give examples to support your conclusion.

49. A, B, and C are nonempty events. If A and B are independent and B and C are independent, are A and C independent? Give examples to support your answer.

50. Show that if E and F are independent events, then E' and F' are independent events. It may be helpful to show that $E' \cap F' = (E \cup F)'$ (use Venn diagrams).

51. Gretchen tosses a fair coin repeatedly, and it turns up heads ten times in a row. What is the probability it turns up heads the eleventh time? Give reasons for your answer.

 52. Many computers perform millions of calculations per second, and we expect them to perform error free for hours. Let's analyze the performance of a computer that performs 1 million calculations per second and has a probability of 0.000001 of making an error on a calculation.

(a) Find the probability the computer runs error free for 1 hour.

(b) Find the probability the computer runs error free for 10 hours.

(c) A computer is expected to run error free for 1 hour with a probability of 0.999. Find the maximum probability of error in one calculation.

(d) A computer is expected to run for 8 hours error free with a probability of 0.9999. Find the maximum probability of error in one calculation.

 53. It is recommended that PC computer users "back up" their hard disk periodically (copy information from the hard disk to another device) so if a malfunction destroys some information on the disk, it can be replaced.

On any given day, the probability that a malfunction will destroy some information on Ron's hard disk is 0.01. Ron plans to back up his hard disk every 30 days. Is that a good decision?

For Angel's PC, the probability of malfunction is 0.0005 on any given date. Determine how long she can go without backup and the probability of malfunction still be less than 0.01.

We deviate a little from the beaten path to discuss how to solve problems like "Show that the probability is 1 that at least two people in New York City have the same number of hairs on their head." We use a simple, and obvious, idea called the **pigeonhole principle**.

Pigeonhole Principle

A flock of pigeons roost in pigeonholes. If there are more pigeons than pigeonholes, then at least one pigeonhole has two or more pigeons roosting in it.

New York City has a population of more than 8 million. The hairiest head has far fewer than 8 million hairs. Thus, we make each person in New York City a "pigeon" and provide one "pigeonhole" for each number 0, 1, 2, . . . up to the maximum number of hairs on a head. Each person is assigned to the pigeonhole corresponding to the number of hairs on his or her head. There are more pigeons than pigeonholes, so at least one pigeonhole must be shared by two or more people having the same number of hairs.

Use the pigeonhole principle to show that the probability = 1 for each event described in Exercises 54 through 58.

54. In a group of eight people, at least two were born on the same day of the week.

55. In a group of 13 people, at least two people have the same birthday month.

56. At least two members of the U.S. House of Representatives share the same birthday. (There are 435 members of the House of Representatives.)

57. A class of 75 students is given a true–false quiz of 6 questions. At least two students answer the questions alike.

58. The KOT Club has 27 members and 9 pledges. To plan and put on the Spring Bash, they assign seven committees of four people each. No one can serve on two committees. Show that some committee has a pledge on it.

7.6 Bayes' Rule

Conditional probability typically deals with the probability of an event when you have information about something that happened earlier. Let's look at a situation that reverses the information. Imagine the following:

A club separates a stack of bills for a drawing and places some in box A and the rest in box B. Each box contains some $50 bills and other bills. At the drawing, a person selects one box and draws a bill from that box. Conditional probability answers questions such as "If box A is selected, what is the probability that a $50 bill is drawn?" Symbolically, this is "What is $P(\$50$ bill, given that box A is selected)?" This question assumes that the first event (selecting a box) is known and asks for the probability of the second event.

Bayes' Rule deals with a reverse situation. It answers a question such as "If the person ends up with a $50 bill, what is the probability that it came from box A?" This assumes that the second event is known and asks for the probability of the first event. Bayes' Rule determines the probability of an earlier event based on information about an event that happened later.

Let's look at an example with some probabilities given to see when and how to use Bayes' Rule. We will then make a formal statement of the rule.

The student body at a college is 60% male and 40% female. The registrar's records show that 30% of the men attended private high schools and 70% attended

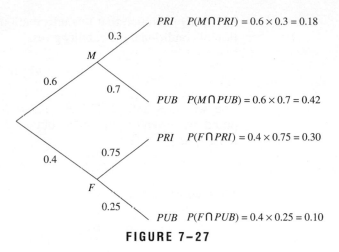

FIGURE 7–27

public high schools. Furthermore, 75% of the women attended private high schools, and 25% attended public schools.

Before we go further, let's summarize this information in probability notation. We use M and F for male and female, and we abbreviate private and public with PRI and PUB. Then, for a student selected at random,

$$P(M) = 0.6, \qquad\qquad P(F) = 0.4$$
$$P(PRI\,|\,M) = 0.3, \qquad P(PUB\,|\,M) = 0.7$$
$$P(PRI\,|\,F) = 0.75, \qquad P(PUB\,|\,F) = 0.25$$

Figure 7–27 is a tree diagram of this information, with its four branches terminating at $M \cap PRI$, $M \cap PUB$, $F \cap PRI$, and $F \cap PUB$. Their respective probabilities are

$$P(M \cap PRI) = P(M)P(PRI\,|\,M) = 0.6 \times 0.3 = 0.18$$
$$P(M \cap PUB) = P(M)P(PUB\,|\,M) = 0.6 \times 0.7 = 0.42$$
$$P(F \cap PRI) = P(F)P(PRI\,|\,F) = 0.4 \times 0.75 = 0.30$$
$$P(F \cap PUB) = P(F)P(PUB\,|\,F) = 0.4 \times 0.25 = 0.10$$

We can find the probability that a randomly selected student attended a private school by locating all branches that terminate in PRI and adding the probabilities. Thus,

$$P(PRI) = 0.18 + 0.30 = 0.48$$
$$P(PUB) = 0.42 + 0.10 = 0.52$$

Notice that in symbolic notation,

$$P(PRI) = P(M \cap PRI) + P(F \cap PRI)$$

and

$$P(PUB) = P(M \cap PUB) + P(F \cap PUB)$$

All the above information was developed for reference throughout the example. Now let's look at a problem in which Bayes' Rule is helpful.

Suppose a student selected at random is known to have attended a private school. What is the probability that the student selected is female; that is, what is

$P(F|PRI)$? Notice that this information is missing from the above. The definition of conditional probability gives

$$P(F|PRI) = \frac{P(F \cap PRI)}{P(PRI)}$$

You will find both $P(F \cap PRI)$ and $P(PRI)$ listed above. They were computed, not given originally. We obtained $P(PRI)$ from the two branches ending in PRI, so

$$P(PRI) = P(M \cap PRI) + P(F \cap PRI)$$

Thus, we can write

$$P(F|PRI) = \frac{P(F \cap PRI)}{P(M \cap PRI) + P(F \cap PRI)}$$

This is one form of Bayes' Rule. We get a more complicated-looking form, but one that uses the given information more directly, when we substitute

$$P(M \cap PRI) = P(M)P(PRI|M)$$

and

$$P(F \cap PRI) = P(F)P(PRI|F)$$

to get

$$P(F|PRI) = \frac{P(F)P(PRI|F)}{P(M)P(PRI|M) + P(F)P(PRI|F)}$$
$$= \frac{0.4 \times 0.75}{0.6 \times 0.3 + 0.4 \times 0.75} = \frac{0.30}{0.18 + 0.30} = \frac{0.30}{0.48} = 0.625$$

This last form has the advantage of using the information that was given directly in the problem.

We need to observe one more fact before making a general statement. We used $P(PRI) = P(M \cap PRI) + P(F \cap PRI)$. The events M and F make up *all* the branches in the first stage of the tree diagram, so $M \cup F$ gives all of the sample space ($M \cup F = S$). Furthermore, M and F are mutually exclusive. These two conditions on M and F are needed for Bayes' Rule.

Now we are ready for the general statement.

NOTE

This means that M and F form a partition of S.

Bayes' Rule

Let E_1 and E_2 be mutually exclusive events whose union is the sample space ($E_1 \cup E_2 = S$). Let F be an event in S, $P(F) \neq 0$. Then,

(a) $P(E_1|F) = \dfrac{P(E_1 \cap F)}{P(F)}$

(b) $P(E_1|F) = \dfrac{P(E_1 \cap F)}{P(E_1 \cap F) + P(E_2 \cap F)}$

(c) $P(E_1|F) = \dfrac{P(E_1)P(F|E_1)}{P(E_1)P(F|E_1) + P(E_2)P(F|E_2)}$

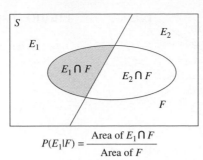

FIGURE 7-28

$$P(E_1|F) = \frac{\text{Area of } E_1 \cap F}{\text{Area of } F}$$

Area of F = Area of $E_1 \cap F$ + Area of $E_2 \cap F$

Let's look at some Venn diagrams to help clarify these rules. The events E_1 and E_2 divide S into two disjoint parts, and except when F lies completely within E_1 or E_2, they divide F into two parts (Figure 7–28).

The conditional probability $P(E_1|F)$ can be viewed as the fraction of F that is covered by $E_1 \cap F$. Rule (a) states this as

$$P(E_1 \mid F) = \frac{P(E_1 \cap F)}{P(F)}$$

We can also view F as the part of F that is in E_1 plus the part in E_2, symbolically stated as

$$F = (E_1 \cap F) \cup (E_2 \cap F)$$

Because these two parts are disjoint,

$$P(F) = P(E_1 \cap F) + P(E_2 \cap F)$$

Making this substitution in the denominator of rule (a) gives rule (b). Rule (c) then follows by replacing each term with its equivalent using the Multiplication Rule.

EXAMPLE 1 ➤

A microchip company has two machines that produce the chips. Machine I produces 65% of the chips, but 5% of its chips are defective. Machine II produces 35% of the chips, and 15% of its chips are defective.

A chip is selected at random and found to be defective. What is the probability that it came from machine I?

SOLUTION

Let I be the set of chips produced by machine I and II be those produced by machine II. We want to find $P(\text{I}|\text{defective})$. We are given $P(\text{I}) = 0.65$, $P(\text{II}) = 0.35$, $P(\text{defective}|\text{I}) = 0.05$, and $P(\text{defective}|\text{II}) = 0.15$. We may use the second form of Bayes' Rule directly:

$$P(\text{I}|\text{defective}) = \frac{P(\text{I})P(\text{defective}|\text{I})}{P(\text{I})P(\text{defective}|\text{I}) + P(\text{II})P(\text{defective}|\text{II})}$$
$$= \frac{0.65 \times 0.05}{0.65 \times 0.05 + 0.35 \times 0.15}$$
$$= \frac{0.0325}{0.085} = 0.38$$

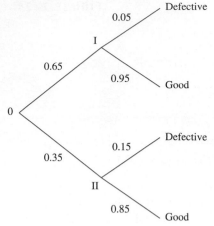

Tree diagram showing the possible situations
of defective and good chips from machines I and II

(a)

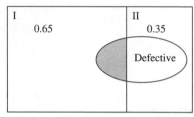

The shaded area represents those chips from
machine I that are defective.

(b)

Locate all of these computations on the tree diagram in Figure 7–29.

■ Now You Are Ready to Work Exercise 11

Bayes' Rule is not restricted to the situation in which just two mutually exclusive events, form all of S. There can be any finite number of mutually exclusive events, as long as their union is the sample space. A more general form of Bayes' Rule is the following:

Bayes' Rule (General)

Let $E_1, E_2, \ldots, E_n$ be mutually exclusive events whose union is the sample space S (they partition S), and let F be any event where $P(F) \neq 0$. Then,

(a) $P(E_i \mid F) = \dfrac{P(E_i \cap F)}{P(F)}$

(b) $P(E_i \mid F) = \dfrac{P(E_i \cap F)}{P(E_1 \cap F) + P(E_2 \cap F) + \cdots + P(E_n \cap F)}$

(c) $P(E_i \mid F) = \dfrac{P(E_i)P(F \mid E_i)}{P(E_1)P(F \mid E_1) + P(E_2)P(F \mid E_2) + \cdots + P(E_n)P(F \mid E_n)}$

As with the Bayes' Rule given earlier, we can represent the general Bayes' Rule with a diagram. We do so with E_1, E_2, and E_3.

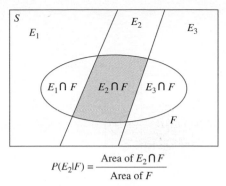

FIGURE 7-30

$$P(E_2|F) = \frac{\text{Area of } E_2 \cap F}{\text{Area of } F}$$

Area of F = Area of $E_1 \cap F$ + Area of $E_2 \cap F$ + Area of $E_3 \cap F$

If we seek $P(E_2 | F)$ for example, we want the ratio

$$\frac{\text{area of } E_2 \text{ in } F}{\text{area of } F}$$

This ratio can be obtained by using the area of all of F, or by finding the area of pieces of F and putting them together (Figure 7–30) as the sum of the parts of F that are in E_1, E_2, and E_3, symbolically stated as

$$F = (E_1 \cap F) \cup (E_2 \cap F) \cup (E_3 \cap F)$$

As these three parts are disjoint,

$$P(F) = P(E_1 \cap F) + P(E_2 \cap F) + P(E_3 \cap F)$$

Making this substitution in the denominator of rule (a) gives rule (b). Rule (c) then follows by replacing each term with its equivalent using the Multiplication Rule.

The next two exercises illustrate some applications of Bayes' Rule.

EXAMPLE 2 ➤

A manufacturer buys an item from three subcontractors, A, B, and C. A has the better quality control; only 2% of its items are defective. A furnishes the manufacturer with 50% of the items. B furnishes 30% of the items, and 5% of its items are defective. C furnishes 20% of the items, and 6% of its items are defective. The manufacturer finds an item defective (D) and would like to know which subcontractor supplied it.

(a) What is the probability that it came from A? (Find $P(A|D)$.)
(b) What is the probability that it came from B? (Find $P(B|D)$.)
(c) What is the probability that it came from C? (Find $P(C|D)$.)
(d) Which subcontractor was the most likely source of the defective item?

SOLUTION

Let

A represent the set of items produced by A.

B represent the set of items produced by B.

C represent the set of items produced by C.

D represent the set of defective items.

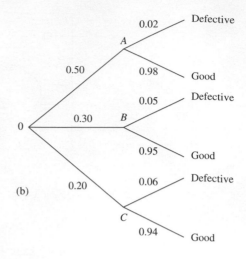

FIGURE 7–31

The following probabilities are given:

$$P(A) = 0.50, \qquad P(B) = 0.30, \qquad P(C) = 0.20$$
$$P(D\,|\,A) = 0.02, \qquad P(D\,|\,B) = 0.05, \qquad P(D\,|\,C) = 0.06$$

Let's use the second form of Bayes' Rule to solve the problem. We need the following probabilities, which are computed by using the Multiplication Rule:

$$P(D\cap A) = P(A)P(D\,|\,A) = 0.50 \times 0.02 = 0.010$$
$$P(D\cap B) = P(B)P(D\,|\,B) = 0.30 \times 0.05 = 0.015$$
$$P(D\cap C) = P(C)P(D\,|\,C) = 0.20 \times 0.06 = 0.012$$

These areas are shown in Figure 7–31a. Then,

(a) $P(A\,|\,D) = \dfrac{0.010}{0.010 + 0.015 + 0.012} = \dfrac{0.010}{0.037} = 0.27$

(b) $P(B\,|\,D) = \dfrac{0.015}{0.010 + 0.015 + 0.012} = \dfrac{0.015}{0.037} = 0.41$

(c) $P(C\,|\,D) = \dfrac{0.012}{0.010 + 0.015 + 0.012} = \dfrac{0.012}{0.037} = 0.32$

(d) The largest probability, $P(B\,|\,D) = 0.41$, suggests that subcontractor B is the most likely source.

Trace the computations on the tree diagram in Figure 7–31b.

■ **Now You Are Ready to Work Exercise 14**

EXAMPLE 3 ➤   Studies show that a pregnant woman who contracts German measles is more likely to bear a child with certain birth defects. In a certain country, the probability that a pregnant woman contracts German measles is 0.2. If a pregnant woman contracts the disease, the probability that her child will have the defect is 0.1. If a pregnant woman does not contract German measles, the probability that her child will have the defect is 0.01. A child is born with this defect. What is the probability that the child's mother contracted German measles while pregnant? Restrict the analysis to pregnant women.

SOLUTION

Let

$M = $ the set of pregnant women who contracted measles

$M' = $ the set of pregnant women who did not contract measles

$S = $ the sample space of pregnant women $(M \cup M')$

$D = $ the set of mothers who bear a child with this defect

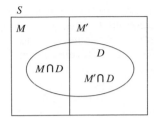

FIGURE 7–32

The Venn diagram of these events is shown in Figure 7-32.

We are given $P(M) = 0.2$, $P(M') = 0.8$, $P(D|M) = 0.1$, and $P(D|M') = 0.01$; and we are to find $P(M|D)$:

$$P(M|D) = \frac{P(M \cap D)}{P(M \cap D) + P(M' \cap D)}$$

$$= \frac{P(M)P(D|M)}{P(M)P(D|M) + P(M')P(D|M')}$$

$$= \frac{0.2 \times 0.1}{0.2 \times 0.1 + 0.8 \times 0.01} = \frac{0.02}{0.028} = 0.71$$

■ **Now You Are Ready to Work Exercise 18**

7.6 EXERCISES

LEVEL 1

1. Draw the tree diagram that shows the following information. A class is composed of 30 girls and 20 boys. On the final exam, 12% of the girls and 9% of the boys made an A.

2. Draw the tree diagram that shows the following information. A psychology class checked a group of students for personality types. The students fell into three categories: science, humanities, and social sciences. Of the group, 25% were in science, 35% in humanities, and 40% in social sciences. The findings are summarized as follows:

Group	Personality Type	
	A	B
Science	72%	28%
Humanities	61%	39%
Social sciences	48%	52%

3. Compute $P(E_1|F)$ for the following:
   (a) $E_1 \cup E_2 = S$, $P(E_1 \cap F) = 0.7$, $P(F) = 0.9$
   (b) $E_1 \cup E_2 = S$, $P(E_1) = 0.75$, $P(E_2) = 0.25$, $P(F|E_1) = 0.40$, $P(F|E_2) = 0.10$

4. E_1 and E_2 partition the sample space S, and F is an event.

   $P(E_1) = 0.3$, $P(E_2) = 0.7$
   $P(F|E_1) = 0.25$, $P(F|E_2) = 0.15$

   Find $P(E_1|F)$.

5. F is an event, and E_1, E_2, and E_3 partition S.

   $$P(E_1) = \frac{5}{12}, \qquad P(E_2) = \frac{4}{12}, \qquad P(E_3) = \frac{3}{12}$$

   $$P(F|E_1) = \frac{2}{5}, \qquad P(F|E_2) = \frac{1}{4}, \qquad P(F|E_3) = \frac{1}{3}$$

   (a) Find $P(E_1 \cap F)$, $P(E_2 \cap F)$, $P(E_3 \cap F)$.
   (b) Find $P(F)$.
   (c) Find $P(E_1|F)$ and $P(E_3|F)$.

6. F is an event, and E_1, E_2, and E_3 partition S.

$$P(E_1) = 0.20, \quad P(E_2) = 0.35, \quad P(E_3) = 0.45$$
$$P(F \mid E_1) = 0.40, \quad P(F \mid E_2) = 0.10, \quad P(F \mid E_3) = 0.25$$

(a) Find $P(E_1 \cap F)$, $P(E_2 \cap F)$, $P(E_3 \cap F)$.
(b) Find $P(F)$. **(c)** Find $P(E_2 \mid F)$.

7. Events E_1 and E_2 partition the sample space S, and F is an event. The number of simple events in each region formed by these events is shown in Figure 7–33.

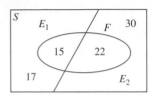

FIGURE 7–33

The outcomes in the sample space S are equally likely. Find:

(a) $P(F)$ **(b)** $P(E_1)$

(c) $P(E_2)$ **(d)** $P(F \mid E_1)$

(e) $P(F \mid E_2)$ **(f)** $P(E_1 \mid F)$

(g) $P(E_1 \cap F)$ **(h)** $\dfrac{P(E_1 \cap F)}{P(F)}$

(i) $\dfrac{P(E_1)P(F \mid E_1)}{P(E_1)P(F \mid E_1) + P(E_2)P(F \mid E_2)}$

8. Plantation Foods has 843 employees. A total of 255 employees have a college degree, and 588 do not have college degrees. Of those with college degrees, 60% are men and 40% are women. Of those who do not have college degrees, 25% are men and 75% are women. The Human Resources Office selects an employee at random to interview about a proposed health insurance change. The person selected is a woman. Find the probability that she does not have a college degree.

9. In the Perri Manufacturing Company, 28% of the employees are sales and administration staff, and 72% are production staff. The company offers a free six-month fitness program for all employees. Forty percent of the sales and administration staff participate, and 64% of the production staff participate. At the end of the program, an employee is selected at random to assess the value of the program. The person is selected from those who participated in the program. Find the probability that the person selected is on the production staff.

10. At a joint meeting, 78 Young Republicans and 97 Young Democrats are present to debate issues in the upcoming campus elections. On the issue of campus parking, 35% of the Young Democrats and 60% of the Young Republicans favor turning a campus green space into a parking lot. A campus newspaper reporter randomly selects a person from the group and finds that the person favors the parking space. Find the probability that the person selected is a Young Democrat.

11. *(See Example 1)* A manufacturer has two machines that produce television picture tubes. Machine I produces 55% of the tubes, and 3% of its tubes are defective. Machine II produces 45% of the tubes, and 4% of its tubes are defective. A picture tube is found to be defective. Find the probability that it was produced by machine II.

12. The incoming freshman class at Winston College shows that 38% come from single-parent homes and 62% come from two-parent homes. The financial aid office reports that 65% of students from single-parent homes receive financial aid and 42% of those from two-parent homes receive financial aid. Find the probability that an incoming freshman who receives financial aid is from a two-parent home.

13. A football team plays 60% of its games at home and 40% away. It typically wins 80% of its home games and 55% of its away games. If the team wins on a certain Saturday, what is the probability that it played at home?

LEVEL 2

14. *(See Example 2)* A distributor receives 100 boxes of transistor radios. Factory A shipped 35 boxes, factory B shipped 40 boxes, and factory C shipped 25 boxes. The probability of a radio being defective is 0.04 from factory A, 0.02 from factory B, and 0.06 from factory C. A box and a radio from that box are both selected at random. The radio is found to be defective. Find the probability that it came from factory B.

15. A mythology class is composed of 10 sophomores, 25 juniors, and 15 seniors. On the first exam, 3 sophomores, 5 juniors, and 6 seniors earned A's. Find the probability that a student who received an A is a junior.

16. The soft drinks in the ice chest at a picnic are 55% Pepsi and 45% RC. The Pepsi drinks are 40% diet, and the RC drinks are 30% diet. A drink is randomly selected and found to be a diet drink. Find the probability that it is an RC drink.

17. At the Campus Coffee Nook, 35% of the customers order regular coffee and 65% order flavored coffee. Of those who order regular coffee, 40% drink it black and 60% use sugar or cream. Of those who drink flavored coffee, 25% drink it black and 75% use sugar or cream.
   (a) Use a tree diagram to show all possible outcomes and their probabilities.
   (b) Find $P(\text{black coffee} \mid \text{regular})$, $P(\text{black coffee and regular})$, and $P(\text{black coffee})$
   (c) Find $P(\text{flavored coffee} \mid \text{cream or sugar})$

18. *(See Example 3)* If a patient is allergic to penicillin, the probability of a reaction to a new drug X is 0.4. If the patient is not allergic to penicillin, the probability of a reaction to drug X is 0.1. Eight percent of the population is allergic to penicillin. A patient is given drug X and has a reaction. Find the probability that the patient is allergic to penicillin.

19. Of the 17 year old persons who hold a driver's license, 80% have had a driver's ed course, and 20% have not. The probability of a 17-year-old student with a driver's license who has had driver's ed being involved in an accident is 0.32 and the probability of a 17-year-old driver with a driver's license who has not had driver's ed being involved in an accident is 0.55. A 17-year-old driver is involved in an accident. Find the probability that the driver has had driver's ed.

20. At a Houston auto inspection station, 68% of the vehicles are automobiles and 32% are pickups. Records show that 14% of the automobiles and 23% of the pickups fail to meet the pollution standards. A vehicle fails the pollution standards. Find the probability that it is an automobile.

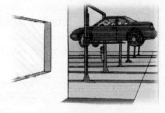

21. In 1995, 18% of Americans age 25 years and older did not hold high school diplomas, and 82% did. Of those who held high school diplomas, 4% were unemployed, compared with 10% unemployed of those with no high school diploma. A randomly selected person is unemployed. Find the probability that the person has no high school diploma.

22. In 1995, for civilians age 25 years and older, we have the following information breakdown:

Category	Percentage in Category	Percentage in Category with H.S. Diploma
Employed	63.5	89.4
Unemployed	3.1	77.4
Not in labor force	33.4	67.1

A person is selected from the population. If the person has
   (a) a high school diploma, find the probability that the person is employed.
   (b) no high school diploma, find the probability that the person is not in the labor force.

23. At a clinic, a preliminary test for hepatitis has been found to be 95% accurate, in that 95% of those with hepatitis have a positive reaction. However, 2% of those without hepatitis also have a positive reaction. Suppose 70% of those examined have hepatitis. Find the probability that a person who has a positive reaction has hepatitis.

24. It has been estimated that each year 1,358,000 new cases of cancer occur, with 594,000 of them women and 764,000 men. Of those cases, 78,000 are lung cancer in women, and 99,000 are lung cancer in men. Find the probability that a person diagnosed with lung cancer is a man.

25. A standardized reading test was given to fourth- and fifth-grade classes at an elementary school. A summary of the results is the following:

	Scoring Below Grade Level	Scoring at Grade Level	Scoring Above Grade Level	Total
4th grade	120	342	216	678
5th grade	105	324	98	527
Total	225	666	314	1205

A student is selected at random. Find the probability that
   (a) the student is in the fourth grade.
   (b) the student is in the fifth grade.

(c) the student tests at grade level.

(d) a student in the fourth grade tests at grade level.

(e) a student in the fifth grade tests at grade level.

(f) the student is a fourth-grade student who tested at grade level.

(g) The student selected tested at grade level. Find the probability the student is in the fourth grade.

26. Profs. Clanton, Duff, and Jonson each have a 9:00 MWF calculus class, so they work together to prepare a joint test. The number of students who made an A or a B on the latest test are the following:

	Clanton	**Duff**	**Jonson**
No. in class	28	24	18
No. making an A	5	6	4
No. making a B	9	6	7

The papers are collected for a joint grading session. One paper is selected at random.

(a) Find the probability that the paper is from Clanton's class.

(b) Find the probability that the paper is from Duff's class.

(c) Find the probability that the paper is from Jonson's class.

(d) The paper is from Clanton's class. Find the probability that it is an A paper.

(e) The paper is from Duff's class. Find the probability that it is a B paper.

(f) The paper is an A paper. Find the probability that it is from Jonson's class.

(g) The paper is a B paper. Find the probability that it is from Duff's class.

LEVEL 3

27. A survey of mathematics students at the college revealed that 43% consistently spent at least 1.5 hours on mathematics homework and 57% spent less. Of those who spent at least 1.5 hours on homework, 78% made an A or B in the course. Of those who spent less than 1.5 hours, 21 % made an A or B. A student made an A or B in the course. Find the probability that the student spent at least 1.5 hours on homework.

28. In a major city, 70% of the drivers are older than 25 years, and 12% of them will have a traffic violation during a 12-month period. The drivers 25 years and younger compose 30% of the drivers, and 28% of them will have a traffic violation in a 12-month period. A driver is charged with a traffic violation. Find the probability that the driver is older than 25.

29. The IRS checks the deduction for contributions to identify fraudulent tax returns. They believe that if a taxpayer claims more than a certain standard amount, there is a 0.20 probability that the return is fraudulent. If the deduction does not exceed the IRS standard, the probability of a fraudulent return reduces to 0.03. About 11 % of the returns exceed the IRS standard.

(a) Estimate the percentage of returns that are fraudulent.

(b) A concerned citizen informs the IRS that a certain return is fraudulent. Find the probability that its deductions exceed the IRS standard.

30. A chain has three stores in Knoxville, A, B, and C. The stores have 60, 80, and 110 employees, respectively, of whom 40%, 55%, and 50% are women. Economic conditions force the chain to lay off some employees. They decide to do this by random selection. The first person selected is a woman. Find the probability that she works at store C.

31. The Business Students Club was composed of 20% sophomores, 35% juniors, and 45% seniors. The Accounting Department was interested to learn that 15% of the sophomores were accounting majors, as were 8% of the juniors and 21 % of the seniors. Find the percentage of accounting majors that were juniors.

32. At a gathering of undergraduate students, 15% were freshmen, 28% sophomores, 46% juniors, and 11% seniors. The percentage who had 8 o'clock classes were the following: freshmen, 38%; sophomores, 24%; juniors, 15%; and seniors, 5%. Find the percentage of those that have 8 o'clock classes that are sophomores.

33. The 106th U.S. Senate was composed of 55 Republicans, and 45 Democrats. An amendment proposed by the president was supported by 80% of the Democrats and by 20% of the Republicans. Find the probability that a person who supported the amendment was a Republican.

34. At the Student Center cafeteria, 30% of those eating lunch are on a diet, and 70% are not. Of those on a

diet, 85% select a low-fat dessert, and 20% of those not on a diet select a low-fat dessert. Find the probability that a person who selects a low-fat dessert is on a diet.

35. A company manufactures integrated circuits on silicon chips at three different plants, X, Y, and Z. Out of every 1000 chips produced, 400 come from X, 350 come form Y, and 250 come from Z. It has been estimated that of the 400 from X, 10 are defective, whereas 5 of those from Y are defective, and only 2 of those from Z are defective. Determine the probability that a defective chip came from plant Y.

EXPLORATIONS

36. Let F be the event that a person has a low-grade fever and let C be the event that a person has a contagious disease. Suppose the following probabilities hold.

$$P(F) = 0.005$$
$$P(C) = 0.010$$
$$P(C \mid F) = 0.80$$
$$P(C \mid F') = 0.001$$

(a) Interpret the meaning of $P(F \mid C)$.
(b) Find $P(F \mid C)$.

37. The Mathematics Department gives a test to entering freshmen who plan to take calculus. If they pass the test, they are advised to proceed into calculus. If they fail the test, they are advised to take precalculus first. The department's experience is that 90% of those who pass the test perform satisfactorily in calculus and 15% of those who fail the test will also perform satisfactorily in calculus.

If 65% of the entering freshmen who take calculus do pass the test, find the probability that a student who performs satisfactorily in calculus actually passed the test.

38. In the practice of medicine, a number of medical tests are used to detect diseases. For example, tests are available to test for prostate cancer, AIDS, tuberculosis, and so on. These tests are not perfect. A probability exists that the test will give a *false positive*; it will indicate the presence of a disease when none exists. It may also give a *false negative,* an indication that the disease is not present when it is in fact present.

Suppose medical research finds a test for the AIDS virus (HIV) that is considered quite accurate. The probability of the test indicating the presence of HIV virus (positive test) when the virus is actually present is 0.993. Thus, the probability of a false negative is 0.007. The test indicates no virus (negative test) when no virus is actually present with probability 0.995. Thus, the probability of a false positive test is 0.005. It is estimated (in 2001) that a randomly selected person in North America has a probability of about 0.001 of having the HIV virus.

(a) Find the probability that a person who tests positive actually has the HIV virus.
(b) Find the probability that a person who tests negative actually does not have the HIV virus.
(c) A routine physical exam includes this test for the HIV virus. If the test is positive, indicating the presence of the HIV virus, would you suggest further testing or would you be convinced that the HIV virus is actually present?
(d) If the test in a routine exam were negative, indicating no HIV virus, would you be convinced no HIV virus was present?

39. Prior to his political party's convention to nominate candidates for president and vice-president, a political analyst showed his estimates of the probabilities of nomination of the leading candidates: Andrews (A) and Belew (B) for president; Cain (C) and Dugas (D) for vice-president.

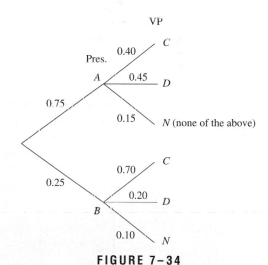

FIGURE 7–34

Find the following probabilities:
(a) $P(A)$ **(b)** $P(D \mid A)$
(c) $P(N \mid B)$ **(d)** $P(N)$
(e) Use information from Figure 7–34 to fill in the probabilities for the branches of the tree diagram at the top of the next page.

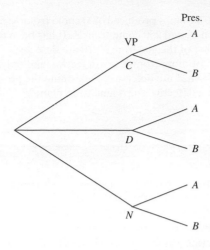

40. In general we do not expect $P(A \mid B)$ and $P(B \mid A)$ to be equal; it is, however, possible. Determine when $P(A \mid B) = P(B \mid A)$.

41. Find x in the tree diagram below so that A and D are independent.

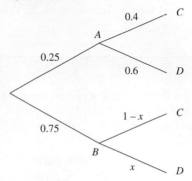

7.1 Markov Chains

- Transition Matrix
- Markov Chain
- Steady State
- Finding the Steady-State Matrix

Transition Matrix

We now study a type of problem that uses probability, matrix operations, and systems of linear equations (you may wish to review those topics). We introduce you to this type of problem with the following simplified example.

The alumni office of a university knows that generally 80% of the alumni who contribute one year will contribute the next year. They also know that 30% of those who do not contribute one year will contribute the next. We want to answer the following kinds of questions:

If 40% of a graduating class contributes the first year, how many will contribute the second year? The fifth year? The tenth year?

Before we solve this problem, let's define some terminology and basic concepts. First, the alumni can be placed in exactly one of two possible categories; they either contribute or do not contribute. These categories are called **states.**

DEFINITION

> A **state** is a category, situation, outcome, or position that a process can occupy at any given time. The states are disjoint and cover all possible outcomes.

For example, the alumni are either in the state of contributors or in the state of noncontributors. A patient is ill or well. A person's emotional state may be happy, angry, or sad. In a Markov process, the system may move from one state to another at any given time. When the process moves from one state to the next, we say that a transition is made from the **present state** to the **next state.** An alumnus can make a transition from the noncontributor state to the contributor state or vice versa.

The information on the proportion of alumni who contribute and who do not contribute can be represented by a **transition matrix.** Let C represent those who contribute and NC represent those who do not:

$$
\begin{array}{cc}
 & \text{Next State} \\
 & \begin{array}{cc} C & NC \end{array} \\
\begin{array}{l} \text{Present} \quad C \\ \text{State} \quad NC \end{array} & \begin{bmatrix} 0.8 & 0.2 \\ 0.3 & 0.7 \end{bmatrix}
\end{array}
$$

The entries in the transition matrix are probabilities that a person will move from one state (present state) to another state (next state) the following year. For example, the probability that a person who now contributes will contribute again next year is 0.8. We get this from the statement "80% of the alumni who contribute one year will contribute the next year."

The headings to the left of the matrix identify the present state, and the headings above the matrix identify the state at the next stage. Each entry in the matrix is interpreted as follows:

0.8 is the probability that a person passes from present state C to state C next; that is, a contributor remains a contributor.

0.2 is the probability that a person passes from present state C to state NC next; that is, a contributor becomes a noncontributor.

0.3 is the probability that a person passes from present state NC to state C next; that is, a noncontributor becomes a contributor.

0.7 is the probability that a person passes from present state NC to state NC next; that is, a noncontributor remains a noncontributor.

Because each row lists the probabilities of going from that state to each of all possible states, the entries in a row always add to 1.

Transition Matrix

A **transition matrix** is a square matrix with each entry a number from the interval 0 through 1. The entries in each row add to 1.

In the alumni example, a row matrix may be used to represent the proportion of people in each state. For example, the matrix [0.40   0.60] indicates that 40% are in state C and 60% are in state NC.

The same row matrix may also be interpreted to indicate that the probability of a person being in state C is 0.40, and the probability of being in state NC is 0.60.

These row matrices are called **state matrices,** or **probability-state matrices.** For each state, they show the probability that a person is in that state.

Why do we put the information in this form? Because we can use matrix operations to provide useful information. Here's how.

Multiply the state matrix and the transition matrix:

$$
\begin{array}{cc}
\begin{array}{cc} C & NC \end{array} \\
\begin{bmatrix} 0.40 & 0.60 \end{bmatrix} \begin{bmatrix} 0.8 & 0.2 \\ 0.3 & 0.7 \end{bmatrix}
\end{array}
$$

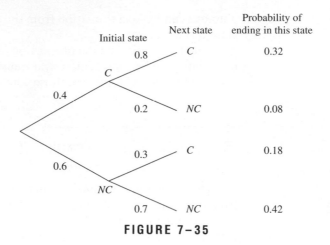

FIGURE 7–35

and the result is

$$[(0.40)(0.8) + (0.60)(0.3) \quad (0.40)(0.2) + (0.60)(0.7)]$$

$$\begin{matrix} C & NC \end{matrix}$$
$$= [0.50 \quad 0.50] \quad \text{(The state matrix at the next stage)}$$

You should interpret this as follows: In one year, the alumni moved from 40% in C and 60% in NC to 50% in each.

Let's look at the tree diagram in Figure 7–35 to help justify this.

The first stage shows the two possible states, C and NC, with the probability of each. The second stage shows the states and the probability of entering those states. At the end of each branch is the probability of terminating there. Notice that two branches terminate in C. We simply add the two probabilities there to obtain the probability of ending in state C. That probability is

$$(0.40)(0.8) + (0.60)(0.3) = 0.32 + 0.18 = 0.50$$

Notice that this is exactly the computation used in the matrix product that gave the first entry in the state matrix for the next stage. The second entry in the state matrix for the next stage is

$$(0.40)(0.2) + (0.60)(0.7)$$

which gives the probability for entering state NC, as obtained from the tree diagram.

This illustrates how the present state matrix and transition matrix can be used to get the next state matrix.

EXAMPLE 1 ➤   Using $[0.40 \quad 0.60]$ as the first year's state matrix (sometimes called the **initial state matrix**) and

$$\begin{bmatrix} 0.8 & 0.2 \\ 0.3 & 0.7 \end{bmatrix}$$

as the transition matrix for the alumni problem, find the percentage of alumni in each category for the second, third, and fourth years.

SOLUTION

As illustrated above, the second-year breakdown is

$$
\begin{array}{cc} C & NC \end{array}
$$
$$
[0.40 \quad 0.60] \begin{bmatrix} 0.8 & 0.2 \\ 0.3 & 0.7 \end{bmatrix} = \begin{array}{cc} C & NC \end{array} [0.50 \quad 0.50]
$$
$$
\text{1st Year} \qquad\qquad\qquad \text{2nd Year}
$$

The third-year state matrix is obtained by multiplying the second-year state matrix (it is now the present matrix) by the transition matrix:

$$
[0.50 \quad 0.50] \begin{bmatrix} 0.8 & 0.2 \\ 0.3 & 0.7 \end{bmatrix} = [(0.50)(0.8) + (0.50)(0.3) \quad (0.50)(0.2) + (0.50)(0.7)]
$$
$$
= [0.40 + 0.15 \quad 0.10 + 0.35]
$$
$$
= [0.55 \quad 0.45]
$$

The fourth-year state matrix is

$$
[0.55 \quad 0.45] \begin{bmatrix} 0.8 & 0.2 \\ 0.3 & 0.7 \end{bmatrix} = [(0.55)(0.8) + (0.45)(0.3) \quad (0.55)(0.2) + (0.45)(0.7)]
$$
$$
= [0.575 \quad 0.425]
$$

This process may be continued for years 5, 6, and so on.

■ **Now You Are Ready to Work Exercise 9**

We apply the same process when there are more than two states.

EXAMPLE 2 ➤

At the end of each fiscal year, the Student Loan Program gathers information on the payment status of the loans. The loans are divided into three categories: payments up to date, with payments made within 15 days of the due date considered current (labeled 0–15); payments that are 16 to 90 days late (labeled 16–90); and payments over 90 days late (labeled 90+). Each year, some of the students change categories because they get behind in payments or catch up. A study of past years gives the following transition matrix showing the fraction of students that change from one category to another or stay in the same category:

		Move to Category		
		0–15	16–90	90+
Move	0–15	0.86	0.08	0.06
from	16–90	0.62	0.29	0.09
Category	90+	0.17	0.37	0.46

One year, the percentage in each category was 0–15, 80%; 16–90, 11%; and 90+, 9%.

(a) Find the percentage in each category the next year.
(b) Find the percentage in each category three years later.

SOLUTION

At the end of each fiscal year, each student loan is in one of three states: paid up (0–15), in arrears 90 days or less (16–90), or over 90 days in arrears (90+). The present state matrix is [0.80   0.11   0.09].

(a) The next-state matrix is

$$[0.80 \quad 0.11 \quad 0.09] \begin{bmatrix} 0.86 & 0.08 & 0.06 \\ 0.62 & 0.29 & 0.09 \\ 0.17 & 0.37 & 0.46 \end{bmatrix}$$

$$= [0.772 \quad 0.129 \quad 0.099]$$

with the entries rounded to three decimals. A year later, there are 77.2% in the 0–15 category, 12.9% in the 16–90 category, and 9.9% in the 90+ category.

(b) Two years later, the state matrix is

$$[0.772 \quad 0.129 \quad 0.099] \begin{bmatrix} 0.86 & 0.08 & 0.06 \\ 0.62 & 0.29 & 0.09 \\ 0.17 & 0.37 & 0.46 \end{bmatrix}$$

$$= [0.760 \quad 0.136 \quad 0.104]$$

Two years later, there are 76.0% in the 0–15 category, 13.6% in the 16–90 category, and 10.4% in the 90+ category.

Three years later, the state matrix is

$$[0.760 \quad 0.136 \quad 0.104] \begin{bmatrix} 0.86 & 0.08 & 0.06 \\ 0.62 & 0.29 & 0.09 \\ 0.17 & 0.37 & 0.46 \end{bmatrix}$$

$$= [0.755 \quad 0.139 \quad 0.106]$$

Three years later, there are 75.5% in the 0–15 category, 13.9% in the 16–90 category, and 10.6% in the 90+ category.

■ **Now You Are Ready to Work Exercise 22**

Markov Chain

Let's summarize the ideas of a Markov chain.

A **Markov chain,** or **Markov process,** is a sequence of experiments with the following properties:

1. An experiment has a finite number of discrete outcomes, called **states.** The process, or experiment, is always in one of these states.
2. With each additional trial, the experiment can move from its present state to any other state or remain in the same state.
3. The probability of going from one state to another on the next trial depends only on the present state and not on past states.
4. The probability of moving from any one state to another in one step is represented in a transition matrix.
   (a) The transition matrix is square, because all possible states are used for rows and columns.

> **(b)** Each entry is between 0 and 1, inclusive.
> **(c)** The entries in each row add to 1.
> **5.** The state matrix times the transition matrix gives the state matrix for the next stage.

Steady State

A study of Markov processes enables us to determine the probability-state matrix for a sequence of trials. Sometimes it helps to know the long-term trends of a population, of the market of a product, or of political processes. A Markov chain may provide some useful long-term information because some Markov processes will tend toward a **steady state,** or **equilibrium.** Here is a simple example.

EXAMPLE 3 ➤

The transition matrix of a Markov process is

$$T = \begin{bmatrix} 0.6 & 0.4 \\ 0.1 & 0.9 \end{bmatrix}$$

and an initial-state matrix is [0.50 0.50].

If we compute a sequence of state matrices for subsequent stages, we obtain the following information:

Step	State Matrix		
Initial	[0.50 0.50]		
1	[0.35 0.65]	= [0.50 0.50]T	
2	[0.275 0.725]	= [0.35 0.65]T	= [0.50 0.50]T^2
3	[0.238 0.762]	= [0.275 0.725]T	= [0.50 0.50]T^3
4	[0.219 0.781]	= [0.238 0.762]T	= [0.50 0.50]T^4
5	[0.209 0.791]	= [0.219 0.781]T	= [0.50 0.50]T^5
6	[0.204 0.796]	= [0.209 0.791]T	= [0.50 0.50]T^6
7	[0.202 0.798]	= [0.204 0.796]T	= [0.50 0.50]T^7
8	[0.201 0.799]	= [0.202 0.798]T	= [0.50 0.50]T^8

It appears that the state matrix is approaching [0.20 0.80] as the sequence of trials progresses. In fact, that is the case. Furthermore, the state matrix [0.20 0.80] has an interesting property, which we can observe when we find the state matrix for the next stage:

$$[0.20 \quad 0.80]\begin{bmatrix} 0.6 & 0.4 \\ 0.1 & 0.9 \end{bmatrix} = [(0.20)(0.6) + (0.80)(0.1) \quad (0.20)(0.4) + (0.80)(0.9)]$$

$$= [0.12 + 0.08 \quad 0.08 + 0.72]$$

$$= [0.20 \quad 0.80]$$

■ **Now You Are Ready to Work Exercise 23**

There is no change in the next state matrix. The process has reached a **steady,** or **equilibrium,** state.

DEFINITION
Steady-State Matrix

A state matrix $X = [p_1 \quad p_2 \ldots p_n]$ is a **steady-state,** or **equilibrium, matrix** for a transition matrix T if $XT = X$.

EXAMPLE 4 ➤

The steady-state matrix for the alumni problem is [0.6 0.4] because an initial state matrix will eventually approach [0.6 0.4] and

$$[0.6 \quad 0.4]\begin{bmatrix} 0.8 & 0.2 \\ 0.3 & 0.7 \end{bmatrix} = [0.48 + 0.12 \quad 0.12 + 0.28]$$

$$= [0.60 \quad 0.40]$$

This indicates that as long as the transition matrix represents the giving practices of the alumni, they will stabilize at 60% contributing and 40% not contributing.

■ **Now You Are Ready to Work Exercise 12**

Finding the Steady-State Matrix

Let's show how to find the steady-state matrix for the alumni problem. Let $X = [x \quad y]$ be the desired, but unknown, steady-state matrix. We want to find x and y so that

$$[x \quad y]\begin{bmatrix} 0.8 & 0.2 \\ 0.3 & 0.7 \end{bmatrix} = [x \quad y]$$

The matrix product on the left gives

$$[0.8x + 0.3y \quad 0.2x + 0.7y] = [x \quad y]$$

so

$$0.8x + 0.3y = x$$
$$0.2x + 0.7y = y$$

which is equivalent to

$$-0.2x + 0.3y = 0$$
$$0.2x - 0.3y = 0$$

As these two equations are equivalent, we can drop one of them.

Because $[x \quad y]$ is a probability matrix, we must have $x + y = 1$. This, with the equation above, gives the system

$$x + \quad y = 1$$
$$-0.2x + 0.3y = 0$$

If we use an augmented matrix to solve this system, we have

$$\begin{bmatrix} 1 & 1 & | & 1 \\ -0.2 & 0.3 & | & 0 \end{bmatrix}$$

which eventually reduces to

$$\begin{bmatrix} 1 & 0 & | & 0.6 \\ 0 & 1 & | & 0.4 \end{bmatrix}$$

so $x = 0.6$, $y = 0.4$ gives [0.6 0.4] as the steady-state matrix. This approach will work in general.

EXAMPLE 5 ➤ Find the steady-state matrix of the transition matrix

$$T = \begin{bmatrix} 0.3 & 0.2 & 0.5 \\ 0.1 & 0.4 & 0.5 \\ 0.4 & 0 & 0.6 \end{bmatrix}$$

SOLUTION

Solve the equation

$$\begin{bmatrix} x & y & z \end{bmatrix} \begin{bmatrix} 0.3 & 0.2 & 0.5 \\ 0.1 & 0.4 & 0.5 \\ 0.4 & 0 & 0.6 \end{bmatrix} = \begin{bmatrix} x & y & z \end{bmatrix}$$

for a probability matrix $\begin{bmatrix} x & y & z \end{bmatrix}$, which is the system

$$\begin{aligned} 0.3x + 0.1y + 0.4z &= x \\ 0.2x + 0.4y &= y \\ 0.5x + 0.5y + 0.6z &= z \end{aligned}$$

plus the equation $x + y + z = 1$. Write this as

$$\begin{aligned} x + y + z &= 1 \\ -0.7x + 0.1y + 0.4z &= 0 \\ 0.2x - 0.6y &= 0 \\ 0.5x + 0.5y - 0.4z &= 0 \end{aligned}$$

Solve this system using the Gauss-Jordan Method. It gives the following sequence of augmented matrices:

$$\begin{bmatrix} 1 & 1 & 1 & | & 1 \\ -0.7 & 0.1 & 0.4 & | & 0 \\ 0.2 & -0.6 & 0 & | & 0 \\ 0.5 & 0.5 & -0.4 & | & 0 \end{bmatrix}$$

Multiply the last three rows by 10 to obtain integer entries and perform row operations indicated

$$\begin{bmatrix} 1 & 1 & 1 & | & 1 \\ -7 & 1 & 4 & | & 0 \\ 2 & -6 & 0 & | & 0 \\ 5 & 5 & -4 & | & 0 \end{bmatrix} \quad \begin{array}{l} 7R1 + R2 \to R2 \\ -2R1 + R3 \to R3 \\ -5R1 + R4 \to R4 \end{array}$$

$$\begin{bmatrix} 1 & 1 & 1 & | & 1 \\ 0 & 8 & 11 & | & 7 \\ 0 & -8 & -2 & | & -2 \\ 0 & 0 & -9 & | & -5 \end{bmatrix} \quad R2 + R3 \to R3$$

$$\begin{bmatrix} 1 & 1 & 1 & | & 1 \\ 0 & 8 & 11 & | & 7 \\ 0 & 0 & 9 & | & 5 \\ 0 & 0 & -9 & | & -5 \end{bmatrix} \quad R3 + R4 \to R4$$

$$\begin{bmatrix} 1 & 1 & 1 & | & 1 \\ 0 & 8 & 11 & | & 7 \\ 0 & 0 & 9 & | & 5 \\ 0 & 0 & 0 & | & 0 \end{bmatrix} \qquad \begin{array}{l} \frac{1}{8}R2 \to R2 \\ \frac{1}{9}R3 \to R3 \end{array}$$

$$\begin{bmatrix} 1 & 1 & 1 & | & 1 \\ 0 & 1 & \frac{11}{8} & | & \frac{7}{8} \\ 0 & 0 & 1 & | & \frac{5}{9} \\ 0 & 0 & 0 & | & 0 \end{bmatrix} \qquad \begin{array}{l} -R3 + R1 \to R1 \\ \\ -\frac{11}{8}R3 + R2 \to R2 \end{array}$$

$$\begin{bmatrix} 1 & 1 & 0 & | & \frac{4}{9} \\ 0 & 1 & 0 & | & \frac{1}{9} \\ 0 & 0 & 1 & | & \frac{5}{9} \\ 0 & 0 & 0 & | & 0 \end{bmatrix} \qquad -R2 + R1 \to R1$$

$$\begin{bmatrix} 1 & 0 & 0 & | & \frac{3}{9} \\ 0 & 1 & 0 & | & \frac{1}{9} \\ 0 & 0 & 1 & | & \frac{5}{9} \\ 0 & 0 & 0 & | & 0 \end{bmatrix}$$

This gives $x = \frac{3}{9}$, $y = \frac{1}{9}$, $z = \frac{5}{9}$ and the steady-state matrix $\begin{bmatrix} \frac{3}{9} & \frac{1}{9} & \frac{5}{9} \end{bmatrix}$.

■ **Now You Are Ready to Work Exercise 25**

EXAMPLE 6 ➤   A sociologist made a regional study of the shift of population between rural and urban areas. The transition matrix of the annual shift from one area to another was found to be

$$\begin{array}{c} & \text{To} \\ & \begin{array}{cc} R & U \end{array} \\ \text{From} \begin{array}{c} R \\ U \end{array} & \begin{bmatrix} 0.76 & 0.24 \\ 0.08 & 0.92 \end{bmatrix} \end{array}$$

indicating that 76% of rural residents remain in rural areas, 24% move from rural to urban areas, 8% of urban residents move from urban to rural areas, and 92% remain in the urban areas. Find the percentage of the population in rural and urban areas when the population stabilizes.

SOLUTION

Let $[x \quad y]$ be the state matrix of the population, with x the proportion in rural areas and y the proportion in urban areas. We want to find the steady-state matrix, that is, the solution to

$$[x \quad y]\begin{bmatrix} 0.76 & 0.24 \\ 0.08 & 0.92 \end{bmatrix} = [x \quad y]$$

This condition, with $x + y = 1$, gives the system

$$\begin{array}{rcl} x + \quad y &=& 1 \\ -0.24x + 0.08y &=& 0 \\ 0.24x - 0.08y &=& 0 \end{array}$$

The solution to the system is $x = 0.25$ and $y = 0.75$, so the steady-state matrix is [0.25 0.75], indicating that the population will stabilize at 25% in rural areas and 75% in urban areas.

■ **Now You Are Ready to Work Exercise 26**

Let's look at the steady-state situation with two states, and thus a 2 × 2 transition matrix, to see how the steady-state solution can be reduced to a linear equation. We use the transition matrix

$$\begin{bmatrix} 0.76 & 0.24 \\ 0.08 & 0.92 \end{bmatrix}$$

from Example 6.

The steady-state solution is found by solving the system

$$x + y = 1$$
$$0.76x + 0.08y = x$$
$$0.24x + 0.92y = y$$

In a steady-state problem with two states, the first equation always appears, and the other equations (like the last two equations here) are always equivalent. Thus, x and y are related with $y = 1 - x$. The steady-state solution can be found by substituting $x = t$ and $y = 1 - t$ into one of the last two equations. Using the middle equation, we have $0.76t + 0.08(1 - t) = t$, which reduces to $0.32t - 0.08 = 0$, which gives $t = 0.25$ and $1 - t = 0.75$. The steady-state solution then is [0.25 0.75]. This procedure can be used when solving a two-state problem with a 2 × 2 transition matrix.

It is sometimes important to know whether a Markov process will eventually reach equilibrium. In Examples 5 and 6 we found the steady state matrix. It happens to be true in those cases that we will eventually reach the steady-state matrix after a sequence of trials, regardless of the initial state matrix. Although this is not true for all transition matrices, there is a rather reasonable property that ensures that a Markov process will reach equilibrium. We call transition matrices with this property **regular.** A regular Markov process will eventually reach a steady state, and its transition matrix has the following property:

DEFINITION
Regular Matrix

A transition matrix T of a Markov process is called **regular** if some power of T has only positive entries.

A regular transition matrix is useful because it defines a Markov process that eventually reaches a steady state.

EXAMPLE 7 ▶

$$T = \begin{bmatrix} 0.3 & 0.7 \\ 0.25 & 0.75 \end{bmatrix}$$

is regular because its first power contains all positive entries.

$$T = \begin{bmatrix} 0 & 1 \\ 0.6 & 0.4 \end{bmatrix}$$

is regular because

$$\begin{bmatrix} 0 & 1 \\ 0.6 & 0.4 \end{bmatrix}^2 = \begin{bmatrix} 0.6 & 0.4 \\ 0.24 & 0.76 \end{bmatrix}$$

has all positive entries.

■ **Now You Are Ready to Work Exercise 34**

EXAMPLE 8 ➤   Find the steady-state matrix of the regular transition matrix

$$T = \begin{bmatrix} 0 & 0.5 & 0.5 \\ 0.5 & 0.5 & 0 \\ 0.5 & 0 & 0.5 \end{bmatrix}$$

(It is regular.)

SOLUTION
The condition

$$[x \quad y \quad z] \begin{bmatrix} 0 & 0.5 & 0.5 \\ 0.5 & 0.5 & 0 \\ 0.5 & 0 & 0.5 \end{bmatrix} = [x \quad y \quad z]$$

with $x + y + z = 1$ yields a system of four equations whose augmented matrix is

$$\begin{bmatrix} 1 & 1 & 1 & | & 1 \\ -1 & 0.5 & 0.5 & | & 0 \\ 0.5 & -0.5 & 0 & | & 0 \\ 0.5 & 0 & -0.5 & | & 0 \end{bmatrix}$$

(Be sure that you can get this matrix.)
    We will not show all the row operations that lead to the solution, but the final matrix is

$$\begin{bmatrix} 1 & 0 & 0 & | & \frac{1}{3} \\ 0 & 1 & 0 & | & \frac{1}{3} \\ 0 & 0 & 1 & | & \frac{1}{3} \\ 0 & 0 & 0 & | & 0 \end{bmatrix}$$

so the steady-state matrix is $\begin{bmatrix} \frac{1}{3} & \frac{1}{3} & \frac{1}{3} \end{bmatrix}$.

■ **Now You Are Ready to Work Exercise 30**

▦ 7.7 EXERCISES

LEVEL 1

1. Which of the following are transition matrices?

(a) $\begin{bmatrix} 0.6 & 0.4 \\ 0.3 & 0.7 \end{bmatrix}$

(b) $\begin{bmatrix} 0.6 & 0 & 0.4 \\ 0.2 & 0.1 & 0.5 \\ 0.3 & 0.4 & 0.3 \end{bmatrix}$

(c) $\begin{bmatrix} 0.5 & 0.5 \\ 0.3 & 0.7 \\ 0.2 & 0.8 \end{bmatrix}$

(d) $\begin{bmatrix} 0.1 & 0 & 0.3 & 0.6 \\ 0.2 & 0.3 & 0 & 0.5 \\ 0.4 & 0.2 & 0.1 & 0.3 \\ 0 & 0.25 & 0.35 & 0.4 \end{bmatrix}$

2. Which of the following are probability-state matrices?
   (a) [0.4 0.3 0.3]   (b) [0.3 0.3 0.3]
   (c) [0.1 0.2 0.3 0.4] (d) [0.6 0.7]
   (e) [0.1 0.2 0 0.3 0.4]

3. The matrix T represents the transition of college students between dorms (D) and apartments (A) at the end of a semester.

$$
\begin{array}{c}
 & \text{To} \\
 & \begin{array}{cc} A & D \end{array} \\
\text{From} \begin{array}{c} A \\ D \end{array} & \begin{bmatrix} 0.9 & 0.1 \\ 0.4 & 0.6 \end{bmatrix}
\end{array}
$$

   (a) What percent of those living in apartments move to a dorm?
   (b) What is the probability that a student will remain in a dorm the next semester?
   (c) What is the probability that an apartment-dwelling student will remain in an apartment?
   (d) What percent of dorm residents move to an apartment?

4. An investment firm invests in stocks, bonds, and mortgages for its clients. One of the partners in the firm analyzed the investment patterns of her clients. She found that during a year they change between types of investments according to the following transition matrix:

$$
\begin{array}{c}
 & \text{To} \\
 & \begin{array}{ccc} S & B & M \end{array} \\
\text{From} \begin{array}{c} S \\ B \\ M \end{array} & \begin{bmatrix} 0.88 & 0.09 & 0.03 \\ 0.15 & 0.75 & 0.10 \\ 0.19 & 0.17 & 0.64 \end{bmatrix}
\end{array}
$$

   (a) What percentage of those investing in bonds move to mortgages?
   (b) What percentage move their investments from stocks to bonds?
   (c) What is the probability that a bond investor will leave his or her investment in bonds?
   (d) What is the probability that a mortgage investor will change to stocks or bonds?

5.
$$
\begin{array}{c}
 & \begin{array}{ccc} A & B & C \end{array} \\
\begin{array}{c} A \\ B \\ C \end{array} & \begin{bmatrix} 0.3 & 0.2 & 0.5 \\ 0.4 & 0.6 & 0 \\ 0.1 & 0.8 & 0.1 \end{bmatrix} = T
\end{array}
$$

From the transition matrix T, find the probability of
(a) moving from state B to state A.
(b) moving from state C to state B.
(c) remaining in state A.

For Exercises 6 through 8, find the state matrix for the next stage from the given present state and transition matrices.

6. $S = [0.2 \quad 0.8], \quad T = \begin{bmatrix} 0.5 & 0.5 \\ 0.8 & 0.2 \end{bmatrix}$

7. $S = [0.45 \quad 0.55], \quad T = \begin{bmatrix} 0.4 & 0.6 \\ 0.9 & 0.1 \end{bmatrix}$

8. $S = [0.2 \quad 0.5 \quad 0.3], \quad T = \begin{bmatrix} 0.3 & 0.5 & 0.2 \\ 0.2 & 0.2 & 0.6 \\ 0.1 & 0.8 & 0.1 \end{bmatrix}$

Initial state and transition matrices are given in Exercises 9 through 11. Find the state matrices for the next two stages.

9. *(See Example 1)*
   $M_0 = [0.65 \quad 0.35], \quad T = \begin{bmatrix} 0.24 & 0.76 \\ 0.36 & 0.64 \end{bmatrix}$

10. $M_0 = [0.3 \quad 0.7], \quad T = \begin{bmatrix} 0.8 & 0.2 \\ 0.2 & 0.8 \end{bmatrix}$

11. $M_0 = [0.25 \quad 0.50 \quad 0.25], \quad T = \begin{bmatrix} 0.3 & 0.4 & 0.3 \\ 0.1 & 0.3 & 0.6 \\ 0.2 & 0.5 & 0.3 \end{bmatrix}$

For Exercises 12 through 15, show that the given state matrix S is the steady-state matrix for the transition matrix T.

12. *(See Example 4)*
   $S = [0.375 \quad 0.625], \quad T = \begin{bmatrix} 0.5 & 0.5 \\ 0.3 & 0.7 \end{bmatrix}$

13. $S = \begin{bmatrix} \dfrac{2}{3} & \dfrac{1}{3} \end{bmatrix}, \quad T = \begin{bmatrix} 0.6 & 0.4 \\ 0.8 & 0.2 \end{bmatrix}$

14. $S = [0.625 \quad 0.375], \quad T = \begin{bmatrix} 0.58 & 0.42 \\ 0.7 & 0.3 \end{bmatrix}$

15. $S = \begin{bmatrix} \dfrac{13}{28} & \dfrac{15}{28} \end{bmatrix}, \quad T = \begin{bmatrix} 0.25 & 0.75 \\ 0.65 & 0.35 \end{bmatrix}$

16. Let T be the transition matrix
   $$T = \begin{bmatrix} 0.25 & 0.75 \\ 0.40 & 0.60 \end{bmatrix}$$
   and $M_0 = [0.2 \quad 0.8]$ be an initial state matrix.
   (a) Compute the state matrix $M_1 = M_0 T$. Then compute the next state matrix from M_1, that is, $M_2 = M_1 T$.
   (b) Compute T^2.
   (c) Compute $M_0 T^2$. Verify that it equals $M_1 T$.

17. For $M = [0.3 \quad 0.3 \quad 0.4]$ and

$$T = \begin{bmatrix} 0.1 & 0.5 & 0.4 \\ 0.2 & 0.6 & 0.2 \\ 0.4 & 0 & 0.6 \end{bmatrix}$$

compute $MT, (MT)T, ((MT)T)T$, and MT^3.

LEVEL 2

18. A department store's charge accounts are either currently paid up or in arrears. The store's records show that
   (i) 90% of accounts that are paid up this month will be paid up next month also.
   (ii) 40% of those in arrears this month will be in arrears next month also.
   (iii) Current accounts are 85% paid up and 15% in arrears.

   (a) Represent this information with a tree diagram.
   (b) Represent this information with a state matrix and a transition matrix.

19. Find x so that $[0.2 \quad x \quad 0.4]$ is a probability-state matrix.

20. Find x, y, z so that

$$\begin{bmatrix} 0.1 & x & 0 \\ y & 0.3 & 0.5 \\ 0.2 & 0.7 & z \end{bmatrix}$$

is a transition matrix.

21. The alumni of State University generally contribute (C) or do not contribute (NC) according to the following pattern: 75% of those who contribute one year will contribute the next year; 15% of those who do not contribute one year will contribute the next. The transition matrix is the following:

		Next Year	
		C	NC
Present	C	0.75	0.25
Year	NC	0.15	0.85

Forty-five percent of last year's graduating class contributed this year. What percent will contribute next year? In two years?

22. *(See Example 2)* The students at Conway College can buy a full-meal plan at the college cafeteria or a one-meal-per-day plan. Each semester students are in one of three "states": the full-meal plan, the one-meal plan, or no plan. Each semester a student may remain in a plan or change to another plan, according to the following transition matrix as estimated from the cafeteria records:

		To Plan		
		Full	One	None
From	Full	0.6	0.2	0.2
Plan	One	0.1	0.8	0.1
	None	0.2	0.3	0.5

If 50% of the students have the full-meal plan, 25% have the one-meal plan and 25% have no plan, what percentage will have each plan
   (a) two semesters later?
   (b) three semesters later?

23. *(See Example 3)* For the initial state matrix $[0.6 \quad 0.4]$ and the transition matrix

$$T = \begin{bmatrix} 0.5 & 0.5 \\ 0.2 & 0.8 \end{bmatrix}$$

find the sequence of the next six state matrices. What appears to be the steady-state matrix?

24. Find the steady-state matrix of

$$\begin{bmatrix} 0.6 & 0.4 \\ 0.2 & 0.8 \end{bmatrix}$$

25. *(See Example 5)* Find the steady-state matrix of the transition matrix

$$T = \begin{bmatrix} 0.6 & 0.2 & 0.2 \\ 0.1 & 0.8 & 0.1 \\ 0.2 & 0.4 & 0.4 \end{bmatrix}$$

26. *(See Example 6)* The transition of college students between dorms and apartments at the end of a semester is given by the following:

		To	
		D	A
From	D	0.9	0.1
	A	0.4	0.6

Find the percentage of the population in dorms and apartments when the population stabilizes.

27. Find the steady-state matrix of the transition matrix given in Exercise 5.

28. Use the transition matrix of Exercise 21 to find the steady-state distribution of State University alumni who contribute and who do not contribute.

29. Use the transition matrix of Exercise 22 to find the proportion of students in each plan when equilibrium is reached.

Find the steady-state matrix of the regular matrices in Exercises 30 through 33.

30. (*See Example 8*)

$$\begin{bmatrix} \frac{1}{3} & \frac{2}{3} \\ \frac{3}{4} & \frac{1}{4} \end{bmatrix}$$

31. $\begin{bmatrix} 0.9 & 0.1 \\ 1 & 0 \end{bmatrix}$

32. $\begin{bmatrix} 0.3 & 0.2 & 0.5 \\ 0 & 0.5 & 0.5 \\ 0.7 & 0.2 & 0.1 \end{bmatrix}$

33. $\begin{bmatrix} \frac{1}{3} & \frac{1}{3} & \frac{1}{3} \\ \frac{1}{2} & \frac{1}{2} & 0 \\ 0 & \frac{1}{4} & \frac{3}{4} \end{bmatrix}$

34. (*See Example 7*) Show that

$$T = \begin{bmatrix} 0 & 0 & 1 \\ 0.2 & 0.3 & 0.5 \\ 0 & 0.3 & 0.7 \end{bmatrix}$$

is regular.

35. There are five points on a circle as shown in Figure 7–36. A particle is on one of the points and moves to an adjacent point in either direction. The probability is $\frac{1}{2}$ that it will move clockwise and $\frac{1}{2}$ that it will move counterclockwise. Write the transition matrix for this process.

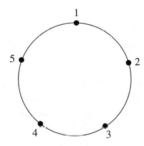

FIGURE 7–36

36. A security guard must check three locations periodically. They are located in a triangle. To relieve the monotony, after arriving at one location, he tosses a coin to determine which of the other two he will check next.
   (a) Form the transition matrix of this process.
   (b) Find the steady-state matrix.

37. A plant with genotype RW can produce red (R), pink (P), or white (W) flowers. When two plants of this genotype are crossed, they produce the three colors according to the following transition matrix:

		Flowers of Offspring	
		R	P	W
Flowers of Parent	R	0.50	0.50	0
	P	0.25	0.50	0.25
	W	0	0.50	0.50

Flowers of this genotype are crossed for successive generations. When the process reaches a steady state, what percentage of the flowers will be red, pink, and white?

38. Assume that a person's profession can be classified as professional, skilled, or unskilled. Sociology studies give the following information about a child's profession as related to his or her parents:

   Professional: Of their children, 80% are professional, 10% are skilled, and 10% are unskilled.

   Skilled: Of their children, 60% are skilled, 20% are unskilled, and 20% are professional.

   Unskilled: Of their children, 35% are skilled, 15% are professional, and 50% are unskilled.

   (a) Set up the transition matrix of this process.
   (b) Find the probability of an unskilled parent having a grandchild who is professional.
   (c) The population will eventually stabilize into fixed proportions of these professions. Find those proportions; that is, find the steady-state population distribution.

39. A country is divided into three geographic regions: I, II, and III. The matrix T is the transition matrix that shows the probabilities of moving from one region to another.

		To		
		I	II	III
From	I	0.9	0.06	0.04
	II	0.15	0.8	0.05
	III	0.3	0.1	0.6

What is the probability of
   (a) moving from region I to region II?
   (b) moving from region II to region III?
   (c) remaining in region II?
   (d) moving from region I to region III?

40. A country is divided into three geographic regions. It is found that each year 5% of the residents move from region I to region II and 5% move from region I to region III. In region II, 15% move to region I, and 10% move to region III. In region III, 10% move to region I, and 5% to region II. Find the steady state population distribution.

■■ IMPORTANT TERMS

7.1

Experiment
Outcome
Trial
Sample Space
Sample Point
Event
Simple Event
Simple Outcome
Probability Assignment
Empirical Probability
Relative Frequency

7.2

Equally Likely
Successes
Failures

Random Selection
Random Outcome

7.3

Compound Event
Union
Intersection
Complement
Mutually Exclusive Events
Disjoint Events

7.4

Conditional Probability
Reduced Sample Space
Multiplication Rule

7.5

Independent Events
Dependent Events

Multiplication Rule for Independent
   Events
Pigeonhole Principle

7.6

Bayes' Rule

7.7

State
Present State
Next State
Transition Matrix
Probability-State Matrix
Initial State Matrix
Markov Chain
Steady State
Equilibrium
Steady-State Matrix
Regular Matrix

■■ REVIEW EXERCISES

1. An experiment has six possible outcomes with the
   following probabilities:

$$P_1 = 0.02, \quad P_2 = 0.3, \quad P_3 = 0.1$$
$$P_4 = 0.0, \quad P_5 = 0.2, \quad P_6 = 0.3$$

Is this a valid probability assignment?

2. An experiment has the sample space
   $S = \{a, b, c, d\}$. Find the probability of each simple
   outcome in S if

$$P(a) = P(b), \quad P(c) = 2 P(b), \quad P(d) = 3 P(c)$$

3. A refreshment stand kept a tally of the number of
   soft drinks sold. One day its records showed the
   following:

Soft Drinks

Size	Number Sold
Small	94
Medium	146
Large	120

Find the probability that a person selected at random will buy a medium-sized soft drink.

4. Three people are selected at random from a group
   of five men and two women.

(a) Find the probability that all three selected are
   men.
(b) Find the probability that two men and one
   woman are selected.
(c) Find the probability that all three selected are
   women.

5. In a class of 30 students, 10 participate in sports,
   12 participate in band, and 5 participate in both.
   If a student is selected at random, find the probability that the student participates in sports
   or band.

6. In a group of 30 schoolchildren, 15 are eight-year-
   olds, 12 are nine-year-olds, and 3 are ten-year-olds.
   Of the eight-year-olds, 10 are boys; of the nine-year-
   olds, 5 are boys; and of the ten-year-olds, 2 are boys.
   One child is selected at random from the group.
   Find the probability that the child is
   (a) an eight-year-old.   **(b)** a boy.
   (c) a nine-year-old.   **(d)** a twelve-year-old.
   (e) a nine-year-old girl.   **(f)** a ten-year-old girl.

7. Mark and Melanie are two of the ten students who
   volunteer to tutor children after school. Four students are selected to tutor at South Elementary.
   Find the probability that Mark and Melanie are
   among the four selected.

8. A die is rolled. Find the probability that an even number or a number greater than 4 will be rolled.

9. A single card is picked from a deck of 52 playing cards. Find the probability that it will be a king or a spade.

10. A coin and a die are tossed. Find the probability of throwing a
    (a) head and a number less than 3.
    (b) tail and an even number.
    (c) head and a 6.

11. A coin is tossed five times. Find the probability that all five tosses will land heads up.

12. A card is selected from a deck of 52 playing cards. Find the probability that it will be a
    (a) red card or a 10.   (b) face card or a spade.
    (c) face card or a 10.

13. An elementary school teacher has a collection of mathematics review questions, including 10 addition, 8 subtraction, and 15 multiplication. A computer randomly selects problems for a student.
    (a) Find the probability that the first problem selected is an addition or subtraction problem.
    (b) Find the probability that the first two are subtraction or multiplication problems.

14. A card is selected at random from a deck of bridge cards. Find the probability that it is not
    (a) an ace.   (b) a face card.

15. A bargain table has 40 books; 10 are romance, 10 are biographies, 10 are crafts, and 10 are historical fiction. If 2 books are selected at random, what is the probability that they are
    (a) the same kind?
    (b) different kinds?

16. A load of lumber contains 40 pieces of birch and 50 pieces of pine. Of the lumber, 5 pieces of birch and 3 pieces of pine are warped. Let F, G, and H be the events of selecting birch, pine, and a warped piece of wood, respectively. Compute and interpret the following probabilities.
    (a) $P(F), P(G), P(H)$ (b) $P(F \cap H)$
    (c) $P(F \cup H)$ (d) $P(F' \cup H)$
    (e) $P((F' \cup H)')$

17. A card is picked from a deck of 52 playing cards. Let F be the event of selecting an even-numbered card and let G be that of selecting a 10. Are G and F independent events?

18. A card is picked from a deck of 52 playing cards. Let R be the event of selecting a red card and let Q be the event of selecting a queen. Are R and Q independent?

19. A die is tossed four times. Find the probability of obtaining
    (a) 1, 2, 3, 4, in that order.
    (b) 1, 2, 3, 4, in any order.
    (c) two even numbers, then a 5, then a number less than 3.

20. Two dice are rolled. Find the probability that the
    (a) sum of the numbers on the dice is 6.
    (b) same number is obtained on each die.

21. A student has four examinations to take. She has determined that the probability of her passing the mathematics examination is 0.8; English, 0.5; history, 0.3; and chemistry, 0.7. Assuming independence of examinations, find the probability of her passing
    (a) mathematics, history, and English but failing chemistry.
    (b) mathematics and chemistry but failing history and English.
    (c) all four subjects.

22. A study of juvenile delinquents shows that 60% come from low-income families (LI), 45% come from broken homes (BH), and 35% come from both ($LI \cap BH$). A juvenile delinquent is selected at random.
    (a) Find the probability that the juvenile is not from a low-income family.
    (b) Find the probability that the juvenile comes from a broken home or a low-income family.
    (c) Find the probability that the juvenile comes from a low-income family, given that the juvenile comes from a broken home.
    (d) Are LI and BH independent?
    (e) Are LI and BH mutually exclusive?

23. A study of the adult population in a midwestern state found the following information on drinking habits:

	Men	Women
Abstain	20%	40%
Infrequent	10%	20%
Moderate	50%	35%
Heavy	20%	5%

The adult population of the state is 55% female and 45% male. An individual selected at random is found to be
    (a) an abstainer. Find the probability that the person is male.
    (b) a heavy drinker. Find the probability that the individual is female.

24. A stock analyst classifies stocks as either blue chip (*BC*) or not (*NBC*). The analyst also classifies stock by whether it goes up (*UP*), remains unchanged (*UC*), or goes down (*D*) at the end of a day's trading. One percent of the stocks are blue chip. The analyst summarizes the performance of stocks as follows:

	Probability of		
	UP	*UC*	*D*
BC	0.45	0.35	0.20
NBC	0.35	0.25	0.40

(a) Show this information with a tree diagram.
(b) A customer selects a stock at random and asks the analyst to buy. Find the probability that the stock is a blue chip stock that goes up the next day.

25. A two-digit number is to be constructed at random from the numbers 1, 2, 3, 4, 5, and 6, repetition not allowed. Find the probability of getting
(a) the number 33   (b) the number 35.
(c) a number the sum of whose digits is 10.
(d) a number whose first digit is greater than its second.
(e) a number less than 24.

26. A finite mathematics professor observed that 90% of the students who do the homework regularly pass the course. He also observed that only 20% of those who do not do the homework regularly pass the course. One semester, he estimated that 70% of the students did the homework regularly. Given a student who passed the course, find the probability that the student did the homework regularly.

27. A sociology class is composed of 10 juniors, 34 seniors, and 6 graduate students. Two juniors, eight seniors, and three graduate students received an A in the course. A student is selected at random and is found to have received an A. What is the probability that the student is a junior?

28. In a certain population, 5% of the men are color-blind and 3% of the women are color-blind. The population is made up of 55% men and 45% women. If a person chosen at random is color-blind, what is the probability that the person is a man?

29. Two National Merit finalists and three semifinalists are seated in a row of five chairs on the stage. In how many ways can they be seated if the
(a) finalists are in the first two chairs and the semifinalists are seated in the last three?
(b) finalists and semifinalists alternate seats?

30. The history department hosts a distinguished visiting historian. Two students are randomly selected

from the history majors — 6 seniors, 5 juniors, and 3 sophomores — to join lunch with the scholar.
(a) Find the probability that the first student selected is a senior and the second is a sophomore.
(b) Find the probability that both are juniors.

31. Ten cards are numbered 1 through 10. A card is drawn. It is then replaced in the deck, and the cards are shuffled. A second card is drawn. Find the probability that
(a) the first card is less than 3 and the second greater than 7.
(b) both cards are less than 4.

32. A college has 1650 female students and 1460 males. The financial aid office reports that 35% of the males and 40% of the females receive financial aid. A student is selected at random. Find the probability that the student
(a) is female.   (b) receives financial aid.
(c) receives financial aid, given that the student is male.
(d) is female, given that the student receives financial aid.

33. A mathematics placement exam is scored high, middle, and low. The performance of these students in calculus is summarized as follows:

	Score			
Grade	**High**	**Middle**	**Low**	**Total**
C or above	98	124	3	225
Below C	12	118	65	195
Total	110	242	68	420

One of the 420 students is selected at random. Find the probability that the student
(a) makes a grade of C or above.
(b) scored low on the placement exam.
(c) made a grade below C, given that the student scored middle on the placement exam.
(d) scored high on the placement exam, given that the student made C or above.

34. A mathematics department compares SAT mathematics scores to performance in calculus. The findings are summarized in the following table:

	SAT Score			
Grade	**Below 550**	**550–650**	**Above 650**	**Total**
A or B	1	55	28	84
C	21	69	25	115
Below C	43	56	2	101
Total	65	180	55	300

A student is selected at random.

(a) Find the probability that the student scored above 650
(b) Find the probability that the student made an A or a B.
(c) Find the probability that the student scored above 650 and made an A or a B.
(d) Find the probability that the student's SAT score was in the 550–650 range.
(e) Find the probability that the student made a C.
(f) Show that "SAT above 650" and "grade of A or B" are dependent.
(g) Show that "SAT is 550–650" and "grade of C" are independent.

35. Reports on 192 accidents showed the following relationship between injuries and using a seat belt:

	Injuries	No Injuries
Seat belt not used	66	28
Seat belt used	14	84

Determine whether the events "seat belt not used" and "injuries" are dependent or independent.

36. A professor gave two forms of an exam to an economics class. The grades by exam form are given in the following table:

Grade	Exam	
	Form I	Form II
A or B	15	20
Below B	24	32

Show that the events "student took Form I" and "student made an A or a B" are independent.

37. A survey of working couples in Davidson County revealed the following information: The probability that the husband is happy with his job is 0.72. The probability that the wife is happy with her job is 0.55. The probability that both are happy with their jobs is 0.35.

A working couple is selected at random. Find the probability that
(a) at least one is happy with his or her job.
(b) neither is happy with his or her job.

Statistics

When the president of the United States submits an annual budget to Congress in the trillion-dollar range, many taxpayers ask, "Where is all that money going?" But few of them wish to be handed a detailed budget a foot thick. They want the information summarized in a few broad categories such as defense, Social Security, education, agriculture, interest on the debt, and so on. Meanwhile, the president might want to know how the voters react to specific budget items such as defense and Social Security budgets. To poll all voters regarding their opinion of, for instance, the defense budget is impractical. However, the president can obtain valuable, although incomplete, information on voter opinion by a sample opinion poll.

"How did your class do on the finite exam?" and "How were the grades on the test?" are questions often asked by faculty and students after an exam. No one expects a response that includes a list of all the students' grades. Faculty want a response that indicates how well the class understood the material, and the students want to compare their performance with others in the class. On other occasions, a faculty member may wish to know if the students understood the concepts appropriate to a specific problem.

In these cases, the relevant information from all of the tests needs to be summarized in a form that conveys the desired information without going into great detail.

The exam summary and the president's budget summary are examples of applying descriptive statistics. **Descriptive statistics** summarize data and describe their more relevant features.

The sample poll, on the other hand, falls into the category of inferential statistics. **Inferential statistics** make generalizations or draw conclusions from representative information. This chapter presents some of the methods used in descriptive and inferential statistics. Sections 8.1 through 8.3 deal with descriptive statistics, and Sections 8.4 through 8.8 deal with inferential statistics.

The discipline of statistics provides methods to collect data, organize them in a meaningful way, and interpret and report conclusions. Because numerous disciplines and organizations depend on statistical analyses, a variety of statistical methods exist. We find statistical specialists in economics, social sciences, sciences, business, medicine, engineering, governmental agencies, and education. To collect, interpret, and report statistical data often requires massive efforts. For example, accurate estimates of unemployment or incidence of crime may be rather easy to determine in a remote county in North Dakota but difficult to determine for the whole country. This chapter is a study of some useful statistical methods, mostly dealing with much simpler situations than those that occur routinely in the larger scheme of things. ■

8.1 Frequency Distributions

- Frequency Table
- Stem-and-Leaf Plots
- Construction of a Frequency Table

- Visual Representations of Frequency Distributions
- Histogram
- Pie Chart

Frequency Table

Opinion polls and population studies use random samples to obtain information. Quite often it helps to organize the information in a tabular or visual form. One tabular form may be obtained by grouping similar observations into **categories** (or **classes**).

For example, the Mathematics Department gives a departmental final exam in calculus. The highest possible score on the exam is 160 points. A complete listing of all exam scores helps the faculty to decide on a grading curve, but a summary like the following table may give a better picture of the students' overall performance on the exam:

Score	Number of Students Making Score
0–20	0
21–40	18
41–60	36
61–80	83
81–100	110
101–120	121
121–140	73
141–160	16
Total	457

We call this kind of summary a **frequency table** or **frequency distribution.** We call the number of observations in a category the **frequency** of that category. This frequency table gives the number of students for each 20-point interval of grades. A **range,** or **interval,** of numbers like 101 to 120 or 61 to 80, determines each category.

The frequency table of grades on the calculus exam does not give complete information. It does not tell us the highest or lowest score made. In fact, it does not tell the number of students who made any score. It does give general information about overall performance on the exam.

In some summaries, the categories might not be numerical. For example, we may summarize the students' majors by subject at a university with a frequency table like the following:

Major	Number of Students
Science	429
Arts	132
Languages	41
Social sciences	631
Engineering	344
Total	1577

This is an example of **qualitative data** where the data are identified by nonnumeric categories. The cases where data are represented by numerical values are **quantitative data.**

EXAMPLE 1 ➤   A mathematics quiz consists of five questions. The professor summarizes the performance of the class of 75 students with the following frequency distribution. Each quiz question determines a different category.

Question	Number of Correct Answers (Frequency)
1	36
2	41
3	22
4	54
5	30

■

 **EXAMPLE 2** ➤ A survey of students reveals that they spent the following amounts of money on books for three courses during a semester:

$ 78	$123	$136	$162	$ 96	$145
$115	$183	$150	$110	$191	$ 88
$157	$137	$122	$172	$165	$119
$105	$127	$148	$170	$131	$118

Make a frequency table to summarize the students' book expenses.

SOLUTION

As the smallest amount is $78 and the largest amount is $191, the numbers cover a range of $191 - 78 = 113$. Let's form five categories of intervals of equal length. Our first estimate of the interval length is $\frac{113}{5} = 22.6$. We round this to 25 and start the first interval at 75, giving the intervals 75–99, 100–124, 125–149, 150–174, and 175–199. Next, we can make a table that places each number in the appropriate category:

Interval	Numbers in the Interval
75–99	78, 96, 88
100–124	123, 115, 110, 122, 119, 105, 118
125–149	136, 145, 137, 127, 148, 131
150–174	162, 150, 157, 172, 165, 170
175–199	183, 191

NOTE

The choice of the number of categories and the dollar range for each category could be made in several different ways.

From this table we have the count of each category and obtain the following frequency table:

Book Expenses	Number of Students (Frequency)
$75–99	3
$100–124	7
$125–149	6
$150–174	6
$175–199	2

■ **Now You Are Ready to Work Exercise 1**

Stem-and-Leaf Plots

To more orderly organize data into categories, a **stem-and-leaf plot** can be used. We do so by breaking the scores into two parts, the *stem,* consisting of the first one or two digits, and the *leaf,* consisting of the other digits.

EXAMPLE 3 ➤ Make a stem-and-leaf plot of the following scores: 21, 13, 17, 24, 48, 7, 31, 46, 44, 39, 9, 15, 10, 41, 46, 33, 24

SOLUTION

We use the first digits 0, 1, 2, 3, 4 for the stems, which will divide the data into intervals 0–9, 10–19, 20–29, 30–39, and 40–49.

Stem	Leaves
0	79
1	3750
2	144
3	193
4	86416

We can now easily count the frequency of each category.

Notice that the digits for the leaves are not in order. They can be listed as they occur in the list.

EXAMPLE 4 ➤ Make a stem-and-leaf plot of the following scores: 3, 2, 6, 12, 14, 0, 11, 8, 2, 5, 7, 6, 6. For the categories, use the intervals 0–4, 5–9, and 10–14.

SOLUTION

Stem	Leaves
0	3202
0	685766
1	241

Notice that the stem 0 occurs twice, the first time for leaves 0 through 4 and the second time for leaves 5 through 9.

EXAMPLE 5 ➤ Make a stem-and-leaf plot of the following data: 10.1, 9.3, 9.7, 11.4, 12.3, 10.8, 10.7, 10.3, 11.7, and 11.9.

SOLUTION

Stem	Leaves
9	37
10	1873
11	479
12	3

The stem-and-leaf plot has the advantage that the original data can be reconstructed from the plot.

■ **Now You Are Ready to Work Exercise 5**

Construction of a Frequency Table

Here are some suggestions for setting up a frequency table.

Construction of a Frequency Table

Construct a frequency table in three steps, as follows:

Step 1.   Choose the categories by which the data will be grouped, for example, the calculus grades in the 81–100 range. Generally, the categories are determined by intervals of equal length.

Step 2.   Place each piece of data in the appropriate category; for example, sort the calculus grades and place them in the appropriate category.

Step 3.   Count the data in each category. This gives the frequency of the category; for example, perhaps 73 students scored in the 121–140 range.

The classification decisions in step 1 make the other two steps mechanical. The determination of categories can be a two-step decision. First, determine the number of categories (e.g., eight) and then the range of values that each category covers (e.g., 61–80, 81–100). No magic formula exists to make these decisions. They depend on the nature of the data and the message you wish to convey. However, some generally accepted rules of thumb might help. Just remember, exceptions are appropriate at times.

Hints on Setting up Category Intervals

1. **Number of categories.**   Usually from five to fifteen. More categories might be unwieldy, and fewer categories might not distinguish between important features. Be sensible: Don't use five categories to summarize a two-category situation, such as the male–female breakdown in enrollment. Generally use a larger number of categories for larger amounts of data and fewer categories for smaller amounts of data.
2. **Range of each category.**   A good guideline for choosing interval length and bounds is the following:
   (a) Find the difference between the largest and smallest observation and divide it by the number of intervals.
   (b) Adjust the length obtained to a relatively simple number.
   (c) Select a number less than, or equal to, the smallest observation for the lower bound of the first category. Drop down to, say, the first multiple of 5 or 10, or whatever number is appropriate for the data.
   　　In summarizing calculus grades on a 160-point test, the interval 80–99 range might be used as the range of one classification. However, a range of 81–100 could be used just as well.
   (d) Be sure to leave no gaps between categories when they might include some data. Don't use intervals like $2.00 to $4.00 and $5.00 to $7.00 when $4.75 is a valid data point.
   (e) Place each piece of data in only one class. The category designations 300–400 and 400–500 leave it unclear where to place 400. Category designations like 300–399 and 400–499 clearly indicate where to place 400.
   (f) Use category intervals that make sense for the situation. To summarize traffic speeds for a traffic study, use intervals like 31–35 and 36–40 rather than 33–38 and 39–42. Although intervals of equal length are generally preferred, sometimes different lengths make sense. For example, we might summarize performance on a test with intervals like 0–59, 60–69, 70–79, 80–89, and 90–100 because they represent letter grades.

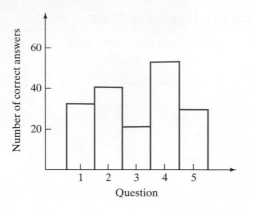

FIGURE 8–1 Summary of correct answers on a mathematics quiz.

Visual Representations of Frequency Distributions

Because a picture conveys a more forceful message than a column of numbers, a visual presentation of a frequency table sometimes provides a better understanding of the data. We will study two common visual methods: the histogram and the pie chart. In each case, the graph shows information obtained from a frequency table.

Histogram

A **histogram** is a bar chart in which each bar represents a category and its height represents the frequency of that category. Figure 8–1 is the histogram of Example 1.

Mark the categories on a horizontal scale and the frequencies on a vertical scale. The bars are of equal width and are centered above the point that designates the category. In this case, a single number forms the category interval. The bars should be of equal width because two bars with the same height and different widths have different areas. This gives an impression of different frequencies. Thus, the *area*, not just the height of the bar, customarily represents the frequency when different bar widths are used. Sometimes a space is left between bars when using a discrete variable. If the category is defined by an interval (such as 1–5, 6–10), the bar is located between the end points of the interval.

The categories in the first two examples are **discrete;** that is, the values in one category are separated from those in another category by a "gap." A summary of the number of correct answers on a quiz uses the possible values 0, 1, 2, 3, 4, and 5 as categories. There is a jump from 3 to 4; a value of 2.65 is not valid. On the other hand, GPA scores are not discrete. Any GPA from 0 through 4.0 is valid. There is no jump from one category to another. In such a case, the data are said to be **continuous.** For two different values of continuous data, all values between them are permissible values of the data. For example, 3.1 and 3.2 are valid values of GPA, as well as any number between, such as 3.135 and 3.1999.

When continuous data are represented by a histogram, the categories are determined by a range of values like 0–0.49, 0.50–1.0, and so on. The next example illustrates the use of a histogram with continuous data.

EXAMPLE 6 ➤

The university registrar selects 100 transcripts at random and records the GPA for each, where all GPAs are rounded to two decimals. The frequency distribution follows:

GPA	Frequency
0–0.49	5
0.5–0.99	9
1.00–1.49	17
1.50–1.99	10
2.00–2.49	18
2.50–2.99	22
3.00–3.49	11
3.50–4.00	8

The histogram representing this information appears in Figure 8–2. Notice that we have rotated the histogram so that the bars are horizontal. We did this because the category labels were too long to fit under a bar.

FIGURE 8–2

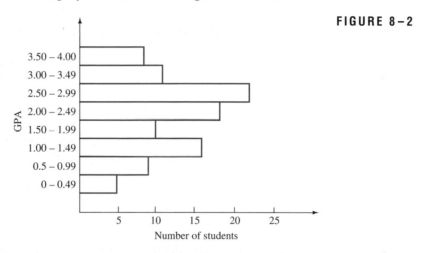

■ **Now You Are Ready to Work Exercise 12**

Sometimes a frequency table is summarized with a histogram that uses relative frequency instead of frequency for the vertical scale. **Relative frequency** counts the fractional part of the data that belong to a category. Compute relative frequency by dividing a frequency by the total number in the distribution. For example, the relative frequency of the 1.00–1.49 category of Example 6 is $\frac{17}{100}$, because 17 is the frequency of the category and 100 is the total of all frequencies. Relative frequency can also be stated as a percentage. If the relative frequency of a category is 0.17, then 17% of the scores are in the category.

EXAMPLE 7 ➤

A question on an economics exam has five possible responses: A, B, C, D, and E. The number of students who gave each response follows:

Response	Frequency
A	6
B	14
C	8
D	22
E	10

Draw a histogram that shows the relative frequency of each response.

SOLUTION

Sixty students answered the question, so the relative frequency is the number responding divided by 60.

Response	Relative Frequency
A	$\frac{6}{60} = 0.10$
B	$\frac{14}{60} = 0.23$
C	$\frac{8}{60} = 0.13$
D	$\frac{22}{60} = 0.37$
E	$\frac{10}{60} = 0.17$

The histogram is shown in Figure 8–3.

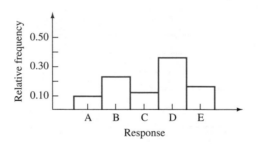

FIGURE 8–3   Relative frequency of responses to a question.

■ **Now You Are Ready to Work Exercise 15**

A histogram sometimes conveys the erroneous impression that there is a natural break between categories. When continuous data are represented (such as the heights of 18-year-old males) it may be difficult to tell in which category a measurement should be placed. For example, if 5′9″ divides two categories, a smooth curve avoids the impression that a person slightly under 5′9″ is distinctly shorter than a person who is 5′9″. A smooth curve conveys the impression of continuous data better than a histogram. You can sketch a smooth curve based on a histogram by drawing it through the midpoints at the top of the bars. (See Figure 8–4.)

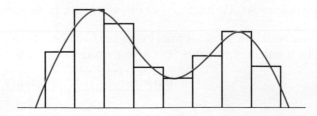

FIGURE 8–4   A curve that smooths the histogram.

EXAMPLE 8 ➤   A frequency table of the heights of male high school seniors in Ponca City is the following:

Height (Inches)	Frequency
61–62.9	10
63–64.9	51
65–66.9	115
67–68.9	200
69–70.9	240
71–72.9	195
73–74.9	104
75–76.9	42
77–78.9	15

Draw a histogram and a smooth curve representing the data. (See Figure 8–5.) ∎

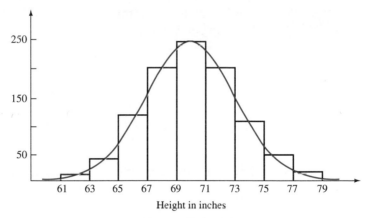

Height in inches

FIGURE 8–5

Notice that Figure 8–5 identifies the boundaries of the bars with 61, 63, 65, 67, 69, and so on. Although this is a simple way to designate category boundaries, you cannot tell, just by looking at the graph, if the score 69, for example, is in the 67–69 or the 69–71 category. We use the convention that a boundary point lies in the bar to its right. Thus, 69 is in the 69–71 category, which is consistent with the frequency table information.

Convention

If one category has boundaries a and b, and the next category has boundaries b and c, $a < b < c$, then point b is placed in the category to the right with boundaries b and c.

Pie Chart

The second visual representation of data, the **pie chart,** emphasizes the proportion of data that falls into each category. You sometimes see a pie chart in the newspaper that represents the division of a budget into parts. The parts are frequently reported as percentages of the total. You may obtain the percentage of the data that falls into each category from a frequency table.

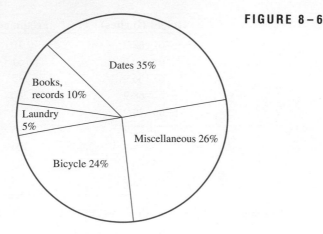

FIGURE 8–6

EXAMPLE 9 ➤ Jim Dandy has $200 for spending money this month. He carefully prepares the following budget:

Category	Amount
Dates	$70
Books and records	$20
Laundry	$10
Bicycle repairs	$48
Miscellaneous	$52

Construct a pie chart that represents this budget.

SOLUTION

First, compute the amount in each category as a percentage of the total by dividing the amount in the category by the total, $200.

Category	Percentage of Total Amount
Dates	35
Books and records	10
Laundry	5
Bicycle repairs	24
Miscellaneous	26

We "cut the pie" into pieces that have areas in the same proportion as the percentages representing the categories. (See Figure 8–6.) A glance at this pie chart tells us the relative share of each category. The angle at the center of each slice determines the size of the slice.

For the purpose of this book, you can sketch a pie chart by estimating the size of slices. If a more accurate drawing is desired, multiply each percentage by 360° to obtain the angle at the center of the slice. Then, use a protractor to mark off the required angle.

For example, the bicycle repairs category in Figure 8–6 accounts for 24% of the budget. The angle used for this category is $0.24 \times 360° = 86°$.

■ **Now You Are Ready to Work Exercise 19**

FIGURE 8–7

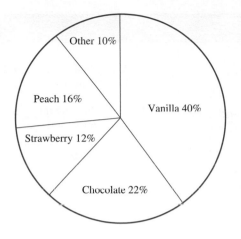

EXAMPLE 10 ➤ An advertising firm asked 150 children their favorite flavor of ice cream. Here is a frequency table of their findings:

Flavor of Ice Cream	Number Who Favor
Vanilla	60
Chocolate	33
Strawberry	18
Peach	24
Other	15
Total	150

Summarize this information with a pie chart.

SOLUTION

Compute the percentages of each category and the size of the angle to be used.

Flavor of Ice Cream	Percentage	Angle in Degrees
Vanilla	$40\left(\text{from } \frac{60}{150}\right)$	$144° = 0.40 \times 360°$
Chocolate	$22\left(\text{from } \frac{33}{150}\right)$	$79° = 0.22 \times 360°$
Strawberry	$12\left(\text{from } \frac{18}{150}\right)$	$43° = 0.12 \times 360°$
Peach	$16\left(\text{from } \frac{24}{150}\right)$	$58° = 0.16 \times 360°$
Other	$10\left(\text{from } \frac{15}{150}\right)$	$36° = 0.10 \times 360°$

(See Figure 8–7.) ■

You might wonder why we have two ways to represent a frequency distribution. The histogram and pie chart give visual representations of the same information. The histogram shows the size of each category in reports on monthly sales, annual gross national product growth for several years, or enrollment in accounting courses. Use the pie chart when you wish to show the proportion of data that falls in each category in cases such as a breakdown of students by home state, a summary of family incomes, and a percentage breakdown of letter grades in a course.

8.1 EXERCISES

1. *(See Example 2)* For the given set of data,

$$-1, 2, 2, 4, 6, 2, -1, 4, 4, 4, 6, 8$$

find the frequency of the following numbers:
(a) -1 **(b)** 2 **(c)** 4 **(d)** 6 **(e)** 8

2. An experiment is repeated 12 times. The results are

$$-3, 2, 4, -3, 4, 8, -3, 2, 8, 8, 8, 9$$

Give the frequency of each result.

3. The daily total number of students who used the state university swimming pool on 40 days during the summer is as follows:

90	98	137	108	128	115	152	122
110	132	149	131	102	109	118	126
121	145	89	149	86	120	97	118
142	139	128	110	105	104	131	159
93	119	107	129	132	129	98	116

Form a frequency table with the classes 85–99, 100–114, 115–129, 130–144, and 145–159.

4. A city police department radar unit recorded the following speeds one afternoon on 17th Street:

30	46	53	28	52	39	34	29
42	27	48	33	37	29	44	42
38	47	31	51	40	31	36	49
41	26	50	39	35	30	45	43
38	41	36	28	52	34	37	43
35	44	49					

Form a frequency table representing these data. Use category intervals of 5 miles per hour.

In Exercises 5 through 8, make a stem-and-leaf plot for the given scores.

5. *(See Example 5)* 26, 31, 53, 23, 57, 44, 27, 38, 41, 59, 33, 40, 39, 48, 41.
   Use the first digit as the stem.

6. 144, 167, 150, 141, 166, 153, 149, 142, 166, 153, 169, 151, 159, 163, 155.
   Use the first two digits as the stem.

7. 31, 22, 40, 20, 41, 33, 27, 35, 37, 48, 29, 26, 34, 45, 36, 46, 23, 38, 43, 36, 21, 41.
   Use the intervals 20–24, 25–29, 30–34, 35–39, 40–44, 45–49 for the categories.

8. 3, 1, 5, 2, 4, 3, 7, 8, 5, 9, 2, 5, 6, 8, 1, 3, 2, 6, 7, 9, 5, 3.
   Use the intervals 1–3, 4–6, 7–9 for the categories.

9. The following table shows the distribution of grades on a mathematics exam:

Grade	Number of Students
0–59	7
60–69	12
70–79	26
80–89	14
90–100	8

Which of the following quantities can be determined from this distribution? If the quantity can be determined, find it.
(a) The number of students who took the test
(b) The number of students who scored below 70
(c) The number of students who scored at least 80
(d) The number of students who scored below 75
(e) The number of students who scored between 69 and 90
(f) The number of students who scored 95

10. The daily totals of students who rode the campus shuttle bus are as follows:

166	172	184	176	181	84	170	198
182	203	210	141	77	93	147	205
164	122	211	137				

Summarize these totals with a frequency table with five categories.

11. A quiz consists of six questions. The following tabulation shows the number of students who received each possible score of 0 through 6:

Score	Number of Students
0	3
1	5
2	12
3	21
4	15
5	8
6	5

Draw a histogram of the data.

12. *(See Example 6)* The student employment office surveyed 150 working students to determine how many hours they worked each week. They found the following information:

Hours per Week	Number of Students
0–4.9	21
5–9.9	34
10–14.9	29
15–19.9	17
20–24.9	27
25–30	22

Draw the histogram of this summary.

13. Prof. Garcia polled his students to determine the number of hours per week they spent on homework for his course. Here are his findings:

3.50	7.50	9.00	2.25	3.50	5.25
4.00	2.75	4.75	1.50	4.50	3.50
3.25	8.50	2.75	5.50	0.25	5.75
2.25	6.25	5.00	6.00	3.50	4.00
0.75	2.25	3.50	3.75	4.00	6.50

Summarize this information with a histogram with five category intervals.

14. A sampling of State University freshmen reveals the following data for mathematics ACT scores:

ACT Score	Number of Students
18	10
19	8
20	15
21	21
22	26
23	30
24	44
25	38
26	50
27	28
28	16
29	12

Using two consecutive scores to form category intervals (18–19, 20–21, and so forth), represent this information with a histogram.

15. *(See Example 7)* Students in Prof. Anderson's class evaluate her on several items. For the item "Stimulated students' thinking," they selected one response from: 1 — Strongly Agree, 2 — Agree, 3 — Neutral, 4 — Disagree, 5 — Strongly Disagree. The following table summarizes the responses.

Response	Frequency
1	9
2	19
3	21
4	8
5	7

Draw a histogram that shows the relative frequency of each response.

16. Students in Prof. Baker's class evaluate him on several items. For the item "Instructor was well prepared," they selected one response from: 1 — Strongly Agree, 2 — Agree, 3 — Neutral, 4 — Disagree, 5 — Strongly Disagree. The following table summarizes the responses:

Response	Frequency
1	22
2	20
3	5
4	3
5	0

Draw a histogram that shows the relative frequency of each response.

17. In the Wild West Bicycle Ride, the cyclists choose from five distances. The number choosing each distance is summarized as follows:

Distance	Frequency
10 miles	160
25 miles	750
50 miles	980
62 miles	1120
100 miles	190
Total	3200

Draw a histogram that shows the relative frequency of each distance category.

18. The grades on a departmental exam in Mathematics 1304 are as follows:

Grade	Frequency
95–100	15
90–94	19
85–89	60
80–84	30
75–79	38
70–74	82
65–69	68
60–64	45
Below 60	18
Total	375

Draw a histogram that shows the relative frequency of each grade category.

19. *(See Example 9)* Draw the pie chart that represents the following frequency distribution:

Brand of Coffee	Number Who Prefer This Brand
Brand X	28
Brand Y	34
Brand Z	18

20. Graph the following information with the most appropriate graph:

Items in Family Budget	Percentage of Income
Food	25
Housing	35
Utilities	22
Clothing	12
Recreation	6

Draw the pie charts for the frequency distributions in Exercises 21 through 25.

21.

Concentration of Ozone in Air of Large City (in parts per billion)	Number of Days
0–40	8
41–80	22
81–120	18
121–160	12

22.

Budget Items of Old Main University	Amount in Budget (dollars)
Instructional	10,500,000
Administrative	1,500,000
Buildings and grounds	2,000,000
Student services	10,100,000
Other	5,700,000

23.

Income of Old Main University	Amount (dollars)
Tuition and fees	11,800,000
Endowment	2,200,000
Gifts	1,500,000
Auxiliary enterprises	10,100,000
Other	2,400,000

24.

Educational Level of Acme Manufacturing Employees	Number of Employees
Less than high school	45
High school graduate	180
College graduate	60
Graduate work	15

25.

GPA of Tech Students	Number of Students
0–0.99	21
1.00–1.99	72
2.00–2.99	98
3.00–4.00	46

EXPLORATIONS

26. Report on the methods used by the *Wall Street Journal* to summarize Dow Jones averages and volume of stocks traded on the New York Stock Exchange.

27. From newspapers or news magazines, collect at least three frequency tables, three histograms, and three pie charts.

Use your graphing calculator or spreadsheet to draw the histograms of data in Exercises 28 through 33.

28. 48, 12, 21, 36, 51, 31, 22, 35, 18, 26, 37, 22, 52, 27, 30, 25, 40, 33, 19, 28, 29, 16, 34, 38, 27
Use five category intervals.

29.

Score	Frequency
3	8
4	5
5	2
6	6
7	12
8	14

Use three category intervals.

30. Use the data in Exercise 11.

31. Use the data in Exercise 14.

32. Use the data in Exercise 18.

33. 11, 12, 14, 14, 16, 17, 19, 20, 20, 21, 23, 23, 23, 24, 27, 28, 29, 29
Use four categories with boundaries 10, 15, 20, 25, 30. Into which category does the program put the value 20? Is this consistent with the convention stated?

34. Represent the following grade distribution with a histogram:

Grade	Frequency
A	12
B	15
C	23
D	8
F	5

35. Summarize the data 6, 7, 3, 8, 10, 3, 14, 12, 1, 3, 13, 7, 9, 11, and 9 in a histogram using the categories 0–2.9, 3.0–5.9, 6.0–8.9, 9.0–11.9, and 12.0–15.0.

36. Summarize the data 3, 5, 4, 6, 8, 9, 2, 1, 12, 15, 9, and 14 in a histogram using four categories.

37. Summarize the data 5.5, 7.6, 8.1, 12.3, 11.1, 18.4, 17.5, 20.3, 21.7, 24.6, 14.9, and 8.8 in a histogram using the categories 5.0–9.9, 10.0–14.9, 15.0–19.9, and 20.0–25.0.

38. Use a pie chart to represent the following survey of 142 students on their usual method of transportation to class.

Method	Frequency
Auto	22
Bicycle	43
Walk	61
Motor scooter	11
Roller blades	5

39. Summarize the data 1, 5, 4, 6, 7, 9, 2, 1, 8, 5, 9, and 4 in a pie chart using the categories 0–2.5, 3–5.5, 6–8.5, and 9–10.

40. Summarize the data 21, 25, 31, 28, 33, 38, 22, 27, 36, 38, 22, 26, and 34 in a pie chart using five categories.

41. Summarize the data 44.2, 51.3, 42.6, 57.5, 58.3, 45.2, 61.3, 42.1, 64.2, 58.9, 47.8, and 55.5 in a pie chart using the categories 40.0–45.5, 45.6–50.5, 50.6–60.00, and 60.1–65.0.

USING YOUR TI-83

HISTOGRAMS

The graph of a histogram may be obtained by the following steps:

1. **Enter data.** Enter the scores in the list L1 and their frequencies in L2.
2. **Set the horizontal and vertical scales.** Set Xmin, Xmax, Xscl, Ymin, Ymax, and Yscl using [WINDOW] in the same way it is used to set the screen for graphing functions.

   Set Xmin and Xmax so that all scores will be in the interval (Xmin, Xmax). This interval will be the *x*-axis of the graph. Xscl is the width of a bar on the histogram and determines the interval length of each category. Ymin and Ymax determine the range of the frequencies.

(continued)

3. **Define the histogram.** Press [STAT PLOT] <1:PLOT1> [ENTER] You will see a screen similar to the following:

Enter a 1 for Freq if the scores are simply a list in L1.

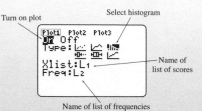

Turn on plot

Select histogram

Name of list of scores

Name of list of frequencies

On that screen select, as shown in the figure, <ON>, histogram for Type, L1 for Xlist (the scores), and L2 for frequencies. When you are finding the histogram for a list of scores, enter the number 1 instead of L2 for Freq. Press [ENTER] after each selection.

4. **Display the histogram.** Press [GRAPH]

EXAMPLE

Draw a histogram with three categories for the data summarized in the frequency table.

Score	Frequency
3	2
4	3
6	7
8	2
9	4
11	5
13	2

We set the range of the scores as 0 through 15 and the length of each category as 5. The frequency range is 0 through 15. This gives the following window settings:

The lists L1 and L2 are

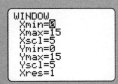

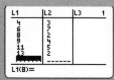

and the window that defines the plot is

giving the histogram

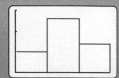

USING EXCEL

HISTOGRAMS AND PIE CHARTS

Both the histogram and the pie chart give a visual representation of data collected in categories. The histogram gives a picture of the number in each category, and the pie chart gives a picture of the percentage in each category. In EXCEL, both use the spreadsheet cells listing the *number* in each category.

EXAMPLE

Draw the histogram and the pie chart for data with four categories where the numbers in each category are 3, 4, 7, and 2.

Solution

Let's call the categories A, B, C, and D and enter their frequencies in cells A2:A5. Because the procedure to draw the histogram and the pie chart have steps in common, we make just one list while showing which steps differ.

- Enter the frequencies in A2:A5.
- Select the cells A2:A5.
- Select the **Wizard** icon from the menu bar.
- Under **Chart type** select **Columns** for a histogram and select **Pie** for a pie chart.
- Under **Chart sub-type,** click on the first one in the top row.
- Click **Next.**
- Click **Next.**
- Select the **Data Labels** tab from the **Chart Option.**

  For a histogram select the **Show label** option.
  For a pie chart select the **Show label and percent** option.

- **Click Finish.**

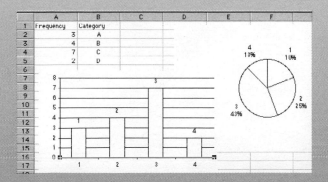

Notice that EXCEL labels the categories 1, 2, 3, and 4 with the labels above the bars in the histogram. The labels can be changed as follows:

For a histogram:
To change the label 1 to A, click on 1 and you will see . Then click on the 1 and hold until you see .

(continued)

Now type A and click somewhere off the histogram. Go to the next label and repeat.

For the pie chart:
To change the label 1 to A, click on 1 and you will see . Then click on 1 and hold until you see .

Now select the 1, and type A. Go to the next label and repeat.

COUNTING FREQUENCIES IN A CATEGORY

When you have a list of scores or a frequency table, you may need to count the number of scores for each category to use EXCEL. We illustrate how to do it.

Given a List, Count the Number of Scores in Each Category
EXAMPLE

For the scores 1, 8, 6, 3, 4, 9, 5, 5, 3, 2, and 5, find the number in each of three categories 1–3, 4–6, and 7–9

Solution

Enter these data in A2:A12. We will place the count for category 1–3 in B2, the count for category 4–6 in B3, and the count for 7–9 in B4 using the COUNTIF function to first count the scores less than or equal ($\leq$) to 3.
   In cell B2, enter =COUNTIF(A2:A12,"<=3").

> ### N O T E
> The symbol <= is used for $\leq$. A2:A12 is the range of the list.

For the frequency of category 4–6, enter into cell B3: =COUNTIF(A2:A12,"<=6")-B2

> ### N O T E
> COUNTIF only counts scores less than (or less than or equal) a number, 6 in this case, so we count scores $\leq$ 6 and subtract the number $\leq$ 3.

For the frequency of category 7–9, enter into cell B4: =COUNTIF(A2:A12,"<=9")-B3-B2.
   Now you can use cells B2:B4 to create the pie chart or histogram.

Given a Frequency Table, Count the Number in Each Category
EXAMPLE

Count the frequencies of the categories 1–6, 7–12, 13–18, and 19–24.

Score	Frequency
2	3
5	4
6	8
7	2
11	3
17	5
18	2
21	3

Solution

Enter the scores in cells A2:A9 and the corresponding frequencies in B2:B9. The category counts will be in cells C2:C5. To count the scores in the 1–6 category, we use the SUMIF function which searches A2:A9 for scores in the 1–6 category and adds their frequencies from B2:B9.

In C2, enter =SUMIF(A2:A9,"<=6",B2:B9).

In C3, enter =SUMIF(A2:A9,"<=12",B2:B9)-C2.

In C4, enter =SUMIF(A2:A9,"<=18",B2:B9)-C3-C2

In C5, enter =SUMIF(A2:A9,"<=24",B2:B9)-C4-C3-C2

 N O T E

The function in C3 counts the scores ≤ 12 and subtracts the number ≤ 6.

Now you can use cells C2:C5 to create the pie chart or histogram.

EXERCISES

1. Draw the histogram and pie chart for the following:

Frequency	Category
6	Adult Male
9	Adult Female
7	Children

2. Draw the histogram and pie chart for the following:

Frequency	Category
10	A
13	B
8	C
18	D

3. Draw the histogram for the following, using the categories 1–4, 5–8, 9–12:

3, 6, 8, 1, 2, 11, 9, 7, 4, 2, 10, 7

4. Draw the pie chart for the following, using the categories 1–5, 6–10, 11–15, 16–20:

8, 4, 19, 3, 11, 17, 20, 1, 5, 6, 13, 9, 16, 13, 7

8.2   Measures of Central Tendency

- The Mean
- The Median
- The Mode
- Which Measure of Central Tendency Is Best?

We have used histograms and pie charts to summarize a set of data. These devices sometimes make it easier to understand the data.

At times, however, we want to be more concise in reporting information, so that comparisons can be easily made or so that some important aspect can be described. You often hear questions like:

"What was the class average on the exam?"

"What kind of gas mileage do you get on your car?"

"What happened to the price of homes from 1977 to 1984?"

You generally expect a response to questions like these to be a single number that is somehow "typical" or at the "center" of the exam grades, distance a car is driven on a certain amount of gas, or price of homes. We will study three ways to obtain such a "central number," which is called a **measure of central tendency.**

A measure of central tendency is associated with a **population,** a collection of objects, such as a collection of exams, a collection of cars, or a collection of homes. Each member of the population has a number associated with it, like a grade on an exam, gas mileage of a car, or the price of a home. We call these numbers **data,** and we find the measure of central tendency of those numbers. We often use subscripted variables to enumerate a list of numbers. Thus, x_1, denotes the first value, x_2 denotes the second value, and x_n denotes the nth value.

A **sample** is a subcollection of a population. Five exam papers form a sample from the population of 37 students in my finite mathematics class. A study of obesity among teenagers could not find data on the weight and height of *all* teenagers. So, estimates of teenagers' weight and height are obtained from representative samples of teenagers. There are times when it is appropriate to find a measure of central tendency of a population, and other times when a sample is used. The measures discussed in this chapter can be applied to both populations and samples.

NOTE

The symbol μ (pronounced *mu*) is the standard notation for the mean of a population. For the mean of a sample, you will usually see the symbol $\bar{x}$ (*x* bar).

The Mean

When we talk about "averages" like test averages, the average price of gasoline, or a basketball player's scoring average, we usually refer to one particular measure of central tendency, the arithmetic **mean.** To compute the mean of a set of numbers, simply add the numbers and divide by how many numbers were used.

DEFINITION
Mean

The population **mean** of n numbers $x_1, x_2, \ldots, x_n$ is denoted by μ (the Greek letter *mu*) and is computed as follows:

$$\mu = \frac{x_1 + x_2 + \cdots + x_n}{n}$$

The **sample mean,** denoted by $\bar{x}$, (*x* bar) is

$$\bar{x} = \frac{x_1 + x_2 + \cdots + x_n}{n}$$

Notice that you calculate the population mean and the sample mean exactly the same. The notation μ and $\bar{x}$ indicate whether the data came from a population or from a sample.

Common terms for the mean are the **average** or the **arithmetic average.**

The Convenience Chain wants to compare sales in its 56 stores during July with sales a year ago, when it had 49 stores. A comparison of total sales for each July may be misleading because the number of stores differs. Sales can be down from a year ago in each of the 49 stores, but total sales can still be up because there are seven additional stores. The mean sales of all stores in each year should better indicate whether sales are improving.

EXAMPLE 1 ➤

The mean of the test grades 82, 75, 96, 74 is

$$\frac{82 + 75 + 96 + 74}{4} = \frac{327}{4} = 81.75$$

■ **Now You Are Ready to Work Exercise 1**

EXAMPLE 2 ➤

Find the mean of the annual salaries $25,000, $14,000, $18,000, $14,000, $20,000, $14,000, $18,000, and $14,000.

SOLUTION

Add the salaries:

$$
\begin{array}{r}
25,000 \\
14,000 \\
18,000 \\
14,000 \\
20,000 \\
14,000 \\
18,000 \\
\underline{14,000} \\
137,000
\end{array}
$$

Divide this total by 8, the number of salaries, to obtain the mean:

$$\text{Mean} = \frac{137,000}{8} = \$17,125 \ ■$$

This mean can be written in this more compact form:

$$\frac{25,000 + 4 \times 14,000 + 2 \times 18,000 + 20,000}{8}$$

where 4 is the frequency of the value 14,000, 2 is the frequency of 18,000, and 25,000 and 20,000 each have a frequency of 1. The divisor, 8, is the sum of the frequencies. This form is useful in cases like the following, where the scores are summarized in a frequency table.

EXAMPLE 3 ➤

Scores are summarized in the following frequency table:

Score (X)	Frequency (f)
3	2
4	1
5	8
6	4
	15

The mean is given by

$$\text{Mean} = \frac{2 \times 3 + 1 \times 4 + 8 \times 5 + 4 \times 6}{15} = \frac{74}{15} = 4.93$$

■ **Now You Are Ready to Work Exercise 27**

The general formula for the mean of a frequency distribution is as follows:

Formula for the Mean of a Frequency Distribution

Given the scores $x_1, x_2, \ldots, x_k$, which occur with frequency $f_1, f_2, \ldots, f_k$, respectively, the mean is

$$\text{Mean} = \frac{f_1 x_1 + f_2 x_2 + \cdots + f_k x_k}{n}$$

where $n = f_1 + f_2 + \cdots + f_k$, the sum of frequencies.

The next example deals with **grouped data.** Scores are combined into categories, and the frequency of that category is given.

EXAMPLE 4 ➤

Professor Tuff gave a 20-question quiz. She summarized the class performance with the following frequency table:

Number of Correct Answers	Number of Students
0–5	8
6–10	14
11–15	23
16–20	10

Estimate the class mean for these grouped data.

SOLUTION

We do not have specific values of the data, only the number in the indicated category. To obtain an *estimate* of the mean, we use the midpoint of each category as the representative value and compute the mean by

$$\text{Mean} = \frac{8 \times 2.5 + 14 \times 8.0 + 23 \times 13.0 + 10 \times 18.0}{55}$$

$$= 11.11$$

where 55 is the sum of the frequencies. (*Note:* The midpoint of an interval is halfway between the left end point and the right end point of the interval. It can be found by taking the mean of the left end point and the right end point.)

■ **Now You Are Ready to Work Exercise 32**

Formula for the Estimated Mean of Grouped Data

If data are grouped in intervals with $m_1, m_2, \ldots, m_k$, the midpoints of each category interval, and $f_1, f_2, \ldots, f_k$ the frequency of each interval, respectively, then, an estimate of the mean is

$$\text{Mean} = \frac{f_1 m_1 + f_2 m_2 + \cdots + f_k m_k}{n}$$

where $n = f_1 + f_2 + \cdots + f_k$, the sum of frequencies.

Notice that the formula for the mean of grouped data looks much like the formula for the mean of a frequency distribution. Why is it repeated? Be sure to notice the difference. First, we can only estimate the mean of grouped data. Second, the *m*'s used in the formula for grouped data are midpoints of the categories. Thus, one number is selected to represent a *range* of numbers in the category. In a frequency distribution, the *x*'s represent the actual scores.

The mean is a useful measure of central tendency because

1. It is familiar to most people.
2. It is easy to compute.
3. It can be computed for any set of numerical data.
4. Each set has just one mean.

The salaries of the Acme Manufacturing Company are

President	$300,000
Vice-president	$ 40,000
Production workers	$ 25,000

The company has 15 production workers. They complain to the president that company salaries are too low. Mr. President responds that the average company salary is about $42,060. He maintains that this is a good salary. The production workers and the vice-president remain unimpressed with this information, because not a single one of them makes this much money. Although the president computed the mean correctly, he failed to mention that a single salary — his — was so large that the mean was in no way typical of all the salaries. This illustrates one of the disadvantages of the mean as a number that summarizes data. One or two **extreme values** can shift the mean, making it a poor representative of the data. ■

When you summarize a set of data with a histogram, you can visualize the mean in the following way. Think of the histogram as being constructed from material of uniform weight, a thickness of plastic or cardboard. If you try to balance the histogram on a point, you will need to position it at the mean. (See Figure 8–8.)

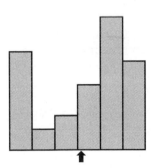

FIGURE 8–8 The mean is where the histogram balances.

The Median

Another measure of central tendency, this one not so easily affected by a few extreme values, is the **median.** When data from an experiment are listed according to size, people tend to focus on the middle of the list. Thus, the median is a useful measure of central tendency.

DEFINITION
Median

The **median** is the middle number after the data have been arranged in order. When there is an even number of data items, the median is the mean of the two middle data items.

Basically, the median divides the data into two equal parts. One part contains the lower half of the data, and the other part contains the upper half of the data.

EXAMPLE 6 ➤

The median of the numbers 3, 5, 8, 13, 19, 22, and 37 is the middle number, 13. Three terms lie below 13, and three lie above.

■ **Now You Are Ready to Work Exercise 13**

EXAMPLE 7 ➤

The median of the numbers 1, 5, 8, 11, 14, and 27 is the mean of the two middle numbers, 8 and 11. The median is

$$\frac{8 + 11}{2} = 9.5$$

Three terms lie below 9.5, and three lie above.

■ **Now You Are Ready to Work Exercise 15**

EXAMPLE 8 ➤

Find the median of the set of numbers 8, 5, 2, 17, 28, 4, 3, and 2.

SOLUTION

First, the numbers must be placed in either ascending or descending order: 2, 2, 3, 4, 5, 8, 17, 28. As there are eight numbers, an even number, there is no middle number. We find the mean of the two middle numbers, 4 and 5, to obtain 4.5 as the median.

■ **Now You Are Ready to Work Exercise 17**

EXAMPLE 9 ➤

The set of numbers 2, 5, 9, 10, and 15 has a mean of 8.2 and a median of 9. If 15 is replaced by 140, the mean changes to 33.2, but the median remains 9. A change in one score of a set may make a significant change in the mean and yet leave the median unchanged. ■

The median is often used to report income, price of homes, and SAT scores.

The Mode

The **mode** is used less often as a measure of central tendency. It is used to indicate which observation or observations dominate the data because of the frequency of their occurrence.

DEFINITION
Mode

The **mode** of a set of data is the value that occurs the largest number of times. If more than one value occurs this largest number of times, those values are also modes. When no value occurs more than once, we say that there is no mode. Thus, a distribution may have one mode, several modes, or no mode.

EXAMPLE 10 ➤

The mode of the numbers 1, 3, 2, 5, 4, 3, 2, 6, 8, 2, and 9 is 2 because 2 appears more often than any other value.

■ **Now You Are Ready to Work Exercise 21**

EXAMPLE 11 ➤

The set of numbers 2, 4, 8, 3, 2, 5, 3, 6, 4, 3, and 2 has two modes, 2 and 3, because they both appear three times, more than any other value. The set of numbers 2, 5, 17, 3, and 4 has no mode because each number appears just once.

■ **Now You Are Ready to Work Exercise 23**

The mode provides useful information about the most frequently occurring categories. The clothing store does not want to know that the mean size of men's shirts is 15.289. However, the store likes to know that it sells more size $15\frac{1}{2}$ shirts than any other size. The mode best represents summaries where the most frequent response is desired, such as the most popular brand of coffee.

Which Measure of Central Tendency Is Best?

With three different measures of central tendency, you probably are curious about which is best. The answer is, "It all depends." It all depends on the nature of the data and the information you wish to summarize.

The mean is a good summary for values that represent magnitudes, like exam scores and price of shoes, if extreme values do not distort the mean. The mean is the best measure when equal distances between scores represent equal differences between the things being measured. For example, the difference between $15 and $20 is $5; the same amount of money is the difference between $85 and $90.

The median is a positional average. It is best used when ranking people or things. In a ranking, an increase or decrease by a fixed amount might not represent the same amount of change at one end of the scale as it does at the other. The difference between the number-one ranked tennis player and the number-two ranked player at Wimbledon may be small indeed. The difference between the tenth- and eleventh-ranked players may be significantly greater. In contests, in student standings in class, and in taste tests, numbers are assigned for ranking purposes. However, this does not imply that the people or things ranked all differ by equal amounts. In such cases, the median better measures central tendency.

The mode is best when summarizing dress sizes or the brands of bread preferred by families. The information desired is the most typical category, the one that occurs most frequently.

EXAMPLE 12 ➤

In Example 5, we used the following data on salaries:

President	$300,000
Vice-president	$ 40,000
15 production workers	$ 25,000 each

We saw that the mean of these salaries, $42,060, was a poor representation of salaries because 16 of the 17 people had lower salaries than the mean. This is an instance of **skewed data,** in which relatively few values of the data distort the location of the central tendency. In cases like this, the median, $25,000, is a better measure of central tendency. ■

8.2 EXERCISES

LEVEL 1

Find the mean of the sets of data in Exercises 1 through 8.

1. *(See Example 1)*   2, 4, 6, 8, 10

2. 3, 8, 2, 14, 21

3. 2.1, 3.7, 5.9

4. 150, 225, 345, 86, 176, 410, 330

5. 6, −4, 3, 5, −8, 2

6. 3, 3, 3, 3, 5, 5, 5, 5

7. 5.9, 2.1, 6.6, 4.7

8. 1525, 1640, 1776, 1492, 2000

9. Find the mean grade of the following exam grades:
   80, 76, 92, 64, 93, 81, 57, 77

10. Six women have the following weights: 106 lb, 115 lb, 130 lb, 110 lb, 120 lb, and 118 lb. What is their mean weight?

11. A radar gun recorded the speeds (in miles per hour) of seven pitches of a baseball pitcher. They were

90.5, 89.2, 78.4, 91.0, 84.2, 73.5, 88.7

Find the average speed of the pitches.

12. One year the rainfall in Central Texas for each month January through May was

1.70 in., 2.05 in., 2.00 in., 3.75 in., 4.70 in.

Find the average monthly rainfall.

In Exercises 13 through 20, find the median of the given set of numbers.

13. *(See Example 6)* 1, 3, 9, 17, 22
14. **(a)** 36, 41, 55, 88, 121, 140, 162
    (b) 12, 4, 8, 3, 1, 10, 6
15. *(See Example 7)* 12, 14, 21, 25, 30, 37

16. 1, 3, 9, 10, 14, 17, 18, 19
17. *(See Example 8)* 6, 8, 1, 29, 15, 9, 14, 22
18. 101, 59, 216, 448, 92, 31
19. 72, 86, 65, 90, 72, 98, 81, 72, 68
20. 379, 421, 202, 598, 148

In Exercises 21 through 26, find the mode(s) of the given set of numbers.

21. *(See Example 10)* 1, 5, 8, 3, 2, 5, 6, 11, 5
22. 3, 2, 4, 6, 5, 4, 1, 6, 8, 4, 1, 6, 8, 4, 4, 13, 6
23. *(See Example 11)* 1, 5, 9, 1, 5, 9, 1, 5, 9, 1, 5
24. 10, 14, 10, 16, 10, 14, 16, 14, 21
25. 5, 4, 9, 13, 12, 1, 2
26. 2, 3, 2, 3, 2, 3, 2, 5, 9

LEVEL 2

27. *(See Example 3)* Prof. Wong had the following grade distribution in her class:

Grade	Frequency
96	2
91	3
85	7
80	13
75	12
70	10
60	8
50	5

What is the class average (mean)?

28. The attendance in a class for a 20-day period was the following:

Attendance	Frequency
19	1
20	2
21	1
22	5
23	8
24	3

Find the average (mean) daily attendance.

29. An inspector of PC computer disks records the number of defective disks found per box of ten

disks. Here is the summary for 1000 boxes inspected:

Number Defective	Frequency
0	430
1	395
2	145
3	25
4	5

Find the mean number of defective disks per box.

30. An avid golf fan recorded the scores of 75 golfers on the ninth hole of the Cottonwood Golf Course as follows:

Score	Frequency
1	1
2	1
3	32
4	27
5	8
6	6

Find the average score.

31. A personnel office gave a typing test to all secretarial applicants. The number of errors for 150 applicants is summarized as follows:

Number of Errors	Frequency
0–5	28
6–10	47
11–20	68
21–30	7

Estimate the mean number of errors.

32. *(See Example 4)* Estimate the mean for the grouped data:

Score	Frequency
0–5	6
6–10	4
11–19	12
20–30	7

33. Prof. Craig posted the following summary of test grades:

Grade	Number of Grades
90–100	8
80–89	15
70–79	22
60–69	11
45–59	5

Estimate the class average.

34. The advertising department of *Food Today* magazine ran a survey to determine the number of ads read by its subscribers. The survey found the following:

Number of Ads Read	Frequency
0–2	45
3–5	75
6–10	35
10–20	15

Estimate the average number of ads read by a subscriber.

35. The daily stock prices of Acme Corporation for one business week were $139.50, $141.25, $140.75, $138.50, and $132.00. What was the average price (mean) for the week?

36. Henry scored 80, 72, 84, and 68 on four exams. What must he make on the fifth exam to have an average of 80?

37. Throughout the 1999 wheat harvest, a farmer sold his wheat each day as it was harvested. The prices he received per bushel were $3.60, $3.57, $3.90, $3.85, $4.00, $4.15, $4.25, and $4.40. Find the mean and median prices.

38. The family income in a depressed area is reported as follows:

Family Income (dollars)	Number of Families
0–2999	3
3000–5999	8
6000–9999	15
10,000–13,999	6
14,000–20,000	4

Estimate the average (mean) family income by using the midpoint of each category as the actual income.

LEVEL 3

39. The scores of 18 players on the eighth hole of the Putt-Putt course are summarized with the following histogram:

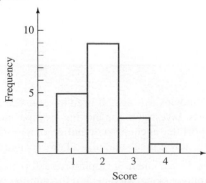

(a) How many made a hole-in-one?
(b) What was the mode score?

(c) What was the mean score?
(d) What was the median score?

40. The results of a 5-point quiz are summarized in the following histogram:

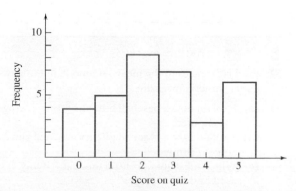

(a) How many scored above 3?
(b) How many scored a 1, 2, or 3?
(c) What is the mode score?
(d) Find the median score.
(e) Find the mean score.

41. Prof. Mitchell gave a quiz to five students. He remembered four of the grades — 72, 88, 81, and 67 — and the mean, 78. What was the other grade?

42. Seven cash registers at a supermarket averaged sales of $2946.38 one day. What were the total sales?

43. A vice-president of a company asked the personnel department for a summary of staff salaries. Personnel provided the following:

21 employees at $22,000

9 employees at $26,000

14 employees at $29,000

26 employees at $32,000

13 employees at $34,000

8 employees at $39,000

The vice-president then decided that she wanted the mean salary. Find it for her.

44. The women taking a physical education course were weighed. The results are summarized below:

Weight (pounds)	Number of Women
105–115	9
116–125	13
126–135	15
136–145	21
146–155	14
156–165	14
166–175	11
176–185	2
186–195	1

In which category does the median fall?

45. The mean salary of 8 employees is $27,450, and the mean salary of 10 others is $31,400. Find the mean salary of the 18 employees.

46. The class average on a test was 73.25. The average of the 16 males in the class was 71.75. There were 20 females tested. What was their average?

47. The median of five test scores is 82. If four of the grades are 65, 93, 77, and 82, what can you determine about the fifth score?

48. If the mean of 10 scores is 13.4, and the mean of 5 other scores is 6.2, find the mean of the 15 scores.

49. The mean price of three sugar-coated cereals is $1.27, and the mean price of two sugar-free cereals is $1.34. Find the mean price of all five cereals.

50. Honest Joe's used-car lot sold an average of 23 cars per month over a 12-month period. Sales for the last 7 months of this period averaged 26 cars per month. What was the average for the first 5 months?

51. The 155 golfers playing in the second round of the 1989 U.S. Open Golf Tournament made the following scores on the sixth hole:

Score	Number of Players
1	4
2	28
3	95
4	24
5	4

Find the mean score for that hole. (Notice that there were four players who made a hole-in-one. This remarkable feat was accomplished by Doug Weaver, Jerry Pate, Nick Price, and Mark Wiebe, all using a 7-iron and all within a period of 1 hour and 50 minutes.)

EXPLORATIONS

52. Find references to the mean and mode in the newspaper or news magazine.

53. **(a)** Explain how a baseball hitter's batting average is computed.
(b) Explain how a baseball pitcher's earned run average is computed.
(c) Which measure of central tendency, if any, is used in these averages?

54. A "moving average" is used in some economic reports. Investigate the moving average and write a report that includes a definition of a moving average, how it is calculated, and gives examples.

55. The Department of Commerce estimates the per capita personal income and the number of people living below poverty level in the United States for 1985 — 2000 is the following:

Year	Per Capita Income ($)	Number Below Poverty Level (in millions)
1985	14,427	33.1
1986	15,122	32.4
1987	15,968	32.2
1988	17,052	31.7
1989	18,176	31.5
1990	19,188	33.6
1991	19,652	35.7
1992	20,576	38.0
1993	21,231	39.3
1994	22,086	38.1
1995	23,562	36.4
1996	24,651	36.5
1997	25,924	35.6
1998	27,203	34.5
1999	28,546	32.3
2000	29,469	31.1

(a) Use the statistics menu on a graphing calculator or spreadsheet to calculate the mean per capita income and the mean number living below the poverty level for these 16 years. Comment on the accuracy of your results given that income is given to the nearest dollar and the number of people is given to the nearest tenth of a million.
(b) Plot the points for per capita income and for poverty level.
(c) Which is increasing faster, per capita income or the number living below poverty level?

Use the mean function on your graphing calculator or spreadsheet to find the mean of the data in Exercises 56 through 60.

56. 3.1, 5.2, 7.8, 12.3, 6.9, 15.5, 11.4, 9.1

57. 54.23, 71.44, 85.41, 99.30, 56.29, 74.12, 44.68

58. The data in Exercise 27.

59. The data in Exercise 28.

60. The data in Exercise 29.

61. Find the mean and median of 3, 2, 5, 6, 4, 7, 8, 9, 3, 1, 5, 7, 9, and 1.

62. Find the mean and median of 21, 25, 34, 28, 37, 23, 33, 34, 35, 22, and 24.

63. Find the mean and median of 42.3, 46.1, 39.7, 34.5, 47.6, 49.9, 36.5, 44.2, and 37.6.

64. Find the mean and median of the number of games won by teams in the Women's National Basketball Association for the 2002 season.

Team	Games Won
Cleveland	10
Charlotte	18
New York	18
Detroit	9
Washington	17
Houston	27
Phoenix	19
Los Angeles	12
Sacramento	8
Utah	8

USING YOUR TI-83

CALCULATING THE MEAN AND MEDIAN
The TI-83 has functions in the LIST menu to find the mean of a list or a frequency table and to find the median of a list. Six lists in the memory (L1, L2, . . . , L6) may be used.

EXAMPLE 1
Find the mean and median of the scores 5, 2, 8, 3, 7, 1, 9.

Solution
Use STAT <4:Clrlist> L1,L2 ENTER to clear lists L1 and L2.
Enter the scores in the list L1.
For the mean, use LIST <MATH> <3:mean(> ENTER L1 ENTER gives mean = 4.33.
For the median, use LIST <MATH> <4:median(> ENTER L1 ENTER gives median = 4.

(continued)

EXAMPLE 2

Find the mean of the frequency table

Score	Frequency
3	2
7	4
8	1

Solution

Enter the scores in L1 and the frequencies in L2. Use [LIST] <MATH> <3:mean(> [ENTER] L1,L2 [ENTER] gives mean =6.

EXERCISES

1. Find the mean and median of the following scores: 3, 4, 7, 14, 2, 8, 7.
2. Find the mean and median of the following scores: 5, 4, 8, 3, 9, 11, 13, 2.
3. Find the mean and median of the following scores: 7.2, 5.6, 2.1, 8.6, 6.9, 12.1, 16.5, 10.4
4. Find the mean of the scores summarized by the following frequency table:

Score	Frequency
4	5
7	3
8	1
10	7
13	4

5. Find the mean of the scores summarized by the following frequency table:

Score	Frequency
1	5
2	7
3	6
4	2

USING EXCEL

CALCULATE THE MEAN AND MEDIAN

You can find the mean and median of a list of numbers using EXCEL. For example, if you have a list of 10 numbers, enter the numbers in cells, say, A1:A10, and enter the formula =AVERAGE(A1:A10) in a cell where you want the mean. Enter =MEDIAN(A1:A10) in the cell where you want the median.

EXERCISES

1. Find the mean and median of the following scores: 3, 4, 7, 14, 2, 8, 7.
2. Find the mean and median of the following scores: 5, 4, 8, 3, 9, 11, 13, 2.
3. Find the mean and median of the following scores: 7.2, 5.6, 2.1, 8.6, 6.9, 12.1, 16.5, 10.4

8.3 Measures of Dispersion: Range, Variance, and Standard Deviation

- Range
- Variance and Standard Deviation
- Measurements of Position
- Application
- Five-Point Summary and Box Plot

A score often has little meaning unless it is compared with other scores. We have used the mean as one comparison. If you know your test score and the class average, you can compare how far you are above or below the class average. However, the average alone does not tell you how many in the class had grades closer to the average than you.

If two bowlers have the same average, do they have the same ability? If two students have the same average in a course, did they learn the same amount? Although the mean gives a rather simple representation of a set of data, sometimes more information is needed about how the scores are clustered about the mean in order to make valid comparisons. Let's use students' class averages to illustrate **measures of dispersion.**

Student A	Student B
80	65
87	92
82	95
92	75
84	98

Each student's mean is 85. However, student A is more consistent. Student B's scores vary more widely. The mean does not distinguish between these two sets of data. As this example shows, two data sets may have the same mean, but in one set the values may be clustered close to the mean, and in the other set the values may be widely scattered. To say something about the amount of clustering as well as the average, we need more information.

Range

One way to measure the dispersion of a set of scores is to find the **range,** the distance between the largest and smallest scores. The range is 12 for student A and 33 for student B, so the range suggests that A's scores are clustered closer to the mean; the mean is thereby a better representation of A's scores than of B's scores. For grouped data, like a frequency table or histogram, the range is the difference between the smaller boundary of the lowest category and the larger boundary of the highest category.

Here are two frequency tables:

TABLE I

Category	Frequency
1–2	10
3–4	5
5–6	3
7 8	6
9–10	16

TABLE II

Category	Frequency
1–2	1
3–4	8
5–6	11
7–8	17
9–10	3

The estimated mean for each is 6.15 (using the grouped data mean), and the range for each is 9. Clearly, the scores in Table II are clustered near the mean, whereas they tend to be more extreme in Table I. The range gives information only about the extreme scores and gives no information on their cluster near the mean.

The **range,** the difference between the largest and smallest scores, is a simple measure of dispersion, but sometimes it helps to know whether the numbers are scattered rather uniformly throughout the range or whether most of them are clustered close together and a few are near the extremes of the range. For example, suppose a company manufactures ball bearings for an automobile company. The automobile company specifies that the bearings should be 0.35 inch in diameter. However, the automobile company and the bearing manufacturer both know that it is impossible to consistently make bearings that are *exactly* 0.35 inch in diameter. Slight variations in the material, limitations on the precision of equipment, and human error will create deviations from the desired diameter. So, the automobile company specifies that the bearings must be 0.35 ± 0.001 inch in diameter. The diameter may deviate as much as 0.001 inch from the desired diameter; that is, the acceptable range of diameters is 0.349 to 0.351. If the manufacturer produces a batch of bearings with all diameters in the range 0.3495 to 0.3508 inch, then there is no problem, because all of them are acceptable. However, if the diameters range from 0.347 to 0.353, there may or may not be a problem. Generally, they expect a few unacceptable bearings. If 90% of the bearings are unacceptable, then major problems exist in the manufacturing process. If fewer than one half of 1% are unacceptable, then the process may be considered satisfactory.

In this situation, a measure of how a batch of bearings varies from the desired diameter, 0.35 inch, provides more useful information than the largest and smallest diameters. With this situation in mind, we now look at two other useful indicators of variation.

Variance and Standard Deviation

A widely used measure of dispersion is the **standard deviation.** A related indicator, the **variance,** is another. These indicators measure the degree to which the scores tend to cluster about a central value, in this case the mean. The variance and standard deviation give measures of the distance of observations from the mean. They are large if the observations tend to be far from the mean and small if they tend to be near the mean. Even when all scores remain within a certain range, the variance and standard deviation will increase or decrease as a score is moved away from or closer to the mean.

Because the computation of the standard deviation is more complicated than that of the mean and the range, we will use the example of student A and student B to go through the steps to compute the variance and standard deviation.

Step 1. Determine the **deviation** of each score from the mean. Compute these deviations by subtracting the mean from each score. The following computations show the deviations for student A and student B using the mean of 85 in each case.

Computation of Standard Deviation

Student A			Student B		
Grade	Deviation	Squared Deviation	Grade	Deviation	Squared Deviation
80	$80 - 85 = -5$	25	65	$65 - 85 = -20$	400
87	$87 - 85 = 2$	4	92	$92 - 85 = 7$	49
82	$82 - 85 = -3$	9	95	$95 - 85 = 10$	100
92	$92 - 85 = 7$	49	75	$75 - 85 = -10$	100
84	$84 - 85 = \underline{-1}$	$\underline{1}$	98	$98 - 85 = \underline{13}$	$\underline{169}$
425		0 88	425		0 818

Notice that the deviations give the distance of the score from the mean a positive value for scores greater than the mean and a negative value for those less than the mean. You should also notice that the deviations add to zero for both students. This always holds. The deviations will *always* add to zero. For this reason, we cannot accumulate the deviations to measure the overall deviation from the mean.

Step 2. Square each of the deviations. These are shown under the heading "squared deviation."

This process of squaring the deviations allows us to accumulate a sum that is large when the scores tend to be far away from the mean. To adjust for cases in which two sets of data have different numbers of scores, we find a mean as in the next step.

Step 3. Find the mean of the squared deviations. For student A, the sum of the squared deviations is 88, and their mean is $\frac{88}{5}$, which equals 17.6. For student B, the sum of the squared deviations is 818, and their mean is $\frac{818}{5}$, which equals 163.6. The number 17.6 is the **variance** for student A, and 163.6 is the **variance** for student B.

Step 4. Find the square root of the means just obtained. For student A, we have $\sqrt{17.6} = 4.20$. For student B, we have $\sqrt{163.6} = 12.79$. These numbers are **standard deviations.** By tradition, the symbol σ (sigma) denotes standard deviation, and σ^2 denotes variance. Thus, for student A, $\sigma = 4.20$, and for student B, $\sigma = 12.79$. The standard deviation measures the spread of the values about their mean. Student B has the larger standard deviation, so her grades are more widely scattered. The grades of student A cluster closer to the mean.

A formal statement of the formula for variance and standard deviation is the following:

Formula for Population Variance and Standard Deviation

Given the n numbers $x_1, x_2, \ldots, x_n$ whose mean is μ, the **population variance,** denoted σ^2, and **population standard deviation, σ,** of these numbers is given by:

$$\sigma^2 = \frac{(x_1 - \mu)^2 + (x_2 - \mu)^2 + \cdots + (x_n - \mu)^2}{n}$$

$$\sigma = \sqrt{\sigma^2}$$

Both the variance and the standard deviation measure the dispersion of data. The variance is measured in the **square** of the units of the original data. The standard deviation is measured in the units of the data, so it is usually preferred as a measure of dispersion.

No practical way exists to enable Congress to survey the total population of the United States regarding proposed legislation on the death penalty. However, valuable information can be obtained from a well-planned, representative sample of the population. From the sample, reasonable estimates can be made regarding the population as a whole. We have already seen that the population mean (when possible) and the sample mean, as an estimate of the population mean, use the same formula. Statisticians have found that variance and standard deviation calculated using $n - 1$ as the divisor for the sample variance and standard deviation gives a better estimate than using n. Consequently, the sample variance and standard deviation use the following formulas:

DEFINITION
Formula for Sample Variance and Standard Deviation

Given the n numbers $x_1, x_2, \ldots, x_n$ whose mean is $\bar{x}$, the **sample variance**, denoted by s^2, and **sample standard deviation**, s, of these numbers is given by

$$s^2 = \frac{(x_1 - \bar{x})^2 + (x_2 - \bar{x})^2 + \cdots + (x_n - \bar{x})^2}{n - 1}$$

$$s = \sqrt{s^2}$$

Unless otherwise stated, variance and standard deviation refer to *population* variance and standard deviation.

EXAMPLE 1 ➤ Compute the mean and standard deviation of the numbers 8, 18, 7, and 10.

SOLUTION

The mean is

$$\frac{8 + 18 + 7 + 10}{4} = \frac{43}{4} = 10.75$$

Scores (x)	Deviation $(x - \mu)$	Squared Deviation $((x - \mu)^2)$
8	$8 - 10.75 = -2.75$	7.56 (rounded)
18	$18 - 10.75 = 7.25$	52.56
7	$7 - 10.75 = -3.75$	14.06
10	$10 - 10.75 = \underline{-0.75}$	$\underline{0.56}$
	0	74.74

$$\sigma = \sqrt{\frac{74.74}{4}} = \sqrt{18.69} = 4.32$$

■ **Now You Are Ready to Work Exercise 1**

You might have noticed that the sum of the deviations is zero in each of the cases shown. This is no accident. The deviations will always add to zero. Thus, the sum of deviations gives no information about the dispersion of scores. We need to use something more complicated, like standard deviation, to determine dispersion.
Let's summarize the procedure for obtaining standard deviation:

Procedure for Computing Standard Deviation

Step 1. Compute the mean of the scores.
Step 2. Subtract the mean from each value to obtain the deviation.
Step 3. Square each deviation.
Step 4. Find the mean of the squared deviations, using n for population data and $n - 1$ for sample data. This gives the variance.
Step 5. Take the square root of the variance. This is the standard deviation.

EXAMPLE 2 ➤ **(a)** Find the population standard deviation of the scores 8, 10, 19, 23, 28, 31, 32, and 41.
(b) Find the sample standard deviation.

SOLUTION

(a) **Step 1.** Find the mean of the scores:

$$\mu = \frac{8 + 10 + 19 + 23 + 28 + 31 + 32 + 41}{8} = \frac{192}{8} = 24$$

Step 2. Compute the deviation from the mean:

Step 3. Square each deviation:

x	$x - \mu$	$(x - \mu)^2$
8	$8 - 24 = -16$	$(-16)^2 = 256$
10	$10 - 24 = -14$	$(-14)^2 = 196$
19	$19 - 24 = -5$	$(-5)^2 = 25$
23	$23 - 24 = -1$	$(-1)^2 = 1$
28	$28 - 24 = 4$	$4^2 = 16$
31	$31 - 24 = 7$	$7^2 = 49$
32	$32 - 24 = 8$	$8^2 = 64$
41	$41 - 24 = \underline{17}$	$17^2 = \underline{289}$
	0	896

Step 4. Find the mean of the squared deviations:

$$\text{Variance} = \frac{896}{8} = 112$$

Step 5.   Take the square root of the result in step 4.

$$\sigma = \sqrt{112} = 10.58$$

(b) Steps 1 through 3 are the same as those above.

Step 4. $\text{Variance} = \dfrac{896}{7} = 128$

Step 5. $s = \sqrt{128} = 11.31$ ∎

The standard deviation of a frequency distribution can be computed similar to the preceding process by keeping in mind the number of times a score is repeated.

EXAMPLE 3 ➤   Find the mean and standard deviation for the following frequency distribution:

Score	Frequency
10	8
15	3
16	13
20	6

SOLUTION

The total of the frequencies is 30, so we have 30 scores. First, compute the mean:

$$\mu = \frac{8 \times 10 + 3 \times 15 + 13 \times 16 + 6 \times 20}{30} = \frac{453}{30} = 15.1$$

Next, compute the variance, using the squares of deviations the number of times the corresponding score occurs:

$$\text{Variance} = \frac{8 \times (10 - 15.1)^2 + 3 \times (15 - 15.1)^2 + 13 \times (16 - 15.1)^2 + 6 \times (20 - 15.1)^2}{30}$$

$$= \frac{8(5.1)^2 + 3(0.1)^2 + 13(0.9)^2 + 6(4.9)^2}{30}$$

$$= \frac{362.7}{30} = 12.09$$

NOTE

We would use a divisor of 29 for the sample standard deviation.

The standard deviation is $\sigma = \sqrt{12.09} = 3.48$.

■ **Now You Are Ready to Work Exercise 11**

You have estimated the mean of grouped data (Example 4, Section 8.2), and we use such a mean in estimating the standard deviation of grouped data.

EXAMPLE 4 ➤   Estimate the standard deviation for the following grouped data:

Score	Frequency
0–5	6
6–10	3
11–19	13
20–25	8

SOLUTION

Use the midpoint of each category to compute the mean. Because we know none of the scores specifically, we use the midpoint of a category as an estimate for each score in the category. By doing so, we convert the grouped data to the following frequency table:

Score (midpoint)	Frequency
2.5	6
8.0	3
15.0	13
22.5	8

We compute the standard deviation of this frequency table to obtain the estimate of the standard deviation of the grouped data.

NOTE

We would use a divisor of 29 for the sample standard deviation.

$$\text{Mean} = \frac{2.5 \times 6 + 8.0 \times 3 + 15.0 \times 13 + 22.5 \times 8}{30} = 13.8$$

The deviations are computed by using the mean and the midpoints of each category.

Deviation	Squared Deviation
$2.5 - 13.8 = -11.3$	$(-11.3)^2 = 127.69$
$8.0 - 13.8 = -5.8$	$(-5.8)^2 = 33.64$
$15.0 - 13.8 = 1.2$	$(1.2)^2 = 1.44$
$22.5 - 13.8 = 8.7$	$(8.7)^2 = 75.69$

To obtain the variance, we need to use each squared deviation multiplied by the frequency of the corresponding category:

$$\text{Variance} = \frac{127.69 \times 6 + 33.64 \times 3 + 1.44 \times 13 + 75.69 \times 8}{30}$$

$$= \frac{1491.30}{30}$$

$$= 49.71$$

$$\text{Standard deviation} = \sqrt{49.71} = 7.05.$$

■ **Now You Are Ready to Work Exercise 15**

You do not expect to compute a standard deviation during your daily activities. Yet on an intuitive level, all of us are interested in standard deviation. For example, you can recall a trip to the grocery store when you selected the shortest checkout line, only to find that it took longer than someone who got in a longer line at the same time. Let's look at the situation from the store's viewpoint. The store's managers are interested in customer satisfaction, so they work to reduce waiting time. They proudly announce that, on the average, they check out a customer in 3.5 minutes. The customer might agree that 3.5 minutes is good service, but an occasional long wait might make the customer question the store's claim. Basically, the customer wants the waiting time to have a small standard deviation. A small standard deviation indicates that you expect the waiting time to be near 3.5 minutes per customer.

A restaurant uses a different system of waiting lines. Customers do not choose a table and line up beside it. All customers wait in a single line and are directed to a table as it becomes available — and for a good reason. Diners do not want someone standing over them waiting for them to finish eating.

If we ignore the psychology of lining up at someone's table, which of these two systems is better? Is it better to let the customer choose one of several lines or to have a single line and allow a customer to go to a station as it becomes available? Both systems are reasonable in a bank. One bank may use a system where the customer chooses the teller, and a line forms at each teller. Another bank may use a system where the customers wait in a single line, and the customer at the head of the line goes to the next available teller. It turns out that the average waiting time is the same under both systems. Thus, from the bank's viewpoint, both systems allow it to be equally efficient. But it also turns out that the single-line concept tends to *decrease the standard deviation* and thereby reduce the extremes in waiting times. More customers will be served near the average waiting time than under the multiple-line system. Thus, from the customer's viewpoint, the single-line concept is better because it helps reduce an occasional long delay. As the bank's efficiency is the same in both cases, management would do well to adopt the single-line concept.

Measurements of Position

Scores are generally meaningless by themselves. "We scored ten runs in our softball game" gives no information about the outcome of the game unless the score of the opponent is known. A grade of 85 is not very impressive if everyone else scored 100.

The mean, median, and mode give a central point of reference for a score. You can compare your score to the mean or median and determine on which side of the central point your score lies.

One familiar measurement of position is the **rank.** A student ranks 15th in a class of 119; a runner places 38th in a field of 420 in the 10-kilometer run; a girl is the fourth runner-up in a beauty contest. Your rank simply gives the position of your score relative to all other scores. It gives no information on the location of the central point or on how closely scores are clustered around it.

If you took the SAT test, you received a numerical score and a **percentile** score, another positional score. A score at the 84th percentile states that 84% of people taking the test scored the same or lower and 16% scored higher. A score at the 50th percentile is a median score: 50% scored higher and 50% scored the same or lower. The percentile score is used as a positional score when a large number of scores is involved. In addition to the 50th percentile (the median), the 25th and 75th percentiles are often used to give information on the dispersion of data.

EXAMPLE 5 ➤
(a) A golfer is ranked 22nd in a field of 114 players. What is his percentile?
(b) A golfer is at the 76th percentile in a field of 88. What is her rank?

SOLUTION

(a) As the golfer is ranked 22nd, there are $114 - 22 = 92$ players ranked lower. Counting the golfer himself, there are 93 ranked the same or lower. The percentile is $\frac{93}{114} \times 100\% = 81.6\%$, which we round down to 81st percentile.

(b) As 76% are ranked the same or lower, there are $0.76 \times 88 = 66.88$ or 67 players ranked the same or lower and 21 ranked higher. We say that she is ranked 22nd.

■ **Now You Are Ready to Work Exercise 32**

Another standard score is the **z-score.** It is more sophisticated than rank or percentile because it uses a central point of the scores (the mean) and a measure of dispersion (the standard deviation). It is useful in comparing scores when two different reference groups are involved. The z-score is

$$z = \frac{\text{score} - \text{mean}}{\text{standard deviation}}$$

EXAMPLE 6 ➤ If $\mu = 85$ and $\sigma = 5$, then the z-score for a raw score of 92 is

$$z = \frac{92 - 85}{5} = \frac{7}{5} = 1.4$$

For a score of 70,

$$z = \frac{70 - 85}{5} = \frac{-15}{5} = -3$$

A negative z-score indicates that the score is below the mean. $z = 1.4$ indicates that the score is 1.4 standard deviations above the mean. A z-score of 0 corresponds to the mean.

■ **Now You Are Ready to Work Exercise 21**

Application

EXAMPLE 7 ➤
A student took both the SAT and ACT tests and made a 480 on the SAT verbal portion and a 22 on the ACT verbal. Which was the better score? (*Note:* You need

to know that for the SAT verbal test, $\mu = 431$ and $\sigma = 111$; and for the ACT verbal, $\mu = 17.8$ and $\sigma = 5.5$.)

SOLUTION

The z-score for each test is

$$\text{SAT } z = \frac{480 - 431}{111} = \frac{49}{111} = 0.44$$

$$\text{ACT } z = \frac{22 - 17.8}{5.5} = \frac{4.2}{5.5} = 0.76$$

The higher z-score, 0.76, indicates that the student performed better on the ACT, because the ACT score was more standard units above the mean.

■ **Now You Are Ready to Work Exercise 37**

Five-Point Summary and Box Plot

We have used range and standard deviation to help understand the dispersion of data. Sometimes, we may want more detail about the dispersion. One method that provides such detail is the **five-point summary** and its visual representation, the **box plot**, also called the **box-and-whisker plot.**

We have already discussed three of the five points used to describe data:

Minimum: The smallest value in the data.

Median: The value in the middle of the data (when they are arranged in order).

Maximum: The largest value in the data.

Two other points that give information on the dispersion of data are the **first quartile, Q_1,** and the **third quartile, Q_3.**

DEFINITION
Median

> The *first quartile*, denoted Q_1, is the median of the values less than the median itself of the data (median itself excluded). It separates the lower 25% of the data from the upper 75%.
>
> The *third quartile*, denoted Q_3, is the median of the values greater than the median of the data (median itself excluded). It separates the upper 25% of the data from the lower 75%.
>
> The *second quartile*, denoted Q_2, is the median of all the data.

EXAMPLE 8 ➤

The five-point summary of 2, 3, 6, 7, 9, 13, 17, 19, and 22 is:

Minimum: 2

Q_1:   4.5

Median: 9

Q_3:   18

Maximum: 22

This can be summarized as {2, 4.5, 9, 18, 22}. We now draw a figure, or box plot, that represents this five-point summary.

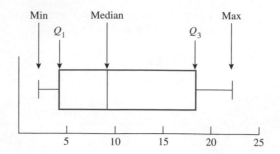

Notice how the box plot is constructed. The ends of the box are Q_1 and Q_3, with the location of the median shown in the box. Whiskers are attached to each end of the box. The whisker on the left extends to the minimum value, and the whisker on the right extends to the maximum value.

■ **Now You Are Ready to Work Exercise 25**

8.3 EXERCISES

LEVEL 1

Determine the population mean, variance, and standard deviation for the sets of data in Exercises 1 through 6.

1. *(See Example 1)*   19, 10, 15, 20

2. 0, 3, 5, 9, 10, 12

3. 4, 8, 9, 10, 14

4. 10, 14, 20, 24

5. 17, 39, 54, 22, 16, 46, 25, 19, 62, 50

6. 1.1, 1.3, 1.5, 1.7

Find the sample mean, variance, and standard deviation for Exercises 7 through 10.

7. $-8, -4, -3, 0, 1, 2$ 8. $-5, 4, 8, 10, 15$

9. $-3, 0, 1, 4, 5, 8, 10, 11$ 10. $-8, -6, 0, 4, 7, 12$

LEVEL 2

11. *(See Example 3)*   Find the population mean, variance, and standard deviation of the following frequency distribution:

x	Frequency
1	2
2	5
3	2
4	3
5	8

Grade	Frequency
65	1
70	2
75	5
78	3
80	7
84	4
85	6
90	2

12. Find the population mean and standard deviation of the following exam grades:

13. Find the sample mean and standard deviation of the following student summer wages:

Wage (per hour)	Frequency
5.65	10
5.90	12
6.00	8
6.24	5

14. Find the sample mean and standard deviation of the following frequency distribution:

x	Frequency
12	3
15	4
20	5
22	4

15. *(See Example 4)* Estimate the population mean and standard deviation of the following grouped data:

Score	Frequency
0–10	5
11–20	12
21–30	8

16. The following is a summary of test grades for a class. Estimate the population mean and standard deviation.

Grades	Frequency
91–100	3
81–90	4
71–80	10
61–70	4
51–60	3

17. Estimate the sample mean and standard deviation of the following grouped data:

Score	Frequency
10–15	14
16–20	18
21–25	18

18. Estimate the mean and standard deviation of the following ACT scores:

Score	Frequency
17–19	25
20–22	35
23–25	45
26–28	25
29–31	10
32–34	10

19. The daily high temperatures (in degrees Fahrenheit) one week in January were 32, 35, 28, 34, 29, 33, and 26. Find the mean and standard deviation of the temperature.

20. **(a)** Calculate the standard deviation of the set 6, 6, 6, 6, 6, and 6.
(b) What are the mean and the standard deviation of a set in which all scores are the same?
(c) What are the median and mode of a set in which all scores are the same?

21. *(See Example 6)* Let $\mu = 160$ and $\sigma = 16$.
(a) Find the z-score for the score 180.
(b) Find the z-score for the score 150.
(c) Find the z-score for the score 160.
(d) Find the score x if the z-score $= 1$.
(e) Find the score x if the z-score $= -0.875$.

22. The following statements give a single-number model of situations. Discuss how our view of the situation might change if we also knew the standard deviation.
(a) The mean monthly rainfall in Waco is 2.51 inches.
(b) The average grade on the last test was 82.
(c) The mean of the temperature on July 11, 1987, was 80 degrees.
(d) "But, Dad, I averaged 50 miles per hour."

23. Find the five-point summary of the following scores: 3, 4, 6, 8, 11, 12, 15, 16, 19, 20.

24. Find the five-point summary of the following scores: 31, 35, 36, 39, 44, 46, 50, 55, 59, 60, 65

25. *(See Example 8)* Draw the box plot that represents the following scores: 3, 4, 7, 8, 11, 12, 15, 17, 20

26. Draw the box plot that represents the following ACT scores: 22, 24, 19, 21, 29, 24, 31, 27, 22, 20, 23

27. Draw the box plot that represents the following scores: 5.6, 7.2, 8.4, 9.1, 4.2, 11.6, 7.5, 9.7, 13.1, 16.4, 18.3, 17.2, 21.5, 23.8, 22.0, 6.1, 8.5, 12.3, 16.4

28. From the box plot shown, determine
(a) the median.
(b) the interval containing the middle 50% of the data.
(c) the maximum value.

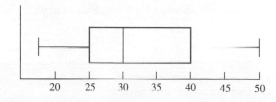

LEVEL 3

29. A machine shop uses two lathes to make shafts for electric motors. A sample of five shafts from each lathe are checked for quality-control purposes. The diameter of each shaft (in inches) is given as follows:

Lathe I	Lathe II
0.501	0.502
0.503	0.497
0.495	0.498
0.504	0.501
0.497	0.502

The mean diameter in each case is 0.500.
(a) Find the standard deviation for each lathe.
(b) Which lathe is more consistent?

30. **(a)** Calculate the mean and the standard deviation of the scores 1, 3, 8, and 12.
(b) Add 2 to each of the scores in part (a) and compute the mean and the standard deviation. How are the mean and the standard deviation related to those in part (a)? Would the same kind of relationship exist if you added another constant instead of 2?

31. A set has 8 numbers. Each one is either 2, 3, or 4. The mean of the set is 3.
(a) If $\sigma = 0.5$, what is the set of numbers?
(b) If $\sigma = 1$, what is the set of numbers?
(c) Can σ be larger than 1?

32. *(See Example 5)* A golfer finished 15th in a field of 90 golfers in a tournament. What was his percentile?

33. A student scored at the 70th percentile on a test taken by 110 students. What was her rank?

34. A student learned that 105 students scored higher and 481 scored the same or lower on a standardized test.
(a) What is the student's rank on the test?
(b) What is the student's percentile on the test?

35. A student scored at the 98th percentile on a verbal test given to 1545 students. How many scored higher?

36. Suppose that $\mu = 140$ and $\sigma = 15$. Find the score 2.2 standard deviations above the mean.

37. *(See Example 7)* Leon took an admissions test at two different universities. At one, where the mean was 72 and the standard deviation 8, he scored 86. At the other, with mean 62 and standard deviation 12, he scored 82. Which one was the better performance?

38. Joe scored 114 on a standardized test that had a mean of 120 and a standard deviation of 8. Josephine scored 230 on a test that had a mean of 250 and a standard deviation of 14. Which one performed better?

39. A runner ran a 3-kilometer race in 19 minutes. The average time of all who finished was 18 minutes with $\sigma = 1$ minute. A cyclist rode a 25-mile race in 64 minutes. The average time of all who finished was 59 minutes with $\sigma = 3$. Which athlete had the better performance?

EXPLORATIONS

40. Can the standard deviation of a data set be zero? Explain.

41. Can the standard deviation of a data set be negative? Explain.

Use the statistics menu on your graphing calculator or spreadsheet to compute both the population and sample mean and standard deviation for the data in Exercises 42 through 49.

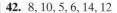
42. 8, 10, 5, 6, 14, 12

43. 14.2, 16.7, 21.3, 18.9, 25.2

44.

Score	Frequency
8	15
10	22
12	19

45. Use the data from Exercise 11.

46. Use the data from Exercise 12.

47. Use the data from Exercise 13.

48. Use the data from Exercise 15.

49. Use the data from Exercise 17.

50. Here are data (in dollars) on taxes collected by the Internal Revenue Service for the years 1992–1996 and 2001:

Year	Total Collected (trillions)	Corporation Income and Profit Tax (billions)	Tax per Capita
1992	1.121	117.9	4374.38
1993	1.177	131.5	4543.33
1994	1.276	154.2	4878.00
1995	1.376	174.4	5216.44
1996	1.478	189.1	5586.00
2001	1.129	186.7	7448.90

Compute the mean and standard deviation for:
(a) Total collected
(b) Corporation taxes
(c) Per capita tax

51. As of May 26, 2002, the top ten Broadway shows, in number of performances, were the following:

Show	Performances
Cats	7485
Les Misérables	6276
A Chorus Line	6137
The Phantom of the Opera	5979
Oh! Calcutta	5959
Miss Saigon	4092
42nd Street	3486
Grease	3388
Beauty and the Beast	3306
Fiddler on the Roof	3242

Compute the mean and standard deviation for the number of performances.

52. The Department of Commerce provides the following data on the number of persons below poverty level during 1986–2000 (in thousands).

Year	Total	White	Black
1986	32,370	22,183	8,938
1987	32,221	21,195	9,520
1988	31,745	20,715	9,356
1989	31,528	20,785	9,302
1990	33,585	22,326	9,837
1991	35,708	23,747	10,242
1992	38,014	25,259	10,827
1993	39,265	26,226	10,877
1994	38,059	25,379	10,196
1995	36,425	24,423	9,872
1996	36,529	24,650	9,694
1997	35,574	24,396	9,116
1998	34,476	23,454	9,091
1999	32,258	21,922	8,360
2000	31,139	21,291	7,901

Compute the mean and standard deviation for:
(a) Total   (b) White   (c) Black

53. In 1997, the top ten college and university libraries (in number of volumes) were the following:

University	Volumes (in millions)
Harvard	13.617
Yale	9.932
Univ. of Illinois–Urbana	9.024
Univ. of Calif.–Berkeley	8.628
Univ. of Texas	7.495
UCLA	7.010
Univ. of Michigan	6.973
Columbia	6.906
Stanford	6.865
Univ. of Chicago	6.117

Compute the mean and standard deviation for the number of volumes.

54. Find the population and sample mean and standard deviation of 2, 3, 5, 6, 7, 9, 11, 14, 18, and 19.

55. Find the population and sample mean and standard deviation of 1, 2, 3, 4, 9, 11, 12, 15, and 19.

56. Find the population and sample mean and standard deviation of 23.4, 25.6, 26.1, 16.6, 19.7, 29.3, 31.7, 32.4, and 34.5.

A general, and conservative, estimate of how data are clustered about the mean is given by Chebyshev's Theorem:

Chebyshev's Theorem

The fraction of scores of *any* set of data lying within k standard deviations of the mean is *always* at least $1 - \dfrac{1}{k^2}$, where k is greater than 1. For example, for $k = 2$, at least $1 - 1/4 = 3/4$ of the scores must lie from 2 standard deviations below the mean to 2 standard deviations above the mean.

57. Find the fraction of scores lying within the following number of standard deviations of the mean.
(a) $k = 3$ (b) $k = 1.5$ (c) $k = 2.5$

58. A set of data has a mean of 75 and a standard deviation of 6. Use Chebyshev's theorem to estimate the fraction of scores in the following intervals:
(a) Between 63 and 87.
(b) Between 57 and 93.
(c) Between 66 and 84.
(d) Between 67 and 83.

59. The following table shows the Department of Labor estimates of the median weekly earnings for 24 selected occupations.

Occupation	Earnings
Food preparation	322
Sales workers	363
Bank tellers	376
Hairdressers	381
Secretaries	479
Cab drivers	487
Farm operators	510
Truck drivers	593
Social workers	644
Mechanics	665
Clergy	699
Electricians	714
Mail carriers	721
Librarians	724
Teachers, other	730
Accountants	773
Psychologists	818
Executives	867
Police	949
Teachers, college	1009
Systems analysts	1100
Pilots	1150
Physicians	1258
Lawyers	1398

The mean for this sample of earnings is $739 and the standard deviation is $290.

(a) Chebyshev's Theorem states that at least 75% of the earnings listed will fall between $739 - 2(290) = 159$ and $739 + 2(290) = 1319$. Count the number in the list that fall in that interval. Is your answer consistent with Chebyshev's Theorem?

(b) Chebyshev's Theorem states that at least 56% of the earnings listed will fall between

$739 - 1.5(290) = 304$ and $739 + 1.5(290) = 1174$. Count the number in the list that fall in that interval. Is your answer consistent with Chebyshev's Theorem?

Another rule that relates the standard deviation to the clustering about the mean is the **empirical rule**. This rule often applies to data that arise from natural causes, such as weights of 10-year-old children, waist sizes of adult men, and IQ scores. The empirical rule applies to these and other data that fall into a bell shaped pattern.

Empirical Rule

For data that are approximately bell shaped:

- About 68% of all scores fall within 1 standard deviation of the mean.
- About 95% of all scores fall within 2 standard deviations of the mean.
- About 99.7% of all scores fall within 3 standard deviations of the mean.

---

60. A set of scores is reasonably bell shaped. The mean is 124, and the standard deviation is 8.
   (a) What percentage of the scores should fall between 116 and 132?
   (b) What percentage of the scores should fall between 108 and 140?

61. In the table of earnings in Exercise 59, count the scores that fall within one standard deviation of the mean. Compare this with the 68% from the empirical rule. Does this suggest that the earnings data are bell shaped?

USING YOUR TI-83

In section 8.2, we used functions from the LIST menu to find the mean and median of data. The STAT menu has functions that calculate the measures of dispersion in this section. The mean and median are also calculated, so you have two routines that calculate mean and median.

CALCULATING THE RANGE

The range of a list of scores in L1 can be found using the **max** and **min** functions in the LIST menu. Here's how.
   [LIST] <MATH> <2:max(> L1 − <1:min(> L1 [ENTER]

CALCULATING THE MEAN AND STANDARD DEVIATION

The TI-83 has a routine in the STAT menu for calculating the mean and standard deviation for a list of numbers or for numbers summarized in a frequency table. The TI-83 has six lists in the memory — L1, L2, ..., L6 — that are used.

Use with $\boxed{\text{STAT}}$ <4:Clrlist> L1, L2 $\boxed{\text{ENTER}}$ to clear lists L1 and L2.

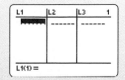

Use one list — say L1 — to enter the list of numbers and use a second list — say L2 — to list the corresponding frequencies if you have a frequency table. $\boxed{\text{STAT}}$ <EDIT> will show a screen of the lists.

Let's illustrate with the following frequency table:

x	Frequency
2	3
5	2
6	4
9	1

With the numbers x in L1 and the frequencies in L2,

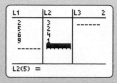

To find the mean and standard deviation of the list L1 only, select $\boxed{\text{STAT}}$ <CALC> <1:1-Var Stats> L1 $\boxed{\text{ENTER}}$

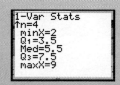

The mean is shown as x-bar $= 5.5$ (the x with a bar over it), $\sigma x = 2.5$ is the population standard deviation, and $Sx = 2.8867$ is the sample standard deviation. Scrolling down the screen, we see the five-point summary, minX $= 2$, $Q_1 = 3.5$, Median $= 5.5$, $Q_3 = 7.5$, and maxX $= 9$.

To find the mean and standard deviation of the frequency table with numbers in L1 and their frequency in L2, select $\boxed{\text{STAT}}$ <CALC> <1:1-Var Stats> L1, L2 $\boxed{\text{ENTER}}$

The mean is shown as x-bar $= 4.9$ (the x with a bar over it); $\sigma x = 2.1656$ is the population standard deviation, and $Sx = 2.2827$ is the sample standard deviation. Scrolling down the screen, we see the five-point summary, minX $= 2$, $Q_1 = 2$, Median $= 5.5$, $Q_3 = 6$, and maxX $= 9$.

(*continued*)

EXERCISES

Find the mean and standard deviation (population and sample) for each of the following sets of data:

1. 3, 5, 1, 6, 8, 2

2. 3.2, 4.1, 6.3, 4.4, 7.5, 5.2

3.

x	Frequency
4	2
5	1
8	3
9	1
11	4

4.

x	Frequency
5.5	3
6.1	2
8.7	1
9.3	3
7.2	1

BOX PLOT

You can draw a box plot of data using your TI-83. Refer back to "Using Your TI-83" in Section 8.1. It describes how to use your TI-83 to draw a histogram. To draw a box plot, follow Steps 1 through 4 for drawing a histogram but substitute *box plot* for *histogram*. Be sure you select the box plot symmbol for Type in Step 3. It is the next to last Type symbol.

EXERCISES

1. Draw the box plot for the following scores: 3, 6, 2, 9, 4, 5, 13, 14, 10.

2. Draw the box plot for the scores summarized by the following frequency table:

Score	Frequency
4	2
7	1
8	4
10	7
12	5
13	3

USING EXCEL

MEASURES OF DISPERSION

EXCEL has routines for finding measures of dispersion for a list of scores. Let's illustrate assuming the data 2, 6, 9, 1, 7, 4, 3, 7, 5, 6 are stored in cells A1:A10.

RANGE

In the cell where you want the range stored, enter the formula =MAX(A1:A10)-MIN(A1:A10), which gives 8.

STANDARD DEVIATION

Sample Standard Deviation: In the cell where you want the sample standard deviation stored, enter the formula =STDEV(A1:A10), which gives 2.4944.

Population Standard Deviation: In the cell where you want the population standard deviation stored, enter the formula =STDEVP(A1:A10), which gives 2.3664.

QUARTILES

First Quartile, Q_1: In the cell where you want the first quartile stored, enter the formula =QUARTILE(A1:A10,1), which gives 3.25.

Third Quartile, Q_3: In the cell where you want the third quartile stored, enter the formula =QUARTILE(A1:A10,3), which gives 6.75.

BOX PLOT

To draw a box plot we need:

   Maximum score, obtained by =MAX(A1:A10), giving 9.

   Minimum score, obtained by =MIN(A1:A10), giving 1.

   Median score, obtained by =MEDIAN(A1:A10), giving 5.5.

From above we have $Q_1 = 3.25$ and $Q_3 = 6.75$, so the five-point summary is {1, 3.25, 5.5, 6.75, 9} and the box plot is

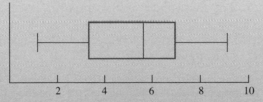

 NOTE

EXCEL does not draw the box plot, so you must sketch it yourself.

EXERCISES

1. Find the range, sample standard deviation, and population standard deviation for the following: 8, 14, 22, 6, 19, 13, 9, 22, 15, 17, 11.

2. Find the range, sample standard deviation, and population standard deviation for the following: 45, 47, 39, 49, 33, 54, 51, 59, 44, 38, 55, 58.

3. Find the five-point summary and sketch the box plot for the following scores: 17, 22, 25, 33, 37, 34, 29, 41, 35, 48.

4. Find the five-point summary and sketch the box plot for the following scores: 1, 5, 7, 3, 11, 15, 9, 18, 20, 11, 8, 22, 14.

8.4 Random Variables and Probability Distributions of Discrete Random Variables

- Random Variables
- Probability Distribution of a Discrete Random Variable

Random Variables

We routinely use numbers to convey information, to help express an opinion or an emotion, and to help evaluate a situation, and to clarify our thoughts. Think how many times you have used phrases such as, "On a scale from 1 to 10, what do you think of . . . ?"; "What is your GPA?"; "The chance of rain this weekend is 40%"; "Which team is first in the American League?"; "How many A's did Professor X give?"; and "What did you make on your math test?"

All of these statements use numbers in some form. Notice that the numbers are not the heart of the statement, they represent a key facet of the message. When he says she is an 8 and she says he is a 3, the numbers are not the emotions that are evoked, the numbers indicate something about the *strength* of the emotions. The product of an examination is an exam paper, but a grade is assigned to the paper to indicate something about the quality of the paper. A trip to the grocery store results in bags of grocery items, but a number — the cost — is usually associated with the purchase.

We associate numbers with outcomes of activities and phenomena because it is easier to compare and analyze numbers. To summarize grades with a GPA is easier than averaging letter grades. The auto driver and the highway patrolman can more nearly agree on the meaning of "drive at 65 mph or less" than they can on "drive at a safe and reasonable speed."

This idea of assigning a number to outcomes of an experiment (activity or process) is an important and useful statistical concept. We call this assignment of numbers a **random variable.**

DEFINITION
Random Variable

> A **random variable** is a rule that assigns a number to each outcome of an experiment.

It is customary to denote the random variable by a capital letter such as X or Y. Let's look at some examples of random variables.

EXAMPLE 1 ➤

If a coin is tossed twice, there are four possible outcomes: HH, HT, TH, and TT. We can assign a number to each outcome by simply giving the number of heads that appear in each case. The values of the random variable X are assigned as follows:

Outcome	X
HH	2
HT	1
TH	1
TT	0

■ **Now You Are Ready to Work Exercise 1**

EXAMPLE 2 ➤

A true–false quiz consists of 15 questions. A random variable may be defined by assigning the total number of correct answers to each quiz. In this case, the random variable ranges from 0 through 15.

■ **Now You Are Ready to Work Exercise 5**

EXAMPLE 3 ➤

The Hardware Store has a box of Super-Special items. Selecting an item from the box is an experiment with each of the items as an outcome. Two possible ways to define a random variable are to

(a) assign the original price to each item (outcome) or
(b) assign the sale price to each item.

■ **Now You Are Ready to Work Exercise 7**

EXAMPLE 4 ➤

(a) Each judge of an Olympic diving contest assigns a number from 1 to 10 to each outcome (dive). We can think of each judge's assignment as a way of defining a random variable.

(b) The housing office gives a congeniality test to incoming students to better assign dormitory roommates. The tests are scored 1 through 5, with a higher number indicating more congeniality. The scores are values of a random variable assigned to levels of congeniality.

(c) An airline baggage claim office records the number of bags claimed by each passenger. The number of bags is a random variable associated with each passenger. ■

EXAMPLE 5 ➤

Three people are selected from a group of five men and four women. A random variable X is the number of women selected. List the possible values of X and the number of different outcomes that can be associated with each value.

SOLUTION

X can assume the value 0, 1, 2, or 3. The number of different outcomes associated with each value is the following:

X	Outcome	Number of Possible Outcomes
0	Three men	$C(5, 3) = 10$
1	One woman, two men	$C(4, 1)C(5, 2) = 40$
2	Two women, one man	$C(4, 2)C(5, 1) = 30$
3	Three women	$C(4, 3) = 4$

■ **Now You Are Ready to Work Exercise 25**

In one sense, you may be quite arbitrary in the way you define a random variable for an experiment. It depends on what is most useful for the problem at hand. Here are several examples:

Experiment	Random Variable, X
A survey of cars entering a mall parking lot	Number of passengers in a car: $1, 2, \ldots, 8$
Rolling a pair of dice	Sum of numbers that turn up: $2, 3, \ldots, 12$
Tossing a coin three times	Number of times tails occurs
Selecting a sample of five tires from an assembly line	Number of defective tires in the sample
Measuring the height of a student selected at random	Observed height
Finding the average life of a brand X tire	Number of miles driven
Selecting a box of Crunchies cereal	Weight of the box of cereal
Checking the fuel economy of a compact car	Distance X the car travels on a gallon of gas

The first four random variables are **discrete variables,** because the values assigned come from a set of distinct numbers and the values between are not permitted as outcomes. For example, the number of auto passengers can take on only the values 1, 2, 3, and so on. The number 2.63 is not a valid assignment for the number of passengers. On the other hand, the last four examples are **continuous**

variables. Assuming an accurate measuring device, 5 feet 4.274 inches is a valid height. There are no distinct gaps that must be excluded as a valid height. Similarly, even though one might expect a car to travel about 23 miles on a gallon of gas, 22.64 miles cannot be excluded as a valid distance. Any distance in a reasonable interval is a valid possibility.

Probability Distribution of a Discrete Random Variable

We now merge two concepts that we studied earlier. We have used the concept of probability to give a measure of the likelihood of a certain outcome or outcomes occurring in an experiment (Section 7.1). At the beginning of this section, we introduced the concept of a random variable, the assigning of a number to each outcome of an experiment. The value assigned to the random variable depends on the outcome that actually occurs. Some outcomes may have a small likelihood of occurring, whereas others may be quite likely to occur. Thus, the associated values of the random variable usually have different levels of likelihood.

We sometimes ask about the chance of observing a certain value of a random variable X (that is, find the *probability* that a certain value occurs). It is no surprise that such a probability closely relates to the probabilities of the outcomes. Rather than look at the probability of just one value of X, we now focus on the probability of each of the possible values of X. We call this assignment of a probability to each value of a discrete random variable a **probability distribution.** Let's look at a simple example.

EXAMPLE 6 ➤ A children's game has 15 cards with 5 colored red, 3 colored black, and 7 colored green. The cards are shuffled, and a child selects 2 cards. If the cards are different colors, no points are given. Five points are given if both cards are green, 10 points are given if both cards are red, and 15 points are given if both cards are black. Because a number (the points) is assigned to each outcome (cards drawn), this determines a random variable X with outcomes and numbers related as follows:

Outcome	Value of X (points)
Red and black	0
Red and green	0
Red and red	10
Black and black	15
Black and green	0
Green and green	5

The probability of drawing two greens is $\dfrac{C(7, 2)}{C(15, 2)} = 0.200$

of two reds $\dfrac{C(5, 2)}{C(15, 2)} = 0.095$

of two blacks $\dfrac{C(3, 2)}{C(15, 2)} = 0.029$

All other outcomes will be different colors, so the probability of different colors is $1 - 0.200 - 0.095 - 0.029 = 0.676$. Now we can give the probability of each value of X as follows.

X	P(X)
0	0.676
5	0.200
10	0.095
15	0.029

■ **Now You Are Ready to Work Exercise 21**

DEFINITION
**Probability
Distribution**

If a discrete random variable has the values

$$x_1, x_2, x_3, \ldots, x_k$$

then a **probability distribution** $P(x)$ is a rule that assigns a probability $P(x_i)$ to each value x_i. More specifically,

(a) $0 \le P(x_i) \le 1$ for each x_i,
(b) $P(x_1) + P(x_2) + \cdots + P(x_k) = 1$

EXAMPLE 7 ➤   A coin is tossed twice. Define a random variable as the number of times heads appears. Compute the probability of zero, one, or two heads in the usual manner to get

$$P(0) = P(\text{TT}) = \frac{1}{4}$$

$$P(1) = P(\text{HT or TH}) = \frac{1}{2}$$

$$P(2) = P(\text{HH}) = \frac{1}{4}$$

We can represent a probability distribution graphically using a histogram. (See Figure 8–9.)

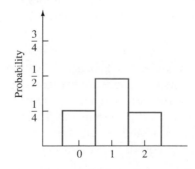

FIGURE 8–9   Probability graph of the number of heads in two tosses of a coin.

> **NOTE**
> The width of a bar of a probability distribution is one unit; therefore, the *area* of the rectangular bar equals the probability of that value of X.

Form a category for each value of the random variable and center the rectangle over the category mark. The probability, $P(x)$, of a value, x, of the random variable is typically represented by the area of the rectangle. When x has only integral values, as in this case, the width of the rectangle is 1, and the height of the rectangle equals the probability, $P(x)$.

■ **Now You Are Ready to Work Exercise 15**

EXAMPLE 8 ➤   An experiment randomly selects two people from a group of five men and four women. A random variable X is the number of women selected. Find the probability distribution of X.

FIGURE 8–10

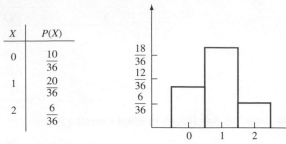

X	$P(X)$
0	$\dfrac{10}{36}$
1	$\dfrac{20}{36}$
2	$\dfrac{6}{36}$

(a) Probability distribution

(b) Graph of probability distribution

SOLUTION

The values of X range over the set $\{0, 1, 2\}$, because 0, 1, or 2 women can be selected. The probability of each value is computed in the usual manner.

$$P(0) = \frac{C(5, 2)}{C(9, 2)} = \frac{10}{36} \qquad \text{(Probability both are men)}$$

$$P(1) = \frac{C(5, 1) \times C(4, 1)}{C(9, 2)} = \frac{20}{36} \qquad \text{(Probability of one man and one woman)}$$

$$P(2) = \frac{C(4, 2)}{C(9, 2)} = \frac{6}{36} \qquad \text{(Probability of two women)}$$

The probability distribution and its graph are shown in Figure 8–10.

■ **Now You Are Ready to Work Exercise 27**

EXAMPLE 9 ➤

Five red cards are numbered 1 through 5, and five black cards are numbered 1 through 5. Cards are drawn, without replacement, until a card is drawn that matches the color of the first card. Let the number of cards drawn be the random variable X. Find the probability distribution of X.

SOLUTION

The possible values of X are 2, 3, 4, 5, 6, and 7, because at least two cards must be drawn in order to have a match. There are five cards that are different from the first color, so the sixth card is the last time a card that is different from the first can be drawn. The seventh card must then be a match. To compute probabilities, designate the color of the first card by C and the different color by D. Then the possible outcomes of the sequence of colors are the following:

$$
\begin{aligned}
X &= 2\text{: CC} \\
X &= 3\text{: CDC} \\
X &= 4\text{: CDDC} \\
X &= 5\text{: CDDDC} \\
X &= 6\text{: CDDDDC} \\
X &= 7\text{: CDDDDDC}
\end{aligned}
$$

We now compute the probability of each sequence.

Note that it doesn't matter which color is drawn first, so the probability the first card is some color equals 1.

$$\text{CC:} \quad P(X = 2) = 1 \times \frac{4}{9} = \frac{4}{9}$$

$$\text{CDC:} \quad P(X = 3) = 1 \times \frac{5}{9} \times \frac{4}{8} = \frac{5}{18}$$

$$\text{CDDC:} \quad P(X = 4) = 1 \times \frac{5}{9} \times \frac{4}{8} \times \frac{4}{7} = \frac{10}{63}$$

$$\text{CDDDC:} \quad P(X = 5) = 1 \times \frac{5}{9} \times \frac{4}{8} \times \frac{3}{7} \times \frac{4}{6} = \frac{5}{63}$$

$$\text{CDDDDC:} \quad P(X = 6) = 1 \times \frac{5}{9} \times \frac{4}{8} \times \frac{3}{7} \times \frac{2}{6} \times \frac{4}{5} = \frac{2}{63}$$

$$\text{CDDDDDC:} \quad P(X = 7) = 1 \times \frac{5}{9} \times \frac{4}{8} \times \frac{3}{7} \times \frac{2}{6} \times \frac{1}{5} \times \frac{4}{4} = \frac{1}{126}$$

■ **Now You Are Ready to Work Exercise 39**

8.4 EXERCISES

LEVEL 1

1. *(See Example 1)* A coin is tossed three times, and the number of heads that occur, X, is assigned to each outcome. List all possible outcomes and the corresponding value of X.

2. A word is selected from the phrase "A stitch in time saves nine." A random variable counts the number of letters in the word selected. List all possible outcomes and the corresponding value of X.

3. Two people are selected from {Ann, Betty, Jason, Tom}, and the number of females selected, X, is assigned to each outcome. List all possible outcomes and the value of X assigned to each.

4. A nickel is tossed. If it turns up heads, a quarter is then tossed once. If the nickel turns up tails, the quarter is then tossed twice. The number of heads that turns up (on both coins) is the random variable X. List all possible outcomes and the corresponding value of X.

Give all possible values of the random variable in each of Exercises 5 through 7.

5. *(See Example 2)* Three people are selected at random from eight men and six women. The random variable is the number of men selected.

6. Four people are selected at random from five women and three men. The random variable is the number of men selected.

7. *(See Example 3)* A mother's purse contains four $1 bills, six $5 bills, and three $10 bills in a random order. A child finds the purse and takes four bills.
   (a) The random variable is the number of $5 bills taken.
   (b) The random variable is the number of $10 bills taken.

8. As a class project, students in the Student Center are asked to report the amount of sleep they had in the previous 24 hours. Then the fraction of the day (time sleeping/24) spent in sleep is recorded for X. What are the possible values for X?

Identify the random variables in Exercises 9 through 12 as discrete or continuous.

9. (a) Number of squirrels in a tree.
   (b) Height of a tree.
   (c) Volume of water in Lake Belton.
   (d) Number of passengers in a car that enters the parking garage.

10. (a) Number of students in each classroom at 9:00 A.M. on Monday.
    (b) The weight of each person enrolling in a diet clinic.
    (c) The runs scored by the home team in a baseball game.
    (d) The speed of a moped passing a certain point on the street.

11. (a) The diameter X of a golf ball.
 (b) From a random selection of ten students, the number X who have type O blood.
 (c) Sixty families live in a remote village. The random variable X is
   (i) the number of children in a family.
   (ii) the amount of electricity used in a month by a family.
   (iii) the monthly income of a family.

12. (a) The number of building permits issued each year in Augusta, Georgia.
 (b) The length of time to play a tennis match.
 (c) The number of cartons of Blue Bell ice cream sold each week in the Food Mart.
 (d) The number of pounds of apples sold each day at Joe's Roadside Fruit Stand.

13. Fifty students are polled on the courses they are taking, and X is the number of courses a student takes. The number of courses ranges from 2 to 6. Is the following a probability distribution of X?

X (number of courses)	Fraction Taking X Courses
2	0.08
3	0.12
4	0.48
5	0.30
6	0.02

14. Customers at Lake Air Mall were asked to state which of three brands of coffee tasted the best. The brands were numbered 1, 2, and 3. The results are summarized as follows:

Brand	Fraction Preferring Brand
1	0.25
2	0.40
3	0.20

If the brand number is a random variable, does this define a probability distribution?

15. *(See Example 7)* A coin is tossed three times. X is the number of heads that appear. Find the probability distribution of X and its graph.

16. A pair of dice is tossed. The random variable X is the sum of the numbers that turn up. Find the probability distribution of X.

17. In the game of SCRABBLE, players randomly draw letters from 100 blocks, which consist of 98 letters and 2 blanks. A player attempts to form a word using letters already played and letters from the player's hand. Each letter is assigned a number, so the player receives a score based on the value of the letters used. The value of letters and the number having that value is given by the following table:

Value	Number Having Value
0	2
1	68
2	7
3	8
4	10
5	1
8	2
10	2
	100

Let the random variable X be the value of a block randomly drawn from the set of 100 blocks. Find the probability distribution of X.

18. A class of 30 students took a 5-point quiz. The results are summarized as follows:

Score	Frequency
0	2
1	4
2	8
3	3
4	12
5	1

Let the random variable X be the score on a randomly selected quiz. Find the probability distribution of X, and draw its graph.

19. A survey of 75 households revealed the following information on the number of television sets they owned:

Number of TV Sets	Frequency
0	8
1	49
2	13
3	4
4	1

Let the random variable X be the number of television sets owned by a randomly selected household. Find the probability distribution of X, and sketch its graph.

20. When Luv That Balloon opened, a record of the daily number of phone calls was kept. The records showed the following data:

Number of Calls per Day	Frequency
1	1
2	3
3	1
4	2
5	14
6	9

Let the number of calls be the random variable X. Use this summary to determine a probability distribution of X.

21. *(See Example 6)* A child's game has ten cards, four colored red and six colored black. The cards are shuffled and two cards drawn. If both cards are red, 10 points are given. If both cards are black, 5 points are given. If the cards are of different colors, no points are given. The random variable is the number of points given. Find the probability distribution of X.

22. In a card game, a player randomly selects 2 cards in succession from a 52-card bridge deck. The player receives a score of 10 if the second card is the same suit as the first, a score of 5 if the second card is the same color and different suit, and a score of 0 if the second card is a different color. Let the random variable X be the score and find the probability distribution of X.

LEVEL 2

23. A word is randomly selected from a newspaper article. A random variable X is defined as the number of letters in that word.
    (a) What are the possible values of X?
    (b) Is X discrete or continuous?

24. An experiment consists of selecting a whole number and finding the remainder after division by 3. What are the possible outcomes?

25. *(See Example 5)* Two people are randomly selected from a group of six men and five women. The random variable X is the number of women selected. List the possible values of X and the number of different outcomes that can be associated with each value.

26. Professor Armitstead has a test bank of questions for his course on the United States presidency. The section on Andrew Jackson has six true–false questions, four discussion questions, and eight multiple-choice questions. The computer randomly selects three questions. Let the random variable X be the number of true–false questions. List the possible values of X and the number of different outcomes associated with each value.

27. *(See Example 8)* Two people are randomly selected from a group of four men and four women. The random variable X is the number of men selected. Find the probability distribution for X and its graph.

28. Three people are randomly selected from a group of five men and six women. The random variable X is the number of women selected. Find the probability distribution of X.

29. In a collection of ten electronic components, three are defective. Two are selected at random, and the number of defective components is noted. Let X be

the number of defective components, and draw the graph of the probability distribution of X.

30. A number is selected at random from the numbers 1, 2, 3, 4, 5, 6, 7, 8, 9, and 10. The random variable X is the remainder when the selected number is divided by 3. Find the probability distribution of X.

31. A jar contains three red jellybeans, two green jellybeans, and one yellow jellybean. As a reward for good behavior, a child receives two randomly selected jellybeans.
    (a) The random variable X is the number of green jellybeans received. Find the probability distribution of X.
    (b) The random variable X is the number of yellow jellybeans received. Find the probability distribution of X.

32. At each meeting of a club, one person is selected to draw a "lucky number." That person gets the amount in dollars of the number drawn. The box contains 30 cards with the number 1, 14 cards with the number 5, 3 cards with the number 10, 2 cards with the number 20, and 1 card with the number 50. Let the random variable X be the numbers on the cards. Determine the probability distribution of X, and draw its graph.

33. A guest at the Downtown Civic Club is allowed to randomly draw 1 bill from a box containing 50 bills: 26 $1 bills, 12 $5 bills, 8 $10 bills, 2 $20 bills, 1 $50 bill, and 1 $100 bill. Let the random variable X be the value of the bill drawn. Find the probability distribution of X.

LEVEL 3

34. Gloria is one of five people who are randomly arranged in five positions 1 through 5. Let the random variable X be the position number where Gloria is placed. List the possible values of X and determine the number of different arrangements associated with each value.

35. The Green and Gold teams meet in a play-off. When a team wins two games, that team wins the play-off. Let the random variable X be the number of games won by the Green team. List the possible values of X and determine the possible number of sequences of games associated with each value.

36. Lindstrom follows the price of Solartech stock each day to see whether it goes up in price or not.
   (a) Let the random variable X be the number of days the price goes up in a certain three-day period. For each value of X, determine the number of different outcomes that are associated with that value.
   (b) Let the random variable X be the number of days the price goes up in a certain five-day period. For each possible value of X, find the number of different outcomes associated with that value.

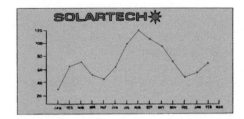

37. A shopping mall contains a number of stores appealing to a variety of customers. The manager of a store views each shopper in terms of what the shopper might buy. The manager of a pet shop might view a shopper in terms of the kind of pet he or she might buy. The manager of a bookstore thinks of the number of books a shopper might buy. You ask the manager of each store in the mall to define a random variable that assigns a number to each shopper. How might the manager of each of the following stores assign a number that reflects the kind of store managed?
   (a) A diet clinic
   (b) A men's clothing store
   (c) A women's shoe shop
   (d) A dental clinic

38. A box contains six green balls and four red balls. A ball is randomly drawn. If it is a red ball, the process stops. If the ball is green, it is replaced and another ball is drawn. The process continues until a red ball is drawn. Let the random variable X be the number of balls drawn. Give the possible values of X.

39. *(See Example 9)* A deck contains four red cards and four black cards. Cards are randomly drawn, without replacement, until a card drawn matches the color of the first card. Let the random variable X be the number of cards drawn. Find the probability distribution of X.

40. Cards are randomly drawn without replacement from a deck of five red and five black cards until a card drawn matches the color of the first card. The number of cards drawn is the random variable X. Find the probability distribution of X.

41. A coin is tossed repeatedly up to three times until a head appears. Even though a head might not appear in the first three times, the process ends with the third toss. Let the number of tosses be the random variable X. Determine the probability distribution of X.

42. A coin is tossed repeatedly until a tail appears. Let the number of tosses be the random variable X. Find the probability distribution of X.

43. Teams A and B meet in a play-off. The first team that wins two games wins the play-off. Let X be the number of games won by team A. For any given game the probability that team A wins is 0.6. Find the probability distribution of X.

44. On any given day, the stock of Hightech Corp. has a probability $\frac{2}{3}$ of going up in price. Two days are selected at random, and the prices are observed. Let the random variable X be the number of days the stock went up. Find the probability distribution of X.

EXPLORATIONS

45. Cards are drawn, without replacement, from a deck of six red and six black cards until two consecutive cards drawn are of the same color. The number of cards drawn is the random variable X. Find the probability distribution of X.

8.5   Expected Value of a Random Variable

- Expected Value
- Variance and Standard Deviation of a Random Variable

If we are honest, most of us will admit to having fantasized about getting rich quick. Some will act on the fantasy by buying lottery tickets, by gambling, or by purchasing merchandise so that we can win a magazine's sweepstakes award. Do you ever ask yourself how much you are willing to pay for the *chance* to become rich quick? Would you pay $1 for the chance to win $1 million? Millions of people would and do by purchasing lottery tickets. Would you pay $5000 for a chance to win $1 million? Far fewer will. In this section, we study one way to judge the worth of a chance to win money, in large or small amounts, and to determine the expected long-term performance of an activity. The measure we study is called **expected value.** We analyze a hypothetical activity to develop the concept of expected value.

Expected Value

Each day a student puts 50¢ in a candy machine to buy a 35¢ candy bar. She observes three possible outcomes:

1. She gets a candy bar and 15¢ change. This happens 80% of the time.
2. She gets a candy bar and no change. This happens 16% of the time.
3. She gets a candy bar and the machine returns her 50¢. This happens 4% of the time.

Over a period of time, what is the average cost of a candy bar?

The three possible costs, 35¢, 50¢, and 0¢, have a mean of 28.3¢. This is *not* the average cost, because it assumes that 35¢, 50¢, and 0¢ occur equally often, whereas, in fact, 35¢ occurs more often than the other two. If the machine behaved this way for 500 purchases, we would expect the following:

Cost (X)	Frequency (per 500)	Probability (P[X])
35¢	400	0.80
50¢	80	0.16
0¢	20	0.04
	500	1.00

This presents the information as a frequency table, so we can compute the mean as we have before with a frequency table (Section 8.3). The mean is

$$\frac{400 \times 35 + 80 \times 50 + 20 \times 0}{500} = \frac{14{,}000 + 4{,}000 + 0}{500} = 36$$

so the average cost is 36¢.

Let's write the mean in another way:

$$\frac{400 \times 35 + 80 \times 50 + 20 \times 0}{500} = \frac{400}{500} \times 35 + \frac{80}{500} \times 50 + \frac{20}{500} \times 0$$
$$= 0.80 \times 35 + 0.16 \times 50 + 0.04 \times 0$$

This last expression is simply the sum obtained by adding each cost of a candy bar times the probability that cost occurred; that is,

$$P(35) \times 35 + P(50) \times 50 + P(0) \times 0$$

This illustrates the procedure used to compute the mean when each value occurs with a specified probability. This mean is called the **expected value;** it represents the long-term mean of numerous trials. Although a few trials likely would not average to the expected value, a larger and larger number of trials will tend to give a mean closer to the expected value.

We emphasize that the term *expected value* does not mean the value we expect in an everyday sense. In the candy bar example, the amount paid for a candy bar is either 0¢, 35¢, or 50¢, whereas the expected value is 36¢. As 36¢ is never the amount paid, you never expect to pay that. However, 36¢ is the average amount paid over a large number of purchases. This long-term average is called the expected value.

We now make a formal definition of expected value in terms of a random variable and probability distribution.

DEFINITION
Expected Value

If X is a random variable with values $x_1, x_2, \ldots, x_n$ and corresponding probabilities $p_1, p_2, \ldots, p_n$, then the **expected value** of X, $E(X)$, is

$$E(X) = p_1 x_1 + \cdots + p_n x_n$$

EXAMPLE 1 ➤

Find the expected value of X, where the values of X and their corresponding probabilities are given by the following table:

x_i	2	5	9	24
p_i	0.4	0.2	0.3	0.1

SOLUTION

$$\begin{aligned} E(X) &= 0.4 \times 2 + 0.2 \times 5 + 0.3 \times 9 + 0.1 \times 24 \\ &= 0.8 + 1.0 + 2.7 + 2.4 \\ &= 6.9 \end{aligned}$$

■ **Now You Are Ready to Work Exercise 1**

EXAMPLE 2 ➤

(a) A contestant tosses a coin and receives $5 if heads appears and $1 if tails appears. What is the expected value of a trial?

(b) A contestant receives $4 if a coin turns up heads and pays $3 if it turns up tails. What is the expected value?

(c) A contestant receives $5 if a coin turns up heads and pays $5 when the coin turns up tails. What is the expected value?

SOLUTION

(a) The probability of receiving $5 is $\frac{1}{2}$ (the probability of tossing a head) and the probability of receiving $1 is $\frac{1}{2}$ (the probability of tossing a tail). Then,

$$E(X) = \frac{1}{2}(5) + \frac{1}{2}(1) = \$3$$

(b) Likewise, the probability of receiving $4 is $\frac{1}{2}$, and the probability of paying $3 is $\frac{1}{2}$. So,

$$E(X) = \frac{1}{2}(4) + \frac{1}{2}(-3) = \$2.00 - \$1.50 = \$0.50$$

(c) Because the probability of receiving $5 is $\frac{1}{2}$ and the probability of paying $5 is $\frac{1}{2}$, the expected value is

$$\frac{1}{2}(5) + \frac{1}{2}(-5) = \$0$$

■ **Now You Are Ready to Work Exercise 5**

We call a game **fair** whenever the expected value is zero. Thus, the game in part (c) is fair and the games in parts (a) and (b) are not fair. In a fair game, a player's expected winnings and losses are equal.

EXAMPLE 3 ➤   An IRS study shows that 60% of all income tax returns audited have no errors; 6% have errors that cause overpayments averaging $25; 20% have minor errors that cause underpayments averaging $35; 13% have more serious errors averaging $500 underpayments; and 1% have flagrant errors averaging $7000 underpayment. If the IRS selects returns at random:

(a) What is the average amount per return that is owed to the IRS, that is, what is the expected value of a return selected at random?

(b) How much should the IRS expect to collect if one million returns are audited at random?

(c) If the budget for the Audit Department is $15 million, how many returns must they examine to collect enough to cover their budget expenses?

SOLUTION

(a) Because an underpayment error eventually results in additional money paid to the IRS and a taxpayer overpayment is money paid back by the IRS, we use positive values for underpayment and negative for overpayment. The probability in each case is the fraction of returns with that type of error.

$$\begin{aligned} E(X) &= 0.60(0) + 0.06(-25) + 0.20(35) + 0.13(500) + 0.01(7000) \\ &= 140.5 \end{aligned}$$

Thus, the IRS expects to collect $140.50 for each return selected.

(b) If one million returns are selected at random, they may expect to collect 140.50×1 million $= \$140.5$ million.

(c) As the average amount they expect to collect is $140.50 per return audited, they must examine

$$15,000,000/140.50 = 106,762 \text{ returns}$$

■ **Now You Are Ready to Work Exercise 9**

The next example illustrates that the expected value need not be an amount of money. It can be any "payoff" associated with each outcome.

EXAMPLE 4 ➤ A tray of electronic components contains nine good components and three defective components. If two components are selected at random, what is the expected number of defective components?

SOLUTION

Let the random variable X be the number of defective components selected. X can have the value 0, 1, or 2. We need the probability of each of those numbers:

$$P(0) = \text{probability of no defective (both good)}$$
$$= \frac{C(9,2)}{C(12,2)} = \frac{36}{66} = \frac{12}{22}$$

$$P(1) = \text{probability of one good and one defective}$$
$$= \frac{C(9,1)C(3,1)}{C(12,2)} = \frac{27}{66} = \frac{9}{22}$$

$$P(2) = \text{probability of two defective}$$
$$= \frac{C(3,2)}{C(12,2)} = \frac{3}{66} = \frac{1}{22}$$

The expected value is

$$E(X) = \frac{12}{22}(0) + \frac{9}{22}(1) + \frac{1}{22}(2) = \frac{11}{22} = \frac{1}{2}$$

so the expected number of components is $\frac{1}{2}$. Clearly, you don't expect to get half of a component. The value $\frac{1}{2}$ simply says that if a large number of selections are made, you will *average* one half each time. You expect to get no defectives a little less than half the time and either one or two the rest of the time, but the average will be one half.

■ **Now You Are Ready to Work Exercise 15**

Variance and Standard Deviation of a Random Variable

The expected value is a "central tendency." As illustrated by the vending machine example at the beginning of this section, the expected value is a long-term average, or mean, of the values of a random variable, taking the probability of their occurrence into consideration. Think back on the mean and median; we shouldn't be surprised that sometimes the dispersion, or spread, of a random variable is needed to give more information. We can find the **variance of a random variable,** and it too measures the dispersion, or spread, of a random variable from the mean (expected value). The greater dispersion gives a greater variance. The histogram of the probability distribution shown in Figure 8–11a has a smaller variance than that shown in Figure 8–11b.

Let's use the candy bar example to illustrate variance and standard deviation. The information is given in the form of a frequency distribution:

Amount Paid	Frequency
35¢	400
50¢	80
0¢	20
	500

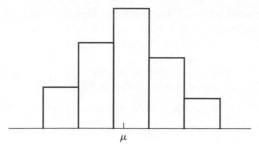

These scores are clustered closer to the mean than those in (b).

(a)

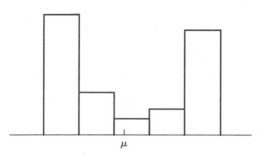

These scores have greater dispersion than those in (a).

(b)

We compute the standard deviation using the method shown in Example 3 of Section 8.3 for frequency distributions:

$$\text{Variance} = \frac{400(35 - 36)^2 + 80(50 - 36)^2 + 20(0 - 36)^2}{500}$$

We rewrite this expression as

$$\begin{aligned}
\text{Variance} &= \frac{400(35 - 36)^2}{500} + \frac{80(50 - 36)^2}{500} + \frac{20(0 - 36)^2}{500} \\
&= \frac{400}{500}(35 - 36)^2 + \frac{80}{500}(50 - 36)^2 + \frac{20}{500}(0 - 36)^2 \\
&= 0.80(35 - 36)^2 + 0.16(50 - 36)^2 + 0.04(0 - 36)^2
\end{aligned}$$

This last expression is of the form

$$\text{Variance} = P(35)(35 - \mu)^2 + P(50)(50 - \mu)^2 + P(0)(0 - \mu)^2$$

where μ is 36. Let's complete the computations for variance and standard deviation:

$$\begin{aligned}
\text{Variance} &= 0.80(-1)^2 + 0.16(14)^2 + 0.04(-36)^2 \\
&= 0.80 + 31.36 + 51.84 \\
&= 84
\end{aligned}$$
$$\text{Standard deviation} = \sigma = \sqrt{84} = 9.165$$

Let's give the formal definition of the variance and standard deviation of a random variable and then give an example showing the computation. We denote the variance by $\sigma^2(X)$ and let the Greek letter μ (mu) represent the mean (expected value).

DEFINITION
Variance, Standard Deviation

If X is a random variable with values $x_1, x_2, \ldots, x_n$, corresponding probabilities $p_1, p_2, \ldots, p_n$, and expected value $E(X) = \mu$, then

$$\text{Variance} = \sigma^2(X) = p_1(x_1 - \mu)^2 + p_2(x_2 - \mu)^2 + \cdots + p_n(x_n - \mu)^2$$
$$\text{Standard deviation} = \sigma(X) = \sqrt{\text{variance}}$$

EXAMPLE 5 ➤

Find the variance and standard deviation for the random variable defined by the following table:

x_i	4	7	10	8
p_i	0.2	0.2	0.5	0.1

SOLUTION

Set up the computations in the following way:

x_i	p_i	$p_i x_i$	$x_i - \mu$	$(x_i - \mu)^2$	$p_i(x_i - \mu)^2$
4	0.2	0.8	-4	16	3.2
7	0.2	1.4	-1	1	0.2
10	0.5	5.0	2	4	2.0
8	0.1	0.8	0	0	0
		$\mu = 8.0$			$\sigma^2(X) = 5.4$

The variance of $\sigma^2(X) = 5.4$. So the standard deviation is

$$\sigma(X) = \sqrt{5.4} = 2.32$$

■ **Now You Are Ready to Work Exercise 11**

8.5 EXERCISES

LEVEL 1

1. *(See Example 1)* Find the expected value of X for the following probability distribution:

x_i	3	8	15	22
p_i	0.3	0.2	0.1	0.4

2. Find $E(X)$ for the following probability distribution:

x_i	2	4	-1	6	10
p_i	0.15	0.08	0.17	0.35	0.25

3. Find $E(X)$ for the following probability distribution:

x_i	150	235	350	410	480
p_i	0.2	0.2	0.2	0.2	0.2

4. Find $E(X)$ for the following probability distribution:

x_i	0	10
p_i	0.7	0.3

5. *(See Example 2)*   A contestant receives $8 if a coin turns up heads and $2 if it turns up tails. What is the expected value of a trial?

6. A contestant rolls a die and receives $5 if it is a 1 or 2, receives $15 if it is a 3, and pays $2 if it is a 4, 5, or 6. What is the expected value of each trial?

7. A game consists of tossing a coin twice. A player who throws the same face, heads or tails, twice wins $1. How much should the organizers charge to enter the game
   (a) if they want to break even?
   (b) if they want to average $1 per person profit?

8. A game involves throwing a pair of dice. The player will win, in dollars, the sum of the numbers thrown. How much should the organizers charge to enter the game if they want to break even?

9. *(See Example 3)*   An auto repair shop's records show that 15% of the cars serviced need minor repairs averaging $20, 65% need moderate repairs averaging $130, and 20% need major repairs averaging $700.
   (a) What is the expected cost of repair of a car selected at random?
   (b) The shop has 125 cars scheduled for repair. What is the total expected repair cost of these cars?

10. The Chamber of Commerce estimates that of the visitors who come to the West Fest, 35% will spend an average of $150, 25% will spend $250, 30% will spend $350, and 10% will spend $500.
    (a) Find the expected expenditure of a visitor.
    (b) If 30,000 visitors attend the West Fest, what is their expected total expenditure?

11. *(See Example 5)*   Compute μ, $\sigma^2(X)$, and $\sigma(X)$ for the random variable defined as follows:

x_i	100	140	210
p_i	0.4	0.5	0.1

12. Compute μ and $\sigma(X)$ for the following random variable:

x_i	-1	2	3	4	10
p_i	0.1	0.2	0.1	0.3	0.3

13. Compute μ and $\sigma(X)$ for the following random variable:

x_i	10	30	50	90
p_i	0.4	0.2	0.3	0.1

14. Compute the expected value (μ), the variance $[\sigma^2(X)]$, and standard deviation $[\sigma(X)]$, for the random variable defined by the following table:

x_i	5	10	14	20
p_i	0.1	0.1	0.5	0.3

LEVEL 2

15. *(See Example 4)*   Of 15 windup toys on a sale table, 4 are defective. If 2 toys are selected at random, find the expected number of defective toys.

16. Community Bank studied the transaction times at the tellers' windows. They found that, rounded to the nearest minute, 15% of the transactions took one minute, 30% took two minutes, 25% took three minutes, 20% took four minutes, and 10% took five minutes.
    (a) Find the expected time of a random transaction.
    (b) If the bank has 150 customers per hour, how many tellers should be available?

17. A group of eight businesspeople meets for lunch every Tuesday. Each one puts $1 in the pot, and at the end of lunch, one of their business cards is drawn and that person wins the pot. Find a person's expected value.

18. Another group of eight businesspeople meets every Wednesday for lunch. Each one puts $1 in the pot, and at the end of lunch, business cards are drawn one at a time until one card is left and that person wins the pot. Find a person's expected value.

19. Herbert plays the lottery at the Riverboat Casino by paying $1 for a ticket and selecting five different numbers from the numbers 1 through 25. If his numbers are the same, not necessarily in the same order, as those drawn by the casino, he wins $1000. Find Herbert's expected value.

20. Ray buys a ticket for the Beach Front Casino's lottery only when his expected value is at least as much as he pays for the ticket. In such a case, he buys a $1 ticket and chooses four different numbers from the numbers 1 through 20. He wins if his numbers are the same, not necessarily in the same order, as the four drawn by the casino. Find the value of the pot necessary for Ray to buy a $1 ticket.

LEVEL 3

21. A company considers two business ventures. The executives believe that venture A has a 0.60 probability of a $50,000 profit, a 0.30 probability of breaking even, and a 0.10 probability of a $70,000 loss. Venture B has a 0.55 probability of $100,000 profit and 0.45 probability of a $60,000 loss. Determine the expected value of each venture. Which venture appears to be more profitable?

22. In one city, an insurance company insures 10,400 houses with wood shingle roofs, 28,700 houses with composition roofs, and 2350 houses with metal roofs. Past experience indicates that a severe hail storm will significantly damage a roof, with the following probabilities:

Type of Roof	Probability of Damage
Wood shingle	0.45
Composition	0.35
Metal	0.15

A severe hail storm strikes the city. Estimate the number of insured houses that will experience significant damage.

23. Air South overbooks as many as five passengers per flight because some passengers with reservations do not show up for the flight. The airline's records indicate the following probabilities that it will be overbooked at flight time:

Number overbooked at flight time	0	1	2	3	4	5
Probability of number overbooked	0.75	0.10	0.06	0.04	0.04	0.01

Find the expected number over for each flight.

24. A mathematics exam consists of six problems. The probability of a student working each problem is known from records kept on similar problems given previously. The point value and probability of working each problem are as follows:

Problem	Points	Probability of Working Problem
1	10	0.90
2	10	0.85
3	15	0.70
4	20	0.75
5	20	0.55
6	25	0.65

Find the expected number of points on the exam.

25. Ten cards of a children's game are numbered with all possible pairs of two different numbers from the set $\{1, 2, 3, 4, 5\}$. A child draws a card, and the random variable is the score of the card drawn. The score is 5 if the two numbers on the card add to 5; otherwise, the score is the smaller number on the card.
  (a) Find the expected score of a card.
  (b) Find the standard deviation of X.

26. In a game you toss a coin until a head turns up, but you get four tosses at most. When the first head occurs, you stop and receive payment as follows:

Toss When First Head Occurs	Payoff
1	$ 2
2	$ 4
3	$ 8
4	$16

If no head occurs in four tosses, you receive nothing.
  (a) Find the expected value of the game.
  (b) If you pay $5 to play this game, what are your expected winnings or losses?

27. Each time Danny, a Little League baseball player, comes to bat, the probability of his getting a hit is 0.4. His dad promises to pay him $2 if he gets one hit in a game, $4 for two hits, $6 for three hits, and $8 for four hits. What are his expected earnings in a game if he bats twice?

28. Four red cards and four green cards are well shuffled. Cards are drawn, without replacement, until the color of a card drawn matches the color of the first card drawn. Find the average number of cards drawn until a match occurs.

29. A plant manufactures microchips, 5% of which are defective. The plant makes a profit of $18 on each good microchip and loses $23 on each defective one.
  (a) What is the expected profit on a microchip?
  (b) The plant plans a production run of 150,000 microchips. What is the expected profit for the run?

EXPLORATIONS

30. Obtain a brochure for a free giveaway contest, such as the *Reader's Digest* Sweepstakes. The fine print should state the chances of winning each prize. Compute the expected value of an entry.

31. A casino plans a game of chance with the following characteristics:
(a) The probability that a player wins is 0.3.
(b) A player pays $5 if he loses.
(c) A player receives an amount, x, if he wins. Find the amount of the payoff, x, if the house averages a profit of $0.20 per player.

32. Hawaiian Airlines conducted a "Dreams in Paradise" promotion. Each person on a Hawaiian Airlines flight received a game card with three symbols hidden under a silver coating. A passenger who found three identical symbols under the coating received an award ranging from a Hawaiian Airlines luggage tag to a condo in Hawaii two weeks a year for 20 years.

Here are the awards, their value, and the number given away:

Award	Value	Number Given
Hawaiian condo	$22,400	4
Honda Civic car	$10,300	2
Trip for two to Tahiti	$ 1,500	2
One-month interisland commuter pass	$ 799	20
West Coast–Honolulu trip	$ 520	100
Interisland round trip	$ 100	200
Eight gallons of Shell gas	$ 12	600
Big Mac or Egg McMuffin	$ 1.99	15,000
Hawaiian Airlines luggage tag	$ 1.00	100,000

The number of game cards distributed was 1,200,000. Find the expected value of a game card.

33. A roulette wheel has 38 spaces of which 18 are red, 18 are black, and 2 are green. Most commonly, a player bets on the red or on the black. A player who makes such a bet pays the "house" $1.00 and upon winning gets the $1.00 back plus another $1.00 from the house. The house keeps the losing player's dollar. So, the net result is that a player who loses a bet pays $1.00 and a player who wins a bet receives $1.00.

(a) Find the expected value of a bet on red or on black.
(b) Does this game favor the house or the player?
(c) Find the amount a winning player should receive for the game to be fair (an expected value of zero).
(d) A player bets $1.00 on green. What should the house pay for a winning bet for the game to be fair?

34. A state lottery draws winning numbers every Wednesday and Saturday nights. Six different numbers are randomly drawn from the numbers 1 through 50. A player selects six different numbers and pays $1.00 for a ticket. If the player's six numbers match the six drawn, the player receives the "pot" or shares it equally with other players who also selected the winning number.

(a) How many possible combinations of six numbers are possible?
(b) What is the probability of selecting the six winning numbers?
(c) If the pot is $1 million, find the expected value of a ticket.
(d) Ray buys a lottery ticket only if the expected value is $1.00 or more. How large should the pot be for Ray to buy a ticket?
(e) If no one selects the winning numbers, the pot accumulates into the next game. After a sequence of games with no winners, the pot rose to $57 million.
  (i) Find the expected value of a ticket.
  (ii) An Australian syndicate decides to buy a ticket for every possible combination of numbers and thereby win the lottery. Because the lottery is drawn twice a week they must buy all the tickets in three days. Assume a person can buy one ticket a second and does so nonstop for three days. How many people will the syndicate need in order to buy all possible tickets in three days?
  (iii) If the syndicate manages to buy all possible combinations of numbers, will they make a profit? Why or why not?

Matrix multiplication can be used to calculate expected value. For example, we illustrate with the following values of the random variable and the corresponding probabilities.

x_i	5	2	4	1
p_i	0.3	0.1	0.2	0.4

Write the x_i values in a 1×4 row matrix, $[5 \quad 2 \quad 4 \quad 1]$, and the probabilities in a column matrix, (4×1),

$$\begin{bmatrix} 0.3 \\ 0.1 \\ 0.2 \\ 0.4 \end{bmatrix}. \text{ The matrix product } [5 \quad 2 \quad 4 \quad 1]\begin{bmatrix} 0.3 \\ 0.1 \\ 0.2 \\ 0.4 \end{bmatrix} = [2.9]$$

is a 1×1 matrix whose entry is the expected value 2.9.

Use matrix multiplication to find the expected value in Exercises 35 through 38.

35.

x_i	4	8	10	6	3
p_i	0.15	0.10	0.25	0.30	0.20

36.

x_i	−5	10	15	−20
p_i	0.40	0.20	0.25	0.15

37.

x_i	140	150	−75	−50	200	−250
p_i	0.12	0.08	0.24	0.16	0.22	0.18

38.

x_i	440	−370	215	144
p_i	0.27	0.19	0.32	0.22

39. Teams A and B compete in a play-off with the winner being the team that wins two out of three games. Team A can win by any of the following sequences, where A denotes a win by team A and B denotes a win by team B:

> Two games: AA
>
> Three games: BAA or ABA

The sequences that indicate how B can win the play-off are obtained by interchanging A and B.

The number of different sequences possible in the playoff is:

> Two-game series, two ways
>
> Three-game series, four ways.

(a) If A and B are evenly matched (the probability that each wins a game is 0.5), find the expected number of games in the play-off.

(b) If A is the better team with a probability of 0.6 of winning each game, find the expected number of games in the play-off.

40. Teams A and B compete in a play-off with the team that wins three out of five games being the winner. Note that the last game played is won by the team that wins the play-off.
(a) List the different sequences of games that are possible when team A wins the play-off.
(b) Find the expected number of games in the play-off if the teams are evenly matched.
(c) Find the expected number of games if A has probability of 0.6 of winning each game.

41. Observe in Exercise 39 that team A can win the play-off (two out of three) by:

> Two games in $C(1, 1)$ ways
>
> Three games in $C(2, 1)$ ways

Team B can win in the same number of ways.
    In Exercise 40, observe that team A can win three games out of five by:

> Three games in $C(2, 2)$ ways
>
> Four games in $C(3, 2)$ ways
>
> Five games in $C(4, 2)$ ways

(a) An NBA play-off is won by the team that wins four out of seven games. Find the number of ways team A can win the play-off.
(b) If teams A and B are evenly matched, find the expected number of games in the play-off.

42. Find the expected value of the probability distribution

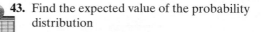

x	4	5	6	7	8
p	0.21	0.15	0.32	0.10	0.22

43. Find the expected value of the probability distribution

x	15	18	21	24
p	0.20	0.25	0.30	0.25

8.6   Binomial Experiments and Binomial Distribution

- Binomial Trials
- Probability of a Binomial Experiment
- Justification of the Binomial Experiment Formula
- Binomial Distribution

Binomial Trials

We have studied the probability of an event occurring in an experiment. We now look at a sequence of experiments to determine probabilities related to the sequence. The experiments involved are not arbitrary; they have specific characteristics like those found in the problems "What is the probability that heads appears seven times in ten tosses of a coin?" or "If you guess at the answers of 15 multiple-choice questions, what are your chances of a passing grade?"

These problems are examples of a certain type of probability problem, **binomial trials.** Such problems involve **repeated trials** of an experiment with only two possible outcomes: heads or tails, right or wrong, yes or no, and so on. We classify the *two outcomes* as success or failure.

To classify an experiment as a binomial trial experiment, several properties must hold.

Binomial Experiment

> **Binomial Experiment with n Trials**
>
> 1. The experiment is repeated a fixed number of times (n times).
> 2. Each trial has only two possible outcomes: success and failure. The possible outcomes are exactly the same for each trial.
> 3. The probability of success remains the same for each trial. (We use p for the probability of success and $q = 1 - p$ for the probability of failure.)
> 4. The trials are independent. (The outcome of one trial has no influence on later trials.)
> 5. We are interested in the total number of successes, not the order in which they occur. There may be $0, 1, 2, 3, \ldots,$ or n successes in n trials.

EXAMPLE 1 ➤

(a) We are interested in the number of times heads occurs when a coin is tossed eight times. Each toss of the coin is a trial, so there are eight repeated trials ($n = 8$). We consider the outcome *heads* a success and *tails* a failure. The probability of success (heads) on each trial is $p = \frac{1}{2}$, and the probability of failure (tails) is $1 - \frac{1}{2} = \frac{1}{2}$. This is an example of a binomial trial.

(b) A student guesses at all the answers on a ten-question multiple-choice quiz (four choices of an answer on each question). This fulfills the properties of a binomial trial because:

1. Each guess is a trial ($n = 10$).
2. There are two possible outcomes: correct and incorrect.
3. The probability of a correct answer is $\frac{1}{4}\left(p = \frac{1}{4}\right)$, and the probability of an incorrect answer is $\frac{3}{4}\left(q = \frac{3}{4}\right)$, on each trial.
4. The guesses are independent because guessing an answer on one question gives no information on other questions.

(c) Suppose eight cards are drawn from a deck and none are replaced. We are interested in the number of spades drawn. This is *not* a binomial trial because the trials (selecting a card) are not independent (the first card drawn affects the possible choices of the second card; consequently, the conditional probability of drawing a spade changes each time a card is removed).

(d) Suppose a card is drawn from a deck, the card is noted and placed back in the deck, and the deck is shuffled. This is repeated eight times. We are interested in the number of times a spade is drawn.

This experiment is a binomial trial because each trial is the same, the trials are independent, and the probability of obtaining a spade remains the same for each trial.

Technically, each trial has 52 outcomes, each card in the deck. However, we reduce the outcomes to the two possible outcomes "success" or "failure" when we collect all spades into the event defined as success and all other cards into the event defined as failure. Then, $p = \frac{13}{52}$ and $q = \frac{39}{52}$. ■

The following example illustrates how a tree diagram may be used to count the number of successes when the number of trials is small.

EXAMPLE 2 ➤   A coin is tossed three times. What is the probability of exactly two heads in the three tosses?

SOLUTION

Look at the paths of Figure 8–12 that begin at 0 and terminate at the end of a branch. There are a total of eight paths, and three of them contain exactly two heads. The probability of terminating at the end of those branches is $\frac{1}{8}$ for each one. (Because the probability of each branch is $\frac{1}{2}$, the probability of taking a sequence of three paths is $\frac{1}{2} \times \frac{1}{2} \times \frac{1}{2} = \frac{1}{8}$.) So, the probability of exactly two heads in three tosses of a coin is $\frac{3}{8}$.

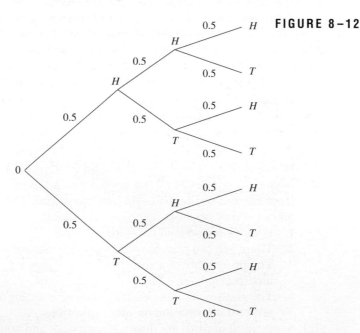

FIGURE 8–12

■ **Now You Are Ready to Work Exercise 25**

Probability of a Binomial Experiment

The tree diagram of a binomial trial problem can become unwieldy with a large number of trials, so an algebraic formula is useful for computing the probability of a specified number of successes. Let's start with a simple example.

EXAMPLE 3 ➤

A quiz has five multiple-choice questions with four possible answers to each. A student wildly guesses the answers. What is the probability that he guesses exactly three correctly?

SOLUTION

As a binomial trial problem, $n = 5$, $p = \frac{1}{4}$, and $q = \frac{3}{4}$. We are about to give a formula that calculates the desired probability, but you need to be aware that you have no reason, at this point, to know why it is true. That will be explained later. We want you to understand the quantities used in the formula so that you can more easily follow the justification. Now for the formula. The probability of exactly three correct answers is

$$P(3 \text{ correct}) = C(5, 3)\left(\frac{1}{4}\right)^3\left(\frac{3}{4}\right)^2$$

$$= 10\left(\frac{1}{64}\right)\left(\frac{9}{16}\right)$$

$$= 0.088 \quad \text{(Rounded to three decimals)}$$

■ **Now You Are Ready to Work Exercise 28**

Let's make some observations about the computation, because they will hold for the general formula.

1. In $C(5, 3)$, 5 is the number of trials, and 3 is the number of successes. (In general, in $C(n, x)$, n is the number of trials, and x is the number of successes.)
2. In $\left(\frac{1}{4}\right)^3$, $\frac{1}{4}$ is the probability of success in a single trial, and 3 is the number of successes in the five trials. (In general, in p^x, p is the probability of success in a single trial, and x is the number of successes in n trials.)
3. In $\left(\frac{3}{4}\right)^2$, $\frac{3}{4}$ is the probability of failure in a single trial, and 2 is the number of failures in five trials. It may seem trivial, but note that $2 = 5 - 3$; the number of failures equals the number of trials minus the number of successes. (In general, in $(1 - p)^{n-x}$, $1 - p$ is the probability of failure in a single trial, and $n - x$ is the number of failures in n trials.)

We now give you the general formula.

Probability of a Binomial Experiment

Given a binomial experiment and

   n independent repeated trials,

   p is the probability of success in a single trial,

   $q = 1 - p$ is the probability of failure in a single trial,

   x is the number of successes ($0 \leq x \leq n$).

   Then, the probability of x successes in n trials is

$$P(x \text{ successes in } n \text{ trials}) = C(n, x)p^x q^{n-x}$$

$P(k \text{ successes in } n \text{ trials})$ may be written $P(X = k)$.

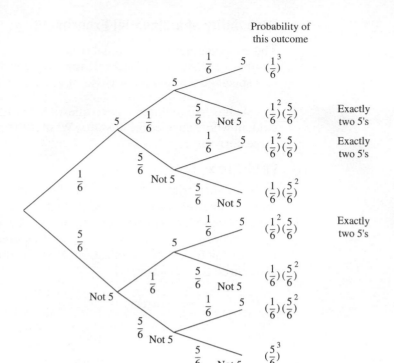

Probability of
this outcome

FIGURE 8-13 Exactly two 5's occurs three ways in three tosses of a die.

EXAMPLE 4 ➤ A single die is rolled three times. Find the probability that a 5 turns up exactly twice.

SOLUTION

$$n = 3, \quad x = 2, \quad p = \frac{1}{6}, \quad q = \frac{5}{6}$$

so

$$P(X = 2) = C(3, 2)\left(\frac{1}{6}\right)^2\left(\frac{5}{6}\right)^1$$
$$= 3\left(\frac{1}{36}\right)\left(\frac{5}{6}\right)$$
$$= 0.0694$$

Notice that in the tree diagram (Figure 8–13) there are three branches with exactly two 5's. This number is $C(3, 2)$, the number of sequences two 5's can occur in three rolls of the die.

The probability of termination at the end of one such branch is

$$\left(\frac{1}{6}\right)^2\left(\frac{5}{6}\right)$$

The notation $P(X = 2)$ is used to indicate the probability of two successes.

■ **Now You Are Ready to Work Exercise 31**

EXAMPLE 5 ➤

A coin is tossed ten times. What is the probability that heads occurs six times?

SOLUTION

In this case, $n = 10$, $x = 6$, $p = \frac{1}{2}$, and $q = \frac{1}{2}$.

$$P(X = 6) = C(10, 6)\left(\frac{1}{2}\right)^6\left(\frac{1}{2}\right)^4$$
$$= 210\left(\frac{1}{64}\right)\left(\frac{1}{16}\right)$$
$$= 0.205$$

■ **Now You Are Ready to Work Exercise 34**

Justification of the Binomial Experiment Formula

Now let's go back to Example 3 and explain how to obtain the expression used to compute the probability of a binomial trial.

Recall the problem: A multiple-choice quiz has five questions with four possible answers to each, one of which is correct. A student guesses the answers. What is the probability that three of the five are correct?

Let's look at how the student succeeds in answering three of the five questions. There are several ways, and the exact number plays a key role in the solution. First, let's list some ways. We will list the sequence of successes (correct answers) and failures (incorrect answers). One sequence is

SSSFF

which indicates that the first three were correct and the last two were incorrect. Some other sequences of three correct and two incorrect are

SSFSF
FSSSF

and so on. Rather than list all the ways in which three correct and two incorrect answers can be given, let's compute the number.

The basic procedure for forming a sequence of three S's and two F's amounts to selecting the three questions with correct answers. The other two answers are automatically incorrect.

In how many different ways can three questions be selected from the five questions? You should recognize this as $C(5, 3)$. As $C(5, 3) = 10$, there are ten possible sequences of three S's and two F's. The student succeeds in passing if his sequence of guesses is any one of the ten sequences.

It helps that the probabilities of all ten sequences are exactly the same. Notice that the probability of SSSFF is

$$\left(\frac{1}{4}\right)\left(\frac{1}{4}\right)\left(\frac{1}{4}\right)\left(\frac{3}{4}\right)\left(\frac{3}{4}\right)$$

by the Multiplication Rule. Also, the probability of SSFSF is

$$\left(\frac{1}{4}\right)\left(\frac{1}{4}\right)\left(\frac{3}{4}\right)\left(\frac{1}{4}\right)\left(\frac{3}{4}\right)$$

Both of these are simply

$$\left(\frac{1}{4}\right)^3\left(\frac{3}{4}\right)^2$$

NOTE

Add the probabilities of each of the ten sequences to obtain the probability that the student's sequence of answers is one of these ten.

each written in different order. In fact, the probability of *any* sequence of three S's and two F's will contain $\frac{1}{4}$ three times and $\frac{3}{4}$ twice, which gives $\left(\frac{1}{4}\right)^3\left(\frac{3}{4}\right)^2$.

When we add up the probabilities of the ten sequences, we are adding $\left(\frac{1}{4}\right)^3\left(\frac{3}{4}\right)^2$ ten times, which is

$$10\left(\frac{1}{4}\right)^3\left(\frac{3}{4}\right)^2$$

This is the probability obtained in Example 3. A similar situation holds in general for other values of n, x, p, and q.

EXAMPLE 6 ➤

Plantation Foods has found that 25% of those hired to load trucks will work more than one week. Find the probability that three of four new hires will work more than one week.

SOLUTION

This is an experiment with four repeated trials. A hire that works more than a week is considered a success.

$$n = 4, \qquad x = 3, \qquad p = 0.25, \qquad q = 0.75$$
$$P(X = 3) = C(4, 3)(0.25)^3(0.75)^1$$
$$= 4(0.0156)(0.75) = 0.0469$$

The probability that three of the four new hires will work more than a week is about 0.047.

■ **Now You Are Ready to Work Exercise 52**

EXAMPLE 7 ➤

Professor Purdue gives a multiple-choice quiz with five questions. Each question has four possible answers. A student guesses all answers. What is the probability that she passes the test if at least three correct answers are needed to pass?

SOLUTION

In this case, $n = 5$, $p = \frac{1}{4}$, and $q = \frac{3}{4}$. The student passes if she gets three, four, or five correct answers. We must find the probability of each of these outcomes and add them:

$$P(X = 3) = C(5, 3)(0.25)^3(0.75)^2$$
$$= 10(0.015625)(0.5625)$$
$$= 0.0879$$
$$P(X = 4) = C(5, 4)(0.25)^4(0.75)^1$$
$$= 5(0.0039062)(0.75)$$
$$= 0.0146$$
$$P(X = 5) = C(5, 5)(0.25)^5(0.75)^0$$
$$= 1(0.0009765)(1)$$
$$= 0.00098 \quad \text{(rounded)}$$

Then the probability of three or more correct, which we indicate by $P(X \geq 3)$, is
$$P(X \geq 3) = 0.0879 + 0.0146 + 0.00098 = 0.10348.$$

■ **Now You Are Ready to Work Exercise 49**

Binomial Distribution

We can use binomial trials to form an important probability distribution. This distribution is called a **binomial distribution.** Define it in the following way.

DEFINITION
Binomial Distribution

For a sequence of binomial experiments of n repeated trials, define a random variable X as the number of successes in n trials. For each value of x, $0 \leq x \leq n$, find the probability of x successes in n trials. The probability distribution obtained is the **binomial distribution.**

EXAMPLE 8 ➤

Form the binomial distribution for the experiment of rolling a die three times and counting the times a 4 appears.

SOLUTION

The random variable X takes on the values 0, 1, 2, and 3, the possible number of successes in three trials. The probability of each value occurring is computed by using binomial trials with $p = \frac{1}{6}$ and $q = \frac{5}{6}$. ($\frac{1}{6}$ is the probability of rolling a 4 in a single trial.)

X	$P(X)$
0	$C(3, 0)(\frac{5}{6})^3 = 0.5787$
1	$C(3, 1)(\frac{1}{6})(\frac{5}{6})^2 = 0.3472$
2	$C(3, 2)(\frac{1}{6})^2(\frac{5}{6}) = 0.0694$
3	$C(3, 3)(\frac{1}{6})^3 = 0.0046$

■ **Now You Are Ready to Work Exercise 19**

The binomial distribution got its name from the binomial $(p + q)^n$. Each term of the expansion of the binomial gives one of the probabilities in the binomial distribution. Notice that

$$(p + q)^3 = p^3 + 3p^2q + 3pq^2 + q^3$$

which can be written

$$C(3, 3)p^3q^0 + C(3, 2)p^2q + C(3, 1)pq^2 + C(3, 0)p^0q^3$$

where each term represents the probability of 3, 2, 1, or 0 successes, respectively, in a binomial experiment with three trials.

EXAMPLE 9 ➤

Find the binomial distribution for $n = 4$ and $p = 0.3$.

SOLUTION

The random variable X may take on the values 0, 1, 2, 3, and 4. The probability distribution is the following.

X	$P(X)$
0	$C(4, 0)(0.7)^4 = 0.2401$
1	$C(4, 1)(0.3)(0.7)^3 = 0.4116$
2	$C(4, 2)(0.3)^2(0.7)^2 = 0.2646$
3	$C(4, 3)(0.3)^3(0.7) = 0.0756$
4	$C(4, 4)(0.3)^4 = 0.0081$

■ **Now You Are Ready to Work Exercise 15**

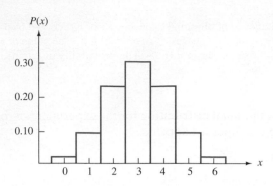

FIGURE 8-14   Probability distribution of the number of heads in six tosses of a coin.

EXAMPLE 10 ➤   Form the binomial distribution of the experiment of tossing a coin six times and counting the number of heads.

SOLUTION

The random variable X takes on the values 0, 1, 2, 3, 4, 5, and 6, the possible number of successes in six tosses. Both p and q are $\frac{1}{2}$. The values of X and the corresponding probabilities computed using binomial trials are the following:

X	$P(X)$
0	$C(6, 0)\left(\frac{1}{2}\right)^6 = 0.0156$
1	$C(6, 1)\left(\frac{1}{2}\right)^1\left(\frac{1}{2}\right)^5 = 0.0938$
2	$C(6, 2)\left(\frac{1}{2}\right)^2\left(\frac{1}{2}\right)^4 = 0.2344$
3	$C(6, 3)\left(\frac{1}{2}\right)^3\left(\frac{1}{2}\right)^3 = 0.3125$
4	$C(6, 4)\left(\frac{1}{2}\right)^4\left(\frac{1}{2}\right)^2 = 0.2344$
5	$C(6, 5)\left(\frac{1}{2}\right)^5\left(\frac{1}{2}\right) = 0.0938$
6	$C(6, 6)\left(\frac{1}{2}\right)^6 = 0.0156$

The histogram of the distribution is given in Figure 8–14.

■ **Now You Are Ready to Work Exercise 21**

8.6 EXERCISES

LEVEL 1

Compute the binomial trial probabilities in Exercises 1 through 12.

1. $C(5, 3)(0.35)^3(0.65)^2$

2. $C(6, 3)(0.45)^3(0.55)^3$

3. $C(10, 4)(0.4)^4(0.6)^6$

4. $C(12, 11)(0.7)^{11}(0.3)$

5. $C(12, 5)(0.65)^5(0.35)^7$

6. $C(9, 2)(0.2)^2(0.8)^7$

7. $n = 8, p = 0.25, x = 3$

8. $n = 7, p = \frac{1}{3}, x = 4$

9. $n = 5, p = 0.1, x = 4$

10. $n = 10, p = \frac{1}{5}, x = 6$

11. $P(X = 3)$ for $n = 5, p = \frac{1}{4}$

12. $P(X = 4)$ for $n = 8, p = \frac{2}{3}$

13. Fill in the following table for a binomial trial experiment with four trials and $p = 0.3$.

X	$P(X \text{ successes})$
0	
1	
2	
3	
4	

14. Fill in the following table for a binomial trial experiment with three trials and $p = 0.5$.

X	$P(X \text{ successes})$
0	
1	
2	
3	

15. *(See Example 9)* Find the binomial distribution for $n = 5$ and $p = 0.3$.

16. Find the binomial distribution for $n = 4$ and $p = 0.6$.

17. Form the binomial distribution of an experiment with $n = 5$ and $p = 0.4$.

18. Draw the probability histogram for the binomial distribution for $n = 4$ and $p = 0.2$.

19. *(See Example 8)* Form the binomial distribution for the experiment of rolling a die four times and counting the number of times a 2 appears.

20. Form the binomial distribution of rolling a die three times and counting the number of times a 6 appears.

21. *(See Example 10)* Form the binomial distribution of the experiment of tossing a coin four times and counting the number of heads. Draw the probability histogram.

22. A coin is tossed seven times, and the number of heads is counted. Determine the binomial distribution of the experiment.

23. Five cards are numbered 1 through 5. The cards are shuffled, and a card is drawn. The number drawn is noted and the card is replaced before the next shuffle. This is done four times, and a count is made of the number of times an odd number appears. Find the probability distribution and draw its histogram.

24. The probability that a new drug will cure a skin rash is 0.85. If the drug is administered to 500 patients with the skin rash, what is the probability that it will cure 400 of them? Set up but do not compute this probability.

25. *(See Example 2)* The probability that a marksman will hit a moving target is 0.7. Use a tree diagram to determine the probability that he will hit the target in two out of three attempts.

26. The probability that an instructor will give a quiz is 0.4. Use a tree diagram to determine the probability that she will give a quiz in two out of four class meetings.

27. A check of autos passing an intersection reveals that 80% of the drivers are wearing seat belts. Find the probability that three of the next five drivers are wearing seat belts.

28. *(See Example 3)* A multiple-choice quiz of four questions is given. Each question has five possible answers. If a student guesses at all the answers, find the probability that three out of four are correct.

29. If a couple, each with genes for both brown and blue eyes, parent a child, the probability that the child has blue eyes is $\frac{1}{4}$. Find the probability that two of the couple's three children will have blue eyes.

30. A certain drug was developed, tested, and found to be effective 70% of the time. Find the probability of successfully administering the drug to at least nine out of ten patients.

31. *(See Example 4)* A single die is rolled six times. Find the probability of a 3 turning up four times.

32. A die is tossed three times. Find the probability of throwing two 6's.

33. A die is thrown four times. Determine the probability of throwing:
    (a) zero 6's.   (b) one 6.
    (c) two 6's.   (d) three 6's.
    (e) four 6's.

34. *(See Example 5)* A coin is tossed seven times. Find the probability of its turning up tails five times.

35. A coin is tossed six times. Determine the probability of its landing heads up four times.

36. A coin is tossed four times. Find the probability of its landing on tails
    (a) zero times.   (b) once.
    (c) twice.   (d) three times.
    (e) four times.

Set up but do not compute the probability for each of the problems in Exercises 37 through 40.

37. The probability of 40 successes, where $n = 90$ and $p = 0.4$

38. The probability of 100 successes, where $n = 500$ and $p = 0.7$

39. $P(X = 35)$, where $n = 75$ and $p = 0.25$

40. $P(40 \leq X \leq 42)$, where $n = 100$ and $p = 0.35$

LEVEL 2

41. Ken is a waiter at the Elite Cafe. His customers tip him 15% or more 80% of the time. Find the probability that four of the next six customers will tip him 15% or more.

42. Five door prizes are to be awarded at a club meeting. There are 100 tickets in the box, and one of them is yours. A ticket is drawn, the prize is awarded, and the ticket is placed back in the box. This continues until all five prizes are awarded.
  (a) Find the probability your ticket is drawn once.
  (b) Find the probability your ticket is drawn twice.

43. JoJo tosses a fair coin five times. Find the probability of $0, 1, 2, \ldots, 5$ heads appearing in the five tosses and fill in the following probability table:

X	$P(X \text{ heads})$
0	
1	
2	
3	
4	
5	

44. Elaine works in the telephone sales department of Davis Publishing. Past records indicate that the probability a phone call will result in a sale is 0.02. If Elaine makes six calls, find the probability of $0, 1, 2, \ldots, 6$ sales and summarize in the following table:

X	$P(X \text{ sales})$
0	
1	
2	
3	
4	
5	
6	

45. The probability that the University Bookstore will buy back a used textbook for more than 40% of the purchase price is 0.4. Meka takes five books back at the end of the semester. Find the probability that she will receive more than 40% of the purchase price for three of them.

46. The most popular color for compact/sports cars is dark green, with 15.2% of owners preferring that color. If seven compact/sports car owners are selected at random, find the probability that four of them prefer dark green.

47. The probability that a new drug raises the HDL cholesterol of a patient by 20% is 0.6. A physician prescribes the drug to five patients. Find the probability that it will raise the HDL by 20% in at least three of the patients.

48. In Portugal, the literacy rate is 87%. If ten people are selected at random, find the probability that nine are literate.

LEVEL 3

49. *(See Example 7)* A single die is rolled five times. Find the probability that a 4 turns up at least three times.

50. In Nigeria, 45% of the population is younger than 15 years of age and 3% are older than 65.
  (a) If eight people are selected at random, find the probability that five are younger than 15.
  (b) If seven people are selected at random, find the probability that two are older than 65.

51. A coin is tossed eight times. Find the probability that it will land on tails at least five times.

52. *(See Example 6)* The Super Store's records show that 70% of checkers hired will remain at work more than one month. Find the probability that three of four newly hired checkers will work more than one month.

53. A true–false quiz has ten questions. A student guesses all the answers.
  (a) Find the probability of getting eight correct.
  (b) Find the probability of getting at least eight correct.
  (c) Find the probability of getting no more than two correct.

54. A coin is tossed four times. Is it true or false that the probability that it turns up heads twice is $\frac{1}{2}$?

55. A psychology student conducts an ESP experiment in which the subject attempts to identify the number that turns up when an unseen die is rolled. The die is rolled six times. On the basis of chance alone, find the probability that the subject identifies the number correctly four of the six times.

56. Brian makes 65% of his free throws in basketball.
  (a) Find the probability that he makes three out of five free throws.
  (b) Find the probability that he makes at least three out of five free throws.

57. A baseball player has a 0.360 batting average. Find the probability that he will have at least two hits in five times at bat.

58. It is estimated that a certain brand of automobile tire has a 0.8 probability of lasting 25,000 miles. Out of four such tires put on a car, find the probability that at least one will have to be replaced before 25,000 miles.

59. Now for a bit of baseball trivia from the distant past. Many baseball experts agree that Joe DiMaggio accomplished the greatest feat in the history of baseball. What was it?

In 1941 "Joltin' Joe" had a batting average of 0.357. (That was not his greatest feat).
  (a) Assume that DiMaggio goes to bat four times in a game. Find the probability that he has at least one hit in the game.
  (b) Find the probability that DiMaggio gets at least one hit in 56 consecutive games. (He did in 1941.) Assume he bats four times each game.

60. The probability that Prof. Ponder will give a quiz on any given day is 0.7. Find the probability of a quiz on six of the next ten classes.

61. Find the probability of exactly five heads in ten tosses of a coin. Is the result what you expected?

62. A single die is rolled 18 times. Find the probability that a 6 turns up exactly three times.

EXPLORATIONS

63. Dan has gotten hooked on playing solitaire on his computer. He finally decides he needs to limit the number of games he plays in a day. He observes that the probability of winning a game is 0.12, and he would like to win at least one game on three-fourths of the days he plays; that is, the probability of winning at least one game on any day is 0.75. Assume solitaire is a binomial trial; how many games should he plan to play per day? (*Hint:* If x = number of games won, $P(X \geq 1) = 1 - P(X = 0)$.)

64. Here is an experiment to compare theoretical probability with an empirical probability. Each student tosses a coin 20 times and records the number of times that the coin turns up heads. Repeat five times.
  (a) Based on the tosses, estimate the probability that a coin turns up heads 5, 6, or 7 times in 20 tosses.
  (b) Compute the binomial probability that a coin turns up heads 5, 6, or 7 times in 20 tosses.

65. Fill in the following tables for the indicated binomial trial experiment.
  (a) Four trials, $p = 0.6$

X	$P(X$ successes in four trials$)$
0	
1	
2	
3	
4	

  (b) Four trials, $p = 0.4$

X	$P(X$ successes in four trials$)$
0	
1	
2	
3	
4	

  (c) Five trials, $p = 0.5$

X	$P(X$ successes in five trials$)$
0	
1	
2	
3	
4	
5	

  (d) Eight trials, $p = 0.3$

X	$P(X$ successes in eight trials$)$
0	
1	
2	
3	
4	
5	
6	
7	
8	

66. Based on the preceding exercise, answer the following.

 (a) A coin is tossed five times. Find the probability that it turns up heads at least four times.

 (b) A coin is tossed five times. Find the probability that it turns up tails two or three times.

 (c) A binomial trial experiment table showing the probability of x successes in n trials for $x = 0, 1, 2, \ldots, n$ is a binomial distribution. How are the binomial distributions for $p = 0.4$, $q = 0.6$ and for $p = 0.6$, $q = 0.4$ related?

 (d) You have a binomial distribution table for a specific n, p, and q. Form a new binomial distribution for the same n, but with p and q interchanged. How is the second table related to the first?

67. The probability that a patient will respond to a medication is 0.65. The medication is administered to 220 patients. Find the probability that 75 respond.

68. One out of nine women in the United States will develop breast cancer during her lifetime. A group of 50 women are randomly selected. Find the probability that at least one will develop breast cancer during her lifetime.

69. One out of five men in the United States will develop prostate cancer during his lifetime. In a group of 30 men selected at random, find the probability that

 (a) at least one will develop prostate cancer during his lifetime.

 (b) at least five will develop prostate cancer during their lifetimes.

Use the **BIOD** program (see "Using Your TI-83") to compute the binomial distributions in Exercises 70 through 74.

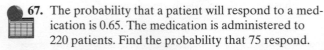

70. $p = 0.5$, $n = 6$

71. $p = 0.5$, $n = 10$

72. $p = 0.4$, $n = 8$

73. $p = 0.25$, $n = 7$

74. $p = 0.15$, $n = 6$

Use the **BIOH** program (see "Using Your TI-83") to find the histogram of the binomial distributions in Exercises 75 through 80.

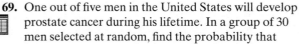

75. $p = 0.5$, $n = 4$

76. $p = 0.6$, $n = 5$

77. $p = 0.4$, $n = 5$

78. $p = 0.3$, $n = 7$

79. $p = 0.7$, $n = 8$

80. $p = 0.8$, $n = 8$

Use a spreadsheet to work Exercises 81 through 86.

81. For a binomial distribution, $p = 0.45$. Find the probability of four successes in five trials.

82. For a binomial distribution, $p = 0.6$. Find the probability of 12 successes in 18 trials.

83. For a binomial distribution, $p = 0.25$. Find the probability of four successes in eight trials.

84. For a binomial distribution, $p = 0.6$. Find the probability of at least three successes in seven trials.

85. For a binomial distribution, $p = 0.3$. Find the probability of at least 8 successes in 13 trials.

86. For a binomial distribution, $p = 0.45$. Find the probability of at least 11 successes in 33 trials.

USING YOUR TI-83

BINOMIAL PROBABILITY DISTRIBUTION

We give you two programs that help with the binomial distribution. The first, **BIOD,** calculates all the probabilities of a binomial distribution. The second, **BIOH,** gives you the histogram of a binomial distribution.

BIOD

```
: Lbl 1                          : C → [A](I, 2)
: Prompt P, N                    : I+1 → I
: {N+1, 2} → dim ([A])           : If X = N
: 0 → X                          : Goto 3
: 1 → I                          : X+1 → X
: Lbl 2                          : Goto 2
: N nCr X → C                    : Lbl 3
: C*(P^X)*(1-P)^(N-X) → C        : Disp "X PR"
: round (C, 4) → C               : Pause [A]
: X → [A](I, 1)                  : Goto 1
                                 : End
```

The BIOD program calculates the binomial probability distribution

$$P(X = x) = C(N, x)p^x(1 - p)^{N-x} \qquad (\text{for } x = 0, 1, 2, \ldots, N)$$

Start the program with $\boxed{\text{PRGM}}$ <BIOD> $\boxed{\text{ENTER}}$ When requested, enter the probability of success (p) and the number of trials (N) and press $\boxed{\text{ENTER}}$. We illustrate with $p = 0.4$ and $N = 5$. The beginning screen is

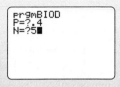

The screen showing the distribution with the value of x in the first column and the corresponding probabilities in the second column is

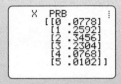

After the probability distribution appears, press $\boxed{\text{ENTER}}$ to go to the start of the program.

EXERCISES

1. Find the binomial probability distribution for $p = 0.35$ and $n = 4$.
2. Find the binomial probability distribution for $p = 0.20$ and $n = 6$.
3. Find the binomial probability distribution for $p = 0.70$ and $n = 5$.
4. Find the binomial probability distribution for $p = 0.5$ and $n = 6$.

HISTOGRAM OF A PROBABILITY DISTRIBUTION

This program calculates the binomial probability distribution

$$P(X = x) = C(N, x)p^x(1 - p)^{N-x} \qquad (\text{for } x = 0, 1, 2, \ldots, N)$$

and draws its histogram. **Warning:** Turn off or clear all y= functions before running the program so that they don't appear on the screen.

BIOH

```
: Lbl 1                    : N nCr X → C
: Prompt P, N              : C*(P^X)*(1-P)^(N-X) → C
: 1 → I                    : 1000*C → C
: 0 → X                    : round (C, 0) → L2 (I)
: ClrDraw                  : I-1 → L1 (I)
: ClrList L1, L2           : I+1 → I
: 0 → Xmin                 : If X=N
: N+1 → Xmax               : Goto 3
: 1 → Xscl                 : X+1 → X
: 0 → Ymin                 : Goto 2
: 800 → Ymax               : Lbl 3
: 100 → Yscl               : Plot 1(Histogram, L1, L2)
: N+1 → dim (L1)           : DispGraph
: N+1 → dim (L2)           : Pause
: Lbl 2                    : Goto 1
                           : End
```

(continued)

Note: Xmin, Xmax, Ymin, and so on are found under VARS <1:Window> Histogram is found under STAT PLOT <TYPE>

Start the program with PRGM <BIOH> ENTER When requested, enter the probability of success (p) and the number of trials (N) and press ENTER. We illustrate with $p = 0.4$ and $N = 5$.

The beginning and final screens are

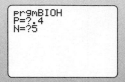
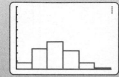

From left to right, each bar represents the probability for $x = 0, 1, 2, \ldots, n$, respectively. The tick marks on the vertical scale represent a probability of $0.1, 0.2, 0.3, \ldots, 0.8$. After the histogram is displayed, press ENTER to go to the beginning of the program.

EXERCISES

1. Draw the histogram of the binomial probability distribution for $p = 0.35$ and $n = 4$.
2. Draw the histogram of the binomial probability distribution for $p = 0.20$ and $n = 6$.
3. Draw the histogram of the binomial probability distribution for $p = 0.70$ and $n = 5$.
4. Draw the histogram of the binomial probability distribution for $p = 0.5$ and $n = 6$.

8.7   Normal Distribution

- Normal Curve
- Area Under a Normal Curve
- The z-Score
- Approximating the Binomial Distribution
- Mean and Standard Deviation of a Binomial Distribution
- Using the Normal Curve to Approximate a Binomial Distribution
- Application

In Section 8.4, we used histograms to give a graphic representation of a probability distribution where the data were **discrete,** that is, the values of the observations had space between that contained no data values. Counting the number of heads in five tosses of a coin gives discrete data because two different values must be at least one unit apart. When we deal with **continuous data,** the values can be arbitrarily close together and for two different values, all numbers between are permissible values of the data. For example, where we measure the heights of male college freshmen, two men could be of different but nearly the same height. Thus, assuming a precise measuring device, any height between 5'9" and 5'10" is a possible height for some male.

Normal Curve

One of the most important probability distributions of data is the **normal distribution.** The normal distribution is a valid representation of the distribution of many populations, such as IQ of 18-year-olds, heights and weights of 12-year-old girls, and scores on standardized tests such as SAT scores. Just as we used histograms in Section 8.4 to graph discrete probability distributions, we use a **normal curve** to graph a normal distribution. Although normal curves vary in size and shape, they all have "bell shapes" similar to that shown in Figure 8–15. The horizontal baseline represents the values of the data, and the height of the curve tells something about the probability of the value.

FIGURE 8–15

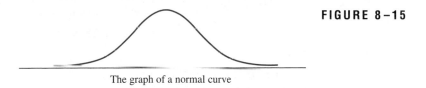

The graph of a normal curve

A histogram can be used to represent continuous data, and in many cases a smooth curve drawn through the tops of the bars gives a normal curve. It is known that the height of 18-year-old males has a mean of 68 inches and a standard deviation of 3 inches, and the normal curve gives a good representation of the heights. Figure 8–16 shows a histogram of heights and a bell-shaped curve obtained by smoothing the histogram.

FIGURE 8–16

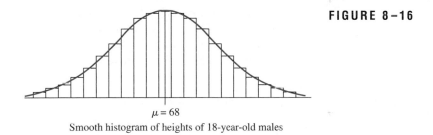

$\mu = 68$

Smooth histogram of heights of 18-year-old males

The normal curve is a smooth curve that peaks at the mean and is perfectly symmetric about the mean. This symmetry means that one half of the area under the curve lies to the left of the mean and the other half lies to the right of the mean. We will use areas under the normal curve to represent probabilities, so the total area is taken as 1, representing the probability of the sample space. The normal curve has the property that the mean and standard deviation of the data determine the shape of the normal curve. Figure 8–17 shows three normal distributions with mean = 0 and three different standard deviations.

FIGURE 8–17

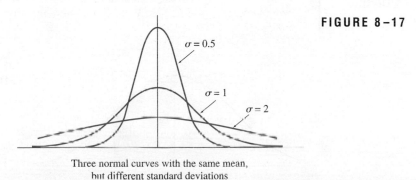

Three normal curves with the same mean, but different standard deviations

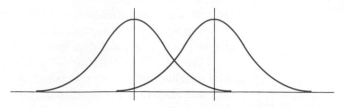

Two normal curves with different means
and equal standard deviations

FIGURE 8–18

Notice that the curves approach the baseline but never touch it. This indicates that fewer and fewer observations are found as you move away from the mean. The normal curve with the smaller standard deviation has a sharper peak, and a larger standard deviation gives a flatter curve.

Figure 8–18 illustrates that two normal distributions with the same standard deviation but different means give normal curves that are the same size and shape. However, the curves are located in different positions.

When dealing with continuous data, like heights of 18-year-old males, remember that few if any males are exactly 70 inches tall. When we say a fellow is 70 inches tall, we usually mean that he is "close to" 70 inches. Depending on how accurately we attempt to measure his height, we may mean he is within $\frac{1}{2}$ inch or $\frac{1}{4}$ inch of 70 inches. When we ask for the probability of a randomly chosen 18-year-old male who is approximately 70 inches tall, we can be more specific and ask for the probability that his height is between 69.5 and 70.5 inches.

Area Under a Normal Curve

The ability to find the area under a portion of the normal curve is important because it gives the probability that an observation is in a specified category or interval. For example, suppose that a normal distribution has a mean of 85. (See Figure 8–19.) The area under the curve between 90 and 95 represents the probability that the observation is between 90 and 95.

Notice that the area under the curve between 85 and 90 is larger than the area between 90 and 95. This indicates a proportionally larger probability of values between 85 and 90.

The Probability for a Normal Distribution

The probability that a value of X lies between two values, x_1 and x_2, is determined by and equals the fraction of the area under the normal curve that lies between x_1 and x_2.

We use the notation $P(x_1 \le X \le x_2)$ to denote the probability that X lies between x_1 and x_2.

FIGURE 8–19

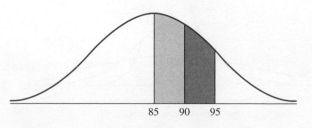

85   90   95

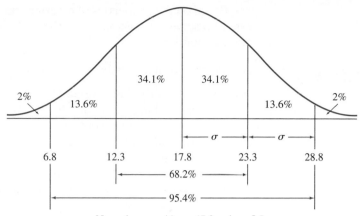

Normal curve with $\mu = 17.8$ and $\sigma = 5.5$

FIGURE 8–20

The normal curve has the unusual property of being completely determined by the mean and standard deviation. This suggests that we need to deal with a different normal curve for each value of the mean and standard deviation. Fortunately, that is not the case. The normal curve can be standardized in a way that allows us to find the area under a portion of a normal curve from one table of values. The distance from the mean, measured in standard deviations, determines the area under the curve in that interval. For example, for all normal curves, about 68% of the scores will lie within 1 standard deviation of the mean, 34% on one side and 34% on the other. Approximately 95% of the scores will lie within 2 standard deviations, and more than 99% will lie within 3 standard deviations. (See Figure 8–20.) Let's apply this idea to the following example.

EXAMPLE 1 ▶

One year, the ACT English test had a mean $\mu = 17.8$ and a standard deviation $\sigma = 5.5$, and the scores formed a normal distribution.

(a) The scores 12.3 and 23.3 are both 1 standard deviation from the mean, so 34% of the area under the normal curve lies between 12.3 and 17.8, and 34% of the area lies between 17.8 and 23.3. Thus, the probability that an ACT English score lies between 12.3 and 17.8 is 0.34. Similarly, the probability that the score lies between 17.8 and 23.3 is 0.34. (See Figure 8–20.)

(b) About 95% of the area lies between 6.8 and 28.8, that is, within 2 standard deviations of the mean, so the probability that an ACT English score lies between 6.8 and 28.8 is 0.95.

(c) Because more than 99% of the area lies within 3 standard deviations of the mean (a distance of 16.5 or less from the mean), you expect less than 1% of the area to lie more than 3 standard deviations away. So, the probability is less than 0.01 that an ACT English score is greater than 34.3 or less than 1.3. ∎

NOTE

The *standard normal curve* is the graph of the function

$$f(x) = \frac{e^{-x^2/2}}{\sqrt{2\pi}}$$

It happens that we can answer questions about normal curves in general by referring to a specific normal curve, the **standard normal curve.** The standard normal curve has mean $\mu = 0$ and $\sigma = 1$.

In the standard normal curve, the letter z is used for the variable, and because $\mu = 0$ and $\sigma = 1$, z also represents the number of standard deviations between the observation and the mean. Thus, $z = 1.6$ indicates 1.6 standard deviations between z and the mean, 0.

Based on the properties of a normal curve that we have already discussed, we can draw some conclusions about the probabilities associated with a standard normal distribution.

EXAMPLE 2 ➤   A standard normal distribution has a mean of zero and a standard deviation of 1.

(a) The probability that a value of z lies between 0 and 1 is about 0.34, and the probability that a value of z lies between -1 and 1 (within 1 standard deviation of the mean) is about 0.68.

(b) The probability that a value of z lies between $z = -2$ and $z = 2$ (within 2σ of the mean) is about 0.95.

(c) The probability that a value of z lies between $z = -3$ and $z = 3$ (within 3 standard deviations of the mean) is about 0.99. ∎

The z-Score

We have already discussed the probability that an observation lies within 1, 2, or 3 standard deviations of the mean in a normal distribution. How do we determine the number of observations that lie within 1.25, 0.63, or 2.50 standard deviations of the mean? Tables such as the standard normal table found inside the back cover of the book enable you to find the area between the mean and an observation. The key is to locate the observation according to the number of standard deviations it lies from the mean. Traditionally, the **z-score** represents the number of standard deviations between an observation and the mean.

Whatever scale is used for x on a normal curve, we can associate a value of z with each value of x.

Here's how we determine a value of z that corresponds to a given value of x. As z represents the number of standard deviations between the mean and the score x, find z by

$$z = \frac{\text{difference between the value of } x \text{ and the mean}}{\text{standard deviation}}$$
$$= \frac{\text{value of } x - \text{mean}}{\text{standard deviation}} = \frac{x - \mu}{\sigma}$$

We use z to find the area under the normal curve between two scores. To do so, we use the standard normal table found inside the back cover of the book. The table does not give area between any two scores; that would require a prohibitively lengthy, cumbersome table. You must understand that the table gives area *between the mean and a z-score* for selected *z-scores*.

EXAMPLE 3 ➤   Compute z for each of the following values of the mean, the standard deviation, and a given value of x:

(a) Mean $= 25$, standard deviation $= 2$, $x = 31$
(b) Mean $= 25$, standard deviation $= 2$, $x = 19$
(c) Mean $= 7.5$, standard deviation $= 1.2$, $x = 10.5$
(d) Mean $= 16.85$, standard deviation $= 2.1$, $x = 14.12$

SOLUTION

(a) Using the formula

$$z = \frac{x - \text{mean}}{\text{standard deviation}}$$

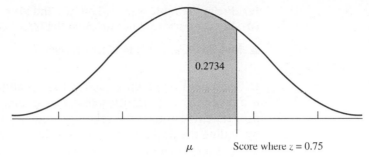

FIGURE 8–21

we obtain

$$z = \frac{31 - 25}{2} = \frac{6}{2} = 3$$

(b) $z = \dfrac{19 - 25}{2} = \dfrac{-6}{2} = -3$

A negative value of z indicates that x is less than the mean.

(c) $z = \dfrac{10.5 - 7.5}{1.2} = \dfrac{3}{1.2} = 2.5$

(d) $z = \dfrac{14.12 - 16.85}{2.1} = \dfrac{-2.73}{2.1} = -1.3$

■ **Now You Are Ready to Work Exercise 1**

EXAMPLE 4 ➤

Use the standard normal table (inside the back cover) to find the fraction of area between the mean and

(a) $z = 0.75$ **(b)** $z = -0.75$
(c) $z = 2.58$ **(d)** $z = -1.92$

SOLUTION

(a) The value for A that corresponds to $z = 0.75$ is 0.2734, the fraction of the area between the mean and $z = 0.75$. (See Figure 8–21.)

(b) Because the curve is symmetric about the mean, the area between the mean and $z = -0.75$ equals the area between the mean and $z = 0.75$, which is 0.2734. (See Figure 8–22.)

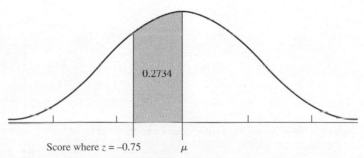

FIGURE 8–22

(c) From the table at $z = 2.58$, we find $A = 0.4951$.
(d) We use $z = 1.92$ to find $A = 0.4726$.

■ **Now You Are Ready to Work Exercise 7**

In Section 7.1 on an introduction to probability, we noted that relative frequency is a probability distribution that we call empirical probability. At times, it is appropriate to estimate the relative frequency with which a value lies in an interval by finding the probability that the value lies in the interval.

For a normal distribution, the same procedure is used to find

(a) the probability that X is between a and b, $P(a \leq X \leq b)$.
(b) an estimate of the fraction of values of X that lie between a and b. (Fraction $= P(a \leq X \leq b)$.)
(c) an estimate of the percentage of values of X that lie between a and b. (Percentage $= 100 \cdot P(a \leq X \leq b)$.)

EXAMPLE 5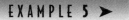

Estimate the fraction of scores between $z = 1.15$ and $z = -1.15$ under the normal curve. (We may also say that we find the area *within* 1.15 standard deviations of the mean.)

SOLUTION

We obtain this area by combining the area between the mean and $z = 1.15$ with the area between the mean and $z = -1.15$, because the table gives only the area between the mean and a score, not between two scores. For the area between the mean and $z = 1.15$, look for the area corresponding to $z = 1.15$. It is 0.3749. The area between the mean and $z = -1.15$ is the same, 0.3749, so the total area is $0.3749 + 0.3749 = 0.7498$. (See Figure 8–23.)

■ **Now You Are Ready to Work Exercise 23**

EXAMPLE 6 ▶

Find the area under the normal curve between $z = -0.46$ and $z = 2.32$.

SOLUTION

The point where $z = -0.46$ lies below the mean, so the area between the mean and where $z = -0.46$ is 0.1772 (look up A for $z = 0.46$). The point where $z = 2.32$ lies above the mean, and the area between the mean and where $z = 2.32$

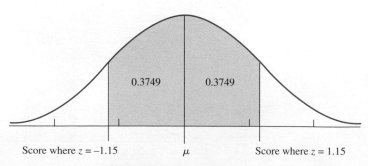

Score where $z = -1.15$ μ Score where $z = 1.15$

FIGURE 8–23

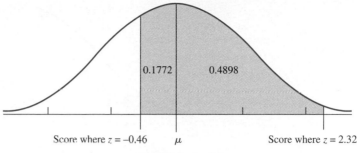

Score where $z = -0.46$ μ Score where $z = 2.32$

FIGURE 8–24

is 0.4898. As the two points lie on opposite sides of the mean, add the two areas found, $0.1772 + 0.4898 = 0.6670$, to find the total area. (See Figure 8–24.)

■ **Now You Are Ready to Work Exercise 31**

EXAMPLE 7 ➤

Find the probability that a score lies to the right of $z = 0.80$.

SOLUTION

Figure 8–25 shows the desired area. The standard normal table gives $A = 0.2881$ for $z = 0.80$. However, this is the area *below* $z = 0.80$ and above the mean, and we want to know the area *above* z. Remember that all the area above the mean is 0.5000 of the area under the curve. Therefore, the area *above* $z = 0.80$ is $0.5000 - 0.2881 = 0.2119$ of the total area. Thus, $P(Z > 0.80) = 0.2119$.

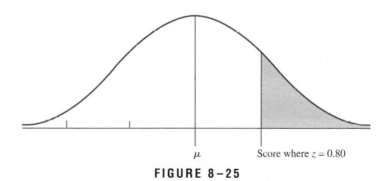

μ Score where $z = 0.80$

FIGURE 8–25

■ **Now You Are Ready to Work Exercise 35**

EXAMPLE 8 ➤

Estimate the fraction of scores that are more than 1.40 standard deviations away from the mean.

SOLUTION

This asks for the area to the right of the score where $z = 1.40$ and the area to the left of the score where $z = -1.40$. (See Figure 8–6.) The area between the mean and a score where $z = 1.40$ is 0.4192, so the area to the right of the score is $0.5000 - 0.4192 = 0.0808$. By symmetry, the area to the left of the score where $z = -1.40$ is also 0.0808. The total area more than 1.40 standard deviations away

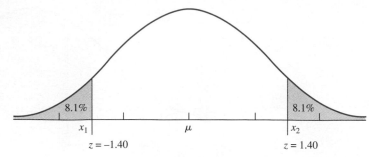

FIGURE 8–26

from the mean is 0.1616, so about 0.1616 of the scores are more than 1.40 standard deviations away from the mean. ■

Generally, you will be given the mean and scores, not values of z. The following examples show how to find the area between two scores.

EXAMPLE 9 ➤   A normal distribution has a mean of 30 and a standard deviation of 7. Find the probability that the value of x is between 30 and 42 $(P(30 \leq X \leq 42))$.

SOLUTION

To use the standard normal table, we must find the z-score that corresponds to 42. It is

$$z = \frac{42 - 30}{7} = \frac{12}{7} = 1.71$$

From the standard normal table, we obtain $A = 0.4564$ when $z = 1.71$. Thus, $P(30 \leq x \leq 42) = 0.4564$. (See Figure 8–27.)

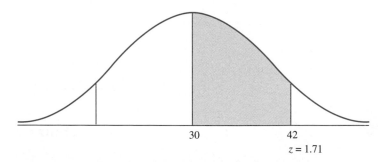

FIGURE 8–27

■ **Now You Are Ready to Work Exercise 43**

When two values of x are on opposite sides of the mean and are different distances from the mean, we add areas to find the area between the values.

EXAMPLE 10 ➤   A normal distribution has a mean $\mu = 50$ and a standard deviation $\sigma = 6$. Estimate the percentage of scores between 47 and 58.

SOLUTION

As the mean is between 47 and 58, we need to find the area under the curve in two steps; that is, we need to find the area from the mean to each score.

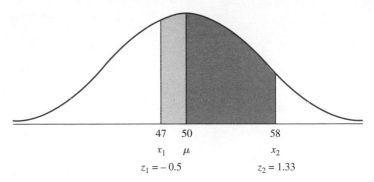

FIGURE 8-28

1. The area between 47 and 50:

$$z_1 - \frac{47 - 50}{6} = \frac{-3}{6} = -0.50$$

From the standard normal table, using that the area between 0 and -0.50 equals the area between 0 and 0.50, this area is $A = 0.1915$.

2. The area between 50 and 58:

$$z_2 = \frac{58 - 50}{6} = \frac{8}{6} = 1.33$$

$$A = 0.4082$$

The total area between 47 and 58 is $0.1915 + 0.4082 = 0.5997$, so 0.5997, or 59.97%, of the scores lie between 47 and 58. (See Figure 8–28.)

■ **Now You Are Ready to Work Exercise 47**

Example 10 illustrates the fundamental concept used to find the probability that a value of x lies in a specified interval of a normal distribution — say, x lies between x_1 and x_2 ($x_1 \le x \le x_2$).

> For the interval $x_1 \le x \le x_2$,
>
> $$P(x_1 \le X \le x_2) = P(z_1 \le Z \le z_2)$$
>
> where z_1, z, and z_2 are the z-values corresponding to x_1, x, and x_2, respectively. $P(z_1 \le Z \le z_2)$ is found using the standard normal table.

Sometimes, we want to find an area between two values of x that lie on the same side of the mean.

EXAMPLE 11 ➤ The Welding Program Department at Paul's Valley Technical School gives an exit test to evaluate the students' skills and knowledge of procedures.

It is designed so that it gives scores that are reasonably close to a normal distribution with a mean of $\mu = 100$ and a standard deviation $\sigma = 8$. A student is selected at random. Find the probability that the student will score between 110 and 120.

SOLUTION

Because we always measure areas from the mean to a score, we can find the area between 100 and 110 and the area between 100 and 120. To find the area between

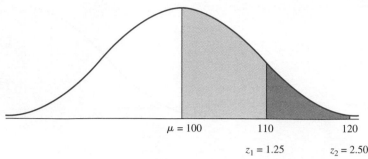

FIGURE 8-29

110 and 120, subtract the area between 100 and 110 from the area between 100 and 120.

For $x_1 = 110$,

$$z_1 = \frac{110 - 100}{8} = \frac{10}{8} = 1.25$$

and $A_1 = 0.3944$.

For $x_2 = 120$,

$$z_2 = \frac{120 - 100}{8} = \frac{20}{8} = 2.5$$

and $A_2 = 0.4938$.

The area between 110 and 120 is then $0.4938 - 0.3944 = 0.0994$, so the probability that a student scores between 110 and 120 $(P(110 \le X \le 120))$ is 0.0944. (See Figure 8–29.)

■ **Now You Are Ready to Work Exercise 51**

EXAMPLE 12 ➤ Students at Flatland University spend an average of 24.3 hours per week on homework, with a standard deviation of 1.4 hours. Assume a normal distribution.

(a) Estimate the percentage of the students who spend more than 28 hours per week on homework.
(b) What is the probability that a student spends more than 28 hours per week on homework?

SOLUTION

(a) The value of z corresponding to 28 hours is

$$z = \frac{28 - 24.3}{1.4} = \frac{3.7}{1.4} = 2.64$$

From the standard normal table, we have $A = 0.4959$ when $z = 2.64$. The value $A = 0.4959$ represents the area from the mean, 24.3, to 28 ($z = 2.64$). All of the area under the curve to the right of the mean is one half of the total area. The area to the right of $z = 2.64$ is $0.5000 - 0.4959 = 0.0041$. Therefore, about 0.41% of the students study more than 28 hours.

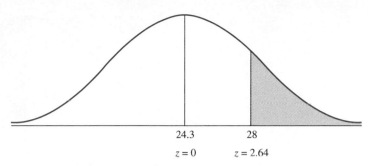

24.3   28

$z = 0$ $z = 2.64$

FIGURE 8–30

(b) The probability that a student studies more than 28 hours is the area that is to the right of 28 — that is, 0.0041. (See Figure 8–30.)

■ **Now You Are Ready to Work Exercise 97**

EXAMPLE 13 ➤

(a) Find the value of z such that an estimated 4% of the scores are to the right of z.
(b) Find the value of z such that 0.04 is the probability that a score lies to the right of z.

SOLUTION

(a) If 4% of the scores are to the right of z, then the other 46% of the scores to the right of the mean are between the mean and z. (Remember that 50% of the scores are to the right of the mean.) Look for $A = 0.4600$ in the standard normal table. It occurs at $z = 1.75$. This is the desired value of z.
(b) This also occurs at $z = 1.75$. (See Figure 8–31.)

■ **Now You Are Ready to Work Exercise 69**

These examples ask for the probability that a score lies in a certain interval (such as $P(50 \le X \le 60)$) or greater than a certain score (such as $P(X \ge 75)$). A natural question is "How do you find the probability of a certain score, such as $P(X = 65)$?" The answer is "The probability is zero." In general, $P(X = c) = 0$ for any value c in a normal distribution. (See Exercise 129.) For example, this states that the probability of randomly selecting an 18-year-old male who is *exactly* 5 feet 10 inches tall is zero.

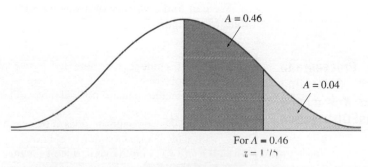

$A = 0.46$

$A = 0.04$

For $A = 0.46$
$z = 1.75$

FIGURE 8–31

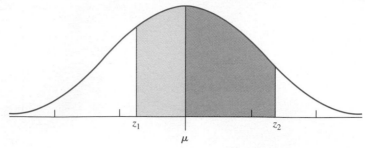

(a) Add the areas for z_1 and z_2 to get the area between z_1 and z_2.

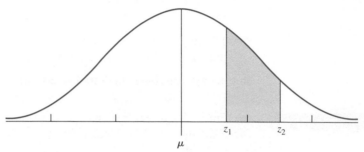

(b) Subtract z_1 area from z_2 area to get the area between z_1 and z_2.

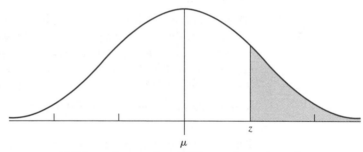

(c) Subtract the z area from 0.500 to get the area beyond z.

FIGURE 8–32

Because the normal curve is symmetric, the standard normal table gives A only for positive values of z. For negative values of z, when the score is to the left of the mean, simply use the value of A for the corresponding positive z. Keep in mind that each z-value determines an area *from the mean* to the z position.

We can find a variety of areas by adding or subtracting areas given from the table. (See Figure 8–32.)

Procedure to Determine $P(c \leq X \leq d)$ **of a Normal Distribution**	**Step 1.** Determine the z-value for $x = c$ and $x = d$ and call them z_1 and z_2, respectively. **Step 2.** From the standard normal table, determine A corresponding to z_1 and to z_2. **Step 3.** **(a)** If c and d are on opposite sides of the mean (z_1 and z_2 have opposite signs), add the values of A corresponding to z_1 and z_2. **(b)** If c and d are on the same side of the mean (z_1 and z_2 have the same signs), subtract the smaller value of A from the larger value.

Procedure to Determine $P(X < c)$ or $P(X > d)$ of a Normal Distribution

Step 1.   Determine the z-value corresponding to c or d.

Step 2.   From the standard normal table, determine the value of A corresponding to z of step 1.

Step 3.

(a) $P(X < c)$: Find the area below c. If c is to the right of the mean (z is positive), add A to 0.5000. If c is to the left of the mean (z is negative), subtract A from 0.5000.

(b) $P(X > d)$: Find the area above d. If d is to the right of the mean (z is positive), subtract A from 0.5000. If d is to the left of the mean (z is negative), add A to 0.5000.

EXAMPLE 14 ➤   A standardized test is given to several hundred thousand junior high students. The mean is 100, and the standard deviation is 10. If a student is selected at random, what is the probability that the student scores in the 114 to 120 interval?

SOLUTION

This question may be answered by using the properties of the normal curve, as it represents standardized test scores well.

The values of z corresponding to 114 and 120 are $z = 1.4$ and $z = 2.0$, respectively. For $z = 1.4$, $A = 0.4192$; for $z = 2.0$, $A = 0.4773$. So, the area between 114 and 120 is $0.4773 - 0.4192 = 0.0581$. The probability that the student's score is in the 114 to 120 interval is 0.0581.

■ **Now You Are Ready to Work Exercise 99**

EXAMPLE 15 ➤   The Quality Cola Bottling Company sells its Quality Cola in the standard size can, 355 milliliters (ml). The manager does not expect every can to contain *exactly* 355 ml of cola but would like to be consistently close. Working with the quality-control manager, she agrees that the quantity can be expected to vary as a normal distribution, but the cans should have a mean of 355 ml, and at least 95% should vary from 355 ml by no more than 5 ml. What value of the standard deviation does this require?

SOLUTION

The company wants 95% of the values to lie between 350 and 360 ml (355 ± 5); that is, $z = -1.96$ at 350 and $z = 1.96$ at 360.

As $5 = 1.96\sigma$,

$$\sigma = \frac{5}{1.96} = 2.55 \quad ■$$

Summary of Properties of a Normal Curve

1.   All normal curves have the same general bell shape.
2.   The curve is symmetric with respect to a vertical line that passes through the peak of the curve.
3.   The vertical line through the peak occurs where the mean, median, and mode coincide.
4.   The area under any normal curve is always 1.
5.   The mean and standard deviation completely determine a normal curve. For the same mean, a smaller standard deviation gives a taller and narrower peak. A larger standard deviation gives a flatter curve.
6.   The area to the right of the mean is 0.5; the area to the left of the mean is 0.5.
7.   About 68.26% of the area under a normal curve is enclosed in the interval formed by the score 1 standard deviation to the left of the mean and the score 1 standard deviation to the right of the mean.

8. If a random variable X has a normal probability distribution, the probability that a score lies between x_1 and x_2 is the area under the normal curve between x_1 and x_2.

9. The probability that a score is less than x_1 equals the probability that a score is less than, or equal to x_1; that is,

$$P(x < x_1) = P(x \le x_1)$$

10. $P(X = c) = 0$

Approximating the Binomial Distribution

EXAMPLE 16 ➤

The probability that a new drug will cure a certain blood disease is 0.7. If it is administered to 100 patients with the disease, what is the probability that 60 of them will be cured?

SOLUTION

You should set up

$$P(X = 60) = C(100, 60)(0.7)^{60}(0.3)^{40}$$

with little difficulty. You then may find it tedious to compute the probability and even more tedious to form the binomial distribution of the experiment. Some calculators have functions that make these computations relatively easy. ■

We do not calculate $P(X = 60)$ in the preceding example because the normal distribution provides a means to avoid this wearisome computation. It can be used to estimate the binomial distribution for large values of n. We will soon show how to do this. First, we need to find the mean and standard deviation of a binomial distribution.

Mean and Standard Deviation of a Binomial Distribution

Recall from Section 8.5 that the computation of the mean (or expected value), the variance, and the standard deviation of a probability distribution can be time-consuming for a random variable with many values.

In Example 9 of Section 8.6, we used the binomial distribution with $n = 4$ and $p = 0.3$. The expected value of the random variable X is

$$
\begin{aligned}
E(X) &= 0.2401(0) + 0.4116(1) + 0.2646(2) + 0.0756(3) + 0.0081(4) \\
&= 0 + 0.4116 + 0.5292 + 0.2268 + 0.0324 \\
&= 1.20
\end{aligned}
$$

The significance of this result is that $1.20 = 4(0.3)$, which is np for this example. This result holds for all binomial distributions, $E(X) = np$. Likewise, the variance and standard deviation of a binomial distribution can be expressed in rather simple terms of n, p, and q, as follows.

<table>
<tr><td>

DEFINITION
Mean, Variance, and Standard Deviation of a Binomial Distribution

</td><td>

Let X be the random variable for a binomial distribution with n repeated trials, with p the probability of success, q the probability of failure, and

$$P(X = x) = C(n, x)p^x(1 - p)^{n-x}$$

Then, the **mean** (expected value), **variance,** and **standard deviation** of X are given by

Mean: $\mu = np$

Variance: $\sigma^2(X) = np(1 - p) = npq$

Standard deviation: $\sigma(X) = \sqrt{np(1 - p)} = \sqrt{npq}$

</td></tr>
</table>

EXAMPLE 17 ➤   **(a)** For the binomial distribution with $n = 20$, $p = 0.35$:

$$\text{Mean} = \mu = 20(0.35) = 7$$
$$\sigma^2(X) = 20(0.35)(0.65) = 4.55$$
$$\sigma(X) = \sqrt{4.55} = 2.133$$

(b) If $n = 160$, $p = 0.21$,

$$\mu = 160(0.21) = 33.6$$
$$\sigma^2(X) = 160(0.21)(0.79) = 26.544$$
$$\sigma(X) = \sqrt{26.544} = 5.152$$

■ **Now You Are Ready to Work Exercise 75**

Using the Normal Curve to Approximate a Binomial Distribution

To compute some binomial probabilities, such as the probability of 40 to 65 successes in 200 trials, can be tedious and subject to mistakes. Fortunately, the normal curve can often be used to obtain a satisfactory **estimate of a binomial probability.**

   To illustrate how the normal curve is used to estimate binomial probabilities, we use the following simple example.

EXAMPLE 18 ➤   Use the normal curve to estimate the probability of

(a) three heads in six tosses of a coin.
(b) three or four heads in six tosses of a coin.

SOLUTION

(a) We first point out that the binomial distribution of this problem, and its histogram, are given in Example 10 of Section 8.6. From it, for example, we see that $P(X = 3) = 0.3125$.

   For the binomial distribution with $n = 6$ and $p = 0.5$, the mean and standard deviation are

$$\mu = 6(0.5) = 3$$
$$\sigma = \sqrt{6(0.5)(0.5)} = \sqrt{1.5} = 1.225$$

Figure 8–33 shows a normal curve with $\mu = 3$ and $\sigma = 1.225$ superimposed on the binomial distribution for $n = 6$ and $p = 0.5$. Notice that a portion of the histogram lies above the curve and that some space under the curve is not filled by the rectangles. It appears that if the portions of the rectangle outside were moved into the empty spaces below the curve, then the area under the

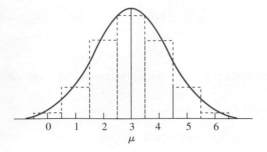

FIGURE 8–33   A normal curve superimposed on a binomial distribution.

curve would pretty well be filled. This states that the total area enclosed by the histogram is "close" to the area under the normal curve.

Let's make another observation. Each bar in the histogram is of width 1, and its height is equal to the probability it represents.

The bar representing $P(X = 3)$ is centered at 3 and is therefore located between 2.5 and 3.5. The area of the bar (1×0.3125) can also be used to represent the probability that $X = 3$. We mention this because it holds the key to using the normal distribution. To find $P(X = 3)$, find the area under the normal curve between 2.5 and 3.5. (See Figure 8–34.)

Here's how we use the normal curve to estimate $P(X = 3)$. Use $\mu = 3$, $\sigma = 1.225$, $x_1 = 2.5$, and $x_2 = 3.5$. Then,

$$z_1 = \frac{2.5 - 3}{1.225} = \frac{-0.5}{1.225} = -0.41$$

$$z_2 = \frac{3.5 - 3}{1.225} = \frac{0.5}{1.225} = 0.41$$

The standard normal table shows that the area under the normal curve between the mean and $z = 0.41$ is $A = 0.1591$. Then, the area between $z_1 = -0.41$ and $z_2 = 0.41$ is $0.1591 + 0.1591 = 0.3182$. This estimates $P(X = 3)$ as 0.3182. This compares to the actual probability of 0.3125 $\big(C(6, 3)(0.5)^3(0.5)^3\big)$. Notice that we approximated the binomial probability $P(X = 3)$ by the normal probability $P(2.5 \le X \le 3.5)$.

(b) To find the probability of three or four heads in six tosses of a coin, $P(3 \le X \le 4)$, we find the total area of the bars for three and four. This amounts to finding the area in the histogram between 2.5 and 4.5. The normal approximation is the area between

$$z_1 = \frac{2.5 - 3}{1.225} \qquad \text{and} \qquad z_2 = \frac{4.5 - 3}{1.225}$$

FIGURE 8–34   To estimate $P(X = 3)$ in a binomial distribution, find the area under the normal curve between $x = 2.5$ and $x = 3.5$.

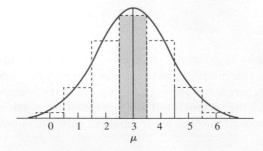

That is,

$$z_1 = -0.41 \quad \text{and} \quad z_2 = 1.22$$

The corresponding areas are

$$A_1 = 0.1591 \quad \text{and} \quad A_2 = 0.3888$$

so the desired probability estimate is $0.1591 + 0.3888 = 0.5479$. The actual probability, from Example 10, Section 8.6, is $0.3125 + 0.2344 = 0.5469$. Notice that we approximated the binomial probability $P(3 \le X \le 4)$ by the normal probability $P(2.5 \le X \le 4.5)$.

■ **Now You Are Ready to Work Exercise 89**

At this point, we must confess that the normal approximation worked rather well in the preceding example because p and q were both $\frac{1}{2}$. Had we used $p = \frac{1}{6}$ and $q = \frac{5}{6}$, the normal approximation would not have worked very well. However, if n is 50 instead of 6, the normal approximation is reasonable. So when is it reasonable to use the normal distribution to approximate the binomial distribution? The answer is, "It depends." It depends on the values of n and p. Several rules of thumb are used by statisticians to judge when the normal approximation is reasonable. We use the following:

Rule

> The normal distribution provides a good estimate of the binomial distribution when both np and nq are greater than or equal to 5.

Figures 8–35, 8–36, and 8–37 show the binomial distribution for $p = 0.60$ and $n = 3, 7$, and 15. Notice that the distribution for $n = 15$ more closely resem-

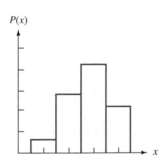

FIGURE 8–35 Binomial distribution for $p = 0.6$ and $n = 3$.

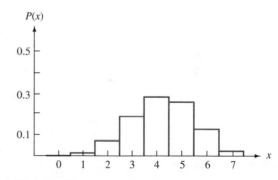

FIGURE 8–36 Binomial distribution for $p = 0.6$ and $n = 7$.

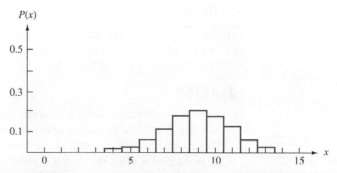

FIGURE 8–37 Binomial distribution for $p = 0.6$ and $n = 15$.

bles a normal distribution than those for $n = 3$ or $n = 7$. We observe that, for $n = 3$, $np = 1.8$ and $nq = 1.2$, both smaller than the desired value of 5 for a good normal approximation. Also, for $n = 7$, $np = 4.2$ and $nq = 2.8$, but for $n = 15$, $np = 9.0$ and $nq = 6.0$. Thus, by our rule, the normal approximation is a reasonable approximation of the binomial distribution for $n = 15$ and $p = 0.6$.

EXAMPLE 19 ➤ For a binomial distribution with $p = 0.6$, find the smallest value of n so that the normal distribution is a reasonable approximation to the binomial distribution.

SOLUTION

For the normal distribution to be a reasonable approximation, both $np \geq 5$ and $nq \geq 5$.

For $np \geq 5$,

$$n(0.6) \geq 5 \quad \text{and} \quad n \geq \frac{5}{0.6} = 8.33$$

For $nq \geq 5$,

$$n(0.4) \geq 5 \quad \text{and} \quad n \geq \frac{5}{0.4} = 12.5$$

Thus, $n \geq 12.5$, so the smallest integer value of n is 13.

■ **Now You Are Ready to Work Exercise 87**

In the next three examples, we estimate some binomial probabilities, using the normal distribution. In each of the three examples, we use the binomial distribution with

$$n = 14, \quad p = 0.4, \quad q = 0.6$$

On the basis of these values, we have

$$\mu = 14(0.4) = 5.6 \quad \text{and} \quad \sigma = \sqrt{14(0.4)(0.6)} = 1.83$$

These values will be used in the three examples.

EXAMPLE 20 ➤ The probability that a gasoline additive increases gasoline mileage in a car is 0.4. In a test conducted by an automotive class at the Rocky Mountain Technical Institute, the additive was used on 14 cars selected at random. Use the normal curve to estimate

(a) the probability that gasoline mileage improves in 3 to 7 cars $(P(3 \leq X \leq 7))$.
(b) the probability that gasoline mileage improves in more than 3 cars and fewer than 7 cars $(P(3 < X < 7))$.

SOLUTION

(a) Figure 8–38 shows the graph of the binomial probability with the bars representing $3 \leq X \leq 7$ shaded and a normal curve superimposed. It shows that the area under the normal curve between $X = 2.5$ and 7.5 approximates the desired probability. To find the area, we need the corresponding z-scores.
    Using $\mu = 5.6$ and $\sigma = 1.83$ we have

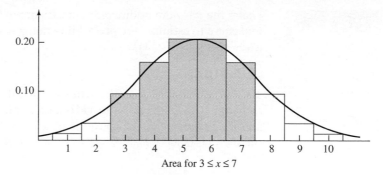

Area for $3 \leq x \leq 7$

FIGURE 8–38

At $X = 2.5$,

$$z = \frac{2.5 - 5.6}{1.83} = -1.69 \quad \text{and} \quad A = 0.4545$$

(from the standard normal table).

At $X = 7.5$,

$$z = \frac{7.5 - 5.6}{1.83} = 1.04 \quad \text{and} \quad A = 0.3508$$

The area under the normal curve between 2.5 and 7.5 is then $0.4545 + 0.3508 = 0.8053$, so $P(3 \leq X \leq 7) = 0.8053$.

(b) To find $P(3 < X < 7)$, we need to find the area under the normal curve between 3.5 and 6.5, which includes the bars for 4, 5, and 6. (See Figure 8–39.)

At $X = 3.5$,

$$z = \frac{3.5 - 5.6}{1.83} = -1.15 \quad \text{and} \quad A = 0.3749$$

At $X = 6.5$,

$$z = \frac{6.5 - 5.6}{1.83} = 0.49 \quad \text{and} \quad A = 0.1879$$

The area under the normal curve between $X = 3.5$ and $X = 6.5$ is then $0.3749 + 0.1879 = 0.5628$, so $P(3 < X < 7) = 0.5628$.

■ **Now You Are Ready to Work Exercise 91**

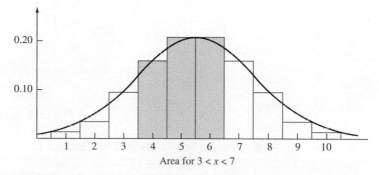

Area for $3 < x < 7$

FIGURE 8–39

EXAMPLE 21 ➤

Using the gasoline additive data in Example 20 ($n = 14$, $p = 0.4$), use the normal curve to estimate the probability that gasoline mileage improves in more than three cars $(P(X > 3))$.

SOLUTION

To approximate $P(X > 3)$, we need to find the area under the normal curve to the right of 3.5. (See Figure 8–40.) From Example 20, we have at $X = 3.5$, $z = -1.15$, and the area from $X = 3.5$ to the mean, 5.6, is 0.3749. As the area to the right of the mean is 0.5000, $P(X > 3) = 0.3749 + 0.5000 = 0.8749$.

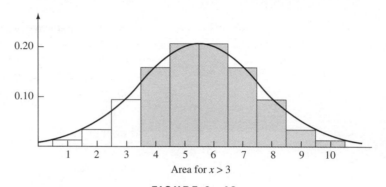

FIGURE 8–40

■ **Now You Are Ready to Work Exercise 93**

EXAMPLE 22 ➤

Using the gasoline additive data in Example 20 ($n = 14$, $p = 0.4$), use the normal curve to estimate the probability that gasoline mileage improves in seven or fewer cars $(P(X \leq 7))$.

SOLUTION

To approximate $P(X \leq 7)$, we need to find the area under the normal curve to the left of 7.5. (See Figure 8–41.) From Example 20, we have at $X = 7.5$, $z = 1.04$, and the area from the mean, 5.6, to $X = 7.5$ is 0.3508. As the area to the left of the mean is 0.5000, $P(X \leq 7) = 0.3508 + 0.5000 = 0.8508$.

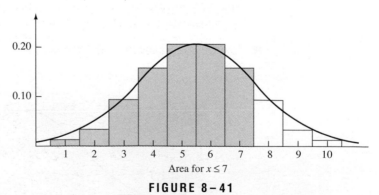

FIGURE 8–41

■ **Now You Are Ready to Work Exercise 95**

In summary, estimate a binomial probability with a normal distribution as follows:

Procedure for Estimating a Binomial Probability

1. If np and nq are both greater than or equal to 5, you may assume that the normal distribution provides a good estimate.

2. Compute $\mu = np$ and $\sigma = \sqrt{npq}$.

3. To estimate $P(X = c)$, find the area under the normal curve between $c - 0.5$ and $c + 0.5$.

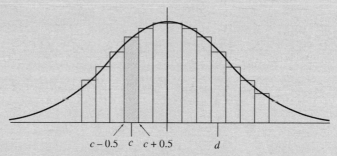

4. To estimate $P(c \le X \le d), c < d$, find the area under the normal curve between $c - 0.5$ and $d + 0.5$.

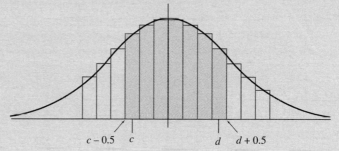

5. To estimate $P(c < X < d), c < d$, find the area under the normal curve between $c + 0.5$ and $d - 0.5$.

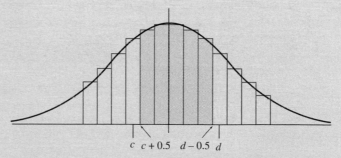

6. To estimate $P(X > c)$, find the area under the normal curve to the right of $c + 0.5$.

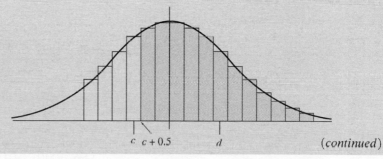

(continued)

7. To estimate $P(X \geq c)$, find the area under the normal curve to the right of $c - 0.5$.

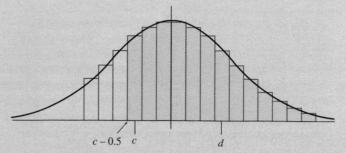

8. To estimate $P(X < c)$, find the area under the normal curve to the left of $c - 0.5$.

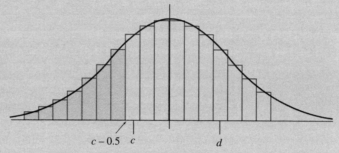

9. To estimate $P(X \leq c)$, find the area under the normal curve to the left of $c + 0.5$.

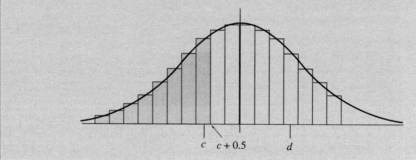

Application

EXAMPLE 23 ➤

The probability that a new drug will cure a certain blood disease is 0.7. It is administered to 100 patients. (You may recognize this as the problem we set up but did not solve in Example 16.) Use a normal curve to estimate

(a) the probability that 60 of them will be cured.
(b) the probability that 60 to 75 of them will be cured.
(c) the probability that more than 75 will be cured.

SOLUTION

Because $n = 100$, $p = 0.7$, and $q = 0.3$, we know that $np = 70$ and $nq = 30$. As both values are greater than 5, a normal curve provides a good estimate, and

$$\mu = 100(0.7) = 70$$

$$\sigma = \sqrt{100(0.7)(0.3)} = \sqrt{21} = 4.583$$

(a) Find the area under the normal curve between 59.5 and 60.5 — that is, between

$$z_1 = \frac{59.5 - 70}{4.583} = \frac{-10.5}{4.583} = -2.29$$

and

$$z_2 = \frac{60.5 - 70}{4.583} = \frac{-9.5}{4.583} = -2.07$$

The corresponding areas are $A_1 = 0.4890$ and $A_2 = 0.4808$. As the z-scores lie on the same side of the mean, we must subtract areas, $0.4890 - 0.4808 = 0.0082$. So, $P(X = 60) = 0.0082$.

(b) To estimate $P(60 \leq X \leq 75)$, we find the area between 59.5 and 75.5. The corresponding z-values and areas are

$$z_1 = \frac{59.5 - 70}{4.583} = -2.29 \qquad \text{and} \qquad z_2 = \frac{75.5 - 70}{4.583} = 1.20$$

$A_1 = 0.4890$, and $A_2 = 0.3849$. As the scores lie on opposite sides of the mean, we add areas to obtain the probability:

$$P(60 \leq X \leq 75) = 0.4890 + 0.3849 = 0.8739$$

(c) To estimate the probability that more than 75 patients will be cured, $P(X > 75)$, find the area under the normal curve that lies to the right of 75.5. (*Note:* To find the probability of 75 or more, $P(X \geq 75)$, find the area to the right of 74.5.)

For a score of 75.5, $z = 1.20$ and $A = 0.3849$. (See part (b).) Because we want the area above $z = 1.20$, we need to subtract $0.5000 - 0.3849 = 0.1151$ to get $P(X > 75) = 0.1151$.

■ **Now You Are Ready to Work Exercise 101**

■■ 8.7 EXERCISES

LEVEL 1

Find the value of z in Exercises 1 through 6.

1. *(See Example 3)* $x = 3.1$, mean $= 4.0$, $\sigma = 0.3$

2. $x = 38.0$, mean $= 22.5$, $\sigma = 6.2$

3. $x = 10.1$, mean $= 10.0$, $\sigma = 2.0$

4. $x = 31$, mean $= 31$, $\sigma = 5$

5. $x = 2.65$, mean $= 0$, $\sigma = 1.0$

6. $x = 192$, mean $= 150$, $\sigma = 7$

In Exercises 7 through 14, find the fraction of the area under the normal curve that lies between the mean and the given z-score.

7. *(See Example 4)* **8.** $z = 1.90$
   $z = 0.50$

9. $z = 0.25$

10. $z = 2.50$

11. $z = 1.10$

12. $z = -0.10$

13. $z = -0.75$

14. $z = -2.25$

In Exercises 15 through 18, find the percentage of the area under the normal curve that lies between the mean and the given number of standard deviations above the mean.

15. 0.46   **16.** 2.17   **17.** 0.38   **18.** 1.30

In Exercises 19 through 22, find the percentage of the area under the normal curve that lies between the mean and the given number of standard deviations below the mean.

19. 1.24   **20.** 0.24   **21.** 2.90   **22.** 0.54

In Exercises 23 through 26, find the fraction of the area under the normal curve between the given values of z.

23. (*See Example 5*) $z = 1.25$ and $z = -1.25$

24. $z = 0.40$ and $z = -0.40$

25. $z = 2.20$ and $z = -2.20$

26. $z = 1.80$ and $z = -1.80$

In Exercises 27 through 30, find the percentage of the area under the normal curve that lies within the given number of standard deviations of the mean.

27. 0.65   **28.** 1.34   **29.** 0.38   **30.** 2.73

In Exercises 31 through 34, find the fraction of the area under the normal curve that lies between the given z-scores.

31. (*See Example 6*) $z = -0.60$ and $z = 1.28$

32. $z = -1.75$ and $z = 1.20$

33. $z = -0.80$ and $z = 2.80$

34. $z = -0.20$ and $z = 2.65$

In Exercises 35 through 38, find the fraction of the area under the normal curve that lies above the given value of z.

35. (*See Example 7*)   **36.** $z = 0.65$
$z = 1.30$

37. $z = 2.40$ **38.** $z = 0.15$

Find the percentage of the area under the normal curve that lies more than the given number of standard deviations away from the mean in Exercises 39 through 42.

39. 0.86   **40.** 1.70   **41.** 1.50   **42.** 2.50

A normal distribution has a mean of 85 and a standard deviation of 5. Estimate the fraction of scores in the intervals given in Exercises 43 through 46.

43. (*See Example 9*)   Between 85 and 98

44. Between 85 and 89   **45.** Between 85 and 80

46. Between 85 and 73

A normal distribution has a mean of 226 and a standard deviation of 12. In Exercises 47 through 50, find the probability that a score is in the interval given.

47. (*See Example 10*)   Between 220 and 235

48. Between 208 and 240   **49.** Between 211 and 241

50. Between 222 and 234

A normal distribution has a mean of 140 and a standard deviation of 8. In Exercises 51 through 54, find the probability that a score is in the interval given.

51. (*See Example 11*)   Between 144 and 152

52. Between 142 and 156   **53.** Between 146 and 156

54. Between 148 and 154

A normal distribution has a mean of 75 and a standard deviation of 5. Estimate the fraction of scores in the intervals indicated in Exercises 55 through 59.

55. Between 80 and 85   **56.** Between 77 and 83

57. Above 76   **58.** Between 68 and 79

59. Either below 70 or above 80

A normal distribution has a mean of 168 and a standard deviation of 10. In Exercises 60 through 68, find the percentage of scores in the interval given.

60. Between 168 and 175   **61.** Between 155 and 169

62. Between 170 and 180   **63.** Less than 172

64. Less than 150   **65.** Larger than 173

66. Larger than 170   **67.** Less than 184

68. Less than 160

Assume normal distributions in the following exercises. Find the value of z in Exercises 69 through 74.

69. (*See Example 13*)   8% of the scores are to the right of z.

70. An estimated 15% of the scores are to the right of z.

71. An estimated 86% of the scores are to the left of z.

72. The probability that a score is to the left of z is 0.96.

73. An estimated 91% of the scores are between z and $-z$.

74. The probability that a score is between z and $-z$ is 0.82.

For each of the binomial distributions in Exercises 75 through 79, compute the mean, variance, and standard deviation.

75. (*See Example 17*)   **76.** $n = 210, p = 0.3$
$n = 50, p = 0.4$

77. $n = 600, p = 0.52$ **78.** $n = 1850, p = 0.24$

79. $n = 470, p = 0.08$

The normal distribution is a good estimate for which of the binomial distributions in Exercises 80 through 86?

80. $n = 30, p = 0.4$ **81.** $n = 50, p = 0.7$

82. $n = 40, p = 0.1$ **83.** $n = 40, p = 0.9$

84. $n = 200$, $p = 0.08$ **85.** $n = 25$, $p = 0.5$

86. $n = 15$, $p = 0.3$

87. *(See Example 19)*   For a binomial distribution with $p = 0.35$, find the smallest value of n so that the normal distribution is a reasonable approximation to the binomial distribution.

88. For a binomial distribution with $p = 0.5$, find the smallest value of n so that the normal distribution is a reasonable approximation to the binomial distribution.

89. *(See Example 18)*   Given the binomial experiment with $n = 50$ and $p = 0.7$, use the normal distribution to estimate
 (a) $P(X = 40)$ **(b)** $P(X = 28)$
 (c) $P(X = 32)$

90. Given the binomial experiment with $n = 12$ and $p = 0.5$, use the normal distribution to estimate
 (a) $P(X = 6)$ **(b)** $P(X = 7)$
 (c) $P(X = 8)$

91. *(See Example 20)*   Given the binomial experiment with $n = 15$ and $p = 0.4$, use the normal distribution to estimate
 (a) $P(4 < X < 8)$ **(b)** $P(4 \le X \le 8)$
 (c) $P(7 \le X \le 8)$

92. Given the binomial experiment with $n = 25$ and $p = 0.3$, use the normal distribution to estimate
 (a) $P(6 < X < 10)$ **(b)** $P(6 \le X \le 10)$
 (c) $P(5 \le X \le 7)$

93. *(See Example 21)*   Given the binomial experiment with $n = 16$ and $p = 0.5$, use the normal distribution to estimate
 (a) $P(X > 5)$ **(b)** $P(X \ge 5)$
 (c) $P(X > 9)$

94. Given the binomial experiment with $n = 30$ and $p = 0.4$, use the normal distribution to estimate
 (a) $P(X > 14)$ **(b)** $P(X > 14)$
 (c) $P(X \ge 9)$

95. *(See Example 22)*   Given the binomial experiment with $n = 24$ and $p = 0.3$, use the normal distribution to estimate
 (a) $P(X < 10)$ **(b)** $P(X \le 10)$
 (c) $P(X < 6)$

96. Given the binomial experiment with $n = 18$ and $p = 0.35$, use the normal distribution to estimate
 (a) $P(X < 4)$ **(b)** $P(X \le 4)$
 (c) $P(X < 9)$

LEVEL 2

97. *(See Example 12)*   Customers at Big Burger spend an average of $3.15 with a standard deviation of $0.75.
 (a) Estimate the percentage of customers who spend more than $3.75
 (b) What is the probability that a customer spends more than $3.75?

98. At Montevallo College the students from Harper County have an average GPA of 2.95 with a standard deviation of 0.50.
 (a) Estimate the percentage who have a GPA below 2.50.
 (b) What is the probability that a student from Harper County has a GPA below 2.50?

99. *(See Example 14)*   One year, the freshmen of all U.S. colleges had a mean average IQ of 110 and a standard deviation of 12. The IQ scores form a normal distribution. If a student is selected at random, find the probability that
 (a) the student has an IQ between 120 and 125.
 (b) the student has an IQ below 100.
 (c) the student has an IQ between 105 and 115.

100. The scores on a standardized test have a mean of 80 and a standard deviation of 5. A test is selected at random. Find the probability that
 (a) the score is between 84 and 90.
 (b) the score is above 88.
 (c) the score is below 74.
 (d) the score is between 75 and 83.

In Exercises 101 through 107, use the normal curve to make the estimates requested.

101. *(See Example 23)*   The probability that a new drug will cure a certain disease is 0.6. It is administered to 100 patients. Find the probability that it will cure
 (a) 50 to 75 of them.   **(b)** more than 75.
 (c) fewer than 50.

102. A coin is tossed 30 times. Find the probability that it will land heads up at least 20 times.

103. A die is rolled 20 times. What is the probability that it turns up a 1 or a 3 six, seven, or eight times?

104. A coin is tossed 100 times. Find the probability of 55 to 60 heads.

105. A coin is tossed 100 times. Find the probability of tossing
  (a) 50 heads.
  (b) 45 to 55 heads, inclusive.
  (c) 49 to 51 heads, inclusive.

106. A die is tossed 36 times. Find the probability of throwing six to eight 4's, inclusive.

107. A certain binomial trial is repeated 64 times. The probability of success in a single trial is $\frac{1}{4}$. Find the probability of success in 20 to 24 of the times, inclusive.

LEVEL 3

108. The IQs of individuals form a normal distribution with mean $= 100$ and $\sigma = 15$.
  (a) Estimate the percentage of the population who have IQs below 85.
  (b) Estimate the percentage who have IQs over 130.
  (c) A college requires an IQ of 120 or more for entrance. Estimate the percentage of the population from which it must draw students.
  (d) In a state with 400,000 high school seniors, estimate how many meet the IQ requirements.
  (e) What is the probability that a student selected at random has an IQ below 90?

109. The grades in a large American history class fall reasonably close to a normal curve with a mean of 66 and a standard deviation of 17. The professor curves the grades.
  (a) If the top 12% receive A's, what is the cutoff score for an A?
  (b) If the top 6% receive A's, what is the cutoff score for an A?

110. The batteries used for a calculator have an average life of 2000 hours and a standard deviation of 200 hours. The normal distribution closely represents the life of the battery. Find the fraction of batteries that can be expected to last the following lengths of time:
  (a) Between 1800 and 2200 hours
  (b) Between 1900 and 2100 hours
  (c) At least 2500 hours
  (d) No more than 2200 hours
  (e) Less than 1500 hours
  (f) Between 1600 and 2300 hours
  (g) Less than 1900 or more than 2100 hours
  (h) Between 2200 and 2400 hours
  (i) What is the probability that a battery lasts longer than 2244 hours?

111. The scores of students on a standardized test form a normal distribution with a mean of 300 and a standard deviation of 40.
  (a) Estimate the fraction of students who scored between 270 and 330.
  (b) Estimate the fraction of students who scored between 210 and 330.
  (c) Estimate the fraction of students who scored less than 300. Less than 326.
  (d) What must a student score to be in the upper 10%?
  (e) A student is selected at random. What is the probability that the student scored between 312 and 324?

112. An automatic lathe makes shafts for a high-speed machine. The specifications call for a shaft with a diameter of 1.800 inches. An inspector closely monitors production for a week and finds that the actual diameters form a normal distribution with a mean of 1.800 inches and a standard deviation of 0.00033 inch. The design engineer will accept a shaft that is within 0.001 inch of the specified diameter of 1.800. If a shaft is selected at random, what is the probability that it is within the specified tolerance, that is, that the diameter is in the interval from 1.799 to 1.801 inches?

113. A standardized test is represented by a normal distribution and has a mean of 120 and a standard deviation of 10. If two students are selected at random, what is the probability that both score below 128?

114. Grades on a sociology test are reasonably close to a normal curve and have a mean of 74 and a standard deviation of 10. The professor wants to curve the grades so that the highest 10% receive A and the lowest 5% receive F. The next 25% below an A receive B, and the next 10% above F receive D. The remainder between the B's and D's receive C. Find the cutoff scores for each letter grade.

115. The manufacturer of an electronics device knows that the length of life of the device is a normal distribution with a mean of 1050 hours and a standard deviation of 50 hours. Find the probability that a device will last at least 1140 hours.

116. The average rainfall for September in Hillsboro is 4.65 inches with a standard deviation of 1.10 inches. Find the probability that next September's rainfall is below 5.2 inches.

117. An airline has 10% of its reservations result in no-shows. It books 270 passengers on a flight that has

250 seats. Find the probability that all passengers who show will have a seat.

118. A drug company has developed a new drug that it believes is 90% effective. It tests the drug on 500 people. Find the probability that at least 90% of the people respond favorably to the drug.

119. A college report states that 30% of its students commute to school. If a random sample of 250 students is taken, find the probability that at least 65 commute.

120. An insurance company estimates that 5% of automobile owners do not carry liability insurance. If 60 cars are stopped at random, what is the probability that less than 5% of them have no liability insurance?

121. A true–false test consists of 90 questions. Three points are given for each correct answer and one point is deducted for each incorrect answer. A student must score at least 98 points to pass.
  (a) How many correct answers are required to pass if a student answers all questions?
  (b) If a student guesses all answers, find the probability of a passing grade.

122. In a certain city of 30,000 people, the probability of a person being involved in a motor accident in any one year is 0.01. Find the probability of more than 250 people having accidents in a year.

In Exercises 123 through 128, use $P(X = x) = C(n, x)\, p^x q^{n-x}$ to compute the binomial probabilities.

123. Given the binomial experiment with $n = 15$ and $p = 0.4$,
  (a) find $P(4 < X < 8) = P(X = 5) + P(X = 6) + P(X = 7)$.

(b) compare your answer with that of Exercise 91 (a).

124. Given the binomial experiment with $n = 15$ and $p = 0.5$,
  (a) find $P(4 < X < 8)$.
  (b) use the normal distribution to estimate $P(4 < X < 8)$.
  (c) compare the results in parts (a) and (b).

125. Given the binomial experiment with $n = 15$ and $p = 0.8$,
  (a) find $P(10 < X < 14)$.
  (b) use the normal distribution to estimate $P(10 < X < 14)$.
  (c) compare the results in parts (a) and (b).

126. Given the binomial experiment with $n = 15$ and $p = 0.8$,
  (a) find $P(7 < X < 11)$.
  (b) use the normal distribution to estimate $P(7 < X < 11)$.
  (c) compare the results in parts (a) and (b).

127. For a binomial distribution $n = 60$ and $p = 0.5$,
  (a) find the mean and standard deviation.
  (b) What values of X (integers) lie within 1 standard deviation of the mean?
  (c) Find the probability that a randomly selected integer, X, lies within 1 standard deviation of the mean.

128. For a binomial distribution $n = 60$ and $p = 0.7$,
  (a) find the mean and standard deviation.
  (b) What values of X (integers) lie within 1 standard deviation of the mean?
  (c) Find the probability that a randomly selected integer, X, lies within 1 standard deviation of the mean.

EXPLORATIONS

129. A probability distribution is a standard normal distribution with a mean of 0 ($\mu = 0$) and a standard deviation of 1 ($\sigma = 1$).
  (a) Find the probability that z is between 0.5 and 1.5, that is, $P(0.5 \le Z \le 1.5)$.
  (b) Find the probability that z is between 0.6 and 1.4, that is, $P(0.6 \le Z \le 1.4)$.
  (c) Find $P(0.7 \le Z \le 1.3)$.
  (d) Find $P(0.8 \le Z \le 1.2)$.
  (e) Find $P(0.9 \le Z \le 1.1)$.
  (f) Find $P(0.95 \le Z \le 1.05)$.
  (g) Find $P(0.99 \le Z \le 1.01)$.
  (h) Notice that as the interval squeezes in on $Z = 1$, the probabilities get smaller. Does it then seem reasonable to say that $P(Z = 1) = 0$? In fact, $P(Z = c) = 0$ for any value of c in a continuous normal distribution.

130. The values in a data set range from 0 through 100, the mean $\mu = 50$, and the median equals 65. Would you expect the data to form a normal distribution?

131. The values in a data set range from 0 to 100 and the mean $\mu = 70$. Would you expect the data to form a normal distribution?

132. Mid-State Utilities made a study of 3159 residential customers' electric bills for one month. They found the bills ranged from $24.75 to $239.60, the mean was $124.30, and the median was $91.80. Based on this information, is it reasonable to conclude that the amounts of the electric bills approximate a normal distribution?

133. The grades on a departmental exam in chemistry range from 25 to 160. The median is 93 and the mean is 92.1. Based on this information alone, do you think it is possible for the grades to approximate a normal distribution?

134. Smooth the histogram shown and conclude whether or not a normal curve provides a fairly reasonable model of the data. Give reasons for your answer.

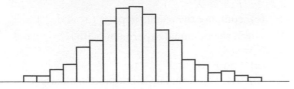

135. Smooth the histogram shown and conclude whether or not a normal curve provides a fairly reasonable model of the data. Give reasons for your answer.

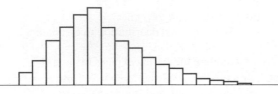

136. Smooth the histogram shown and conclude whether or not a normal curve provides a fairly reasonable model of the data. Give reasons for your answer.

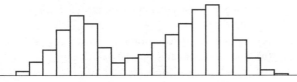

137. The frequency table shown gives the 360 grades on the precalculus departmental final at Baylor University for fall 1994.

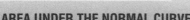

(a) Find the mean, median, the 25th percentile, and the 75th percentile of the data.

(b) Find the standard deviation.

(c) Find the intervals, $\pm\sigma$, $\pm2\sigma$, and $\pm3\sigma$ from the mean, and the percentage of grades in each interval. Are the percentages obtained reasonably consistent with the percentages expected in a normal distribution?

(d) Make a histogram using 10-point intervals for the categories.

138. Graph the histogram for each of the following binomial distributions. In each case, tell whether or not you think a normal curve is a reasonable representation of the distribution. Use the BIOD or BIOH program to compute the distributions.

(a) $n = 8$, $p = 0.5$, $q = 0.5$

(b) $n = 8$, $p = 0.3$, $q = 0.7$

(c) $n = 8$, $p = 0.1$, $q = 0.9$

In general, is it reasonable to approximate a binomial distribution with a small value of n with a normal curve?

Grade	Fre-quency	Grade	Fre-quency	Grade	Fre-quency
7	1	49	4	75	10
11	1	50	3	76	10
13	1	51	2	77	14
16	1	52	3	78	12
18	1	53	1	79	12
19	2	55	1	80	9
20	1	56	3	81	8
21	1	57	6	82	14
24	1	59	1	83	8
25	1	60	8	84	9
29	2	61	10	85	7
30	1	62	7	86	7
32	2	63	4	87	4
33	3	64	3	88	2
34	2	65	9	89	5
35	1	66	13	90	2
37	2	67	9	91	9
38	3	68	13	92	9
39	1	69	8	93	4
43	3	70	7	94	6
44	1	71	11	95	3
45	1	72	10	96	3
46	4	73	6	97	1
47	4	74	6	98	1
48	2				

USING YOUR TI-83

AREA UNDER THE NORMAL CURVE

The standard normal table enables you to find the area under a normal curve for limited values of z. We have a program for the TI-83 that finds the area for other z-values. This program calculates the area under the normal curve between two scores.

NORML

: Lbl 1	: Input U
: Disp "MEAN"	: (L-M)/S → W
: Input M	: (U-M)/S → Z
: Disp "STD DEV"	: fnInt((e^(-X²/2))/√ (2π),X,W,Z) → A
: Input S	: Disp "AREA="
: Disp "LOW LIMIT"	: Pause A
: Input L	: Goto 1
: Disp "UP LIMIT"	: End

Note: fnInt is found in the [MATH] <MATH> menu.

   Input: The mean and standard deviation of the normal distribution, the scores that define the lower and upper limits of the area. If the area above a specified score is desired, use a score that is 3.5 standard deviations above the mean for the upper limit. Similarly, a score 3.5 standard deviations below the mean is used as the lower limit when the area below a specified score is desired.

   Output: The area under the normal curve bounded by the given upper and lower limit scores.
   We give two illustrations.

(a) A data set has a mean of 26 and a standard deviation of 1.5. Find the area under the normal curve between 23 and 30. The first screen shows the mean, standard deviation, and limits entered. The second screen shows the area between limits, A = .9734.

(b) For the data set in (a), find the area above 24. Notice in the first screen that the upper limit is 32 because it is more than 3.5 standard deviations above the mean, 26. ($3.5 \times 1.5 = 5.25$, and 32 is 6 above the mean.) The second screen shows the area under the normal curve above 24, A = 0.9087.

 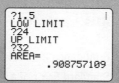

   Note: This program may give slightly different answers than those obtained using the standard normal table because z in the table is rounded to two decimals and A is rounded to four decimals. The program carries more decimal places in its calculations.

EXERCISES

1. A data set has a mean of 48 and a standard deviation of 3. Find the area under the normal curve between 46 and 53.

2. A data set has a mean of 120 and a standard deviation of 10. Find the area under the normal curve between 125 and 135.

3. A data set has a mean of 65 and a standard deviation of 4.2. Find the area under the normal curve above 62.

4. A data set has a mean of 76 and a standard deviation of 5. Find the area under the normal curve below 73.5.

USING EXCEL

EXCEL has a function that finds the area under the normal curve that is *below* a score x. This differs from the standard normal table we have used, that finds the area between the mean and the score. Here's how to use it.

Store the value of x in A2, the mean in C2, and the standard deviation in D2. Let's put the area under the normal curve below x in cell A3 by entering the formula =NORMDIST(A2,C2,D2,1) in A3. For example, if we enter $x = 35$ in A2, mean $= 30$ in C2, and standard deviation $= 5$ in D2, then the formula returns 0.84134474 in A3.

To find the area between two scores, we enter the smaller score, say 23, in A2, the larger score, say 35, in B2, the mean, 30, in C2, and the standard deviation, 5, in D2. We obtain the area between the two scores in A4 by subtracting the area below the smaller score from the area below the larger score. In A4, enter the formula =NORMDIST(B2,C2,D2,1)-NORMDIST(A2,C2,D2,1). In this case, it returns the area 0.76058803.

To find the area above the score in A2, use the formula =1-NORMDIST(A2,C2,D2,1).

EXERCISES

In the following exercises, find the area (to 4 decimals) under the normal curve with the given mean and standard deviation.

1. Mean $= 45$, standard deviation $= 3$. Find the area below $x = 49$.
2. Mean $= 45$, standard deviation $= 3$. Find the area below 40.
3. Mean $= 130$, standard deviation $= 8$. Find the area between 125 and 136.
4. Mean $= 130$, standard deviation $= 8$. Find the area between 120 and 128.
5. Mean $= 130$, standard deviation $= 8$. Find the area between 134 and 138.
6. Mean $= 92$, standard deviation $= 6$. Find the area above 95.
7. Mean $= 92$, standard deviation $= 6$. Find the area above 85.

8.8   Estimating Bounds on a Proportion

- Confidence Intervals
- Standard Error of a Proportion
- Computing Error Bounds for a Proportion
- Quality Control

Mr. Alexander Quality, President of Quality Cola Company, grew tired of Quality Cola. He wanted a cola with more zest and a new taste. Like his father before him, he had vowed never to drink his competitor's cola. Thus, he must develop a new quality cola or resign himself to the traditional taste. He discussed the problem with departmental heads. The head of the research division agreed they could develop a new formula, but it would cost thousands of dollars. The chief accountant insisted that they should recover the development cost and make a profit. The marketing manager hesitated to put a new product on the market unless she was confident that it would succeed.

Mr. Quality, an astute executive, agreed that his managers had valid points, so he instructed them to develop a new formula, find out if the public liked it, and if so, pour money into advertising it. After weeks of work, the research division developed a formula that both they and Mr. Quality liked. Now the marketing manager wanted to know if the public liked it. She quickly determined that it was quite unrealistic and prohibitive in cost to give everyone in the country a taste test. So, she asked the company statistician to help her. She told the statistician that she was confident that the new cola would be successful if 40% or more of the population liked it. The statistician outlined the following plan:

1. Select, by a random means, 500 people throughout the country.
2. Give each one a taste test.
3. Find the proportion of the sample that like the new cola.
4. Use the sample proportion as an estimate of the proportion of the total population that like the new cola.

It took the statistician several weeks to select and survey the sample. When the information was in and tabulated, it showed that 43% of the people in the sample test liked the new cola. At first the marketing manager was elated. Enough people liked the new product to make it successful. Then, she had second thoughts. What about the millions of people who did not participate in the taste test? They were the ones who would determine the success of the new cola, so she called the statistician.

"Can I depend on 43% of everyone liking the new cola? Perhaps you just happened to pick the few people who like it."

"I cannot guarantee that precisely 43% of the general public will like it. I told you this was an estimate."

"How good is the estimate? If the estimate is off 2 or 3 percentage points, we are O.K. If the proportion for the entire population is actually only 20%, we are in real trouble. Can you put some bounds on how much the estimate might be in error?"

We interrupt this saga to give some background of how this analysis works. The market analysis in this story involves a population and a sample. The **population** consists of all the people who are potential customers, millions of them, perhaps. The **sample** consists of the 500 people selected for the taste test. The results of the taste test showed that the **proportion** of the sample who liked Quality Cola was 43%. We denote the **sample proportion** by $\bar{p}$ and $\bar{p} = 0.43$. (In this case, we use the decimal form rather than the percentages.)

For the entire population, some proportion will like Quality Cola, and we denote that **population proportion** with p. For all practical purposes, the population proportion, p, is impossible to determine even though the marketing manager would dearly love to know it.

Let's use the Quality Cola story to describe an important relationship between the proportion, p, of an arbitrary population and the proportion, $\bar{p}$, of a sample taken from the population.

The proportion of the population who like Quality Cola is unknown, but let's suppose that somehow we know it is $p = 0.45$ (45% like Quality Cola). In such a case, the sample proportion $\bar{p} = 0.43$ yields a reasonably good estimate.

However, if we took other random samples of 500 people, we would not expect to find the same proportion who like Quality Cola. We might find $\bar{p} = 48\%$ in one sample, 39% in another, 41% in another, or maybe even as much as 65% in a sample.

Now here is the interesting, and useful, relationship. If a large number of random samples are taken, and the percentage who like Quality Cola is recorded in each case, then a data set of sample proportions is obtained. As a set of numbers they have a mean and a standard deviation.

If the sample size is large enough, then the proportions obtained from a large number of samples form what is reasonably close to a normal distribution. An amazing result occurs when the mean of all possible sample proportions is computed. This mean equals the population proportion, p.

Let's illustrate this with a simple example.

EXAMPLE 1 ➤

A deck of cards contains a large number of cards (thousands or maybe even millions of cards). Each card in the deck is numbered with a 1, 2, 3, or 4. Each of the numbers appears on one fourth of the cards.

The population in this example is the deck of cards. The proportion of cards that contain a 1 is $p = \frac{1}{4}$. Now suppose someone who doesn't know the makeup of the deck wants to find the proportion of cards that contain a 1. They proceed by randomly selecting two cards from the deck and noting the proportion of 1's, that is, they select a sample of two cards and note $\bar{p}$ for the sample. The cards are replaced and other random samples are taken. For each sample taken, three kinds of outcomes are possible:

(a) Neither card is a 1 $(\bar{p} = 0)$.
(b) One card is a 1 and the other card is not $(\bar{p} = \frac{1}{2})$.
(c) Both cards are 1's $(\bar{p} = 1)$.

After a number of samples, a list of numbers consisting of 0's, $\frac{1}{2}$'s, and 1's is obtained. The mean of these numbers estimates the population proportion. If we did the impractical and took *all* possible samples of two cards, the theory indicates that the mean of the sample proportions would equal the population proportion. Rather than trying the impractical, let's find the mean of the proportions by finding the *expected value*. We need the probability of the three possible values of $\bar{p}$:

(a) $\bar{p} = 0$ when neither card is a 1. The probability that a single card drawn is not a 1 is $\frac{3}{4}$. Thus, the probability the first card is not a 1, and the second card is not a 1 is $\left(\frac{3}{4}\right)\left(\frac{3}{4}\right) = \frac{9}{16}$.

(b) $\bar{p} = \frac{1}{2}$ when one card is a 1 and the other isn't. This occurs when the first card is a 1 and the second isn't, or the first card isn't a 1 and the second is. The probability of this occurring is $\frac{1}{4}\left(\frac{3}{4}\right) + \frac{3}{4}\left(\frac{1}{4}\right) = \frac{6}{16}$.

(c) $\bar{p} = 1$ when both cards are 1's. The probability of this occurring is $\frac{1}{4}\left(\frac{1}{4}\right) = \frac{1}{16}$.

We have $P(\bar{p} = 0) = \frac{9}{16}$, $P(\bar{p} = \frac{1}{2}) = \frac{6}{16}$, and $P(\bar{p} = 1) = \frac{1}{16}$. So, the expected value of $\bar{p}$ is

$$0\left(\frac{9}{16}\right) + \frac{1}{2}\left(\frac{6}{16}\right) + 1\left(\frac{1}{16}\right) = \frac{4}{16} = \frac{1}{4}$$

Thus, the mean of *all* sample proportions is equal to the population proportion $p = \frac{1}{4}$.

Notice that *none* of the sample proportions could possibly equal the population proportion, but the mean of the sample proportions does. ■

It is unrealistic to take all possible samples of a large population, so one possible strategy is similar to that taken at a nearby university. Here's the situation.

EXAMPLE 2 ➤

The dean of students received a student petition requesting a change in the date of spring break. The dean had no desire to consider a change unless there was strong student support beyond the relatively small number who signed the petition.

The dean approached Professor Turner, a statistics professor, for help. Professor Turner had just taught the chapter on sampling to the class, so it seemed like a good project to assign to the students. Each of the 15 students was to make a random sample of 60 students and determine the proportion of the sample that favored a change in spring break.

The dean was somewhat dismayed when the sample proportions were all different. Which one should be used?

Because the mean of all sample proportions equals the population proportion, it seems reasonable that the mean of the 15 samples should provide a better estimate of the population proportion than any sample alone. So, Professor Turner recommended that the dean use the mean of the 15 sample proportions as the estimate of the population proportion. ■

The procedure of taking several samples and averaging the $\bar{p}$'s is not the method generally used because it is often a major effort to find just one random sample. So, we go to the other extreme and take just one sample. The proportion obtained is used as the estimate for the population proportion. When only one sample is used, we cannot take a small sample like Example 1 where it was impossible to obtain a sample proportion near the population proportion. The sample size must be large enough to allow a sample proportion close to the population proportion. Realizing the sample proportion is likely to be different from the population proportion, it is important to know how close the sample proportion is to the population proportion. The fact that the sample proportions form a normal distribution helps answer that.

We remind ourselves that we are dealing with

1. A population of which a proportion, p, have a characteristic of interest (such as liking Quality Cola).
2. Random samples of the population are taken and the proportion, $\bar{p}$, of the sample that have the characteristic of interest is obtained. The set of possible $\bar{p}$'s has a mean equal to the population proportion p.
3. If the sample size is large, the distribution of the $\bar{p}$'s is approximately normal.

Because the $\bar{p}$'s form a normal distribution, they tend to be clustered fairly close to the population proportion, p.

Confidence Intervals

In the Quality Cola example, the marketing manager would feel comfortable if she knew the sample proportion was within 3% of the actual, but unknown, population proportion. (See Figure 8–42a.)

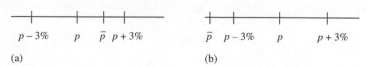

(a)   (b)

FIGURE 8–42 The Quality Cola marketing manager would like $\bar{p}$ to be in the interval $p - 3\%$ to $p + 3\%$ [as in part (a)], not outside the interval [as in part (b)].

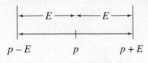

A confidence interval for *p*.

FIGURE 8–43

Because the population proportion, p, is unknown and unattainable, we use a sample proportion, $\bar{p}$, as an estimate of the population proportion, even though $\bar{p}$ rarely turns out to be p. Still, we expect the sample proportion $\bar{p}$ to be close to p in the sense that $\bar{p}$ is in some interval centered at p, say in the interval from $p - E$ to $p + E$. We call E the **maximum error of the proportion.** (In the Quality Cola example, $E = 3\%$.) The longer this interval, the more confident we are that it contains p. (See Figure 8–43.)

When we state how close p is to $\bar{p}$, we consider the interval $\bar{p} - E$ to $\bar{p} + E$ and discuss the probability that p lies in this interval.

It is reasonable to expect a larger value of E, and thus a longer interval, to be more likely to contain p than a shorter interval. We say we are more *confident* the longer interval contains the population proportion p. We would like to do more than say we are more or less confident. We want to measure the confidence. This leads to the idea of a confidence interval and confidence level. We call the degree of confidence a **confidence level,** and we express it as a percentage. When we say that we are 100% confident that p lies in the interval from $\bar{p} - E$ to $\bar{p} + E$, we are saying we are certain. When we say that we are 90% confident p lies in the interval $\bar{p} - E$ to $\bar{p} + E$ ($\bar{p} - E \le p \le \bar{p} + E$), we are saying that if we take a sample, the probability that p for the population lies in the confidence interval is 0.90.

DEFINITION
**Confidence Interval,
Confidence Level**

Let $\bar{p}$ be the proportion of any random sample from a population and let E be a positive number.

The numbers between $\bar{p} - E$ and $\bar{p} + E$ form an interval called the **confidence interval** for the population proportion.

The **confidence level** associated with this interval is the probability $P(\bar{p} - E \le p \le \bar{p} + E)$ expressed as a percentage.

We now proceed to discuss how we determine E.

Standard Error of a Proportion

We introduce the term **standard error,** abbreviated S.E. The standard error is the term used when referring to the standard deviation of the $\bar{p}$'s, and it is rather simple to compute:

$$\text{S.E.} = \sqrt{\frac{p(1 - p)}{n}}$$

where n is the sample size and p is the mean of the $\bar{p}$'s and is also the population proportion.

Unfortunately, we don't know the value of p, which is the very thing we are attempting to find. Because we are going to need S.E. to estimate a confidence interval, we use

$$\text{S.E.} \approx \sqrt{\frac{\bar{p}(1 - \bar{p})}{n}}$$

which is considered a reasonable estimate.

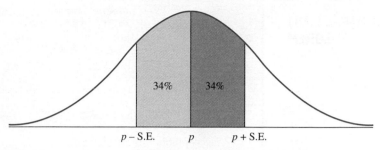

FIGURE 8–44 Because $\bar{p}$ forms a normal distribution, about 68% of the $\bar{p}$'s lie within 1 S.E. of the mean of the $\bar{p}$'s (mean $= p$).

DEFINITION
Estimate of Standard Error

$$\text{S.E.} \approx \sqrt{\frac{\bar{p}(1 - \bar{p})}{n}}$$

The properties of a normal distribution tell us, for example, that about 68% of the $\bar{p}$'s lie within 1 S.E. of the p. (See Figure 8–44.)

We also say that the probability that $\bar{p}$ lies within 1 S.E. of p is 0.68, and consequently, the probability that p lies within 1 S.E. of $\bar{p}$ is 0.68. The interval $\bar{p} - \text{S.E.}$ to $\bar{p} + \text{S.E.}$ is a 68% confidence interval.

We are primarily interested in three confidence levels: 90%, 95%, and 99%, with the 95% confidence level being the one most used in practice.

Computing Error Bounds for a Proportion

Let's look at how we find a 95% confidence interval. We want to find the interval $(\bar{p} - E, \bar{p} + E)$ so that the probability that p is in that interval is 0.95; that is, $P(\bar{p} - E \leq p \leq \bar{p} + E) = 0.95$. As $\bar{p}$ forms a normal distribution, we are seeking the area under the normal curve such that 95% of the area is between $\bar{p} - E$ and $\bar{p} + E$, 47.5% to the left of $\bar{p}$ and 47.5% to the right of $\bar{p}$.

From the standard normal table, $z = 1.96$ for $A = 0.475$, so we conclude that $E = 1.96\,(\text{S.E.})$. (See Figure 8–45.)

The procedure for estimating p by using the proportion of a random sample, $\bar{p}$, is justified by a fundamental theorem in statistics, called the Central Limit Theorem.

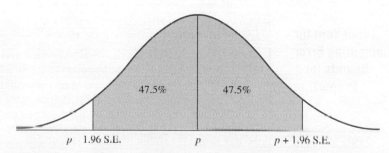

FIGURE 8–45 95% of all $\bar{p}$'s lie within 1.96 S.E.'s of the population proportion, p.

CENTRAL LIMIT THEOREM

Let $\bar{p}$ represent the proportion of a sample of size n that has a certain property and let p represent the proportion of the entire population that has the same property. When the sample size is large $(n\bar{p} \geq 5$ and $n(1 - \bar{p}) \geq 5)$, then a normal distribution is a good representation of the distribution of $\bar{p}$ with

Mean of the $\bar{p}$'s $= \mu_{\bar{p}} = p =$ the population mean

Standard error of the $\bar{p}$'s $=$ S.E. $= \sqrt{\dfrac{p(1 - p)}{n}}$

An acceptable estimate for S.E. is

$$\text{S.E.} \approx \sqrt{\dfrac{\bar{p}(1 - \bar{p})}{n}}$$

The Central Limit Theorem allows us to use the properties of a normal distribution in the study of a distribution that is not normal. Even if a population does not have normal distribution properties, the sample proportions, $\bar{p}$, are normally distributed. When we use sufficiently large samples, generally over 50, we can make inferences about the population based on the properties of a normal distribution.

Now let's go back to the Quality Cola story and find the 95% confidence interval for the sample of 500 that gave $\bar{p} = 0.43$.

EXAMPLE 3 ▶

$$n = 500 \quad \text{and} \quad \bar{p} = 0.43$$

Find the 95% confidence interval.

SOLUTION

$$\text{S.E.} = \sqrt{\dfrac{0.43(0.57)}{500}} = 0.0221$$

Then the maximum error $E = 1.96(0.0221) = 0.043$. The 95% confidence interval is from $0.43 - 0.043 = 0.387$ to $0.43 + 0.043 = 0.473$.

Thus, we say $0.387 \leq p \leq 0.473$ with 95% confidence. The Quality Cola marketing manager should be pleased because the lower end of the confidence interval is very nearly the 40% favorable response she desired. ■

In general, we use $E = z \times$ S.E., where z depends on the confidence level desired. For a 90% confidence level, find z in the standard normal table that corresponds to $A = 0.45$ (one-half of 0.90). You will find $z = 1.65$. For a 99% confidence level, use $z = 2.58$ because it corresponds to $A = 0.495$ (one-half of 0.99).

Procedure for Computing Error Bounds for a Proportion

Let n be the sample size and $\bar{p}$ the proportion of the sample that respond favorably.

1. Decide on the confidence level to be used and write it as a decimal c.
2. Compute $A = \frac{c}{2}$. This corresponds to the area under the standard normal curve between $z = 0$ and the z-score that corresponds to an area equal to A. (A is the same as that found in the normal distribution table.)
3. Find the value of z in the standard normal table that corresponds to A.
4. Compute the standard error estimate S.E. $\approx \sqrt{\dfrac{\bar{p}(1 - \bar{p})}{n}}$.

5. Compute the maximum error $E = z \times \text{S.E.} = z \times \sqrt{\dfrac{\bar{p}(1 - \bar{p})}{n}}$.

6. Compute the upper and lower limits

$$\bar{p} + E \quad \text{and} \quad \bar{p} - E$$

7. Then c, the confidence level, is the probability that the proportion of the *total* population, p, lies in the interval

$$\bar{p} - E \le p \le \bar{p} + E$$

EXAMPLE 4 ➤

Compute the error bounds for the proportion $\bar{p} = 0.55$ obtained from a sample of size $n = 120$. Use the 95% confidence level.

SOLUTION

The steps for the procedure give

1. $n = 120$, $\bar{p} = 0.55$, and $c = 0.95$.
2. $A = \dfrac{0.95}{2} = 0.475$.
3. The value of z that corresponds to $A = 0.475$ is $z = 1.96$.
4. $\text{S.E.} = \sqrt{\dfrac{(0.55)(0.45)}{120}} = \sqrt{0.0020625} = 0.045$.
5. $E = z \times \text{S.E.} = 1.96 \times 0.045 = 0.0882$.
6. The upper and lower bounds of the proportion are

$$0.55 + 0.0882 = 0.6382$$
$$0.55 - 0.0882 = 0.4618$$

and the 95% confidence interval is

$$0.4618 < p < 0.6382$$

■ **Now You Are Ready to Work Exercise 2**

EXAMPLE 5 ➤

A marketing class made a random selection of 150 shoppers at a shopping mall to participate in a taste test of different brands of coffee. They found that 54 shoppers preferred brand X. Find, at the 95% confidence level, the error bounds of the proportion of shoppers who prefer brand X.

SOLUTION

For this problem, $n = 150$, $\bar{p} = \frac{54}{150} = 0.36$, and $c = 0.95$. Then,

$$A = \frac{0.95}{2} = 0.475$$

$z = 1.96$ corresponds to $A = 0.475$ in the standard normal table

$$\text{S.E.} = \sqrt{\frac{(0.36)(0.64)}{150}} = \sqrt{0.001536} = 0.0392$$

$$E = 1.96(0.0392) = 0.0768$$

The bounds are

$$0.36 + 0.0768 = 0.4368 \quad \text{and} \quad 0.36 - 0.0768 = 0.2832$$

and the 95% confidence interval is

$$0.2832 < p < 0.4368$$

The marketing class is 95% confident that between 28.32% and 43.68% of all shoppers prefer brand X.

■ **Now You Are Ready to Work Exercise 9**

It should not escape your attention that the error bounds in this example give a rather large confidence interval. One way to maintain the same confidence level and reduce the size of the confidence interval is to increase the sample size.

EXAMPLE 6 ➤ Suppose that the sample in Example 5 was $n = 500$ in size but the proportion remained the same. Compute the error bounds of the proportion.

SOLUTION

Now, $n = 500$, $\bar{p} = 0.36$, and $c = 0.95$. We still have $z = 1.96$, but

$$\text{S.E.} = \sqrt{\frac{(0.36)(0.64)}{500}} = \sqrt{0.0004608} = 0.0215$$

Then, $E = 1.96(0.0215) = 0.0421$, and the upper and lower bounds are

$$0.36 + 0.0421 = 0.4021 \qquad \text{and} \qquad 0.36 - 0.0421 = 0.3179$$

and the confidence interval is

$$0.3179 < p < 0.4021$$

■ **Now You Are Ready to Work Exercise 15**

Notice that this confidence interval is smaller, so the sample proportion is a better estimate of the total population. This is generally true; a larger sample size reduces the maximum error. The sample size that will keep the maximum error to a specified level can be determined. (See Example 9.)

EXAMPLE 7 ➤ A random sample of 200 people shows that 46 of them use No-Plaque toothpaste. Based on a 98% confidence level, estimate the proportion of the general population that uses the toothpaste.

SOLUTION

For this sample,

$$n = 200$$
$$\bar{p} = \frac{46}{200} = 0.23$$
$$c = 0.98$$
$$A = 0.49$$
$$z = 2.33$$
$$\text{S.E.} = \sqrt{\frac{(0.23)(0.77)}{200}} = 0.02976$$
$$E = 2.33(0.02976) = 0.0693$$

Then, the interval that contains the proportion of the general population is from $0.23 - 0.0693 = 0.1607$ to $0.23 + 0.0693 = 0.2993$. We conclude, with 98% confidence, that about 16% to 30% of the population use No-Plaque toothpaste. Notice that the higher confidence level, 98% in this case, requires a larger confidence interval.

■ **Now You Are Ready to Work Exercise 19**

EXAMPLE 8 ➤

A random sample of 25 shoppers showed that 24% shopped at Cox's Department Store. Find a 90% confidence interval of the proportion of the general population that shop there.

SOLUTION

$n = 25$, $\bar{p} = 0.24$, $c = 0.90$, $A = 0.45$, $z = 1.65$, S.E. $= \sqrt{0.007296} = 0.0854$, and $E = 0.1409$. (Be sure you check these computations.) So, the upper and lower bounds are

$$0.24 + 0.1409 = 0.3809 \qquad \text{and} \qquad 0.24 - 0.1409 = 0.0991$$

So, the 90% confidence interval is $0.0991 < p < 0.3809$. Notice that this small sample yields a wide confidence interval.

■ **Now You Are Ready to Work Exercise 21**

EXAMPLE 9 ➤

KWTX television station wants an estimate of the proportion of the population that watches its late movie. The station wants the estimate correct within 5% at the 95% confidence level. How big a sample should it select?

SOLUTION

Basically, the television station wants a maximum error of 5%, written 0.05 in our computations, at the 95% confidence level.

Look at the computations to obtain E. They are

$$E = z\sqrt{\frac{\bar{p}(1 - \bar{p})}{n}}$$

We are given $E = 0.05$, and we know that $z = 1.96$ for the 95% confidence level. We need to find n so that

$$0.05 = 1.96\sqrt{\frac{\bar{p}(1 - \bar{p})}{n}}$$

We face a dilemma. We need to know the value $\bar{p}$ so we can solve for n. However, we find $\bar{p}$ from the sample. Thus, it appears that we need $\bar{p}$ before we know the size of the sample, and we need the sample to find $\bar{p}$. There is a way out of this vicious circle. It can be shown that the largest possible value of $\bar{p}(1 - \bar{p})$ is 0.25 and occurs when $\bar{p} = 0.5$. (See Exercise 43.) So, if we use 0.25 for $\bar{p}(1 - \bar{p})$, the value of E may be a little too high and the resulting confidence interval a little larger than necessary, but we have erred on the safe side. We proceed using 0.25. Then,

$$0.05 = 1.96\sqrt{\frac{0.25}{n}}$$

Squaring both sides, we get

$$0.0025 = (1.96)^2 \frac{0.25}{n} = \frac{0.9604}{n}$$

Then,

$$0.0025n = 0.9604$$
$$n = \frac{0.9604}{0.0025} = 384.16$$

A sample size of 385 will be sufficient to provide the desired maximum error.

■ **Now You Are Ready to Work Exercise 23**

Quality Control

Manufacturing companies are concerned with maintaining the quality of their products. Defective items can occur because of a random glitch or because a machine may become worn or out of adjustment or a new employee may not understand the procedures. A manufacturer usually recognizes that some defective items caused by random glitches will occur, but if the defective items become excessive, production is halted and corrective measures are taken.

A quality-control manager monitors the production process for defective items by taking random samples and using the proportion of defective items in the sample to estimate the proportion of defective items overall. As with opinion polls, the sample proportion is only an estimate, and error bounds can be quite useful. Let's look at a simple example.

EXAMPLE 10 ➤

The quality-control policy of Arita China Company specifies that production must continue as long as the quality-control manager is 95% confident that defects occur in less than 1% of the pieces produced. Otherwise, production is halted, and the equipment is adjusted and calibrated. The quality-control manager periodically takes a random sample of 450 pieces of china and carefully examines them for defects. On February 15, she found two defective pieces in the sample of 450. Should she allow production to continue?

SOLUTION

The sample estimate of the proportion of defective pieces is

$$\bar{p} = \frac{2}{450} = 0.0044 = 0.44\%$$

$$\text{S.E.} = \sqrt{\frac{.0044(.9956)}{450}} = 0.00312$$

giving the error bounds

$$0.0044 + 1.96(.00312) = 0.0105 \quad \text{and} \quad 0.0044 - 1.96(.00312) = -0.00172$$

So, with 95% confidence the population proportion of defective items is in the interval $0 < x < 0.0105$. (0 was used because the proportion is never negative.)

The accepted interval for proportion of defective items is $0 < x < 0.01$. As the computed interval $0 < x < 0.0105$ is so close to the desired interval, the quality-control manager should not halt production.

Had the computed interval been $0 < x < 0.0195$, the quality-control manager would have been justified in requesting that production be halted to calibrate the equipment.

■ **Now You Are Ready to Work Exercise 39**

8.8 EXERCISES

LEVEL 1

1. Compute the standard error for the given sample sizes and proportions:
  (a) $n = 50, \bar{p} = 0.45$ **(b)** $n = 100, \bar{p} = 0.32$
  (c) $n - 400, \bar{p} - 0.20$

Find the error bounds for the proportion given in Exercises 2 through 8.

2. *(See Example 4)* $\bar{p} = 0.64, n = 140, 95\%$ confidence level

3. $\bar{p} = 0.30$, sample size $= 50, 90\%$ confidence level

4. $\bar{p} = 0.30$, sample size $= 50, 95\%$ confidence level

5. $\bar{p} = 0.30$, sample size $= 50, 98\%$ confidence level

6. $\bar{p} = 0.40, n = 300, 95\%$ confidence level

7. $\bar{p} = 0.5, n = 400, 95\%$ confidence level

8. $\bar{p} = 0.1, n = 1000, 95\%$ confidence level

LEVEL 2

9. *(See Example 5)* A survey of 300 people showed that 45% preferred brand X cola. Find error bounds on the proportion of the population who prefer brand X, using a
  (a) 90% confidence level.
  (b) 95% confidence level.

10. A survey of 450 people showed that 35% thought that inflation would decrease the next year. Find error bounds on the proportion of the population who think inflation will decrease, using a
  (a) 95% confidence level.
  (b) 98% confidence level.

11. The Hillsboro High school paper reported that 243 students out of 300 surveyed cruise Main Street on Saturday nights. Based on a 98% confidence level, estimate the proportion of the high school population that cruise Main Street.

12. A survey of 100 adults showed that 60 of them drank coffee daily. Estimate the proportion of the adult population that drinks coffee daily and find error bounds at the 95% confidence level.

13. To estimate the proportion of patients who respond to an experimental drug, a medical researcher administered the drug to a random sample of 195 patients. She found that 144 patients responded to the drug.

(a) Estimate the proportion of the population that will respond to the drug.
(b) Find the 95% confidence interval for the proportion.
(c) Find the 99% confidence interval for the proportion.

14. Creative Education Institute took a random sample of 800 schoolchildren to estimate the proportion of children who have reading difficulties. In their sample, 108 children had reading difficulties.
  (a) Estimate the population proportion of children who have reading difficulties.
  (b) Find the 99% confidence interval of the proportion.

15. *(See Example 6)* A consulting firm found that 27% of a sample of voters favored a flat-rate income tax.
  (a) Find the 95% error bounds if the sample size was 300.
  (b) Find the 95% error bounds if the sample size was 1000.
  (c) What effect does a larger sample size have on the confidence interval?

16. A random sample of books in the Marlin Public Library revealed that 15% had not been checked out during the last 12 months.
    (a) Find the 95% error bounds if the sample size was 500.
    (b) Find the 95% error bounds if the sample size was 1600.

(c) What effect does a larger sample size have on the confidence interval?

17. A radar unit on an interstate found that 28% of the vehicles exceeded 60 mph. Find the error bounds on the proportion at the 95% confidence level if the number of vehicles checked was
    (a) 200.   (b) 400.

LEVEL 3

18. A random sample of freshmen transcripts revealed that 21% made lower than a C average in their first semester. Find error bounds on the proportion at the 95% confidence level if the sample consisted of
    (a) 275 freshmen.   (b) 500 freshmen.

19. *(See Example 7)* A health class found that 144 of 320 people surveyed were overweight by 5 pounds or more. Based on a 95% confidence level, estimate the proportion of the population that is overweight by 5 pounds or more.

20. A random check of 90 students at Classic College revealed that 40 of them did not know the location of the Reserve Room in the Library. Find error bounds on the proportion of Classic students who do not know where the Reserve Room is located. Use a 98% confidence level.

21. *(See Example 8)* A random sample of 60 shoppers showed that 54% shopped at Nate Chadrow's Department Store. Find a 90% confidence interval of the proportion of the general population that shop there.

22. A survey of 2000 families showed that 18% owned at least one handgun. Find the error bounds of this proportion using a 95% confidence level.

23. *(See Example 9)* The ARA food service wants to estimate the proportion of students who like their food. They want the estimate correct within 6% at the 95% confidence level. How big a sample should they select?

24. A candidate for Falls County commissioner wants an estimate of the proportion of voters who will vote for her. She wants the estimate to be accurate within 1% (0.01) at the 95% confidence level.
    (a) How big a sample should be used?
    (b) If a 99% confidence level is desired, how big should the sample be?

25. A state highway department needs an estimate of the percentage of vehicles that exceed 60 mph on an interstate. How large a sample is needed if the maximum error of the estimate is no more than 2% at the 95% confidence level?

26. The manufacturer of brand Y coffee wants to estimate the proportion of adults who prefer its coffee. How large a sample should the manufacturer survey to estimate the proportion with an error of 5% or less? They want a 95% confidence level.

27. Reading Skills, Inc., plans to test a computerized reading comprehension program. They speculate that 65% of children would benefit from the program. To compare confidence intervals (95% level), they assumed several sample sizes. Help the company by computing the 95% level confidence intervals for the 0.65 proportion benefiting from the program for the following samples.
    (a) Sample size = 300   (b) Sample size = 360
    (c) Sample size = 400   (d) Sample size = 460
    (e) Sample size = 500   (f) Sample size = 1000

28. Shortly after the United States launched air strikes against terrorist sites in Afghanistan in 2001, ABC News polled 506 randomly selected adults. They were asked if they supported the air strikes, the sending of humanitarian aid, and related questions. About 70% favored sending humanitarian aid. Using a 95% confidence level, find the error bounds of this proportion.

29. How big a sample should be taken to ensure a maximum error of 0.0785 in the estimate of the proportion of adults who smoke? Use a 95% confidence level.

30. A marketing class took a random sample to estimate the proportion of people in the city that watch the educational TV channel. At the 95% confidence level they reported the confidence interval $0.624 < p < 0.656$. Find the
    (a) sample proportion, $\bar{p}$.   (b) sample size, n.

31. A statistics student reported that the proportion of students who read the campus newspaper lies in the 90% confidence interval.

$$0.5007 < p < 0.5393$$

Find $\bar{p}$ and n for the sample.

32. At State University, 42% of the faculty are tenured. The president randomly selects a committee of 100 to study merit pay. Find the probability that the committee contains from 35% to 49% tenured faculty.

33. The student body is split 60–40 in favor of assessing a student fee to help finance a recreation center. If a random sample of 300 students is polled, find the probability that less than 54% of the sample favor the fees.

34. An insurance company knows that 30% of the vehicles insured by the company are equipped with air bags. If the company takes a random sample of 1800 of their insured vehicles, find the probability that a sample proportion of vehicles with air bags will be between 0.27 and 0.33.

35. At Broadman College, 45% of the students are from out of state. Find the probability that in a random sample of 500 students, at least 50% are from out of state.

36. A 1994 NCAA report stated that 57% of scholarship athletes who entered Division I institutions in 1987 graduated in six years. Suppose a random sample of these athletes is obtained. Find the probability that the sample proportion of those athletes who graduate is between 54% and 60% if the sample size is
    (a) 100.   **(b)** 300.   **(c)** 500.   **(d)** 1000.

37. A news magazine reported that 46% of a sample of 1498 people indicated they were better off now than they were five years ago. The margin of error was stated at ± 3%. Find the confidence level used.

38. A dental publication reported that 33% of 3650 children studied in a random sample could benefit from orthodontics treatment. The margin of error was reported as ± 2%. Find the confidence level.

39. *(See Example 10)* The Allied Canning Corp. has a policy that when more than 0.7% of the labels on canned food are torn or crooked, the machine that pastes labels on cans is halted and corrective adjustments are made. One day the quality-control manager found 6 cans with crooked labels in a random sample of 1600 cans. With 95% confidence, determine if the label machine should be shut down for adjustment.

40. The Ball Bearing Company requires that its production line be shut down for corrective calibration when more than 0.5% of the bearings are defective. The quality-control manager finds 6 defective bearings in a random sample of 1500 bearings. With 95% confidence, determine if production should be halted in order to calibrate the equipment.

EXPLORATIONS

41. It is known that 68% of the licensed drivers in a certain state carry liability insurance. If a large number of random samples of size 4500 licensed drivers are taken, then the sample proportion $\bar{p}$ of those who carry liability insurance, would exceed _____ only about 8.08% of the time.

42. Suppose a sample proportion $\bar{p} = 1.00$. Tell what this suggests about the population and S.E.

43. Show that $\frac{1}{4} \geq \bar{p}(1 - \bar{p})$ and the largest value occurs when $\bar{p} = \frac{1}{2}$. The steps that follow are an outline of the proof. Fill in the details and be sure you understand the steps.
    (i) $(1 - 2\bar{p})^2 \geq 0$ **(ii)** $1 - 4\bar{p} + 4\bar{p}^2 \geq 0$
    (iii) $\frac{1}{4} \geq \bar{p}(1 - \bar{p})$
    (iv) When $\bar{p} = \frac{1}{2}$, $\bar{p}(1 - \bar{p})$ achieves its largest value, $\frac{1}{4}$.

44. Report on a news article reporting the estimate of a population proportion based on a random sample. Give $\bar{p}$, n, the error bounds, and the confidence level.

45. Report on the 1948 presidential election between Thomas Dewey and Harry Truman. Discuss the erroneous prediction by opinion polls of Harry Truman's defeat.

46. Each member of the class is to select a 3-inch section of a column in a newspaper and count the frequency of the letters of the alphabet in that section. Find the total frequencies by combining everyone's finding.
    (a) Draw a histogram of the percentage of time each letter occurs.
    (b) Which is the most frequently occurring vowel?
    (c) Which are the five most frequently occurring consonants?
    (d) For the grand prize on *Wheel of Fortune,* the contestant is given the letters R, S, T, L, N, and E and is allowed to choose four consonants and one vowel. Why were the consonants R, S, T, L, and N selected for the contestant? If you were the contestant, which consonants and vowel would you choose?

IMPORTANT TERMS

8.1

Descriptive Statistics
Inferential Statistics
Categories
Frequency Table
Frequency Distribution
Qualitative Data
Quantitative Data
Histogram
Discrete Data
Continuous Data
Relative Frequency
Pie Chart
Stem-and-Leaf Plot

8.2

Measure of Central Tendency
Population Sample
Mean
Average (Arithmetic Average)
Grouped Data
Median
Mode
Skewed Data

8.3

Measures of Dispersion
Range

Standard Deviation
Variance
Deviation
Squared Deviation
Rank
Percentile
z-Score
Quartile
Box Plot
Five-Point Summary

8.4

Random Variable
Discrete Variable
Continuous Variable
Probability Distribution

8.5

Expected Value
Variance of a Random Variable
Standard Deviation of a Random
    Variable
Fair Game

8.6

Binomial Trials
Repeated Trials

Binomial Experiment
Binomial Distribution

8.7

Discrete Data
Continuous Data
Normal Distribution
Normal Curve
Standard Normal Curve
z-Score
Mean of a Binomial Distribution
Variance of a Binomial
    Distribution
Standard Deviation of a Binomial
    Distribution
Estimate of a Binomial
    Probability

8.8

Sample Proportion
Population Proportion
Maximum Error of the
    Proportion
Confidence Interval
Confidence Level
Standard Error
Error Bounds

REVIEW EXERCISES

1. Draw a histogram and a pie chart based on the following frequency table:

New Accounts Opened	Frequency
Monday	17
Tuesday	31
Wednesday	20
Thursday	14
Friday	8

2. Find the mean of 4, 6, −5, 12, 3, 2, and 9.

3. Find the median of
   (a) 8, 12, 3, 5, 6, 3, and 9.
   (b) 4, 9, 16, 12, 3, 22, 1, and 95.
   (c) 3, −2, 6, 1, 4, and −3.

4. Find the mean for the following quiz data:

Score on Quiz	Frequency
0	2
1	3
2	6
3	9
4	2
5	4

5. Estimate the mean for the number of passengers in cars arriving at a play.

Number of Passengers in a Car	Frequency
1–2	54
3–4	32
5–6	12

6. Find the mean, median, and mode for the numbers 2, 8, 4, 3, 2, 9, 6, 2, and 7.

7. A shopper paid a total of $90.22 for his purchases. The mean price was $3.47. How many items did he purchase?

8. Find the variance and standard deviation for the numbers 8, 18, 10, 16, 3, and 11.

9. A professor selects three students each day. Let X be the random variable that represents the number who completed their homework. Give all possible values of X.

10. Prof. Delgado asked her students to evaluate her teaching at the end of the semester. Their response to the statement "The tests were a good measure of my knowledge" were marked on a scale of 1 to 5 as follows:

Outcome	Random Variable X	Responses
Strongly agree	1	10
Agree	2	35
Neutral	3	30
Disagree	4	20
Strongly disagree	5	15

Find the probability distribution of X and sketch its graph.

11. A store has a special on bread with a five-loaf limit. The probability distribution for sales is listed in the following table. Find the average number of loaves per customer.

Number of Loaves per Customer	Probability
0	0.05
1	0.20
2	0.15
3	0.20
4	0.25
5	0.15

12. Find the expected value and variance for the following probability distribution:

X	$P(X)$
1	0.14
2	0.06
3	0.22
4	0.15
5	0.36
6	0.07

13. An instructor summarized his student ratings (scale of 1 to 5) in the following probability distribution:

Response	X	Probability
Excellent	1	0.20
Good	2	0.32
Average	3	0.21
Fair	4	0.15
Poor	5	0.12

What is his expected average rating?

14. A normal distribution has a mean of 80 and a standard deviation of 6. Find the fraction of scores **(a)** between 80 and 88.   **(b)** between 70 and 84.   **(c)** greater than 90.

15. A distributor averages sales of 350 mopeds per month with a standard deviation of 25. Assume that sales follow a normal distribution. What is the probability that sales will exceed 400 during the next month?

16. A construction company contracts to build an apartment complex. The total construction time follows a normal distribution with an average time of 120 days and a standard deviation of 15 days.
(a) The company will suffer a penalty if construction is not completed within 140 days. What is the probability that it will be assessed the penalty?
(b) The company will be given a nice bonus if it completes construction in less than 112 days. What is the probability that it will receive the bonus?
(c) What is the probability that the construction will be completed in 115 to 130 days?

17. Find the binomial distribution for $n = 4$ and $p = 0.25$.

18. A binomial distribution has $n = 22$ and $p = 0.35$. Find the mean and standard deviation of the distribution.

19. The probability that a new drug will cure a certain disease is 0.65. If it is administered to 80 patients, estimate the probability that it will cure
(a) more than 50 patients.
(b) 65 patients.
(c) more than 55 and fewer than 60 patients.

20. For a set of scores, the mean = 240 and $\sigma = 10$. Find
(a) z for the score 252.   **(b)** z for the score 230.
(c) the score corresponding to $z = -2.3$

21. In a cross-country ski race, a skier came in 43rd in a field of 216 skiers. What was her percentile?

22. A brother and sister took two different standardized tests. He scored 114 on a test that had a mean of 100 and a standard deviation of 18. She scored 85 on a test that had a mean of 72 and a standard deviation of 12. Which one had the better score?

23. The housing office measures the congeniality of incoming students with a test scored 1–5. A higher score indicates a higher level of congeniality. A summary of 800 tests is the following:

X (score)	Number Receiving Score
1	20
2	160
3	370
4	215
5	35

On the basis of this information, determine a probability distribution of X.

24. A number is drawn at random from $\{1, 2, 3, 4, 5, 6, 7\}$. A random variable X has the value $X = 0$ when the number drawn is even and $X = 1$ when the number is odd. Find the probability distribution of X.

25. Three papers in a creative writing class were graded A, and five were graded B. The professor selects three of them to read to the class. Let the random variable, X, be the number of A papers selected. List the possible values of X and give the number of possible outcomes that can be associated with each value.

26. A chemistry teacher gives a challenge problem for each meeting of a class of 19 students. A random variable, X, is the number of students who work the challenge problem. What are the possible values of X?

27. A die is rolled, and the player receives, in dollars, the number rolled on the die or $3, whichever is smaller. Let X be the number of dollars received. Determine the number of different ways in which each can occur.

28. A store ran a special on six-pack cartons of cola, with a maximum of four allowed. A cola representative recorded the number purchased by each customer. The results are summarized as follows:

Number of Cartons	Frequency
0	85
1	146
2	268
3	204
4	122

Find the mean number of cartons purchased.

29. Estimate the mean score of the following grouped data:

Score	Frequency
0–6	8
7–10	13
11–14	6
15–20	15

30. Roy bought four textbooks at a mean price of $34.60, and Rhonda bought three textbooks at a mean price of $29.70. Find the mean price of the seven books.

31. Compute the standard error for the given sample sizes and proportions:
(a) $n = 60$, $\bar{p} = 0.35$ (b) $n = 700$, $\bar{p} = 0.64$
(c) $n = 950$, $\bar{p} = 0.40$

32. Compute the error bounds for the following:
(a) $n = 50$, $\bar{p} = 0.45$, 95% confidence level
(b) $n = 100$, $\bar{p} = 0.30$, 98% confidence level

33. A survey of 30 individuals revealed that 22 watched Monday Night Football. Find, at the 95% confidence level, the error bounds of the proportion who watch Monday Night Football.

34. A manufacturer wants an estimate of the proportion of customers who will respond favorably to its new product. The manufacturer wants the estimate to be correct within 3% at the 95% confidence level. How big a sample should it select?

Game Theory

People of all ages like to play games. The games may range from simple children's games (hide and seek), to games of luck (roulette), to games requiring special skills (baseball) or strategies (chess). In some games, each opponent tries to anticipate the other's actions and act accordingly. For example, the defense in a football game may anticipate a pass and set up a blitz while the offense, in turn, tries to anticipate the defense's strategy. The study of game theory is not restricted to our usual concept of a game. It analyzes strategies used when the goals of "competing" parties conflict. Such a situation could arise when labor and management meet to discuss a contract, when airlines vie for a certain route, or when supermarkets compete against one another. ■

9.1 Two-Person Games

• Strictly Determined Games

This chapter focuses on a simple situation with exactly two sides, or **players,** involved. These two players compete for a **payoff** that one player pays to the other. Let's begin with a simple example using a coin-matching game.

Two friends, Rob and Chad, play the following game. Each one has a coin, and each decides which side to turn up. They show the coins simultaneously and make payments according to the following.

If both show heads, Rob pays Chad 50¢. If both show tails, Rob pays Chad 25¢. If a head and a tail show, Chad pays Rob 35¢. The following figure shows these payoffs for each possible outcome:

	Chad	
	Heads	**Tails**
Heads	Rob pays 50¢ Chad gets 50¢	Rob gets 35¢ Chad pays 35¢
Tails	Rob gets 35¢ Chad pays 35¢	Rob pays 25¢ Chad gets 25¢

Rob

This game is called a **two-person game** because exactly two people participate. Notice that the amount won by a player is exactly the same as the amount lost by the opponent. Whenever this is the case, we call the game a **zero-sum game.** We can represent the payoffs in a two-person zero-sum game by a **payoff matrix.** The payoffs indicated by the figure can be simplified to the following matrix:

$$\begin{array}{c} \textbf{Chad} \\ \begin{array}{cc} \text{Heads} & \text{Tails} \end{array} \\ \textbf{Rob} \begin{array}{c} \text{Heads} \\ \text{Tails} \end{array} \begin{bmatrix} -50¢ & 35¢ \\ 35¢ & -25¢ \end{bmatrix} \end{array}$$

The matrix shows all possible payoffs for the way heads and tails are paired. An entry represents the amount Rob receives for that pair. Consequently, a negative entry indicates that Rob pays Chad.

A player's plan of action against the opponent is called a **strategy.** Game theory attempts to determine the best strategy so that each player will maximize his payoff. It is assumed that each player knows all strategies available to himself and to his opponent, but each player selects a strategy without the opponent knowing which strategy is selected. In this coin-matching game, the choice of a strategy simply amounts to selecting heads or tails. Notice that when Rob selects the strategy *heads,* he has selected the first row of the payoff matrix. Chad's selection of a strategy is equivalent to selecting a column of the matrix.

EXAMPLE 1 ➤ Rob and Chad match quarters. When the coins match, Rob receives 25¢. When the coins differ, Chad receives 25¢. The payoff matrix is

$$
\begin{array}{cc}
 & \textbf{Chad} \\
 & \begin{array}{cc} H & T \end{array} \\
\textbf{Rob} \begin{array}{c} H \\ T \end{array} & \left[\begin{array}{cc} 25\cancel{c} & -25\cancel{c} \\ -25\cancel{c} & 25\cancel{c} \end{array} \right]
\end{array}
$$

■ **Now You Are Ready to Work Exercise 1**

EXAMPLE 2 ➤ The payoff matrix for a game is

$$
\begin{array}{cc}
 & C \\
 & \begin{array}{cc} c_1 & c_2 \end{array} \\
R \begin{array}{c} r_1 \\ r_2 \end{array} & \left[\begin{array}{cc} 14 & -3 \\ -6 & -5 \end{array} \right]
\end{array}
$$

(a) What are the possible payoffs to R if strategy r_1 is selected?

(b) What are the possible payoffs to C if strategy c_2 is selected?

(c) What is the payoff when R selects r_2 and C selects c_1?

SOLUTION

(a) R receives 14 if C selects c_1 and R pays 3 if C selects c_2.

(b) C receives 3 or 5, depending on whether R selects r_1 or r_2.

(c) C receives 6, and R pays 6.

■ **Now You Are Ready to Work Exercise 5**

Strictly Determined Games

We begin our analyses of the best strategy with some simple games that have fixed strategies.

In a competitive situation, we might ask if a strategy exists that will improve our chances of winning or that will increase our payoff. Some simple games have a fixed strategy or strategies that result in the best payoff possible when the competitors both have knowledge of the payoff associated with each strategy. Other games have no fixed best strategy. We first analyze the best strategy for games that have fixed strategies. We call them **strictly determined games.**

Two players, R and C, play a game in which R can take two alternative actions (strategies), called r_1 and r_2, and C can take two actions, c_1 and c_2. Assume that each player chooses a strategy with no foreknowledge of the other's likely strategy. Because there are two possible strategies for R and two for C, a 2×2 payoff matrix represents the outcome, or payoff, corresponding to each pair of strategies selected. Here is an example of a payoff matrix:

$$
\begin{array}{cc}
 & C \\
 & \begin{array}{cc} c_1 & c_2 \end{array} \\
R \begin{array}{c} r_1 \\ r_2 \end{array} & \left[\begin{array}{cc} 4 & -9 \\ 6 & 8 \end{array} \right]
\end{array}
$$

Following the convention that the entries represent the payoff to the row player, R, we interpret the matrix as follows: If R chooses strategy r_1 and C chooses c_1, then C pays 4 to R. If R chooses strategy r_1 and C chooses c_2, then R pays 9 to C, and so on. Notice that the sum of the amounts won by R and C is zero whatever strategies are selected. For example, if R selects r_2, and C selects c_1, then

R wins 6 and C loses 6 (or wins -6), so the total is zero. Therefore, this is called a *zero-sum game*.

We emphasize that a matrix entry represents an amount paid by player C to player R. To indicate the situation where C receives payment from R, enter a negative amount.

R and C each wish to gain as much as they can (or lose as little as possible). Which strategies should they select? Let's analyze the situation. First, observe that for R to gain as much as possible, R's strategy attempts to select the *maximum* entry. As a gain for C is represented by a negative number, C attempts to select the *minimum* entry.

R should select r_2, because R then stands to gain the most, 6 or 8, regardless of C's strategy.

At first glance, it appears that C's strategy should be c_2, because that is C's only chance of winning. However, we assume that both parties know all possible strategies, so C knows that R's best strategy is r_2. Thus, C expects R to choose r_2, so C can choose only between losing 8 or losing 6. C should select strategy c_1, because C then risks giving the least away. With the strategies r_2 and c_1, C pays 6 to R. This entry of the matrix is called the **value of the game;** its location in the matrix is called a **saddle point;** and the pair of strategies leading to the value is called a **solution.** This game has a value of 6; the $(2, 1)$ location is the saddle point; and the solution is the pair of strategies r_2 and c_1.

Let's use this example and analyze the general approach that determines whether a solution exists in a two-person game and how we can find a solution, if a solution exists.

Several courses of action may be available to both players, and there need not be the same number for each.

The strategies available to R and C in a two-person game can be represented by a **payoff matrix** A:

$$
\begin{array}{c}
 \\
\boldsymbol{R} \quad
\begin{array}{c}
r_1 \\ r_2 \\ \vdots \\ r_m
\end{array}
\end{array}
\overset{\displaystyle \boldsymbol{C}}{
\overset{\rule{6cm}{0.4pt}}{
\begin{array}{c}
\begin{array}{cccc}
c_1 & c_2 & \cdots & c_n
\end{array} \\
\left[
\begin{array}{cccc}
a_{11} & a_{12} & \cdots & a_{1n} \\
a_{21} & a_{22} & \cdots & a_{2n} \\
\vdots & \vdots & & \vdots \\
a_{m1} & a_{m2} & \cdots & a_{mn}
\end{array}
\right]
\end{array}
}} = A
$$

Note that the number of strategies available to R is m and the number of strategies available to C is n. The entry a_{ij} represents the payoff that R receives when R adopts strategy r_i and C adopts c_j. The following are the conditions of the game:

Underlying Assumptions of a Strictly Determined Game

1. Each player aims to choose the strategy that will enable the player to obtain as large a payoff as possible (or to lose as little as possible).
2. It is assumed that neither player has any prior knowledge of what strategy the other will adopt.
3. Each player makes a choice of strategy under the assumption that the opponent is an intelligent person adopting an equally rational approach to the game.

Let's first approach the game from the viewpoint of R who is trying to maximize winnings. R scans the rows of A trying to decide which strategy, $r_1, r_2, \ldots, r_m$, to adopt. For any given strategy (row), what is the least possible payoff R can expect? It is the *minimum* entry of that row. In the example with

$$
\begin{array}{c}
\textbf{\textit{C}} \\
\begin{array}{cc}
c_1 & c_2
\end{array} \\
\textbf{\textit{R}}\ \ \begin{array}{c} r_1 \\ r_2 \end{array}
\begin{bmatrix}
4 & -9 \\
6 & 8
\end{bmatrix}
\end{array}
$$

as the payoff matrix, the minimum row entries are -9 for row 1 and 6 for row 2.

R is aware of playing an intelligent opponent who aims to hold R to a small payoff. When R chooses a row, R is guaranteed at least the minimum entry of that row. R then selects the row containing the largest of these minima. R is then guaranteed that payoff, and it is the largest payoff that R can hope for against C's best counterstrategy.

We now look at the situation from the viewpoint of C, who wants the minimum payoff because a gain for C is represented by a negative number. C scans the columns of A trying to decide which strategy, $c_1, c_2, \ldots, c_n$, to adopt. C marks the maximum element in each column, C's least possible gain (or largest possible loss) for that strategy. In the example given, this is 6 in column 1 and 8 in column 2. C then selects the strategy that has the smallest of those maxima. The smallest of the maxima represents C's greatest gain if it is negative and C's smallest loss if it is positive. This is the least payoff that C need make against R's strategy.

Thus, the approaches that R and C should adopt are the following:

Strategy for a Two-Person Zero-Sum Game

> R marks off the minimum element in each row, and selects the row that has the largest of these minima.
>
> C marks off the maximum element in each column, and selects the column that has the smallest of these maxima.
>
> If the largest of the row minima occurs in the same location as the smallest of the column maxima, that payoff is the **value** of the game.
>
> The two strategies leading to the value, the **solution**, will be the ones that should be adopted.
>
> When the largest row minimum and the smallest column maximum are the same, the game is said to be **strictly determined.** The location of this element is the **saddle point** of the game. (There are games that are not strictly determined. We discuss strategies for such games in the following section.)

In the example, R selects the largest of the row minima, 6. C selects the smallest of the column maxima, 6. Because these are the same, the game is strictly determined, and the $(2, 1)$ location is the saddle point of the game. This game is strictly determined because each player should always play the same strategy. If R wants to maximize average earnings, R should always play r_2. If C wants to minimize average payoff, C should always play c_1. It is in this sense that the game is strictly determined.

EXAMPLE 3 ➤ The following payoff matrix shows the strategies of a game played by players R and C, with R having three strategies and C having two strategies:

$$
\begin{array}{c}
\textbf{\textit{C}} \\
\begin{array}{cc}
c_1 & c_2
\end{array} \\
\textbf{\textit{R}}\ \ \begin{array}{c} r_1 \\ r_2 \\ r_3 \end{array}
\begin{bmatrix}
1 & 2 \\
3 & 4 \\
7 & 5
\end{bmatrix}
\end{array}
$$

Determine the value of the game, if it exists.

SOLUTION

First, analyze R's strategies by selecting the minimum entry from each row and write it on the right of the matrix. Next, analyze C's strategies by finding the maximum entry in each column and write it below the column. Then select the largest of the minima and the smallest of the maxima.

$$
\begin{array}{c}
& & \text{Row} \\
& & \text{Minima} \\
\begin{bmatrix} 1 & 2 \\ 3 & 4 \\ 7 & 5 \end{bmatrix} & & \begin{array}{l} 1 \\ 3 \\ \text{⑤} \quad \text{Largest of Minima} \end{array}
\end{array}
$$

$$
\left. \begin{array}{c} \text{Column} \\ \text{Maxima} \end{array} \right\} \quad 7 \quad \text{⑤}
$$

Smallest of Maxima

Observe that the smallest of the column maxima and the largest of the row minima are the same and both occur in the $(3, 2)$ location. Thus, the game is strictly determined with value 5. The saddle point is $(3, 2)$ and the strategies r_3 for R and c_2 for C form the solution.

When the players adopt these strategies, as they should, player R will receive 5 from player C.

■ **Now You Are Ready to Work Exercise 6**

Perhaps this game seems unfair because C has no chance of winning. The best C can do is to lose 5 each time the game is played. Unfortunately, a game like this may correspond to a real situation. Sometimes a business manager may have several options for the business, but economic conditions are such that each option results in a financial loss. The best strategy is then to minimize losses.

What happens when one player chooses another strategy? Suppose C adopts c_1 in the hope of reducing losses to 1. If R keeps to strategy r_3, then C loses 7, rather than the loss of 5 by choosing c_2.

The following example illustrates that it is possible for a two-person game to have more than one saddle point.

EXAMPLE 4 ➤

Look at the analysis of the following two-person game:

$$
\begin{array}{cccccc}
& & & C & & \text{Row} \\
& & c_1 & c_2 & c_3 & \text{Minima} \\
& r_1 & \begin{bmatrix} 3 & 4 & 11 \\ & & & \end{bmatrix} & & & 3 \\
\boldsymbol{R} \quad & r_2 & \begin{bmatrix} -5 & 2 & -3 \end{bmatrix} & & & -5 \\
& r_3 & \begin{bmatrix} 6 & 6 & 9 \end{bmatrix} & & & \text{⑥ Max.}
\end{array}
$$

$$
\left. \begin{array}{c} \text{Column} \\ \text{Maxima} \end{array} \right\} \quad \begin{array}{ccc} \text{⑥} & \text{⑥} & 11 \\ \text{Min.} & \text{Min.} \end{array}
$$

Each 6 in this matrix occurs at a saddle point, $(3, 1)$ and $(3, 2)$. (Verify this.) The best strategy for R is r_3, whereas C has the option of either strategy c_1 or c_2. In either case, the payoff is 6. When more than one saddle point occurs, the entries in the matrix at the saddle points are the same, being the value of the game.

■ **Now You Are Ready to Work Exercise 12**

A game may have no saddle point, and thereby is not strictly determined. Here is an example to illustrate this type of situation.

EXAMPLE 5 ➤ Here is the payoff matrix of a two-person game:

$$
\begin{array}{cc}
 & C \\
 & \begin{array}{cc} c_1 & c_2 \end{array}
\end{array}
$$

$$
\boldsymbol{R} \begin{array}{c} r_1 \\ r_2 \end{array}
\begin{bmatrix} 3 & -2 \\ 5 & 4 \end{bmatrix}
\begin{array}{l} \text{Row} \\ \text{Minima} \\ \boxed{-2} \;\text{Max.} \\ -5 \end{array}
$$

Column Maxima} $\quad \boxed{3} \quad 4$
Min.

For a game to be strictly determined, the largest of the row minima and the smallest of the column maxima must be the same and occur in the same location in the matrix. That does not occur in this game because the largest of the row minima is -2 and the smallest of the column maxima is 3, and they appear in different locations. Thus, there is no saddle point.

■ **Now You Are Ready to Work Exercise 13**

The next section discusses strategies each player should take in such a game.

EXAMPLE 6 ➤ Two major discount stores, A-Mart and B-Mart, compete for the business of the same customer base. A-Mart has 55% of the business and B-Mart 45%. Both companies are considering building new superstores to increase their market share. If both build, or neither builds, they expect their market share to remain the same. If A-Mart builds and B-Mart doesn't, A-Mart's share increases to 65%. If B-Mart builds and A-Mart doesn't, then A-Mart's share drops to 50%. Determine which strategy, to build or not build, each company should take.

SOLUTION

Set up a payoff matrix showing A-Mart's share in all possible decisions. Let B represent the decision to build a superstore and NB the decision not to build.

$$
\begin{array}{cc}
 & \text{B-Mart} \\
 & \begin{array}{cc} B & NB \end{array}
\end{array}
$$

A-Mart $\begin{array}{c} B \\ NB \end{array}
\begin{bmatrix} 55 & 65 \\ 50 & 55 \end{bmatrix}$

The row minima and the column maxima give

$$
\begin{array}{cc}
 & \text{B-Mart} \\
 & \begin{array}{cc} B & NB \end{array}
\end{array}
\begin{array}{l} \text{Row} \\ \text{Minima} \end{array}
$$

A-Mart $\begin{array}{c} B \\ NB \end{array}
\begin{bmatrix} 55 & 65 \\ 50 & 55 \end{bmatrix}
\begin{array}{l} \boxed{55} \;\text{Max.} \\ 50 \end{array}$

Column Maxima} $\quad \boxed{55} \quad 65$
Min.

Each store should adopt the build strategy in which case A-Mart retains 55% market share and B-Mart retains 45%. ■

■■ 9.1 EXERCISES

LEVEL 1

1. *(See Example 1)* Write the payoff matrix for the coin-matching game between two players. If both coins are heads, R pays C $1. If both coins are tails, C pays R $2. If the coins differ, R pays C 50¢.

2. Two players, R and C, each have two cards. R has one black card with the number 5 written on it and one red card with the number 3. C has a black card with a 4 written on it and a red card with a 2. They each select one of their cards and simultaneously show the cards. If the cards are the same color, R gets, in dollars, the difference of the two numbers shown. If the cards are different colors, C gets, in dollars, the smaller of the two numbers shown. Write the payoff matrix of this game.

3. Two-Finger Morra is a game in which two players each hold up one or two fingers. The payoff, in dollars, is the total number of fingers shown. R receives the payoff if the total is even, and C receives the payoff if the total is odd. Write the payoff matrix.

4. Player R has a $1 bill, a $10 bill, and a $50 bill. Player C has a $5 bill and a $20 bill. They simultaneously select one of their bills at random. The one with the larger bill collects the bill shown by the other. Write the payoff matrix for this game.

5. *(See Example 2)* Given the following payoff matrix:

$$
\begin{array}{c}
 \\
R \quad
\begin{array}{c}
r_1 \\ r_2 \\ r_3
\end{array}
\end{array}
\begin{array}{c}
\begin{array}{ccc}
c_1 & c_2 & c_3
\end{array} \\
\left[\begin{array}{ccc}
5 & -3 & 4 \\
0 & 10 & -7 \\
-8 & 4 & 0
\end{array}\right]
\end{array}
$$

(a) What is the payoff when strategies r_2 and c_2 are selected?

(b) What is the largest gain possible for the row player?

(c) What is the greatest loss possible for the column player?

(d) What is the largest gain possible for the column player?

Exercises 6 through 11 are payoff matrices for two-person games. Decide whether the games are strictly determined. Find the saddle point, value, and solution for each strictly determined game.

6. *(See Example 3)*
$$\left[\begin{array}{cc} 5 & 2 \\ -4 & 1 \end{array}\right]$$

7. $$\left[\begin{array}{cc} -3 & -5 \\ 2 & -1 \end{array}\right]$$

8. $$\left[\begin{array}{cc} 5 & 9 \\ 7 & 0 \end{array}\right]$$

9. $$\left[\begin{array}{ccc} 1 & 2 & 3 \\ 4 & -5 & 1 \\ 2 & 6 & 3 \end{array}\right]$$

10. $$\left[\begin{array}{ccc} -4 & 3 & 0 \\ -1 & 2 & 5 \\ -3 & 4 & -2 \end{array}\right]$$

11. $$\left[\begin{array}{ccc} 2 & 0 & 1 \\ 0 & -3 & 4 \\ 3 & -2 & 0 \end{array}\right]$$

12. *(See Example 4)* Find the saddle points and values of the payoff matrices.

(a) $$\left[\begin{array}{ccc} 3 & 3 & 8 \\ 1 & -2 & -3 \\ 3 & 3 & 9 \end{array}\right]$$

(b) $$\left[\begin{array}{cccc} 6 & 7 & 6 & 8 \\ 3 & 6 & 5 & 12 \\ 6 & 9 & 6 & 11 \end{array}\right]$$

13. *(See Example 5)* For the following payoff matrices, find the saddle points, values, and solutions, if they exist.

(a) $$\left[\begin{array}{ccc} -3 & 2 & 1 \\ 0 & 2 & 3 \\ -1 & -4 & 2 \end{array}\right]$$

(b) $$\left[\begin{array}{ccc} -1 & 2 & 3 \\ 4 & -1 & 0 \\ 0 & 1 & -1 \end{array}\right]$$

(c) $$\left[\begin{array}{ccc} 1 & -2 & 1 \\ 5 & 7 & 3 \\ -1 & 3 & -4 \end{array}\right]$$

(d) $$\left[\begin{array}{ccc} -1 & -2 & 3 \\ 5 & 0 & 2 \\ 4 & -1 & 1 \end{array}\right]$$

LEVEL 2

14. Two coffeehouses near the campus, The Coffee Club and The Rendezvous Room, compete for the same customers. To attract more customers, The Coffee Club is considering adding a blues musical group or a cheesecake counter. There is room to add only one of these. At the same time, The Rendezvous Room is considering adding a jazz musical group or a bagel counter. They too have room to add only one.

The payoff matrix shows the percentage increase in business for the Rendezvous Room for each op-

tion adopted by the coffeehouses. Their options are no change, add music, and add food.

		Coffee Club		
		No Change	Add Blues	Add Cheese-cake
Rendezvous Room	No Change	0	-10	-15
	Add Jazz	15	2	5
	Add Bagels	10	-5	6

Which option should each adopt? Find the resulting value.

15. Two television networks, Century and ReMark, are competing for a viewer audience in prime time. Century is considering a new talk show, a game show, and an educational documentary show. ReMark is considering a sports highlights show, a mystery drama, and a variety show. The following payoff matrix shows the estimated gain by ReMark in audience ratings for the pairings of shows.

		Century		
		Talk	Game	Educa-tional
ReMark	Sports	10	−5	25
	Mystery	-20	−15	10
	Variety	−10	−10	5

Find the strategy each network should adopt and the value.

16. Crofford College and Round Rock Tech are neighboring schools that are considering tuition changes. Some argue that tuition should be increased to obtain more revenue, and others argue that tuition should remain the same or even be lowered to recruit more students. The decisions will cause some students to change from one school to the other. The following matrix describes student movement, with a positive entry indicating the number of students moving from Crofford to Round Rock and a negative number indicating movement from Round Rock to Crofford.

Determine if there is a saddle point. If so, what strategy should each school adopt?

		Crofford		
		Lower	Same	Raise
Round Rock	Lower	125	200	320
	Same	−175	80	175
	Raise	−400	−220	0

LEVEL 3

17. The farming occupation may be considered a game between the farmer, who has a choice of crops to plant, and nature, which has a "choice" of types of weather. For the type of soil and climate in his area, a farmer has the choice of planting crops of milo, corn, or wheat. The matrix shows the expected gross income per acre for each crop and weather condition for the growing season.

		Nature		
		Dry Season	Normal Season	Wet Season
Farmer	Milo	85	120	150
	Corn	60	165	235
	Wheat	70	150	175

(a) Based on this "two-person" game, which crop strategy should the farmer adopt?

(b) The strategy of a two-person game assumes that each player is an intelligent person using a rational approach to the game. Do you think that game theory is a valid approach to determine which crop strategy is best?

(c) Weather records show that dry, normal, and wet seasons do not occur with equal frequency. Dry and wet seasons each occur about 20% of the time, and normal seasons occur about 60% of the time. For each crop, find the income over a five-year period assuming one year is dry, one is wet, and three are normal. Based on your findings, which is the best crop strategy for the long term? Is it the same as that found in part (a)?

9.2 Mixed-Strategy Games

- Games That Are Not Strictly Determined
- Should I Have the O
- Fair Games

Games That Are Not Strictly Determined

In the preceding section, we analyzed which each person had one strate game. We also saw that some ...ined two-person games in ...der the conditions ..tly determined, ...tions of the

The following game is not strictly determined because the largest of the row minima is 10 and the smallest column maxima is 16.

$$
\begin{array}{cc}
 & \begin{array}{cc} & C \\ c_1 & c_2 \end{array} & \begin{array}{c} \text{Row} \\ \text{Minima} \end{array}
\end{array}
$$

$$
R \begin{array}{c} r_1 \\ r_2 \end{array} \left[\begin{array}{cc} 10 & 16 \\ 17 & 8 \end{array} \right] \quad \begin{array}{c} 10 \\ 8 \end{array}
$$

$$
\left. \begin{array}{c} \text{Column} \\ \text{Maxima} \end{array} \right\} \quad 17 \quad 16
$$

Thus, the game has no saddle point. Notice that C's best strategy is c_1 if R plays r_1 and is c_2 if R plays r_2. The best strategy for R is r_1 when C plays c_2 and is r_2 when C plays c_1. Which strategy should each player adopt?

Games are often played a number of times, not just once. For example, a football team has a variety of offensive strategies (plays) available and the opposing team has several defensive strategies. Although the offensive team may have a play that is successful more consistently than others, they do not choose it every time because the opponents would then use the defensive strategy that works best against that play. The best offensive strategy uses a variety of plays in a way that maximizes overall efforts. Similarly, the defensive team uses a variety of strategies that they hope will keep the offensive gains to a minimum. In the simple example

$$
\begin{array}{cc}
 & \begin{array}{cc} & C \\ c_1 & c_2 \end{array}
\end{array}
$$

$$
R \begin{array}{c} r_1 \\ r_2 \end{array} \left[\begin{array}{cc} -17 & 16 \\ 10 & 8 \end{array} \right]
$$

players must conceal their choice of strategy or else the other players can select the strategy that is most beneficial to them. If R realizes that C selects strategy c_2, then R will select r_1 to maximize payoff. If C selects c_1, then R knows to select r_2.

If each player keeps the strategy secret, with a mixture of strategies, then the best strategy for each is to select a strategy in a random manner but with a frequency that provides the greatest benefit. Thus, R should select r_1 part of the time in a random manner and select r_2 the rest of the time. In a similar manner, C selects c_1 part of the time and c_2 the rest of the time.

Player R would like to adopt a sequence of strategies that would maximize R's long-term average payoff from C, and C would like to minimize the long-term average payoff to R. How often should R adopt r_1 and how often should C adopt c_1 for the most beneficial payoff to each? We use probability theory to control the randomness of each player's strategy.

We use the following payoff matrix for a general two-person game:

$$
\begin{array}{cc}
 & \begin{array}{cc} & C \\ c_1 & c_2 \end{array}
\end{array}
$$

$$
R \begin{array}{c} r_1 \\ r_2 \end{array} \left[\begin{array}{cc} a_{11} & a_{12} \\ a_{21} & a_{22} \end{array} \right]
$$

We use the notation p_1 for the **probability** that R will adopt strategy r_1 and p_2 for the probability that R will adopt strategy r_2. For player C, we use q_1 for the probability that C will adopt strategy c_1, and q_2 represents the probability that c_2 be adopted. In practice, over a sequence of a number of games, player R uses strategy r_1 the fraction p_1 of the time and uses r_2 the fraction p_2 of the time. adopts c_1 the fraction q_1 of the time and adopts c_2 the fraction q_2 of the time. Whenever the players adopt strategies in a random manner with the fraction of times indicated, then the probability of the payoff a_{11} is $p_1 q_1$ — that is, the probability that R selects r_1 and C selects c_1. Similarly, the probability of payoff a_{12} is $p_1 q_2$, the probability of a_{21} is $p_2 q_1$, and the probability of a_{22} is $p_2 q_2$.

With each possible pair of strategies the following probabilities and payoffs hold:

Strategies	$r_1 c_1$	$r_1 c_2$	$r_2 c_1$	$r_2 c_2$
Probability	$p_1 q_1$	$p_1 q_2$	$p_2 q_1$	$p_2 q_2$
Payoff	a_{11}	a_{12}	a_{21}	a_{22}

When each player randomly selects a strategy according to these probabilities, the average long-term payoff is the expected value of the payoff. (Note that this is the expected value as defined in Section 8.5.) We call this the **expected payoff** of the game, and it is

$$\text{Expected payoff} = p_1 q_1 a_{11} + p_1 q_2 a_{12} + p_2 q_1 a_{21} + p_2 q_2 a_{22}$$

The expected payoff denotes the average payoff to R when the game is played a large number of times.

We can represent the strategy probability with the matrices

$$P = [p_1 \quad p_2] \quad \text{and} \quad Q = \begin{bmatrix} q_1 \\ q_2 \end{bmatrix}$$

and denote the payoff matrix by A and the expected payoff associated with these probabilities by $E(P, Q)$. Then, in matrix form $E(P, Q)$ becomes

$$E(P, Q) = PAQ$$

We call the matrices P and Q the **strategies adopted** by R and C, respectively.

We verify this by performing the matrix multiplication

$$[p_1 \quad p_2] \begin{bmatrix} a_{11} & a_{12} \\ a_{21} & a_{22} \end{bmatrix} \begin{bmatrix} q_1 \\ q_2 \end{bmatrix} = [p_1 a_{11} + p_2 a_{21} \quad p_1 a_{12} + p_2 a_{22}] \begin{bmatrix} q_1 \\ q_2 \end{bmatrix}$$

$$= (p_1 a_{11} + p_2 a_{21})q_1 + (p_1 a_{12} + p_2 a_{22})q_2$$

$$= a_{11} p_1 q_1 + a_{21} p_2 q_1 + a_{12} p_1 q_2 + a_{22} p_2 q_2$$

which is the expected payoff.

EXAMPLE 1 ➤

Here is the payoff matrix of a mixed strategy game:

$$\begin{array}{c} \\ R \begin{array}{c} r_1 \\ r_2 \end{array} \end{array} \overset{\displaystyle \overset{C}{\overline{}}}{\overset{\displaystyle \begin{array}{cc} c_1 & c_2 \end{array}}{\begin{bmatrix} 15 & 75 \\ 45 & -30 \end{bmatrix}}}$$

Let's compute the expected payoff using the strategy $[\frac{2}{3} \quad \frac{1}{3}]$ for R and the strategy $[\frac{2}{5} \quad \frac{3}{5}]$ for C. This means that $\frac{2}{3}$ of the time R will adopt strategy r_1 and $\frac{1}{3}$ of the time will adopt strategy r_2, whereas $\frac{2}{5}$ of the time C will adopt strategy c_1 and $\frac{3}{5}$ of the time C will adopt strategy c_2

$$E(P, Q) = [\tfrac{2}{3} \quad \tfrac{1}{3}] \begin{bmatrix} 15 & 75 \\ 45 & -30 \end{bmatrix} \begin{bmatrix} \tfrac{2}{5} \\ \tfrac{3}{5} \end{bmatrix} = [25 \quad 40] \begin{bmatrix} \tfrac{2}{5} \\ \tfrac{3}{5} \end{bmatrix} = 34$$

R can expect an average payoff of 34 if the game is played a large number of times provided that the players select strategies in a random manner with the relative frequency given by P and Q. If the players choose to vary the relative frequency of their strategies, then the expected payoff will change.

■ **Now You Are Ready to Work Exercise 1**

We have seen that a strictly determined game has a solution that provides an optimal strategy for both players. In the case of a mixed-strategy game, no such single strategy exists. However, player R would like a strategy P that gives the best payoff against player C's best strategy. We call such a strategy the **optimal strategy for player R.**

Likewise, C wants to adopt a strategy Q that yields the least payoff to R against R's best strategy. We call it the **optimal strategy for player C.**

Both players in a mixed-strategy game have the option of selecting their strategy probabilities. Because different strategies generally yield different expected payoffs, a player would like to know how to determine the optimal strategy. Mathematicians have determined how to find optimal strategies from the 2×2 payoff matrix of a two-person game.

For the payoff matrix

$$
\begin{array}{c}
\quad C \\
R \quad
\begin{bmatrix}
a_{11} & a_{12} \\
a_{21} & a_{22}
\end{bmatrix}
\end{array}
$$

the optimal strategy for player R is $P = [p_1 \quad p_2]$ where

$$
p_1 = \frac{a_{22} - a_{21}}{a_{11} + a_{22} - a_{12} - a_{21}} \quad \text{and} \quad p_2 = 1 - p_1
$$

The optimal strategy for player C is $Q = [q_1 \quad q_2]$ where

$$
q_1 = \frac{a_{22} - a_{12}}{a_{11} + a_{22} - a_{12} - a_{21}} \quad \text{and} \quad q_2 = 1 - q_1
$$

The expected payoff for the optimal strategies is

$$
E = PAQ = \frac{a_{11}a_{22} - a_{12}a_{21}}{a_{11} + a_{22} - a_{12} - a_{21}}
$$

The values of $p_1, p_2, q_1,$ and q_2 are meaningless when $a_{11} + a_{22} - a_{12} - a_{21} = 0$. However, this never occurs when the game is not strictly determined.

EXAMPLE 2 ➤   Now let's find the optimal strategies, and the value associated with the strategies, for the game with payoff matrix

$$
\begin{array}{c}
\quad C \\
R \quad
\begin{bmatrix}
15 & 75 \\
45 & -30
\end{bmatrix}
\end{array}
$$

SOLUTION

$$
p_1 = \frac{-30 - 45}{15 - 30 - 75 - 45} = \frac{-75}{-135} = \frac{5}{9}, \quad p_2 = \frac{4}{9}
$$

$$
q_1 = \frac{-30 - 75}{-135} = \frac{7}{9}, \quad q_2 = \frac{2}{9}
$$

The optimal strategy for R is $\left[\frac{5}{9} \quad \frac{4}{9}\right]$ and the optimal strategy for C is $\left[\frac{7}{9} \quad \frac{2}{9}\right]$. These strategies yield the expected payoff

$$E = \frac{15(-30) - 75(45)}{-135} = \frac{-3825}{-135} = 28.3$$

For a large number of games, R can expect, on the average, to receive about 28.3 from C provided that both players adopt their optimal strategies.

■ **Now You Are Ready to Work Exercise 6**

EXAMPLE 3 ➤ A two-person game with payoff matrix

$$
\begin{array}{cc}
 & C \\
R & \begin{bmatrix} 4 & 6 \\ 2 & -5 \end{bmatrix}
\end{array}
$$

is strictly determined with value 4. R's strategy is r_1, and C's strategy is c_1, which we can represent in matrix form as $P = [1 \quad 0]$ for R and $Q = [1 \quad 0]$ for C.

If we use the formulas for optimal strategies and expected payoff, we obtain for R the strategy $P = \left[\frac{7}{9} \quad \frac{2}{9}\right]$ and for C the strategy $Q = \left[\frac{11}{9} \quad \frac{-2}{9}\right]$ with $E = \frac{32}{9}$.

This result is erroneous because neither $\frac{11}{9}$ nor $\frac{-2}{9}$ is a valid probability and because $E = 4$ is the optimal value, not $\frac{32}{9}$. ■

 CAUTION

If a game is strictly determined, the preceding method of computing the value is not correct.

This example emphasizes that a mixed-strategy analysis does not apply to a strictly determined game.

In some instances, a payoff matrix can be simplified to a smaller matrix. Here is an example.

EXAMPLE 4 ➤ Here is a payoff matrix in which each player has the option of three strategies. The game is not strictly determined.

$$
\begin{array}{cc}
 & C \\
 & \begin{array}{ccc} c_1 & c_2 & c_3 \end{array} \\
R \begin{array}{c} r_1 \\ r_2 \\ r_3 \end{array} & \begin{bmatrix} 50 & 20 & -10 \\ 0 & 15 & 5 \\ -15 & 10 & -5 \end{bmatrix}
\end{array}
$$

Notice that each entry in row 2 is greater than the corresponding entry in row 3. Thus, R should never adopt strategy r_3 because, regardless of the strategy adopted by C, the strategy r_2 yields a better value for R. As C wants to minimize the payoff to R, C should never select strategy c_2 because strategy c_3 yields a smaller payoff to R regardless of the strategy R adopts. In a game like this, we say that strategy r_2 **dominates** strategy r_3, strategy c_3 **dominates** strategy c_2, and strategies r_3 and c_2 can be removed from consideration.

■ **Now You Are Ready to Work Exercise 9**

DEFINITION
Dominant Row or Column

Row i of a payoff matrix **dominates** row j if every entry of row i is greater than or equal to the corresponding entry of row j.

Column i of a payoff matrix **dominates** column j if every entry of column i is less than or equal to the corresponding entry of column j.

Whenever row i dominates row j, then row j may be removed from the payoff matrix without affecting the analysis.

Whenever column i dominates column j, then column j may be removed from the payoff matrix without affecting the analysis.

Because row 2 dominates row 3 and column 3 dominates column 2 of the payoff matrix of the example, we can remove row 3 and column 2 to obtain the payoff matrix

$$A = \begin{bmatrix} 50 & -10 \\ 0 & 5 \end{bmatrix}$$

From this 2×2 matrix, we can now find the optimal strategy for R, $P = [p_1 \ \ p_2 \ \ 0]$. The zero occurs because r_3 will never be used.

$$p_1 = \frac{5 - 0}{50 + 5 - (-10) - 0} = \frac{5}{65} = \frac{1}{13} \quad \text{and} \quad p_2 = \frac{12}{13}$$

Thus, $P = \left[\frac{1}{13} \ \ \frac{12}{13} \ \ 0 \right]$.

For C, the optimal strategy is $Q = [q_1 \ \ 0 \ \ q_3]$. $q_2 = 0$ because c_2 is never used.

$$q_1 = \frac{5 - (-10)}{65} = \frac{15}{65} = \frac{3}{13} \quad \text{and} \quad q_3 = \frac{10}{13}$$

Thus, $Q = \left[\frac{3}{13} \ \ 0 \ \ \frac{10}{13} \right]$.

The expected payoff is

$$\frac{50(5) - (-10)(0)}{65} = \frac{250}{65} = 3.85$$

Thus, in the long term, R will receive an average value of 3.85 from C.

Fair Games

If a game is not strictly determined, it is called a **fair game** when its value is zero. This means that the long-term average gain of each player is zero; that is, the losses of each player equal his gains, so his net gain is zero.

EXAMPLE 5 ▶

The game

$$\begin{bmatrix} 3 & -1 \\ 2 & 4 \end{bmatrix}$$

is not strictly determined, and its value is

$$\frac{3(4) - 2(-1)}{3 + 4 + 1 - 2} = \frac{14}{6}$$

so it is not a *fair* game.

The game

$$\begin{bmatrix} 2 & -4 \\ -3 & 6 \end{bmatrix}$$

is not strictly determined, and its value is

$$\frac{2(6) - (-3)(-4)}{2 + 6 + 4 + 3} = \frac{0}{15} = 0$$

so this game is fair.

■ **Now You Are Ready to Work Exercise 13**

Generally, a strictly determined game is not fair.

Should I Have the Operation?

Medical diagnoses have an element of uncertainty. For two people with the same symptoms, one may have the disease, and the other may not. Because of the uncertainty involved, further tests or a second opinion is often recommended. Some medical procedures involve nontrivial risks. Should a patient undergo surgery when, historically, there is a relatively small probability a tumor is malignant?

Let's look at the decision to have surgery or not as a game between the patient and nature. To illustrate, we use the example in which the expected years of life remaining for the patient are

1. 3 years if the disease is present and no surgery is performed.
2. 15 years if the disease is present and surgery is performed.
3. 30 years if no disease is present and no surgery is performed.
4. 25 years if no disease is present and surgery is performed.

Thinking of this as a two-person game, nature has the options of disease (D) and no disease (ND), whereas the patient has the options of surgery (S) or no surgery (NS).

The matrix representing these options is

$$
\begin{array}{c}
 \\
\text{Patient}
\end{array}
\begin{array}{cc}
 & \text{Nature} \\
 & \overline{\begin{array}{cc} D & ND \end{array}} \\
\begin{array}{c} S \\ NS \end{array} &
\begin{bmatrix} 15 & 25 \\ 3 & 30 \end{bmatrix}
\end{array}
$$

where the patient's strategy is $P = \lfloor p_1 \quad p_2 \rfloor$ and nature's strategy is $Q = [q_1 \quad q_2]$. If nature's strategy is $[1 \quad 0]$, the disease is present; then the patient's strategy is clearly $[1 \quad 0]$, have surgery. However, because the patient knows only the probability of nature's options, we ask how the probability should influence the patient's decision. Suppose the probability that the disease is present is 0.20. Then, $Q = [0.20 \quad 0.80]$. If the patient opts for surgery, then $P = [1 \quad 0]$ and the expected number of years remaining is

$$
[1 \quad 0]\begin{bmatrix} 15 & 25 \\ 3 & 30 \end{bmatrix}\begin{bmatrix} 0.20 \\ 0.80 \end{bmatrix} = [15 \quad 25]\begin{bmatrix} 0.20 \\ 0.80 \end{bmatrix} = 23
$$

If the patient opts for no surgery, then the expected survival time is

$$
[0 \quad 1]\begin{bmatrix} 15 & 25 \\ 3 & 30 \end{bmatrix}\begin{bmatrix} 0.20 \\ 0.80 \end{bmatrix} = [3 \quad 30]\begin{bmatrix} 0.20 \\ 0.80 \end{bmatrix} = 24.6
$$

In this case, no surgery appears to be a slightly better option.

The following example, using the preceding payoff matrix, illustrates how we can estimate when surgery is a better option and when it is not.

EXAMPLE 6 ➤

The payoff matrix for nature's options of disease or no disease and a patient's options of surgery or no surgery is

$$
\begin{array}{c}
 \\
\text{Patient}
\end{array}
\begin{array}{cc}
 & \text{Nature} \\
 & \overline{\begin{array}{cc} D & ND \end{array}} \\
\begin{array}{c} S \\ NS \end{array} &
\begin{bmatrix} 15 & 25 \\ 3 & 30 \end{bmatrix}
\end{array}
$$

For what values of Q, the probability of disease, is surgery a better option?

SOLUTION

We determine if surgery is the better option by considering the expected survival time for both the surgery and no surgery options and using $Q = [q_1 \quad 1 - q_1]$.

For surgery:

$$\text{Expected survival} = [1 \quad 0]\begin{bmatrix} 15 & 25 \\ 3 & 30 \end{bmatrix}\begin{bmatrix} q_1 \\ 1 - q_1 \end{bmatrix}$$

$$= [15 \quad 25]\begin{bmatrix} q_1 \\ 1 - q_1 \end{bmatrix}$$

$$= 15q_1 + 25(1 - q_1)$$

$$= 25 - 10q_1$$

For no surgery:

$$\text{Expected survival} = [0 \quad 1]\begin{bmatrix} 15 & 25 \\ 3 & 30 \end{bmatrix}\begin{bmatrix} q_1 \\ 1 - q_1 \end{bmatrix} = [3 \quad 30]\begin{bmatrix} q_1 \\ 1 - q_1 \end{bmatrix}$$

$$= 3q_1 + 30(1 - q_1)$$

$$= 30 - 27q_1$$

Surgery is the better option when

$$25 - 10q_1 > 30 - 27q_1$$
$$17q_1 > 5$$
$$q_1 > \frac{5}{17} = 0.294$$

Thus, surgery is the better option when the probability that the disease is present is greater than 0.294.

■ **Now You Are Ready to Work Exercise 17**

You might question the use of game theory to help make decisions about medical treatment because of these reasons:

1. Game theory assumes both players are intelligent people who know all strategies available to both players and who will adopt a rational approach to the game.
2. Data on life expectancies and the probability that a treatment is successful give information that may be reliable for large groups of people but doesn't predict the outcome for a particular individual very well. When we are at risk, we tend to want to be treated and treated aggressively.

Although nature is not an intelligent person, nature tends to "behave" in a random manner, often with known probability. We can play a mixed-strategy game with nature *as if* nature were an intelligent person who decided to choose options randomly with a known probability. The medical researcher might find such a game useful in determining research actions.

9.2 EXERCISES

1. *(See Example 1)*   Find the expected payoff for the payoff matrix

$$R \begin{array}{c} C \\ \begin{bmatrix} 20 & 12 \\ 8 & 30 \end{bmatrix} \end{array}$$

using $P = \begin{bmatrix} \frac{2}{3} & \frac{1}{3} \end{bmatrix}$ for R's strategy and using $Q = \begin{bmatrix} \frac{1}{4} & \frac{3}{4} \end{bmatrix}$ for C's strategy.

2. Find the expected payoff for the payoff matrix

$$A = \begin{bmatrix} 3 & -1 \\ -2 & 4 \end{bmatrix}$$

using the row strategy [0.3 0.7] and the column strategy [0.6 0.4].

3. Find the expected payoff for the payoff matrix

$$A = \begin{bmatrix} 3 & -6 \\ -2 & 4 \end{bmatrix}$$

using the row strategy $\begin{bmatrix} \frac{2}{5} & \frac{3}{5} \end{bmatrix}$ and the column strategy $\begin{bmatrix} \frac{2}{3} & \frac{1}{3} \end{bmatrix}$.

4. Find the expected payoff for the payoff matrix

$$R \begin{array}{c} C \\ \begin{bmatrix} -12 & 18 & 6 \\ 8 & -10 & -4 \\ -6 & 14 & -8 \end{bmatrix} \end{array}$$

using $P = \begin{bmatrix} 0 & \frac{1}{2} & \frac{1}{2} \end{bmatrix}$ for R's strategy and using $Q = \begin{bmatrix} \frac{1}{2} & 0 & \frac{1}{2} \end{bmatrix}$ for C's strategy.

5. Here is a payoff matrix for a two-person game:

$$R \begin{array}{c} C \\ \begin{bmatrix} -15 & 40 & 25 \\ 25 & -10 & -5 \\ 45 & 20 & -15 \end{bmatrix} \end{array}$$

Find the expected payoff for each of the pairs of strategies.

(a) $P = [1 \ 0 \ 0]$ $Q = [0 \ 1 \ 0]$
(b) $P = \begin{bmatrix} \frac{1}{2} & \frac{1}{2} & 0 \end{bmatrix}$ $Q = \begin{bmatrix} \frac{1}{2} & 0 & \frac{1}{2} \end{bmatrix}$
(c) $P = \begin{bmatrix} \frac{1}{5} & \frac{2}{5} & \frac{2}{5} \end{bmatrix}$ $Q = \begin{bmatrix} \frac{1}{3} & \frac{1}{3} & \frac{1}{3} \end{bmatrix}$
(d) $P = [0.3 \ 0.1 \ 0.6]$ $Q = [0.2 \ 0.2 \ 0.6]$

The payoff matrices in Exercises 6 through 8 describe mixed-strategy games. Determine the optimal strategies for each game and the resulting value of the game.

6. *(See Example 2)* $\begin{bmatrix} 4 & 9 \\ 10 & 3 \end{bmatrix}$

7. $\begin{bmatrix} 15 & 10 \\ -5 & 20 \end{bmatrix}$ **8.** $\begin{bmatrix} 12 & -4 \\ -6 & 14 \end{bmatrix}$

9. *(See Example 4)*   Reduce the following matrices by deleting rows or columns that are dominated by other rows or columns.

(a) $\begin{bmatrix} 6 & -5 & 1 & 9 \\ 2 & 0 & 4 & 7 \\ -3 & -6 & 0 & 3 \end{bmatrix}$ (b) $\begin{bmatrix} 5 & 2 & -2 & 8 \\ 3 & 1 & 13 & 22 \\ 1 & 3 & 6 & 11 \end{bmatrix}$

10. Reduce the following matrices by deleting rows or columns that are dominated by other rows or columns.

(a) $\begin{bmatrix} 6 & 2 & 3 \\ 2 & 0 & 4 \\ 7 & 2 & 1 \\ -3 & -8 & 3 \end{bmatrix}$ (b) $\begin{bmatrix} 6 & 2 & 2 & 1 \\ 4 & 1 & 9 & 2 \\ 2 & 4 & 6 & 8 \\ 1 & 3 & 6 & 1 \end{bmatrix}$

11. Determine the optimal strategy and the resulting value for this mixed-strategy game:

$$\begin{bmatrix} 6 & 7 & 3 \\ 4 & 5 & 2 \\ 5 & 6 & 8 \end{bmatrix}$$

12. Determine the optimal strategy and the resulting value for this mixed-strategy game:

$$\begin{bmatrix} 4 & 6 & 5 \\ 6 & -1 & 7 \\ 8 & 2 & 9 \end{bmatrix}$$

13. *(See Example 5)*   Determine whether or not the following games are fair.

(a) $\begin{bmatrix} 6 & -8 \\ -3 & 4 \end{bmatrix}$ (b) $\begin{bmatrix} 5 & 2 \\ 1 & 3 \end{bmatrix}$

(c) $\begin{bmatrix} 3 & -9 \\ 2 & 6 \end{bmatrix}$

14. Determine whether or not the following games are fair.

(a) $\begin{bmatrix} 3 & 2 \\ -1 & 5 \end{bmatrix}$ (b) $\begin{bmatrix} 5 & -10 \\ -2 & 4 \end{bmatrix}$

(c) $\begin{bmatrix} 7 & 8 \\ 2 & 6 \end{bmatrix}$

LEVEL 2

15. Two television networks, Century and ReMark, are competing for a viewer audience in prime time. Century is considering a new talk show, a game show, and an educational documentary show. ReMark is considering a sports highlights show, a mystery drama, and a variety show. The following payoff matrix shows the estimated gain by ReMark in audience ratings for the pairings of shows.

		Century		
		Talk	Game	Educational
ReMark	Sports	10	−5	25
	Mystery	−20	15	10
	Variety	−10	−10	5

The networks are prepared to air their shows in a random manner. Find the optimal strategy of each network and the expected value.

16. Baseball has classic one-on-one competition between the pitcher and the batter. Let's analyze the strategy of the pitcher, Perez, and the batter, Blakemore. Perez has two pitches, a fastball and a curve. When Blakemore is set for a fastball and Perez throws a fastball, Blakemore's batting average is 0.450. If, instead, Perez throws a curve, Blakemore's batting average is 0.200. When Blakemore is set for a curve and Perez throws a curve, Blakemore's batting average is 0.350. If Perez throws a fastball instead, Blakemore's batting average is 0.150. Thus, the payoff matrix is

		Perez Throws	
		F	C
Blakemore	F	0.450	0.200
Expects	C	0.150	0.350

Find the optimal strategy for each player and the expected batting average.

LEVEL 3

17. *(See Example 6)*   A patient has symptoms that suggest a disease that requires exploratory surgery to determine if the disease exists. The expected survival times based on whether the disease exists or not (D or ND) and on whether or not surgery is performed or not (S or NS) are given in the matrix

		Nature	
		D	ND
Patient	S	25	30
	NS	5	35

The probability that the disease exists is $q_1 = 0.60$.
(a) Find the expected survival time with surgery.
(b) Find the expected survival time without surgery.
(c) For what value of q_1 is no surgery the better option?

18. A patient faces surgery because of a suspected disease. The expected years of survival are given in the matrix

		Nature	
		D	ND
Patient	S	15	35
	NS	1	40

The probability that the disease actually exists is $q_1 = 0.35$.
(a) Find the expected survival time if surgery is done.
(b) Find the expected survival time if surgery is not done.
(c) Find the probability q_1 for which surgery is the better option.

19. A physician tells her patient that symptoms suggest a tumor that requires surgery for further diagnosis and treatment. The following matrix gives expected survival time:

		Nature	
		D	ND
Patient	S	20	22
	NS	3	25

The physician indicates the probability that the tumor is malignant is considered to be about 0.70, but it can vary from that. Assuming $q_1 = 0.70$,
(a) find the expected survival with surgery.
(b) find the expected survival without surgery.
(c) Because $q_1 = 0.70$ is somewhat questionable, find the range of q_1 such that surgery is the better option.

20. A patient who has a tumor that is suspected to be malignant has three treatment options: surgery, radiation–chemotherapy, or both. The probability that the tumor is malignant is 0.80. The survival time of the patient under treatment or no treatment options is given for each option.

Surgery

	Nature	
	D	ND
Patient T	20	22
NT	3	25

Radiation

	Nature	
	D	ND
Patient T	12	25
NT	2	30

Both

	Nature	
	D	ND
Patient T	18	20
NT	2	30

Based on expected survival, rank the treatments in order of effectiveness.

EXPLORATIONS

21. Show that $PAQ = \dfrac{a_{11}a_{22} - a_{12}a_{21}}{a_{11} + a_{22} - a_{12} - a_{21}}$ where $A = \begin{bmatrix} a_{11} & a_{12} \\ a_{21} & a_{22} \end{bmatrix}$ and P and Q are the optimal strategy probabilities of the two players.

▟ IMPORTANT TERMS

9.1

Payoff
Two-Person Game
Strategy
Strictly Determined Game
Zero-Sum Game
Value of the Game

Saddle Point
Solution
Payoff Matrix

9.2

Mixed-Strategy Game
Probability of a Strategy

Expected Payoff
Optimal Strategy
Dominant Row or Column
Fair Game

▟ REVIEW EXERCISES

1. The following represent payoff matrices for two-person games. For each one, determine whether the game is strictly determined. If a game is strictly determined, find the saddle point and value.

(a) $\begin{bmatrix} 5 & -1 \\ 2 & 4 \end{bmatrix}$

(b) $\begin{bmatrix} 1 & 3 & 9 \\ 7 & 4 & 8 \\ -5 & 3 & 4 \end{bmatrix}$

(c) $\begin{bmatrix} 140 & 210 \\ 300 & 275 \end{bmatrix}$

(d) $\begin{bmatrix} -6 & 2 & 9 & 1 \\ 5 & -4 & 0 & 2 \\ 4 & 2 & 8 & 3 \end{bmatrix}$

2. R and C are competing retail stores. Each has two options for an advertising campaign. The results of the campaign are expected to result in the following payoff matrix:

$$ R \begin{bmatrix} 40 & 35 \\ -25 & 30 \end{bmatrix} $$

The entries represent the amount in sales that R will gain from C. What strategy should each adopt?

3. The following are payoff matrices that describe the payoff (in yards gained) of possible strategies of the offense and defense in a football game. Determine the solutions, if they exist.

(a) $\begin{bmatrix} 3 & 15 & -4 \\ 2 & -1 & 6 \\ 4 & 2 & 7 \end{bmatrix}$

(b) $\begin{bmatrix} -5 & 8 & 7 \\ 4 & 5 & 25 \\ 3 & -2 & 2 \end{bmatrix}$

4. Two players X and Y play the following game. X gives Y $5 before the game starts. They each write 1, 2, or 3 on a slip of paper. Then, Y gives X an amount determined by the numbers written. The following payoff matrix gives the amount paid to X in each possible case:

$$
\begin{array}{c}
 & & Y \\
 & & \begin{array}{ccc} 1 & 2 & 3 \end{array} \\
X & \begin{array}{c} 1 \\ 2 \\ 3 \end{array} & \left[\begin{array}{ccc} 2 & 3 & 7 \\ 4 & 0 & 6 \\ 5 & 4 & 9 \end{array} \right]
\end{array}
$$

What strategy should each player use?

5. Determine the expected payoff of the following games:

(a) The strategy of X is [0.3 0.7]; the strategy of Y is [0.6 0.4]; the payoff matrix is

$$
\begin{array}{c}
 & Y \\
X & \left[\begin{array}{cc} 5 & 9 \\ 11 & 2 \end{array} \right]
\end{array}
$$

(b) The strategy of X is [0.5 0.5]; the strategy of Y is [0.1 0.9]; the payoff matrix is

$$
\begin{array}{c}
 & Y \\
X & \left[\begin{array}{cc} -2 & 6 \\ 3 & 9 \end{array} \right]
\end{array}
$$

(c) The strategy of X is [0.1 0.4 0.5]; the strategy of Y is [0.2 0.2 0.6]; the payoff matrix is

$$
\begin{array}{c}
 & Y \\
X & \left[\begin{array}{ccc} -3 & 2 & 1 \\ 4 & -2 & 5 \\ 3 & 1 & 2 \end{array} \right]
\end{array}
$$

6. Determine the optimal strategy and its resulting value for each of the following games:

(a) $\left[\begin{array}{cc} 2 & 9 \\ 6 & 3 \end{array} \right]$ (b) $\left[\begin{array}{cc} -2 & 5 \\ 8 & 4 \end{array} \right]$

(c) $\left[\begin{array}{ccc} 3 & 5 & 7 \\ 1 & 4 & 6 \\ 6 & 7 & 5 \end{array} \right]$

7. A truck-garden farmer raises vegetables in the field and in a greenhouse. The weather determines where they should be planted. The following payoff matrix describes the strategies available:

$$
\begin{array}{c c}
 & \begin{array}{cc} \text{Wet} & \text{Dry} \end{array} \\
\begin{array}{c} \text{Field} \\ \text{Greenhouse} \end{array} & \left[\begin{array}{cc} 250 & 140 \\ 175 & 210 \end{array} \right]
\end{array}
$$

What percentage of the crop should be planted in the field and what percentage in the greenhouse?

Logic

Communication is an important activity of the human race, but it is often difficult. We sometimes have difficulty expressing our thoughts. We make statements that another person interprets in a way we did not intend. To be sure that no question arises about the meaning, we sometimes ask lawyers to draw up a document to convey the precise intention of the information in the document. Even so, a lawsuit may arise when two parties disagree on the meaning and intent of the document.

Because problems in clear and precise communication do exist, mathematicians have sought to study these problems in order to clarify them and make some areas of communication more precise. In this chapter, we introduce you to this area of mathematics, logic. ■

10.1 Statements

- Notation
- Compound Statements
- Conjunction

- Disjunction
- Negation

In logic, we study sentences and relationships between certain kinds of sentences. In fact, we limit the study to just a portion of sentences used in your daily speech. If we examine the sentences that we use in daily life, we find they can be classified into several categories, which include the following:

1. Open your book to page 93. (a command)
2. Did you enjoy the concert? (a question)
3. What an exciting game! (an exclamation)
4. The sun rises in the east. (a true declarative statement)
5. Three plus two is nine. (a false declarative statement)
6. That was a good movie. (an ambiguous sentence)

The last sentence is ambiguous because there may be no agreement on what makes a movie good.

We restrict our study of logic to declarative sentences that are unambiguous and that can be classified as true or false but not both. We call such sentences **statements.** Only sentences 4 and 5 above are statements.

EXAMPLE 1 ➤ Classify each of the following sentences as a statement or not a statement.

(a) When is your next class?
(b) George Washington was the first president of the United States.
(c) Andrew Jackson was a great president.
(d) In 2010, February will have 29 days.
(e) That was a hard test!
(f) Three plus five is eight.

SOLUTION

(a) This is a question, not a declarative sentence, so it is not a statement.
(b) This is a true declarative sentence, so it is a statement.
(c) This is an ambiguous sentence, because there is no uniform understanding of the meaning of a "great" president. Therefore, it is not a statement.
(d) This is a false declarative sentence, so it is a statement.
(e) This is not a statement because it is an exclamation, not a declarative sentence. It is also ambiguous.
(f) This is a true declarative sentence, so it is a statement.

■ **Now You Are Ready to Work Exercise 1**

Notation

We use the letters $p, q, r, \ldots$ to denote statements. Thus, we might use p to represent a specific statement such as "Mathematics is required for my degree." In many cases, p or q is used to represent a general or unspecified statement, just as John Doe or Jane Doe is used to represent a general or unspecified person. For example, we might let p represent an arbitrary statement taken from *The Story of My Life* by Helen Keller.

Because we deal only with statements that can be classified as "true" or "false," we can assign a **truth value** to a statement p. We use T to represent the value "true" and F to represent the value "false" (as you do on a true–false quiz).

Compound Statements

We all frequently use statements in daily conversation that can be constructed by combining two or more simple statements. We call these **compound statements.** Compound statements can be long, complicated, and difficult to understand. Their meaning depends on the meaning of the component statements and how these components are put together into a single statement. We will look at some compound statements and how to analyze them.

Conjunction

The first type of compound statement that we study is a statement like "Ted is taking art, and Susan is taking history" that can be formed by connecting the statement "Ted is taking art" to the statement "Susan is taking history" using the word "and." If we use the p, q notation, this becomes:

Let p be the statement "Ted is taking art."

Let q be the statement "Susan is taking history."

Then "p and q" represents the statement "Ted is taking art, and Susan is taking history."

We use the notation **"$p \wedge q$"** to denote "p and q." A statement of the form $p \wedge q$ is called a **conjunction.**

When we form a statement by combining statements that we know to be true or false, can we decide on the truth of the compound statement? The answer is yes. In particular, the truth value of a conjunction is determined by the truth value of the statements making up the conjunction. The conjunction "Ted is taking art, and Susan is taking history" is true when both statements "Ted is taking art" and "Susan is taking history" are true. If either one or both statements are false, then the conjunction is false. In general, we can summarize such a situation with a **truth table.** When we form a compound statement from two statements, the two statements could be both true, both false, or one true and one false. The truth table below gives the truth value of $p \wedge q$ using the four possible combinations of truth values for p and for q:

p	q	$p \wedge q$
T	T	T
T	F	F
F	T	F
F	F	F

EXAMPLE 2 ➤ The statement "1992 was a leap year, and the Fourth of July is a national holiday in the United States" is true because both statements "1992 was a leap year" and "The Fourth of July is a national holiday in the United States" are true. The statement "1992 was a leap year, and February has 30 days" is false because the statement "February has 30 days" is false. ∎

Disjunction

Two statements can be connected with the word "or" to form a **disjunction.** The statement "We advertise in the Sunday paper, or we buy time on TV" is the disjunction of the statement "We advertise in the Sunday paper" and the statement "We buy time on TV." The notation $p \vee q$ represents the disjunction of statement p with statement q. We may read $p \vee q$ as "p or q."

EXAMPLE 3 ➤ Identify each of the following statements as conjunction, disjunction, or neither:

(a) The toast is burned, and the eggs are cold.
(b) The cafeteria opens for lunch at 11:00 A.M.
(c) Students in English 102 are required to analyze two poems, or they are required to analyze two short stories.

SOLUTION

(a) Conjunction   **(b)** Neither   **(c)** Disjunction

∎ **Now You Are Ready to Work Exercise 4**

Again, the truth value of $p \vee q$ is determined by the truth values of p and of q. The statement $p \vee q$ is true when one or both of p and q are true; otherwise, it is false. The following truth table summarizes the truth value of $p \vee q$:

p	p	$p \vee q$
T	T	T
T	F	T
F	T	T
F	F	F

One common use of the connective p or q is interpreted as "either p or q but not both." We call this **exclusive or.** Unless otherwise specified, mathematicians use p or q to mean "either p or q or both." This is called the **inclusive or.**

EXAMPLE 4 ➤ Determine the truth value of each of the following statements:

(a) June has 30 days, or September has 31 days.
(b) There are 24 hours in a day, or there are seven days in a week.
(c) Thomas Jefferson was the first U.S. president, or there are 65 states in the United States.

SOLUTION

(a) This is true because the statement "June has 30 days" is true.
(b) This is true because both statements forming the disjunction are true.
(c) This statement is false because both statements "Thomas Jefferson was the first U.S. president" and "there are 65 states in the United States" are false.

∎ **Now You Are Ready to Work Exercise 10**

Negation

Sometimes we want to make a statement that means the opposite of a given statement. We call such a statement the **negation** of the given statement. We write the negation of statement p by writing "It is not true that p" (denoted by $\sim p$, which is read "not p"). The negation of "March is a summer month" is "It is not true that March is a summer month." You would probably express this rather awkward sentence as "March is not a summer month." We define the negation of a true statement as false and the negation of a false statement as true. The following table summarizes the truth values of p and $\sim p$ for the possible truth values of p:

p	$\sim p$
T	F
F	T

EXAMPLE 5 ➤

Write the negation of the following statements:

(a) Thanksgiving is in November.   **(b)** $2 + 2 = 7$
(c) My car won't start.

SOLUTION

(a) Thanksgiving is not in November.   **(b)** $2 + 2 \neq 7$
(c) My car will start.

■ **Now You Are Ready to Work Exercise 14**

EXAMPLE 6 ➤

Let p be the statement "Susan's dog is a poodle" and let q be the statement "Jake has a black cat." Write the following statements:

(a) $p \vee q$ **(b)** $\sim q$ **(c)** $p \wedge (\sim q)$

SOLUTION

(a) Susan's dog is a poodle, or Jake has a black cat.
(b) Jake does not have a black cat.
(c) Susan's dog is a poodle, and Jake does not have a black cat.

■ **Now You Are Ready to Work Exercise 16**

EXAMPLE 7 ➤

Determine the truth value of each of the following:

(a) February has 30 days, or Franklin Roosevelt was not the first president of the United States.
(b) A quart is not larger than a liter, and water does not run uphill.

SOLUTION

(a) This statement is of the form $p \vee (\sim q)$, where p is false and q is false (so $\sim q$ is true). Therefore, the statement is true.
(b) This statement is of the form $(\sim p) \wedge (\sim q)$, where p is false ($\sim p$ is true) and q is false ($\sim q$ is true). As both $\sim p$ and $\sim q$ are true, the statement is true. ■

EXAMPLE 8 ➤

Find the truth value of $(p \vee \sim q) \wedge q$ for all possible truth values of p and q.

SOLUTION

The following table shows all values of p and q and the corresponding values of the parts making up the given statement.

p	q	$p \vee \sim q$	$(p \vee \sim q) \wedge q$
T	T	T	T
T	F	T	F
F	T	F	F
F	F	T	F

■ **Now You Are Ready to Work Exercise 20**

In the next example, you are to find truth values of a statement with variables p, q, and r. You can construct the table of all possible values of p, q, and r as follows:

1. List all possible values of p and q and for each case enter T under the r column.
2. Again list all possible values of p and q under the part of the table just formed and for each case enter F in the r column. Going from a table with two variables (four rows) to one with three variables (eight rows), double the number of rows in the table. Each time you add another variable, double the number of rows needed to list all possible values of the variables.

EXAMPLE 9 ➤

Find the truth value of $p \vee (q \wedge r)$ for all possible truth values of p, q, and r.

SOLUTION

We list all possible combinations of T and F for the statements and fill in the following truth table:

p	q	r	$q \wedge r$	$p \vee (q \wedge r)$
T	T	T	T	T
T	F	T	F	T
F	T	T	T	T
F	F	T	F	F
T	T	F	F	T
T	F	F	F	T
F	T	F	F	F
F	F	F	F	F

■ **Now You Are Ready to Work Exercise 22**

▪▪ 10.1 EXERCISES

1. *(See Example 1)* Determine whether each of the following sentences is a statement:
   (a) A day is 24 hours long.
   (b) An hour is 65 minutes long.
   (c) Are there more than 30 seconds in a minute?
   (d) I think 50 minutes is too little time for a history class.

2. Determine whether each of the following sentences is a statement:
   (a) *Gone with the Wind* is a movie about Civil War times and Reconstruction.
   (b) Pick up a video of *Man from Snowy River* on your way home.
   (c) Clark Gable was a great actor in *Gone with the Wind*.
   (d) *Gone with the Wind* was not produced until 1987.

3. Determine whether each of the following sentences is a statement:
   (a) $2x + 4 = 2(x + 2)$
   (b) $6 + 3 = 17$
   (c) $5 + 5 = 10$
   (d) Find x such that $2x = 10$.
   (e) Prime numbers are interesting.

4. *(See Example 3)* Identify each of the following as conjunction, disjunction, or neither:
   (a) Jim has on a brown coat, and Jules has on a blue coat.
   (b) My car has a flat tire, and my brother's car won't start.
   (c) I ride to work with Mary on Tuesdays, or I am late for work.
   (d) The weather forecast calls for scattered showers.

5. Identify each of the following as conjunction, disjunction, or neither:
   (a) What a beautiful day!
   (b) You must drive within the speed limit, or you will receive a ticket.
   (c) I did my homework every day, and I received an A on the exam.
   (d) I need to find a job, or I will not be able to pay my rent.

6. Let p be the statement "Coffee contains caffeine" and q the statement "Sugar contains 16 calories per teaspoon." Write out the following statements:
   (a) $p \wedge q$ (b) $p \vee q$
   (c) $\sim p$ (d) $\sim p \wedge q$

7. Let p be the statement "Betty has blonde hair" and q the statement "Angela has dark hair." Write out the following statements:
   (a) $p \vee q$ (b) $\sim q$
   (c) $p \wedge q$ (d) $\sim p \wedge \sim q$

8. Convert the following statements into symbolic form using p for "Joe made a B on the exam" and q for "Al did not take the exam."
   (a) Joe made a B on the exam, and Al did not take the exam.
   (b) Al took the exam.
   (c) Al did not take the exam, or Joe made a B on the exam.

9. Convert the following statements into symbolic form, letting p be "$2 + 3 = 5$" and q be "$17 > 12$."
   (a) $2 + 3 = 5$ and $17 > 12$.
   (b) $2 + 3 = 5$ or $17 > 12$.
   (c) $2 + 3 \neq 5$ and $17 > 12$.

10. *(See Example 4)* Determine whether the following statements are true or false:
    (a) There are 60 seconds in a minute and 60 minutes in an hour.
    (b) $2 + 2 = 4$ and $2 \times 5 = 25$.
    (c) The year 2000 had 400 days, and the Moon is made of green cheese.

11. Determine whether the following statements are true or false:
    (a) $3^2 = 9$ and $4^2 = 15$.
    (b) Lyndon Johnson was the first president of the United States, and algebra is a subject in mathematics.
    (c) $2 \times 5 = 10$ and June has 30 days.

12. Determine the truth value of each of the following statements:
    (a) There are 24 hours in a day, or there are 366 days in a leap year.
    (b) The Sun rises in the east, or the Sun rises in the north.
    (c) Water runs uphill or there is no water in the ocean.

13. Determine the truth value of each of the following statements:
    (a) $3 + 5 = 17$ or 19 is greater than 20.
    (b) A triangle has three sides, or a square has equal sides.
    (c) A triangle has three sides, or a square has five sides.

14. *(See Example 5)* Write the negation of each of the following statements:
    (a) The metric system is the official system of measurement in the United States.
    (b) The Sun did not shine yesterday.
    (c) Josie's history paper is 37 pages long.

15. Write the negation of the following statements:
    (a) I have six $1 bills in my wallet.
    (b) Roy can name all 50 states.
    (c) A quorum was not present for the meeting.

16. *(See Example 6)* Let p represent the statement "My bookcase is full of books" and let q represent "My desk drawer is full of paper." Write the following statements:
    (a) $p \wedge (\sim q)$ (b) $(\sim p) \vee (\sim q)$
    (c) $(\sim p) \wedge (\sim q)$

17. Let p represent the statement "I drink coffee at breakfast," let q represent "I eat salad for lunch,"

and let r represent "I like a dessert after dinner." Write the following statements:

(a) $p \wedge q \wedge r$ (b) $p \vee (q \wedge r)$

(c) $p \wedge q \wedge (\sim r)$ (d) $(p \wedge q) \vee (p \wedge r)$

18. Let p represent a statement having a truth value T, q represent a statement having a truth value F, and r represent a statement having a truth value T. Find the truth value of each of the following:

(a) $p \wedge (\sim q)$ (b) $p \wedge q \wedge r$

(c) $p \wedge (q \vee r)$ (d) $\sim (p \wedge q)$

(e) $(\sim p) \vee (\sim q)$

19. Find the truth value of each of the following when p, q, and r all have truth values T:

(a) $p \vee (q \wedge r)$ (b) $(\sim p) \wedge (q \vee r)$

(c) $(\sim p) \wedge p$ (d) $(\sim p \vee q) \wedge (\sim r)$

(e) $(p \vee q) \vee (r \wedge \sim q)$

20. *(See Example 8)* Use a truth table to find the value of $\sim (p \vee q)$ for all possible truth values of p and q.

21. Use a truth table to find the value of $\sim p \wedge \sim q$ for all possible truth values of p and q.

22. *(See Example 9)* Use a truth table to find the truth value of $(p \vee q) \wedge (p \vee r)$ for all possible values of p, q, and r.

23. (a) Write the following statement in symbolic form. "Jane brings the chips and Tony brings the drinks and Hob brings the cookies, or Ingred brings the chips and Alex or Hester bring the drinks and Alice or Jenn bring the cookies."

(b) Determine if the statement is true or false if the following occurs.

Jane brings chips.

Tony does not bring drinks but brings cookies.

Hob brings drinks and does not bring cookies.

Ingred brings chips.

Alex and Hester both bring drinks.

Neither Alice nor Jenn bring cookies.

24. Exactly two of the following statements are true, and one is false.

(a) Heidi works at Auto Parts and Scott works at the Pizza Parlor.

(b) Heidi works at Auto Parts, or Scott works at the Pizza Parlor.

(c) Scott works at the Pizza Parlor.

What information does this give about where Heidi and Scott work?

10.2   Conditional Statements

- Conditional
- Converse, Inverse, and Contrapositive of a Conditional
- Biconditional

Conditional

We sometimes make sentences that contain a condition rather than an outright assertion. For example, the sentence "If the Sun shines, I will cut the grass" contains a condition (If the Sun shines) regarding the cutting of grass. We call such sentences **conditional** or **implication statements.**

DEFINITION
Conditional (Implication) Statement

A **conditional (implication) statement** is a statement of the form

"If p, then q."

We denote it by $p \rightarrow q$.

In an implication, $p \rightarrow q$, we call p the **hypothesis** and q the **conclusion.**

EXAMPLE 1 ▶

Let p be the statement "I study four hours" and q the statement "I can make an A on the exam."

(a) Write the statement $p \rightarrow q$.   (b) Write the statement $q \rightarrow p$.

(c) Write the statement with p as the hypothesis and q as the conclusion.

SOLUTION

(a) If I study four hours, then I can make an A on the exam.
(b) If I can make an A on the exam, then I study four hours.
(c) If I study four hours, then I can make an A on the exam.

As the statements (a) and (c) in the solution are the same statements, this illustrates that the notation in part (a) describes the same statement as does part (c).

■ **Now You Are Ready to Work Exercise 1**

We usually think that the truth value of the hypothesis of a conditional has some effect on the conclusion. For example, the amount of time spent studying has some effect on the exam grade. However, we can construct implications where the hypothesis and conclusion have no relationship. The sentence "If June has 30 days, then Humpty Dumpty sat on the wall" is a true conditional even though the number of days in June has nothing to do with Humpty Dumpty. Although such a conditional might appear to be somewhat nonsensical, and you might not use this form in conversation, we still want to assign it a truth value. The following table shows how the truth values of p and q determine the truth value of $p \rightarrow q$:

p	q	$p \rightarrow q$
T	T	T
T	F	F
F	T	T
F	F	T

To assign T as the truth value of $p \rightarrow q$ in the last two lines may not seem natural. You might prefer a value of F, or you may think it doesn't make much sense to put either T or F. Granted, we most often use a conditional statement in situations where the first two lines apply. However, in logic we insist that statements be either true or false, so we must assign values to the last two cases. The truth table given is the definition accepted by mathematicians.

Even so, we do not arbitrarily assign true values. The following examples help us to understand this.

Your professor tells you, "If you make an A on the final exam, I will give you an A in the course." You make an A on the final exam, and the professor gives you a B in the course. Isn't your conclusion that the professor made a false statement? (Or in less nice terms, the professor lied.) Thus, assigning the value F to $p \rightarrow q$ when p is T and q is F is consistent with our usual interpretation.

Dick has just received a driver's license, and his parents impose some rules to help him become a responsible driver. One of the rules is "If you speed, you will be grounded." This rule is of the form $p \rightarrow q$, a conditional. When Dick returns one evening, he is asked, "Did you speed?" He had not sped and truthfully answered, "No." His parents do not ground Dick. Thus, p and q both have the truth value F. Even though both parts of the rule are false, the rule has not been violated, so we should assign the value T to the conditional in this case.

Another evening Dick returns quite late and he again is asked "Did you speed?" His truthful answer was. "No." Dick was grounded because he stayed out past the time to be home. Here p has the value F and q the value T. Has the rule been violated? The rule does not state that speeding is the only reason for grounding. Thus, the rule has not been violated, and we should assign the value T to the rule.

EXAMPLE 2 ➤

Determine the truth of the following conditionals:

(a) If $2 \times 3 = 6$, then $15 - 4 = 11$.
(b) If $3 + 3 = 6$, then $2 = 1$.
(c) If there are 8 days in a week, then June has 30 days.
(d) If there are 8 days in a week, then August has 79 days.

SOLUTION

(a) True because the hypothesis and conclusion are true
(b) False because the hypothesis is true and the conclusion is false
(c) True because the hypothesis is false and the conclusion is true
(d) True because the hypothesis and conclusion are false

■ **Now You Are Ready to Work Exercise 3**

Notice that a conditional is always true when the hypothesis is false, regardless of whether or not the conclusion is true. A true conclusion always makes a conditional true.

Converse, Inverse, and Contrapositive of a Conditional

We have just discussed how to form a conditional statement from two given statements. Actually, more than one conditional can be formed. For example, the statements "My car has a flat tire" and "I want to ride to the game with you" can be used to form the conditional "If my car has a flat tire, then I want to ride to the game with you." They can also be combined to form the conditional "If I want to ride to the game with you, then my car has a flat tire." Two other conditionals, which are commonly formed by using the negation of the given statements, are "If my car does not have a flat tire, then I do not want to ride to the game with you" and "If I do not want to ride to the game with you, then my car does not have a flat tire." Because the four conditionals of these forms occur frequently in mathematics, we give them names: the **original conditional,** the **converse,** the **inverse,** and the **contrapositive.**

DEFINITION
**Converse, Inverse,
Contrapositive**

From a given conditional "If p then q" we can form the following conditionals:

    Converse: "If q, then p." $(q \rightarrow p)$

    Inverse: "If not p, then not q." $(\sim p \rightarrow \sim q)$

    Contrapositive: "If not q, then not p." $(\sim q \rightarrow \sim p)$

EXAMPLE 3 ➤

Write the converse, inverse, and contrapositive of "If it does not rain, then we want to go on a picnic."

SOLUTION

    Converse: "If we want to go on a picnic, then it does not rain."

    Inverse: "If it docs rain, thcn wc do not want to go on a picnic."

    Contrapositive: "If we do not want to go on a picnic, then it does rain."

■ **Now You Are Ready to Work Exercise 5**

Now let's compare the truth tables of a conditional and its converse, inverse, and contrapositive. We can summarize them in one table as follows:

p	q	$p \to q$	$q \to p$	$\sim p \to \sim q$	$\sim q \to \sim p$
T	T	T	T	T	T
T	F	F	T	T	F
F	T	T	F	F	T
F	F	T	T	T	T

Notice that $p \to q$ and $q \to p$ have different truth values when p is true and q is false or vice versa. This indicates that a conditional and its converse cannot be interchanged because a converse might not be true even though the conditional is. It is a common error for some people to think that a conditional and its converse have exactly the same meaning. You certainly would not accept that the statement "If I live in Chicago, then I live in Illinois" has the same meaning as the statement "If I live in Illinois, then I live in Chicago."

Also notice that $p \to q$ (a statement) and $\sim q \to \sim p$ (its contrapositive) have exactly the same truth values in all cases. This means that one statement can be substituted for the other. Similarly, $q \to p$ (the converse of $p \to q$) and $\sim p \to \sim q$ (the inverse of $p \to q$) also have the same truth values. For example, the statements "If I go home, I will have some home cooking" and "If I do not have some home cooking, then I do not go home" have the same meaning. Also the statements "If I have home cooking, then I go home" and "If I do not go home, then I do not have home cooking" have the same meaning.

Biconditional

Another fundamental compound statement is the **biconditional** statement, denoted $p \leftrightarrow q$, and read "p if and only if q." You can also think of the biconditional, $p \leftrightarrow q$, as $(p \to q) \land (q \to p)$. We develop the truth table for the biconditional as follows:

p	q	$p \to q$	$q \to p$	$(p \to q) \land (q \to p)$	$p \leftrightarrow q$
T	T	T	T	T	T
T	F	F	T	F	F
F	T	T	F	F	F
F	F	T	T	T	T

We summarize this table with the following:

p	q	$p \leftrightarrow q$
T	T	T
T	F	F
F	T	F
F	F	T

Notice that $p \leftrightarrow q$ is true exactly when p and q have the same truth values.

EXAMPLE 4 ➤

Determine the truth value of the following biconditionals:

(a) $2 + 2 = 4$ if and only if $5 \times 2 = 10$.
(b) The English alphabet contains 36 letters if and only if Mexico is north of New York.
(c) $x + x = 2x$ if and only if $2 = 1$.

SOLUTION

(a) The biconditional is true because both $2 + 2 = 4$ and $5 \times 2 = 10$ are true.
(b) This is true because the component statements have the same truth value, false.
(c) This is false because the statements $x + x = 2x$ and $2 = 1$ have different truth values, true and false, respectively.

■ **Now You Are Ready to Work Exercise 7**

We can combine the basic compound and simple statements to form other compound statements. Using the truth tables of the basic compound statements, we can find the truth values of other compound statements.

EXAMPLE 5 ➤

(a) Find the truth table for $\sim p \rightarrow (p \wedge q)$.
(b) Find the truth table for $(p \wedge q) \rightarrow r$.

SOLUTION

(a) It is easier to analyze $\sim p \rightarrow (p \wedge q)$ if we include columns for $\sim p$ and $p \wedge q$ in the truth table:

p	q	$\sim p$	$p \wedge q$	$\sim p \rightarrow (p \wedge q)$
T	T	F	T	T
T	F	F	F	T
F	T	T	F	F
F	F	T	F	F

■ **Now You Are Ready to Work Exercise 9**

(b) Because there are three statements $p, q,$ and r, eight rows are required in the truth table to allow for all the possible combinations of the truth values of $p, q,$ and r:

p	q	r	$p \wedge q$	$(p \wedge q) \rightarrow r$
T	T	T	T	T
T	T	F	T	F
T	F	T	F	T
T	F	F	F	T
F	T	T	F	T
F	T	F	F	T
F	F	T	F	T
F	F	F	F	T

■ **Now You Are Ready to Work Exercise 17**

■■ 10.2 EXERCISES

1. *(See Example 1)* Let p be the statement "I have $5.00" and q the statement "I can rent a video."
   (a) Write the statement $p \to q$.
   (b) Write the statement $q \to p$.

2. Let p be the statement "Monday is a holiday" and q the statement "I can sleep late."
   (a) Write the statement $p \to q$.
   (b) Write the statement $q \to p$.

3. *(See Example 2)* Determine the truth of the following conditionals:
   (a) If a is a vowel of the English alphabet, then bees make honey.
   (b) If k is a vowel of the English alphabet, then bees make honey.
   (c) If New Year's Day is the first day of the year, then the Fourth of July is the last day of the year.

4. Determine the truth of the following conditionals:
   (a) If 5 is larger than 10, then a penny is worth more than $1.00.
   (b) If 1 foot equals 12 inches, then z is the last letter of the alphabet.
   (c) If 60 seconds equal 1 minute, there are 30 hours in a day.

5. *(See Example 3)* Write the converse, inverse, and contrapositive of "If I live in Denver, then I live in Colorado."

6. Write the converse, inverse, and contrapositive of "If I carry an umbrella, then I do not get wet."

7. *(See Example 4)* Determine the truth values of the following biconditionals:
   (a) $8 + 7 = 15$ if and only if 5 is larger than 25.
   (b) a is a vowel in the English alphabet if and only if i is a vowel in the English alphabet.
   (c) April has 30 days if and only if July has 31 days.

8. Determine the truth of the following biconditionals:
   (a) Twelve inches equals 1 foot if and only if 3 feet equals 1 yard.
   (b) California is on the West Coast if and only if the year 2001 is a leap year.
   (c) p is a vowel of the English alphabet if and only if 2 divides 14.

Construct the truth tables for the statements expressed in Exercises 9 through 24.

9. *[See Example 5(a)]* $\sim (p \wedge q)$

10. $\sim (\sim p)$

11. $p \wedge (\sim q)$

12. $(\sim p) \vee q$

13. $\sim p \vee \sim q$

14. $p \to (q \vee \sim p)$

15. $\sim p \to (q \vee p)$

16. $(p \wedge q) \to (p \vee q)$

17. *[See Example 5(b)]* $(p \vee q) \wedge r$

18. $[(p \vee q) \wedge r] \wedge \sim r$

19. $(p \to q) \wedge (q \to r)$

20. $(p \wedge q) \leftrightarrow (p \vee q)$

21. $(p \vee q) \leftrightarrow r$

22. $(p \to q) \leftrightarrow (q \to p)$

23. $\sim (\sim p) \leftrightarrow p$

24. $(p \wedge q) \vee (p \wedge r)$

25. The following statements appeared in IRS 1040 tax return instructions. Identify simple statements within each one and symbolically represent them with $p, q, r, \ldots$ and the connectives $\wedge, \vee, \to$.
   (a) If your insurance company paid the provider directly for part of your expenses, and you paid only the amount that remained, include on line 1 only the amount you paid.
   (b) If you leave line 65 blank, the IRS will figure the penalty and send you the bill.
   (c) If you changed your name because of marriage, divorce, etc., and you made estimated tax payments using your former name, attach a statement to the front of Form 1040 explaining all the payments you and your spouse made in 1997, the service center where you made the payments, and the name and SSN under which you made the payments.

10.3 Equivalent Statements

Two compound statements are **logically equivalent,** or **equivalent,** if they have exactly the same truth values. This means that a statement can be substituted for its equivalent because when one is true, the other is also true, and when one is false, the other is false. Two equivalent statements will have the same truth values for any choice of truth values of the component parts of the compound statements.

EXAMPLE 1 ➤

Show that $p \rightarrow q$ and $\sim p \vee q$ are logically equivalent statements.

SOLUTION

We form the truth table for each statement and see whether they are the same. We can form the table as follows:

p	q	$\sim p$	$p \rightarrow q$	$\sim p \vee q$
T	T	F	T	T
T	F	F	F	F
F	T	T	T	T
F	F	T	T	T

Note that for each choice of p and q (each row), the truth values of $p \rightarrow q$ and $\sim p \vee q$ are the same. Therefore, the statements are equivalent.

■ **Now You Are Ready to Work Exercise 1**

EXAMPLE 2 ➤

Show that $p \rightarrow q$ and $q \rightarrow p$ are not logically equivalent.

SOLUTION

Form the truth table and compare truth values:

p	q	$p \rightarrow q$	$q \rightarrow p$
T	T	T	T
T	F	F	T
F	T	T	F
F	F	T	T

Note that the truth values of $p \rightarrow q$ and $q \rightarrow p$ differ in rows 2 and 3, so the statements are not equivalent.

A difference of truth values in just one row prevents two statements from being equivalent.

■ **Now You Are Ready to Work Exercise 6**

EXAMPLE 3 ➤

Determine if the two following statements are equivalent or not:

(a) If you check the box, you cannot take the standard deduction.
(b) If you can take the standard deduction, you do not check the box.

SOLUTION

Let p represent the statement "You check the box" and let q represent the statement "You can take the standard deduction."

Then statement (a) is $p \rightarrow \sim q$ and statement (b) is $q \rightarrow \sim p$. We form a truth table and compare the truth values of $p \rightarrow \sim q$ and $q \rightarrow \sim p$.

p	q	$\sim p$	$\sim q$	$p \to \sim q$	$q \to \sim p$
T	T	F	F	F	F
T	F	F	T	T	T
F	T	T	F	T	T
F	F	T	T	T	T

As $p \to \sim q$ and $q \to \sim p$ have identical truth values, they are equivalent.

■ **Now You Are Ready to Work Exercise 11**

10.3 EXERCISES

Use truth tables to determine whether or not the pairs of statements in Exercises 1 through 10 are logically equivalent.

1. *(See Example 1)* $p \to q$ and $\sim q \to \sim p$

2. $p \leftrightarrow q$ and $(p \to q) \wedge (q \to p)$

3. $\sim p \vee (p \wedge q)$ and $\sim p \vee q$

4. **(a)** $\sim (p \wedge q)$ and $\sim p \vee \sim q$
   (b) $\sim (p \vee q)$ and $\sim p \wedge \sim q$
   These are called DeMorgan's laws.

5. $\sim p \vee (q \wedge r)$ and $p \to (q \wedge r)$

6. *(See Example 2)* $\sim (p \wedge q)$ and $\sim p \wedge \sim q$

7. $\sim (p \to q)$ and p

8. $(p \to q) \wedge (q \to r)$ and $p \to r$

9. $p \wedge (q \vee r)$ and $(p \wedge q) \vee r$

10. $p \vee (q \wedge r)$ and $(p \vee q) \wedge (p \vee r)$

11. *(See Example 3)* Determine if the two following statements are equivalent:

    (a) If the exceptions above do not apply, then use form 2210.
    (b) The exceptions above apply or use form 2210.

12. Determine if the two following statements are equivalent:
    (a) Sara did not come to work, and Tom is on vacation.
    (b) Sara came to work, and Tom is not on vacation.

13. Determine if the following statements are equivalent:
    (a) I will buy a jacket, and I will buy a shirt or tie.
    (b) I will buy a jacket and a shirt, or I will buy a jacket and a tie.

14. For the statement "I will go to a movie, or I will go to a ball game," which of the following is equivalent to its negation?
    (a) I will not go to a movie, or I will not go to a ball game.
    (b) I will go to a movie, and I will not go to a ball game.
    (c) I will not go to a movie, and I will not go to a ball game.

10.4 Valid Arguments

We all like to win an argument or prove a point. Unfortunately, false information or intimidation, not logic, sometimes wins arguments. Although we like to win arguments, it is more important that we draw a valid conclusion from information given. We would like a jury to reach a verdict that is supported by evidence. We do not like to see a person's reputation ruined because an invalid conclusion was drawn from information given. A business executive can harm the corporation if a decision is not supported by the facts relevant to the decision.

Some arguments are generally recognized as invalid. For example, those who accept the premise "If you live in Omaha, then you live in Nebraska" will not accept the conclusion "If you live in Nebraska, then you live in Omaha." However, it appears that a significant number of people who accept the premise "If you are homosexual, then you will acquire AIDS" will accept the conclusion "If you acquire AIDS, then you are homosexual." Actually, this AIDS argument and the Omaha argument are of the same form. Each one has a premise of the form $p \rightarrow q$ and a conclusion of the form $q \rightarrow p$. We would like an argument to be valid or invalid based on whether or not the premises support the conclusion, not on which statements are used. We are interested in **arguments** or **proofs** that are logically sound; we call them **valid arguments.**

An argument consists of a statement, called the **conclusion,** that follows from one or more statements, called the **premises.** If the conclusion follows logically from the premises, we say the argument is **valid.** You might ask what we mean by "the conclusion follows logically from the premises." Here is what we mean. Suppose an argument has two statements for its premises: call them p and q. Call the conclusion r. The conclusion, r, follows logically from the premises, p and q, if whenever all the premises, p and q, are true, then the conclusion, r, is also true. An argument may have any number (one or more) premises.

**DEFINITION
Valid Argument**

> An argument with premises p_1, p_2, p_3, ..., p_n and a conclusion r is **valid** if r is true when *all* of p_1, p_2, p_3, ..., p_n are true.

Here is a simple example of a valid argument.

EXAMPLE 1 ▶

Premises:	If Jenny has a job, she saves money.
	Jenny has a job.
Conclusion:	Jenny saves money.

Let's put this in symbolic form. Write

p: Jenny has a job.
q: Jenny saves money.

Now the argument can be written in the notation

Premises: $p \rightarrow q$
p
Conclusion: q

where the premises are written above the line and the conclusion below. The argument is valid if q is true whenever both $p \rightarrow q$ and p are true. Check the following truth table:

p	q	$p \rightarrow q$
T	T	T
T	F	F
F	T	T
F	F	T

Notice that when $p \rightarrow q$ and p are both true (first line only), q is also true, so the argument is valid.

The entries in the truth table would be the same if statements other than "Jenny has a job" and "Jenny saves money" are used for p and q. Therefore, for any statements p and q the argument

$$\begin{array}{rl} \text{Premises:} & p \rightarrow q \\ & \underline{p} \\ \text{Conclusion:} & q \end{array}$$

is a valid argument. This is sometimes called the **Law of Detachment.**

■ **Now You Are Ready to Work Exercise 1**

DEFINITION
Law of Detachment

For any pair of statements p and q, the argument

$$\begin{array}{rl} \text{Premises:} & p \rightarrow q \\ & \underline{p} \\ \text{Conclusion:} & q \end{array}$$

is a valid argument.

EXAMPLE 2 ➤

By the Law of Detachment, the following arguments are valid:

(a) Premises: If my car has a flat tire, I will ride to the game with Sam.
               My car has a flat tire.
───
Conclusion: I will ride to the game with Sam.

(b) Premises: If I make an A on the final, I will make an A in the course.
               I made an A on the final.
───
Conclusion: I will make an A in the course.   ■

We can determine the validity of an argument in a way a little different from the definition by observing that if all the premises $p_1, p_2, p_3, \ldots, p_n$ are true, then

$$p_1 \wedge p_2 \wedge \ldots \wedge p_n \rightarrow r$$

is true only when r is also true. If any one of $p_1, p_2, \ldots, p_n$ is false, then $p_1 \wedge p_2 \wedge \ldots \wedge p_n$ is false. In this case, the conditional $p_1 \wedge p_2 \wedge \ldots \wedge p_n \rightarrow r$ is true.

All of this makes for a situation where an argument with premises $p_1, p_2, \ldots, p_n$ and conclusion r is valid provided the conditional $p_1 \wedge p_2 \wedge \ldots \wedge p_n \rightarrow r$ is true for all possible values of the premises and conclusion. (A statement that is always true is called a **tautology.**) This approach will be used to test the validity of arguments.

THEOREM
Valid Argument

An argument with premises $p_1, p_2, p_3, \ldots, p_n$ and a conclusion r is **valid,** if

$$p_1 \wedge p_2 \wedge \ldots \wedge p_n \rightarrow r$$

is true for all possible truth values of $p_1, p_2, p_3, \ldots, p_n$ and r.

EXAMPLE 3 ➤

Show that the following argument is valid:

$$\begin{array}{rl} \text{Premises:} & p \wedge q \\ & \underline{q} \\ \text{Conclusion:} & p \end{array}$$

SOLUTION

By the theorem above we can prove the argument valid by showing $(p \wedge q) \wedge q \to p$ is true for all possible values of p and q. We check the truth table for

p	q	$p \wedge q$	$(p \wedge q) \wedge q$	$(p \wedge q) \wedge q \to p$
T	T	T	T	T
T	F	F	F	T
F	T	F	F	T
F	F	F	F	T

The argument is valid because only T appears in the last column. ∎

EXAMPLE 4 ➤

Determine whether the following argument is valid:

$$\text{Premises:} \quad p \to q$$
$$\underline{\qquad\qquad q \to r}$$
$$\text{Conclusion:} \quad p \to r$$

This argument is called a **syllogism.**

SOLUTION

Check $[(p \to q) \wedge (q \to r)] \to (p \to r)$:

p	q	r	$p \to q$	$q \to r$	$p \to r$	$(p \to q) \wedge (q \to r)$	$[(p \to q) \wedge (q \to r)] \to (p \to r)$
T	T	T	T	T	T	T	T
T	F	T	F	T	T	F	T
F	T	T	T	T	T	T	T
F	F	T	T	T	T	T	T
T	T	F	T	F	F	F	T
T	F	F	F	T	F	F	T
F	T	F	T	F	T	F	T
F	F	F	T	T	T	T	T

The argument is valid.

■ **Now You Are Ready to Work Exercise 3**

EXAMPLE 5 ➤

Determine whether the following argument is valid:

Premises: If Mike jogs daily, he keeps his weight under control.
             If Mike keeps his weight under control, he can wear last
             year's clothes.

Conclusion: If Mike jogs daily, he can wear last year's clothes.

SOLUTION

Write the arguments in symbolic form.
    Let

p: Mike jogs daily.
q: Mike keeps his weight under control.
r: Mike can wear last year's clothes.

The argument is of the form

$$\begin{array}{ll} \text{Premises:} & p \to q \\ & \underline{q \to r} \\ \text{Conclusion:} & p \to r \end{array}$$

Example 4 shows that this argument (a syllogism) is valid. ■

EXAMPLE 6 ➤

Determine whether the following argument is valid:

Premises: Marci will mow the lawn, or Lee will mow the lawn.
<u>Marci cannot mow the lawn.</u>

Conclusion: Lee will mow the lawn.

SOLUTION

Let

p: Marci will mow the lawn.
q: Lee will mow the lawn.

The argument is of the form

$$\begin{array}{ll} \text{Premises:} & p \vee q \\ & \underline{\sim p} \\ \text{Conclusion:} & q \end{array}$$

Check the following truth table for $[(p \vee q) \wedge (\sim p)] \to q$:

p	q	$p \vee q$	$\sim p$	$(p \vee q) \wedge (\sim p)$	$[(p \vee q) \wedge (\sim p)] \to q$
T	T	T	F	F	T
T	F	T	F	F	T
F	T	T	T	T	T
F	F	F	T	F	T

The last column indicates that the argument is valid. ■

This argument is called a **disjunctive syllogism.**

EXAMPLE 7 ➤

Show that the following argument is not valid:

$$\begin{array}{ll} \text{Premises:} & p \to q \\ & \underline{\sim p} \\ \text{Conclusion:} & \sim q \end{array}$$

SOLUTION

Check $[(p \to q) \wedge (\sim p)] \to (\sim q)$:

p	q	$p \to q$	$\sim p$	$\sim q$	$(p \to q) \wedge (\sim p)$	$[(p \to q) \wedge (\sim p)] \to \sim q$
T	T	T	F	F	F	T
T	F	F	F	T	F	T
F	T	T	T	F	T	F
F	F	T	T	T	T	T

The argument is not valid because an F appears in the last column.

■ **Now You Are Ready to Work Exercise 5**

EXAMPLE 8 ➤   Show that the following argument is valid:

$$p \rightarrow q$$
$$\underline{\sim q}$$
$$\sim p$$

SOLUTION

Here is the truth table:

p	q	$p \rightarrow q$	$\sim q$	$(p \rightarrow q) \wedge \sim q$	$[(p \rightarrow q) \wedge \sim q] \rightarrow \sim p$
T	T	T	F	F	T
T	F	F	T	F	T
F	T	T	F	F	T
F	F	T	T	T	T

The last column indicates that the argument is valid. ■

This is called **indirect reasoning.**
We now summarize the valid arguments discussed:

Four Valid Arguments

Law of Detachment

$$p \rightarrow q$$
$$\underline{p}$$
$$q$$

Syllogism

$$p \rightarrow q$$
$$\underline{q \rightarrow r}$$
$$p \rightarrow r$$

Disjunctive Syllogism

$$p \vee q$$
$$\underline{\sim p}$$
$$q$$

Indirect Reasoning

$$p \rightarrow q$$
$$\underline{\sim q}$$
$$\sim p$$

◨ 10.4 EXERCISES

Determine whether the arguments in Exercises 1 through 10 are valid or not.

1. *(See Example 1)*

Premise: If you eat your beans, you may have dessert.
You ate your beans.

Conclusion: You may have dessert.

2. Premise: If you are absent more than three times, your average will be lowered 2 points.
You are absent more than three times.

Conclusion: Your average will be lowered 2 points.

3. *(See Example 4)*

Premise: If you do not study, you cannot do the homework.
If you cannot do the homework, you cannot pass the course.

Conclusion: If you do not study, you cannot pass the course.

4. Premise: If the word processor is working, I will write a paper for extra credit.
If I write a paper for extra credit, I will receive a B in the course.

Conclusion: If the word processor is working, I will receive a B in the course.

5. *(See Example 7)*

Premise: If you eat your beans, you may have dessert.
You did not eat your beans.

Conclusion: You may not have dessert.

6. Premise: If the ice is 6 inches thick, Shelley will go skating.
The ice is not 6 inches thick.

Conclusion: Shelley will not go skating.

7. Premise: If the ice is 6 inches thick, Shelley will go skating.
Shelley did not go skating.

Conclusion: The ice is not 6 inches thick.

8. Premise: If Andy smokes, he will likely get lung cancer.
Andy smokes.

Conclusion: Andy will likely get lung cancer.

9. Premise: If inflation increases, the price of new cars will increase.
If the price of new cars increases, more people will buy used cars.

Conclusion: If inflation increases, more people will buy used cars.

10. Premise: If I don't clean my room, I'll be in trouble with my mother.
I cleaned my room.

Conclusion: I will not be in trouble with my mother.

Determine whether the arguments in Exercises 11 through 20 are valid or not.

11. Premise $p \to q$
$q \wedge r$
Conclusion: $p \vee r$

12. Premise: $\sim p \to q$
p
Conclusion: $\sim q$

13. Premise: $p \wedge q$
$p \to \sim q$
Conclusion: $p \wedge \sim q$

14. Premise: $p \to q$
$\sim p \to r$
Conclusion: $q \vee r$

15. Premise: $q \to r$
$\sim p \vee q$
p
Conclusion: r

16. Premise: $p \to q$
$q \to r$
$r \to s$
Conclusion: $p \to s$

17. Premise: $p \to q$
$p \to r$
Conclusion: $q \wedge r$

18. Premise: $p \to q$
$\sim q \to \sim r$
Conclusion: $p \to \sim r$

19. Premise: $p \to q$
$q \to r$
$\sim q$
Conclusion: $\sim r$

20. Premise: $p \to (q \wedge r)$
$\sim q$
Conclusion: $\sim p$

Identify the arguments in Exercises 21 through 33 as Law of Detachment, syllogism, disjunctive syllogism, or indirect reasoning.

21. Premise: If we have a test Wednesday, I cannot go to the movies.
I can go to the movies.

Conclusion: We do not have a test Wednesday.

22. Premise: If I make an A on the final, I will make an A in the course.
I made an A on the final.

Conclusion: I made an A in the course.

23. Premise: If I take calculus, then I will need a graphing calculator.
If I need a graphing calculator, then I will need money to buy it.

Conclusion: If I take calculus, then I will need money to buy a graphing calculator.

24. Premise: If I take calculus, then I will need a graphing calculator.
I do not need a graphing calculator.

Conclusion: I am not taking calculus.

25. Premise: If the KOTs have a party Friday night, I will go.
The KOTs are having a party Friday night.

Conclusion: I will go to the party.

26. Premise: For spring break, I will go home, or I will stay here and help build a Habitat for Humanity house.
I did not go home.

Conclusion: I stayed here and helped build a Habitat house.

27. Premise: If I trim the hedge, I may go to the movies.
I trimmed the hedge.

Conclusion: I may go to the movies.

28. Premise: If rain is forecast, I will take my umbrella.
I did not take my umbrella.

Conclusion: Rain was not forecast.

29. Premise: You will enjoy the book, or you will enjoy the movie version.
You did not enjoy the book.

Conclusion: You enjoyed the movie version.

30. Premise: I will take calculus next semester, or I will take physics.
I will not take calculus.

Conclusion: I will take physics.

31. Premise: If you read this book, you will enjoy the movie version.
You did not enjoy the movie version.

Conclusion: You did not read this book.

32. Premise: If I hear that joke one more time, I will scream.
I did not scream.

Conclusion: I did not hear the joke one more time.

33. The class votes for an oral exam, or the class votes for a take-home exam. The class does not vote for an oral exam. Therefore, the class votes for a take-home exam.

■■ IMPORTANT TERMS

10.1

Statement
Truth Value
Compound Statement
Conjunction
Truth Table
Disjunction
Negation
Exclusive Or
Inclusive Or

10.2

Conditional (Implication)
Converse
Inverse
Contrapositive
Biconditional

10.3

Equivalent Statements

10.4

Valid Argument
Proof
Premise
Valid Conclusion
Law of Detachment
Tautology
Syllogism
Disjunctive Syllogism
Indirect Reasoning

■■ REVIEW EXERCISES

1. Determine whether each of the following sentences is a statement.
(a) It rained on March 30.
(b) Did you see him make a hole-in-one?
(c) Harriet's painting is the best work of art in the exhibit.
(d) February 1995 had five Fridays.

2. Identify each of the following as a conjunction, disjunction, or neither.
(a) The wind blew over the house plant, and rain soaked the carpet.
(b) We will attend the Brazos River Festival, or we will drive out to see the wildflowers.
(c) My antique car won't start today.

(d) We plan to watch a documentary on TV or go to a concert.

3. Let p be the statement "Rhonda is sick today" and let q be the statement "Rhonda has a temperature." Write out the following statements.
(a) $\sim p$ **(b)** $p \wedge q$
(c) $\sim p \wedge \sim q$ **(d)** $p \vee q$

4. Convert the following statements into symbolic form using p for "Angela is in biology lab today" and q for "Pete has a history exam tomorrow."
(a) Angela is in biology lab today, and Pete does not have a history exam tomorrow.
(b) Angela is not in biology lab today, or Pete does not have a history exam tomorrow.

5. Determine whether the following statements are true or false.
(a) Water freezes at 0°C, and water boils at 100°C.
(b) Water freezes at 0°C, or George Washington was the first president of the United States.
(c) The year 2001 was a leap year, and January has 31 days.
(d) The year 2001 was not a leap year, or February has 30 days.

6. Write the negation of the following statements.
(a) This sentence is false.
(b) Dillard's operates a chain of department stores.
(c) Halloween is not a national holiday.

7. Find the truth value of each of the following where p and q have truth values T and r has the truth value F:
(a) $p \wedge (q \vee r)$ **(b)** $(p \wedge q) \vee (p \wedge r)$
(c) $p \wedge q \wedge \sim r$ **(d)** $\sim [p \vee (q \wedge r)]$

8. Let p be the statement "Monty has a laptop computer" and q be the statement "Tanya has a graphing calculator."
(a) Write the statement $p \rightarrow q$.
(b) Write the statement $q \rightarrow p$.

9. Determine the truth of the following conditionals.
(a) If ten dimes make a dollar, then 10% of $3.50 is $0.35.
(b) If New Year's Day is January 1, then Thanksgiving Day is in June.

10. Determine the truth value of the following conditionals:
(a) If Florida is the northernmost state of the United States, then New York City is on the West Coast.

(b) If William Shakespeare was a French novelist, then Franklin Roosevelt was president of the United States.

11. Write the inverse, converse, and contrapositive of the statement "If I turn my paper in late, then I will be penalized."

12. Determine the truth value of the following biconditionals:
(a) Water freezes at 0°C if and only if water boils at 100°C.
(b) Ten dimes make a dollar if and only if 10% of $5.00 is $0.65.

Construct truth tables for Exercises 13 and 14.

13. $\sim p \rightarrow (p \wedge q)$

14. $(p \wedge \sim q) \leftrightarrow (\sim p \vee q)$

15. Use truth tables to determine if $p \wedge (p \vee q)$ and p are logically equivalent.

16. Use truth tables to determine if $p \wedge (q \vee r)$ and $(p \wedge q) \vee (p \wedge r)$ are logically equivalent.

17. Determine if the following argument is valid:

Premise: If you study logic, mathematics is easy. Mathematics is not easy.

Conclusion: You did not study logic.

18. Determine if the following argument is valid:

Premise: If Sally studies music, then she is artistic. Sally is not artistic.

Conclusion: Sally does not study music.

19. Determine if the following argument is valid:

Premise: If there is money in my account, then I will pay my rent. If I pay my rent, then I will not be evicted.

Conclusion: If there is money in my account, then I will not be evicted.

20. Determine if the following argument is valid: If wheat prices are steady, then exports will rise. Wheat prices are steady. Therefore, exports will rise.

21. Ms. Lopez will request a transfer out of state, or she will start a business locally. Ms. Lopez did not request a transfer out of state. Therefore, she will start a business locally.

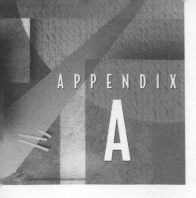
Review Topics

This appendix contains basic algebra topics that are necessary for the materials in this book. You are encouraged to study those topics for which you need review and skip those topics with which you are familiar. ■

A.1   Properties of Real Numbers

- The Real Number Line
- The Arithmetic of Real Numbers

The Real Number Line

The most basic numbers in our study of mathematics are the **natural numbers,**

$$1, 2, 3, 4, \ldots$$

These are in fact the numbers that we use to count. They can be represented graphically as points on a line, as in Figure A–1.

FIGURE A–1

Begin with a starting point on the line, which we label 0, and a convenient unit scale. Starting at 0, mark off equal lengths to the right. These marks represent $1, 2, 3, \ldots$.

We can mark off lengths in the opposite direction from 0 and let them represent numbers, the negatives of the natural numbers (Figure A–2). This collection of numbers represented by the marks on the line is written

$$\ldots, -5, -4, -3, -2, -1, 0, 1, 2, 3, 4, 5, \ldots$$

and is called the set of **integers.** We call the number represented by the symbol 0 zero. It is neither positive nor negative.

FIGURE A–2

Other numbers, the fractions, can be represented with points between the integers. In Figure A–3, point A, halfway between 1 and 2, represents $1\frac{1}{2}$; point B, one-quarter of the way from 3 to 4, represents $3\frac{1}{4}$. Point C is three-quarters of the way from -4 to -5, so it represents $-4\frac{3}{4}$. These numbers together with the integers are called **rational numbers.**

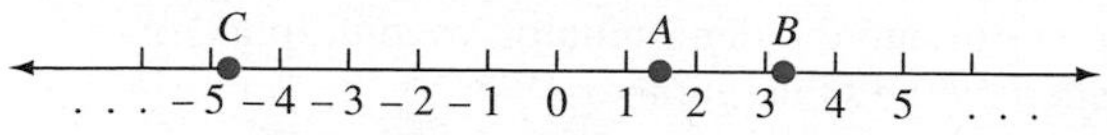

FIGURE A-3

All these points can be expressed in terms of finite or infinite decimals. Point A is the point 1.5, B the point 3.25, and C the point -4.75. Every rational number can be expressed either as a finite decimal that terminates, such as A, B, and C above, or as a decimal that repeats infinitely. For example, $5\frac{1}{3}$ is a rational number that can be written in decimal form as $5.333\ldots$; the 3 repeats endlessly.

There are, however, certain numbers called **irrational numbers** that do not have any pattern of repetition in their decimal form. One such number is $\sqrt{2}$. Its decimal form is $1.414213\ldots$.

The set of all rational and irrational numbers is called the set of **real numbers.** One way to visualize the set of real numbers is to think of each point on the line as representing a real number.

The Arithmetic of Real Numbers

There are four useful operations on the set of real numbers: addition, subtraction, multiplication, and division. The following tables summarize some rules that govern these operations:

Rule		Example
Division by zero is not allowed.		$\frac{5}{0}$ and $\frac{0}{0}$ have no meaning.

Rules of Operations		Examples
$a + b = b + a$	Numbers can be added in either order.	$4 + 9 = 9 + 4$
$ab = ba$	Numbers can be multiplied in either order.	$3 \times 11 = 11 \times 3$
$ab + ac = a(b + c)$	A common number can be factored from each term in a sum.	$12a + 15b = 3(4a + 5b)$

Rules of Signs		Examples
$-a = (-1)a$	The negative of a number a is the product $(-1)a$.	$-12 = (-1)12$
$-(-a) = a$		$-(-7) = 7$
$(-a)b = a(-b) = -ab$	The product of a positive and a negative number is a negative number.	$(-4)8 = -32$
		$5(-3) = -15$
$(-a)(-b) = ab$	The product of two negative numbers is a positive number.	$(-2)(-7) = 14$
$(-a) + (-b) = -(a + b)$	The sum of two negative numbers is a negative number.	$(-4) + (-5) = -9$
$-(a - b) = (-a) + b$		$-(8 - 3) = -8 + 3$
$\dfrac{a}{b} = \dfrac{a}{-b}$	Division using a negative number and a positive number is a negative number.	$\dfrac{-18}{3} = -6$
$\quad = -\left(\dfrac{a}{b}\right)$		$\dfrac{22}{(-11)} = -2$

Arithmetic of Fractions		**Examples**
$\dfrac{ac}{bc} = \dfrac{a}{b}$	The value of a fraction is unchanged if both the numerator and the denominator are multiplied or divided by the same number.	$\dfrac{10}{15} = \dfrac{2}{3}$ $\dfrac{7}{4} = \dfrac{14}{8}$
$\dfrac{a}{b} + \dfrac{c}{b} = \dfrac{a+c}{b}$	To add two fractions with the same denominators, add the numerators and keep the same denominator.	$\dfrac{5}{3} + \dfrac{2}{3} = \dfrac{7}{3}$
$\dfrac{a}{b} + \dfrac{c}{d} = \dfrac{ad}{bd} + \dfrac{bc}{bd}$ $= \dfrac{ad + bc}{bd}$ $(b, d \neq 0)$	To add two fractions with different denominators, convert them to fractions with the same denominators by multiplying the numerator and denominator of each fraction by the denominator of the other.	$\dfrac{3}{7} + \dfrac{2}{5}$ $= \dfrac{3(5)}{7(5)} + \dfrac{7(2)}{7(5)}$ $= \dfrac{3(5) + 7(2)}{7(5)}$ $= \dfrac{15 + 14}{35}$ $= \dfrac{29}{35}$
$\dfrac{a}{b} \times \dfrac{c}{d} = \dfrac{ac}{bd}$	To multiply two fractions, multiply their numerators and multiply their denominators.	$\dfrac{3}{4} \times \dfrac{6}{11} = \dfrac{18}{44} = \dfrac{9}{22}$
$\dfrac{a}{b} \div \dfrac{c}{d} = \dfrac{a}{b} \times \dfrac{d}{c} = \dfrac{ad}{bc}$ Division may also be written in the form $\dfrac{\frac{a}{b}}{\frac{c}{d}} = \dfrac{a}{b} \div \dfrac{c}{d}$ $= \dfrac{a}{b} \times \dfrac{d}{c} = \dfrac{ad}{bc}$ $(b, c, d \neq 0)$ $\dfrac{a}{b} = \dfrac{c}{d}$ if and only if $ad = bc$	To divide by a fraction, invert the divisor and multiply.	$\dfrac{2}{3} \div \dfrac{5}{8} = \dfrac{2}{3} \times \dfrac{8}{5}$ $= \dfrac{16}{15}$ $\dfrac{\frac{5}{3}}{\frac{4}{7}} = \dfrac{5}{3} \div \dfrac{4}{7}$ $= \dfrac{5}{3} \times \dfrac{7}{4}$ $= \dfrac{35}{12}$ $\dfrac{3}{7} = \dfrac{9}{21}$ because $3(21) = 7(9)$ $\dfrac{5}{9} \neq \dfrac{3}{4}$ because $5(4)$ $\neq 9(3)$

◫ A.1 EXERCISES

Evaluate the expressions in Exercises 1 through 49.

1. $(-1)13$

2. $(-1)(-7)$

3. $-(-23)$

4. $(-10)(-4)$

5. $(-5)(6)$

6. $(-2)(-4)$

7. $5(-7)$

8. $(-6) + (-11)$

9. $-(7 - 2)$

10. $\dfrac{-10}{5}$

11. $\dfrac{21}{-3}$

12. $\dfrac{5 \times 3}{7 \times 3}$

13. $(-4) + (-6)$

14. $(-3)(-2)$

15. $(-4)2$

16. $\dfrac{4}{9} + \dfrac{2}{9}$

17. $\dfrac{5}{3} + \dfrac{4}{3}$

18. $\dfrac{4}{11} - \dfrac{2}{11}$

19. $\dfrac{12}{5} - \dfrac{3}{5}$

20. $\dfrac{6}{10} - \dfrac{13}{10}$

21. $\dfrac{2}{3} + \dfrac{3}{4}$

22. $\dfrac{5}{8} - \dfrac{1}{3}$

23. $\dfrac{5}{6} - \dfrac{7}{4}$

24. $\dfrac{5}{12} - \dfrac{1}{6}$

25. $\dfrac{2}{5} + \dfrac{1}{4}$

26. $(-3) + 6$

27. $\dfrac{4}{7} - \dfrac{3}{5}$

28. $\dfrac{2}{3} \times \dfrac{4}{5}$

29. $\dfrac{\frac{3}{4}}{\frac{9}{8}}$

30. $\dfrac{3}{8} + \dfrac{2}{5}$

31. $\dfrac{\frac{2}{7}}{\frac{4}{5}}$

32. $\left(\dfrac{4}{3}\right)\left(\dfrac{6}{7}\right)$

33. $\left(\dfrac{1}{3}\right)\left(\dfrac{1}{5}\right)$

34. $6 \times (-3)$

35. $\dfrac{2}{5} \times \dfrac{4}{3}$

36. $\left(-\dfrac{2}{3}\right)\left(\dfrac{1}{9}\right)$

37. $\left(-\dfrac{3}{5}\right)\left(-\dfrac{4}{7}\right)$

38. $\dfrac{4}{5} \div \dfrac{2}{15}$

39. $\dfrac{3}{11} + \dfrac{1}{3}$

40. $\dfrac{-\frac{4}{9}}{\frac{5}{2}}$

41. $\dfrac{5}{7} \div \dfrac{15}{28}$

42. $\dfrac{\frac{4}{9}}{\frac{16}{3}}$

43. $\dfrac{5}{8} \div \dfrac{1}{3}$

44. $\left(\dfrac{1}{2} - \dfrac{1}{3}\right)\left(\dfrac{5}{7}\right)$

45. $\left(\dfrac{3}{4} + \dfrac{1}{5}\right) \div \left(\dfrac{2}{9}\right)$

46. $5(4a + 2b)$

47. $-2(3a + 11b)$

48. $2(a - 3b)$

49. $-5(2a + 10b)$

A.2   Solving Linear Equations

Numerous disciplines including science, technology, the social sciences, business, manufacturing, and government find mathematical techniques essential in day-to-day operations. They depend heavily on mathematical equations that describe conditions or relationships between quantities.

In an equation such as

$$4x - 5 = 7$$

the symbol x, called a **variable,** represents an arbitrary, an unspecified, or an unknown number just as John Doe and Jane Doe often denote an arbitrary, unspecified, or unknown person.

The equation $4x - 5 = 7$ may be true or false depending on the choice of the number x. If we substitute the number 3 for x in

$$4x - 5 = 7$$

both sides become equal and we say that $x = 3$ is a **solution** of the equation. If 5 is substituted for x, then both sides are *not* equal, so $x = 5$ is *not* a solution.

It may help a sales representative to know that the expression $0.20x + 11$ describes the daily rental of a car, where x represents mileage. Furthermore, the solution of the equation $0.20x + 11 = 40$ answers the question "You paid \$40 for car rental, how many miles did you drive?" Solutions of equations can sometimes help one to make a decision or gain needed information.

One basic procedure for solving an equation is to obtain a sequence of equivalent equations with the goal of isolating the variable on one side of the equation and the appropriate number on the other side.

The following two operations help to isolate the variable and find the solution.

1. The same number may be added to or subtracted from both sides of an equation.
2. Both sides of an equation may be multiplied or divided by a nonzero number.

Either of these operations yields another equation that is equivalent to the first, in other words, a second equation that has the same *solution* as the first.

EXAMPLE 1 ➤

Solve the equation $3x + 4 = 19$.

SOLUTION

We begin to isolate x by subtracting 4 from both sides:

$$3x + 4 - 4 = 19 - 4$$
$$3x = 15$$

Next, divide both sides by 3:

$$\frac{3x}{3} = \frac{15}{3}$$
$$x = 5$$

is the solution. We can check our answer by substituting $x = 5$ into the original equation,

$$3(5) + 4 = 15 + 4 = 19$$

so the solution checks.

■ **Now You Are Ready to Work Exercise 3**

EXAMPLE 2 ➤

Solve $4x - 2 = 2x + 12$.

SOLUTION

$$4x - 2 = 2x + 12 \qquad \text{(First, add 2 to both sides.)}$$
$$4x - 2 + 2 = 2x + 12 + 2$$
$$4x = 2x + 14 \qquad \text{(Next, subtract } 2x \text{ from both sides.)}$$
$$4x - 2x = 2x + 14 - 2x$$
$$2x = 14 \qquad \text{(Now divide both sides by 2.)}$$
$$x = 7$$

Check: $4(7) - 2 = 28 - 2 = 26$ (left-hand side) and $2(7) + 12 = 14 + 12 = 26$ (right-hand side), so it checks.

■ **Now You Are Ready to Work Exercise 7**

EXAMPLE 3 ➤

Solve $7x + 13 = 0$.

SOLUTION

$$7x + 13 = 0 \qquad \text{(Subtract 13 from both sides.)}$$
$$7x = -13 \qquad \text{(Divide both sides by 7.)}$$
$$x = -\frac{13}{7}$$

■ **Now You Are Ready to Work Exercise 9**

The previous examples all use **linear equations.**

DEFINITION
Linear Equation

A **linear equation in one variable,** x, is an equation that can be written in the form

$$ax + b = 0 \quad \text{where} \quad a \neq 0$$

A **linear equation in two variables** is an equation that can be written in the form

$$y = ax + b \quad \text{where} \quad a \neq 0$$

EXAMPLE 4 ➤

Solve $\dfrac{3x - 5}{2} + \dfrac{x + 7}{3} = 8$.

SOLUTION

We show two ways to solve this. First, use rules of fractions to combine the terms on the left-hand side:

$$\frac{3x - 5}{2} + \frac{x + 7}{3} = 8 \quad \text{(Convert fractions to the same denominator.)}$$

$$\frac{3(3x - 5)}{6} + \frac{2(x + 7)}{6} = 8 \quad \text{(Now add the fractions.)}$$

$$\frac{3(3x - 5) + 2(x + 7)}{6} = 8$$

$$\frac{9x - 15 + 2x + 14}{6} = 8$$

$$\frac{11x - 1}{6} = 8 \quad \text{(Now multiply both sides by 6.)}$$

$$11x - 1 = 48$$

$$11x = 49$$

$$x = \frac{49}{11}$$

An alternative, and simpler, method is the following:

$$\frac{3x - 5}{2} + \frac{x + 7}{3} = 8$$

Multiply through by 6 (the product of the denominators):

$$3(3x - 5) + 2(x + 7) = 48$$

$$9x - 15 + 2x + 14 = 48$$

$$11x - 1 = 48$$

$$11x = 49$$

$$x = \frac{49}{11}$$

■ **Now You Are Ready to Work Exercise 13**

EXAMPLE 5 ➤

Rent-A-Car charges $0.21 per mile plus $10 per day for car rental. Thus, the daily fee is represented by the equation

$$y = 0.21x + 10$$

where x is the number of miles driven and y is the daily fee.

(a) Determine the rental fee if the car is driven 165 miles during the day.
(b) Determine the rental fee if the car is driven 420 miles during the day.
(c) The rental fee for one day is $48.64. Find the number of miles driven.

SOLUTION

(a) $x = 165$, so $y = 0.21(165) + 10 = 44.65$. The fee is $44.65.
(b) $x = 420$, so $y = 0.21(420) + 10 = 98.2$. The fee is $98.20.
(c) $y = 48.64$, so x is the solution of the equation.

$$0.21x + 10 = 48.64$$
$$0.21x = 48.64 - 10$$
$$0.21x = 38.64$$
$$x = \frac{38.64}{0.21} = 184$$

The car was driven 184 miles.

■ **Now You Are Ready to Work Exercise 17**

A.2 EXERCISES

Determine which of the following values of x are solutions to the equations in Exercises 1 and 2. Use $x = 1, 2, -3, 0, 4,$ and -2.

1. $2x - 4 = -10$

2. $3x + 1 = x + 5$

Solve the equations in Exercises 3 through 16.

3. *(See Example 1)*
$2x - 3 = 5$

4. $-4x + 2 = 6$

5. $4x - 3 = 5$

6. $7x - 4 = 0$

7. *(See Example 2)*
$7x + 2 = 3x + 4$

8. $2x - 4 = -5x + 2$

9. *(See Example 3)*
$12x + 21 = 0$

10. $5 - x = 8 + 3x$

11. $3(x - 5) + 4(2x + 1) = 9$

12. $6(4x + 5) + 7 = 2$

13. *(See Example 4)*
$\dfrac{2x + 3}{3} + \dfrac{5x - 1}{4} = 2$

14. $\dfrac{4x + 7}{6} + \dfrac{2 - 3x}{5} = 5$

15. $\dfrac{12x + 4}{2x + 7} = 4$

16. $\dfrac{x + 1}{x - 1} = \dfrac{3}{4}$

17. *(See Example 5)* The U-Drive-It Rental Company charges $0.20 per mile plus $112 per week for car rental. The weekly rental fee for a car is represented by the linear equation

$$y = 0.20x + 112$$

where x is the number of miles driven and y is the weekly rental charge.
(a) Determine the rental fee if the car is driven 650 miles during the week.
(b) Determine the rental fee if the car is driven 1500 miles.
(c) The weekly rental fee is $302. How many miles were driven?

18. Joe Cool has a summer job selling real estate in a subdivision development. He receives a base salary of $100 per week plus $50 for each lot sold. Therefore, the equation

$$y = 50x + 100$$

represents his weekly income, where x is the number of lots sold.
(a) What is his weekly income if he sells 7 lots?
(b) What is his weekly income if he sells 15 lots?
(c) If he receives $550 one week, how many lots did he sell?

19. A Girl Scout troop collects aluminum cans for a project. The recycling center weighs the cans in a container that weighs 8 pounds, so the scouts are paid according to the equation

$$y = 0.42(x - 8)$$

where x is the weight in pounds given by the scale and y is the payment in dollars.
(a) How much money do the Girl Scouts receive if the scale reads 42 pounds?

(b) How much do they receive if the scale reads 113 pounds?

(c) The scouts received $22.26 for one weekend's collection. What was the reading on the scale?

20. The tuition and fees paid by students at a local junior college are given by the equation

$$y = 27x + 85$$

where x is the number of hours enrolled and y is the total cost of tuition and fees ($85 fixed fees and $27 per hour tuition).

(a) How much does a student pay who is enrolled in 13 hours?

(b) A student who pays $517 is enrolled in how many hours?

A.3   Coordinate Systems

We have all seen a map, a house plan, or a wiring diagram that shows information recorded on a flat surface. Each of these uses some notation unique to the subject to convey the desired information. In mathematics, we often use a flat surface called a **plane** to draw figures and locate points. We place a reference system in the plane to record and communicate information accurately. The standard mathematical reference system consists of a horizontal and a vertical line (called **axes**). These two perpendicular axes form a **Cartesian,** or **rectangular, coordinate system.** They intersect at a point called the **origin.**

We name the horizontal axis the **x-axis,** and we name the vertical axis the **y-axis.** The origin is labeled O.

Two numbers are used to describe the location of a point in the plane, and they are recorded in the form (x, y). For example, $x = 3$ and $y = 2$ for the point $(3, 2)$. The first number, 3, called the **x-coordinate,** or **abscissa,** represents the horizontal distance from the y-axis to the point. The second number, 2, called the **y–coordinate** or **ordinate,** represents the vertical distance measured from the x-axis to the point. The point $(3, 2)$ is shown as point P in Figure A–4. Points located to the right of the y-axis have positive x-coordinates; those to the left have negative x-coordinates. The y-coordinate is positive for points located above the x-axis and negative for those located below.

Figure A–4 shows other examples of points in this coordinate system: Q is the point $(-4, 3)$, and R is the point $(-3, -2.5)$. The origin O has coordinates $(0, 0)$.

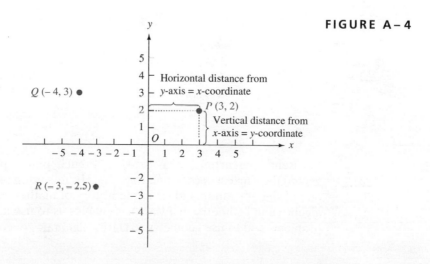

FIGURE A–4

FIGURE A-5

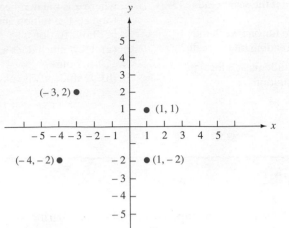

Figure A–5 shows the points $(-3, 2)$, $(-4, -2)$, $(1, 1)$, and $(1, -2)$ plotted on the Cartesian coordinate system.

The coordinate axes divide the plane into four parts called **quadrants.** The quadrants are labeled I, II, III, and IV as shown in Figure A–6. Point A is located in the first quadrant, where x and y are both positive; B, in the second quadrant, where x is negative and y is positive; C in the third quadrant, where both x and y are negative; and D is in the fourth quadrant, where x is positive and y is negative. Points A and E lie in the same quadrant.

FIGURE A-6

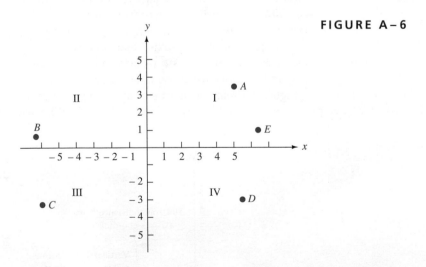

René Descartes (1596–1650), a French philosopher–mathematician, invented the Cartesian coordinate system. His invention of the coordinate system is one of the outstanding ideas in the history of mathematics because it combined algebra and geometry in a way that enables us to use algebra to solve geometry problems and to use geometry to clarify algebraic concepts.

A.3 EXERCISES

1. The following are the coordinates of points in a rectangular Cartesian coordinate system. Plot these points.

$$(-5, 4), (-2, -3), (-2, 4), (1, 5), (2, -5)$$

2. What are the coordinates of the points $P, Q, R,$ and S in the coordinate system in Figure A–7?

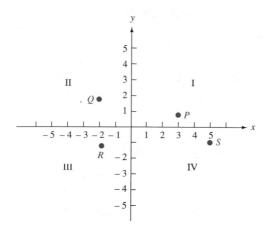

FIGURE A–7

3. Locate the following points on a Cartesian coordinate system:

$$(-2, 5), (3, -2), (0, 4), (-2, 0), \left(\tfrac{7}{2}, 2\right),$$
$$\left(\tfrac{2}{3}, \tfrac{9}{4}\right), (-4, -2), (0, -5), (0, -2), (-6, -3)$$

4. Give the coordinates of $A, B, C, D, E,$ and F, in the coordinate system shown in Figure A–8.

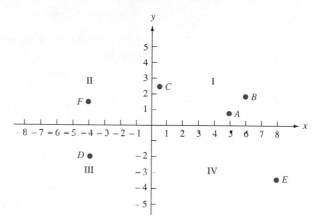

FIGURE A–8

5. Note that all points in the first quadrant have positive x-coordinates and positive y-coordinates. What are the characteristics of the points in:
   (a) the second quadrant?
   (b) the third quadrant?
   (c) the fourth quadrant?

6. For each case shown in Figure A–9, find the property the points have in common.

7. An old map gives these instructions to find a buried treasure: Start at giant oak tree. Go north 15 paces, then east 22 paces to a half-buried rock. The key to the treasure chest is buried at the spot that is 17 paces west and 13 paces north of the rock. From the place where the key is buried, go 32 paces west and 16 paces south to the location of the buried treasure. Use a coordinate system to represent the location of the oak tree, the rock, the key, and the treasure.

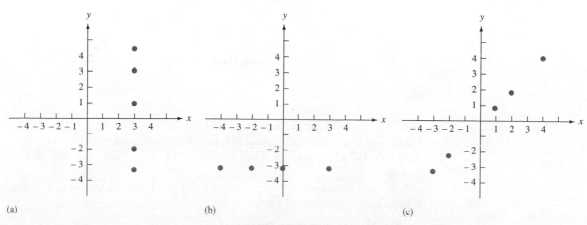

FIGURE A–9

A.4 Linear Inequalities and Interval Notation

- Solving Inequalities
- Interval Notation

We frequently use inequalities in our daily conversation. They may take the form "Which store has the lower price?" "Did you make a higher grade?" "Our team scored more points." "My expenses are greater than my income." Statements such as these basically state that one quantity is greater than another. Statements using the terms "greater than" or "less than" are called **inequalities.** Our goal is to solve inequalities. First, we give some terminology and notation.

The symbol $<$ means "less than," and $>$ means "greater than." Just remember that each of these symbols points to the smaller quantity. The notations $a > b$ and $b < a$ have exactly the same meaning. We may interpret the definition of $a > b$ in three equivalent ways. At times, one may be more useful than the other, so choose the most appropriate one.

DEFINITION
$a > b$

If a and b are real numbers, the following statements have the same meaning.

(a) $a > b$ means that a lies to the right of b on a number line.
(b) $a > b$ means that there is a positive number p such that $a = b + p$.
(c) $a > b$ means that $a - b$ is a positive number p.

The positive numbers lie to the right of zero on a number line, and the negative numbers lie to the left.

We will also use the symbols $<$ (less than), $\geq$ (greater than or equal to), and $\leq$ (less than or equal to).

DEFINITION
$a < b, a \geq b, a \leq b$

$a < b$ means $b > a$.
$a \geq b$ means $a = b$ or $a > b$.
$a \leq b$ means $a = b$ or $a < b$.

EXAMPLE 1 ➤ The numbers 5, 8, 17, -2, -3, and -15 are plotted on a number line in Figure A–10.

FIGURE A–10

Notice the following:

1. **(a)** 17 lies to the right of 5.   **(b)** $17 = 5 + 12$ **(c)** $17 - 5 = 12$
   Each of these three statements is equivalent to saying that $17 > 5$.
2. $8 > -3$ because $8 - (-3) = 8 + 3 = 11$ (by part (c) of the above definition).
3. $-2 > -15$ because $-2 = -15 + 13$ (by part (b) of the above definition, where $p = 13$).

■ **Now You Are Ready to Work Exercise 1**

Solving Inequalities

By the **solution of an inequality** like

$$3x + 5 > 23$$

we mean the value or values of x that make the statement true. The method for solving inequalities is similar to that for solving equations. We want to operate on an inequality in a way that gives an equivalent inequality but that enables us to determine the solution.

Here are some simple examples of useful properties of inequalities.

1. Because $18 > 4$, $18 + 6 > 4 + 6$, that is, $24 > 10$ (6 added to both sides).
2. Because $23 > -1$, $23 - 7 > -1 - 7$; that is, $16 > -8$ (7 subtracted from both sides).
3. Because $6 > 2$, $4(6) > 4(2)$; that is, $24 > 8$ (both sides multiplied by 4).
4. Because $10 > 3$, $-2(10) < -2(3)$; that is, $-20 < -6$ (both sides multiplied by -2).
5. Because $-15 > -21$, $\dfrac{-15}{3} > \dfrac{-21}{3}$; that is, $-5 > -7$. (Divide both sides by 3.)
6. Because $20 > 6$, $\dfrac{20}{-2} < \dfrac{6}{-2}$; that is, $-10 < -3$. (Divide both sides by -2.) ∎

These examples illustrate basic properties that are useful in solving inequalities.

CAUTION

The inequality symbol reverses when we multiply each side by a negative number.

CAUTION

The inequality symbol reverses when dividing each side by a negative number.

Properties of Inequalities

For real numbers a, b, and c, the following are true:

1. Adding a number to both sides of an inequality leaves the direction of the inequality unchanged.

$$\text{If } a > b, \text{ then } a + c > b + c.$$

2. Subtraction of a number from both sides of an inequality leaves the direction of the inequality unchanged.

$$\text{If } a > b, \text{ then } a - c > b - c.$$

3. Multiply both sides of an inequality by a nonzero number:
   (a) If $a > b$ and c is positive, then $ac > bc$.
   (b) If $a > b$ and c is negative, then $ac < bc$. (Notice the change from $>$ to $<$.)
4. Divide both sides of an inequality by a nonzero number:
   (a) If $a > b$ and c is positive, then $\dfrac{a}{c} > \dfrac{b}{c}$.
   (b) If $a > b$ and c is negative, then $\dfrac{a}{c} < \dfrac{b}{c}$. (Notice the change from $>$ to $<$.)

(*Note:* All these properties hold if $>$ is replaced by $<$ and vice versa; and if $>$ is replaced by $\geq$ and $<$ is replaced by $\leq$.)

We use these properties to solve an inequality — that is, to find the values of x that make the inequality true. In general, we proceed by finding equivalent inequalities that will eventually isolate x on one side of the inequality and the appropriate number on the other side.

EXAMPLE 3 ➤

Solve the inequality $3x + 5 > 14$.

SOLUTION

Begin with the given inequality:

$$3x + 5 > 14 \quad \text{(Next, subtract 5 from each side (Property 2))}$$
$$3x > 9 \quad \text{(Now divide each side by 3 (Property 4))}$$
$$x > 3$$

Thus, all x greater than 3 make the inequality true. This solution can be graphed on a number line as shown in Figure A–11. The empty circle indicates that $x = 3$ is omitted from the solution, and the heavy line indicates the values of x included in the solution.

FIGURE A–11 $x > 3$.

■ **Now You Are Ready to Work Exercise 3**

EXAMPLE 4 ➤

Solve the inequality $5x - 17 > 8x + 14$ and indicate the solution on a graph.

SOLUTION

Start with the given inequality:

$$5x - 17 > 8x + 14 \quad \text{(Now add 17 to both sides (Property 1))}$$
$$5x > 8x + 31 \quad \text{(Now subtract 8x from both sides (Property 2))}$$
$$-3x > 31 \quad \text{(Now divide both sides by } -3 \text{ (Property 4))}$$
$$x < -\frac{31}{3} \quad \text{(This reverses the inequality symbol)}$$

Thus, the solution consists of all x to the left of $-\frac{31}{3}$. See Figure A–12.

FIGURE A–12 $x < \dfrac{-31}{3}$.

■ **Now You Are Ready to Work Exercise 9**

EXAMPLE 5 ➤

Solve and graph $2(x - 3) \le 3(x + 5) - 7$.

SOLUTION

$$2(x - 3) \le 3(x + 5) - 7 \quad \text{(First perform the indicated multiplications)}$$
$$2x - 6 \le 3x + 15 - 7$$
$$2x - 6 \le 3x + 8 \quad \text{(Now add 6 to both sides (Property 1))}$$
$$2x \le 3x + 14 \quad \text{(Subtract 3x from both sides (Property 2))}$$
$$-x \le 14 \quad \text{(Multiply both sides by } -1 \text{ (Property 3))}$$
$$x \ge -14$$

Because the solution includes -14 and all numbers greater, the graph shows a solid circle at -14 (see Figure A–13).

FIGURE A–13 $x \ge -14$.

■ **Now You Are Ready to Work Exercise 13**

The next example illustrates a problem that involves two inequalities.

EXAMPLE 6 ➤

Solve and graph $3 < 2x + 5 \le 13$.

SOLUTION

This inequality means both $3 < 2x + 5$ *and* $2x + 5 \le 13$. Solve it in a manner similar to the preceding examples except that you try to isolate the x in the middle.
    Begin with the given inequality:

$$3 < 2x + 5 \le 13 \quad \text{(Subtract 5 from all parts of the inequality.)}$$
$$-2 < 2x \le 8 \quad \text{(Divide each part by 2.)}$$
$$-1 < x < 4$$

The solution consists of all numbers between -1 and 4, including 4 but not including -1. The graph of the solution (see Figure A–14) shows an empty circle at -1 because -1 is not a part of the solution. It shows a solid circle at 4 because 4 is a part of the solution. The solid line between -1 and 4 indicates that all numbers between -1 and 4 are included in the solution.

FIGURE A–14 $-1 < x \le 4$.

■ **Now You Are Ready to Work Exercise 17**

Interval Notation

The solution of an inequality can be represented by yet another notation, the **interval notation.** Identify the portion of the number line that represents the solution of an inequality by its end points; brackets or parentheses indicate whether or not the end point is included in the solution. A parenthesis indicates that the end point is not included, and a bracket indicates that the end point is included. For example, the notation $(-1, 4]$ means $-1 < x \le 4$ and indicates the set of all numbers between -1 and 4 with -1 excluded and 4 included in the set. The notation $(-1, 4)$ means $-1 < x < 4$ and indicates that both -1 and 4 are excluded from the set. The notation $(-1, \infty)$ denotes the set of all numbers greater than -1, $x > -1$. The symbol ∞ denotes infinity and indicates that there is no upper bound to the interval.
    Table A–1 shows the variations of the interval notation.

TABLE A–1

Inequality Notation		Interval Notation		Graph of Interval
General	Example	General	Example	
$a < x < b$	$-1 < x < 4$	(a, b)	$(-1, 4)$	
$a \le x < b$	$-1 \le x < 4$	$[a, b)$	$[-1, 4)$	
$a < x \le b$	$-1 < x \le 4$	$(a, b]$	$(-1, 4]$	
$a \le x \le b$	$-1 \le x \le 4$	$[a, b]$	$[-1, 4]$	
$x < b$	$x < 4$	$(-\infty, b)$	$(-\infty, 4)$	
$x \le b$	$x \le 4$	$(-\infty, b]$	$(-\infty, 4]$	
$a < x$	$-1 < x$	(a, ∞)	$(-1, \infty)$	
$a \le x$	$-1 \le x$	$[a, \infty)$	$[-1, \infty)$	

EXAMPLE 7 ➤

Solve $1 \le 2(x - 5) + 3 < 5$.

SOLUTION

$$
\begin{aligned}
1 &\le 2(x - 5) + 3 < 5 \quad &&\text{(Multiply to remove parentheses.)} \\
1 &\le 2x - 10 + 3 < 5 \\
1 &\le 2x - 7 < 5 \quad &&\text{(Add 7 throughout.)} \\
8 &\le 2x < 12 \quad &&\text{(Divide through by 2.)} \\
4 &\le x < 6
\end{aligned}
$$

The solution consists of all values of x in the interval $[4, 6)$. The graph is shown in Figure A–15.

FIGURE A–15 $4 \le x < 6$.

■ **Now You Are Ready to Work Exercise 25**

EXAMPLE 8 ➤

The total points on an exam given by Professor Passmore are 20 points plus 2.5 points for each correct answer. A total score in $[70, 80)$ is a C. If Scott made a C on the exam, how many questions did he answer correctly?

SOLUTION

The score on an exam is given by $20 + 2.5x$, where x is the number of correct answers. Thus the condition for a C is

$$70 \le 20 + 2.5x < 80$$

Solve for x to obtain the number of correct answers:

$$
\begin{aligned}
70 &\le 20 + 2.5x < 80 \\
50 &\le 2.5x < 60 \\
\frac{50}{2.5} &\le x < \frac{60}{2.5} \\
20 &\le x < 24
\end{aligned}
$$

In this case, only whole numbers make sense, so Scott got 20, 21, 22, or 23 correct answers.

■ **Now You Are Ready to Work Exercise 31**

▪▪ A.1 EXERCISES

LEVEL 1

1. *(See Example 1)*   The following inequalities are of the form $a > b$. Verify the truth or falsity of each one by the property $a > b$ means $a - b$ is a positive number.

(a) $9 > 3$ (b) $4 > 0$ (c) $-5 > 0$

(d) $-3 > -15$ (e) $\dfrac{5}{6} > \dfrac{2}{3}$

2. Plot the numbers $10, 2, 5, -4, 3,$ and -2 on a number line. Verify the truth or falsity of each of the following by the property $a > b$ if a lies to the right of b.

(a) $10 > 5$ (b) $-4 > 2$ (c) $10 > 2$

(d) $5 > 10$ (e) $-4 > -2$ (f) $-2 > -4$

Solve the inequalities in Exercises 3 though 8. State your solution using inequalities.

3. *(See Example 3)*
$3x - 5 < x + 4$

4. $12 > 1 - 5x$

5. $5x - 22 \le 7x + 10$

6. $13x - 5 \le 7 - 4x$

7. $3(2x + 1) < 9x + 12$

8. $14 - 5x \ge 6x - 15$

15. $3(2x + 1) < -1(3x - 10)$

16. $-2(3x + 4) > -3(1 - 6x) - 17$

17. *(See Example 6)* $-16 < 3x + 5 < 22$

18. $124 > 5 - 2x \ge 68$

19. $14 < 3x + 8 < 32$

20. $-9 \le 3(x + 2) - 15 < 27$

Solve the inequalities in Exercises 9 through 20. Graph the solution.

9. *(See Example 4)*
$3x + 2 \le 4x - 3$

10. $3x + 2 < 2x - 3$

11. $6x + 5 < 5x - 4$

12. $78 < 6 - 3x$

13. *(See Example 5)* $3(x + 4) < 2(x - 3) + 14$

14. $4(x - 2) > 5(2x + 1)$

Solve the inequalities in Exercises 21 through 26. Give the solution in interval form.

21. $3x + 4 \le 1$

22. $5x - 7 > 3$

23. $-7x + 4 \ge 2x + 3$

24. $-3x + 4 < 2x - 6$

25. *(See Example 7)*
$-45 < 4x + 7 \le -10$

26. $16 > 2x - 10 \ge 4$

LEVEL 2

Solve the inequalities in Exercises 27 through 30.

27. $\dfrac{6x + 5}{-2} \ge \dfrac{4x - 3}{5}$

28. $\dfrac{2x - 5}{3} < \dfrac{x + 7}{4}$

29. $\dfrac{2}{3} < \dfrac{x + 5}{-4} \le \dfrac{3}{2}$

30. $\dfrac{3}{4} < \dfrac{7x + 1}{6} < \dfrac{5}{2}$

LEVEL 3

31. *(See Example 8)*   Prof. Tuff computes a grade on a test by $35 + 5x$, where x is the number of correct answers. A grade in the interval $[75, 90)$ is a B. If a student receives a B, how many correct answers were given?

32. A professor computes a grade on a test by $25 + 4x$ where x is the number of correct answers. A grade in the interval $[70, 79]$ is a C. What number of correct answers can be obtained to receive a C?

33. On a final exam, any grade in the interval $[85, 100]$ was an A. The professor gave 3 points for each correct answer and then adjusted the grades by adding 25 points. If a student made an A, how many correct answers were given?

34. A sporting goods store runs a special on jogging shoes. The manager expects to make a profit if the number of pairs of shoes sold, x, satisfies $32x - 4230 > 2x + 480$. How many pairs of shoes need to be sold to make a profit?

▦ IMPORTANT TERMS

A.1

Natural Numbers
Integers
Rational Numbers
Irrational Numbers
Real Numbers

A.2

Variable
Solution
Linear Equation

A.3

Axes
Cartesian Coordinate System
Rectangular Coordinate System
Origin
x-Axis
y-Axis
Abscissa
Ordinate
Quadrants

A.4

Inequalities
$>, <, \ge, \le$
Solution of an Inequality
Properties of Inequalities
Interval Notation

Using a TI-83 Graphing Calculator

The purpose of this appendix is to provide a summary of some of the key instructions that are useful in applying the graphing calculator to this course. This appendix is not intended to replace the instruction guidebook. It covers only the TI-83 graphing calculator. Students are free to use other graphing calculators and computers in working the Technology Explorations. ∎

 N O T E

A key may be used to make two or more selections. For example, the notation "TEST A" appears above the $\boxed{\text{MATH}}$ key. When two names appear above a key, the one on the left is selected by the $\boxed{\text{2nd}}$ key, and the one on the right is selected by the $\boxed{\text{ALPHA}}$ key. Press $\boxed{\text{2nd}}$ $\boxed{\text{MATH}}$ and the TEST menu will appear. Press $\boxed{\text{ALPHA}}$ $\boxed{\text{MATH}}$ and the letter A will appear. When only one name appears above a key, use the $\boxed{\text{2nd}}$ key, to select it. For example, INS appears above the $\boxed{\text{DEL}}$ key, and is accessed with $\boxed{\text{2nd}}$ $\boxed{\text{DEL}}$. INS is used to insert characters/numbers in an exisisting string of characters/numbers. The use of the $\boxed{\text{2nd}}$ and $\boxed{\text{ALPHA}}$ with other keys will give the menus or letters indicated above the key.

Thus, when you see a notation like $\boxed{\text{TEST}}$ in this material, you are to press $\boxed{\text{2nd}}$ $\boxed{\text{MATH}}$.

Notation

We will use the following notation:

Symbols enclosed in a rectangle, like

$$\boxed{\text{A}}, \boxed{5}, \boxed{\text{Y=}}, \boxed{\text{MATRX}}, \text{ and } \boxed{\text{2nd}}$$

refer to keys to be pressed.

Symbols enclosed with <>, like

$$<\text{OPS}>, <\text{NUM}>, \text{ and } <\text{PROB}>$$

refer to commands selected from a menu.

Arithmetic Operations

Arithmetic calculations are done much like the calculations on a nongraphing calculator. Here are a few basic hints that might be helpful:

1. The $\boxed{\text{ENTER}}$ key plays the same role as "equals" in calculation.
2. The $\boxed{-}$ key denotes subtraction.
3. The $\boxed{(-)}$ key denotes "negative."

To calculate $2 - 5$, you key $\boxed{2}$ $\boxed{-}$ $\boxed{5}$

To calculate $-2 \cdot 8$, you key $\boxed{(-)}$ $\boxed{2}$ $\boxed{\times}$ $\boxed{8}$

4. The $\boxed{\wedge}$ key is used to indicate that an exponent follows.

$$3^4 \text{ is keyed as } \boxed{3}\ \boxed{\wedge}\ \boxed{4}$$

5. The parentheses keys are used in the usual manner to group calculations:

$$3(5 + 8) \text{ is keyed } \boxed{3}\ \boxed{(}\ \boxed{5}\ \boxed{+}\ \boxed{8}\ \boxed{)}$$

Graphing

The TI-83 graphing calculator will graph functions written in the form $y = f(x)$. It will not graph functions written as $3x + 2y = 27$ or $x^2 + y^2 = 5$. To graph $3x + 2y = 27$, you must solve for y and graph $y = -\frac{3}{2}x + \frac{27}{2}$ or $y = -1.5x + 13.5$ or $y = (27 - 3x)/2$.

To enter the equation to be graphed, display the <y=> screen. It is obtained by pressing $\boxed{Y=}$.

To enter an equation like $y = -2x + 3$, key the numbers and symbols of the equation. The "x" in the equation is entered with the key $\boxed{X, T, \theta, n}$ on the TI-83. To initiate graphing press the $\boxed{GRAPH}$ key.

The Range

From time to time, you will want to change the portion of the x-y plane that shows on the screen. Here's how. Press $\boxed{WINDOW}$ and you will obtain a menu that includes something like

WINDOW
Xmin=10
Xmax=10
Xscl=1
Ymin=−10
Ymax=10
Yscl=1
Yres=1

xMin and xMax specify the range of x-values that will show on the screen, xScl specifies the spacing of the tick marks on the x-axis. yMin, yMax, and yScl do the same for y.

The Standard Screen

The screen that shows x from -10 to 10 and y from -10 to 10 is called the **standard screen.** For the standard screen, xMin and the rest of the range settings can be set automatically with $\boxed{ZOOM}$ <6:ZStandard>.

The Square Window

The screen on the TI-83 graphing calculator is 1.5 times as wide as high. Therefore, the standard window with ranges of -10 to 10 for both x and y results in a different scale for the two axes. A graph is actually distorted somewhat. To obtain a graph with no distortion, the axes must be scaled the same. This is done by selecting the **square window** option. Here's how to do it: $\boxed{\text{ZOOM}}$ <5:ZSquare>.

The TRACE Function

The TRACE function has the useful feature that it moves the cursor along a curve and gives the coordinates of the point where the cursor is located.

1. To initiate the TRACE function press $\boxed{\text{TRACE}}$.
2. To trace, move the cursor toward the left with the $\boxed{<}$ key and to the right with the $\boxed{>}$ key. As you do so, the x- and y-coordinates show at the bottom of the screen.

Evaluating a Function

1. You can evaluate a function for a given value of x by tracing the graph until the x-coordinate reaches the desired value. The corresponding function value shows as the y-coordinate.

    The TRACE function is useful in locating a value of x that corresponds to a given y-value. Trace the graph until the given value of y shows as the y-coordinate and read the value of the x-coordinate.

    Because the cursor moves in discrete steps when using TRACE, it may skip over a value of x that you want to use in evaluating the function. By zooming in, you reduce the step size and thus can obtain values of x closer to, if not actually equal to, the desired value.
2. Using **value** you can evaluate a function more accurately for a given value of x. To illustrate how, suppose you have entered $y1 = 2x^2 + 3$ in the <y=> window. For a value of x—say, $x = 4.2$—you obtain $y1 = 2(4.2)^2 + 3$ with the sequence $\boxed{\text{CALC}}$ <1:value> $\boxed{\text{ENTER}}$. When <x=> shows on the screen, enter 4.2 $\boxed{\text{ENTER}}$. The value $y = 38.28$ will show on the screen.

ZOOM

The ZOOM feature zooms in or out at the location of the cursor and gives a magnified or reduced view. The location of the cursor determines the center of the new area. The ranges of x and y are reduced by one fourth or enlarged by a factor of 4.

To zoom in and magnify a part of the area:

1. On the graph, move the cursor near the point you want to be the center of the new area and press $\boxed{\text{ENTER}}$.
2. Now zoom in by $\boxed{\text{ZOOM}}$ <2:Zoom In> $\boxed{\text{ENTER}}$.

Finding the Intersection of Two Graphs

The TRACE command can be used to estimate the point of intersection of two graphs. Locate the cursor on one of the graphs and trace toward the point of intersection. You may zoom in to obtain a more accurate estimate.

Press the $\boxed{\lor}$ or $\boxed{\land}$ key to move the cursor from one graph to the other. As you move from one graph to the other, compare the y-coordinates of points on the graph to help determine the accuracy of the estimates.

You can locate the point of intersection of two graphs more accurately using the **Intersect** command. Graph the two functions with their intersection or intersections showing on the graph. Then follow these steps:

1. Select $\boxed{\text{CALC}}$ <5:intersect>. The screen will display <First curve?>. The equation of the curve selected will show in the upper left corner of the screen.
2. Press $\boxed{\lor}$ or $\boxed{\land}$, if necessary, to move the cursor to one of the curves, then press $\boxed{\text{ENTER}}$. The screen will display <Second curve?>.
3. Press $\boxed{\lor}$ or $\boxed{\land}$, if necessary, so that the cursor moves to the second curve. Press $\boxed{\text{ENTER}}$. If the display still shows <Second curve?>, press the other of $\boxed{\lor}$ or $\boxed{\land}$.
4. The screen will display <Guess?>. Move the cursor near the point of intersection and press $\boxed{\text{ENTER}}$. Then the screen will display the x- and y-coordinates of the point of intersection.

Constructing a Table

(See also Sections 1.1 and 8.3) The **Table** function allows you to specify a list of values of x and then compute the corresponding values of a function, or functions, as defined in the $\boxed{\text{Y=}}$ menu. You may let the calculator generate a list of equally spaced values of x or you may list them one by one.

Setting Up the x-List

(a) Let the calculator generate the list.
  - Select $\boxed{\text{TBLSET}}$ and on the screen that appears enter the starting value of x, say 5, at **TblStart.**
  - Enter the amount by which you want to increment the values of x, say 3, at **ΔTbl.**
  - Select **Auto** for **Indpnt.**

The calculator will generate a list for x beginning with 5 and increasing by 3 throughout the list.

(b) The user enters the x values.
  - Select $\boxed{\text{TBLSET}}$ and on the screen that appears you can ignore **TblStart** and **ΔTbl.**
  - Select **Ask** for **Indpnt.**

Calculating the Formulas for y

Open the $\boxed{\text{Y=}}$ screen and enter the desired formula for Y1. If other formulas are needed, enter them in Y2, Y3, …

To view the table, press $\boxed{\text{TABLE}}$ and the screen will show the table of x and y values. Only two columns of y values show. If you have entered three or more formulas for y, scroll to the right to view those columns. If you have entered the formulas for y before entering the values of x, the calculated values of y will apear as you enter the values of x.

The table shows as many as six digits of the numbers. If a value of y is more than six digits, select the number and the 12-digit form shows at the bottom of the screen.

Matrices

Ten matrices are allowed on the TI-83 and, depending on memory available, may have up to 99 rows or columns.

Matrix Names

The names of matrices may be obtained from the [MATRIX] menu, [A] through [J].

Entering a Matrix into Memory

To enter a matrix into memory, select [MATRIX] <EDIT>, select the name you wish to use, then press [ENTER]. Enter the size of the matrix and the matrix entries, row by row. To return to the home screen, press [QUIT].

Displaying a Matrix

To display a matrix A, display the name of the matrix on the screen and press [ENTER].

Storing a Matrix

To store a matrix in a second matrix, display the name of the first matrix on the screen, press [STO>], display the name of the second matrix on the screen, press [ENTER].

Performing Matrix Operations

Here are the commands to perform the basic matrix operations.

Adding Matrices [A] and [B]   [[A]] [+] [[B]] [ENTER]

Multiplying Matrices [A] and [B]   [[A]] [×] [[B]] [ENTER] or [[A]] [[B]] [ENTER]

Multiplying All Entries of a Matrix [A] by Scalar such as 5   [5] [×] [[A]] [ENTER] or [5] [[A]] [ENTER]

Finding the Inverse of a Matrix [A]   [[A]] [x^{-1}] [ENTER]

Entering the Powers of a Matrix A (Say, A^5)   [[A]] [^] [5] [ENTER]

Using Row Operations to Reduce a Matrix.   The row operations are used on an augmented matrix to solve a system of equations. Think through the steps used to manually solve a system. Use the corresponding row operation from the menu.

Row Swap.   This interchanges two rows of a matrix. Here's how to interchange rows 2 and 3 of a matrix named [B].

   [MATRX] <MATH> <C:rowSwap(> [ENTER]

The screen will display <rowSwap(>. Complete the command as <rowSwap ([B], 2, 3)> [ENTER]

Multiplying All Entries in a Row by a Constant.   To multiply row 1 of a matrix named [B] by 5:

MATRX <MATH> <E:*row(> ENTER

Complete the command with <*row(5, [B], 1)> ENTER

Adding Two Rows of a Matrix.   This adds two rows and places the result in the second row named.

To add row 2 to row 3 of a matrix named [B] and to place the result in row 3:

MATRX <MATH> <D:row+(> ENTER

Complete the command as <row+([B], 2, 3)>.

Multiplying a Row by a Constant and Adding to Another Row.   To multiply row 2 by −5 and add the result to row 4 and then replace row 4 with the result:

MATRX <MATH> <F:*row+(> ENTER

When <*row+(> appears, complete the command with <row+(−5, [B], 2, 4)>.

A Sequence of Row Operations on a Matrix.   If you perform a sequence of row operations on matrix [B] then, at any stage, the answer is the last operation performed on the *original* matrix. For example, the following sequence

$$*row(2, [B], 1)$$
$$row+([B], 1, 3)$$
$$*row+(−4, [B], 2, 3)$$

will give the same final answer as

$$*row+(−4, [B], 2, 3)$$

gives. Each operation in the sequence operates on the original matrix B and saves the result in a temporary location, ANS, not in B.

On the other hand, the sequence

$$*row(2, [B], 1)$$
$$row+(ANS, 1, 3)$$
$$*row+(−4, ANS, 2, 3)$$

will give the answer that carries through each operation in the sequence with the final result stored in ANS.

A row operation on a matrix — say, [B] — does not modify [B] itself unless other action is taken. You can record the effect of a row operation on [B] by a store command.

The row operations can be carried through the sequence with the following commands:

$$*row(2, [B], 1) \text{ STO> } [B]$$
$$row+([B], 1, 3) \text{ STO> } [B]$$
$$*row+(−4, [B], 2, 3) \text{ STO> } [B]$$

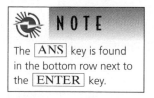

NOTE

The ANS key is found in the bottom row next to the ENTER key.

Row Echelon Form of a Matrix.   A system of equations can be solved by reducing the augmented matrix to the reduced echelon form through a sequence of row operations. This can be quite tedious for larger systems of equations.

The basic procedure of reducing a matrix to solve a system uses a series of row operations in an attempt to reduce the columns (except for the last one) to columns containing a single entry of 1 and all other entries 0.

rref.   The TI-83 has a command <rref> that reduces a matrix to its reduced echelon form. Here is the procedure for using <rref>:

Display <rref(> on the screen by: MATRX <MATH> <B:rref(>. Then fill in the matrix name, say [A], giving <rref([A])>. Press ENTER and the reduced echelon matrix will appear on the screen.

Statistics

The data for statistical analyses are entered in a list, or lists, and the analysis is done on the lists.

Entering Data Into a List

You may use up to six lists with names $L_1, L_2, \ldots, L_6$ and the names appear on the keyboard above the keys for $1, 2, \ldots, 6$.

Clearing Lists

To clear the lists L_1 and L_2:

STAT <EDIT> <4:CLRLST> ENTER L1 , L2 ENTER .

Entering Data

To enter data in L_1 (scores) and L_2 (frequency): STAT <EDIT> ENTER . Enter a score in L_1 and press ENTER . Repeat until all scores are entered. Press > to move to L_2. Enter frequencies in a similar manner; then press QUIT .

Statistical Calculations

The TI-83 has two menus, the STAT and LIST menus, for calculating some of the measures of central tendency and dispersion. In both cases, the data are entered in list L1 for a single list of scores. For a frequency table, list the scores in L1 and the corresponding frequencies in L2.

STAT Menu

To obtain the statistical calculations for a single list in L1 use

STAT <CALC><1:1-Var stats> ENTER

The screen shows the values for:

$\bar{x}$ (Mean)

Sx (Sample standard deviation)

σx (Population standard deviation)

$\min X$ (Smallest score)

Q_1 (First quartile)

Med   (Median)

Q_3 (Third quartile)

$\max X$ (Largest score)

These values give you the five-point summmary and both forms of the standard deviation. For a frequency table with scores in L1 and frequencies in L2 use

$\boxed{\text{STAT}}$ <CALC><1:1-Var stats> L1,L2 $\boxed{\text{ENTER}}$

to obtain the same list of values.

LIST Menu

Using the $\boxed{\text{LIST}}$ <MATH> menu, you can obtain the mean, maximum, minimum, median, sample variance, and standard deviation from a single list in L1 or a frequency table in L1 and L2. Unlike the STAT menu, these values do not appear all at the same time. They must be obtained separately. For example, the mean is obtained with

$\boxed{\text{LIST}}$ <MATH><3:mean(L1)

for a single list and with

$\boxed{\text{LIST}}$ <MATH><3:mean(L1,L2)

for a frequency table. Although this is less convenient than using STAT, it has the advantage that the calculations can be used in other formulas. One example is the calculation of the range.

Range

$\boxed{\text{LIST}}$ <MATH><2:max> L1-<1:min> L1 $\boxed{\text{ENTER}}$

Permutations and Combinations

Find $P(9, 7)$, $C(9, 7)$, and 6!

$P(9, 7)$: $\boxed{9}$ $\boxed{\text{MATH}}$ <PRB> <2:nPr> $\boxed{\text{ENTER}}$ $\boxed{7}$ $\boxed{\text{ENTER}}$

$C(9, 7)$: $\boxed{9}$ $\boxed{\text{MATH}}$ <PRB> <3:nCr> $\boxed{\text{ENTER}}$ $\boxed{7}$ $\boxed{\text{ENTER}}$

6!: $\boxed{6}$ $\boxed{\text{MATH}}$ <PRB> <4:!> $\boxed{\text{ENTER}}$

Histograms

In order to obtain the graph of a histogram, follow these steps:

1. Enter data into appropriate lists.
2. Clear or turn off all functions in the <y=> screen so that they will not appear on the graph of the histogram.
3. Turn off other plots in the PLOT screen so that they will not appear.
4. Clear any previous drawings that remain.

5. Set the viewing window so that the appropriate range of the variable and the frequencies will appear on the screen.
6. Define the histogram.
7. Display the graph.

Now for more detail of the steps:

1. **Enter data.** See earlier section.
   In the examples, the list name L_1 is used for the variable list and L_2 is used for the frequency list. For a single list of data, use 1 for each frequency.
2. **Clear <y=> screen.**
   Select the <y=> screen and clear or turn off the functions.
3. **Turn off other plots,** say PLOT 3 with:
   $\boxed{\text{STAT PLOT}}$ <PLOT 3> <OFF> $\boxed{\text{ENTER}}$
4. **Clear drawings**
   Turn off Plot and <y=> functions.
   $\boxed{\text{DRAW}}$ <1:CLRDRAW> $\boxed{\text{ENTER}}$ $\boxed{\text{ENTER}}$
5. **Set the viewing window.**
   Set XMIN, XMAX, XSCL, YMIN, YMAX, YSCL using the window screen used to set the screen when graphing functions. Set XMIN and XMAX so all scores are in the interval (XMIN, XMAX). This interval will be the x-axis on the graph. XSCL is the width of a bar on the histogram and determines the interval length of each category. YMIN and YMAX determine the range of frequencies (the y-scale).
6. **Define the histogram.**
   Three plots are allowed. This example will use PLOT 1.
   Press $\boxed{\text{STAT PLOT}}$ <1:PLOT1> $\boxed{\text{ENTER}}$
   You will see a screen similar to the following.

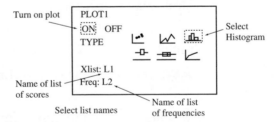

On that screen select as shown in the figure: <ON>, histogram for type, L_1 for Xlist, and L_2 for Freq. Press $\boxed{\text{ENTER}}$ after each selection.
7. **Display the Histogram.**
   Go to home screen and press $\boxed{\text{GRAPH}}$.

Box Plot. You can graph the box plot of data. Follow the steps for a histogram except for step 6. Substitute the following in step 6:

6. Select the box plot type. The symbol is ⊢▭⊣ instead of histogram.

Graphing Calculator Programs

Some problems, such as finding the reduced echelon form of a matrix or the simplex solution of a linear programming problem, can be lengthy and tedious. Programs are available in which the calculator performs the calculations. The programs are listed in the Graphing Calculator Manual or the Instructor's Manual.

Some of the programs are included at the end of sections in the text. In the list below, the section where the program appears in the text is given after the name of the program. The programs available include the following:

PIVOT (Section 2.3).   The program pivots on the matrix entry specified by the user. This program is useful in reducing a matrix or performing the pivots in a simplex solution.

SMPLX (Section 4.2).   The user specifies the pivot row and column in a simplex tableau, and the program finds the next tableau.

BINOM.   For a given N, p, and x, this calculates the binomial probability

$$P(X = x) = C(N, x)p^x(1 - p)^{N-x}$$

ANNA.   This calculates the amount of an annuity when the periodic payment, periodic interest rate, and number of periods are given.

PAYANN.   This computes the periodic payment of an annuity that will yield a specified amount at a future date.

PVAL.   This computes the present value of an annuity with specified periodic payments.

LNPAY.   This finds the monthly payments required to amortize a loan.

ANNGRO (Section 5.3).   This shows the growth of an annuity by computing the amount of annuity period by period.

AMLN (Section 5.4).   This computes the amortization schedule of a loan, that is, the interest paid, principal paid, and balance of the loan for each month.

BIOD (Section 8.6).   This computes the binomial probability distribution:

$$P(X = x) = C(N, x)p^x(1 - p)^{N-x} \quad (\text{for } x = 0, 1, 2, \ldots, N)$$

BIOH (Section 8.6).   This computes the binomial probability distribution and displays it as a histogram.

NORML (Section 8.7).   This finds the area under a normal curve between specified limits.

Using EXCEL

Instructions for using EXCEL can be found at the end of some sections to show how EXCEL may be applied. Here is a summary of the topics and the section in which they occur. ■

Answers to Selected Odd-Numbered Exercises

CHAPTER 1

Section 1.1

1. $y = 15x + 20$. Domain is number of hours worked. Range is number of dollars of fee.   **3. (a)** \$23.75
(b) \$14.25   **5. (a)** 1   **(b)** -11 **(c)** -1 **(d)** $4a - 3$ **7. (a)** $\frac{3}{2}$ **(b)** $\frac{5}{7}$ **(c)** -1 **(d)** $\dfrac{2c + 1}{2c - 1}$
9. (a) $p(1995) = 43.9$ thousand, $p(2010) = 63.7$ thousand   **(b)** In 2018   **11. (a)** 540 calories   **(b)** 83.3 minutes
13. $y = 2.40x + 25$ **15.** $y = x - 0.20x$ or $y = 0.80x$ **17.** $y = 0.60x + 12$ **19.** $y = 3500x + 5,000,000$
21. $y = 0.88x$ **23. (a)** \$328.13   **(b)** \$356.25   **(c)** 8.5 hours   **25. (a)** $A = \pi r^2$ is a function.   **(b)** Domain:
positive numbers. Range: positive numbers.   **27. (a)** $p = $ price per pound times w is a function.   **(b)** Domain: positive numbers. Range: positive numbers.   **29. (a)** $y = x^2$ is a function.   **(b)** Domain: all real numbers. Range: all
nonnegative numbers.   **31.** y is not a function of x. There can be more than one person with a given family name. x is
a function of y.   **33.** Not a function because two classes can have the same number of boys, but the combined weights
different.   **35.** Not a function because two families with the same number of children can have a different number
of boys.   **37.** The domain is the set of numbers in the interval $[-2, 4]$. The range is the set of numbers in the interval
$[-1, 3]$.   **39.** The domain is the set of numbers in the intervals $[0, 4]$ or $[7, 12]$ and the range is the set of numbers in
the interval $[-4, 8]$.   **43. (a)** $p = 0.40(220 - x)$ for age x **(b)** $p = 0.70(220 - x)$ for age x **45. (a)** 83 ft.
(b) 264 ft.   **47. (a)** 12.47   **(b)** 23.93   **(c)** 20.99   **49. (a)** 57.6625   **(b)** 32.4205   **(c)** 258.776
51. (a) 68.1%   **(b)** 40.5%   **(c)** 75.0%   **(d)** 81.9%   **53. (a)** 274.4 million   **(b)** 154.3 million   **(c)** 70.0 million
(d) 19.6 million   **(e)** 3.0 million   **(f)** 329.1 million   **(g)** 424.4 million   **(h)** 400 million about 2042; 500 million about
2071.   **55.** $y = 6, 3, -6, -18, -45$ **57.** $y = 39.28, 46.48, 104.24$ **59.** $y = 16, 19, 22, 25, 28$

Using Your TI-83, Section 1.1

1. $y = 3, 27, 51$ **3.** $y = 9, 9, 14, 30$ **5.** $y = 0, -0.25, -1.6,$ and 8.9231

Using EXCEL, Section 1.1

1. =A4+B4   **3.** −C4+C5   **5.** =B2*B3   **7.** =(B1+B2)/2   **9.** =2.1*A5-1.8

Section 1.2

1.

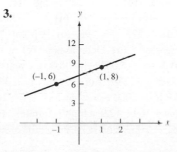

3.

5.

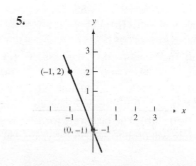

7. Slope $= 7$, y-intercept $= 22$ **9.** Slope $= -\frac{2}{5}$, y-intercept $= 6$ **11.** Slope $= -\frac{2}{5}$, y-intercept $= \frac{3}{5}$
13. Slope $= \frac{1}{3}$, y-intercept $= 2$ **15.** 1   **17.** $-\frac{4}{3}$ **19.** Negative   **21.** Positive   **23.** $y = -2$
25. $y = 0$ **27.**   **29.**   **31.** $x = 3$ **33.** $x = 10$

35.   **37.**

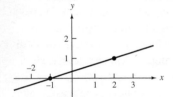

39. $y = 4x + 3$ **41.** $y = -x + 6$ **43.** $y = \frac{1}{2}x$ **45.** $y = -4x + 9$ **47.** $y = \frac{1}{2}x + \frac{3}{2}$
49. $y = 7x - 2$ **51.** $y = \frac{1}{5}x + \frac{21}{5}$
53. $y = \frac{1}{3}x + \frac{1}{3}$ **55.** $y = 2x$ **57.** $y = 4$

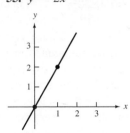

59. $y = -5$, y-intercept   **61.** $y = -5$, y-intercept
     $x = 3$, x-intercept $x = 12.5$, x-intercept

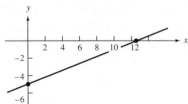

63. The lines are parallel.   **65.** The lines are not parallel.   **67.** The lines are parallel.   **69.** The lines are not
parallel.   **71.** Perpendicular   **73.** Not perpendicular   **75.** $y = 3x + 8$ **77.** $y = -\frac{5}{7}x + 8$
79. For Exercise 76, $3x + 2y = 18$. For Exercise 77, $5x + 7y = 56$. For Exercise 78, $5x - 2y = 26$.   **81.** $\frac{11}{3}$
83. $3x - 5y = -9$ **85. (a)** A point and slope   **(b)** $y = -3x + 238$ **(c)** 238 pounds   **87.** $y = 1.4x + 640$
89. (a) $y = -6$ **(b)** $x = \frac{3}{4}$ **(c)** $y = \frac{21}{4}$ **(d)** $x = -15$ **91. (a)** $y = 836.78x + 13{,}892$ **(b)** The estimated
average cost for 2005 is \$25,607.   **93.** $y = 50x + 375$ **95. (a)** A point and slope   **(b)** $y = 0.078x + 5$
97. (a) Increases 4   **(b)** Decreases 3   **(c)** Increases $\frac{2}{3}$ **(d)** Decreases $\frac{1}{2}$ **(e)** $y = -\frac{2}{3}x + \frac{4}{3}$, so it decreases $\frac{2}{3}$.   **(f)** No
change   **99.** $y = 0.11x + 22$ **101. (a)** $y = -15{,}000x + 536{,}000$, x years after 1997. **(b)** $y = 4500x + 536{,}000$,
x years after 1997.   **103.** $y = 0.27(x - 27{,}950) + 3892.5$ or $y = 0.27x - 3654$ **105.** $y = 0.15(x - 12{,}000)$
$+ 1200$ or $y = 0.15x - 600$. This equation holds for $12{,}000 \le x \le 46{,}700$.   **107. (a)** $y = -0.167x + 13.7$ for x years
after 1980.   **(b)** The birth rate for 1985 is estimated to be 12.9, a high estimate.   **(c)** This function estimates that Japan's
birth rate will drop to zero in the year $1980 + 82 = 2062$. This is based on the assumption that birth rates will drop in a lin-

car manner at the same rate they dropped in 1980–2002. It is unrealistic to expect that no babies will be born in an entire year, so the linear function is not a valid long-term estimate.   **109. (a)** $y = -45x + 410$ **(b)** 95 **(c)** \$3.55 **(d)** An admission of \$9.11, or more, would result in no attendance. **(e)** 410   **111.** Approximately 249 pounds per square inch.   **115.** About 4:30 A.M.   **117.** All have y-intercepts of 4, but they are not parallel.   **119.** All go through the origin.   **121.** $-\frac{10}{3}$ **123.** 1.240   **125.** $y = \frac{12}{5}x - 8$ **127.** $y = 0.351x + 5.821$

Using EXCEL, Section 1.2

1. 0.25   **3.** 2.73   **5.** $y = -0.67x + 4.33$ **7.** $y = -0.56x + 4.24$
1. $y = 1.4x + 0.8$ **3.** $y = 1.25x + 1.35$

Section 1.3

1. (a) \$10,040 **(b)** $x = 223$ bikes **(c)** Unit cost is \$43; fixed cost is \$2300.   **3. (a)** Fixed cost is \$400; unit cost is \$3. **(b)** For 600 units, \$2200; for 1000 units, \$3400   **5. (a)** $R(x) = 32x$ **(b)** \$2496 **(c)** 21 pairs   **7. (a)** $R(x) = 3.39x$ **(b)** \$2827.26   **9. (a)** $C(x) = 57x + 780$ **(b)** $R(x) = 79x$ **(c)** The break-even number is 36 coats.   **11.** $C(x) = 4x + 500, C(800) = \$3700$ **13. (a)** $C(x) = 2x + 200$ **(b)** \$200 **(c)** \$2   **15. (a)** $C(x) = 649x + 1500$ **(b)** $R(x) = 899x$ **(c)** \$25,513 **(d)** \$33,263 **(e)** $x = 6$ computers   **17. (a)** $BV = -50x + 425$ **(b)** \$50 **(c)** \$275   **19. (a)** $BV = -1575x + 9750$ **(b)** \$1575 **(c)** $BV(2) = \$6600$, $BV(5) = \$1875$ **21.** At least 2160 cookies must be sold to make a profit.   **23.** At least 94 ties per week   **25.** At least 3683 bagels per week   **27.** Company A is the better deal when the weekly mileage is less than 1400 miles.   **29.** At least 2154 copies must be sold to make a profit.   **31.** $C(x) = 12.65x + 2140$ **33. (a)** 183 tickets must be sold to break even. **(b)** 368 tickets must be sold to clear \$700. **(c)** 264 tickets must be sold to clear \$700.   **35. (a)** $R(x) = 35x$ **(b)** \$43,330 **(c)** 17   **37. (a)** $BV = -1425x + 18,450$ **(b)** \$1425 **(c)** \$18,450   **39. (a)** $R(x) = 12x$ **(b)** $C(x) = 6.5x + 1430$ **41. (a)** $R(x) = 22.5x$ **(b)** $C(x) = 16.7x + 1940$ **(c)** 335 for break-even quantity   **43. (a)** The two options cost the same for 265 minutes per month. **(b)** The first plan is less costly when calls total more than 265 minutes a month.   **45.** Plan 2 is better when the number of checks written in a month is more than 62 and less than 125.   **47.** The new process is more economical when the plant produces more than 282 knives per day.   **49. (b)** Shortage **(c)** Surplus **(d)** When the demand is 20, the price is $-\$40$, so people must be paid to purchase. It is unrealistic to expect a demand of 20.   **51.** 595   **53.** Plan 1 is better when sales are less than \$2941. Plan 2 is better when sales are between \$2941 and \$3750. Plan 3 is better when sales are greater then \$3750.   **55. (a)** 220 tickets **(b)** 223 tickets **(c)** The Ferrell Center is more profitable when more than 233 tickets are sold; otherwise, the Convention Center is more profitable.   **57. (a)** For 10 trips, plan 2 is better at a cost of \$1600. For 15 trips, plan 2 is better at a cost of \$2250. **(b)** Plan 3 is better for less than 9 trips. Plan 2 is better for 9 through 16 trips. Plan 1 is better for more than 16 trips.   **59. (a)** \$5600 **(b)** \$7925   **61.** For $x = 10$, \$840 loss. For $x = 30$, \$240 profit. For $x = 45$, \$1050 profit. For $x = 62$, \$1968 profit.   **63.** $(27.2, 587.52)$ **65. (a)** $(64, 5088)$ **(b)** $x = 121.14$ **(c)** $x = 47.23$

Using Your TI-83, Section 1.3

1. $(4, 7)$ **3.** $(4.27, 0.91)$ **5.** $(2, 6)$

Using EXCEL, Section 1.3

1. $-35, 215, 590, 1015,$ and 1490 **3.** $-1396; 16,709; 52,919; 77,059; 127,753$ **5.** $(25, 1220)$

Review Exercises, Chapter 1

1. (a) 16 **(b)** 2 **(c)** 12.5 **(d)** $\dfrac{7b - 3}{2}$ **3.** 22   **5. (a)** \$4.20 **(b)** 2.75 pounds
7. (a) $f(x) = 29.95x$ **(b)** $f(x) = 1.25x + 40$

9. (a) 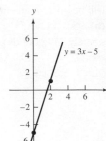   **(b)** 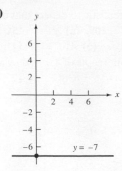   **(c)**    **(d)**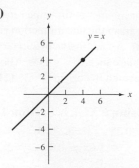

11. (a) Slope is -2, y-intercept is 3.   **(b)** Slope is $\frac{2}{3}$, y-intercept is -4.   **(c)** Slope is $\frac{5}{4}$, y-intercept is $\frac{3}{2}$.   **(d)** Slope is $-\frac{6}{7}$, y-intercept is $-\frac{5}{7}$.   **13. (a)** $-\frac{6}{5}$ **(b)** 3 **(c)** $\frac{5}{2}$ **15. (a)** $y = -\frac{3}{4}x + 5$ **(b)** $y = 8x - 3$ **(c)** $y = -2x + 9$
(d) Horizontal line, $y = 6$ **(e)** $y = -\frac{1}{6}x + \frac{23}{6}$ **(f)** Vertical line, $x = -2$ **(g)** $y = \frac{4}{3}x + \frac{13}{3}$ or $4x - 3y = -13$
17. (a) $y = 2$ **(b)** $x = -4$ **(c)** $x = 5$ **(d)** $y = 6$ **19.** The lines are not parallel.   **21.** The lines are not
parallel.   **23.** The lines are not parallel.   **25.** $C(x) = 36x + 12{,}800$ **27. (a)** \$4938 **(b)** 655 bags
29. (a) $R(x) = 11x$ **(b)** $C(x) = 6.5x + 675$ **(c)** $x = 150$ **31.** 365 watches   **33. (a)** $BV = -2075x + 17{,}500$
(b) \$2075 **(c)** \$7125   **35.** $BV = -296x + 1540$ **37.** $k = 5$ **39.** $BV = -3800x + 22{,}000$
41. $C(x) = 0.67x + 480$ **43.** The second plan is better when sales exceed 20,000 items.   **45.** $k = 26$
47. \$1015   **49.** Plan 1 is better when sales are less than \$2667. Plan 2 is never better. Plan 3 is better when sales are
larger than \$2667.   **51. (a)** $y = 120x + 8400$ **(b)** During the 14th month   **53.** $y = 500x - 1700$

CHAPTER 2

Section 2.1

1. $(2, 3)$ **3.** $(-3, 0)$ **5.** $(-3, -15)$ **7.** $(5, 3)$ **9.** $(6, -8)$ **11.** $(3.5, 4)$ **13.** $\left(\frac{2}{3}, -\frac{5}{3}\right)$
15. $(-2.5, -12.5)$ **17.** $(6, -1)$ **19.** $\left(\frac{5}{2}, -3\right)$ **21.** $(-2, 3)$ **23.** $(-14, 19)$ **25.** $(-2.4, 5.1)$
27. $(0.06, 0.13)$ **29.** No solution   **31.** Infinite number   **33.** Infinite number   **35.** $(4, 3)$ **37.** $(30, 10)$
39. $(15, 8)$ **41.** $(25, 85.5)$ **43.** $(2.5$ oranges, 3.25 apples)   **45. (a)** 38,554.2 CDs at break even, so we
round up ro 38,555 **(b)** A loss of \$77,000 occurs. **(c)** A profit of \$47,500 occurs.   **47.** 43 nickels, 122 dimes
49. Mix 225 ounces of the first drink and 375 of the second drink   **51.** 595 at McGregor, 905 at Ennis   **53.** 231 boxes
of oranges and 271 boxes of grapefruit   **55.** 11 cases of Golden, 14 cases of Light Punch   **57.** 7 two-place tables
and 13 four-place tables   **59.** \$23,500 in tax-free, \$26,500 in money market   **61.** Federal tax is \$38,000, state tax
is \$8000   **63. (a)** \$2370 to Habitat for Humanity, \$1050 to the Family Abuse Center **(b)** \$1580 for their church
67. Their deposits will be equal in a little less than 11 years.   **69. (a)** $x = 4.54, y = 7.74$ **(b)** $x = 0.75, y = 9.25$
(c) $x = 1.26, y = 4.80$ **(d)** $x = 3.02, y = -0.72$ **71.** 20.55 tricycles with a price of \$117.53. If we round to 21
tricycles, the supply price is \$118.80, and the demand price is \$115.50.   **73. (a)** Equilibrium occurs at $(27, 126)$.
(b) Shortage of about 24 TVs **(c)** Surplus of about 14 TVs   **75.** $(5.32, 2.26)$ **77.** $(2.5, 6.25)$

Section 2.2

1. $(3, 2)$ **3.** $(2, -1)$ **5.** $(2, -1, 2)$ **7.** $(7, 5, 3)$ **9.** $(3, 1, 2)$
11. (a) $a_{11} = 2, a_{22} = 3, a_{33} = 6, a_{43} = 11$ **(b)** $(2, 3)$ **(c)** $a_{12} = 4, a_{32} = 0, a_{41} = 9$

13. Coefficients: $\begin{bmatrix} 5 & -2 \\ 3 & 1 \end{bmatrix}$, augmented matrix: $\begin{bmatrix} 5 & -2 & | & 1 \\ 3 & 1 & | & 7 \end{bmatrix}$

15. Coefficients: $\begin{bmatrix} 1 & 1 & -1 \\ 3 & 4 & -2 \\ 2 & 0 & 1 \end{bmatrix}$, augmented matrix: $\begin{bmatrix} 1 & 1 & -1 & | & 14 \\ 3 & 4 & -2 & | & 9 \\ 2 & 0 & 1 & | & 7 \end{bmatrix}$

17. Coefficients: $\begin{bmatrix} 1 & 5 & -2 & 1 \\ 1 & -1 & 2 & 4 \\ 6 & 3 & -11 & 1 \\ 5 & -3 & -7 & 1 \end{bmatrix}$, augmented matrix: $\begin{bmatrix} 1 & 5 & -2 & 1 & 12 \\ 1 & -1 & 2 & 4 & -5 \\ 6 & 3 & -11 & 1 & 14 \\ 5 & -3 & -7 & 1 & 22 \end{bmatrix}$

19. $5x_1 + 3x_2 = -2$
$-x_1 + 4x_2 = 4$

21. $5x_1 + 2x_2 - x_3 = 3$
$-2x_1 + 7x_2 + 8x_3 = 7$
$3x_1 \qquad + x_3 = 5$

23. $3x_1 \qquad + 2x_3 + 6x_4 = 4$
$-4x_1 + 5x_2 + 7x_3 + 2x_4 = 2$
$x_1 + 3x_2 + 2x_3 + 5x_4 = 0$
$-2x_1 + 6x_2 - 5x_3 + 3x_4 = 4$

25. $\begin{bmatrix} 1 & 2 & -4 & 6 \\ 4 & 2 & 5 & 7 \\ 1 & -1 & 0 & 4 \end{bmatrix}$

27. $\begin{bmatrix} 1 & 3 & 2 & -4 \\ 0 & -7 & -1 & 13 \\ 0 & -6 & -10 & 19 \end{bmatrix}$

29. $\begin{bmatrix} 1 & -3 & 2 & -6 \\ 0 & 1 & -2 & 4 \\ 0 & 4 & 3 & 8 \end{bmatrix}$

31. $(1, 1)$ **33.** $(5, 2)$ **35.** $\left(\frac{3}{2}, -\frac{5}{2}\right)$ **37.** $(3, 1, 2)$ **39.** $(9, -4, 2)$ **41.** $(4, -3, -1)$
43. $(3, 0, 2)$ **45.** $(0, 4, 2)$ **47.** $(5, -4, 1)$ **49.** $(4, 1, -3, 2)$ **51.** $(1, 2, 3, 4)$
53. Let x_1 = number cases of Regular, x_2 = number cases of Premium, x_3 = number cases of Classic.
$4x_1 + 4x_2 + 5x_3 = 316$ (Apple juice)
$5x_1 + 4x_2 + 2x_3 = 292$ (Pineapple juice)
$x_1 + 2x_2 + 3x_3 = 142$ (Cranberry juice)
55. Let x_1 = number of student tickets, x_2 = number of faculty tickets, x_3 = number of general public tickets.
$3x_1 + 5x_2 + 8x_3 = 2542$
$x_1 = 3x_2$
$x_3 = 2x_1$
57. Let x = number shares of X, y = number shares of Y, z = number shares of Z.
$44x + 22y + 64z = 20{,}480$
$42x + 28y + 62z = 20{,}720$
$42x + 30y + 60z = 20{,}580$
59. A six-pack cost $2.40, a bag of chips cost $2.00, and a package of cookies cost $2.70
61. $22,000 in stock A, $3000 in stock B, $15,000 in stock C   **63.** 540 to high school students, 1230 to college students, 680 to adults   **65.** 24 cases of Regular, 35 cases of Premium, 16 cases of Classic   **67.** 123 students, 41 faculty, 246 general public   **69.** 90 shares of X, 140 shares of Y, 210 shares of Z
71. Let x_1 = number of days A operates, x_2 = number of days B operates, x_3 = number of days C operates.
$300x_1 + 700x_2 + 400x_3 = 39{,}500$
$500x_1 + 900x_2 + 400x_3 = 52{,}500$
$200x_1 + 100x_2 + 800x_3 = 12{,}500$
75. Madeline is correct. The solution is $(7, -4, -4, 3)$.

77. $(3, 10)$ **79.** $(2.5, 1)$ **81. (a)** $\begin{bmatrix} 1.1 & 1.2 & -1.3 & 29.76 \\ 3.5 & 4.1 & -2.2 & 81.94 \\ 2.3 & 0 & 1.4 & 17.70 \end{bmatrix}$ **(b)** $(12.2, 5.6, -7.4)$ **83. (a)** $(5.149, -0.990)$

Using Your TI-83, Section 2.2
1. $(1, -2, 3)$ **3.** $(6, -1, 5)$

Using EXCEL, Section 2.2
1. $(1, -2, 3)$ **3.** $(6, -1, 5)$ **5.** $(2, -3, 5, 7)$

Section 2.3
1. Reduced echelon form   **3.** Not in reduced echelon form because column 3 does not contain a zero in row 2
5. Reduced echelon form   **7.** Not in reduced echelon form because the leading 1 in row 3 is to the left of the leading 1 in row 2, and the 3 in row 3 should be 0

9. $\begin{bmatrix} 1 & 0 & 0 & | & 1 \\ 0 & 1 & 0 & | & -8 \\ 0 & 0 & 1 & | & 2 \end{bmatrix}$ **11.** $\begin{bmatrix} 1 & 0 & -1 & | & -8 \\ 0 & 1 & 2 & | & 6 \\ 0 & 0 & 5 & | & -3 \end{bmatrix}$ **13.** $\begin{bmatrix} 1 & 0 & -13 & 10 & | & -19 \\ 0 & 1 & 4 & -2 & | & 7 \\ 0 & 0 & 2 & 3 & | & 5 \end{bmatrix}$ **15.** $\begin{bmatrix} 1 & 0 & 0 & | & -8 \\ 0 & 1 & 0 & | & 8 \\ 0 & 0 & 1 & | & 2 \end{bmatrix}$

17. $\begin{bmatrix} 1 & 0 & \frac{7}{5} & | & \frac{14}{5} \\ 0 & 1 & \frac{1}{5} & | & -\frac{3}{5} \\ 0 & 0 & 0 & | & 0 \end{bmatrix}$ **19.** $\begin{bmatrix} 1 & -1 & 0 & 0 & | & 0 \\ 0 & 0 & 1 & 0 & | & 0 \\ 0 & 0 & 0 & 1 & | & 0 \\ 0 & 0 & 0 & 0 & | & 1 \end{bmatrix}$ **21.** $x_1 = 3$ The solution is $(3, -2, 5)$.
$x_2 = -2$
$x_3 = 5$

23. $x_1 + 3x_3 = 4$
$ x_2 + x_3 = -6$
$ x_4 = 2$
There are an infinite number of solutions
of the form $(4 - 3k, -6 - k, k, 2)$.

25. $x_1 = 0$
$x_2 = 0$
$0 = 1$
No solution

27. $x_1 + 3x_4 = 0$
$ x_2 - 2x_4 = 0$
$ x_3 + 7x_4 = 0$
$ 0 = 0$
There are an infinite number of solutions
of the form $(-3k, 2k, -7k, k)$

29. No solution   **31.** $x_1 = 4, x_2 = -\frac{1}{2}, x_3 = -\frac{3}{2}$

33. $x_1 = -32 - 2x_3, x_2 = 45 + x_3, x_4 = 16$ or $(-32 - 2k, 45 + k, k, 16)$ **35.** No solution
37. No solution   **39.** $x_1 = -59 + 11x_3, x_2 = 23 - 5x_3$ or $(-59 + 11k, 23 - 5k, k)$
41. $x_1 = 3 + x_3 + x_4, x_2 = 6 + x_3 + x_4 - x_5$ or $(3 + k + r, 6 + k + r - s, k, r, s)$ **43.** No solution
45. No solution   **47.** $x = 2, y = -3$ **49.** $x = -5, y = 2$ **51.** $x_1 = 2, x_2 = 1, x_3 = 3$
53. No solution   **55.** No solution   **57.** $x_1 = \frac{9}{2}, x_2 = -\frac{5}{2}, x_3 = 0, x_4 = \frac{1}{2}$ **59.** $(-9, 6, 2, 1)$
61. $x_1 = -2x_2 + 3x_3 - x_5 + 3x_6, x_4 = -2x_5 - x_6$ or $(-2k + 3r - s + 3t, k, r, -2s - t, s, t)$
63. $15,000 in stocks, $18,000 in bonds, $12,000 in money market
65. She may bike from 0 to 20 minutes, then should jog for twice that time, and play handball for the rest of the 60 minutes.
67. (a) Let $x_1 =$ hours of math, $x_2 =$ hours of English,   (b) 14 hours for math, 7 hours for English, 17.5 hours
$x_3 =$ hours of chemistry, $x_4 =$ hours of history.   for chemistry, and 3.5 hours for history
$x_1 + x_2 + x_3 + x_4 = 42$
$x_1 + x_2 = 21$
$x_1 = 2x_2$
$x_2 = 2x_4$
69. The federal tax is $19,200, the state tax is $7200, and the city tax is $3200   **71.** Infinite number of solutions
73. (a) Let   (b) $x_5 = 5, x_6 = 5$ yields $x_1 = 0, x_2 = 15, x_3 = 25,$
$x_1 =$ number supplied by Sweats-Plus to Spirit Shop 1 $x_4 = 15. x_5 = 10, x_6 = 5$ yields $x_1 = 5, x_2 = 10, x_3 = 25,$
$x_2 =$ number supplied by Sweats-Plus to Spirit Shop 2 $x_4 = 10. x_5 = 0, x_6 = 15$ yields $x_1 = 5, x_2 = 20, x_3 = 15,$
$x_3 =$ number supplied by Sweats-Plus to Spirit Shop 3 $x_4 = 10.$ Other ways are possible.
$x_4 =$ number supplied by Imprint-Sweats to Spirit Shop 1   (c) x_5 represents the number of sweatshirts supplied by
$x_5 =$ number supplied by Imprint-Sweats to Spirit Shop 2   Imprint-Sweats to Spirit Shop 2. Because $x_2 = 20 - x_5,$
$x_6 =$ number supplied by Imprint-Sweats to Spirit Shop 3 x_5 must be no greater than 20, or else x_2 would be a
negative number. Thus, $x_5 \leq 20$.

The system of equations is   The solution is   (d) If Imprint-Sweats supplies no sweatshirts to Spirit
$x_1 + x_4 = 15$ $x_1 = -10 + x_5 + x_6$ Shop 2 and Spirit Shop 3, then $x_5 = 0$ and $x_6 = 0$ so
$x_2 + x_5 = 20$ $x_2 = 20 - x_5$ $x_1 = -10.$ Thus, the order cannot be filled.
$x_3 + x_6 = 30$ $x_3 = 30 - x_6$ (e) Because $x_1 = -10 + x_5 + x_6,$ in order for $x_1 \geq 0,$
$x_1 + x_2 + x_3 = 40$ $x_4 = 25 - x_5 - x_6$ then $x_5 + x_6 \geq 10.$ Imprint-Sweats must supply a total of
$x_4 + x_5 + x_6 = 25$ 10 or more to Spirit Shops 2 and 3.

(f) Because $x_2 = 20 - x_5$, and $0 \le x_5 \le 20$, then $x_2 \le 20$. Thus, Sweats-Plus supplies 20 or less to Spirit Shop 2.
75. The last equation should be $3x + y + 4z = 5$.   **81.** The system has infinitely many solutions.
83. (a) Label the traffic flow on each block by $x_1, x_2, \ldots, x_7$ as shown. Then the condition that incoming traffic equals outgoing traffic at each intersection gives:

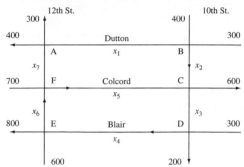

At A: $x_1 + x_7 = 700$
At B: $x_1 + x_2 = 700$
At C: $x_2 + x_5 = 600 + x_3$
At D: $x_3 + 300 = 200 + x_4$
At E: $600 + x_4 = 800 + x_6$
At F: $x_6 + 700 = x_5 + x_7$

(b) The solution to the above system is

$x_1 = 700 - x_7$
$x_2 = x_7$
$x_3 = 100 + x_6$
$x_4 = 200 + x_6$
$x_5 = 700 + x_6 - x_7$

(c) x_3 represents the traffic flow on 10th St. between Colcord and Blair. Because $x_3 = 100 + x_6$, $x_3 \ge 100$. The minimum traffic flow is 100 vehicles per hour when $x_6 = 0$. The other traffic flows then become

$x_1 = 700 - x_7$
$x_2 = x_7$
$x_4 = 200$
$x_5 = 700 - x_7$

85. $c \ne 14$ **87.** $x_1 = 0.2 + 1.6x_3$, $x_2 = -0.2 + 1.4x_3$ **89.** $(-19, 28, -10)$
91. (a) $x_1 = 0.5 + 3.5x_4$ **(b)** $x_2 = 0.5 - 1.83x_4$ implies $0.5 - 1.83x_4 > 0$, so $x_4 < \frac{0.5}{1.83} = 0.273$. Thus,
$\qquad x_2 = 0.5 - 1.83x_4$ $0 < x_4 < 0.273$. $x_3 = 2.33x_4$ implies $0 < x_3 < 0.636$. $x_1 = 0.5 + 3.5x_4$ implies $x_1 > 0.5$
$\qquad x_3 = 2.33x_4$ and the maximum value of x_1 occurs when x_4 is maximum. Thus, $x_1 < 0.5 + 3.5(0.273) =$
$\qquad$ 1.456 so $0.50 < x_1 < 1.456$. $x_2 = 0.5 - 1.83x_4$ implies $x_2 < 0.5$ and x_2 is minimum, 0,
$\qquad$ when x_4 reaches 0.273 so $0 < x_2 < 0.5$.

Using Your TI-83, Section 2.3

1. $(2, -3, 4)$ **3.** $(3, 4, 6)$ **1.** $\begin{bmatrix} 1 & 0 & 0 & 1.3 & 2.5 \\ 0 & 1 & 0 & 0.7 & 0.5 \\ 0 & 0 & 1 & -1.2 & 0.5 \end{bmatrix}$ **3.** $(-2, 5, -3)$

Using EXCEL, Section 2.3

1. $(-5, 2, 6)$ **3.** $(6, 2, 7)$

Section 2.4

1.

	Alpha	Beta
Salvation Army	50	65
Boy's Club	85	32
Girl Scouts	68	94

3.

	Joe	Jane	Judy
Checking	12	11	5
Savings	15	18	8
Boxes	8	9	21

5. 2 by 2   **7.** 3 by 3
9. 4 by 1   **11.** 2 by 4
13. 2 by 3   **15.** 2 by 2
17. Not equal   **19.** Equal

21. Not equal   **23.** $\begin{bmatrix} 3 & 3 & 2 \\ 7 & 4 & 3 \end{bmatrix}$ **25.** $\begin{bmatrix} 2 \\ 58 \end{bmatrix}$ **27.** Cannot add them.   **29.** $\begin{bmatrix} 6 & 5 & 8 \\ 5 & 1 & 3 \\ 5 & -3 & 4 \end{bmatrix}$ **31.** $\begin{bmatrix} 12 & 3 \\ 6 & 15 \end{bmatrix}$

33. $\begin{bmatrix} 20 \\ 15 \\ 5 \\ 10 \end{bmatrix}$ **35.** $\begin{bmatrix} -12 & 6 & -15 \end{bmatrix}$ **37.** $\begin{bmatrix} 0 & 0 \\ 0 & 0 \end{bmatrix}$

39. (a) $3A = \begin{bmatrix} 3 & 12 \\ -6 & 9 \end{bmatrix}$, $-2B = \begin{bmatrix} 0 & -4 \\ -8 & -2 \end{bmatrix}$, $5C = \begin{bmatrix} 5 & -10 \\ 5 & -15 \end{bmatrix}$ **(b)** $\begin{bmatrix} 2 & 2 \\ -1 & 0 \end{bmatrix}$ **(c)** $\begin{bmatrix} 3 & 8 \\ -14 & 7 \end{bmatrix}$

(d) $\begin{bmatrix} 6 & -10 \\ -5 & -14 \end{bmatrix}$

41.

	I	II	III
PC	23	17	20
Printer	19	22	11
Disk	151	151	105

43. $x = 3$ **45.** $\frac{17}{8}$

47.

A B C	
$\begin{bmatrix} \frac{65}{12} & \frac{55}{6} & \frac{20}{3} \\ \frac{30}{4} & \frac{45}{4} & 5 \\ \frac{25}{4} & \frac{28}{3} & 7 \end{bmatrix}$	Small Regular Giant

49.

	S.D.	N.O.	P.M.
12-month totals: Fairfield	2760	1080	1680
Tyler	3120	1380	1992

51.

A	89.7
B	68.0
C	75.3
D	82.0
E	75.7

53. (a) The change in the data is found by subtracting:

$$Y2002 - Y1994 = \begin{bmatrix} 199 & 7990 & 905 \\ 67 & 5744 & 1867 \\ -913 & 7180 & 1740 \\ 123 & 4387 & 963 \\ -324 & 4664 & 1284 \\ 23 & 4710 & 2470 \end{bmatrix}$$

(b) Bowdoin had the largest increase in tuition, \$7990.   **(c)** Marquette had a decrease of 913 students, and Samford had a decrease of 324 students.

57. (a) $\begin{bmatrix} 5 & 1 & 9 \\ 10 & 4 & 10 \\ 3 & 6 & 3 \end{bmatrix}$ **(b)** $\begin{bmatrix} 1 & 1 & -1 \\ -6 & -4 & -6 \\ -1 & -4 & -5 \end{bmatrix}$ **(c)** $\begin{bmatrix} 6 & 2 & 8 \\ 4 & 0 & 4 \\ 2 & 2 & -2 \end{bmatrix}$ **(d)** $\begin{bmatrix} 18 & 4 & 31 \\ 32 & 12 & 32 \\ 10 & 19 & 8 \end{bmatrix}$

59. (a) $\begin{bmatrix} 9 & 0 & 8 \\ 6 & 2 & 7 \\ 4 & 8 & 8 \end{bmatrix}$ **(b)** $\begin{bmatrix} -1.4 & 8.3 \\ 7.7 & 6.6 \end{bmatrix}$ **(c)** $\begin{bmatrix} 5.4 & 3.0 & 3.1 \\ 3.8 & 4.2 & 1.6 \\ 3.1 & 6.2 & -5.0 \end{bmatrix}$ **(d)** $\begin{bmatrix} 3 & 15 & 3 \\ 6 & 3 & 6 \\ 9 & 0 & 12 \end{bmatrix}$ **(e)** $\begin{bmatrix} 16.20 & 11.88 \\ 32.94 & 39.42 \\ -7.56 & -13.50 \end{bmatrix}$

(f) $\begin{bmatrix} 6 & 9 & 0 \\ 9 & 2 & 13 \\ 0 & 13 & -8 \end{bmatrix}$

Using Your TI-83, Section 2.4

1. (a) $\begin{bmatrix} 4 & 1 & 3 \\ 10 & 13 & 2 \end{bmatrix}$ **(b)** $\begin{bmatrix} 2 & 6 & 4 \\ 8 & 10 & 14 \end{bmatrix}$ **(c)** $\begin{bmatrix} -2 & 5 & 1 \\ -2 & -3 & 12 \end{bmatrix}$ **(d)** $\begin{bmatrix} 9 & 5 & 8 \\ 24 & 31 & 11 \end{bmatrix}$

Using EXCEL, Section 2.4

1. $\begin{bmatrix} 9 & 5 & -2 \\ 2 & 14 & 11 \\ -2 & 9 & 7 \end{bmatrix}$ **3.** $\begin{bmatrix} 4 & 12 & -8 \\ 20 & 36 & 28 \\ -16 & 9 & 24 \end{bmatrix}$ **5.** $\begin{bmatrix} 52 & 24 & -8 \\ 2 & 66 & 52 \\ -4 & 54 & 30 \end{bmatrix}$

Section 2.5

1. 14   **3.** 12   **5.** 14   **7.** \$6.25   **9.** $\begin{bmatrix} -5 & 11 \\ 0 & 14 \end{bmatrix}$ **11.** $\begin{bmatrix} 30 & 2 \\ 39 & -3 \end{bmatrix}$

13. (a) Not possible   **(b)** Possible, a 2 × 3 matrix   **(c)** Not possible   **15. (a)** Possible, a 3 × 5 matrix   **(b)** Not possible

17. $\begin{bmatrix} -7 & 7 \\ -1 & 6 \end{bmatrix}$ **19.** $\begin{bmatrix} 15 \\ -2 \end{bmatrix}$ **21.** Multiplication is not possible.   **23.** Multiplication is not possible.

25. $\begin{bmatrix} -4 & 4 \\ 3 & 27 \end{bmatrix}$ **27.** Multiplication is not possible.   **29.** $\begin{bmatrix} 5 & 9 & 12 \\ 5 & 17 & 21 \end{bmatrix}$ **31.** $\begin{bmatrix} 8 \\ 13 \\ 7 \end{bmatrix}$

33. $AB = \begin{bmatrix} 1 & 8 \\ -1 & 2 \end{bmatrix}$, $BA = \begin{bmatrix} 3 & 10 \\ -1 & 0 \end{bmatrix}$ **35.** $AB = \begin{bmatrix} -3 & 10 \\ -2 & 5 \end{bmatrix}$, $BA = \begin{bmatrix} -1 & 2 \\ -4 & 3 \end{bmatrix}$

37. $AB = \begin{bmatrix} 6 & 2 & 13 \\ -11 & -6 & -4 \end{bmatrix}$; BA not possible   **39.** $AB = [27 \quad 38]$; BA not possible

41. $AB = BA = \begin{bmatrix} 10 & 6 \\ -9 & 10 \end{bmatrix}$ **43.** $\begin{bmatrix} -19 & 16 \\ -4 & 6 \end{bmatrix}$ **45.** $\begin{bmatrix} 8 & 1 & -2 & -14 \\ 5 & 3 & 9 & -11 \\ 8 & 2 & 12 & 2 \end{bmatrix}$ **47.** $\begin{bmatrix} 4 & 4 \\ 10 & 10 \end{bmatrix}$

49. $\begin{bmatrix} 3 & 4 \\ 1 & 2 \end{bmatrix}$ **51.** $\begin{bmatrix} 2 & -10 \\ 3 & 7 \end{bmatrix}$ **53.** $\begin{bmatrix} 3x + y \\ 2x + 4y \end{bmatrix}$ **55.** $\begin{bmatrix} x_1 + 2x_2 - x_3 \\ 3x_1 + x_2 + 4x_3 \\ 2x_1 - x_2 - x_3 \end{bmatrix}$

57. $\begin{bmatrix} x_1 + 3x_2 + 5x_3 + 6x_4 \\ -2x_1 + 9x_2 + 6x_3 + x_4 \\ 8x_1 \quad\quad + 17x_3 + 5x_4 \end{bmatrix}$ **59. (a)** R1 + R2 → R2   **(b)** 2R1 → R1, R2 + R3 → R3   **(c)** R1 − R3 → R1
(d) R1 + R2 + R3 → R2   **(e)** R1 + 4R3 → R1

61. (a) $\begin{bmatrix} 2 & 10 & 6 & 18 \\ -2 & 7 & 4 & 11 \\ 27 & 0 & -6 & -15 \\ 6 & 3 & 3 & 2 \end{bmatrix}$ **(b)** $\begin{bmatrix} 3 & 15 & 9 & 27 \\ -2 & 7 & 4 & 11 \\ 5 & 14 & 10 & 27 \\ 6 & 3 & 3 & 2 \end{bmatrix}$ **(c)** $\begin{bmatrix} 1 & 5 & 3 & 9 \\ 0 & 17 & 10 & 29 \\ 9 & 0 & 2 & 5 \\ 6 & 3 & 3 & 2 \end{bmatrix}$ **(d)** $\begin{bmatrix} 1 & 5 & 3 & 9 \\ -2 & 7 & 4 & 11 \\ 0 & -45 & -25 & -76 \\ 6 & 3 & 3 & 2 \end{bmatrix}$

(e) $\begin{bmatrix} 1 & 5 & 3 & 9 \\ 0 & 17 & 10 & 29 \\ 0 & -45 & -25 & -76 \\ 0 & -27 & -15 & -52 \end{bmatrix}$ **63.** 3675 hours assembly time, 1200 hours checking

65. $18,740 on Monday, $18,840 on Wednesday, $19,000 on Friday   **67.**

	A	B$_1$	B$_2$	C	
	1490	0.56	0.85	67.5	I
	2480	0.52	1.70	6	II

69. 705 pounds of regular coffee, 244 pounds of High Mountain coffee, and 41 pounds of chocolate

71. (a) $\begin{bmatrix} 82.0 \\ 84.0 \\ 73.0 \\ 90.0 \\ 95.3 \\ 73.0 \end{bmatrix}$ **(b)** $\begin{bmatrix} 82.6 \\ 83.9 \\ 73.2 \\ 89.8 \\ 94.9 \\ 74.3 \end{bmatrix}$ **73. (a)** $AB =$

	Male	Female
Well	104,750	102,000
Sick	42,000	40,000
Carrier	13,250	13,000

(b) 42,000 sick males
(c) 102,000 well females

This matrix gives the number of males and the number of females who are well, sick, or carriers.

75. (a)

	SEA	LA	DEN	KC	SLC
SEA	0	1	0	0	1
LA	1	0	1	0	0
DEN	0	1	0	1	1
KC	0	0	1	0	1
SLC	1	0	1	1	0

$= A$

(b) The desired matrix is A^2:

$A^2 =$

	SEA	LA	DEN	KC	SLC
SEA	2	0	2	1	0
LA	0	2	0	1	2
DEN	2	0	3	1	1
KC	1	1	1	2	1
SLC	0	2	1	1	3

77. Number of columns of A equals number of rows of B. Number of rows of A equals number of columns of B.

79. BA may not exist. BA may exist and may or may not equal AB.

81. $AB = \begin{bmatrix} 99.9 & 97.2 \\ 71.4 & 65.6 \\ 133.0 & 144.0 \end{bmatrix}$ This gives the total shipping costs by department for the two companies.

83. (a) $AB = \begin{bmatrix} 55,600 & 9,550 \\ 29,550 & 5,310 \\ 34,850 & 5,970 \end{bmatrix}$ This gives the total salaries and benefits by school.

(b) $CD = \begin{bmatrix} 55,600 & 29,550 & 34,850 \\ 9,550 & 5,310 & 5,970 \end{bmatrix}$ **(c)** $FE = [631.50 \quad 1,967.20]$ This gives the total income tax and FICA.

85. $AB = \begin{bmatrix} 1 & 10 \\ -2 & 22 \end{bmatrix}$ **87.** $AB = \begin{bmatrix} 4 & -7 & 7 \\ 23 & 10 & 35 \\ 12 & 6 & 18 \end{bmatrix}$, $BA = \begin{bmatrix} 19 & 9 & 13 \\ -4 & -1 & -12 \\ 14 & 6 & 14 \end{bmatrix}$ **89.** $AB = \begin{bmatrix} 16 & -13 \\ 6 & -8 \end{bmatrix}$

91. (a) Every three years, the age-group populations are equal, but increasing by 50%.

(b) $A^{15}B = \begin{bmatrix} 7593.75 \\ 7593.75 \\ 7593.75 \end{bmatrix}$, $A^{16}B = \begin{bmatrix} 91,125.0 \\ 3796.875 \\ 1898.4375 \end{bmatrix}$, $A^{17}B = \begin{bmatrix} 22,781.25 \\ 45,562.5 \\ 949.21875 \end{bmatrix}$, $A^{18}B = \begin{bmatrix} 11,390.625 \\ 11,390.625 \\ 11,390.625 \end{bmatrix}$

93. The population repeats in a cycle of three years.

Using Your TI-83, Section 2.5

1. $\begin{bmatrix} 2 & 10 \\ 29 & -11 \end{bmatrix}$ **3.** $\begin{bmatrix} 13 & 7 \\ 12 & -10 \end{bmatrix}$ **5.** $A^2 = \begin{bmatrix} 2 & 3 & 2 \\ 6 & 7 & 6 \\ 8 & 6 & 8 \end{bmatrix}$ $A^3 = \begin{bmatrix} 10 & 9 & 10 \\ 26 & 25 & 26 \\ 28 & 30 & 28 \end{bmatrix}$ $A^3 = \begin{bmatrix} 38 & 39 & 38 \\ 102 & 103 & 102 \\ 116 & 114 & 116 \end{bmatrix}$

Using EXCEL, Section 2.5

1. $\begin{bmatrix} 12 & 15 \\ 22 & 29 \end{bmatrix}$ **3.** $\begin{bmatrix} -3 & 7 & 17 \\ 8 & 7 & 13 \end{bmatrix}$

Section 2.6

1. $25^{-1} = 0.04$, $\left(\frac{2}{3}\right)^{-1} = \frac{3}{2}$, $(-5)^{-1} = -\frac{1}{5}$, $(0.75)^{-1} = \frac{4}{3}$, $11^{-1} = \frac{1}{11}$ **3.** Yes **5.** No **7.** Yes

9. $\begin{bmatrix} -5 & 2 \\ 3 & -1 \end{bmatrix}$ **11.** $\begin{bmatrix} 3 & -2 \\ -4 & 3 \end{bmatrix}$ **13.** $\begin{bmatrix} -\frac{5}{4} & \frac{9}{4} & \frac{3}{4} \\ 3 & -6 & -1 \\ -\frac{3}{4} & \frac{7}{4} & \frac{1}{4} \end{bmatrix}$ **15.** $\begin{bmatrix} -\frac{7}{5} & -\frac{8}{5} & \frac{13}{5} \\ \frac{3}{10} & \frac{1}{5} & -\frac{1}{5} \\ \frac{1}{10} & \frac{2}{5} & -\frac{2}{5} \end{bmatrix}$ **17.** No inverse

19. No inverse **21.** No inverse **23.** $\begin{bmatrix} \frac{3}{2} & -\frac{1}{2} \\ -2 & 1 \end{bmatrix}$ **25.** $\begin{bmatrix} -\frac{3}{7} & \frac{5}{7} & -\frac{11}{7} \\ \frac{2}{7} & -\frac{1}{7} & -\frac{2}{7} \\ \frac{2}{7} & -\frac{1}{7} & \frac{5}{7} \end{bmatrix}$

27. (a) $\begin{bmatrix} 3 & 4 & -5 & | & 4 \\ 2 & -1 & 3 & | & -1 \\ 1 & 1 & -1 & | & 2 \end{bmatrix}$ **(b)** $\begin{bmatrix} 3 & 4 & -5 \\ 2 & -1 & 3 \\ 1 & 1 & -1 \end{bmatrix}$ **(c)** $\begin{bmatrix} 3 & 4 & -5 \\ 2 & -1 & 3 \\ 1 & 1 & -1 \end{bmatrix} \begin{bmatrix} x_1 \\ x_2 \\ x_3 \end{bmatrix} = \begin{bmatrix} 4 \\ -1 \\ 2 \end{bmatrix}$

29. (a) $\begin{bmatrix} 4 & 5 & | & 2 \\ 3 & -2 & | & 7 \end{bmatrix}$ **(b)** $\begin{bmatrix} 4 & 5 \\ 3 & -2 \end{bmatrix}$ **(c)** $\begin{bmatrix} 4 & 5 \\ 3 & -2 \end{bmatrix} \begin{bmatrix} x \\ y \end{bmatrix} = \begin{bmatrix} 2 \\ 7 \end{bmatrix}$ **31.** $\begin{bmatrix} 1 & 3 \\ 2 & -1 \end{bmatrix} \begin{bmatrix} x_1 \\ x_2 \end{bmatrix} = \begin{bmatrix} 5 \\ 6 \end{bmatrix}$

33. $\begin{bmatrix} 1 & 2 & -3 & 4 \\ 1 & 1 & 0 & 1 \\ 3 & 2 & 1 & 2 \end{bmatrix} \begin{bmatrix} x_1 \\ x_2 \\ x_3 \\ x_4 \end{bmatrix} = \begin{bmatrix} 0 \\ 5 \\ 4 \end{bmatrix}$ **35.** No inverse **37.** $\begin{bmatrix} \frac{1}{9} & \frac{1}{3} & \frac{5}{9} \\ \frac{1}{3} & 0 & -\frac{1}{3} \\ -\frac{2}{9} & \frac{1}{3} & -\frac{1}{9} \end{bmatrix} \begin{bmatrix} 2 \\ 0 \\ 1 \end{bmatrix} = \begin{bmatrix} \frac{7}{9} \\ \frac{1}{3} \\ -\frac{5}{9} \end{bmatrix}$

39. $\begin{bmatrix} -\frac{1}{4} & \frac{3}{4} & \frac{5}{12} & -\frac{1}{12} \\ 0 & -1 & -\frac{1}{3} & \frac{2}{3} \\ \frac{1}{4} & \frac{1}{4} & \frac{1}{4} & -\frac{1}{4} \\ \frac{3}{4} & -\frac{1}{4} & -\frac{7}{12} & -\frac{1}{12} \end{bmatrix}\begin{bmatrix} 4 \\ 6 \\ 3 \\ 9 \end{bmatrix} = \begin{bmatrix} 4 \\ -1 \\ 1 \\ -1 \end{bmatrix}$

41. (a) $\begin{bmatrix} -16 & 4 & -13 \\ 5 & -1 & 4 \\ -12 & 3 & -10 \end{bmatrix}\begin{bmatrix} 1 \\ 5 \\ 2 \end{bmatrix} = \begin{bmatrix} -22 \\ 8 \\ -17 \end{bmatrix}$ **(b)** $\begin{bmatrix} -16 & 4 & -13 \\ 5 & -1 & 4 \\ -12 & 3 & -10 \end{bmatrix}\begin{bmatrix} -1 \\ 3 \\ 1 \end{bmatrix} = \begin{bmatrix} 15 \\ -4 \\ 11 \end{bmatrix}$ **(c)** $\begin{bmatrix} -16 & 4 & -13 \\ 5 & -1 & 4 \\ -12 & 3 & -10 \end{bmatrix}\begin{bmatrix} 0 \\ 1 \\ 2 \end{bmatrix} = \begin{bmatrix} -22 \\ 7 \\ -17 \end{bmatrix}$

43. (a) $\begin{bmatrix} -5 & 2 \\ 3 & -1 \end{bmatrix}\begin{bmatrix} 3 \\ 8 \end{bmatrix} = \begin{bmatrix} 1 \\ 1 \end{bmatrix}$ **(b)** $\begin{bmatrix} -5 & 2 \\ 3 & -1 \end{bmatrix}\begin{bmatrix} 4 \\ 9 \end{bmatrix} = \begin{bmatrix} -2 \\ 3 \end{bmatrix}$ **(c)** $\begin{bmatrix} -5 & 2 \\ 3 & -1 \end{bmatrix}\begin{bmatrix} 3 \\ 7 \end{bmatrix} = \begin{bmatrix} -1 \\ 2 \end{bmatrix}$

45. (a) Vitamin C intake $= 32x + 24y$ and Vitamin A intake $= 900x + 425y$. So, $\begin{bmatrix} 32 & 24 \\ 900 & 425 \end{bmatrix}\begin{bmatrix} x \\ y \end{bmatrix} = \begin{bmatrix} b_1 \\ b_2 \end{bmatrix}$ where b_1 = vitamin C intake and b_2 = vitamin A intake. **(b)** 162.4 mg of C, 3942.5 IU of A **(c)** 120 mg of C, 2625 IU of A **(d)** 1.25 units of A, 2.8 units of B **(e)** 2.2 units of A, 1.4 units of B

47. (a) 340 children, 560 adults **(b)** 435 children, 565 adults **(c)** 110 children, 640 adults

49. (a)

$A =$	Early R.	After D.	Deluxe	
	0.80	0.75	0.50	Regular
	0.20	0.20	0.40	High Mt.
	0	0.05	0.10	Chocolate

(b) $A^{-1} = \begin{bmatrix} 0 & 5 & -20 \\ 2 & -8 & 22 \\ -1 & 4 & -1 \end{bmatrix}$

(c) The amount of each blend that could be produced is

Day	1	2	3
Early Riser	350	420	320
After Dinner	200	460	300
Deluxe	150	170	180

51. (a) $\begin{bmatrix} \frac{1}{4} & 0 \\ 0 & \frac{1}{5} \end{bmatrix}$ **(b)** $\begin{bmatrix} \frac{1}{a} & 0 \\ 0 & \frac{1}{b} \end{bmatrix}$ **(c)** $\begin{bmatrix} \frac{1}{a} & 0 & 0 \\ 0 & \frac{1}{b} & 0 \\ 0 & 0 & \frac{1}{c} \end{bmatrix}$

55. $\begin{bmatrix} 1 & -1 \\ -2 & 3 \end{bmatrix}$ **57.** $\begin{bmatrix} 0.50 & 0 & 0 \\ 0 & 0.25 & 0 \\ 0 & 0 & 0.20 \end{bmatrix}$ **59.** $\begin{bmatrix} 1 & -1 & 0 \\ -2 & 5 & 1 \\ -1 & 3 & 1 \end{bmatrix}$ **61.** $\begin{bmatrix} 0.4 & 0.8 & -0.2 & -0.8 \\ -0.4 & 0.2 & 0.2 & -0.2 \\ 0 & -0.1429 & 0.1429 & 0 \\ -0.1 & 0.4857 & 0.0857 & 0.7 \end{bmatrix}$

63. (a) $AB = \begin{bmatrix} 1 & 0 & 1 \\ 0 & 0 & 1 \\ 1 & -1 & 1 \end{bmatrix}, A^{-1} = \begin{bmatrix} 1 & -1 & 0 \\ 0 & 1 & -1 \\ 0 & 0 & 1 \end{bmatrix}, B^{-1} = \begin{bmatrix} 1 & 0 & 0 \\ 1 & 1 & 0 \\ 0 & 1 & 1 \end{bmatrix}, (AB)^{-1} = \begin{bmatrix} 1 & -1 & 0 \\ 1 & 0 & -1 \\ 0 & 1 & 0 \end{bmatrix},$

$A^{-1}B^{-1} = \begin{bmatrix} 0 & -1 & 0 \\ 1 & 0 & -1 \\ 0 & 1 & 1 \end{bmatrix}, B^{-1}A^{-1} = \begin{bmatrix} 1 & -1 & 0 \\ 1 & 0 & -1 \\ 0 & 1 & 0 \end{bmatrix}$ **(b)** $(AB)^{-1} = B^{-1}A^{-1}$ **65.** $(AB)^{-1} = B^{-1}A^{-1}$

67. (a) $A^{-1} = \begin{bmatrix} 2 & -1.5 & 2.5 \\ 0.5 & -0.25 & 0.75 \\ 0 & -0.25 & 0.25 \end{bmatrix}$

The reduced echelon form of A is $\begin{bmatrix} 1 & 0 & 0 \\ 0 & 1 & 0 \\ 0 & 0 & 1 \end{bmatrix}$.

(b) The graphing calculator gives an error when trying to compute A^{-1}, indicating that no inverse exists. The reduced echelon form of A is $\begin{bmatrix} 1 & 0 & 2 \\ 0 & 1 & 0 \\ 0 & 0 & 0 \end{bmatrix}$.

69. When the reduced form of A is the identity matrix, A^{-1} exists. When the reduced form of A contains a row of zeros, A^{-1} does not exist.

71. $\begin{bmatrix} 107 \\ 12 \\ 44 \end{bmatrix}$

Using Your TI-83, Section 2.6

1. $\begin{bmatrix} 0.4 & 1.4 & -1 \\ -0.8 & -0.8 & 1 \\ -0.2 & -1.2 & 1 \end{bmatrix}$ **3.** No inverse

Using EXCEL, Section 2.6

1. $\begin{bmatrix} -17 & -24 & 15 \\ -26 & -37 & 23 \\ 41 & 58 & -36 \end{bmatrix}$ **3.** $\begin{bmatrix} -85 & -120 & 75 \\ -26 & -37 & 23 \\ 32.8 & 46.4 & -28.8 \end{bmatrix}$ **5.** A has no inverse.

Section 2.7

1. $\begin{bmatrix} 2.16 \\ 4.8 \end{bmatrix}$ **3.** $\begin{bmatrix} 3.06 \\ 2.90 \\ 1.40 \end{bmatrix}$ **5.** $\begin{bmatrix} 1.4 & 0.6 \\ 0.4 & 1.6 \end{bmatrix}$ **7.** $\begin{bmatrix} 21.33 \\ 15.17 \end{bmatrix}$

9. The output levels required to meet the demands $\begin{bmatrix} 20 \\ 28 \end{bmatrix}$ and $\begin{bmatrix} 15 \\ 11 \end{bmatrix}$ are $\begin{bmatrix} 63 \\ 76 \end{bmatrix}$ and $\begin{bmatrix} 37.25 \\ 37.00 \end{bmatrix}$, respectively.

11. $I - A = \begin{bmatrix} 0.8 & -0.2 & -0.2 \\ -0.1 & 0.4 & -0.2 \\ -0.1 & -0.1 & 0.6 \end{bmatrix}$ and $(I - A)^{-1} = \begin{bmatrix} \frac{22}{15} & \frac{14}{15} & \frac{4}{5} \\ \frac{8}{15} & \frac{46}{15} & \frac{6}{5} \\ \frac{1}{3} & \frac{2}{3} & 2 \end{bmatrix}$.

The values of X required to meet the demands $\begin{bmatrix} 30 \\ 24 \\ 42 \end{bmatrix}$ and $\begin{bmatrix} 60 \\ 45 \\ 75 \end{bmatrix}$ are $\begin{bmatrix} 100 \\ 140 \\ 110 \end{bmatrix}$ and $\begin{bmatrix} 190 \\ 260 \\ 200 \end{bmatrix}$, respectively.

13. (a) $AX = \begin{bmatrix} 23.5 \\ 28.5 \end{bmatrix}$ **(b)** $X - AX = \begin{bmatrix} 16.5 \\ 21.5 \end{bmatrix}$ **15. (a)** $AX = \begin{bmatrix} 21.6 \\ 64.8 \\ 36.0 \end{bmatrix}$ **(b)** $X - AX = \begin{bmatrix} 14.4 \\ 7.2 \\ 0 \end{bmatrix}$

17. (a) $\begin{array}{c} \\ A = \begin{array}{c} A \\ N \end{array} \end{array} \begin{array}{cc} A & N \\ \begin{bmatrix} 0.1 & 0.3 \\ 0.6 & 0.4 \end{bmatrix} \end{array}$ **(b)** Agriculture internal consumption = $1.91 million, leaving $1.59 million for export. Nonagriculture consumption = $4.18 million, leaving $1.02 million for export.
(c) To export $2 million of each kind of product requires production of $5 million of agriculture and $8.3 million of nonagriculture products.   **(d)** For $2 million of agriculture and $3 million of nonagriculture exports, $5.83 million of agriculture and $10.83 million of nonagriculture products are required.

19. (a) $\begin{array}{c} A = \begin{array}{c} P \\ E \end{array} \end{array} \begin{array}{cc} P & E \\ \begin{bmatrix} 0.10 & 0.40 \\ 0.20 & 0.20 \end{bmatrix} \end{array}$ **(b)** $15.3 million worth of plastics and $11.4 million worth of electronics are used internally.
(c) The corporation must produce $72.5 million worth of plastics and $73.125 million worth of electronics to have $36 million worth of plastics and $44 million worth of electronics available for external sales.

21. (a) $A = \begin{array}{c} C \\ M \\ US \end{array} \begin{array}{ccc} C & M & US \\ \begin{bmatrix} 0.2 & 0.1 & 0.3 \\ 0.2 & 0.4 & 0 \\ 0.4 & 0 & 0.3 \end{bmatrix} \end{array}$ **(b)** Canada uses $8.3 million worth of components; Mexico, $9.2 million; and United States, $8.5 million.
(c) Canada must produce $63.12 million worth of vehicles; Mexico, $71.04 million; and Unites States $64.64 million.
23. A negative entry indicates that a negative cost is associated with producing a product. An entry greater than 1 indicates it costs more than $1.00 to produce $1.00 worth of a product.   **25.** The corporation must produce $7,000,000 of service and $9,000,000 of retail in order to have $200,000 of service and $100,000 of retail available to the consumer.

27. (a) The output that produces no grain and equal values of lumber and energy is any scalar multiple of $\begin{bmatrix} 6 \\ 25 \\ 10 \end{bmatrix}$.

(b) The output that yields no lumber and energy twice the value of grain is any scalar multiple of $\begin{bmatrix} 8 \\ 20 \\ 10 \end{bmatrix}$.

29. (a) When $X = \begin{bmatrix} 10 \\ 30 \\ 20 \end{bmatrix}$, $\begin{bmatrix} 13 \\ 27 \\ 20 \end{bmatrix}$ is the amount consumed internally, and $\begin{bmatrix} -3 \\ 3 \\ 0 \end{bmatrix}$ is the amount left for consumers.

More of the first industry is used internally than is produced. When $X = \begin{bmatrix} 20 \\ 40 \\ 30 \end{bmatrix}$, $\begin{bmatrix} 20 \\ 40 \\ 30 \end{bmatrix}$ is the amount consumed

internally, and $\begin{bmatrix} 0 \\ 0 \\ 0 \end{bmatrix}$ is the amount left for consumers. All production is used internally.

(b) When $X = \begin{bmatrix} 30 \\ 30 \\ 40 \end{bmatrix}$, $\begin{bmatrix} 25 \\ 45 \\ 34 \end{bmatrix}$ is the amount consumed internally, and $\begin{bmatrix} 5 \\ -15 \\ 6 \end{bmatrix}$ is the amount left for consumers.

When $X = \begin{bmatrix} 10 \\ 20 \\ 20 \end{bmatrix}$, $\begin{bmatrix} 12 \\ 23 \\ 17 \end{bmatrix}$ is the amount consumed internally, and $\begin{bmatrix} -2 \\ -3 \\ 3 \end{bmatrix}$ is the amount left for consumers.

31. $\begin{bmatrix} 1.254 & 0.226 & 0.167 \\ 0.743 & 1.593 & 0.700 \\ 1.087 & 0.883 & 1.604 \end{bmatrix}$

33. $\begin{bmatrix} 1,333,000 \\ 937,560 \\ 764,890 \\ 687,110 \end{bmatrix}$, production for year 1 $\begin{bmatrix} 1,432,740 \\ 996,770 \\ 753,510 \\ 728,430 \end{bmatrix}$, production for year 2 $\begin{bmatrix} 1,401,910 \\ 1,034,920 \\ 747,140 \\ 754,830 \end{bmatrix}$, production of year 3

Section 2.8

1. Regression line: $y = 1.20x + 1.60$

3. Regression line: $y = -0.90x + 11.86$

5. (a) $y = 7.51x + 23.28$ **(b)** 83.36% **(c)** When $x = 10.2$, in the year 2007   **7. (a)** $y = 0.23x + 66.79$
(b) The life expectancy of females born in 2010 is about 83 years.   **(c)** About 144 years after 1940, namely, in 2084, females born that year are estimated to have a life expectancy of 100 years.

9. (a)

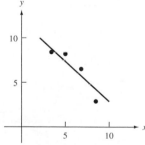

(b) Regression line:
$y = 8.65x + 16.87$

(c)

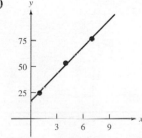

The regression line
fits the data well.

(d) For October, $x = 10$ and $y = 103.4$ estimates the average temperature to be over 103 degrees.   **(e)** The estimate is not realistic because the average temperature does not rise in a straight line. It peaks in the summer and then declines.   **11. (a)** Israel: $y = -0.31x + 26.92$,   U.S.: $y = 0.026x + 15.53$ **(b)** The birth rates will be the same, about 14.5, during the year 2015.   **13. (a)** $f(x) = 26.66x + 161.05$ **(b)** $f(35) = 1094.15$, $f(55) = 1627.35$
15. (a) $y = 0.085x + 1.09$ **(b)** 4.07   **(c)** By the year 2080   **17. (a)** $f(x) = 701.3x + 7499.9$
(b) $f(28) = 27{,}136.3$ **(c)** In the year $1980 + 24.95$, by the year 2005   **19. (a)** $f(x) = 1697.3x + 38015.2$
(b) About \$32,900 for 1985   **(c)** About \$58,400 for 2000   **(d)** In the year $1988 + 21.8$, by the year 2009

Using Your TI-83, Section 2.8

1. $y = 1.64x - 2.71$ **3.** $y = 0.50x + 4.13$

Using EXCEL, Section 2.8

1. $y = 1.2x + 0.8$ **3.** $y = 0.99x - 24.14$

Review Exercises, Chapter 2

1. $\left(\frac{1}{4}, \frac{17}{8}\right)$ **3.** $(6, -4)$ **5.** $\left(2, 0, \frac{1}{3}\right)$ **7.** $(-6, 2, -5)$ **9.** No solution   **11.** $x = -z$, $y = 1 - z$
13. $x_1 = -56 + 29x_4$, $x_2 = 23 - 12x_4$, $x_3 = -13 + 8x_4$ **15.** $x_1 = 3 + \left(\frac{1}{2}\right)x_3$, $x_2 = \frac{2}{3} + \left(\frac{4}{3}\right)x_3$ **17.** $\frac{3}{4}$

19. $\begin{bmatrix} -3 & -2 \\ 6 & 7 \end{bmatrix}$ **21.** $\begin{bmatrix} 11 & -3 \\ 7 & -1 \\ 3 & 0 \end{bmatrix}$ **23.** $[3]$ **25.** Cannot multiply them.   **27.** $\begin{bmatrix} -\frac{5}{2} & 3 \\ \frac{7}{2} & -4 \end{bmatrix}$

29. $\begin{bmatrix} -2 & -6 & 15 \\ 0 & -1 & 2 \\ 1 & 2 & -5 \end{bmatrix}$ **31.** $\begin{bmatrix} 6 & 4 & -5 & 10 \\ 3 & -2 & 0 & 12 \\ 1 & 1 & -4 & -2 \end{bmatrix}$ **33.** $\begin{bmatrix} 1 & 0 & -\frac{3}{5} & 0 \\ 0 & 1 & \frac{9}{5} & 0 \\ 0 & 0 & 0 & 1 \end{bmatrix}$

35. 5 free throws, 11 two-pointers, and 3 three-pointers   **37.** \$20,000 in bonds, \$30,000 in stocks   **39.** 120 of High-Tech, 68 of Big Burger   **41.** 630 at plant A and 270 at plant B   **43.** 7400   **45.** $y = 2.3x + 4.1$

CHAPTER 3

Section 3.1

1. $(1, -1)$, $(3, 1)$, and $(2, 3)$ are solutions.

3.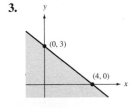

5.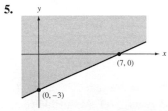

7.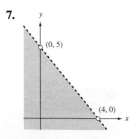

9.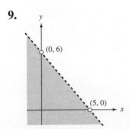

11.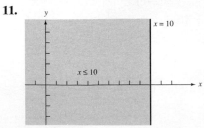

13.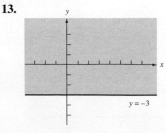

15.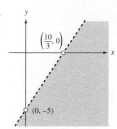

17.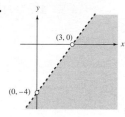

19.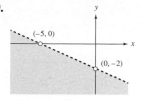

21. x = number of air conditioners, y = number of fans.
$3.2x + 1.8y \leq 144$

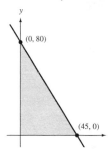

23. (a) $4x + 6y \geq 500$. x = number of members, y = number of pledges.

(b)

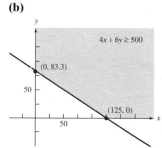

25. x = number of TV ads, y = number of newspaper ads. x and y must satisfy $900x + 830y \leq 75,000$. The graph is

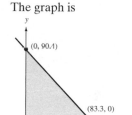

27. Let x = number of acres of strawberries
y = number of acres of tomatoes
$9x + 6y \leq 750$

31. Let x_1 = number of Ham and Egg
x_2 = number of Roast Beef
$360x_1 + 300x_2 \leq 2000$ (Calories)
$14x_1 + 5x_2 \leq 65$ (Total Fat)
$4x_1 + 2x_2 \leq 20$ (Saturated Fat)

29. Let x = number of days at Glen Echo
y = number of days at Speegleville Road
(a) $200x + 300y \geq 2400$ (paperback)
(b) $300x + 200y \geq 2100$ (hardback)
33. Let x represent the number of servings of milk.
Let y represent the number of servings of bread:
$12x + 15y \geq 50$

Section 3.2

1.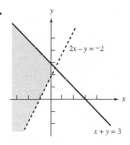

3.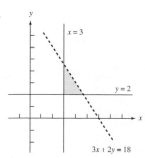

5.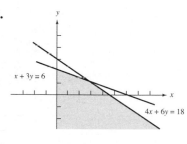

7.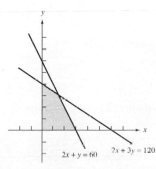

9.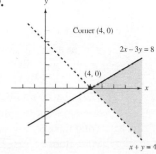

11.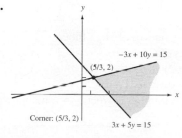

13.

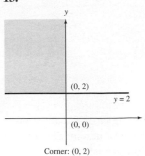

Corner: $(0, 2)$

15.

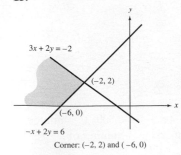

Corner: $(-2, 2)$ and $(-6, 0)$

17. No feasible region

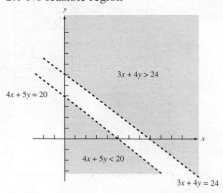

19. No feasible region

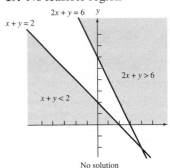

No solution

21.

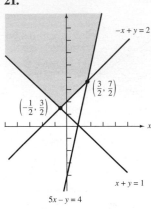

Corners $\left(\frac{3}{2}, \frac{7}{2}\right)$ and $\left(-\frac{1}{2}, \frac{3}{2}\right)$

23. Corners: $(-4, 1), (1, 6), (8, 0),$ and $(2, -4)$

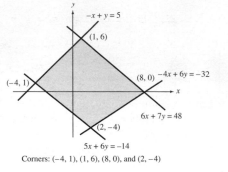

Corners: $(-4, 1), (1, 6), (8, 0),$ and $(2, -4)$

25. Bounded feasible region

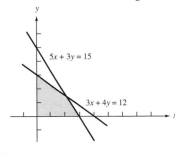

27. Unbounded feasible region

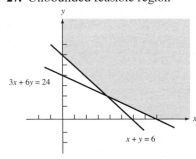

29. Corners $\left(\frac{1}{5}, \frac{12}{5}\right), \left(\frac{7}{4}, -\frac{9}{4}\right), \left(\frac{1}{2}, -\frac{7}{2}\right), \left(-\frac{5}{3}, -\frac{4}{3}\right)$

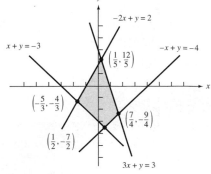

31. Let x = ounces of High Fiber,
y = ounces of Corn Bits
$0.25x + 0.02y \geq 0.40$
$0.04x + 0.10y \geq 0.25$

35. Let x = number correct, y = number wrong
$x + y \geq 60$
$4x - y \geq 200$

33. Let x = number of student tickets,
y = number of adult tickets
$x + y \geq 500$
$6x + 12y \geq 5400$

37. Let x = number of balcony tickets, y = number of main-floor tickets
Total tickets $\qquad x + y \geq 3000$
Main floor $\qquad\qquad\quad y \geq 1200$
Sales $\qquad\qquad 15x + 25y \geq 60,000$

39. Corner $(1.85, 2.69)$ **41.** Corners: $(0, 0)$ $(0, 5)$ $(1.43, 4.29)$ $(2.86, 2.57)$ $(4, 0)$
43. Corners: $(0, 9.5)$ $(1.33, 5.55)$ $(7.5, 0)$ **45.** Corners: $(0, 0)$ $(0, 5.47)$ $(1.99, 4.55)$ $(4.49, 1.22)$ $(5.00, 0)$

Using Your TI-83, Section 3.2

1. Corners: $(0, 0), (8, 2), (3, 6), (0, 7), (9.33, 0)$ **3.** Corners: $(0, 0), (0, 11), (10, 9), (21, 4), (27, 0)$

Using EXCEL, Section 3.2

1. Intersection at $(7, 5)$ **3.** The first two intersect at $(10, 12)$ The first and third intersect at $(14, 4)$ The second and third intersect at $(5, 10)$ **5.** Intersect at $(2.3, 6.7)$

Section 3.3

1. Let x = number of style A,
y = number of style B
$x + 2y \leq 110$ (Labor restriction)
$x + y \leq 80$ (Space restriction)
$x \geq 0, y \geq 0$
Maximize $z = 50x + 40y$

5. At $(2.4, 17.2), z = 432$
At $(21.6, 15.4), z = 542.4$
At $(25.2, 4.7), z = 314.4$
Maximum $z = 542.4$ at $(21.6, 15.4)$
Minimum $z = 314.4$ at $(25.2, 4.7)$

3. At $(0, 0), z = 0$
At $(0, 8), z = 120$
At $(7, 6), z = 132$
At $(10, 0), z = 60$
Maximum $z = 132$ at $(7, 6)$
Minimum $z = 0$ at $(0, 0)$

7. At $(0, 9), z = 108$
At $(4, 3), z = 116$
At $(5, 0), z = 100$
At $(0, 0), z = 0$
So, maximum z is 116 at $(4, 3)$.

9. 66 at $(4, 15)$ **11.** 32 at $(16, 0)$ **13.** 343.5 at $(16.5, 15)$ **15.** 630 at $(18, 9)$ **17.** $z = 28,200$ at $(75, 30)$ **19.** 52 at $(8, 4)$ **21.** 66 at $(6, 12)$ **23.** Maximum z is 380 at $(10, 6)$. Minimum z is 40 at $(2, 0)$.
25. Maximum $z = 245$ at $(40, 7.5)$. Minimum $z = 90$ at $(0, 15)$.   **27.** Maximum z is 90 at $(3, 5), (0, 10)$, and all points on the line segment between.   **29.** Minimum $z = 180$ at $(4, 24), \left(\frac{180}{11}, \frac{60}{11}\right)$ and all points on the line segment joining them.   **31.** Minimum $z = 500$ at $(40, 30)$ and $(60, 20)$ and all points on the line segment joining these points.   **33.** Unbounded feasible region, no maximum   **35.** No feasible solution   **37.** No feasible region, no maximum

39. Let x = number of standard VCR
y = number of deluxe VCR
Maximize $z = 39x + 26y$, subject to
$8x + 9y \leq 2200$
$115x + 136y \leq 18,000$
$x \geq 35, y \geq 0$

41. Let x = number shipped to A
y = number shipped to B
Minimize $z = 13x + 11y$, subject to
$x + y \geq 250$
$x \leq 110$
$y \leq 147$
$x \geq 0, y \geq 0$

43. Let x = number of cartons of regular
y = number of cartons of diet
Maximize $z = 0.15x + 0.17y$, subject to
$x + 1.20y \leq 5400$
$x + y \leq 5000$
$x \geq 0, y \geq 0$

45. Let x = number of desk lamps
y = number of floor lamps
Maximize $z = 2.65x + 3.15y$, subject to
$0.8x + 1.0y \leq 1200$
$4x + 3y \leq 4200$
$x \geq 0, y \geq 0$
Maximum profit is \$3828.75 for 375 desk lamps and 900 floor lamps.

47. Let x = number of sandwiches and
y = number bowls of soup
Maximize $z = 9x + 5y$, subject to
$400x + 200y \leq 2000$
$12x + 8.5y \leq 65$
$x \geq 0, y \geq 0$
Maximum fiber is 46 grams with
4 sandwiches and 2 bowls of soup.

49. Let x = amount of food I
y = amount of food II
Minimize $z = 0.03x + 0.04y$, subject to
$0.4x + 0.6y \geq 10$
$0.5x + 0.2y \geq 7.5$
$0.06x + 0.04y \geq 1.2$
$x \geq 0, y \geq 0$
Minimum cost is \$0.72 for 16 grams of food I and 6 grams of food II.

51. Let x = number of standard gears
    y = number of heavy duty gears
Minimize $z = 15x + 22y$, subject to
$8x + 10y \geq 12{,}000$
$3x + 10y \geq 8400$
$x \geq 0, y \geq 0$
Minimum cost = $24,528 for 720
standard and 624 heavy duty gears.

55. (a) Maximum profit is $790 with 3000
cartons of regular and 2000 cartons of diet.

(b)   **(i)** The optimal solution changes to
the corner$(5000, 0)$ and the maximum
profit increases to $1000.

    (ii) The optimal solution changes to
the corner $(0, 4500)$ and the maximum
profit increases to $855.

    (iii) The optimal solution remains
at the corner $(3000, 2000)$ but
the maximum profit decreases to $770.

59. Let x = number of days at Glen Echo plant
    y = number of days at Speegleville plant
Minimize $z = x + y$, subject to
$200x + 300y \geq 2400$
$300x + 200y \geq 2100$
Minimum number of days is nine with three at
Glen Echo and six at Speegleville.

63. The minimum cost is $235 and it occurs when
supplier A ships 10 to Emporia and none to Ardmore.
Supplier B ships 15 to Emporia and 20 to Ardmore.

67. Corner B **69.** Corner C **71. (a)** Corner B
touched as the objective function line M moves upward.
75. The maximum z is 31.68 at $(0, 6.6)$

53. Let x = number of SE
    y = number of LE
Minimize $z = 2700x + 2400y$, subject to
$16{,}000x + 20{,}000y \leq 160{,}000$
    $x + y \geq 9$
    $x \geq 0, y \geq 0$
Minimum operating costs = $23,100 per year for five SE and
four LE.

57. Let x = number of square feet of type A
    y = number of square feet of type B
(a) Minimize $z = x + 0.25y$, subject to
    $x + y \geq 4000$
$0.80x + 1.20y \leq 4500$
    $x \geq 0, y \geq 0$
Minimum conductance = 1562.50 BTU with 750 square
feet of type A and 3250 square feet of type B.
(b) Minimize $z = 0.80x + 1.20y$, subject to
    $x + y \geq 4000$
$x + 0.25y \leq 2200$
    $x \geq 0, y \geq 0$
Minimum cost is $4160 with 1600 square feet of type A
and 2400 square feet of type B.

61. The minimum cost is $13,025 and occurs when
supplier A ships 10 to Raleigh and 190 to Greensboro.
Supplier B ships 155 to Raleigh and none to Greensboro.

65. In a maximization problem, the profit line is moved
away from the origin until a corner point is the only point
it touches in the feasible region. To find the optimal
integer solution move the profit line outward until it
touches the last point in the feasible region with integer
coordinates. In this case that is the point $(7, 0)$; so the
optimal profit occurs when $x = 7$ and $y = 0$. If neither x
nor y can be zero, the last point is $(6, 1)$.

(b) Corner B is still the last corner of the feasible region
73. The maximum z is 60.74 at $(0, 8.32)$

Section 3.4

1. Let x_1 = number of acres of onions
    x_2 = number of acres of carrots
    x_3 = number of acres of lettuce
Maximize profit $z = 65x_1 + 70x_2 + 50x_3$, subject to
    $x_1 + x_2 + x_3 \leq 75$ (Acreage available)
$250x_1 + 300x_2 + 325x_3 \leq 225{,}000$ (Production costs)
    $x_1 \geq 0, x_2 \geq 0, x_3 \geq 0$

3. Let x_1 = number of chests
    x_2 = number of desks
    x_3 = number of silverware boxes
Maximize profit $z = 180x_1 + 300x_2 + 45x_3$, subject to
    $270x_1 + 310x_2 + 90x_3 \leq 5000$ (Cost limitation)
    $7x_1 + 18x_2 + 1.5x_3 \leq 1500$ (Space limitations)
        $x_1 + x_2 + x_3 \leq 200$ (Maximum number
    $x_1 \geq 0, x_2 \geq 0, x_3 \geq 0$ allowed)

5. Let x_1 = number pounds of Lite
x_2 = number pounds of Trim
x_3 = number pounds of Regular
x_4 = number pounds of Health Fare
Maximize profit $z = 0.25x_1 + 0.25x_2 + 0.27x_3 + 0.32x_4$, subject to

$0.75x_1 + 0.50x_2 + 0.80x_3 + 0.15x_4 \leq 2400$ (Wheat)
$0.20x_1 + 0.25x_2 + 0.20x_3 + 0.50x_4 \leq 1400$ (Oats)
$0.05x_1 + 0.20x_2 + 0.25x_4 \leq 700$ (Raisins)
$0.05x_2 + 0.10x_4 \leq 250$ (Nuts)
$x_1 \geq 0, x_2 \geq 0, x_3 \geq 0, x_4 \geq 0$

9. The following chart summarizes the given information.

Plantation

	P–1	P–2	P–3	
Plant A	Cost = 50 No. = x_1	Cost = 65 No. = x_2	Cost = 58 No. = x_3	Processes at least 540
Plant B	Cost = 40 No. = x_4	Cost = 55 No. = x_5	Cost = 69 No. = x_6	Processes at least 450
	Produces 250	Produces 275	Produces 310	

11. Let x_1 = number of days Red Mountain operates
x_2 = number of days Cahaba operates
x_3 = number of days Clear Creek operates

13. Let x_1 = number of Mini Packets
x_2 = number of Mid Packets
x_3 = number of Maxi Packets
Minimize costs $z = 19x_1 + 30x_2 + 45x_3$, subject to
$8x_1 + 7x_2 + 6x_3 \geq 260$ (Number of "A" tickets)
$2x_1 + 7x_2 + 14x_3 \geq 175$ (Number of "B" tickets)
$x_1 \geq 0, x_2 \geq 0, x_3 \geq 0$

17. Let x_1 = percentage invested in A bonds
x_2 = percentage invested in AA bonds
x_3 = percentage invested in AAA bonds
Maximize return $z = 0.072x_1 + 0.068x_2 + 0.065x_3$, subject to
$x_1 + x_2 + x_3 \leq 100$ (Investment Total)
$x_1 + x_2 \leq 65$ (Invested in A and AA)
$x_2 + x_3 \geq 50$ (Invested in AA and AAA)
$x_1 \geq 0, x_2 \geq 0, x_3 > 0$

7. The following figure shows the possible two-day and three-day periods with the periods labeled $x_1, x_2, \ldots, x_5$, which also represent the number of computers needed for that time period.

	2-day rental		3-day rental	
Day 1	x_1			
Day 2		x_2	x_4	
Day 3			x_3	x_5
Day 4				

Minimize rent $z = 80x_1 + 80x_2 + 80x_3 + 100x_4 + 100x_5$, subject to

$x_1 + x_4 \geq 12$ (Number needed first day)
$x_1 + x_2 + x_4 + x_5 \geq 15$ (Number needed second day)
$x_2 + x_3 + x_4 + x_5 \geq 18$ (Number needed third day)
$x_3 + x_5 \geq 20$ (Number needed fourth day)
$x_1 \geq 0, x_2 \geq 0, x_3 \geq 0, x_4 \geq 0, x_5 \geq 0$

The linear programming problem is the following:
Minimize shipping costs
$z = 50x_1 + 65x_2 + 58x_3 + 40x_4 + 55x_5 + 69x_6$, subject to
$x_1 + x_2 + x_3 \geq 540$ (Plant A)
$x_4 + x_5 + x_6 \geq 450$ (Plant B)
$x_1 + x_4 \geq 250$ (P–1 production)
$x_2 + x_5 \geq 275$ (P–2 production)
$x_3 + x_6 \geq 310$ (P–3 production)
$x_1 \geq 0, x_2 \geq 0, x_3 \geq 0, x_4 \geq 0, x_5 \geq 0, x_6 \geq 0$

Minimize cost $z = 8{,}000x_1 + 14{,}000x_2 + 12{,}000x_3$, subject to
$105x_1 + 295x_2 + 270x_3 \geq 1800$ (Amount of low-grade)
$90x_1 + 200x_2 + 85x_3 \geq 1350$ (Amount of high-grade)
$x_1 \geq 0, x_2 \geq 0, x_3 > 0$

15. Let x_1 = number of servings of milk
x_2 = number of servings of vegetables
x_3 = number of servings of fruit
x_4 = number of servings of bread
x_5 = number of servings of meat
Maximize carbohydrates $z = 12x_1 + 7x_2 + 11x_3 + 15x_4$, subject to
$8x_1 + 2x_2 + 2x_4 + 7x_5 \leq 35$ (Amount of protein)
$10x_1 + x_4 + 5x_5 \leq 40$ (Amount of fat)
$x_1 \geq 0, x_2 \geq 0, x_3 \geq 0, x_4 \geq 0, x_5 \geq 0$

19. Let x_1 = number of sheets of cutting plan 1,
x_2 = number of sheets of cutting plan 2, etc.
The amount to be minimized as waste equals
$x_2 + 3x_3 + 2x_4 + 12x_5 + 14x_7$, and the constraints are
$3x_2 + 9x_3 + 6x_4 + 6x_6 \geq 165$ (No. of 15-inch doors)
$9x_1 + 6x_2 + 3x_4 + 3x_7 \geq 200$ (No. of 16-inch doors)
$6x_1 + 3x_6 + 3x_7 \geq 85$ (No. of 18-inch doors)
$x_1 \geq 0, x_2 \geq 0, x_3 \geq 0, x_4 \geq 0, x_5 \geq 0, x_6 \geq 0, x_7 \geq 0$

Review Exercises, Chapter 3

1.

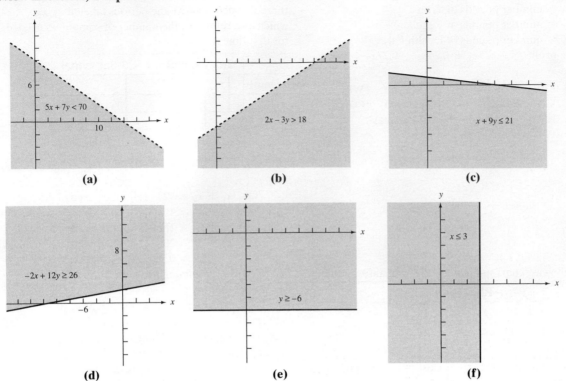

(a) $5x + 7y < 70$

(b) $2x - 3y > 18$

(c) $x + 9y \le 21$

(d) $-2x + 12y \ge 26$

(e) $y \ge -6$

(f) $x \le 3$

3. Corners are $(-9, -5)$, $(-1, -5)$, and $(3, -1)$.

5. Corners are $(-4, 2)$, $(4, 8)$, $(5, 0)$, and $(-2, 0)$.

7. Corners are $\left(-\frac{4}{3}, -\frac{2}{3}\right)$, $(0, 2)$, $(2, 2)$, and $(2, 1)$.

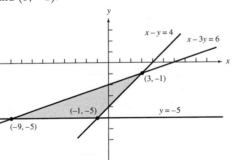

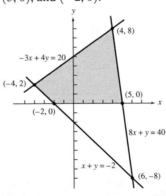

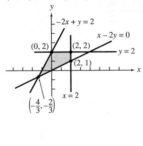

9. Maximum z is 22 at $(2, 3)$. in between.

11. (a) Minimum z is 34 at $(2, 6)$.   **(b)** Minimum z is 92 at $(2, 6)$, $(8, 1)$ and points

13. Maximum $z = 290$ at $(20, 30)$.

15. (a) $65x + 105y \le 700$ **(b)**

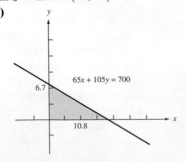

$65x + 105y = 700$

17. Let x = number of adult tickets
y = number of children's tickets
$x + y \le 275$
$4.50x + 3.00y \ge 1100$
$x \ge 0, y \ge 0$

19. 640 bars of standard and 280 bars of premium yield a revenue of $85,600.

21. Let x_1 = number of A-teams
x_2 = number of B-teams
x_3 = number of C-teams
Maximize
number of inoculations $z = 175x_1 + 110x_2 + 85x_3$,
subject to
$x_1 + x_2 + x_3 \le 75$ (Number of doctors)
$3x_1 + 2x_2 + x_3 \le 200$ (Number of nurses)
$x_1 \ge 0, x_2 \ge 0, x_3 \ge 0$
Maximize profit $z = 48x_1 + 45x_2 + 55x_3 + 65x_4$,
subject to
$40x_1 + 25x_2 + 30x_3 + 45x_4 \le 1250$ (Number tulips)
$25x_1 + 50x_2 + 40x_3 + 45x_4 \le 1600$ (Number daffodils)
$6x_1 + 4x_2 + 8x_3 + 2x_4 \le 195$ (Number boxwood)
$x_1 \ge 0, x_2 \ge 0, x_3 > 0, x_4 \ge 0$

23. Let x_1 = number of Type I pattern
x_2 = number of Type II pattern
x_3 = number of Type III pattern
x_4 = number of Type IV pattern

CHAPTER 4

Section 4.1

1. $2x_1 + 3x_2 + s_1 = 9$
$x_1 + 5x_2 + s_2 = 16$

3. $x_1 + 7x_2 - 4x_3 + s_1 = 150$
$5x_1 + 9x_2 + 2x_3 + s_2 = 435$
$8x_1 - 3x_2 + 16x_3 + s_3 = 345$

5. $2x_1 + 6x_2 + s_1 = 9$
$x_1 - 5x_2 + s_2 = 14$
$-3x_1 + x_2 + s_3 = 8$
$-3x_1 - 7x_2 + z = 0$

7. $6x_1 + 7x_2 + 12x_3 + s_1 = 50$
$4x_1 + 18x_2 + 9x_3 + s_2 = 85$
$x_1 - 2x_2 + 14x_3 + s_3 = 66$
$-420x_1 - 260x_2 - 50x_3 + z = 0$

9.
$$\left[\begin{array}{ccccc|c} 4 & 5 & 1 & 0 & 0 & 10 \\ 3 & 1 & 0 & 1 & 0 & 25 \\ -3 & -17 & 0 & 0 & 1 & 0 \end{array}\right]$$

11.
$$\left[\begin{array}{ccccccc|c} 16 & -4 & 9 & 1 & 0 & 0 & 0 & 128 \\ 8 & 13 & 22 & 0 & 1 & 0 & 0 & 144 \\ 5 & 6 & -15 & 0 & 0 & 1 & 0 & 225 \\ -20 & -45 & -40 & 0 & 0 & 0 & 1 & 0 \end{array}\right]$$

13. Let x_1 = number of cartons of screwdrivers
x_2 = number of cartons of chisels
x_3 = number of cartons of putty knives

$$\left[\begin{array}{cccccc|c} 3 & 4 & 5 & 1 & 0 & 0 & 2200 \\ 15 & 12 & 11 & 0 & 1 & 0 & 8500 \\ -5 & -6 & -5 & 0 & 0 & 1 & 0 \end{array}\right]$$

15. Let x_1 = amount of salad
x_2 = amount of potatoes
x_3 = amount of steak

$$\left[\begin{array}{cccccc|c} 20 & 50 & 56 & 1 & 0 & 0 & 1000 \\ 1.5 & 3 & 2 & 0 & 1 & 0 & 35 \\ -0.5 & -1 & -9 & 0 & 0 & 1 & 0 \end{array}\right]$$

17. Let x_1 = pounds of Lite
x_2 = pounds of Trim
x_3 = pounds of Health Fare

$$\left[\begin{array}{ccccccc|c} 0.75 & 0.50 & 0.15 & 1 & 0 & 0 & 0 & 2320 \\ 0.25 & 0.25 & 0.60 & 0 & 1 & 0 & 0 & 1380 \\ 0 & 0.25 & 0.25 & 0 & 0 & 1 & 0 & 700 \\ -0.25 & -0.25 & -0.32 & 0 & 0 & 0 & 1 & 0 \end{array}\right]$$

19. Let x_1 = number of military trunks
x_2 = number of commercial trunk
x_3 = number of decorative trunks

$$\left[\begin{array}{ccccccc|c} 4 & 3 & 2 & 1 & 0 & 0 & 0 & 4900 \\ 1 & 2 & 4 & 0 & 1 & 0 & 0 & 2200 \\ 0.1 & 0.2 & 0.3 & 0 & 0 & 1 & 0 & 210 \\ -6 & -7 & -9 & 0 & 0 & 0 & 1 & 0 \end{array}\right]$$

21. Let x_1 = number of Majestic Clocks, x_2 = number of Traditional Clocks, x_3 = number of Wall Clocks.

$$\left[\begin{array}{ccccccc|c} 4 & 2 & 1 & 1 & 0 & 0 & 0 & 120 \\ 3 & 2 & 1 & 0 & 1 & 0 & 0 & 80 \\ 1 & 1 & 0.5 & 0 & 0 & 1 & 0 & 40 \\ -400 & -250 & -160 & 0 & 0 & 0 & 1 & 0 \end{array}\right]$$

23. Let x_1 = number of pounds of Early Riser
x_2 = number of pounds of Coffee Time
x_3 = number of pounds of After Dinner
x_4 = number of pounds of Deluxe

Maximize profit $z = 1.00x_1 + 1.10x_2 + 1.15x_3 + 1.20x_4$, subject to

$0.80x_1 + 0.75x_2 + 0.75x_3 + 0.50x_4 \leq 260$ (Regular)
$0.20x_1 + 0.23x_2 + 0.20x_3 + 0.40x_4 \leq 90$ (High Mountain)
$0.02x_2 + 0.05x_3 + 0.10x_4 \leq 20$ (Chocolate)
$x_1 \geq 0, x_2 \geq 0, x_3 \geq 0, x_4 \geq 0$

$$\begin{bmatrix} 0.80 & 0.75 & 0.75 & 0.50 & 1 & 0 & 0 & 0 & 260 \\ 0.20 & 0.23 & 0.20 & 0.40 & 0 & 1 & 0 & 0 & 90 \\ 0 & 0.02 & 0.05 & 0.10 & 0 & 0 & 1 & 0 & 20 \\ \hline -1.00 & -1.10 & -1.15 & -1.20 & 0 & 0 & 0 & 1 & 0 \end{bmatrix}$$

27. (a) $12x + 10y + s_1 = 120$
$3x + 10y + s_2 = 60$
$7x + 12y + s_3 = 84$

(b) The point (x, y, s_1, s_2, s_3) may not lie in the feasible region. The point (x, y) satisfies the constraint $12x + 10y \leq 120$, so it lies in the half plane that is the solution to this constraint.

(c) The point (x, y) lies on the boundary line $12x + 10y = 120$. It may or may not be in the feasible region depending on where it is on the line.

25. Let x_1 = amount invested in stocks
x_2 = amount invested in treasury bonds
x_3 = amount invested in municipal bonds
x_4 = amount invested in corporate bonds

Maximize $z = 0.10x_1 + 0.06x_2 + 0.07x_3 + 0.08x_4$, subject to

$x_1 + x_2 + x_3 + x_4 \leq 10{,}000$
$0.60x_1 - 0.40x_2 - 0.40x_3 - 0.40x_4 \leq 0$
$-0.15x_1 + 0.85x_2 - 0.15x_3 - 0.15x_4 \leq 0$
$-0.30x_1 - 0.30x_2 + 0.70x_3 - 0.30x_4 \leq 0$
$-0.25x_1 - 0.25x_2 - 0.25x_3 + 0.75x_4 \leq 0$
$x_1 \geq 0, x_2 \geq 0, x_3 \geq 0, x_4 \geq 0$

$$\begin{bmatrix} 1 & 1 & 1 & 1 & 1 & 0 & 0 & 0 & 0 & 0 & 10{,}000 \\ 0.60 & -0.40 & -0.40 & -0.40 & 0 & 1 & 0 & 0 & 0 & 0 & 0 \\ -0.15 & 0.85 & -0.15 & -0.15 & 0 & 0 & 1 & 0 & 0 & 0 & 0 \\ -0.30 & -0.30 & 0.70 & -0.30 & 0 & 0 & 0 & 1 & 0 & 0 & 0 \\ -0.25 & -0.25 & -0.25 & 0.75 & 0 & 0 & 0 & 0 & 1 & 0 & 0 \\ \hline -0.10 & -0.06 & -0.07 & -0.08 & 0 & 0 & 0 & 0 & 0 & 1 & 0 \end{bmatrix}$$

(d) The point (x, y) is not in the feasible region when s_1 is negative.

(e) The point $(x, y, 0, s_2, 0)$ satisfies the first equation and $s_1 = 0$, so the point is on the "line" $12x + 10y + s_1 = 120$. Similarly, $(x, y, 0, s_2, 0)$ satisfies the third line and $s_3 = 0$, so the point lies on the "line" $7x + 12y + s_3 = 84$. Thus, the point (x, y) is the intersection of the lines $12x + 10y = 120$ and $7x + 12y = 84$.

29. (a) Enter the matrix

$$\begin{bmatrix} 6 & 3 & 1 & 0 & 0 & 18 \\ 5 & 2 & 0 & 1 & 0 & 27 \\ -40 & -22 & 0 & 0 & 1 & 0 \end{bmatrix}$$

(b) Enter the matrix

$$\begin{bmatrix} 10 & 14 & 1 & 0 & 0 & 0 & 73 \\ 6 & 21 & 0 & 1 & 0 & 0 & 67 \\ 15 & 8 & 0 & 0 & 1 & 0 & 48 \\ -134 & -109 & 0 & 0 & 0 & 1 & 0 \end{bmatrix}$$

(c) Enter the matrix

$$\begin{bmatrix} 1 & 1 & 1 & 1 & 0 & 0 & 0 & 24 \\ 3 & 1 & 4 & 0 & 1 & 0 & 0 & 37 \\ 2 & 5 & 3 & 0 & 0 & 1 & 0 & 41 \\ -15 & -23 & -34 & 0 & 0 & 0 & 1 & 0 \end{bmatrix}$$

(d) Enter the matrix

$$\begin{bmatrix} 7 & 4 & 1 & 2 & 1 & 0 & 0 & 0 & 435 \\ 5 & 3 & 6 & 1 & 0 & 1 & 0 & 0 & 384 \\ 2 & 8 & 4 & 5 & 0 & 0 & 1 & 0 & 562 \\ -24 & -19 & -15 & -33 & 0 & 0 & 0 & 1 & 0 \end{bmatrix}$$

(e) Enter the matrix

$$\begin{bmatrix} 4.7 & 3.2 & 1.58 & 1 & 0 & 0 & 0 & 40.6 \\ 2.14 & 1.82 & 5.09 & 0 & 1 & 0 & 0 & 61.7 \\ 1.63 & 3.44 & 2.84 & 0 & 0 & 1 & 0 & 54.8 \\ -12.9 & -11.27 & -23.85 & 0 & 0 & 0 & 1 & 0 \end{bmatrix}$$

Section 4.2

1. Basic: $x_1 = 8, s_2 = 10$; nonbasic: $x_2 = 0, s_1 = 0$ **3.** Basic $x_2 = 86, s_1 = 54, s_3 = 39$; nonbasic: $x_1 = x_3 = s_2 = 0$ **5. (a) (i)** 7 variables **(ii)** 3 basic variables **(iii)** 4 nonbasic variables. **(b) (i)** 7 variables **(ii)** 4 basic variables **(iii)** 3 nonbasic variables. **(c) (i)** 9 variables **(ii)** 4 basic variables **(iii)** 5 nonbasic variables. **7.** Pivot element = 4 in row 1, column 2 **9.** Pivot element = 4 in row 2, column 3 **11.** Pivot element = either the 4 in row 2, column 2 or the 5 in row 3, column 2 **13.** Pivot element = either the 8 in row 2, column 2 or the 6 in row 1, column 3 **15.** Pivot element = 4 in row 2, column 1

17.
$$\begin{bmatrix} 0 & -1 & 1 & -2 & 0 & 0 & 0 \\ 1 & 2 & 0 & 1 & 0 & 0 & 6 \\ 0 & 1 & 0 & -2 & 1 & 0 & 8 \\ 0 & 5 & 0 & 4 & 0 & 1 & 24 \end{bmatrix}$$

19.
$$\begin{bmatrix} \frac{14}{3} & \frac{29}{3} & 0 & 1 & 0 & \frac{4}{3} & 0 & -270 \\ -\frac{7}{3} & -\frac{34}{3} & 0 & 0 & 1 & -\frac{8}{3} & 0 & 580 \\ \frac{1}{3} & \frac{1}{3} & 1 & 0 & 0 & -\frac{1}{3} & 0 & 130 \\ 0 & -40 & 0 & 0 & 0 & -10 & 1 & 3900 \end{bmatrix}$$

21. $x_1 = 6, x_2 = 4, z = 16$

23. $x_1 = 1$, $x_2 = 2$, $z = 14$ **25.** $x_1 = 180$, $x_2 = 0$, $z = 1440$ **27.** $x_1 = 0$, $x_2 = 100$, $x_3 = 50$, $z = 22{,}500$
29. $x_1 = 0$, $x_2 = 75$, $x_3 = 15$, $z = 450$ **31.** $x_1 = 168$, $x_2 = 0$, $x_3 = 6$, $z = 2610$
33. $x_1 = 24$, $x_2 = 0$, $z = 792$ **35.** $x_1 = 20$, $x_2 = 20$, $x_3 = 20$, $z = 1200$ **37.** 316.7 cartons of screwdrivers,
312.5 of chisels, and no putty knives, for profit of $3458.33. It is reasonable to round this to 317 cartons of screwdrivers,
312 cartons of chisels, and no putty knives.   **39.** The maximum profit is $568.60 using 116 of Pack I, 63 of Pack II,
and 186 of Pack III.   **41.** Maximum profit is $380 using 100 pounds Early Riser, 200 pounds After Dinner, and
50 pounds Deluxe.   **43.** Maximum revenue is $386 making 10 packages of TV Mix, 55 packages of Party Mix,
and 15 packages of Dinner Mix.   **45.** Maximum revenue is $14,110 for 16 Majestic Clocks, 27 Traditional Clocks,
and 6 Wall Clocks.

47. Using row 1 as the pivot row in the first step, the sequence of corner points was $(0, 0)$, $(40, 0)$, and $(30, 20)$ to attain maximum $z = 230$. Using row 3 as the pivot row in the first step, the sequence of corner points was $(0, 0)$, $(40, 0)$, $(40, 0)$, and $(30, 20)$.

49. Using row 3 as the pivot row in the second step, maximum $z = 445$ at $(35, 25)$. The sequence of corner points used was $(0, 0)$, $(0, 50)$, $(10, 45)$, and $(35, 25)$. Using row 4 as the pivot row in the second step, maximum $z = 445$ at $(35, 25)$. The sequence of corner points used was $(0, 0)$, $(0, 50)$, $(10, 45)$, $(10, 45)$, and $(35, 25)$.

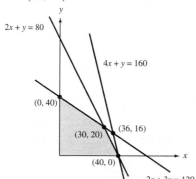

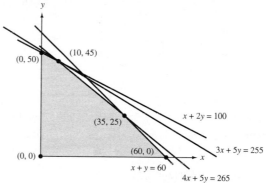

51. (a) A negative appears in the last column, indicating that you are outside the feasible region.   **(b)** It goes to the corner $(6.58, 0, 0)$ instead of $(0, 0, 5)$. If the pivoting continues, it will eventually reach the optimal solution.

53.
$$\begin{bmatrix} 0.67 & 0 & 1 & 0.67 & 0 & -0.33 & 0 & 33.33 \\ -1.67 & 0 & 0 & -0.67 & 1 & -0.67 & 0 & -8.33 \\ 0.67 & 1 & 0 & -0.33 & 0 & 0.67 & 0 & 58.33 \\ 3.33 & 0 & 0 & 5.33 & 0 & 7.33 & 1 & 1266.67 \end{bmatrix}$$

55.
$$\begin{bmatrix} 0.75 & 1 & 0 & -0.13 & -0.13 & 0 & 0 & 9 \\ -0.17 & 0 & 1 & 0.21 & 0.13 & 0 & 0 & 17 \\ 3.83 & 0 & 0 & -1.79 & -0.88 & 1 & 0 & -55 \\ 3.67 & 0 & 0 & 2.42 & 0.25 & 0 & 1 & 658 \end{bmatrix}$$

57. The solution is $(0, 6.81, 17.96, 12.02)$ with a profit of $2075.71. However, the solution must be an integer number of patterns, so we might round the solution to $(0, 7, 18, 12)$, which gives a maximum profit of $2085. Unfortunately, this scheme requires 1255 tulips, 1610 daffodils, and 196 boxwood, while there are only 1250 tulips, 1600 daffodils and 195 boxwood available. We can round down to $(0, 6, 18, 12)$ with a profit of $2040 and remain within the constraints.

Using EXCEL, Section 4.2

1. Maximum $z = 117$ at $(6, 5)$ **3.** Maximum $z = 696$ at $(12, 12, 24)$

Section 4.3

1.
$$\begin{bmatrix} 2 & 4 \\ 1 & 0 \\ 3 & 2 \end{bmatrix}$$
3.
$$\begin{bmatrix} 4 & 1 & 6 & 2 \\ 3 & 8 & -7 & 4 \\ 2 & -2 & 1 & 6 \end{bmatrix}$$

5. (a)
$$\begin{bmatrix} 6 & 5 & 30 \\ 8 & 3 & 42 \\ 25 & 30 & 1 \end{bmatrix}$$
(b)
$$\begin{bmatrix} 6 & 8 & 25 \\ 5 & 3 & 30 \\ 30 & 42 & 1 \end{bmatrix}$$
(c)
$$\begin{bmatrix} 6 & 8 & 1 & 0 & 0 & 25 \\ 5 & 3 & 0 & 1 & 0 & 30 \\ -30 & -42 & 0 & 0 & 1 & 0 \end{bmatrix}$$

7. (a) $\begin{bmatrix} 22 & 30 & 110 \\ 15 & 40 & 95 \\ 20 & 35 & 68 \\ 500 & 700 & 1 \end{bmatrix}$ **(b)** $\begin{bmatrix} 22 & 15 & 20 & 500 \\ 30 & 40 & 35 & 700 \\ 110 & 95 & 68 & 1 \end{bmatrix}$ **(c)** $\begin{bmatrix} 22 & 15 & 20 & 1 & 0 & 0 & 500 \\ 30 & 40 & 35 & 0 & 1 & 0 & 700 \\ -110 & -95 & -68 & 0 & 0 & 1 & 0 \end{bmatrix}$

9. Minimum $z = 40$ at $(6, 4)$.   **11.** Minimum $z = 510$ at $(12, 10, 0)$.   **13.** Minimum $z = 30$ at $(6, 2)$.
15. Minimum $z = 108$ at $(6, 3, 0)$.   **17.** Minimum $z = 279$ at $(18, 27, 0)$.   **19.** 15 at Dallas, 32 at New Orleans for cost of $714,000   **21.** The minimum operating costs are $900,000 when the Chicago plant operates 30
days and the Detroit plant 20 days.   **23.** Elizabeth should select 2.5 cereals, 3 cheeseburgers, and no fries for a
total of 1675 calories.

Section 4.4

1. Minimum z is -200 at $(0, 40)$.   **3.** Minimum $z = -158$ at $\left(0, 2, \frac{56}{3}\right)$.

5. $\begin{bmatrix} 9 & 7 & 10 & 1 & 0 & 0 & 0 & 154 \\ -3 & -5 & -8 & 0 & 1 & 0 & 0 & -106 \\ -6 & -12 & -1 & 0 & 0 & 1 & 0 & -98 \\ -5 & -3 & -8 & 0 & 0 & 0 & 1 & 0 \end{bmatrix}$ **7.** $\begin{bmatrix} 15 & 23 & 9 & 1 & 0 & 0 & 0 & 0 & 85 \\ -7 & -9 & -15 & 0 & 1 & 0 & 0 & 0 & -48 \\ 1 & 3 & 5 & 0 & 0 & 1 & 0 & 0 & 27 \\ -1 & -3 & -5 & 0 & 0 & 0 & 1 & 0 & -27 \\ 7 & 5 & 8 & 0 & 0 & 0 & 0 & 1 & 0 \end{bmatrix}$

9. Maximum $z = 60$ at $(12, 0)$.   **11.** Maximum is $z = 1050$ at $(70, 0)$.   **13.** Maximum $z = 450$ at $(0, 22.5)$.
15. Maximum $z = 5000$ at $(50, 0, 150)$.   **17.** Minimum $z = 168$ at $(8, 6)$.   **19.** Minimum $z = 30$ at $(5, 0, 10)$.
21. Maximum $z = 120$ at $(12, 6)$.   **23.** Maximum $z = 120$ at $(10, 15)$.   **25.** Maximum $z = 640$ at $(0, 18, 8)$.
27. Minimum z is 166 at $(14, 8)$.   **29.** Minimum $z = 63$ at $\left(\frac{5}{2}, \frac{5}{2}, 9\right)$.

31. Let $x =$ number of Custom,
$y =$ number of Executive.

$\begin{bmatrix} -1 & -1 & 1 & 0 & 0 & 0 & -100 \\ -90 & -120 & 0 & 1 & 0 & 0 & -10{,}800 \\ 4 & 5 & 0 & 0 & 1 & 0 & 800 \\ 70 & 80 & 0 & 0 & 0 & 1 & 0 \end{bmatrix}$

33. Let $x =$ number of A, $y =$ number of B,
$z =$ number of C.

$\begin{bmatrix} -1 & -1 & -1 & 1 & 0 & 0 & 0 & 0 & -6{,}600 \\ 20 & 25 & 15 & 0 & 1 & 0 & 0 & 0 & 133{,}000 \\ 1 & 3 & 2 & 0 & 0 & 1 & 0 & 0 & 13{,}600 \\ 8 & 10 & 15 & 0 & 0 & 0 & 1 & 0 & 73{,}000 \\ -6 & -9 & -6 & 0 & 0 & 0 & 0 & 1 & 0 \end{bmatrix}$

35. $\begin{bmatrix} -1 & -1 & -1 & 1 & 0 & 0 & 0 & 0 & 0 & -6{,}800 \\ 20 & 25 & 15 & 0 & 1 & 0 & 0 & 0 & 0 & 133{,}000 \\ 1 & 3 & 2 & 0 & 0 & 1 & 0 & 0 & 0 & 13{,}600 \\ 8 & 10 & 15 & 0 & 0 & 0 & 1 & 0 & 0 & 80{,}000 \\ 0 & -1 & 0 & 0 & 0 & 0 & 0 & 1 & 0 & -2{,}000 \\ -6 & -9 & -6 & 0 & 0 & 0 & 0 & 0 & 1 & 0 \end{bmatrix}$

37. Let $x_1 =$ minutes jogging
  $x_2 =$ minutes playing handball
  $x_3 =$ minutes swimming.

(a) $\begin{bmatrix} -13 & -11 & -7 & 1 & 0 & 0 & 0 & 0 & -660 \\ 1 & 0 & -1 & 0 & 1 & 0 & 0 & 0 & 0 \\ -1 & 0 & 1 & 0 & 0 & 1 & 0 & 0 & 0 \\ 2 & -1 & 0 & 0 & 0 & 0 & 1 & 0 & 0 \\ 1 & 1 & 1 & 0 & 0 & 0 & 0 & 1 & 0 \end{bmatrix}$ **(b)** $\begin{bmatrix} 1 & 1 & 1 & 1 & 0 & 0 & 0 & 0 & 90 \\ 1 & 0 & -1 & 0 & 1 & 0 & 0 & 0 & 0 \\ -1 & 0 & 1 & 0 & 0 & 1 & 0 & 0 & 0 \\ 2 & -1 & 0 & 0 & 0 & 0 & 1 & 0 & 0 \\ -13 & -11 & -7 & 0 & 0 & 0 & 0 & 1 & 0 \end{bmatrix}$

39. Let x_1 = number of pounds of Lite
x_2 = number of pounds of Trim
x_3 = number of pounds of Regular
x_4 = number of pounds of Health Fare

$$\begin{bmatrix} 0.75 & 0.50 & 0.80 & 0.15 & 1 & 0 & 0 & 0 & 0 & 0 & 2400 \\ 0.20 & 0.25 & 0.20 & 0.50 & 0 & 1 & 0 & 0 & 0 & 0 & 1400 \\ 0.05 & 0.20 & 0 & 0.25 & 0 & 0 & 1 & 0 & 0 & 0 & 700 \\ 0 & 0.05 & 0 & 0.10 & 0 & 0 & 0 & 1 & 0 & 0 & 250 \\ 0 & -0.05 & 0 & -0.10 & 0 & 0 & 0 & 0 & 1 & 0 & -200 \\ -0.25 & 0.25 & -0.27 & -0.32 & 0 & 0 & 0 & 0 & 0 & 1 & 0 \end{bmatrix}$$

41. Let x_1 = number units of food A
x_2 = number units of food B
x_3 = number units of food C

$$\begin{bmatrix} -15 & -10 & -23 & 1 & 0 & 0 & 0 & 0 & -80 \\ -20 & -30 & -11 & 0 & 1 & 0 & 0 & 0 & -95 \\ -500 & -400 & -200 & 0 & 0 & 1 & 0 & 0 & -1200 \\ 8 & 3 & 6 & 0 & 0 & 0 & 1 & 0 & 35 \\ 1.40 & 1.65 & 1.95 & 0 & 0 & 0 & 0 & 1 & 0 \end{bmatrix}$$

43. The manager should order 40 Custom and 60 Executive for a minimum cost of $7600.   **45.** Maximum profit is $46,800 when 2000 of A, 2400 of B, and 2200 of C are ordered.   **47.** Maximum profit is $47,040 when ordering 2080 of A, 2000 of B, and 2760 of C.   **49. (a)** 0 minutes jogging, 60 minutes playing handball, 0 minutes swimming. **(b)** No time jogging or swimming, all 90 minutes playing handball

51.

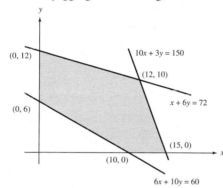

Corner	z	Feasible?
$(0,0)$	0	No
$(0,6)$	60	Yes
$(10,0)$	80	Yes
$(15,0)$	120	Yes
$(12,10)$	196	Yes

The simplex method starts at $(0,0)$, which is not feasible, then moves to $(0,6)$, which is feasible. From there on each pivot moves to an adjacent corner that is in the feasible region.

53. (a)

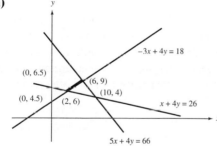

(b) The feasible region is the portion on the line $-3x + 4y = 18$ that lies in the region bounded by $5x + 4y \le 66$, and by $x + 4y \ge 26$. This is simply the line segment joining $(2,6)$ and $(6,9)$.
(c) The feasible region is a segment of the line $a_1x + b_1y = k_1$ where it intersects the feasible region defined by
$a_2x + b_2y \le k_2$
$a_3x + b_3y \ge k_3$
$x \ge 0, y \ge 0$

55. Let the variables be the following:
x_1, x_2, and x_3 = quantity shipped from Cleveland to Chicago, Dallas, and Atlanta, respectively.
x_4, x_5, and x_6 = quantity shipped from St. Louis to Chicago, Dallas, and Atlanta, respectively.
x_7, x_8, and x_9 = quantity shipped from Pittsburgh to Chicago, Dallas, and Atlanta, respectively.
(a) The constraints are

$x_1 + x_2 + x_3 \le 4200$	(Quantity shipped from Cleveland)
$x_4 + x_5 + x_6 \le 4800$	(Quantity shipped from St. Louis)
$x_7 + x_8 + x_9 \le 3700$	(Quantity shipped from Pittsburgh)
$x_1 + x_4 + x_7 = 5400$	(Quantity shipped to Chicago)
$x_2 + x_5 + x_8 = 3800$	(Quantity shipped to Dallas)
$x_3 + x_6 + x_9 = 2700$	(Quantity shipped to Atlanta)

with all variables nonnegative.
Because we are to minimize shipping costs, the objective function is
Minimize $z = 35x_1 + 64x_2 + 60x_3 + 37x_4 + 59x_5 + 51x_6 + 49x_7 + 68x_8 + 57x_9$

(b) The initial simplex tableau is

x_1	x_2	x_3	x_4	x_5	x_6	x_7	x_8	x_9	s_1	s_2	s_3	s_4	s_5	s_6	s_7	s_8	s_9	z	
1	1	1	0	0	0	0	0	0	1	0	0	0	0	0	0	0	0	0	4200
0	0	0	1	1	1	0	0	0	0	1	0	0	0	0	0	0	0	0	4800
0	0	0	0	0	0	1	1	1	0	0	1	0	0	0	0	0	0	0	3700
1	0	0	1	0	0	1	0	0	0	0	0	1	0	0	0	0	0	0	5400
-1	0	0	-1	0	0	-1	0	0	0	0	0	0	1	0	0	0	0	0	-5400
0	1	0	0	1	0	0	1	0	0	0	0	0	0	1	0	0	0	0	3800
0	-1	0	0	-1	0	0	-1	0	0	0	0	0	0	0	1	0	0	0	-3800
0	0	1	0	0	1	0	0	1	0	0	0	0	0	0	0	1	0	0	2700
0	0	-1	0	0	-1	0	0	-1	0	0	0	0	0	0	0	0	1	0	-2700
35	64	60	37	59	51	49	68	57	0	0	0	0	0	0	0	0	0	1	0

57. Maximum $z = 308$ at the corner $(14, 30, 0)$.   **59.** The maximum value of z is 616 and it occurs at $(8, 7, 5, 10)$

Section 4.5

1. The -20 in the last column with nonnegative entries in the rest of the first row indicates no feasible solution.
3. There is an unbounded feasible region, so no solution, because all entries in column 4 are negative.
5. Maximum of 210 at $(5, 9)$ and $(10, 4)$ and points between.   **7.** Maximum $= 40$ at $(5, 5, 2.5)$, $\left(\frac{20}{3}, \frac{20}{3}, 0\right)$ and
points between.   **9.** Unbounded feasible region because all entries in column 1 become negative after a pivot.
11. Unbounded feasible region because all entries in column 2 become negative after a pivot.   **13.** Pivoting leads
to a row with a negative entry in the last column and all other entries in the row are not negative. Thus, there is no feasible solution.   **15.** Pivoting leads to a row with a negative entry in the last column and all other entries in the row
are not negative. Thus, there is no feasible solution.   **17.** Unbounded feasible region because all entries in column 1
become negative after a pivot.   **19.** Pivoting leads to a row with a negative entry in the last column and all other
entries in the row are not negative. Thus, there is no feasible solution.   **21.** Pivoting leads to a row with a negative
entry in the last column and all other entries in the row are not negative. Thus, there is no feasible solution.
23. Pivoting leads to a row with a negative entry in the last column and all other entries in the row are not negative.
Thus, there is no feasible solution.   **25.** Multiple solutions, minimum $z = 51$ at $\left(\frac{51}{4}, \frac{15}{2}, 3\right)$ and $(24, 0, 3)$ and points
between.   **27.** Unbounded feasible region because all entries in column 1 become negative after a pivot.
29. Pivoting leads to a row with a negative entry in the last column and all other entries in the row are not negative.
Thus, there is no feasible solution.   **31.** Column 3 contains all negative entries that indicate an unbounded feasible
region, so there is no maximum expected return.   **33.** In row 3 the last entry is negative and all other entries are
nonnegative. This indicates no feasible solution.

Section 4.6

1. (a) For $(0, 0)$, $s_1 = 17$. For $(2, 2)$, $s_1 = 3$. For $(5, 10)$, $s_1 = -33$. For $(2, 1)$, $s_1 = 6$. For $(2, 3)$, $s_1 = 0$.
(b) $(0, 0)$, $(2, 2)$, $(2, 1)$, and $(2, 3)$ **(c)** $(2, 3)$ **3.** $(6, 0)$ **5.** For $(0, 0, 0)$, $s_1 = 40$. For $(1, 2, 3)$, $s_1 = 26$.
For $(0, 10, 0)$, $s_1 = 0$. For $(4, 2, 7)$, $s_1 = 13$.

7.

Point	s_1	s_2	s_3	Is Point on Boundary?	Is Point in Feasible Region?
$(5, 10)$	15	32	18	No	Yes
$(8, 10)$	12	14	-3	No	No
$(5, 13)$	0	29	0	Yes	Yes
$(11, 13)$	-6	-7	-42	No	No
$(10, 12)$	0	0	-29	No	No
$(15, 11)$	0	-29	-58	No	No

9. (i) (a) $x_1 = 0$, $x_2 = 0$, $s_1 = 900$, $s_2 = 2800$, $z = 0$
    (b) Intersection of $x_1 = 0$ and $x_2 = 0$.
 (ii) (a) $x_1 = 180$, $x_2 = 0$, $s_1 = 0$, $s_2 = 1360$, $z = 540$
    (b) Intersection of $5x_1 + 2x_2 = 900$ and $x_2 = 0$.
 (iii) (a) $x_1 = 100$, $x_2 = 200$, $s_1 = 0$, $s_2 = 0$, $z = 700$
    (b) Intersection of $5x_1 + 2x_2 = 900$ and
    $8x_1 + 10x_2 = 2800$.

11. (a) $(3, 5)$ is not on the boundary line.   **(b)** No, because $s_1 = -14$ at $(3, 5)$.   **13. (a)** $\frac{36}{7}$ **(b)** 6

15.

Corner (x, y)	Value of z	Increase in z	Slack (s_1, s_2, s_3, s_4)
$(0, 0)$	0		$(50, 90, 105, 20)$
$(0, 5)$	100	100	$(0, 65, 80, 20)$
$(10, 12)$	410	310	$(0, 0, 5, 10)$
$(15, 9)$	435	25	$(65, 0, 0, 5)$
$(20, 5)$	440	5	$(140, 5, 0, 0)$

17.

Corner (x_1, x_2, x_3)	Value of z	Increase in z	Slack (s_1, s_2, s_3)
$(0, 0, 0)$	0		$(36, 45, 46)$
$(0, 6, 0)$	144	144	$(0, 9, 10)$
$(2.5, 5.583, 0)$	159	15	$(0, 4, 0)$
$(3, 5, 1)$	163	4	$(0, 0, 0)$

19.

Corner (x_1, x_2, x_3)	Value of z	Increase in z	Slack (s_1, s_2, s_3)
$(0, 0, 0)$	0		$(330, 330, 132)$
$(0, 0, 132)$	396	396	$(66, 66, 0)$
$(0, 11, 154)$	484	88	$(11, 0, 0)$
$(6, 6, 156)$	486	2	$(0, 0, 0)$

21. The value of z did not change with the pivot because the corner point did not change.

Section 4.7

1. (a) The objective function is not maximum at the intersection.   **(b)** The objective function attains its maximum value at the intersection.   **(c)** The objective function attains its maximum value at the intersection.   **3.** $-\frac{18}{5} \leq m \leq -\frac{2}{3}$. A slope of -2 serves as one example.   **5. (a)** z does not reach a maximum at $(0, 28)$.   **(b)** z reaches a maximum at $(0, 28)$.   **(c)** z does not reach a maximum at $(0, 28)$.   **7. (a)** The objective function does not reach a maximum at $(30, 0)$.   **(b)** The objective function reaches a maximum at $(30, 0)$.   **(c)** The objective function reaches a maximum at $(30, 0)$.   **9.** $m \leq -\frac{6}{7}$. A slope of -3 serves as one example.   **11.** The coefficient of x must lie between 8 and 22.5.   **13.** B lies between 3 and 7.5.   **15.** Maximum $z = 3120$ at $(60, 0)$.   **17. (a)** $(20 + \frac{5}{22}D, 12 - \frac{4}{22}D)$ **(b)** $z = 248 + \frac{34}{22}D$ **(c)** Each unit change in D creates a change of 1.55 units in z.   **(d)** D is restricted to the interval $-88 \leq D \leq 66$.

Review Exercises, Chapter 4

1.
$$6x_1 + 4x_2 + 3x_3 + s_1 = 220$$
$$x_1 + 5x_2 + x_3 + s_2 = 162$$
$$7x_1 + 2x_2 + 5x_3 + s_3 = 139$$

3.
$$6x_1 + 5x_2 + 3x_3 + 3x_4 + s_1 = 89$$
$$7x_1 + 4x_2 + 6x_3 + 2x_4 + s_2 = 72$$

5.
$$10x_1 + 12x_2 + 8x_3 + s_1 = 24$$
$$7x_1 + 13x_2 + 5x_3 + s_2 = 35$$
$$-20x_1 - 36x_2 - 19x_3 + z = 0$$

7.
$$3x_1 + 7x_2 + s_1 = 14$$
$$9x_1 + 5x_2 + s_2 = 18$$
$$x_1 - x_2 + s_3 = 21$$
$$-9x_1 - 2x_2 + z = 0$$

9.
$$x_1 + x_2 + x_3 + s_1 = 15$$
$$2x_1 + 4x_2 + x_3 + s_2 = 44$$
$$-6x_1 - 8x_2 - 4x_3 + z = 0$$

11. (a) 6 in row 2, column 2   **(b)** 5 in row 2, column 4

13. (a) $x_1 = 0, x_2 = 80, s_1 = 0, s_2 = 42, z = 98$ **(b)** $x_1 = 73, x_2 = 42, x_3 = 15, s_1 = 0, s_2 = 0, s_3 = 0, z = 138$

15. (a)
$$\begin{bmatrix} 11 & 5 & 3 & 1 & 0 & 0 & 0 & 142 \\ -3 & -4 & -7 & 0 & 1 & 0 & 0 & -95 \\ 2 & 15 & 1 & 0 & 0 & 1 & 0 & 124 \\ -3 & -5 & -4 & 0 & 0 & 0 & 1 & 0 \end{bmatrix}$$

(b)
$$\begin{bmatrix} 7 & 4 & 1 & 0 & 0 & 28 \\ -1 & -3 & 0 & 1 & 0 & -6 \\ 14 & 22 & 0 & 0 & 1 & 0 \end{bmatrix}$$

17. (a)
$$\begin{bmatrix} -15 & -8 & 1 & 0 & 0 & 0 & 0 & -120 \\ 10 & 12 & 0 & 1 & 0 & 0 & 0 & 120 \\ 15 & 5 & 0 & 0 & 1 & 0 & 0 & 75 \\ -15 & -5 & 0 & 0 & 0 & 1 & 0 & -75 \\ -5 & -12 & 0 & 0 & 0 & 0 & 1 & 0 \end{bmatrix}$$

(b)
$$\begin{bmatrix} 14 & 9 & 1 & 0 & 0 & 0 & 0 & 126 \\ -10 & -11 & 0 & 1 & 0 & 0 & 0 & -110 \\ -5 & 1 & 0 & 0 & 1 & 0 & 0 & 9 \\ 5 & -1 & 0 & 0 & 0 & 1 & 0 & -9 \\ 3 & 2 & 0 & 0 & 0 & 0 & 1 & 0 \end{bmatrix}$$

19. Maximum $z = 47$ at $(13, 0, 4)$.   **21.** Unbounded feasible region, no maximum.

23. Maximum $z = 456$ at $(60, 132, 0)$.   **25. (a)** x_3 **(b)** x_2 **27.** $\begin{bmatrix} 3 & 4 & 5 \\ 1 & 0 & 7 \\ -2 & 6 & 8 \end{bmatrix}$ $\begin{bmatrix} 4 & -5 \\ 3 & 0 \\ 2 & 12 \\ 1 & 9 \end{bmatrix}$

29. Multiple solutions, minimum $z = 64$ at $(0, 6, 1)$ and $(0, 8, 0)$.   **31.** Maximum $z = 92$ at $(4, 4, 3)$.
33. Maximum $z = 22$ at $(2, 3, 1)$.   **35.** Unbounded feasible region, no maximum.   **37.** Minimum $z = 252$
at $(6, 4)$.   **39.** Maximum $z = 1050$ at $(30, 60)$.   **41.** Maximum $z = \frac{17}{4}$ at $\left(\frac{3}{4}, \frac{11}{4}\right)$.
43. Let $x_1 =$ number of hunting jackets   **45.** A can be any value from 6 through 20.
         $x_2 =$ number of all-weather jackets
         $x_3 =$ number of ski jackets
Maximize $z = 7.5x_1 + 9x_2 + 11x_3$, subject to
$3x_1 + 2.5x_2 + 3.5x_3 \leq 3200$
$26x_1 + 20x_2 + 22x_3 \leq 18{,}000$
$x_1 \geq 0, x_2 \geq 0, x_3 \geq 0$

$$\begin{bmatrix} 3 & 2.5 & 3.5 & 1 & 0 & 0 & 3{,}200 \\ 26 & 20 & 22 & 0 & 1 & 0 & 18{,}000 \\ -7.5 & -9 & -11 & 0 & 0 & 1 & 0 \end{bmatrix}$$

CHAPTER 5

Section 5.1

1. $66   **3.** $90   **5.** $31.66   **7.** $I = \$18, A = \$318$ **9.** $90   **11.** $52.50   **13.** $116.38
15. $660   **17.** $1100   **19.** 6.5%   **21.** 8.5%   **23.** $2862   **25.** $6909.50   **27.** $1651.38
29. $D = \$83.25, PR = \$1766.75$ **31.** $D = \$27.65, PR = \$457.35$ **33.** $402.50   **35.** $1198.75
37. (a) $361.25 **(b)** $21,675   **39.** 3 years   **41.** 4.5 months   **43.** $650   **45.** $150 toward interest,
$29.95 toward principal   **47.** 9.3%   **49.** 10.53%   **51.** 10.1% discount is $104.52, and simple interest
is $103.00. So, simple interest is better.   **53.** $49.5 billion   **55.** $162,500   **57. (a)** $17,187.50
(b) $275,000   **59.** $982,500   **61.** 7.2%   **63.** $1,991,458   **65.** Quarterly interest $= \$3675$; total
interest $= \$88{,}200$

Section 5.2

1. (a) 3.3% **(b)** 1.65% **(c)** 0.55%   **3. (a)** $6944.86 **(b)** $2444.86   **5. (a)** $55,287.81 **(b)** $24,287.81
7. (a) $6553.98 **(b)** $1553.98   **9.** First quarter, $812.00; second quarter, $824.18; third quarter, $836.54; fourth
quarter, $849.09   **11.** $1965.65   **13. (a)** $15,972 **(b)** $16,081.15 **(c)** $16,138.67   **15.** $13,468.55
17. $472.56   **19.** $720.74   **21.** $242.20   **23.** $10,299.67   **25.** $2919.63   **27.** 5.51%   **29.** 5.2%
31. 7.2%   **33.** About 10% annual rate   **35.** $281.48   **37.** Ken had $1294.76, Barb had $1291.61. Ken had
$3.07 more.   **39. (a)** $11,070 **(b)** $15,347.07 Compound interest gives $4277.07 more.   **41.** The effective rate
for 6.8% is 6.9156%. The effective rate for 6.6% is 6.7651%, so 6.8% compounded semiannually is better.   **43.** The
effective rate of 6.2% compounded quarterly is 6.345%, so 6.4% is better.   **45.** The effective rate of 5.0% compounded quarterly is 5.094%, which is not better than 5.1% compounded annually.   **47.** $22,138.80   **49.** 8%
51. 6% per year   **53.** $6000 will yield $16,110.38, so it is better   **55.** 5.88%   **57.** $161,513.12
59. (a) 6.91% **(b)** $6970.32   **61.** Over $20.5 million   **63.** 5.7%

65.

Interest Rate	Years to Double	Actual Value of $1
3%	24	2.03
5%	14.4	2.02
6%	12	2.01
7%	10.3	2.01
8%	9	2.00
9%	8	1.99
10%	7.2	1.99

67.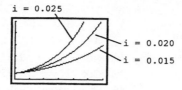

69.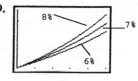

71. For 5% interest

Years	Amount
50	$ 52,750.04
100	$ 604,905.79
150	$ 6,936,696.48
200	$ 79,545,871.75

For 4.5%
200 years, the amount is $30,620,756.86
For 5.5%
200 years, the amount is $205,707,325.80

73. 0.09

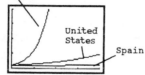

0.07

As compounding occurs more frequently, the effective rate increases but tends to level off.

75. (a), (b)

$y = e^{0.08}$

(c) As x increases and the frequency of compounding increases, the amount of $1 increases with a limiting value of $e^{0.08}$.

77. (a) $162.89
(b) 6,728,970
(c) Kenya

United States

Spain

79. $P = 1000, r = 0.052.$

Year	Lauren	Anna
1	1052.00	1052
2	1106.70	1104
3	1164.25	1156
4	1224.79	1208
5	1288.48	1260
6	1355.48	1312
7	1425.97	1364
8	1500.12	1416
9	1578.13	1468
10	1660.19	1520

Using Your TI-83, Section 5.2

1.

Year	Amount
0	200.00
1	210.20
2	220.92
3	232.19
4	244.03
5	256.47
6	269.55
7	283.30
8	297.75
9	312.94
10	328.89

3. About 16 years

Using EXCEL, Section 5.2

1.

Year	Amount
0	500.00
1	530.00
2	561.80
3	595.51
4	631.24
5	669.11
6	709.26

5. Just over 3 years

3.

Year	6%	5.50%
0	2000.00	2000.00
1	2120.00	2110.00
2	2247.20	2226.05
3	2382.03	2348.48
4	2524.95	2477.65
5	2676.45	2613.92

Section 5.3

1.

Year Deposited	Value at End of Four Years
1	$ 694.58
2	$ 661.50
3	$ 630.00
4	$ 600.00
Final Value	$2586.08

3. $402,064.35   **5.** $5504.75   **7.** $4403.80   **9.** $106,394.86
11. $13,658.37   **13.** $567.26   **15.** $539.63   **17.** $766.80
19. $4086.63   **21.** $796.76   **23.** $2513.71   **25.** $4802.44
27. $306.72   **29.** $59,889.94   **31.** $20,969.37   **33.** $18,205.56
35. These deposits will accumulate to $22,548.37, less than the required amount.   **37. (a)** $14,486.56   **(b)** The annual interest of $1158.92 will make the $1000 payment.   **39.** $3812.16   **41.** $198.88
43. $13,597.20   **45. (a)** The value of Evelyn's fund at retirement was $16,387.93. The value of Esther's fund at retirement was $23,102.04. The value of Lois's fund at retirement was $25,112.88.   **(b)** Start your savings program early for the best return.
47. $177,735   **49. (a)** $159,473   **(b)** $173,309   **(c)** $188,130   **(d)** $204,007   **(e)** $221,015   **(f)** $239,234
51. (a) 309,524   **(b)** $222,870   **(c)** $109,729   **(d)** $46,552   **55.** $x = 0.006238$ per month $= 7.49\%$ annual
57. (a) In about 55.5 quarters, when their amount is $15,000   **(b)** In about 71.2 quarters, when their amount is $20,480
(c) Cutter's investment never catches up; Andrew's is increasing faster.   **59.** $3419.87

Using Your TI-83, Section 5.3

1.

Month	Amount
1	200.0
2	401.50
3	604.51
4	809.05

3.

Year	Amount
1	1000.00
2	2072.00
3	3221.18
4	4453.11

Using EXCEL, Section 5.3

1.

Month	Amount
1	1,500.00
2	3,085.50
3	4,761.37
4	6,532.77
5	8,405.14
6	10,384.23
7	12,476.13
8	14,687.27
9	17,024.45
10	19,494.84

3.

Month	Amount
1	75.00
2	150.34
3	226.01
4	302.03
5	378.39
6	455.09
7	532.14
8	609.54
9	687.28
10	765.37
11	843.82
12	922.61

Section 5.4

1. $23,885.19   **3.** $16,800   **5.** $21,042.03   **7.** $5468.07   **9.** $2400   **11.** $8650
13. $2310.76 annually   **15.** $205.44 quarterly   **17.** $439.47 monthly   **19.** $794.63 monthly
21. $304.72   **23.** $434.29   **25. (a)** $198,115.20   **(b)** $123,115.20   **27. (a)** $391 interest, $85.28 to principal   **(b)** $142,884   **29.** $42,861.47   **31.** $8600   **33.** $574.49   **35.** $124,622.10
37. $533.47   **39.** $1326.61 annual payments   **41.** $10,345.11   **43.** $188.88   **45.** $378,640.20
47. (a) $79,156.54   **(b)** $917.73
49. Costs of owning the house:

Down payment	$10,000.00
Payment of balance	78,131.19
Monthly payments	16,112.64
Real estate agent	5,400.00
Closing costs	1,350.00
Total	110,993.83
Income from sale of house	90,000.00
Net cost	20,993.83

Average monthly cost is $874.74.

55. (a) $79.05 **(b)** $185.50   **57. (a)** $114,528.83 **(b)** $965.11 **(c)** $42,794.40   **63.** 7.2%
65. About $26   **69. (a)** $560,476 **(b)** $455,396 **(c)** $300,746 **(d)** $317,402 **(e)** $299,010 **(f)** $302,756
(h) $298,557   **71.** This represents the present value of a $1 periodic payment for x time periods at 1.5% per period. As x approaches 250, the value of y tends to level off and approaches 65. If x continues beyond 250, the value of y will approach, but never exceed, 66.67. Thus an investment of $66.67 at 1.5% per payment period will provide $1 per period forever.   **73. (a)** For 6%, $R = \$599.55$. For 8.1%, $R = \$740.75$. For 10.2%, $R = \$892.39$.   **(c)** Months required to reduce the balance to 75%, 50%, and 25% of the original loan:

Balance	Interest Rate		
	6%	8.1%	10.2%
75%	163 mo.	189 mo.	212 mo.
50%	252 mo.	269 mo.	283 mo.
25%	313 mo.	321 mo.	328 mo.

(d) As the interest rate increases, it takes longer to reduce the balance to the 75%, 50%, and 25% levels.
75. It will take about 252 months, or 21 years, to repay the loan.   **77.** It will take about 100 months to pay off the credit card bill.   **79.** APR is about 9.88%.   **81.** 1.13% per month = an annual rate of 13.56%
83. $x = 0.01515$ per month = 18.18% annual   **85.** $x = 0.00616$ per month = 7.39% annual

Using Your TI-83, Section 5.4

1.

Month	Interest	Principal Repaid	Balance
1	15.00	159.90	1840.10
2	13.80	161.10	1679.00
3	12.59	162.31	1516.69
4	11.38	163.52	1353.17

3.

Month	Interest	Principal Repaid	Balance
1	552.50	250.02	84,749.98
2	550.07	251.65	84,498.33
3	549.24	253.28	84,245.05
4	547.59	254.93	83,990.12

Review Exercises, Chapter 5

1. $90   **3.** $960   **5.** $1350   **7.** $D = \$1530$, $PR = \$6970$ **9.** $1125.22   **11.** After 10 years, the value is $A = 1000(1.017)^{40} = 1962.63$, so it will not double in value.   **13.** 6.09%   **15.** 7.442%
17. Effective rate of 5.6% = 5.72%. So, 5.6% compounded quarterly is better.   **19. (a)** $10,099.82 **(b)** $2099.82
21. (a) $7063.49 **(b)** $2063.49   **23.** $16,695.38   **25.** $33,648.57   **27.** The investment will not reach
$3500 in 6 years.   **29.** $5637.09   **31.** $6942.02   **33.** $7245.15   **35.** $160,500   **37.** $3736.29
39. $4411.16   **41.** $33,648.57   **43.** After 5 years, the total payment is $297,289.48, so the term should be less than 5 years.   **45.** $276.76   **47.** $1183.24   **49.** $573,937.65   **51.** $17,706.09   **53.** Amount after 5 years, $1360.19; amount after 10 years, $1868.42   **55.** $3023.65   **57.** $39,443.62   **59.** $1157.29

CHAPTER 6

Section 6.1

1. (a) True **(b)** False **(c)** False **(d)** False **(e)** True **(f)** True **(g)** False **(h)** True **(i)** False
3. $B = \{M, I, S, P\}$ **5.** $C = \{16, 18, 20, 22, \ldots\}$ **7.** Not equal   **9.** Equal   **11.** Not equal
13. Subset   **15.** Not a subset   **17.** Subset   **19.** Subset   **21. (a)** $\varnothing$, $\{-1\}$, $\{2\}$, $\{4\}$, $\{-1, 2\}$, $\{-1, 4\}$, $\{2, 4\}$, $\{-1, 2, 4\}$ **(b)** $\varnothing$, $\{4\}$ **(c)** $\varnothing$, $\{-3\}$, $\{5\}$, $\{6\}$, $\{8\}$, $\{-3, 5\}$, $\{-3, 6\}$, $\{-3, 8\}$, $\{5, 6\}$, $\{5, 8\}$, $\{6, 8\}$, $\{-3, 5, 6\}$, $\{-3, 5, 8\}$, $\{-3, 6, 8\}$, $\{6, 5, 8\}$, $\{-3, 5, 6, 8\}$ **23.** $\varnothing$ **25.** Not empty   **27.** $\varnothing$ **29.** $\varnothing$
31. $\varnothing$ **33.** $\{1, 2, 4, 6, 7\}$ **35.** $\{a, b, c, d, x, y, z\}$ **37.** $\{9, 12\}$ **39.** $\{17, 18, 19\}$

41. $\{11, 12, 13\}$ **43. (a)** $\{-1, 0, 12\}$ **(b)** $\{1, 12, 13\}$ **(c)** $\{12\}$ **(d)** $\{-1, 0, 1, 12, 13\}$ **45.** $\{1, 2, 3\}$
47. $\{1, 2, 3, 6, 9\}$ **49.** $\{2, 3\}$ **51.** $\varnothing$ **53.** $\{1, 2, 3, 6, 9\}$ **55.** $\{13, 22, 33, \dots\}$
57. $\{1, 2, 3, 4\}$ **59. (a)** $\subseteq$ **(b)** $\nsubseteq$ **(c)** $=$ **(d)** $\nsubseteq$ **(e)** $\subseteq$ **61.** $\{x \mid x$ is an integer that is a multiple of 35$\}$
63. $A \cup B = \{5, 6, 10, 12, 15, 18, 20, 24, 25, 30\}$ **65.** Disjoint   **67.** $A \cap B$ is the set of students at Miami Bay
University who are taking both finite math and American history.   **69. (a)** $A \cap B$ is the set of students at Winfield
College who had a 4.0 GPA for both the fall 1997 and spring 1998, semesters.   **(b)** A is not necessarily a subset of B
because some students may have a 4.0 GPA in the fall semester and not in the spring semester.   **(c)** $A \cup B$ is the set of
students who had a 4.0 GPA for the fall semester, the spring semester, or both.

Section 6.2

1. (a) 10   **(b)** 5   **(c)** 4   **(d)** 11   **3.** 180   **5.** 7   **7.** 19   **9.** 35   **11. (a)** 42   **(b)** 60   **(c)** 84
(d) 57   **(e)** 39   **(f)** 81   **(g)** 15   **(h)** 15   **(i)** 81   **13. (a)** 35   **(b)** 9   **(c)** 6   **(d)** 38   **(e)** 38   **(f)** 12
15.   **17.** 225   **19. (a)** 14   **(b)** 24   **(c)** 38   **(d)** 44   **21. (a)** 11   **(b)** 14   **(c)** 9

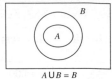

(d) 16   **23. (a)** 21   **(b)** 10   **(c)** 8   **(d)** 56   **(e)** 29   **(f)** 62   **25. (a)** 28
(b) 9   **(c)** 20   **(d)** 7   **27.** 69   **29.** It must be the case that
$n(A \cup B) \geq n(A)$, but it is not.   **31. (a)** 30   **(b)** 20   **33.** Yes, because these
numbers total only 133, not 135   **35.** Total number of people $= 60$, which is not
consistent with the reported 56 total

37. (a) $n(A \cap B) = 0$ **(b)** $A \cap B =$ empty set.   **39.** When $A \subset B$, the Venn diagram is
41. 23   **43.** 1045

$A \cup B = B$

Section 6.3

1.

Shirt	Ties
Blue	Red / Green
Yellow	Red / Green
White	Red / Green

3.

H → H, T
T → H, T

5. 12   **7.** 45   **9.** 42   **11.** 28

13. (a) 169   **(b)** 28,561   **15.** 462   **17.** 90   **19.** 22   **21.** 4096   **23.** 8,000,000
25. (a) 1296   **(b)** 360   **27.** 26,000   **29.** 56   **31.** 18
33.

Allen — Jones — Cooper
Allen — Cooper — Jones
Jones — Allen — Cooper
Jones — Cooper — Allen
Cooper — Allen — Jones
Cooper — Jones — Allen

35. 576   **37.** 20   **39. (a)** 210   **(b)** 30   **(c)** 60   **(d)** 60
(e) 120   **41.** 384   **43.** 144   **45. (a)** 360   **(b)** 750
47. 1440   **49.** 128

51.

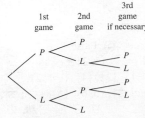

1st game, 2nd game, 3rd game if necessary

53. There are $26 \times 26 \times 26 = 17{,}576$ different sets of three initials, so it is not
possible to index all slides with three initials. If four initials are used, there are
$26^4 = 456{,}976$ possibilities, which is more than adequate for the 22,500 slides.

55. (a) 1,000,000,000 **(b)** This provides 308,915,776 possibilities, less than the current system. **(c)** If all ten digits and all 26 letters are used, there are 2,176,782,336 possible Social Security numbers, more than the current system. If the 26 letters of the alphabet and the eight digits 2–9 are used, there are 1,544,804,416 possibilities, more than the current system. **57.** About 217 billion

Section 6.4

1. 6 **3.** 120 **5.** 720 **7.** 840 **9.** 95,040 **11.** 360 **13.** 970,200 **15.** 840
17. 120 **19.** 2520 **21.** 336 **23. (a)** 5040 **(b)** 210 **25.** 6 **27.** 9240
29. (a) 24 **(b)** 840 **31.** 210 **33.** 120 **35.** 24 **37. (a)** 24 **(b)** 64 **39.** 658,627,200
41. 20,160 **43.** 120 **45. (a)** 10,080 **(b)** 34,650 **(c)** 151,200 **(d)** 151,200 **47. (a)** 35 **(b)** 21
49. (a) 35 **(b)** 35 **(c)** 70 **51.** 5040 **53.** 19,656 **55. (a)** $P(n, 2) = n(n-1)$
(b) $P(n, 3) = n(n-1)(n-2)$ **(c)** $P(n, 1) = n$ **(d)** $P(n, 5) = n(n-1)(n-2)(n-3)(n-4)$
57. 1,423,656 **59.** 32,760 **61. (a)** 35 **(b)** 56 **69. (a)** 5,814 **(b)** 1,320 **(c)** 1,716

Using Your TI-83, Section 6.4

1. 95,040 **3.** 201,600

Using EXCEL, Section 6.4

1. 120 **3.** 1.15585 E+29

Section 6.5

1. 15 **3.** 286 **5.** 126 **7.** 1 **9.** $\{a, b\}, \{a, c\}, \{a, d\}, \{b, c\}, \{b, d\}, \{c, d\}$ **11.** $\{a, b, c, d\}$,
$\{a, b, c, e\}, \{a, b, d, e\}, \{a, c, d, e\}, \{b, c, d, e\}$ **13.** $x^5 + 5x^4y + 10x^3y^2 + 10x^2y^3 + 5xy^4 + y^5$
15. $x^{10} + 10x^9y + 45x^8y^2 + 120x^7y^3 + 210x^6y^4 + 252x^5y^5 + 210x^4y^6 + 120x^3y^7 + 45x^2y^8 + 10xy^9 + y^{10}$
17. 32 **19.** 15 **21.** 455 **23. (a)** 35 **(b)** 840 **25.** 5600 **27.** 4200 **29.** 249
31. 55 **33.** 63 **35.** 119 **37.** 181 **39.** 826 **41.** 187,200 **43.** 27,720 **45.** 3276
47. 15,504 **49. (a)** {Alice}, {Bianca}, {Cal}, {Dewayne} **(b)** {Alice, Bianca}, {Alice, Cal}, {Alice, Dewayne}, {Bianca, Cal}, {Bianca, Dewayne}, {Cal, Dewayne} **(c)** {Alice, Bianca, Cal}, {Alice, Bianca, Dewayne}, {Bianca, Cal, Dewayne}, {Alice, Cal, Dewayne} **(d)** {Alice, Bianca, Cal, Dewayne} **(e)** ∅ **(f)** 16 **51. (a)** 24
(b) 144 **53. (a)** 11,760 **(b)** 6885 **55. (a)** 4960 days **(b)** 11 days, with two students being invited the last day **59. (a)** 20 at $x = 3$ **(b)** 35 at $x = 3$ and 4 **(c)** 70 at $x = 4$ **(d)** 126 at $x = 4$ and 5 **(e)** 252 at $x = 5$
(f) $x = 25$ makes $C(50, x)$ largest. $x = 100$ makes $C(200, x)$ largest. **(g)** $x = 27$ and 28 make $C(55, x)$ largest.
$x = 150$ and 151 make $C(301, x)$ largest. **63. (a)** 25,827,165 **(b)** Height = 23.978 miles using 17 pennies per inch. **65.** 65,780 **69.** 6.35×10^{11} **71. (a)** The number is $P(45, 6) = 5,864,443,200$. **(b)** $C(45, 6) = 8,145,060$

Using Your TI-83, Section 6.5

1. 3003 **3.** 2,882,880 **5.** 72.833

Using EXCEL, Section 6.5

1. 20 **3.** 1.80535 E+13

Section 6.6

1. 44 **3.** 56 **5.** 6720 **7.** About 9.128×10^{11} **9.** 78 **11.** 120 **13.** 120
15. 8! = 40,320 **17.** 45 **19.** 225,225 **21. (a)** 8,000,000 **(b)** 483,840 **(c)** 7,516,160 **23.** 146
25. 870 **27.** 1,078,377,300 **29.** 793 **31.** 1281

Section 6.7

1. 369,600   **3.** 35   **5.** 1260   **7.** 15   **9.** 1680   **11.** 168,168   **13.** 17,153,136   **15.** 630,630
17. 3150   **19.** 252   **21.** 17,153,136   **23.** 280   **25.** 5775   **27.** 3,783,780   **29.** 840,840
31. $\dfrac{22!}{3!(2!)^3 4!(4!)^4}$ **33.** 1260   **35.** $\dfrac{50!}{10!(5!)^{10}}$

Review Exercises, Chapter 6

1. (a) True   **(b)** False   **(c)** False   **(d)** False   **(e)** True   **(f)** False   **(g)** True   **(h)** False   **(i)** True   **(j)** False
(k) True   **(l)** False   **(m)** True   **(n)** False   **(o)** False   **3. (a)** Equal   **(b)** Not equal   **(c)** Equal, both are
empty   **5.** 14
7.

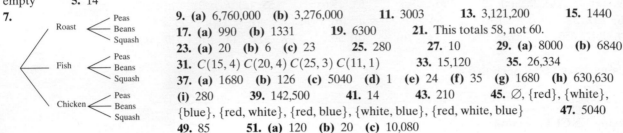

9. (a) 6,760,000   **(b)** 3,276,000   **11.** 3003   **13.** 3,121,200   **15.** 1440
17. (a) 990   **(b)** 1331   **19.** 6300   **21.** This totals 58, not 60.
23. (a) 20   **(b)** 6   **(c)** 23   **25.** 280   **27.** 10   **29. (a)** 8000   **(b)** 6840
31. $C(15, 4)\,C(20, 4)\,C(25, 3)\,C(11, 1)$ **33.** 15,120   **35.** 26,334
37. (a) 1680   **(b)** 126   **(c)** 5040   **(d)** 1   **(e)** 24   **(f)** 35   **(g)** 1680   **(h)** 630,630
(i) 280   **39.** 142,500   **41.** 14   **43.** 210   **45.** ∅, {red}, {white},
{blue}, {red, white}, {red, blue}, {white, blue}, {red, white, blue}   **47.** 5040
49. 85   **51. (a)** 120   **(b)** 20   **(c)** 10,080

CHAPTER 7

Section 7.1

1. (a) {true, false}   **(b)** {vowel, consonant} or the letters {a, b, c, …, x, y, z}   **(c)** {1, 2, 3, 4, 5, 6}   **(d)** {HHH, HHT, HTH, THH, TTH, THT, HTT, TTT}   **(e)** {Mon., Tue., Wed., Thurs., Fri., Sat., Sun.}   **(f)** {Grand Canyon wins, Bosque wins}, or {Bosque wins, Bosque loses}   **(g)** {pass, fail}, {A, B, C, D, F}   **(h)** {Susan and Leah, Susan and Dana, Susan and Julie, Leah and Dana, Leah and Julie, Dana and Julie}   **(i)** {Mon., Tue., Wed., Thur., Fri., Sat., Sun.}, or {1, 2, 3, 4, …, 31}   **(j)** {male, female}, {passing student, failing student}   **3. (a)** Male of normal weight, female of normal weight   **(b)** Female underweight, female of normal weight, female overweight   **(c)** Male underweight, female underweight, male overweight, female overweight   **5. (a)** Valid   **(b)** Valid   **(c)** Invalid: $P(T) + P(D) + P(H) = 0.95$ **(d)** Invalid: $P(2) + P(4) + P(6) + P(8) + P(10) = 1.15$ **(e)** Valid   **(f)** Invalid $P(\text{Utah}) = -0.4 < 0$ **(g)** Valid   **(h)** Invalid: $P(\text{maybe}) = 1.1 > 1$ **(i)** Invalid: $P(\text{True}) + P(\text{False}) = 0$
7. (a) 0.3   **(b)** 0.6   **(c)** 0.8   **(d)** 1   **9. (a)** 0.65   **(b)** 0.35   **11. (a)** 0.21   **(b)** 0.43   **(c)** 0.64   **(d)** 0.27
13. (a)

	Busby	Butler	Hutchison
Calculator	0.160	0.136	0.210
No Calculator	0.185	0.198	0.111

(b) 0.506   **(c)** 0.185   **(d)** 0.334

15. $P(A) = 0.20, P(B) = 0.30, P(C) = 0.40, P(D) = 0.10$ **17.** $P(C) = 0.25, P(B) = 0.20, P(A) = 0.55$
19. $P(\text{Mini}) = \frac{140}{800} = 0.175, P(\text{Burger}) = \frac{345}{800} = 0.431, P(\text{Big B}) = \frac{315}{800} = 0.394$ **21.** $P(\text{below 40}) = 0.089,$
$P(40\text{–}49) = 0.150, P(50\text{–}65) = 0.569, P(\text{over 65}) = 0.192$ **23. (a)** 0.46   **(b)** 0.54   **(c)** 0.55   **25. (a)** $\frac{1}{13}$
(b) $\frac{2}{13}$ **(c)** $\frac{1}{4}$ **(d)** $\frac{1}{2}$

Section 7.2

1. $\frac{4}{15} = 0.267$ **3.** $\frac{22}{36} = 0.611$ **5.** $\frac{1}{6}$ **7.** $\frac{45}{136} = 0.33$ **9.** 0.0045   **11. (a)** $\frac{5}{9}$ **(b)** $\frac{4}{9}$ **(c)** 1
13. $\frac{3}{4}$ **15.** $\frac{594}{1330} = 0.447$ **17. (a)** 0.274   **(b)** 0.095   **(c)** 0.516   **19. (a)** $\frac{1}{1365} = 0.00073$ **(b)** $\frac{44}{1365} = 0.0322$

(c) $\frac{22}{91} = 0.242$ (d) $\frac{22}{91} = 0.242$ **21. (a)** $\frac{15}{728} = 0.0206$ **(b)** $\frac{15}{364} = 0.0412$ **23. (a)** $\frac{1}{70} = 0.0143$
(b) $\frac{3}{35} = 0.0857$ **25.** $\frac{12}{336} = 0.0357$ **27.** $\frac{8}{21} = 0.381$ **29.** $\frac{5}{9} = 0.555$ **31. (a)** $\frac{1}{120}$ **(b)** $\frac{1}{10}$
33. $\frac{48}{143} = 0.336$ **35. (a)** $\frac{3}{14}$ **(b)** $\frac{3}{28}$ **(c)** $\frac{5}{14}$ **37. (a)** 0.462 **(b)** 0.118 **39. (a)** $\frac{4}{5}$ **(b)** $\frac{1}{5}$
41. (a) $\frac{8}{65} = 0.123$ **(b)** $\frac{28}{195} = 0.144$ **(c)** $\frac{2}{91} = 0.022$ **43.** $\frac{324}{570} = 0.568$ **45.** $\frac{1287}{2,598,960} = 0.000495$
47. $\frac{720}{6^6} = 0.0154$ **49. (a)** $3{,}838{,}380$ **(b)** $\frac{1}{3,838,380}$ **51.** 50 **53. (a)** $\frac{1}{40} = 0.025$ **(b)** $\frac{1}{8} = 0.125$
(c) $\frac{1}{6} = 0.167$ **55.** The calculated probability indicates that two pennies dated before 2000 should occur about 22% of the time compared to the 25% found by the class.   **57. (a)** $\frac{1}{25,827,165} = 0.0000000387$ **(b)** $\frac{288}{25,827,165} = 0.0000112$

Section 7.3

1. $E \cup F = \{1, 2, 3, 4, 5, 7, 9\}, E \cap F = \{1, 3\}, E' = \{2, 4, 6, 8, 10\}$ **3.** $E \cup F =$ the set of students who are passing English or failing chemistry (or both). $E \cap F =$ the set of students who are passing English and failing chemistry. $E' =$ the set of students who are failing English.   **5.** $\frac{2}{5}$ **7.** 0.3 **9.** 0.7 **11.** Mutually exclusive
13. Not mutually exclusive   **15.** Mutually exclusive   **17. (a)** $\frac{1}{270,725}$ **(b)** $\frac{2}{270,725}$ **19.** 0.600
21. (a) $\frac{21}{22} = 0.955$ **(b)** $\frac{37}{44} = 0.841$ **23.** $\frac{221}{980} = 0.226$ **25. (a)** $\frac{3}{10} = 0.30$ **(b)** $\frac{11}{20} = 0.55$ **(c)** $\frac{11}{80} = 0.138$
(d) $\frac{285}{400} = 0.713$ **(e)** $\frac{165}{400} = 0.413$ **(f)** $\frac{235}{400} = 0.588$ **27. (a)** $\frac{4}{35}$ **(b)** $\frac{4}{7}$ **29.** 0.81 **31.** 0.51 **33. (a)** 0.90
(b) 0.10 **35. (a)** 0.84 **(b)** 0.16 **37. (a)** $\frac{1}{2}$ **(b)** $\frac{1}{3}$ **(c)** $\frac{1}{2}$ **(d)** $\frac{2}{3}$ **39.** Mutually exclusive
41. Mutually exclusive   **43. (a)** $\frac{5}{6}$ **(b)** $\frac{1}{6}$ **(c)** $\frac{1}{2}$ **(d)** $\frac{1}{3}$ **(e)** $\frac{1}{3}$ **45. (a)** $\frac{1}{10}$ **(b)** $\frac{1}{10}$ **(c)** $\frac{1}{2}$ **(d)** $\frac{1}{2}$ **(e)** $\frac{1}{2}$
(f) $\frac{4}{5}$ **(g)** $\frac{9}{10}$ **47.** $\frac{16}{52} = \frac{4}{13}$ **49.** 0.49 **51.** 10% **53.** $\frac{2}{3}$ **55. (a)** $12{,}341/246{,}905 = 0.050$
(b) $43(2016)/246{,}905 = 0.351$ **(c)** $22{,}016/246{,}905 = 0.089$ **57.** 0.97 **59. (a)** $P(\text{At least one 6 in 4 rolls}) = 0.518$ **(b)** $P(\text{Two sixes at least once in 26 rolls}) = 0.519$. The two events are essentially equally likely.

Section 7.4

1. (a) $\frac{3}{4}$ **(b)** $\frac{1}{2}$ **3. (a)** $\frac{16}{45}$ **(b)** $\frac{7}{24}$ **(c)** $\frac{9}{16}$ **5. (a)** $\frac{3}{8}$ **(b)** $\frac{7}{10}$ **7.** $P(K\,|\,F) = \frac{1}{10}, P(F\,|\,K) = 1$
9. (a) $\frac{7}{15}$ **(b)** $\frac{3}{10}$ **(c)** $\frac{3}{14}$ **11.** $P(E\,|\,F) = \frac{3}{7}, P(F\,|\,E) = \frac{1}{2}$ **13.** $P(E\,|\,F) = \frac{3}{5}, P(F\,|\,E) = \frac{2}{5}$
15. (a) $\frac{4}{9} = 0.444$ **(b)** $\frac{4}{15} = 0.267$ **17. (a)** 0.0825 **(b)** 0.523 **(c)** 0.292 **(d)** 0.177 **(e)** 0.400 **(f)** 0.440
19. $\frac{4}{663} = 0.006$ **21. (a)** $\frac{1}{169}$ **(b)** $\frac{1}{16}$ **(c)** $\frac{1}{4}$ **23.** $\frac{2}{55}$ **25.** $\frac{15}{91} = 0.165$ **27.** $\frac{9}{1024} = 0.0088$
29. (a) $\frac{16}{225}$ **(b)** $\frac{12}{225}$ **31. (a)** $\frac{16}{375}$ **(b)** $\frac{1}{27}$ **33.** $\frac{2}{10}$ **35. (a)** 0.556 **(b)** 0.350 **(c)** 0.400 **(d)** 0.378
(e) 0.588 **37. (a)** 0.189 **(b)** 0.261 **(c)** 0.521 **39. (a)** $\frac{17}{462} = 0.037$ **(b)** $\frac{1}{154} = 0.006$ **(c)** $\frac{34}{77} = 0.442$
(d) $1 - 0.006 = 0.994$.
41. $P(\text{forged}) = \frac{1}{10,000}$, $P(\text{postdated}) = 0.05$,

$$P(\text{postdated}\,|\,\text{forged}) = 0.80. \text{ Find}$$
$$P(\text{forged}\,|\,\text{postdated}) = \frac{P(\text{forged and postdated})}{P(\text{postdated})}$$

As $P(\text{postdated}\,|\,\text{forged}) = \frac{P(\text{forged and postdated})}{P(\text{forged})}$, then
$$0.80 = \frac{P(\text{forged and postdated})}{0.0001}$$
$$P(\text{forged and postdated}) = 0.00008$$
$$P(\text{forged}\,|\,\text{postdated}) = \frac{0.00008}{0.05} = 0.0016$$

43. 0.27 **45. (a)** $\frac{1}{729}$ **(b)** $\frac{6}{729}$ **47. (a)** $\frac{1}{10}$ **(b)** $\frac{4}{5}$ **(c)** $\frac{9}{10}$ **49. (a)** 0.438 **(b)** 0.563 **(c)** 0.778
51. (a) 0.0375 **(b)** 0.178
55. (a) A wager is good if the probability of winning is greater than one half.
(b)   (i) $P(2 \text{ different colors}) = \frac{5}{9}$. This is a good wager.
(ii) $P(2 \text{ different numbers}) = \frac{8}{9}$. This is a good wager.
(c)   (i) $P(3 \text{ different colors}) = 0.275$. This is not a good wager.
(ii) $P(3 \text{ different numbers}) = 0.593$. This is a good wager.
(d)   (i) $P(4 \text{ different colors}) = 0.129$. This is not a good wager.
(ii) $P(4 \text{ different numbers}) = 0.264$. This is not a good wager.

57. (a) 0.63 **(b)** 0.18 **(c)** 0.17 **(d)** 0.54 **(e)** 0.36

Section 7.5

1. **(a)** **(i)** Not mutually exclusive
(ii) Independent, as $P(E) = \frac{1}{4} = P(E\,|\,F)$
(b) **(i)** Not mutually exclusive
(ii) Dependent, as $P(E) = \frac{1}{4}$ and $P(E\,|\,F) = \frac{1}{5}$
(c) **(i)** Not mutually exclusive
(ii) Independent, as $P(E) = \frac{1}{2} = P(E\,|\,F)$
(d) **(i)** Mutually exclusive
(ii) Dependent, as $P(E \cap F) = 0 \neq P(E)P(F) = \frac{1}{16}$
(e) **(i)** Mutually exclusive
(ii) Dependent, as $P(E) = \frac{1}{4}$ and $P(E\,|\,F) = 0$
(f) **(i)** Mutually exclusive
(ii) Independent, as $P(E) = 0 = P(E\,|\,F)$

3. Independent, as $(0.3)(0.5) = 0.15$ **5.** Dependent, as $(0.3)(0.7) \neq 0.20$
7. **(a)** $P(F\,|\,E) = 0.60$ $P(E \cup F) = 0.74$ **(b)** E and F are independent.   **9.** 0.86   **11.** **(a)** Not mutually exclusive; a ring may have both a diamond and ruby.   **(b)** Independent   **13.** Dependent   **15.** $\frac{12}{49}$
17. $\frac{1}{20}$ **19.** **(a)** 0.06 **(b)** 0.56 **(c)** 0.38   **21.** **(a)** 0.24 **(b)** 0.04   **23.** **(a)** 0.65 **(b)** 0.68 **(c)** 0.76
25. **(a)** 0.92 **(b)** 0.52   **27.** 0.93   **29.** $P(\text{male})P(\text{mathematics major}) = \frac{6}{30} = P(\text{male and mathematics major})$ so "male" and "mathematics major" are independent.   **31.** 0.20
33. **(a)**

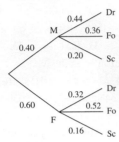

(b) 0.456 **(c)** 0.176 **(d)** Dependent
35. Independent   **37.** **(a)** $(0.5)^{12} = 0.000244$ **(b)** $(0.90)^{12} = 0.282$ **39.** $\frac{1}{5}$
41. $P(\text{met verbal}) = 0.89 \neq P(\text{met verbal}\,|\,\text{met math}) = 0.94$, so dependent.
43. $P(\text{female}) = 0.4545 = P(\text{female}\,|\,\text{science})$, so independent.   **45.** $P(\text{at most 1 head}) = \frac{5}{16}$ and $P(\text{at most 1 head}\,|\,\text{at least 1 head and 1 tail}) = \frac{4}{14}$, so the events are dependent.   **53.** For Ron, $P(\text{no malfunction for 30 days}) = 0.74$. Angela can go 20 days with the probability of malfunction less than 0.01.   **55.** There are more people than months, so at least two must share the same month.   **57.** There are $2^6 = 64$ ways to answer the questions. Because there are more students than ways to answer the questions, at least 2 students must give the same set of answers.

Section 7.6

1.

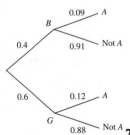

3. **(a)** 0.778 **(b)** 0.923   **5.** **(a)** $P(E_1 \cap F) = \frac{1}{6}$, $P(E_2 \cap F) = \frac{1}{12}$, $P(E_3 \cap F) = \frac{1}{12}$
(b) $\frac{1}{3}$ **(c)** $P(E_1\,|\,F) = \frac{1}{2}$, $P(E_3\,|\,F) = \frac{1}{4}$

7. **(a)** $P(F) = \frac{37}{84}$ **(b)** $P(E_1) = \frac{32}{84} = \frac{8}{21}$ **(c)** $P(E_2) = \frac{52}{84} = \frac{13}{21}$ **(d)** $P(F\,|\,E_1) = \frac{15}{32}$
(e) $P(F\,|\,E_2) = \frac{22}{52} = \frac{11}{26}$
(f) $P(E_1\,|\,F) = \frac{15}{37}$ **(g)** $P(E_1 \cap F) = \frac{15}{84} = \frac{5}{28}$ **(h)** $\dfrac{P(E_1 \cap F)}{P(F)} = \dfrac{\frac{15}{84}}{\frac{37}{84}} = \frac{15}{37}$ **(i)** $\frac{15}{37}$ **9.** 0.804   **11.** 0.522

13. 0.686   **15.** 0.357   **17.** **(a)**

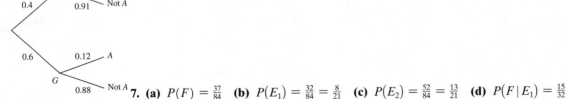

(b) $P(\text{black coffee}\,|\,\text{regular}) = 0.40$
$P(\text{black coffee and regular}) = 0.14$
$P(\text{black coffee}) = 0.303$
(c) $P(\text{flavored coffee}\,|\,\text{cream or sugar}) = 0.699$

19. 0.699   **21.** 0.354   **23.** 0.991   **25. (a)** $P(\text{4th grade}) = \frac{678}{1205} = 0.563$ **(b)** $P(\text{5th grade}) = \frac{527}{1205} = 0.437$
(c) $P(\text{At grade level}) = \frac{666}{1205} = 0.553$ **(d)** $P(\text{At grade level} \mid \text{4th grade}) = \frac{342}{678} = 0.504$ **(e)** $P(\text{At grade level} \mid \text{5th}$
grade$) = \frac{324}{527} = 0.615$ **(f)** $P(\text{4th grade and at grade level}) = \frac{342}{1205} = 0.284$ **(g)** $P(\text{4th grade} \mid \text{At grade level}) =$
$\frac{342}{666} = 0.514$ **27.** 0.737   **29. (a)** 4.87%   **(b)** 0.452   **31.** 18.4%   **33.** 0.234   **35.** 0.294
37. 0.918   **39. (a)** $P(A) = 0.75$ **(b)** $P(D \mid A) = 0.45$ **(c)** $P(N \mid B) = 0.10$ **(d)** $P(N) = 0.1375$
(e) $P(C) = 0.475; P(D) = 0.3875; P(A \mid C) = 0.632; P(B \mid C) = 0.368; P(A \mid D) = 0.871; P(B \mid D) = 0.129;$
$P(A \mid N) = 0.818; P(B \mid N) = 0.182$

41. A and D are independent when
$$P(D \mid B) = 0.6.$$

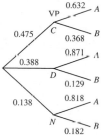

Section 7.7

1. (a) A transition matrix   **(b)** Not a transition matrix   **(c)** Not a transition matrix   **(d)** A transition matrix
3. (a) 10%   **(b)** 0.60   **(c)** 0.90   **(d)** 40%   **5. (a)** 0.4   **(b)** 0.8   **(c)** 0.3   **7.** [0.675 0.325]
9. $M_1 = M_0 T = [0.282 \quad 0.718]$ **11.** $M_1 = M_0 T = [0.175 \quad 0.375 \quad 0.45]$
$M_2 = M_1 T = [0.282 \quad 0.718]T = [0.32616 \quad 0.67384]$ $M_2 = M_1 T = [0.18 \quad 0.4075 \quad 0.4125]$

13. $ST = \begin{bmatrix} \frac{2}{3} & \frac{1}{3} \end{bmatrix} \begin{bmatrix} 0.6 & 0.4 \\ 0.8 & 0.2 \end{bmatrix} = \begin{bmatrix} \frac{2}{3} & \frac{1}{3} \end{bmatrix}$ **15.** $ST = \begin{bmatrix} \frac{13}{28} & \frac{15}{28} \end{bmatrix} \begin{bmatrix} 0.25 & 0.75 \\ 0.65 & 0.35 \end{bmatrix} = \begin{bmatrix} \frac{13}{28} & \frac{15}{28} \end{bmatrix} = S$

17. $MT = [0.25 \quad 0.33 \quad 0.42]$
$(MT)T = [0.259 \quad 0.323 \quad 0.418]$
$((MT)T)T = [0.2577 \quad 0.3233 \quad 0.419]$
$MT^3 = [0.2577 \quad 0.3233 \quad 0.419]$

19. 0.4   **21.** 42% will contribute next year. 40.2% will contribute in two years.
23. [0.38 0.62], [0.314 0.686], [0.2942 0.7058], [0.28826 0.71174], [0.286478 0.713522], [0.2859434 0.7140566] It
appears that [0.286 0.714] is the steady-state matrix. The steady-state matrix is actually $\begin{bmatrix} \frac{2}{7} & \frac{5}{7} \end{bmatrix}$.   **25.** $\begin{bmatrix} \frac{4}{17} & \frac{10}{17} & \frac{3}{17} \end{bmatrix}$
27. $\begin{bmatrix} \frac{18}{57} & \frac{29}{57} & \frac{10}{57} \end{bmatrix}$ **29.** Steady-state matrix $\begin{bmatrix} \frac{7}{29}, \frac{16}{29}, \frac{6}{29} \end{bmatrix}$ so at steady state, 24.14% are in full-meal, 55.17% are in one-
meal, 20.69% are in no plan.   **31.** $\begin{bmatrix} \frac{10}{11} & \frac{1}{11} \end{bmatrix}$ **33.** $\begin{bmatrix} \frac{3}{11} & \frac{4}{11} & \frac{4}{11} \end{bmatrix}$

35.

		To			
	1	2	3	4	5
1	0	$\frac{1}{2}$	0	0	$\frac{1}{2}$
2	$\frac{1}{2}$	0	$\frac{1}{2}$	0	0
From 3	0	$\frac{1}{2}$	0	$\frac{1}{2}$	0
4	0	0	$\frac{1}{2}$	0	$\frac{1}{2}$
5	$\frac{1}{2}$	0	0	$\frac{1}{2}$	0

37. When the process reaches a steady
state, 25% of the flowers will be red, 50% of
the flowers will be pink, and 25% of the
flowers will be white.

39. (a) 0.06   **(b)** 0.05
(c) 0.80   **(d)** 0.04

Review Exercises, Chapter 7

1. No   **3.** $\frac{146}{360} = 0.406$ **5.** $\frac{17}{30} = 0.567$ **7.** $\frac{2}{15}$ **9.** $\frac{4}{13}$ **11.** $\frac{1}{32}$ **13. (a)** $\frac{6}{11} = 0.545$
(b) $\frac{133}{528} = 0.252$ **15. (a)** $\frac{3}{13} = 0.231$ **(b)** 0.769   **17.** Because $P(\text{even})P(10) = \frac{5}{169} \neq P(\text{even} \cap 10) = \frac{4}{52}$,
even and 10 are not independent events.   **19. (a)** $\frac{1}{1296}$ **(b)** $\frac{24}{1296} = \frac{1}{54}$ **(c)** $\frac{1}{72}$ **21. (a)** 0.036   **(b)** 0.196
(c) 0.084   **23. (a)** 0.290   **(b)** 0.234   **25. (a)** 0   **(b)** $\frac{1}{40}$ **(c)** $\frac{2}{40}$ **(d)** $\frac{15}{40}$ **(e)** $\frac{7}{30}$ **27.** 0.154
29. (a) 12   **(b)** 12   **31. (a)** 0.06   **(b)** 0.09   **33. (a)** $\frac{225}{420} = 0.536$ **(b)** $\frac{68}{420} = 0.162$ **(c)** $\frac{118}{242} = 0.488$
(d) $\frac{98}{225} = 0.436$

35. P(seat belt not used and injuries) = 0.344
P(seat belt not used)P(injuries) = 0.204
Dependent

37. (a) 0.92   **(b)** 0.08

CHAPTER 8

Section 8.1

1. (a) 2   **(b)** 3   **(c)** 4   **(d)** 2   **(e)** 1

3.

Category	Frequency
85–99	7
100–114	8
115–129	13
130–144	7
145–159	5

5.

Stem	Leaves
2	637
3	1839
4	41081
5	379

7.

Stem	Leaves
2	2031
2	796
3	134
3	57686
4	0131
4	856

9. (a) 67   **(b)** 19   **(c)** 22   **(d)** Cannot determine   **(e)** 40   **(f)** Cannot determine

11.
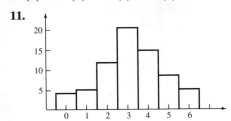

13.
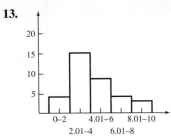

15.
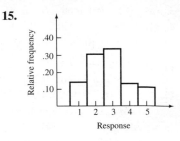

17.
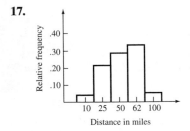

19.

Brand of coffee preferred

21.
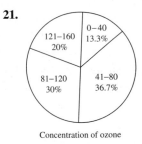

Concentration of ozone

23.
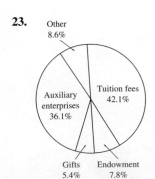

Income of Old Main University

25.
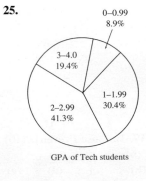

GPA of Tech students

29.
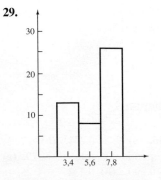

31.

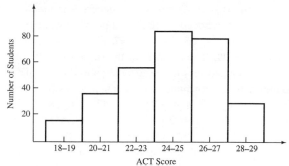

33.

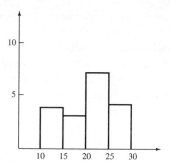

35.

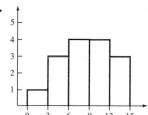

37.

39.

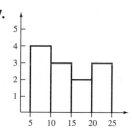

41.

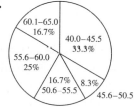

Section 8.2

1. 6   **3.** 3.9   **5.** 0.667   **7.** 4.825   **9.** 77.5   **11.** 85.07 mph   **13.** 9   **15.** 23
17. 11.5   **19.** 72   **21.** 5   **23.** 1 and 5   **25.** No mode   **27.** 73.83   **29.** 0.780   **31.** 11.19
33. 76   **35.** $138.40   **37.** Mean = $3.965, median = $3.95   **39. (a)** 5 **(b)** 2 **(c)** 2 **(d)** 2
41. 82   **43.** $29,538.46   **45.** $29,644.44   **47.** It is 82 or larger.   **49.** $1.30   **51.** 2.974
55. (a) Mean per capita income = $21,427; mean number below poverty = 34.5 million   **57.** 69.353
59. 22.3   **61.** Mean = 5; median = 5   **63.** Mean = 42.04; median = 42.3

Using Your TI-83, Section 8.2

1. Mean = 6.43, median = 7   **3.** Mean = 8.675, median = 7.9   **5.** 2.25

Using EXCEL, Section 8.2

1. Mean = 6.43, median = 7   **3.** Mean = 8.675, median = 7.9

Section 8.3

1. Mean = 16; variance = 15.5; $\sigma = \sqrt{15.5} = 3.94$ **3.** Mean = 9; variance = 10.4; $\sigma = \sqrt{10.4} = 3.22$
5. Mean = 35; variance = 266.2; $\sigma = 16.32$ **7.** Mean = -2; variance = 14; $s = 3.74$ **9.** Mean = 4.5;
variance = 24.68; standard deviation = 4.99   **11.** Mean = 3.5; variance = 2.15; standard deviation = 1.47
13. Mean = 5.90; variance = 0.0377; standard deviation = 0.194   **15.** Mean = 16.6; variance = 52.84;
$\sigma = 7.27$ **17.** Mean = 18.26; standard deviation = 4.21   **19.** Mean = 31; standard deviation = 3.12
21. (a) 1.25 **(b)** $\frac{10}{16} = 0.625$ **(c)** 0 **(d)** 176 **(e)** 146   **23.** Minimum = 3, $Q_1 = 6$, median = 11.5,
$Q_3 = 16$, maximum = 20
25. The five-point summary is $\{3, 5.5, 11, 16, 20\}$ **27.** The five-point summary is $\{4.2, 7.5, 11.6, 17.2, 23.8\}$

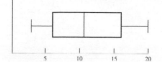

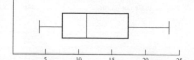

29. (a) For lathe I, $\sigma = 0.00346$; for lathe II, $\sigma = 0.00210$.   **(b)** Lathe II is more consistent because σ is smaller.
31. (a) Six 3's, one 2, and one 4   **(b)** Four 2's, no 3's, and four 4's   **(c)** No, because the largest possible value of the sum of squared deviations is 8, for which $\sigma = 1$.   **33.** 34 out of 110   **35.** 31   **37.** For the score of 86, $z = 1.75$; for the score of 82, $z = 1.67$, so the score of 86 is better.   **39.** The z-score for the runner was 1; the z-score for the cyclist was 1.67. The runner's time was one z-score higher (slower) than the mean, and the cyclist's time was 1.67 z-scores above the mean (slower). So, the runner had the better performance.   **43.** $\mu = 19.26$; $\sigma = 3.79$; $s = 4.23$ **45.** $\mu = 3.5$; $\sigma = 1.47$; $s = 1.50$ **47.** $\mu = 5.9$; $\sigma = 0.191$; $s = 1.94$ **49.** $\mu = 18.26$; $\sigma = 4.17$; $s = 4.21$ **51.** $\mu = 4935$; $\sigma = 1503.6$ **53.** $\mu = 8.26$ million; $\sigma = 2.11$ million   **55.** Mean $= 8.44$, population standard deviation $= 5.96$, sample standard deviation $= 6.33$.   **57. (a)** About 89%   **(b)** About 56%   **(c)** About 84%   **59. (a)** 23 of the 24, about 96%, fall in the interval. This is consistent because it is more than 75%.   **(b)** 22 of the 24, about 92%, fall in the interval. This is consistent because it is more than 56%.   **61.** 16 scores, 75%, fall in the 449–1029 interval. This is 7% higher than the Empirical Rule, but this suggests the data might be roughly bell shaped.

Using Your TI-83, Section 8.3

1. Mean $= 4.17$, population standard deviation $= 2.41$, sample standard deviation $= 2.64$.
3. Mean $= 8.18$, population standard deviation $= 2.66$, sample standard deviation $= 2.79$.
1. Five-point summary $= \{2, 3.5, 6, 11.5, 14\}$

Using EXCEL, Section 8.3

1. Range $= 16$, sample standard deviation $= 5.4556$, population standard deviation $= 5.2017$.
3. $\{17, 25, 33.5, 37, 48\}$

Section 8.4

1.

Outcome	X
HHH	3
HHT	2
HTH	2
THH	2
TTH	1
THT	1
HTT	1
TTT	0

3.

Outcome	X
Ann, Betty	2
Ann, Jason	1
Ann, Tom	1
Betty, Jason	1
Betty, Tom	1
Jason, Tom	0

5. 0, 1, 2, 3   **7. (a)** 0, 1, 2, 3, 4   **(b)** 0, 1, 2, 3   **9. (a)** Discrete   **(b)** Continuous   **(c)** Continuous   **(d)** Discrete   **11. (a)** Continuous   **(b)** Discrete   **(c) (i)** Discrete **(ii)** Continuous **(iii)** Discrete   **13.** Yes

15.

X	P(X)
0	$\frac{1}{8}$
1	$\frac{3}{8}$
2	$\frac{3}{8}$
3	$\frac{1}{8}$

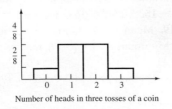

Number of heads in three tosses of a coin

17.

X	P(X)
0	0.02
1	0.68
2	0.07
3	0.08
4	0.10
5	0.01
8	0.02
10	0.02

19.

X	P(X)
0	$\frac{8}{75}$
1	$\frac{49}{75}$
2	$\frac{13}{75}$
3	$\frac{4}{75}$
4	$\frac{1}{75}$

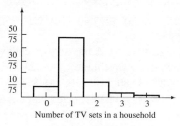

Number of TV sets in a household

21.

X	P(X)
0	0.533
5	0.333
10	0.133

23. (a) 1, 2, 3, 4, …   (b) Discrete

25.

X	No.
0	15
1	30
2	10

27.

X	P(X)
0	$\frac{3}{14}$
1	$\frac{8}{14}$
2	$\frac{3}{14}$

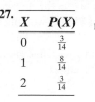

Number of men selected

29.

X	P(X)
0	$\frac{7}{15}$
1	$\frac{7}{15}$
2	$\frac{1}{15}$

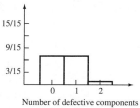

Number of defective components

31. (a)

X	P(X)
0	$\frac{6}{15}$
1	$\frac{8}{15}$
2	$\frac{1}{15}$

(b)

X	P(X)
0	$\frac{2}{3}$
1	$\frac{1}{3}$

33.

X	P(X)
1	$\frac{26}{50} = \frac{13}{25}$
5	$\frac{12}{50} = \frac{6}{25}$
10	$\frac{8}{50} = \frac{4}{25}$
20	$\frac{2}{50} = \frac{1}{25}$
50	$\frac{1}{50}$
100	$\frac{1}{50}$

35.

X	No.
0	1
1	2
2	3

37. (a) Weight of shoppers   (b) Shirt size, suit size, etc.
(c) Shoe size   (d) Number of cavities

39.

X	P(X)
2	$\frac{3}{7}$
3	$\frac{2}{7}$
4	$\frac{6}{35}$
5	$\frac{3}{35}$
6	$\frac{1}{35}$

41.

X	P(X)
1	$\frac{1}{2}$
2	$\frac{1}{4}$
3	$\frac{1}{4}$

43.

X	P(X)
0	0.160
1	0.192
2	0.648

45.

X (no. of cards drawn)	P(X)
1	0
2	$\frac{5}{11}$
3	$\frac{3}{11}$
4	$\frac{4}{33}$
5	$\frac{5}{66}$
6	$\frac{5}{154}$
7	$\frac{5}{231}$
8	$\frac{2}{231}$
9	$\frac{1}{154}$
10	$\frac{1}{462}$
11	$\frac{1}{462}$
12	$\frac{1}{462}$

The last possibility is that the cards alternate colors throughout, with probability $\frac{1}{462}$.

Section 8.5

1. 12.8 **3.** 325 **5.** $5.00 **7. (a)** $0.50 **(b)** $1.50 **9. (a)** $227.50 **(b)** $28,437.50

11. $\mu = 131; \sigma^2(X) = 1049; \sigma(X) = 32.39$ **13.** $\mu = 34; \sigma(X) = 24.98$

15.

X	P(X)
0	$\frac{11}{21}$
1	$\frac{44}{105}$
2	$\frac{2}{35}$

Expected number of defective = 0.533

17. $1 **19.** Almost 2 cents **21.** Venture A has an expected value of $23,000, and venture B $28,000, so B is apparently more profitable. **23.** 0.55

25. (a) $\mu = 2.7$ **(b)** $\sigma = 1.487$ **27.** $0.80 **29. (a)** $15.95 **(b)** $2,392,500

31. The payoff for winning is $11. **33. (a)** The expected value is about −5 cents per play, so a player can expect to lose an average of about 5 cents per play. **(b)** As the expected value is negative to the player, the game favors the house. **(c)** A payoff of $1.11 would make the game fair. **(d)** A payoff of $18 for winning a green bet makes the game fair. **35.** The expected value is 6.3. **37.** The expected value is 1.8.

39. (a) 2.5 **(b)** 2.48 **41. (a)** Four games in 1 way; five games in 4 ways; six games in 10 ways; seven games in 20 ways **(b)** 5.8 games **43.** 19.8

Section 8.6

1. 0.181 **3.** 0.251 **5.** 0.059 **7.** 0.208 **9.** 0.00045 **11.** 0.0879

13.

X	P(X successes)
0	0.2401
1	0.4116
2	0.2646
3	0.0756
4	0.0081

15.

X	P(X)
0	0.1681
1	0.3602
2	0.3087
3	0.1323
4	0.0284
5	0.0024

17.

X	P(X)
0	0.0778
1	0.2592
2	0.3456
3	0.2304
4	0.0768
5	0.0102

19.

X	P(X)
0	0.4823
1	0.3858
2	0.1157
3	0.0154
4	0.0008

21.

X	P(X)
0	0.0625
1	0.2500
2	0.3750
3	0.2500
4	0.0625

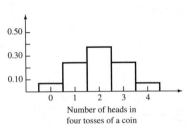

Number of heads in four tosses of a coin

23.

X	P(X)
0	0.0256
1	0.1536
2	0.3456
3	0.3456
4	0.1296

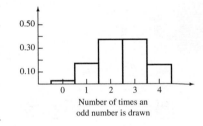

Number of times an odd number is drawn

25.

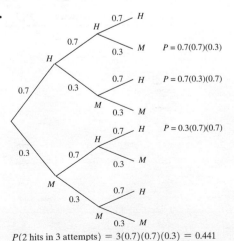

$P = 0.7(0.7)(0.3)$

$P = 0.7(0.3)(0.7)$

$P = 0.3(0.7)(0.7)$

$P(2 \text{ hits in } 3 \text{ attempts}) = 3(0.7)(0.7)(0.3) = 0.441$

27. 0.2048 **29.** 0.141 **31.** 0.0080 **33. (a)** 0.482 **(b)** 0.386 **(c)** 0.116 **(d)** 0.0154 **(e)** 0.00077 **35.** 0.234

37. $C(90, 40)(0.4)^{40}(0.6)^{50}$ **39.** $C(75, 35)(0.25)^{35}(0.75)^{40}$

41. 0.246 **43.** $n = 5$ and $p = 0.5; q = 0.5$

X	P(X heads)
0	0.0313
1	0.1563
2	0.3125
3	0.3125
4	0.1563
5	0.0313

45. 0.230 **47.** 0.683 **49.** 0.0355 **51.** 0.363

53. (a) 0.0439 **(b)** 0.0547 **(c)** 0.0547 **55.** 0.0080

57. 0.591 **59. (a)** 0.83 **(b)** $(0.83)^{56} = 0.000029$

61. 0.246 **63.** 11 games per day

65. (a) $n = 4; p = 0.6$

x	P(x successes)
0	0.0256
1	0.1536
2	0.3456
3	0.3456
4	0.1296

(b) $n = 4; p = 0.4$

x	P(x successes)
0	0.1296
1	0.3456
2	0.3456
3	0.1536
4	0.0256

(c) $n = 5; p = 0.5$

x	P(x successes)
0	0.0313
1	0.1563
2	0.3125
3	0.3125
4	0.1563
5	0.0313

(d) $n = 8; p = 0.3$

x	P(x successes)
0	0.0576
1	0.1977
2	0.2965
3	0.2541
4	0.1361
5	0.0467
6	0.0100
7	0.0012
8	0.0001

67. $C(220, 75)(0.65)^{75}(0.35)^{145} = 8.3 \times 10^{-21}$, practically zero   **69.** (a) 0.9988 (b) 0.745

71.

X	P(X)
0	0.0010
1	0.0098
2	0.0439
3	0.1172
4	0.2051
5	0.2461
6	0.2051
7	0.1172
8	0.0439
9	0.0098
10	0.0010

73.

X	P(X)
0	0.1335
1	0.3115
2	0.3115
3	0.1730
4	0.0577
5	0.0115
6	0.0013
7	0.0001

75.

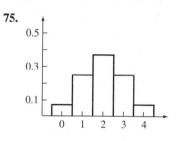

77.

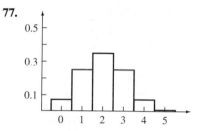

79.

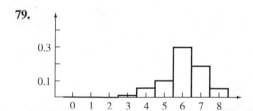

81. 0.1128   **83.** 0.0865   **85.** 0.0183

Using Your TI-83, Section 8.6

1.

X	P(X)
0	0.1785
1	0.3845
2	0.3105
3	0.1115
4	0.0150

3.

X	P(X)
0	0.0024
1	0.0284
2	0.1323
3	0.3087
4	0.3602
5	0.1681

Section 8.7

1. −3.0   **3.** 0.05   **5.** 2.65   **7.** 0.1915   **9.** 0.0987   **11.** 0.3643   **13.** 0.2734   **15.** 17.72%
17. 14.80%   **19.** 39.25%   **21.** 49.81%   **23.** 0.7888   **25.** 0.9722   **27.** 48.44%   **29.** 29.60%
31. 0.6255   **33.** 0.7855   **35.** 0.0968   **37.** 0.0082   **39.** 38.98%   **41.** 13.36%   **43.** 0.4953
45. 0.3413   **47.** 0.4649   **49.** 0.7888   **51.** 0.2417   **53.** 0.2039   **55.** 0.1360   **57.** 0.4207
59. 0.3174   **61.** 44.30%   **63.** 65.54%   **65.** 30.85%   **67.** 94.52%   **69.** 1.41   **71.** 1.08
73. 1.70   **75.** $\mu = 20; \sigma^2(X) = 12; \sigma(X) = 3.46$ **77.** $\mu = 312; \sigma^2(X) = 149.76; \sigma(X) = 12.24$
79. $\mu = 37.6; \sigma^2(X) = 34.592; \sigma(X) = 5.88$ **81.** $np = 35; nq = 15$, so the normal distribution is a good estimate.   **83.** $np = 36; nq = 4$, so normal distribution is not a good estimate.   **85.** $np = 12.5; nq = 12.5$, so
normal distribution is a good estimate.   **87.** The smallest integer value of n is 15.   **89. (a)** 0.0377   **(b)** 0.0118
(c) 0.0805   **91.** $\mu = 6; \sigma = 1.90$ **(a)** 0.5704   **(b)** 0.8132   **(c)** 0.3040   **93.** $\mu = 8; \sigma = 2$ **(a)** 0.8944
(b) 0.9599   **(c)** 0.2266   **95.** $\mu = 7.2; \sigma = 2.24$ **(a)** 0.8485   **(b)** 0.9292   **(c)** 0.2236   **97. (a)** 21.19%
(b) 0.2119   **99. (a)** 0.0977   **(b)** 0.2033   **(c)** 0.3256   **101. (a)** 0.9830   **(b)** 0.0008   **(c)** 0.0162
103. 0.5167   **105. (a)** 0.0796   **(b)** 0.7286   **(c)** 0.2358   **107.** 0.1493   **109. (a)** 86   **(b)** 93
111. (a) 0.5468   **(b)** 0.7612   **(c)** 0.5000, 0.7422   **(d)** 352 or higher   **(e)** 0.1079   **113.** 0.621   **115.** 0.0359
117. 0.9357   **119.** 0.9265   **121. (a)** 47   **(b)** 0.3745   **123. (a)** 0.5696   **(b)** The normal distribution gives
a slightly higher estimate.   **125. (a)** 0.6686   **(b)** 0.6680   **(c)** The normal distribution estimate is quite close.
127. (a) $\mu = 30; \sigma = 3.873$ **(b)** Integers in the interval $27 \le x \le 33$ **(c)** 0.6338   **129. (a)** 0.2417   **(b)** 0.1934
(c) 0.1452   **(d)** 0.0968   **(e)** 0.0484   **(f)** 0.0242   **(g)** 0.0049   **131.** No, the mean should be at the midpoint of a
normal distribution.   **133.** Yes, the mean and median are approximately equal and near the midpoint.
135. As a whole, the curve does not have the symmetry needed for a normal distribution. The portion obtained by deleting
the 7 rightmost bars is somewhat normal in shape, so the data in that region might be considered approximately normal.
137. (a) Mean $= 69.91$, median $= 73$, 25th percentile $= 62$, 75th percentile $= 82$.   **(b)** $\sigma = 17.37$ **(c)** The interval
$69.91 \pm \sigma$ (52.54, 87.28) contains 72% of the scores, which is fairly consistent with 68% in a normal distribution. The interval $69.91 \pm 2\sigma$ (35.17, 104.65) contains 94% of the scores, which is consistent with 95.4% in a normal distribution. The
interval $69.91 \pm 3\sigma$ (17.8, 122.02) contains 98.9% of the scores, which is consistent with 99.7% in a normal distribution.
(d)

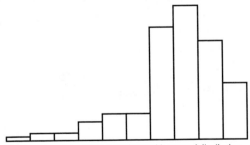

The percentages in part (c) are consistent with a normal distribution.
However, the mean should be approximately 50, and the scores should
extend up to 125.

Using Your TI-83, Section 8.7

1. 0.6997   **3.** 0.7625

Using EXCEL, Section 8.7

1. 0.9088   **3.** 0.5074   **5.** 0.1499   **7.** 0.8783

Section 8.8

1. (a) 0.070   **(b)** 0.047   **(c)** 0.020   **3.** $0.1931 < x < 0.4069$ **5.** $0.1490 < x < 0.4510$
7. $0.451 < x < 0.549$ **9. (a)** $0.4026 < x < 0.4974$ **(b)** $0.3937 < x < 0.5063$ **11.** $0.7573 < x < 0.8627$
13. (a) $0.738 = 73.8\%$ **(b)** $0.6763 < x < 0.7997$ **(c)** $0.6567 < x < 0.8193$ **15. (a)** Error bounds: 0.2198 and
0.3202   **(b)** Error bounds: 0.2425 and 0.2975   **(c)** The larger sample yields a smaller confidence interval.

17. (a) $0.2179 < x < 0.3421$ **(b)** $0.2360 < x < 0.3240$ **19.** $0.3955 < x < 0.5045$ **21.** $0.4339 < x < 0.6461$
23. 267   **25.** 2401   **27. (a)** $0.5960 < x < 0.7040$ **(b)** $0.6007 < x < 0.6993$ **(c)** $0.6033 < x < 0.6967$
(d) $0.6064 < x < 0.6936$ **(e)** $0.6082 < x < 0.6918$ **(f)** $0.6204 < x < 0.6796$ **29.** 156   **31.** $\bar{p} = 0.52$;
$n = 1825$ students   **33.** The probability that less than 54% of the sample favor the fees is 0.017.   **35.** The
probability that at least 50% are from out of state is 0.0122.   **37.** 98% confidence level   **39.** The proportion
of defective labels is in the interval $0.000755 < p < 0.00675$ with probability 0.95, so the machine should not be shut
down.   **41.** The sample proportion would exceed 0.6897 only about 8.08% of the time.

Review Exercises, Chapter 8

1.

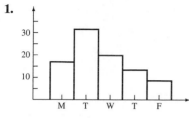
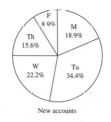

New accounts

3. (a) 6   **(b)** 10.5   **(c)** 2   **5.** 2.64   **7.** 26

9. 0, 1, 2, 3   **11.** 2.85   **13.** 2.67   **15.** 0.0227   **17.**

X	P(X)
0	0.3164
1	0.4219
2	0.2109
3	0.0469
4	0.0039

19. (a) 0.6368   **(b)** 0.0009   **(c)** 0.1669   **21.** 81st percentile

23.

X	P(X)
1	0.025
2	0.200
3	0.4625
4	0.2688
5	0.0438

25.

X	Number of Outcomes
0	10
1	30
2	15
3	1

27.

X	Number of Ways x Can Occur
1	1
2	1
3	4

29. 11.238   **31. (a)** 0.0616   **(b)** 0.0181   **(c)** 0.0159   **33.** $0.575 < x < 0.891$

CHAPTER 9

Section 9.1

1.
$$
\begin{array}{c}
 & C \\
 & \begin{array}{cc} H & T \end{array} \\
\begin{array}{c} R \end{array} \begin{array}{c} H \\ T \end{array} & \begin{bmatrix} -1 & -0.5 \\ -0.5 & 2 \end{bmatrix}
\end{array}
$$
3.
$$
\begin{array}{c}
 & C \\
 & \begin{array}{cc} 1 & 2 \end{array} \\
\begin{array}{c} R \end{array} \begin{array}{c} 1 \\ 2 \end{array} & \begin{bmatrix} 2 & -3 \\ -3 & 4 \end{bmatrix}
\end{array}
$$
5. (a) R receives 10 from C.   **(b)** 10   **(c)** 10   **(d)** 8   **7.** Strictly determined, the saddle point is $(2, 2)$,
value $= -1$, and the solution is row 2 and column 2.   **9.** Not strictly determined   **11.** Strictly determined,
the saddle point is $(1, 2)$, value $= 0$, and the solution is row 1 and column 2.   **13. (a)** Strictly determined, saddle
point $(2, 1)$, value $= 0$, solution row 2, column 1   **(b)** Not strictly determined because the largest row minimum does
not equal the smallest column maximum   **(c)** Strictly determined, saddle point $(2, 3)$, value $= 3$, solution row 2 and
column 3   **(d)** Strictly determined, saddle point $(2, 2)$, value $= 0$, solution row 2, column 2   **15.** ReMark should

choose option 1, the sports highlights, and Century should choose option 2, the game show. The value is -5. ReMark should expect to lose 5 points in the ratings.

17. (a) The saddle point is $(1, 1)$, so the farmer should plant milo.

(c) Total income over five years:

Milo: $85 + 3(120) + 150 = 595$

Corn: $60 + 3(165) + 235 = 790$

Wheat: $70 + 3(150) + 175 = 695$

In the long term, corn appears to be the best and wheat second.

Section 9.2

1. 17.5   **3.** 0   **5. (a)** 40 **(b)** 7.5 **(c)** $\frac{34}{3}$ **(d)** 8.4   **7.** Row strategy $= \begin{bmatrix} \frac{5}{6} & \frac{1}{6} \end{bmatrix}$ and column

strategy $= \begin{bmatrix} \frac{1}{3} & \frac{2}{3} \end{bmatrix}$, $E = 11.67$ **9. (a)** $\begin{bmatrix} -5 \\ 0 \end{bmatrix}$ **(b)** $\begin{bmatrix} 5 & 2 & -2 \\ 3 & 1 & 13 \\ 1 & 3 & 6 \end{bmatrix}$

11. $P = \begin{bmatrix} \frac{1}{2} & 0 & \frac{1}{2} \end{bmatrix}$; $Q = \begin{bmatrix} \frac{5}{6} & 0 & \frac{1}{6} \end{bmatrix}$; $E = \frac{11}{2}$ **13. (a)** $E = 0$. This is a fair game. **(b)** $E = \frac{13}{5}$. This is not a fair game. **(c)** $E = \frac{36}{16}$. This is not a fair game.   **15.** ReMark should show the sports highlights 70% of the time, the mystery drama 30% of the time, and drop the variety show. Century should air the talk show 40% of the time, the game show 60% of the time, and drop the educational documentary. Using these strategies, ReMark can expect an average gain of 1 rating point.   **17. (a)** 27 years **(b)** 17 years **(c)** No surgery is the better option when $q_1 < 0.20$.   **19. (a)** 20.6 years **(b)** 9.6 years **(c)** Surgery is the better option when the probability of malignancy is greater than 0.15.

Review Exercises, Chapter 9

1. (a) This game is not strictly determined. **(b)** This game is strictly determined. The $(2, 2)$ location is the saddle point. The value of the game is 4. **(c)** This game is strictly determined. The $(2, 2)$ location is the saddle point. The value of the game is 275. **(d)** This game is strictly determined. The $(3, 2)$ location is the saddle point. The value of the game is 2.   **3. (a)** This game is not strictly determined, so there is no solution. **(b)** Offense adopts strategy 2, and the defense adopts strategy 1.   **5. (a)** 7.16 **(b)** 6.8 **(c)** 2.4   **7.** The farmer should plant 24% of the crop in the field and 76% in the greenhouse.

CHAPTER 10

Section 10.1

1. (a) Statement. It is a true declarative sentence. **(b)** Statement. It is a false declarative sentence. **(c)** Not a statement. It is a question. **(d)** Not a statement. It is an opinion.   **3. (a)** Statement. It is a true declarative sentence. **(b)** Statement. It is a false declarative sentence. **(c)** Statement. It is a true declarative sentence. **(d)** Not a statement. It is a command. **(e)** Not a statement. It is an opinion.   **5. (a)** Neither **(b)** Disjunction **(c)** Conjunction **(d)** Disjunction   **7. (a)** Betty has blonde hair, or Angela has dark hair. **(b)** Angela does not have dark hair. **(c)** Betty has blonde hair, and Angela has dark hair. **(d)** Betty does not have blonde hair, and Angela does not have dark hair.   **9. (a)** $p \wedge q$ **(b)** $p \vee q$ **(c)** $(\sim p) \wedge q$ **11. (a)** False because $4^2 = 15$ is false **(b)** False because the first part is false **(c)** True because both statements are true   **13. (a)** False because both parts are false **(b)** True because both parts are true **(c)** True because the first part is true   **15. (a)** I do not have six $1 bills in my wallet. **(b)** Roy cannot name all 50 states. **(c)** A quorum was present for the meeting.   **17. (a)** I drink coffee at breakfast and I eat salad for lunch and I like a dessert after dinner. **(b)** I drink coffee at breakfast, or I eat salad for lunch and I like a dessert after dinner. **(c)** I drink coffee at breakfast and I eat salad for lunch, and I do not like a dessert after dinner. **(d)** I drink coffee at breakfast and I eat salad for lunch, or I drink coffee at breakfast and I like dessert after dinner.   **19. (a)** T **(b)** F **(c)** F **(d)** F **(e)** T

21.

p	q	~p	~q	~p ∧ ~q
T	T	F	F	F
T	F	F	T	F
F	T	T	F	F
F	F	T	T	T

23. (a) p: Jane brings chips.
q: Tony brings drinks.
r: Hob brings cookies.
s: Ingred brings chips.
t: Alex brings drinks.
u: Hester brings drinks.
v: Alice brings cookies.
w: Jenn brings cookies.
The statement is: $(p \wedge q \wedge r) \vee (s \wedge [t \vee u] \wedge [v \vee w])$

(b) The statements have the truth values:
$p = T, q = F, r = F, s = T, t = T, u = T, v = F, w = F$
which gives
$(T \wedge F \wedge F) \vee (T \wedge (T \vee T) \wedge (F \vee F))$
which is F.

Section 10.2

1. (a) If I have $5.00, then I can rent a video. **(b)** If I can rent a video, then I have $5.00.   **3. (a)** True because the hypothesis and the conclusion are true **(b)** True because the hypothesis is false **(c)** False because the hypothesis is true and the conclusion is false   **5.** *Converse*: If I live in Colorado, then I live in Denver. *Inverse*: If I do not live in Denver, then I do not live in Colorado. *Contrapositive*: If I do not live in Colorado, then I do not live in Denver.
7. (a) False because the components have different truth values, true and false, respectively **(b)** True because both components are true **(c)** True because both components are true

9.

p	q	p ∧ q	~(p ∧ q)
T	T	T	F
T	F	F	T
F	T	F	T
F	F	F	T

11.

p	q	~q	p ∧ ~q
T	T	F	F
T	F	T	T
F	T	F	F
F	F	T	F

13.

p	q	~p	~q	~p ∨ ~q
T	T	F	F	F
T	F	F	T	T
F	T	T	F	T
F	F	T	T	T

15.

p	q	q ∨ p	~p	~p → (q ∨ p)
T	T	T	F	T
T	F	T	F	T
F	T	T	T	T
F	F	F	T	F

17.

p	q	r	p ∨ q	(p ∨ q) ∧ r
T	T	T	T	T
T	F	T	T	T
F	T	T	T	T
F	F	T	F	F
T	T	F	T	F
T	F	F	T	F
F	T	F	T	F
F	F	F	F	F

19.

p	q	r	p → q	q → r	(p → q) ∧ (q → r)
T	T	T	T	T	T
T	F	T	F	T	F
F	T	T	T	T	T
F	F	T	T	T	T
T	T	F	T	F	F
T	F	F	F	T	F
F	T	F	T	F	F
F	F	F	T	T	T

21.

p	q	r	p ∨ q	(p ∨ q) ↔ r
T	T	T	T	T
T	F	T	T	T
F	T	T	T	T
F	F	T	F	F
T	T	F	T	F
T	F	F	T	F
F	T	F	T	F
F	F	F	F	T

23.

p	$\sim p$	$\sim(\sim p)$	$\sim(\sim p) \leftrightarrow p$
T	F	T	T
F	T	F	T

25. (a) p: Your insurance company paid the provider directly for part of your expenses.
q: You paid only the amount that remained.
r: Include on line 1 only the amount you paid.
The statement is $(p \wedge q) \rightarrow r$.
(b) p: You leave line 65 blank.
q: The IRS will figure the penalty.
r: The IRS will send you the bill.
The statement is $p \rightarrow (q \wedge r)$.

(c) p: You changed your name because of marriage, divorce, etc.
q: You made estimated tax payments using your former name.
r: Attach a statement to the front of Form 1040, explaining all the payments you and your spouse made in 1997.
s: Attach the name of the service center where you made the payments.
t: Attach the name and SSN under which you made the payments.
The statement is $(p \wedge q) \rightarrow (r \wedge s \wedge t)$.

Section 10.3

1.

p	q	$p \rightarrow q$	$\sim q \rightarrow \sim p$
T	T	T	T
T	F	F	F
F	T	T	T
F	F	T	T

Equivalent

3.

p	q	$\sim p$	$p \wedge q$	$\sim p \vee (p \wedge q)$	$\sim p \vee q$
T	T	F	T	T	T
T	F	F	F	F	F
F	T	T	F	T	T
F	F	T	F	T	T

Equivalent

5.

p	q	r	$\sim p$	$q \wedge r$	$\sim p \vee (q \wedge r)$	$p \rightarrow (q \wedge r)$
T	T	T	F	T	T	T
T	F	T	F	F	F	F
F	T	T	T	T	T	T
F	F	T	T	F	T	T
T	T	F	F	F	F	F
T	F	F	F	F	F	F
F	T	F	T	F	T	T
F	F	F	T	F	T	T

Equivalent

7.

p	q	$p \rightarrow q$	$\sim(p \rightarrow q)$	p
T	T	T	F	T
T	F	F	T	T
F	T	T	F	F
F	F	T	F	F

Not equivalent

9.

p	q	r	$q \vee r$	$p \wedge q$	$p \wedge (q \vee r)$	$(p \wedge q) \vee r$
T	T	T	T	T	T	T
T	F	T	T	F	T	T
F	T	T	T	F	F	T
F	F	T	T	F	F	T
T	T	F	T	T	T	T
T	F	F	F	F	F	F
F	T	F	T	F	F	F
F	F	F	F	F	F	F

Not equivalent

Some of the programs are included at the end of sections in the text. In the list below, the section where the program appears in the text is given after the name of the program. The programs available include the following:

PIVOT (Section 2.3).   The program pivots on the matrix entry specified by the user. This program is useful in reducing a matrix or performing the pivots in a simplex solution.

SMPLX (Section 4.2).   The user specifies the pivot row and column in a simplex tableau, and the program finds the next tableau.

BINOM.   For a given N, p, and x, this calculates the binomial probability

$$P(X = x) = C(N, x)p^x(1 - p)^{N-x}$$

ANNA.   This calculates the amount of an annuity when the periodic payment, periodic interest rate, and number of periods are given.

PAYANN.   This computes the periodic payment of an annuity that will yield a specified amount at a future date.

PVAL.   This computes the present value of an annuity with specified periodic payments.

LNPAY.   This finds the monthly payments required to amortize a loan.

ANNGRO (Section 5.3).   This shows the growth of an annuity by computing the amount of annuity period by period.

AMLN (Section 5.4).   This computes the amortization schedule of a loan, that is, the interest paid, principal paid, and balance of the loan for each month.

BIOD (Section 8.6).   This computes the binomial probability distribution:

$$P(X = x) = C(N, x)p^x(1 - p)^{N-x}   \text{(for } x = 0, 1, 2, \dots, N)$$

BIOH (Section 8.6).   This computes the binomial probability distribution and displays it as a histogram.

NORML (Section 8.7).   This finds the area under a normal curve between specified limits.

Instructions for using EXCEL can be found at the end of some sections to show how EXCEL may be applied. Here is a summary of the topics and the section in which they occur. ∎

11. Let p represent the statement "The exceptions above apply" and let q represent the statement "Use Form 2210." Then, statement (a) can be represented by $\sim p \to q$, and statement (b) can be represented by $p \vee q$. Make a truth table and compare the truth values of $\sim p \to q$ and $p \vee q$.

p	q	$\sim p$	$\sim p \to q$	$p \vee q$
T	T	F	T	T
T	F	F	T	T
F	T	T	T	T
F	F	T	F	F

As the truth values of $\sim p \to q$ and $p \vee q$ are identical, the statements are equivalent.

13. p: I will buy a jacket.
q: I will buy a shirt.
r: I will buy a tie.
Then, statement (a) is $p \wedge (q \vee r)$ and statement (b) is $(p \wedge q) \vee (p \wedge r)$.
    We form a truth table and compare the truth values of the two statements.

p	q	r	$q \vee r$	$p \wedge (q \vee r)$	$p \wedge q$	$p \wedge r$	$(p \wedge q) \vee (p \wedge r)$
T	T	T	T	T	T	T	T
T	F	T	T	T	F	T	T
F	T	T	T	F	F	F	F
F	F	T	T	F	F	F	F
T	T	F	T	T	T	F	T
T	F	F	F	F	F	F	F
F	T	F	T	F	F	F	F
F	F	F	F	F	F	F	F

Equivalent.

Section 10.4

1. p: Eat your beans.
  q: You can have dessert.
  $p \to q$
  $\underline{p}$
  q
  Valid, law of detachment

3. p: You do not study.
  q: You cannot do the homework.
  r: You cannot pass the course.
  $p \to q$
  $\underline{q \to r}$
  $p \to r$
  Valid, syllogism

5. p: You can eat your beans.
  q: You can have dessert.
  $p \to q$
  $\underline{\sim p}$
  $\sim q$
  Not valid; see Example 5.

7. p: The ice is six inches thick.
  q: Shelley will go skating.
  $p \to q$
  $\underline{\sim q}$
  $\sim p$
  Valid: indirect reasoning

9. p: Inflation increases.
  q: The price of new cars increases.
  r: More people will buy used cars.
  $p \to q$
  $\underline{q \to r}$
  $p \to r$
  Valid, syllogism

11. Check $[(p \to q) \wedge (q \wedge r)] \to (p \vee r)$:

p	q	r	$p \to q$	$q \wedge r$	$p \vee r$	$[(p \to q) \wedge (q \wedge r)] \to (p \vee r)$
T	T	T	T	T	T	T
T	F	T	F	F	T	T
F	T	T	T	T	T	T
F	F	T	T	F	T	T
T	T	F	T	F	T	T
T	F	F	F	F	T	T
F	T	F	T	F	F	T
F	F	F	T	F	F	T

Valid.

13. Check $[(p \wedge q) \wedge (p \to \sim q)] \to (p \wedge \sim q)$:

p	q	$p \wedge q$	$p \to \sim q$	$p \wedge \sim q$	$[(p \wedge q) \wedge (p \to \sim q)] \to (p \wedge \sim q)$
T	T	T	F	F	T
T	F	F	T	T	T
F	T	F	T	F	T
F	F	F	T	F	T

Valid.

15. Check $[(q \to r) \wedge (\sim p \vee q) \wedge p] \to r$:

p	q	r	$q \to r$	$\sim p \vee q$	$[(q \to r) \wedge (\sim p \vee q) \wedge p] \to r$
T	T	T	T	T	T
T	F	T	T	F	T
F	T	T	T	T	T
F	F	T	T	T	T
T	T	F	F	T	T
T	F	F	T	F	T
F	T	F	F	T	T
F	F	F	T	T	T

Valid.

17. Check $[(p \to q) \wedge (p \to r)] \to (q \wedge r)$:

p	q	r	$p \to q$	$p \to r$	$q \wedge r$	$[(p \to q) \wedge (p \to r)] \to (q \wedge r)$
T	T	T	T	T	T	T
T	F	T	F	T	F	T
F	T	T	T	T	T	T
F	F	T	T	T	F	F
T	T	F	T	F	F	T
T	F	F	F	F	F	T
F	T	F	T	T	F	F
F	F	F	T	T	F	F

Not valid.

19. Check $[(p \to q) \wedge (q \to r) \wedge \sim q] \to \sim r$:

p	q	r	$p \to q$	$q \to r$	$\sim q$	$\sim r$	$[(p \to q) \wedge (q \to r) \wedge \sim q] \to \sim r$
T	T	T	T	T	F	F	T
T	F	T	F	T	T	F	T
F	T	T	T	T	F	F	T
F	F	T	T	T	T	F	F
T	T	F	T	F	F	T	T
T	F	F	F	T	T	T	T
F	T	F	T	F	F	T	T
F	F	F	T	T	T	T	T

Not valid.

21. This argument is of the form

Premise: $\quad p \to q$

$\qquad\qquad \dfrac{\sim q}{}$

Conclusion: $\sim p$

so it is indirect reasoning.

27. *p:* I trim the hedge.

q: I may go to the movie.

The argument can be represented as

Premise: $\quad p \to q$

$\qquad\qquad \dfrac{\sim p}{}$

Conclusion: $\sim q$

Valid, Law of Detachment.

29. $p \lor q$

$\dfrac{\sim p}{q}$

Valid, disjunctive syllogism.

23. This argument is of the form

Premise: $\quad p \to q$

$\qquad\qquad \dfrac{q \to r}{}$

Conclusion: $p \to r$

so it is a syllogism.

31. $p \to q$

$\dfrac{\sim q}{\sim p}$

Valid, indirect reasoning.

25. This argument is of the form

Premise: $\quad p \to q$

$\qquad\qquad \dfrac{p}{}$

Conclusion: q

so it is the Law of Detachment.

33. This argument is of the form $p \lor q$

$\dfrac{\sim p}{q}$

Valid, disjunctive syllogism.

Review Exercises, Chapter 10

1. (a) Statement   **(b)** Not a statement   **(c)** Not a statement   **(d)** Statement   **3. (a)** Rhonda is not sick today.
(b) Rhonda is sick today, and she has a temperature.   **(c)** Rhonda is not sick today, and Rhonda does not have a temperature.   **(d)** Rhonda is sick today, or she has a temperature.   **5. (a)** True   **(b)** True   **(c)** False   **(d)** True
7. (a) T **(b)** T **(c)** T **(d)** F **9. (a)** True   **(b)** False   **11.** *Inverse*: "If I do not turn in my paper late, then I will not be penalized." *Converse*: "If I will be penalized, then I will turn in my paper late." *Contrapositive*: "If I am not penalized, then I did not turn in my paper late."

13.

p	q	$\sim p$	$p \land q$	$\sim p \to (p \land q)$
T	T	F	T	T
T	F	F	F	T
F	T	T	F	F
F	F	T	F	F

15.

p	q	$p \lor q$	$p \land (p \lor q)$
T	T	T	T
T	F	T	T
F	T	T	F
F	F	F	F

As p and $p \land (p \lor q)$ have the same truth values, they are logically equivalent.

17. Valid by indirect reasoning   **19.** Valid by syllogism   **21.** Valid by disjunctive syllogism.

APPENDIX A

Section A.1

1. -13 **3.** 23 **5.** -30 **7.** -35 **9.** -5 **11.** -7 **13.** -10 **15.** -8 **17.** 3
19. $\frac{9}{5}$ **21.** $\frac{17}{12}$ **23.** $-\frac{11}{12}$ **25.** $\frac{13}{20}$ **27.** $-\frac{1}{35}$ **29.** $\frac{2}{3}$ **31.** $\frac{5}{14}$ **33.** $\frac{1}{15}$ **35.** $\frac{8}{15}$
37. $\frac{12}{35}$ **39.** $\frac{20}{33}$ **41.** $\frac{4}{3}$ **43.** $\frac{15}{8}$ **45.** $\frac{171}{40}$ **47.** $-6a - 22b$ **49.** $-10a - 50b$

Section A.2

1. -3 **3.** 4 **5.** 2 **7.** $\frac{1}{2}$ **9.** $-\frac{7}{4}$ **11.** $\frac{20}{11}$ **13.** $\frac{15}{23}$ **15.** 6 **17. (a)** $242   **(b)** $412
(c) 950 miles   **19. (a)** $14.28   **(b)** $44.10   **(c)** 61 pounds

Section A.3

1.

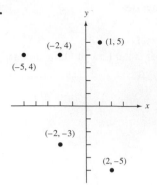

3.

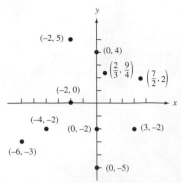

5. (a) x is negative and y is positive.   **(b)** x and y are both negative.   **(c)** x is positive and y is negative.

7.

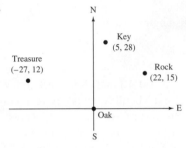

Section A.4

1. (a) True because $9 - 3$ is positive   **(b)** True because $4 - 0$ is positive   **(c)** False because $-5 - 0$ is not positive
(d) True because $-3 - (-15)$ is positive.   **(e)** True because $\frac{5}{6} - \frac{2}{3} = \frac{1}{6}$ is positive   **3.** $x < \frac{9}{2}$ **5.** $x \geq -16$

7. $x > -3$ **9.** ⟶ 5   **11.** ⟵ -9 **13.** ⟵ -4

15. ⟵ 7/9   **17.** -7 —o o— 17/3   **19.** 2 —o o— 8   **21.** $(-\infty, -1]$

23. $\left(-\infty, \frac{1}{9}\right]$ **25.** $\left(-13, -\frac{17}{4}\right]$ **27.** $x \leq -\frac{1}{2}$ **29.** $-\frac{23}{3} > x \geq -11$ **31.** 8, 9, or 10 correct answers
33. 20, 21, 22, 23, 24, or 25

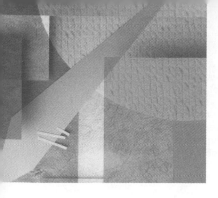

Index

Area Under the Standard Normal Curve

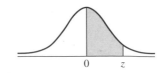

z	A	z	A	z	A	z	A	z	A	z	A	z	A
0.00	0.0000	0.50	0.1915	1.00	0.3413	1.50	0.4332	2.00	0.4773	2.50	0.4938	3.00	0.4987
0.01	0.0040	0.51	0.1950	1.01	0.3438	1.51	0.4345	2.01	0.4778	2.51	0.4940	3.01	0.4987
0.02	0.0080	0.52	0.1985	1.02	0.3461	1.52	0.4357	2.02	0.4783	2.52	0.4941	3.02	0.4987
0.03	0.0120	0.53	0.2019	1.03	0.3485	1.53	0.4370	2.03	0.4788	2.53	0.4943	3.03	0.4988
0.04	0.0160	0.54	0.2054	1.04	0.3508	1.54	0.4382	2.04	0.4793	2.54	0.4945	3.04	0.4988
0.05	0.0199	0.55	0.2088	1.05	0.3531	1.55	0.4394	2.05	0.4798	2.55	0.4946	3.05	0.4989
0.06	0.0239	0.56	0.2123	1.06	0.3554	1.56	0.4406	2.06	0.4803	2.56	0.4948	3.06	0.4989
0.07	0.0279	0.57	0.2157	1.07	0.3577	1.57	0.4418	2.07	0.4808	2.57	0.4949	3.07	0.4989
0.08	0.0319	0.58	0.2190	1.08	0.3599	1.58	0.4430	2.08	0.4812	2.58	0.4951	3.08	0.4990
0.09	0.0359	0.59	0.2224	1.09	0.3621	1.59	0.4441	2.09	0.4817	2.59	0.4952	3.09	0.4990
0.10	0.0398	0.60	0.2258	1.10	0.3643	1.60	0.4452	2.10	0.4821	2.60	0.4953	3.10	0.4990
0.11	0.0438	0.61	0.2291	1.11	0.3665	1.61	0.4463	2.11	0.4826	2.61	0.4955	3.11	0.4991
0.12	0.0478	0.62	0.2324	1.12	0.3686	1.62	0.4474	2.12	0.4830	2.62	0.4956	3.12	0.4991
0.13	0.0517	0.63	0.2357	1.13	0.3708	1.63	0.4485	2.13	0.4834	2.63	0.4957	3.13	0.4991
0.14	0.0557	0.64	0.2389	1.14	0.3729	1.64	0.4495	2.14	0.4838	2.64	0.4959	3.14	0.4992
0.15	0.0596	0.65	0.2422	1.15	0.3749	1.65	0.4505	2.15	0.4842	2.65	0.4960	3.15	0.4992
0.16	0.0636	0.66	0.2454	1.16	0.3770	1.66	0.4515	2.16	0.4846	2.66	0.4961	3.16	0.4992
0.17	0.0675	0.67	0.2486	1.17	0.3790	1.67	0.4525	2.17	0.4850	2.67	0.4962	3.17	0.4992
0.18	0.0714	0.68	0.2518	1.18	0.3810	1.68	0.4535	2.18	0.4854	2.68	0.4963	3.18	0.4993
0.19	0.0754	0.69	0.2549	1.19	0.3830	1.69	0.4545	2.19	0.4857	2.69	0.4964	3.19	0.4993
0.20	0.0793	0.70	0.2580	1.20	0.3849	1.70	0.4554	2.20	0.4861	2.70	0.4965		
0.21	0.0832	0.71	0.2612	1.21	0.3869	1.71	0.4564	2.21	0.4865	2.71	0.4966		
0.22	0.0871	0.72	0.2642	1.22	0.3888	1.72	0.4573	2.22	0.4868	2.72	0.4967		
0.23	0.0910	0.73	0.2673	1.23	0.3907	1.73	0.4582	2.23	0.4871	2.73	0.4968		
0.24	0.0948	0.74	0.2704	1.24	0.3925	1.74	0.4591	2.24	0.4875	2.74	0.4969		
0.25	0.0987	0.75	0.2734	1.25	0.3944	1.75	0.4599	2.25	0.4878	2.75	0.4970		
0.26	0.1026	0.76	0.2764	1.26	0.3962	1.76	0.4608	2.26	0.4881	2.76	0.4971		
0.27	0.1064	0.77	0.2794	1.27	0.3980	1.77	0.4616	2.27	0.4884	2.77	0.4972		
0.28	0.1103	0.78	0.2823	1.28	0.3997	1.78	0.4625	2.28	0.4887	2.78	0.4973		
0.29	0.1141	0.79	0.2852	1.29	0.4015	1.79	0.4633	2.29	0.4890	2.79	0.4974		
0.30	0.1179	0.80	0.2881	1.30	0.4032	1.80	0.4641	2.30	0.4893	2.80	0.4974		
0.31	0.1217	0.81	0.2910	1.31	0.4049	1.81	0.4649	2.31	0.4896	2.81	0.4975		
0.32	0.1255	0.82	0.2939	1.32	0.4066	1.82	0.4656	2.32	0.4898	2.82	0.4976		
0.33	0.1293	0.83	0.2967	1.33	0.4082	1.83	0.4664	2.33	0.4901	2.83	0.4977		
0.34	0.1331	0.84	0.2996	1.34	0.4099	1.84	0.4671	2.34	0.4904	2.84	0.4977		
0.35	0.1368	0.85	0.3023	1.35	0.4115	1.85	0.4678	2.35	0.4906	2.85	0.4978		
0.36	0.1406	0.86	0.3051	1.36	0.4131	1.86	0.4686	2.36	0.4909	2.86	0.4979		
0.37	0.1443	0.87	0.3079	1.37	0.4147	1.87	0.4693	2.37	0.4911	2.87	0.4980		
0.38	0.1480	0.88	0.3106	1.38	0.4162	1.88	0.4700	2.38	0.4913	2.88	0.4980		
0.39	0.1517	0.89	0.3133	1.39	0.4177	1.89	0.4706	2.39	0.4916	2.89	0.4981		
0.40	0.1554	0.90	0.3159	1.40	0.4192	1.90	0.4713	2.40	0.4918	2.90	0.4981		
0.41	0.1591	0.91	0.3186	1.41	0.4207	1.91	0.4719	2.41	0.4920	2.91	0.4982		
0.42	0.1628	0.92	0.3212	1.42	0.4222	1.92	0.4726	2.42	0.4922	2.92	0.4983		
0.43	0.1664	0.93	0.3238	1.43	0.4236	1.93	0.4732	2.43	0.4925	2.93	0.4983		
0.44	0.1700	0.94	0.3264	1.44	0.4251	1.94	0.4738	2.44	0.4927	2.94	0.4984		
0.45	0.1736	0.95	0.3289	1.45	0.4265	1.95	0.4744	2.45	0.4929	2.95	0.4984		
0.46	0.1772	0.96	0.3315	1.46	0.4279	1.96	0.4750	2.46	0.4931	2.96	0.4985		
0.47	0.1808	0.97	0.3340	1.47	0.4292	1.97	0.4756	2.47	0.4932	2.97	0.4985		
0.48	0.1844	0.98	0.3365	1.48	0.4306	1.98	0.4762	2.48	0.4934	2.98	0.4986		
0.49	0.1879	0.99	0.3389	1.49	0.4319	1.99	0.4767	2.49	0.4936	2.99	0.4986		